# Clean Energy Conversion

Reinhold Kneer · Heinz Pitsch ·
Andreas Dreizler · Viktor Scherer ·
Anna Maßmeyer · Stefan Pielsticker
Editors

# Clean Energy Conversion

## Oxy-fuel Combustion of Solid Fuels

Springer

*Editors*
Reinhold Kneer
Institute of Heat and Mass Transfer
RWTH Aachen University
Aachen, Germany

Heinz Pitsch
Institute for Combustion Technology
RWTH Aachen University
Aachen, Germany

Andreas Dreizler
Institute of Reactive Flows and Diagnostics
Technische Universität Darmstadt
Darmstadt, Germany

Viktor Scherer
Department of Energy Plant Technology
Ruhr-Universität Bochum
Bochum, Germany

Anna Maßmeyer
Institute of Heat and Mass Transfer
RWTH Aachen University
Aachen, Germany

Stefan Pielsticker
Institute of Heat and Mass Transfer
RWTH Aachen University
Aachen, Germany

ISBN 978-3-031-98955-1 ISBN 978-3-031-98956-8 (eBook)
https://doi.org/10.1007/978-3-031-98956-8

This work was supported by RWTH Aachen University.



This Springer imprint is published by the registered company Springer Nature Switzerland AG
The registered company address is: Gewerbestrasse 11, 6330 Cham, Switzerland

# Preface

The *German Environment Agency* stated in 2023 [1], based on the recent report from the *International Panel on Climate Change* (IPCC), that "Expectations about the role of carbon capture and storage (CCS) in meeting the global temperature target of 1.5 °C are high". This statement emphasises the critical role of CCS to meet climate targets. Of the various approaches to CCS, this book focuses on one route, oxy-fuel combustion. It presents the achievements and findings of a 12-year research initiative, funded by the *German Research Foundation* (DFG), on solid fuel combustion in an $O_2/CO_2/H_2O$ atmosphere (oxy-fuel combustion). The primary aim of the research was to understand the fundamental processes and mechanisms involved in order to determine the decisive differences arising from the replacement of air as the oxidising agent by an $O_2/CO_2$ mixture.

Research and development activities led by the industry are focused primarily on system-level aspects demonstrating the feasibility of the concepts and to enable rapid implementation. *Vattenfall* for example, driving large scale oxy-fuel combustion, operated a 30 $MW_{th}$ boiler from 2008 to 2014 to specifically prove the viability of flue gas cleaning, the combustion process itself and the related controls. Due to changes in the political landscape, the focus of the power sector in Germany has shifted away from oxy-fuel to a mix of renewables and hydrogen-fueled power plants.

Nowadays, the pace in oxy-fuel combustion development is therefore being set by industries that face the additional problem of unavoidable product-related $CO_2$ emissions, such as the cement and lime industry. These industries are under particularly intense pressure to achieve future $CO_2$ reduction targets. For example, a large-scale oxy-fuel-fired pilot cement plant is scheduled to begin operations in Mergelstetten (Germany) in 2026 and oxy-fuel lime shaft kilns are already offered on the market.

With larger scale operations and feasibility studies, fundamental aspects often remain unexplored, yet they are crucial for a reliable design and the detection of failures in case of damage. Therefore, to address open questions around oxy-fuel combustion of solid fuels, a research consortium (CRC/TRR129 "Oxyflame") between *RWTH Aachen University, Ruhr-Universität Bochum*, and *Technische Universität Darmstadt* was established and funded by the DFG. *Oxyflame* got started in October 2013 and ended in June 2025.

This book is of interest for scientists, advanced students interested in combustion, and development engineers in the industry. Readers are encouraged to refresh their understanding of the fundamental principles of oxy-fuel combustion.

Reinhold Kneer
Heinz Pitsch
Andreas Dreizler
Viktor Scherer
Anna Maßmeyer
Stefan Pielsticker

## Reference

1. Umweltbundesamt. *Carbon Capture and Storage in the Global Climate Debate*. 2023.

# Contents

# The Collaborative Research Centre "Oxyflame"

1

Reinhold Kneer, Viktor Scherer, Heinz Pitsch, Andreas Dreizler, Anna Maßmeyer, Stefan Pielsticker and Benjamin Gövert

## Contents




R. Kneer (✉) · A. Maßmeyer · S. Pielsticker · B. Gövert
Institute of Heat and Mass Transfer, RWTH Aachen University, Aachen, Germany
e-mail: kneer@wsa.rwth-aachen.de

A. Maßmeyer
e-mail: massmeyer@wsa.rwth-aachen.de

S. Pielsticker
e-mail: pielsticker@wsa.rwth-aachen.de

B. Gövert
e-mail: goevert@wsa.rwth-aachen.de

V. Scherer
Department of Energy Plant Technology, Ruhr-Universität Bochum, Bochum, Germany
e-mail: scherer@leat.rub.de

H. Pitsch
Institute for Combustion Technology, RWTH Aachen University, Aachen, Germany
e-mail: h.pitsch@itv.rwth-aachen.de

A. Dreizler
Institute of Reactive Flows and Diagnostics, Technische Universität Darmstadt, Darmstadt, Germany
e-mail: dreizler@rsm.tu-darmstadt.de



R. Kneer et al. (eds.), *Clean Energy Conversion*,
https://doi.org/10.1007/978-3-031-98956-8_1

**Abstract**

To combat climate change, it is imperative to reduce anthropogenic $CO_2$ emissions to net-zero by approximately 2055, especially in solid fuel combustion processes. In the present book, the focus is placed on the detailed investigation of the oxy-fuel process, where the combustion air is replaced with a mixture of oxygen, carbon dioxide, and water vapour, facilitating near carbon-neutral combustion. Research conducted within the presented research centre *Oxyflame* encompasses a diverse array of fuel types, spanning from pulverised fossil fuels such as bituminous coal and lignite to renewable biomass sources including walnut shells, beech wood or miscanthus. Substantial advancements in the experimental characterisation of reaction kinetics, particle dynamics, flame behaviour, and radiation effects in combustion environments of different scales are presented. The results yield into the development of robust models for combustion and transport phenomena as well as efficient simulation methodologies for designing combustion chambers, paving the way for efficient oxy-fuel systems capable of effectively reducing greenhouse gas emissions in alignment with global climate protection goals.

## 1.1 Introduction

To limit global warming to 1.5 °C above pre-industrial levels, as outlined in the *Paris Agreement* [1], anthropogenic carbon dioxide ($CO_2$) emissions must be reduced to net zero by roughly 2055 [2]. At present, 79% of global $CO_2$ emissions are caused by the power generation and industry [2–4]. Therefore, the major contribution to reduce $CO_2$ emissions has to come from these sectors [2,3].

Noting that current predictions for primary energy demand do not see a rapid decline in the consumption of fossil fuels—be it solid, liquid, or gaseous—up to 2040 [4], an essential part of technical solutions to reduce $CO_2$ emissions will be the application of carbon capture, utilisation, and storage (CCUS) technologies. A pre-requisite for achieving CCUS technologies is the transformation of existing energy conversion processes for large-scale power generation or in industrial applications. One advanced method to achieve CCUS in combustion systems is the oxy-fuel process.

Here, the term "oxy-fuel" refers to a combustion atmosphere consisting of oxygen ($O_2$), carbon dioxide ($CO_2$), and water vapour ($H_2O$), which replaces air in conventional combustion systems. The major advantage of this process is that the flue gas is composed almost exclusively of $CO_2$ and $H_2O$. Water vapour is easily removed from the exhaust stream by condensation, leaving $CO_2$ in high concentrations. This enables $CO_2$ to be efficiently captured for storage and use. The combination of the oxy-fuel process with CCUS enables near carbon-neutral combustion of fossil fuels. In addition, using biomass as a fuel for CCUS (BECCUS) can create a net carbon sink [5,6].

Due to the different properties of the combustion atmosphere compared to air, the chemical reactions as well as heat and mass transfer phenomena occurring on and around the particles can no longer be described using the approaches developed for air combustion. Chemical reactions are particularly influenced by the high partial pressure of $CO_2$, which significantly affects both, the heterogeneous and gas-phase reactions. In heterogeneous reactions, the gasification of the remaining carbon solids (char) or diffusion inside the char pores can alter reaction dynamics. In homogeneous gas-phase reactions, high $CO_2$ concentrations increase the importance of the reverse of the main radical-producing reaction $O_2 + H \rightarrow OH + O$. Further, heat and mass transfer in a $CO_2$-containing atmosphere are affected in multiple ways. The $CO_2$ has a significantly higher molar heat capacity than the main air component $N_2$, thereby withdrawing more energy from the reaction zones. At high temperatures, $CO_2$ (and $H_2O$) are highly radiative molecules. The changed heat transfer directly impacts the heat-up process of the fuel particles, which affects the release of combustible gas species (pyrolysis) and the ignition behaviour. In contrast to $O_2$ and $N_2$, which have an effective Lewis number (defined as the ratio of thermal conductivity to diffusion coefficient) of approximately one, $CO_2$ has a significantly higher Lewis number. This can be particularly relevant for small-scale processes near the particle surface. The underlying coupled reaction-transport processes, particularly in the boundary layer of solid fuel particles, are currently only partially understood and, consequently, not fully considered in current state-of-the-art modelling approaches.

To close the existing knowledge gaps, the oxy-fuel combustion of solid fuels was investigated in the Collaborative Research Centre 129 "Oxyflame" funded by the *German Research Foundation* (DFG). The *Oxyflame* team comprises researchers from the fields of solid fuel and gas combustion, chemistry, thermodynamics, and fluid mechanics from the three participating locations: *RWTH Aachen University*, *Ruhr-Universität Bochum*, and *Technische Universität Darmstadt*.

## 1.2 Goals of Oxyflame

The effective and reliable design of large-scale oxy-fuel combustion systems requires a detailed understanding of the relevant chemical and physical processes across multiple orders of magnitude, i.e. from single molecular reactions over particles up to the entire combustion chamber. The need for this detailed understanding defines the scientific goals of *Oxyflame*:

- Development of a comprehensive understanding of all relevant mechanisms involved in the combustion of solid, pulverised fuels in an oxy-fuel atmosphere primarily composed of $O_2$, $CO_2$, and $H_2O$.
- Development of valid generalised models for combustion and transport phenomena in an oxy-fuel atmosphere, while accounting for all relevant effects at each scale.
- Development of physics-based, efficient simulation methods for the layout of oxy-fuel combustion chambers to allow for model-based design, optimisation, and sensitivity analysis.

As mentioned above, replacing air by a mixture of $O_2$, $CO_2$, and $H_2O$ significantly alters combustion behaviour. This can result in modified chemical conversion rates, flame instabilities, or even local flame extinction and reignition. This also affects all relevant transport processes, which occur across length scales ranging from atomic scale $\mathcal{O}\left(10^{-10}\,\mathrm{m}\right)$ to typical furnace dimensions $\mathcal{O}\left(10^{1}-10^{2}\,\mathrm{m}\right)$. Related time scales range from chemical reactions $\mathcal{O}\left(10^{-9}\,\mathrm{s}\right)$ to the fuel residence time in large boilers, which is of the order of several seconds $\mathcal{O}(1\,\mathrm{s})$. To identify the dominant mechanisms underlying the coupled reaction-transport processes, this range of scales has been addressed within *Oxyflame*, both through generic laboratory-scale experiments and through larger-scale validation experiments of pulverised solid fuel combustion. Modelling approaches covering different scales have been considered ranging from molecular dynamics simulations over approaches that partially or even fully resolve turbulence all the way to a multi-physics and multi-scale description at the system-scale with large eddy simulation (LES).

## 1.3 Examined Fuels

To consider the effect of fuel type in experiments and model development within *Oxyflame*, fuels of different rank and elemental composition have been selected. The van Krevelen diagram in Fig. 1.1 depicts the fuel range by comparing the molar H/C ratio against the O/C ratio. The fuel spectrum covers coals as well as biomass in raw and torrefied state.

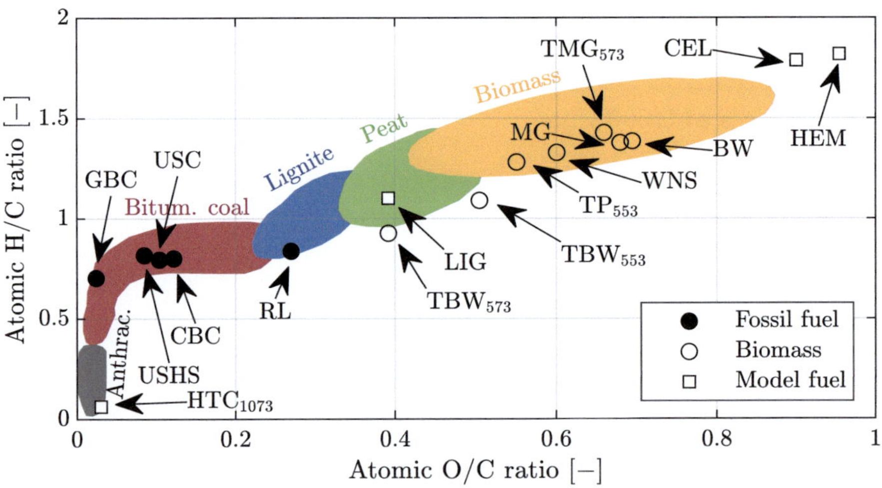


**Fig. 1.1** Van Krevelen diagram of fuels used within *Oxyflame*. Fuel codes are listed in Tab. 1.1

Fossil fuels from various origins and ages served as a starting point in the experimental and modelling cycles and a reference to already existing research data. Two bituminous coals from the German Prosper-Haniel colliery (GBC) on the northern edge of the Ruhr area and the Minas Norte region in Colombia (CBC) serve as representatives of the oldest geological age. One US reference coal (USC) has been added for targeted comparisons and one high-sulphur fuel (USHS) allowed the investigation of pollutant formation. Rhenish lignite (RL) from the Hambach opencast mine is a representative of the younger geological age.

The selected raw biogenic fuels include one woody biomass (beech wood (BW)), one fast-growing energy plant (miscanthus × giganteus (MG)) and one agricultural residue (walnut shells (WNS)), all cultivated in Europe. These fuels exhibit diverse structural characteristics (woody or straw-like) and significantly different proportions of biomass structural components. Selected raw biomasses have been pre-treated by torrefaction at various temperatures and residence times. Torrefied biomasses possess properties comparable to "young" fossil fuels (e.g. Rhenish lignite), making them a suitable renewable alternative.

The fuel spectrum is supplemented by artificial fuels including extracted biomass components (cellulose (CEL), hemicellulose (HEM), and lignin (LIG)) and model fuels produced by hydrothermal carbonisation. The latter were used in experiments to study the catalytic activity of minerals by doping the base fuel with constituents such as Fe, K, and Na. Properties from proximate and ultimate analysis are listed in Table 1.1 for all fuels.

As the focus of *Oxyflame* was on pulverised solid fuel combustion, all fuels were milled and, in most cases, sieved to multiple size fractions typically spanning the overall range of 90–200 μm. Two exemplary particle size analyses for the fuels RL and WNS are provided in Fig. 1.2.

It is worth noting that research on waste-based fuels and larger particles was part of an *Oxyflame* spin-off. However, these findings are not included in this book but are documented in other sources [7–9].

**Table 1.1** Properties of fuels investigated in *Oxyflame*. All values in mass-%

| Code | Fuel name | Proximate analysis* | | | Elemental analysis• | | | | | |
|---|---|---|---|---|---|---|---|---|---|---|
| | | Ash | Volat. | Fixed C◇ | C | H | N | S | Cl | O◇ |
| CBC | Colombian bit. coal (Minas Norte) | 7.0 | 40.5 | 52.5 | 78.7 | 5.26 | 2.11 | 1.10 | 0.01 | 12.8 |
| GBC | German bit. coal (Prosper-Haniel) | 8.5 | 31.2 | 60.2 | 89.6 | 5.26 | 1.73 | 0.37 | 0.06 | 2.9 |
| RL | Rhenish lignite (Hambach) | 5.8 | 52.1 | 42.1 | 68.9 | 4.81 | 1.28 | 0.27 | 0.02 | 24.8 |
| USHS | US high sulphur bit. coal | 9.3 | 37.3 | 53.4 | 80.4 | 5.48 | 1.38 | 3.32 | 2.80 | 9.1 |
| USC | US reference bit. coal (Pittsburgh #8) | 12.3 | 37.5 | 50.2 | 81.3 | 5.39 | 1.64 | 0.38 | – | 11.3 |
| $TBW_{553}$ | Torrefied beech wood (553 K) | 0.7 | 71.2 | 28.1 | 56.5 | 5.13 | 0.32 | 0.00 | – | 38.0 |
| $TBW_{573}$ | Torrefied beech wood (573 K) | 1.0 | 58.6 | 40.5 | 62.3 | 4.81 | 0.35 | 0.01 | – | 32.6 |
| $TMG_{573}$ | Torrefied miscanthus (573 K) | 4.1 | 63.0 | 32.9 | 49.9 | 5.94 | 0.21 | 0.00 | – | 43.9 |
| $TP_{553}$ | Torrefied poplar wood (553 K) | 0.6 | 78.1 | 21.3 | 54.2 | 5.78 | 0.21 | 0.02 | 0.00 | 39.8 |
| BW | Beech wood | 0.4 | 85.3 | 14.3 | 48.9 | 5.64 | 0.23 | 0.02 | – | 45.2 |
| MG | Miscanthus × giganteus | 1.3 | 83.4 | 15.3 | 49.3 | 5.66 | 0.38 | 0.00 | – | 44.7 |
| WNS | Walnut shells | 0.4 | 81.1 | 18.5 | 52.1 | 5.77 | 0.28 | 0.02 | 0.29 | 41.8 |
| CEL | Cellulose | 0.3 | 95.2 | 4.5 | 42.4 | 6.32 | 0.30 | 0.00 | – | 50.9 |
| HEM | Hemicellulose | 3.7 | 84.7 | 11.6 | 41.1 | 6.23 | 0.21 | 0.10 | – | 52.3 |
| LIG | Lignin | 1.3 | 68.2 | 30.5 | 60.7 | 5.57 | 1.22 | 0.81 | – | 31.7 |
| $HTC_{1073}$ | Hydrothermally carbonised cellulose (1073 K) | 0.0 | 3.0 | 97.0 | 95.6 | 0.50 | 0.00 | 0.00 | – | 3.9 |

*: dry, •: dry and ash-free; ◇: by difference

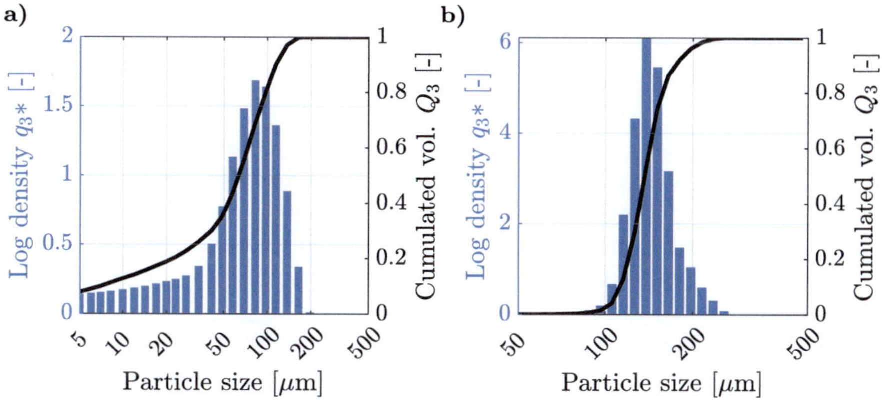


**Fig. 1.2** Exemplary particle size analyses for **a**) unsieved Rhenish lignite (RL) and **b**) walnut shells (WNS) sieved by the manufacturer

## 1.4 Major Results of Oxyflame

The major results of *Oxyflame* are reported in this book, with a very brief overview given here, also outlining the chapter-wise content:

**Parts I and II: Reaction Kinetics Measurements and Modelling**

Flame and flow field development in the combustion chamber depend largely on volatile release and char combustion characteristics of the utilised solid fuel. To improve model approaches for oxy-fuel kinetics of pyrolysis and char conversion, a novel and extensive experimental database has been established. This database includes data gathered under realistic high heating rates for a wide variety of fuels and a broad range of different oxy-fuel atmospheres. Data comprise results from solid sampling and gas-phase analysis as well as data from optical measurements, such as single particle temperatures, size and shape evolution, for both ignited and non-ignited particles. The influence of catalytically active minerals such as Fe, K, and Na on pyrolysis-gas yields and char conversion was investigated. The findings confirmed that the catalytic activity of these minerals is relevant for the high heating rates under oxy-fuel conditions. Experimental analyses also included the study of sorption processes occurring on the fuel particles, leading to the development of a novel pore-structure-dependent sorption model that goes beyond current approaches. Regarding system-relevant corrosive effects, the release of minority species such as sulphur and chlorine was also examined.

The reactions and processes taking place in the carbon matrix were described on an atomic and molecular basis, particularly for the char burnout phase. It was shown that this allows to provide kinetic data that are based on defined molecular compositions and, thus, transferable between different solid fuels independent of their origins. For a knowledge-based design, fast and predictive models that depend on the chemical structure of the fuels are required for CFD simulations. Hence, the well-known CPD model for pyrolysis has been transferred into a modular object-oriented formulation that allows for flexible coupling with different particle and reactor models. The modified CPD model was able to replicate all boundary conditions of the experimental test rigs within *Oxyflame*. It also has been successfully tested with data sets of extracted biomass components. Furthermore, the popular CCK code has been extended by a novel model for thermal annealing of biomass. To finally provide a seamless model for LES that covers pyrolysis and char burnout kinetics simultaneously, a more holistic approach was pursued, which is based on the CRECK-S model from *Politecnico di Milano*. This model has been calibrated based on the experimental results of *Oxyflame*.

For the gas-phase kinetics, detailed measurements have been performed in counterflow-diffusion flames using time-of-flight mass spectrometry. Different fuel blends representing volatile gases have been investigated in air and oxy-fuel atmosphere. Furthermore, the impact of different oxygenated and nitrogen-containing aromatic compounds on combustion and pollutant formation was studied in the same setup as tar representatives with particular relevance for biomass combustion. These data were used to develop well-validated detailed and reduced kinetic models for the gas-phase kinetics of volatiles from coal and biomass.

**Part III: Particle Flow Interactions**

For examining the coupled reaction-transport processes occurring on the meso-scale, the particle-laden two-phase flow assuming typical conditions of solid fuel flames was investigated using both, canonical experiments and DNS modelling. The DNS modelling for non-reacting conditions using up to 70,000 fully resolved particles in isotropic turbulence gave new insights into the interaction of flow field and the motion of spherical and elliptical particles. These results were used, e.g. to derive drag correlations for point-particle approaches. Experimental studies of individual particles and small particle groups in a laminar flow field revealed the changes in flame topology when progressing from individual particle combustion to particle group combustion. For measurements on particles moving in turbulent jets, an advanced high-speed measurement technique was developed and successfully used as a counterpart to respective DNS studies. On the modelling side, DNS for single particles in air and $O_2/CO_2$ environments showed that elevated $CO_2$ concentrations consume H radicals during devolatilisation, leading to a slowdown in chain-branching reactions and consequently to a destabilisation of the combustion process.

**Part IV: Flame Studies**

Regarding the investigations of flames and respective flow fields in large-scale combustion chambers, multiple unique data sets were compiled by means of cascaded experiments. Parameters for these investigations were oxidiser atmosphere, oxygen concentration, and fuel. Detailed investigation of the near-burner environment in a fully optically accessible combustor, operated with a supporting gas flame to compensate for the associated heat losses due to the large windows, produced unique data sets that are useful not only for validating solid fuel flame models but also for supporting the interpretation of results obtained in larger combustors with less optical access. Using similar burners in a pilot-scale combustion chamber, and operating it purely with solid fuel, the effects of various $O_2/CO_2$ mixtures on the flame shape could be investigated in detail. To bring the results closer to typical industrial scales, the burner design has been scaled up and was employed for an application-oriented validation of the results, using a combustion chamber with a thermal output of up to 1 $MW_{th}$.

**Part V: Radiation**
Another physical effect that comes into play mostly for larger combustors is thermal radiation, especially when the gas phase contains high concentrations of $H_2O$ and $CO_2$. For this reason, the gas phase radiation has been modelled using correlated-k approaches, including particle radiation. This modelling approach was supported by detailed numerical investigations of the effect of particle shape and composition as well as on particle volume concentration. Various measurement setups were developed to compensate for the lack of biomass refractive index data. In addition, measurements of particle emissivity changes during char burnout provided valuable input for the particle radiation modelling.

**Part VI: Full-Scale Modelling**
The primary research objective of *Oxyflame* is to investigate, understand, and predict oxy-fuel combustion, spanning from small-scale laboratory configurations to semi-industrial applications. To achieve a comprehensive combustion modelling of various solid fuels, the scientific challenges can be categorised into four key areas: 1) solid fuel conversion, 2) gas-phase turbulence-chemistry interaction, 3) turbulent flow, mixing, and particle dynamics, and 4) radiation. Groundbreaking advancements have been achieved in these fields, both in depth and breadth.

Building on this understanding, the 3-D computational fluid dynamics (CFD) framework *OxySim-129* has been developed and is the focus of Chap. 19. *OxySim-129* extends the open-source CFD code OpenFOAM®, offering unprecedented predictive capabilities for the combustion of pulverised carbonaceous solid fuels under both oxy-fuel and standard air conditions at various reactor scales. Advanced detailed kinetic models for the solid phase have been seamlessly coupled with the turbulent reacting gas phase. Additionally, novel flamelet models have been developed and integrated into large eddy simulations (LES) of turbulent multiphase gas flows. The framework also incorporates fully coupled radiation models for the gas phase, particles, and their interactions, ensuring a comprehensive and highly accurate predictive tool.

Committed to the principles of open science, *Oxyflame* follows the FAIR data principles: Research data and software must be findable, accessible, interoperable, and reusable. In alignment with this philosophy, the *OxySim-129* software package will be released as open-source, providing access to both the scientific community and interested industrial partners.

**Acknowledgements** This work has been funded by the Deutsche Forschungsgemeinschaft (DFG, German Research Foundation)—215035359—SFB/TRR 129 'Oxyflame'.

## References

1. United Nations Framework Convention on Climate Change. *Paris Agreement.* United Nations. 2015.
2. Intergovernmental Panel on Climate Change. *Global Warming of* 1.5 °C: *An IPCC Special Report on the Impacts of Global Warming of* 1.5 °C *Above Pre-Industrial Levels and Related Global Greenhouse Gas Emission Pathways, in the Context of Strengthening the Global Response to the Threat of Climate Change, Sustainable Development, and Efforts to Eradicate Poverty*. 2018.
3. Intergovernmental Panel on Climate Change. *Climate Change 2023: Synthesis Report: A Report of the Intergovernmental Panel on Climate Change*. Intergovernmental Panel on Climate Change (IPCC), 2023. https://doi.org/10.59327/IPCC/AR6-9789291691647.
4. International Energy Agency. *World Energy Outlook 2019.* 2019.
5. Gough, C., Thornley, P., Mander, S., Vaughan, N. and Falano, T., eds. *Biomass Energy and Carbon Capture and Storage (BECCS): Unlocking Negative Emissions*. Hoboken, New Jersey, USA: John Wiley & Sons, 2018.
6. Tanzer, S. E. and Ramírez, A. "When Are Negative Emissions Negative Emissions?" *Energy & Environmental Science* 12.4 (2019), 1210–1218. https://doi.org/10.1039/C8EE03338B.
7. Streier, R., Veckenstedt, I., Deck, T., Lampe, K. and Scherer, V. "Analysis of the Impact of RDF Combustion on Cement Clinker Quality in a Superstoichiometric Oxy-Fuel Atmosphere by CFD Simulations". *Combustion Science and Technology* online (2023), 1–19. https://doi.org/10.1080/00102202.2023.2239463.
8. Streier, R., Wirtz, S., Aleksandrov, K., Gehrmann, H.-J., Stapf, D., Zhang, M., Vogelbacher, M., Matthes, J. and Scherer, V. "Determination of the Statistical Distribution of Drag and Lift Coefficients of Refuse Derived Fuel by Computer Vision". *Fuel* 352 (2023), 128847. https://doi.org/10.1016/j.fuel.2023.128847.
9. Solana Gómez, R., Streier, R., Pielsticker, S., Scherer, V. and Kneer, R. "Experimentelle Bestimmung der Pyrolysekinetik von Ersatzbrennstoffen". *31. Deutscher Flammentag*. Berlin, Germany, 2023.

# Reaction Kinetics Experiments

# Pyrolysis Products and Kinetics

# 2

Stefan Pielsticker, Erik Freisewinkel, Jannik Böttger, Francesca Cerciello, Osvalda Senneca, Dominik König, Viktor Scherer, Martin Muhler, Bernd Epple and Reinhold Kneer

## Contents



S. Pielsticker (✉) · R. Kneer
Institute of Heat and Mass Transfer, RWTH Aachen University, Aachen, Germany
e-mail: pielsticker@wsa.rwth-aachen.de

R. Kneer
e-mail: kneer@wsa.rwth-aachen.de

E. Freisewinkel · V. Scherer
Department of Energy Plant Technology, Ruhr-Universität Bochum, Bochum, Germany
e-mail: freisewinkel@leat.rub.de

V. Scherer
e-mail: scherer@leat.rub.de

J. Böttger · M. Muhler
Laboratory of Industrial Chemistry, Ruhr-Universität Bochum, Bochum, Germany
e-mail: jannik.boettger@rub.de

M. Muhler
e-mail: martin.muhler@rub.de

F. Cerciello · O. Senneca
Istituto di Scienze e Tecnologie per l'Energia e la Mobilità Sostenibili, Naples, Italy
e-mail: francesca.cerciello@stems.cnr.it





R. Kneer et al. (eds.), *Clean Energy Conversion*,
https://doi.org/10.1007/978-3-031-98956-8_2

**Abstract**

In oxy-fuel combustion, pyrolysis is crucial for solid fuel conversion. This chapter presents experimental data on pyrolysis products from various biomasses (including minority species), forming one of the most comprehensive biomass pyrolysis data sets. Multiple setups (TG, fluidised bed, drop tube, heated strip reactors) cover a wide range of conditions, with design optimisations and discussions of advantages (e.g. reduced secondary reactions) and limitations (kinetics vs. transport). Since biomass is often modeled as cellulose, hemicellulose, and lignin, single first-order kinetics for these components under flash pyrolysis complement TG data. With walnut shells, superposition works only at slow heating rates, while higher rates demand extended modelling. Temperature strongly influences primary gas products, and residence time affects secondary distribution. The data also support more sophisticated kinetic schemes that describe the devolatilisation of biomass on the basis of its detailed chemical structure.

## 2.1 Introduction

Pyrolysis describes the thermochemical decomposition of organic materials in the absence of oxygen. The supply of energy causes chemical bonds to break and to form a variety of different products. The pyrolysis of biogenic or fossil solid fuels produces light gases, liquid products (referred to as oil or tar), and a solid carbon-rich residue (char). The light gases mainly include $H_2O$, CO, $CO_2$, $H_2$, small hydrocarbons $C_xH_y$ with up to approximately four carbon atoms as well as different pollutant precursors, e.g. HCN, $NH_3$, $NO_x$ or $SO_2$, COS, and $CS_2$. Tars are defined as volatile products that are condensable under ambient temperature conditions and include mainly

O. Senneca
e-mail: osvalda.senneca@stems.cnr.it

D. König · B. Epple
Institute for Energy Systems and Technology, Technische Universität Darmstadt, Darmstadt, Germany
e-mail: dominik.koenig@est.tu-darmstadt.de

B. Epple
e-mail: bernd.epple@est.tu-darmstadt.de

oxygenated hydrocarbons such as alcohols, acids, and esters and long-chained and (poly)aromatic hydrocarbons, e.g. naphthalene. The remaining char particle is characterised by an increased carbon content and internal porosity. The char morphology (e.g. surface functional groups, specific surface, ...) strongly affects the char reactivity in the subsequent char conversion processes.

The product distribution and reaction kinetics are influenced by the biomass itself, i.e. the proportions of cellulose, hemicellulose and lignin, as well as the external boundary conditions of the pyrolysis process (temperature, pressure, heating rate, particle size, residence time, atmosphere, ...) [1,2]. The proportions of cellulose and hemicellulose mainly influence the quantity and composition of the gas and tar fraction, while the proportion of lignin essentially determines the amount of remaining char. Temperature is the factor that most strongly influences the quantity and composition of the individual fractions, as the degree of decomposition of the individual structural components is highly temperature-dependent. In general, a higher temperature leads to a higher conversion, which is synonymous with a decreasing char yield and increasing gas and tar yields. A high temperature favours both the primary pyrolysis processes and the secondary reactions in the gas phase. High temperatures cause the splitting of long-chain tar molecules into short-chain light gas molecules and thus shift the proportions of the two volatile fractions. A maximum yield of condensable pyrolysis products can be realised at moderate temperatures between 673 and 873 K and the shortest possible residence times. In addition to the cracking reactions, the residence time can also influence the recondensation of tars on the char matrix and thus increase the char yield at moderate temperatures and long residence times. A higher absolute pressure causes an increase in the char fraction at the expense of the tar fraction due to the increased required evaporation temperature of the metaplastic fragments [3,4]. Furthermore, the morphology of the char is strongly influenced by the pyrolysis conditions, especially temperature and heating rate [5–7]. During flash pyrolysis, the particles undergo plastic deformation, often becoming spherical and highly porous [8,9]. Especially, the porosity parameters are important in single particle combustion models [10]. However, details on the effects on single gas species or the reaction kinetics under flash pyrolysis conditions (temperatures above 750 K and heating rates larger than $10^3\,\mathrm{K\,s^{-1}}$) are still open questions. Especially for biomass, the question if, and under which conditions, the reaction behaviour of natural biomass can be described as a superposition of the reaction behaviours of three basic components is controversially discussed [11–13].

Within the present chapter, first, typical experimental setups (Sect. 2.2) for investigating the standalone pyrolysis process on particle scale are illustrated, limitations discussed, and specific improvements presented. These setups are then used to focus on different aspects of biomass pyrolysis, e.g. the difference to fossil fuels, important influencing factors, or the possibility to map the pyrolysis process of natural biomass by the superposition of the three reference components cellulose, hemicellulose and lignin. After focussing on the characterisation of volatile kinetics in Sect. 2.3 and volatile products in Sect. 2.4 (including minority pecies and tar components), Sect. 2.5 addresses the morphology of formed char particles. All those data build important references for the development of pyrolysis models with different complexity, presented in Chap. 8, e.g. the bio-chemical percolation devolatilisation (Bio-CPD) model or the CRECK-S-B model.

## 2.2 Experimental Setups

For the experimental determination of pyrolysis products and kinetics, a wide variety of setups exists that have different characteristics and thus offer advantages and disadvantages to different investigation objectives. The most frequently used reactor concepts include thermogravimetry (TG), fluidised bed reactors (FBR), drop tube reactors (DTR), and heated strip reactors (HSR). Table 2.1 provides an overview of the characteristics, typical boundary conditions and analysis methods of the four reactor concepts used in the present study. The following results show the importance of using complementary experimental setups to take into account different thermal boundary conditions in particular and to minimise the influence of the selected measurement methods as far as possible.

**Table 2.1** Typical boundary conditions and used measurement techniques for pyrolysis experiments

| | TG | FBR | DTR | HSR |
|---|---|---|---|---|
| Fuel feed | Batch-wise | Batch-wise | Continuously | Batch-wise |
| Sample mass (flow) | 50 mg | 15–50 mg | 0.5–5 $g\,min^{-1}$ | 100 mg |
| Reaction condition | Constant heat-up | Nearly isothermal | Non-isothermal | Non-isothermal |
| Temperature | Up to 1373 K | 573–1073 K | 950–1600 K | Up to 2300 K |
| Heating rate | 1–20 $K\,min^{-1}$ | $\approx 10^4\,K\,s^{-1}$ | $10^3$–$10^4\,K\,s^{-1}$ | $\approx 10^4\,K\,s^{-1}$ |
| Atmosphere | $N_2$, $O_2$ or $CO_2$ | $N_2$ | $N_2$, $CO_2$, Ar or $H_2O$ | $N_2$ or $CO_2$ |
| Gas flow rate (STP) | 0.1 $L\,min^{-1}$ | 5 $L\,min^{-1}$ | 20–50 $L\,min^{-1}$ | – |
| Particle residence time | Unlimited | Unlimited | 10–400 ms | Unlimited |
| Gas residence time (hot) | $\approx$ 60 s | 2–10 s | 10–400 ms | – |
| Measurement techniques | | | | |
| Char | Mass balance | Only by burning | Ash/mineral tracer | Near-infrared camera |
| Tars | – | Low moleculars via FTIR | FTIR/tar trap | External via GC/MS |
| Light gases | MS | FTIR | FTIR/GC | – |
| Time resolution | 0.1 s (balance) | 0.1 s (FTIR) | $\approx$ 40 ms (reaction dist.) | 3 ms (camera) |
| References | [14,15] | [16–18] | [19–21] | [22,23] |

### 2.2.1 Thermogravimetry

Thermogravimetry (TG) is a standardised characterisation method to provide insights into material composition, decomposition behaviour, and reactivity within pyrolysis, oxidation, or gasification processes. Within the analysis, a fuel sample with known initial sample mass $m_0$ is heated under controlled conditions in a high-sensitive balance, where the mass over time $m(t)$ and the corresponding temperature $T$ are recorded. From this data and the final mass $m_\infty$, the normalised first derivative of the TG curve is calculated:

$$\mathrm{DTG} = \frac{-\left(\frac{\mathrm{d}m}{\mathrm{d}T}\right)}{m_0 - m_\infty} \tag{2.1}$$

Typically, the sample is subjected to a temperature-time programme, characterised by a constant heating rate and a final temperature. However, also isothermal experiments are possible. Before performing the measurement, a blank measurement without a sample is needed with the same experimental conditions to compensate for buoyancy and flow forces due to changing gas density with increasing temperature. The laminar flow, on the one hand, controls the reaction conditions (pyrolysis, oxidation, gasification) by the gas composition and, on the other hand, removes reaction products preventing secondary charring reactions on the sample and allowing further analysis of the reaction products via gas-phase analysis (e.g. mass or FTIR spectroscopy).

### 2.2.2 Fluidised Bed Reactor

The fluidised bed reactor (FBR) approximates the concept of a perfectly mixed reactor, which provides constant and homogeneous reaction boundary conditions for the reacting fuel particles in terms of space and time. The main advantage of the FBR concept is the combination of high particle heating rates with an unlimited particle residence time at reaction conditions. Furthermore, the concept allows for the independent adjustment of temperature and reaction gas atmosphere resulting in flexible boundary conditions.

**Design**

The reactor system consists of three main parts as illustrated in Fig. 2.1: a gas supply from bottles, the fluidised bed reactor itself, and an ex-situ gas analyser using Fourier-transform infrared spectroscopy (FTIR). Its details are described in detail by Pielsticker et al. [16].

The gas supply is realised from bottles. Using mass flow controllers, any mixing ratios of $N_2$, $O_2$, and $CO_2$ can be set to provide not only reaction conditions for pyrolysis but also for char combustion and gasification as shown in Sects. 3.2 and 3.2.2.

The reactor comprises two coaxially mounted ceramic pipes fixed to a stainless steel head. The previously set gas flow passes from top to bottom through the annular gap between these pipes, where it is heated to the desired reactor temperature

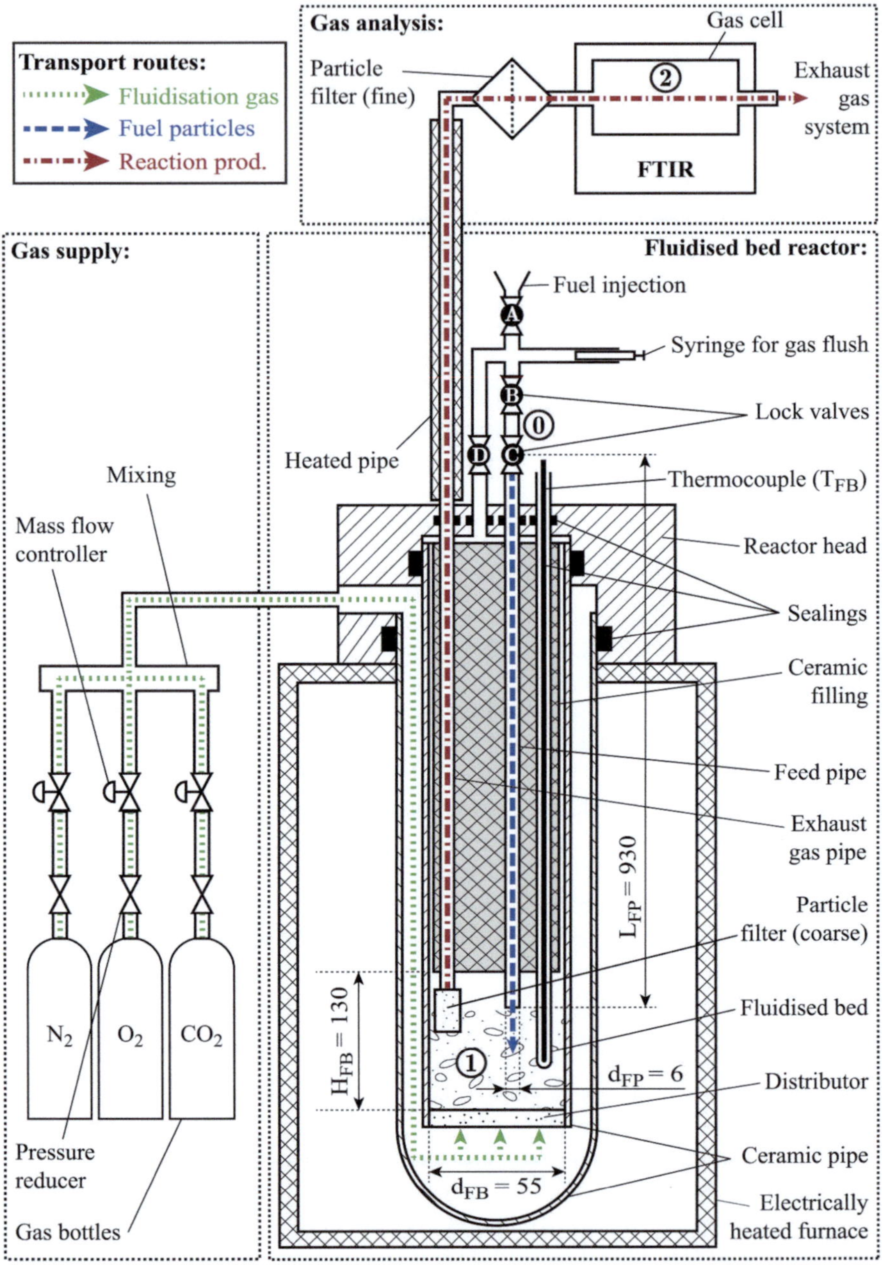


**Fig. 2.1** Scheme of fluidised bed reactor for reaction kinetic experiments: Batch samples are injected from the fuel lock (position ⓪) to the reaction zone (position ①), while the gaseous reaction products are analysed ex situ (position ②) reprinted from Pielsticker et al. [16], licensed under CC BY 4.0)

controlled by the electrically heated furnace. The flow enters the reaction zone from the bottom of the inner pipe through a porous quartz glass plate (distributor). The bed consists of inert $Al_2O_3$ particles and provides continuously well-stirred and thereby homogeneous boundary conditions for the investigated fuel particles. The pulverised fuel ($d_P \approx 50 - 200\,\mu m$) is injected batch-wise ($m_{Fuel} = 1 - 50$ mg) into the fluidised bed. A fuel lock system minimises the influence of contamination from ambient air. A flush-assisted feeding system ensures that particles do not stick to the feed pipe walls and results in quick fuel delivery to the bed with particle heating rates in the order of $10^4\,K\,s^{-1}$. Thereby, the onset of any reaction in the feed pipe before reaching the bed is minimised. Once reaching the bed, the particles react, and the released product gases mix with the fluidising gas flow and all gases leave the reactor through the exhaust gas pipe, fine and coarse filters, and a heated pipe before entering the FTIR gas cell. The volumes of sampling lines, filters, and the freeboard above the bed are reduced to a minimum to shorten the response time between the reaction zone in the bed and the FTIR gas cell [24].

Depending on the investigated fuel type, gas volume fractions $\varphi_i$ of up to 22 different gas species including CO, $CO_2$, $H_2O$, hydrocarbons ranging from $CH_4$ to $C_{10}H_8$, as well as multiple nitrogen- and sulphur-containing species are determined simultaneously [18] and are subject to a filtering routine to eliminate species erroneously predicted from the measurement noise [25].

**Evaluation Procedure**

With the known fluidisation volume flow $\dot{V}_{Fl}$ and the injected fuel mass $m_{Fuel}$, the mass-specific release rate can be calculated for each gas species from the measured volume fraction $\varphi_i$ obtained in the FTIR gas cell according to (2.2):

$$\frac{dy_i(t)}{dt} = \frac{\rho_{N_2}\, M_i}{M_{N_2}} \frac{\dot{V}_{Fl}}{m_{Fuel}} \frac{\varphi_i(t)}{1 - \sum_{j=1}^{N} \varphi_j(t)} \tag{2.2}$$

Figure 2.2a) shows exemplarily the contributions of $H_2O$, $CO_2$, CO, and other hydrocarbons ($C_xH_yO_z$) to the overall volatile release rate for one single pyrolysis experiment of cellulose (CEL) at a reactor temperature $T_{FB} = 823$ K.

To obtain kinetic parameters for fuel-specific empiric pyrolysis reaction models, the difference between experimental data and model predictions is minimised in the least-square fit, whereby characteristic parameters (e.g. the reaction rate) are adjusted. Figure 2.2b) shows the obtained model predictions with optimised parameters for two kinetic models, i.e. the single first-order reaction (SFOR) model and a 2-step model with the formation of an intermediate and subsequent decomposition as explained in detail by Pielsticker et al. [16]. The ex-situ obtained volatile release rates do not directly reflect the particle decomposition rate stemming from the pyrolysis reaction. During the gas transport from the reaction zone (see position ① in Fig. 2.1) to the gas analyser (position ②), dispersion and mixing take place that smears the observed signal compared to the original signal. However, this behaviour can be captured by a transfer function that is based on the considerations

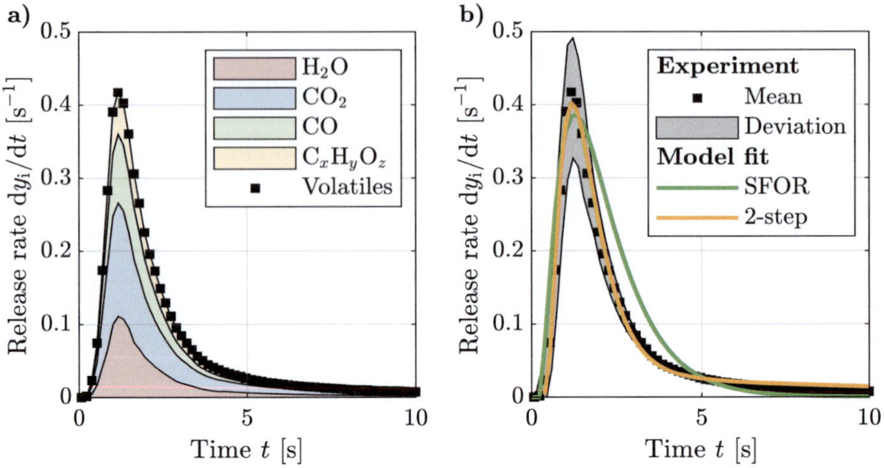


**Fig. 2.2** Single pyrolysis experiment of pulverised cellulose (CEL, $m_{\text{Fuel}} = 15$ mg) at $T_{\text{FB}} = 823$ K and with $\dot{V}_{\text{Fl}} = 300\,\text{L h}^{-1}$: **a)** contribution of different gases to the total volatile release rate and **b)** fit of empirical reaction kinetic models to the experimental data

by Abad et al. [26]. A detailed description of how this procedure is implemented for the present reactor system is given by Pielsticker et al. [17]. Alternatively, a 1-D gas transport model can be used that additionally considers secondary decomposition reactions (e.g. tar cracking) [27]. Results are shown in Sect. 8.3.2.

**Operational Limits**

The fluidised bed reactor system has two main control variables: the set temperature of the furnace, which controls the bed temperature $T_{\text{FB}}$ and the fluidisation volume flow $\dot{V}_{\text{Fl}}$. The technical, fluid mechanical, heat transfer, and gas transport limitations define the appropriate field of application for the reaction kinetics analysis of pulverised fuel particles as illustrated in Fig. 2.3.

The operation of the furnace is limited to a temperature $T_{\text{max}} = 1573$ K. No limit exists for a minimum operating temperature. The small ratio of sample to bed mass ensures that the bed temperature does not change significantly due to the released or absorbed reaction enthalpy throughout a single experiment. The volume flow, instead, is technically limited by the operating range of the mass flow controllers ($\dot{V}_{\text{t}} = 1.25$–$500\text{L h}^{-1}$). However, to achieve the desired movement behaviour of the bed particles, the fluidised bed regime must be maintained. The respective limiting volume flows for the fixed bed regime and the transition to the pneumatic transport regime can be estimated via the minimum fluidisation and the free-fall velocity, respectively. Suitable correlations for the present setup have been provided by Gövert [28].

The solid conversion process inside the fluidised bed is controlled by two mechanisms, the intrinsic reaction kinetics and the heat transfer from the gas phase to the particle as well as inside the particle. To directly derive reaction kinetics from the

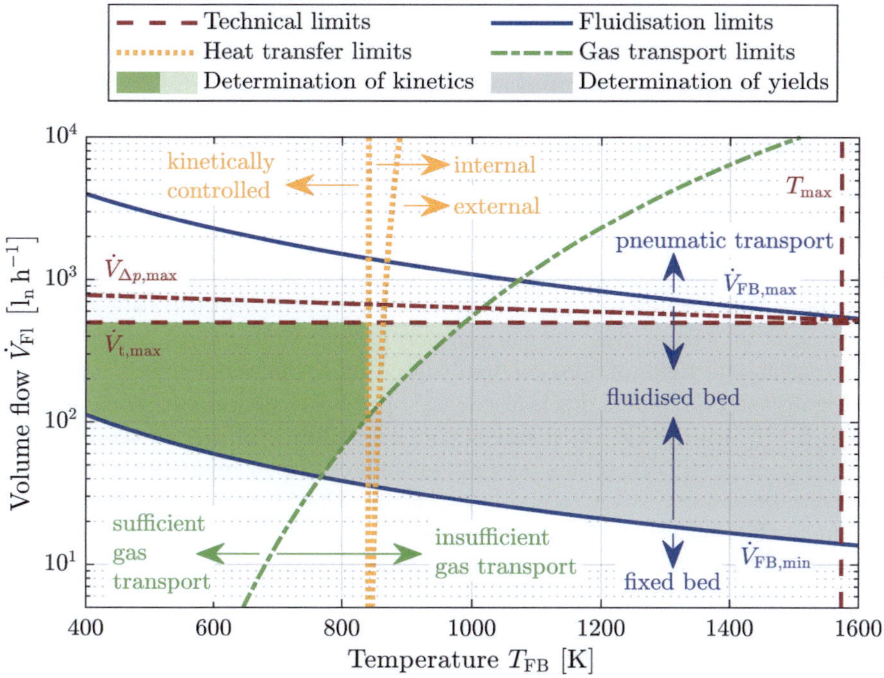


**Fig. 2.3** Operating range of the fluidised bed reactor taking into account technical, fluid mechanical, heat transfer, and gas transport limitations for the determination of reaction kinetic parameters and integrally released pyrolysis products: The fuel-specific reactivity is described with the SFOR model and kinetic parameters for beech wood (BW) from Pielsticker et al. [17] assuming a particle with $d_P = 100\,\mu m$

measured profiles, the particle must react purely in the kinetically controlled regime without any heat and mass transfer limitations. The regime can be estimated by the dimensionless pyrolysis numbers

$$Py_{int} = \frac{4\lambda_P}{\rho_P\, c_P\, d_P^2\, r} \quad \text{and} \quad Py_{ext} = \frac{2\alpha}{\rho_P\, c_P\, d_P\, r} \tag{2.3}$$

that give the ratio of characteristic time scales of the reaction process compared to the heat transfer process inside the particle and externally, respectively [29]. Herein, the thermophysical properties of the particle are described with thermal conductivity $\lambda_P$, the density $\rho_P$ and the heat capacity $c_P$, the heat transfer is characterised by the convective heat transfer coefficient $\alpha$ (correlation derived by Gunn [30]), and the reactivity is represented via the characteristic reaction rate $r$ of the SFOR model. By neglecting radiative heat flows, this represents a conservative approximation. Figure 2.3 reveals that for temperatures below 850 K, the assumption of purely kinetic control is fulfilled. However, for larger particles or higher reactivities, the regime of purely reaction kinetic control shifts towards lower temperatures and vice versa.

The last limitation for the determination of reaction kinetics is the gas transport from the reaction zone to the FTIR gas cell. To be able to clearly determine a characteristic reaction rate from the measured profile, this profile must differ significantly from profiles of other reaction rates. For high reaction rates, the shape of the measured profile is no longer determined by the rate of the reaction, but by the velocity of the gas transport. Different reaction rates are then no longer distinguishable from each other, taking into account the measurement noise, and consequently, no reliable kinetic data can be extracted under these conditions [24]. The green dashed line in Fig. 2.3 splits the field of operation into two areas: In the green area, the gas transport is sufficiently faster than the reaction and thus kinetic data can be derived. By contrast, in the grey area, the gas transport is much slower than the reaction and thus only integral yields of released pyrolysis products can be determined here. Like the heat transfer limitation, this is also a fuel-specific limitation, which is estimated in Fig. 2.3 by using the reaction kinetic parameters for beech wood from the study by Pielsticker et al. [17]. Within the light green area, the gas transport is sufficient, however, the reaction is limited by heat transfer. This is seen in the experimental data in an effective decrease in the reaction rate compared to the expected intrinsic reaction rate [16] and must be taken into account, e.g. by a particle-resolved model.

### 2.2.3 Drop Tube Reactor

The laminar drop tube reactor (DTR) for pyrolysis experiments with temperatures up to 1573 K and heating rates in the range $10^4$–$10^5\,\mathrm{K\,s^{-1}}$ approximates conditions typical for pulverised fuel combustion. The DTR allows for variation of the reactor atmosphere to collect pyrolysis products (solid, liquid, gaseous) at different pyrolysis stages. The DTR has already been used in servals studies for coal [31,32] and biomass [33,34].

**Design**

The reactor system of the DTR consists of three main parts that are shown in Fig. 2.4. The first part is the gas supply system, the second one the DTR and the third one the pyrolysis product analysis system.

The gas supply system provides $CO_2$, $N_2$, Ar and $H_2O$ via gas bottles and mass flow controllers. It also delivers the gas ($N_2$ and $CO_2$) to feed the particles to the reactor and the particle quenching gas ($N_2$) for the sampling probe. In addition, steam can be dosed into the reactor via a connected vaporiser.

The drop tube reactor block consists mainly of fireclay blocks in which the hot gas is supplied through a canal from a microwave to the injector and the reaction zone. Eight silicon carbide heating rods around the fireclay blocks control the reactor wall temperature. The blocks are enclosed in a casing made of stainless steel plates, which is used for atmospheric sealing and for fastening all equipment to the reactor. The particles fall from the feeding system (0.5–5.0 g/min) through the air-cooled injector to the reaction zone. The feeding system is enclosed in a gas-tight acrylic glass housing and consists of a silo, a screw conveyor and a vibrating chute (reduces

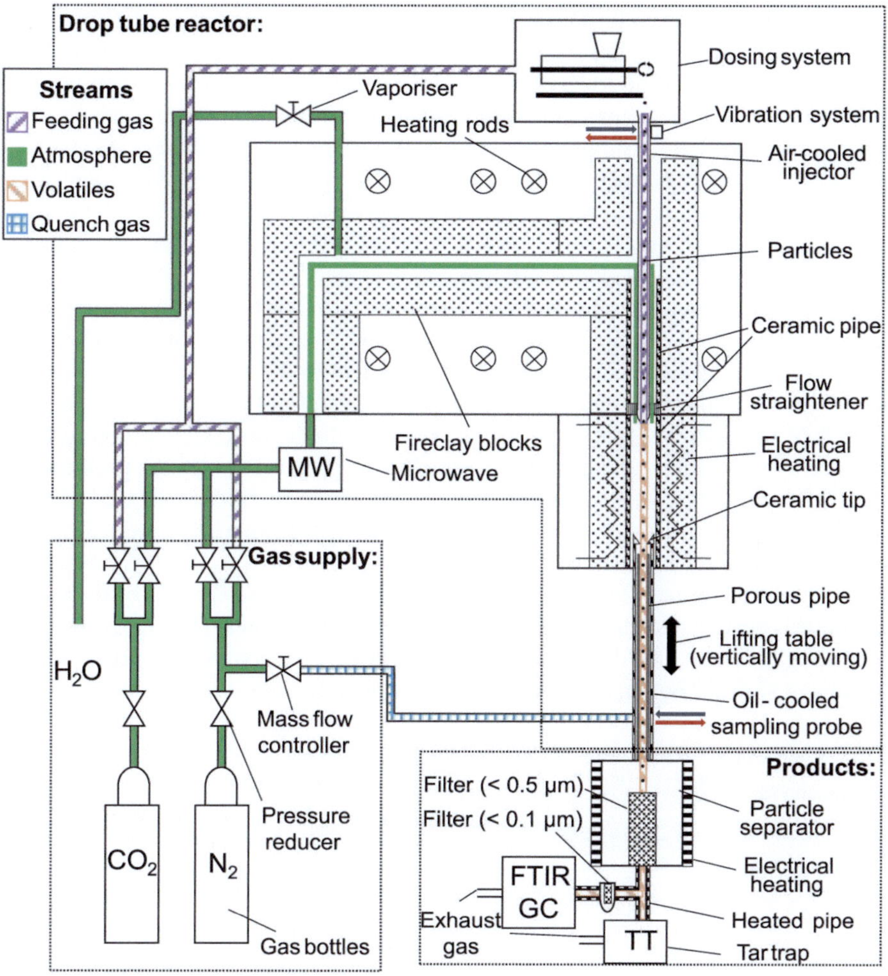


**Fig. 2.4** Scheme of the drop tube reactor with all necessary and optional gas flows

the flow fluctuations of the screw conveyor). A gas volume flow (experimental atmosphere) is supplied via a flow regulator, which inertises the housing and conveys the particles through the injector. To prevent particles from clogging, two vibration units are attached at the top of the injector. The injector is 750 mm long and cooled with compressed air to approximately 475 K, controlled by a thermocouple at the tip of the injector. The cooling prevents the fuel from heating up to pyrolysis temperatures before entering the reaction zone. In the electrically heated reaction zone, the particles start to pyrolyse and are collected by a sampling probe.

The sampling probe is part of the pyrolysis product analysis system. The multi-walled probe is mounted on a lifting table and can be adjusted vertically in height to set the reaction distance, with a maximum distance of 320 mm. In the sampling probe, the pyrolysis products (char, tar, gas) are quenched with $N_2$. A particle separator is connected with the sampling probe, which is heated to approximately 423 K via a jacket heater during the experiments to prevent the condensation of tars. A filter (membrane consisting of $Al_2O_3$ fibres and SiC particles with a separation limit of 0.5 μm) is fitted in the centre and separates the char from the gaseous products. The particle-free gas flow is supplied via a pipe to a heated adapter, where the exhaust gas passes through a further filter (glass fibre) to the gas analysers and through a tar trap to the exhaust gas system. Pumps and rotameters are connected downstream of the gas analysers and the tar trap, which are used to force the process gas into the probe and regulate the gas volume flows.

Compared to coal, the physical and chemical properties of biomass complicate the collection with the sampling probe. Due to their lower density and lower sphericity, biomass falls less linearly and is more sensitive to the gas flow in the reactor. Furthermore, biomass particles are sticky due to metaplast formation between 50 % to 90 % of the pyrolysis mass loss. Upon contact with the oil-cooled probe surface, the pyrolysed particles tend to solidify and clog the probe. To solve the problems of unstable flight characteristics and sticking at the probe tip, its design was adapted (see Fig. 2.5 comparison of old sampling probe for coal experiments [31,32] (left) and the new one for biomass experiments [33,34] (right)).

To reduce the probability of particles hitting the tip and inner walls of the probe, the probe is designed to have the largest possible inner diameter. The disadvantage of increasing the inner tube diameter is the higher amount of hot gas aspiration and the associated slower cooling of the collected products. In addition, to achieve a more

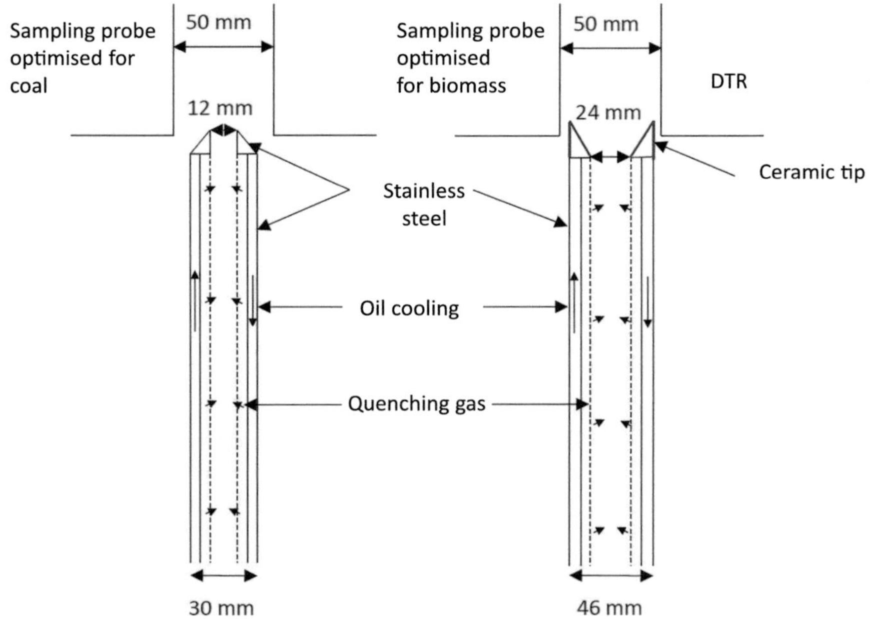


**Fig. 2.5** Modification of the old sampling probe for coal experiments (left) to the new sampling probe (right), optimised for biomass experiments

even flow of the gas and to reduce the contact of the fuel with the tip, the tip of the sampling probe was modified. The new tip is made of ceramic and can be placed loosely on the probe and heats up via the radiation from the reactor walls and the hot gas. The funnel-shaped inlet of the ceramic tip is intended to reduce the contact of the volatile components with the sampling probe.

Due to the design of the reactor, there are some operational limits which must be taken into account. The maximum reaction distance is set to 320 mm, which limits the achievable residence times to less than 1 s. The maximum particle size is limited by the diameter larger than the injector (6 mm). Elongated particles with a high aspect ratio and a particle diameter larger than 400 μm (sieve fraction) are almost impossible to feed. The maximum operating temperature of the reactor is limited to 1600 K. Lastly, the experimental conditions of the DTR do not allow a reliable evaluation of any tars with a molecular weight above 92 g mol$^{-1}$, because they cannot be detected by the FTIR gas analysers, nor can they be collected in the tar trap. Some tars have boiling points above the temperature of the inner probe tube and condense in the sampling probe or in the particle filter.

**Evaluation Procedure**

The volatile yield indicates how far pyrolysis has progressed and is defined as the initial mass of the material that is converted into volatile compounds during the pyrolysis process. The volatile yield is calculated by the collected chars at different reactor distances. As already mentioned, the reactor was initially used for coal and the widely practised ash tracer method was used to calculate the volatile yield. This method is not trivial for biomass because of the low ash content and high volatility of some inorganic species. A further development of the ash tracer was needed. The mineral tracer method aims to consider only thermally stable elements of the ash as tracers. Elements that have already been used as tracers in the literature [35,36] are: titanium, silicon, aluminium, calcium, magnesium and iron. For the lignocellulosic biomass, calcium is suitable as a tracer, as its concentration in the raw biomass is higher compared to titanium and magnesium. With calcium as a tracer, the ash-free volatile yield is calculated as follows:

$$y_V = \frac{1 - \frac{w_{Ca,raw}}{w_{Ca,py}}}{1 - w_{ash,raw}} \tag{2.4}$$

with the calcium mass fractions $w_{Ca}$ of raw and pyrolysed samples and the ash content $w_{ash,raw}$ of the raw biomass. The calcium content is detected with a *NITON XL2 air plus* X-ray fluorescence spectrometer, while the ash content is determined from the proximate analysis. This method has already been successfully used for various biomass samples [33,34].

### 2.2.4 Heated Strip Reactor

The heated strip reactor (HSR) is designed for conversion experiments under high heating rates. Figure 2.6 shows a schematic drawing of the setup including the gas supply and the reactor itself. The main advantage of the HSR setup is that the reaction products are released from the hot particles in a cold environment and thus secondary pyrolysis reactions are largely suppressed. This qualifies the HSR in particular for detailed tar analyses (see Sect. 2.2.5).

**Design**

The core of the setup is a graphite strip located between two copper blocks at both ends. By passing electric power through the graphite strip, it is heated according to the Joule effect reaching up to a heating rate of up to $10^4$ K s$^{-1}$ [22]. Temperature monitoring is performed through quartz windows. The maximum detectable temperature of 1800–1900 K slightly depends on the equipment used: either a two-colour *LAND* infrared pyrometer (1.5–5 μm) in combination with a high-speed near-infrared camera [22] or an infrared *thermoMETER CT* pyrometer from *Micro-Epsilon* [23]. Due to the high heating rate, the strip can be considered isothermal throughout the pyrolysis process. However, as the particle heating typically occurs only via the contact points to the strip, an inhomogeneous temperature profile within the particle is the result. Thus, experiments from the HSR are not considered for the determining reaction kinetics but for detailed product analysis only.

**Experimental Procedure**

Before initiating the heating process, the stainless-steel vessel (volume 2 L) is flushed with either inert gas (He/$N_2$) or $CO_2$ to eliminate any residual air, ensuring a controlled atmosphere during the experiment. Then samples, typically in the form of

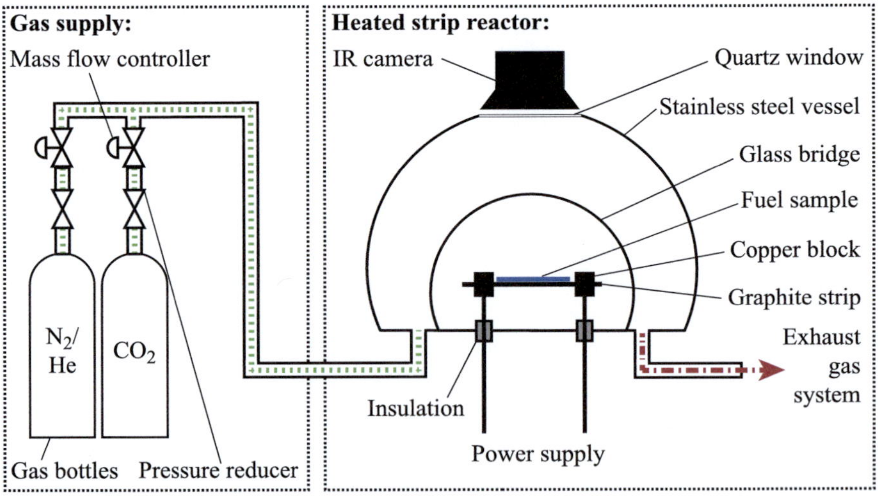


**Fig. 2.6** Scheme of the heated strip reactor

solid fuel particles, are placed directly on the graphite strip. Tars formed during the heating process are collected using a borosilicate glass bridge (Pyrex) positioned above the strip. The glass, being nearly transparent to thermal radiation, remains cool, allowing for tar condensation on its surface. These tars are subsequently dissolved for external analysis as decribed in Sect. 2.2.5.

### 2.2.5 Analysis of Captured Tars

Both, the DTR and HSR setups are equipped with collection devices for the released tars. Before being subject to various analysis techniques, the condensed products of pyrolysis are dissolved, partly separated, and classified into different product groups: hydrocarbons with low molecular weight (light tar), larger hydrocarbons (heavy tar), and fine carbonaceous particles (soot) of nano and even micrometric size for coal samples. Figure 2.7 summarises analytical methods for the chemical and structural characterisation of captured tar samples. Solid phase analysis of soot particles is not described in the present chapter but can be found in different literature studies [37, 38].

While the light tars from coal or biomass pyrolysis can be effectively detected by gas chromatography combined with mass spectrometry (GC-MS), a quantitative analysis of heavy tars turned out to be susceptible to errors, especially when the techniques originally developed for fossil fuels-derived products are applied to the oxygen-rich biomass products. Size exclusion chromatography (SEC) coupled with a UV-VIS absorption detector is applied to both coal and biomass heavy tars, providing evidence of components with molecular weights of thousands of u, similar to incipient soot. In some coal pyrolysis products, carbon particles reach the micrometric scale,

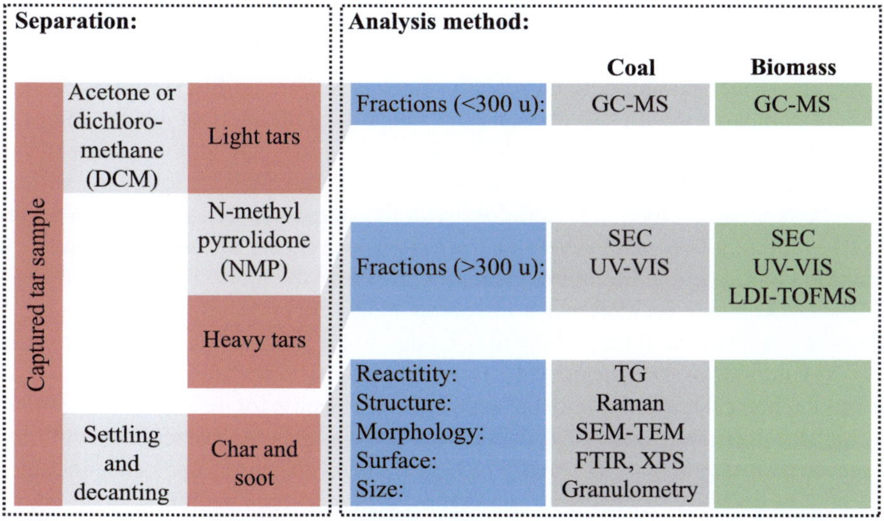


**Fig. 2.7** Tar separation and analysis techniques for coal and biomass

so that ad hoc solvent-based separation procedures are developed to distinguish soot from char fines [37,38]. In biomass pyrolysis products, carbon-rich particles remain confined to the nano-size range.

The molecular mass and size of the heaviest species in biomass pyrolysis products are better assessed by SEC and a UV-VIS fluorescence detector. However, the technique still requires refinement due to the lack of SEC calibration data valid for all tar species. Laser desorption ionisation time of flight mass spectrometry (LDI TOFMS) turned out to be a valid support to evaluate the molecular range of such ill-defined products.

## 2.3 Reaction Kinetics

Reaction kinetics describe the time-dependent reaction progress. Kinetic data are derived for the extracted and separately investigated biomass components cellulose, hemicellulose, and lignin as well as naturally available biomass, e.g. walnut shells.

### 2.3.1 Extracted Biomass Components

Reaction kinetics of biomass components can be found in the literature for low heating rates obtained in TG experiments (e.g. by Orfão et al. [39] or Yu et al. [40]). They are either determined individually for extracted biomass components or calculated with a parallel reaction model approach from natural biomass. The study by Pielsticker et al. [16] investigated the kinetic behaviour of the extracted components under flash pyrolysis conditions realised with the FBR and related them to the TG-derived data. These kinetic data are the basis for the analysis of the superposition of individual component kinetics to describe the reaction behaviour of natural biomass in Sect. 2.3.2.

Figure 2.8 depicts the derived reaction rates assuming an SFOR model for the FBR in comparison with TG data and the range of literature data representing the detailed data out of the study by Pielsticker et al. [16]. The data obtained in the study fit well into the overall picture of the literature data, with the following noticeable characteristics emerging. Cellulose with its comparatively simple linear structure has already been extensively characterised and shows the smallest spread of reaction kinetics in the literature. Generally, it shows the highest activation energy ($E_a = 124\,\mathrm{kJ\,mol^{-1}}$ for FBR) across all components. For hemicellulose, the literature data show a less clear trend, which may be due to the different hemicellulose types (e.g. xylan, mannan, or glucan) [16]. Especially at low temperatures, hemicellulose shows the highest reactivity and thus mostly is responsible for the onset of reactions. At higher temperatures, the FBR data reveal two distinct reaction regimes with different activation energies (77 and 37 kJ $\mathrm{mol^{-1}}$). As the lignin molecule includes the highest diversity of chemical bond types, also reported reactions kinetics show the highest spread in the literature, which is most likely related to the biological origin of the lignin material. However, the SFOR model in most cases is incapable

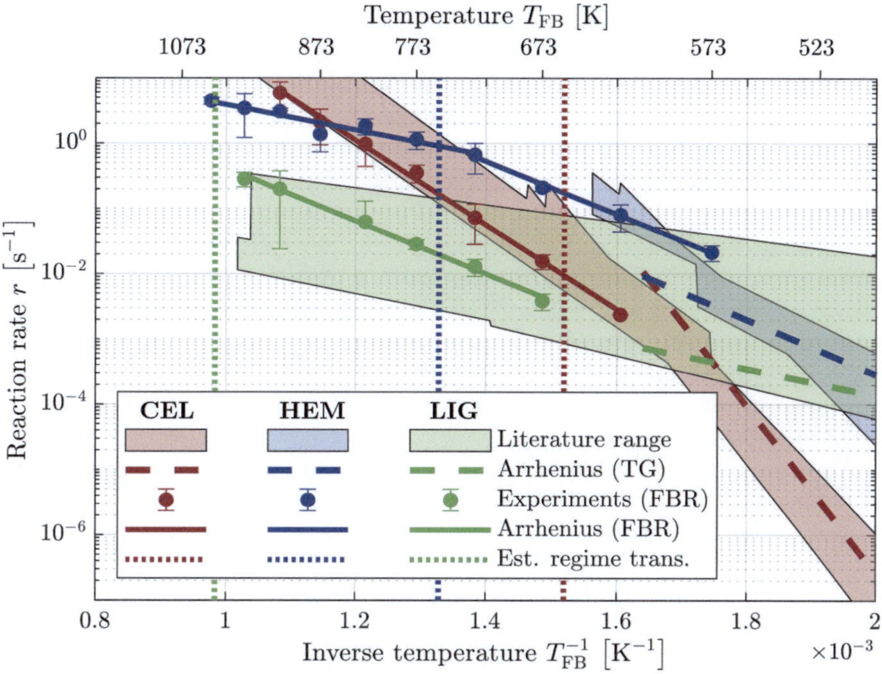


**Fig. 2.8** Arrhenius diagram of pyrolysis kinetics of extracted biomass components derived from experiments in the fluidised bed reactor. The literature range represents the detailed data provided by Pielsticker et al. [16]. Reaction rates are derived assuming an SFOR model. The expected regime transition between reaction kinetic control and heat transfer limitation is calculated with the Py numbers from (2.3)

of capturing the diversity of decomposition reactions taking place in parallel and subsequently and thus the observed kinetic parameters ($E_a = 78\,\mathrm{kJ\,mol^{-1}}$ for FBR) from the SFOR model should only be used for a rough estimation of affected reaction time scales [16]. The reaction kinetics of cellulose (CEL) and hemicellulose (HEM) show different regimes characterised by different activation energies. The transition temperatures are $T \approx 600\,\mathrm{K}$ for cellulose and $T \approx 750\,\mathrm{K}$ for hemicellulose, respectively. Both transitions can be traced back to the transition from the reaction kinetics-controlled regime to the heat transfer limitation, characterised by the pyrolysis numbers from (2.3). For lignin, by contrast, the transition temperature is not exceeded in the experiments and consequently, no change in the activation energy is visible.

### 2.3.2 Natural Biomass—Walnut Shells

In contrast to the extracted biomass components, natural biomass consists of a mixture of the three components and additional extractives. The coexistence of the components may lead to interactions during the conversion process. So far, this analysis has been carried out via thermogravimetric analysis (e.g. in the studies from Zhu and Zhong [11], Yu et al. [40], or Zhao et al. [13]). While, e.g. Zhu and Zhong [11] observed a suitable superposition for synthetically mixed samples, other studies (e.g. Zhao et al. [13]) identified significant interactions between the components in the mixture. Intraparticular gas-phase reactions with metaplastic fragments and the effect of catalytically active minerals were named as possible causes by Couhert et al. [12].

The study by Ontyd et al. [21] investigates the pyrolysis kinetics of pulverised walnut shells (WNS) under various thermal boundary conditions in a drop tube reactor, a fluidised bed reactor, and a TG. The results are compared with the kinetics obtained in Sect. 2.3.1 for the extracted components to evaluate the relevance of interactions and to assess the superpositionability for different thermal boundary conditions. Figure 2.9 shows the experimentally obtained reaction rate assuming an SFOR approach.

All measured reaction rates fit well into the range of available literature data for walnut shell pyrolysis kinetics obtained in different reactor systems. A direct comparison of the kinetics of the three reactors shows a trend where activation

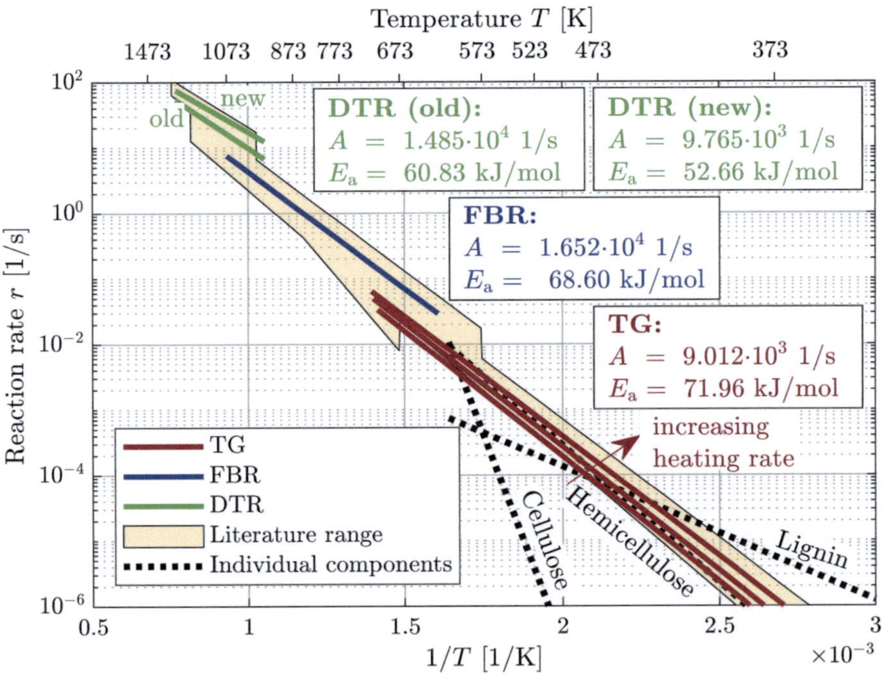


**Fig. 2.9** Arrhenius diagram of pyrolysis kinetics of pulverised walnut shells (WNS) derived with different reactor setups. The literature range represents the detailed data provided by Ontyd et al. [21]. Reaction rates are derived assuming an SFOR model. The individual component kinetics belong to the TG experiments by Pielsticker et al. [16]. Compared to the old data set [21], the new data set [33] takes into account additional experiments with longer residence times so that the pyrolysis process can be fully captured for all temperatures

### 2.4.1 Effect of Temperature and Residence Time

The effect of temperature and residence time are exemplarily investigated for walnut shell particles pyrolysed in the fluidised bed and the drop tube reactor. Within the FBR, the temperature varies between 623 and 1073 K, while the DTR is operated with two different wall temperatures to investigate five different reaction times each. Figure 2.11 shows the mass-specific composition of the volatiles categorised in single gas species ($H_2O$, $CO_2$, CO) and lumped components ($C_xH_y$ for light hydrocarbons, $C_xH_yO_z$ for oxygen-containing hydrocarbons, and heavy tars for molecules with molecular mass $M > 70\,\mathrm{g\,mol^{-1}}$.).

At low temperatures, the product spectrum is dominated by the small molecules $H_2O$ and $CO_2$, which stem from the depolymerisation reactions taking place in the cellulose and hemicellulose molecules. With increasing temperature, the decomposition of the resulting fragments (oligomers/monomers, e.g. levoglucosan in the case of cellulose) is responsible for the release of oxygenated hydrocarbons including e.g. aldehydes, ketones and furans [2]. By further increasing the temperature, the fragments are further decomposed and form small hydrocarbon light gases $C_xH_y$, carbon monoxide and hydrogen. This decomposition is accelerated by an increase in both temperature (as seen in the FBR experiments) and gas residence time (DTR experiments). The more than ten times higher gas residence time in the FBR compared to the DTR (compare Table 2.1) leads to significantly stronger secondary pyrolysis reactions, especially indicated by the absence of $C_xH_yO_z$ and heavy tars.

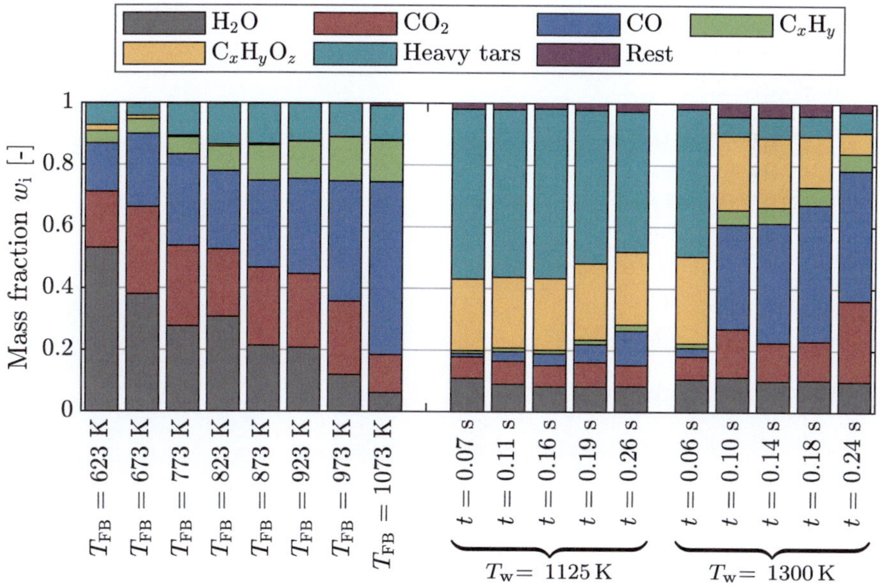


**Fig. 2.11** Mass-specific composition of volatiles from walnut shell (WNS) pyrolysis at different bed temperatures $T_{FB}$ in the FBR [21] and at different residence times for two wall temperatures $T_w$ in the DTR [33]

The importance of including secondary tar cracking reactions in the FBR modelling is shown in Sect. 8.3.2.

### 2.4.2 Effect of Fuel Type

Fig. 2.12 shows pyrolysis yields obtained experimentally in the fluidised bed reactor at a bed temperature $T = 973$ K for a large variety of different fuel types spanning from natural biomass over pretreated biomass to fossil fuels. Basic analysis data are given in Table 1.1.

Of the three natural biomass, beech wood (BW) has the highest yield of volatiles, while walnut shell (WNS) shows the lowest yield. While the high levels of cellulose in beech wood and hemicellulose in miscanthus are responsible for a high volatile yield, the high lignin content in walnut shells preferentially promotes char formation [41]. The comparably long residence times of released volatiles in contact with the hot bed particles lead to the detection of relatively small quantities of long-chain or aromatic hydrocarbons (tars, included in the „other" category) for all fuels as these components undergo secondary cracking reactions.

The pretreatment of the raw natural biomass via torrefaction leads to a partial decomposition of some of the biomass components in the pretreatment process. Torrefaction at 553 K ($TBW_{553}$) mainly decomposes the hemicellulose fraction, while torrefaction at 573 K ($TBW_{573}$) leads additionally to a partial decomposition of cellulose. This is reflected in the pyrolysis yields, which decrease with increasing degree of torrefaction. The study by Pielsticker et al. [25] further revealed decreasing CO,

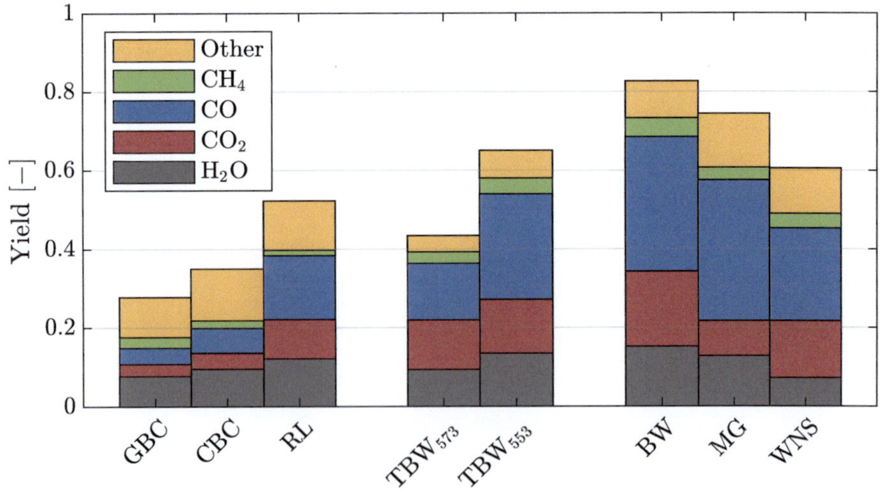


**Fig. 2.12** Obtained volatiles during pyrolysis experiments in the fluidised bed reactor at 973 K for different fuel types: German bituminous coal (GBC) [42], Colombian bituminous coal (CBC) [18], Rhenish lignite (RL) [43], torrefied beech wood (TBW, subscript indicates torrefaction temperature, 30 min residence time) [25], beech wood (BW) [25], miscanthus giganteus (MG) [25] and walnut shells (WNS) [21]

$C_2H_x$ and tar fractions. However, the composition of the condensable fraction was not significantly influenced. The torrefaction pretreatment shifts the pyrolysis behaviour towards those of fossil fuels, especially those of Rhenish lignite (RL). This in combination with improved properties concerning energy density, grindability and storage makes torrefied biomass a good alternative for replacing fossil fuels (especially lignite) in pulverised coal combustion power plants [43]. To achieve similar behaviour to that of bituminous coals, significantly higher degrees of torrefaction would be required, which would be associated with an economically too high mass loss during torrefaction.

### 2.4.3 Formation of Minority Species

In the previous section, the effect of temperature on the species contributing to the main mass loss is discussed. In addition to these species, so-called minority species are released. Although they do not account for a significant fraction of the volatile mass, they are regarded as important precursors for the formation of pollutants and include sulphur-, nitrogen-, and chlorine-containing species.

To experimentally characterise the release behaviour of those minority species, an electrically heated entrained flow reactor (in the concept similar to the DTR described in Sect. 2.2.3) with a total length $L = 5.92\,\mathrm{m}$ is used [44]. At fixed distances—reflecting different residence times—gas samples are taken and analysed with an FTIR to quantify the nitrous species and a mass spectrometer to derive concentrations of sulphur and chlorine species. The element-specific yield gives the contribution of species $i$ to the total injected mass of element $e$ (N, S, Cl) and is calculated by (2.5)

$$y_{\mathrm{i,e}} = \frac{\dot{V}\,\dfrac{p}{R\,T}\,\dfrac{1}{1-\varphi_{\mathrm{N_2}}}\,N_{\mathrm{e,i}}\,M_{\mathrm{e}}\,\varphi_{\mathrm{i}}}{\dot{m}_{\mathrm{Fuel}}\,w_{\mathrm{e}}}, \tag{2.5}$$

wherein $\varphi_{\mathrm{i}}$ is the measured volume fraction, $N_{\mathrm{e,i}}$ is the number of atoms of the element $e$ in the molecule and $M_{\mathrm{e}}$ is the elemental molar mass. The yield gives the fraction of exiting mass (calculated with volume flow $\dot{V}$, pressure $p$, temperature $T$ and universal gas constant $R$) to the injected elemental mass (calculated with fuel mass flow $\dot{m}_{\mathrm{Fuel}}$ and elemental fraction $w_{\mathrm{e}}$).

Figure 2.13 shows the nitrous, sulphur and chlorine species release for pyrolysis experiments in $N_2$ and $CO_2$ atmospheres with walnut shells (WNS) as fuel at different reactor temperatures for a fixed residence time of $t \approx 0.5\,\mathrm{s}$ [44].

Regarding the N-containing species, nitric oxides (NO and $NO_2$) are favourably formed at lower temperatures. In comparison, higher temperatures lead to the formation of $NH_3$ and HCN. Additionally, the overall amount of detected N in the four species decreases with increasing temperature. Most likely, non-detectable $N_2$ is formed. The switch from an $N_2$ to a $CO_2$ pyrolysis atmosphere generally leads to a reduction of the nitrous species yields, whereby especially the NO yield drops more than 28 %. The lower NO yields in the $N_2$ atmosphere can be explained with

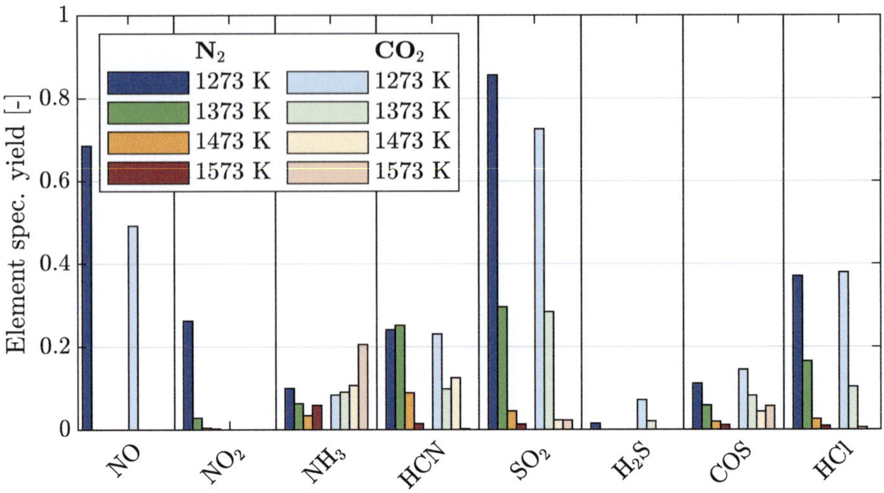


**Fig. 2.13** Release of nitrogenous, sulphur and chloride species from walnut shells (WNS) pyrolysed at different temperatures in $N_2$ and $CO_2$ atmospheres (based on the publication by Yildiz et al. [44])

the mechanism given by Wang et al. [45], where NO and $NO_2$ react with carbon from the char to form $N_2$ according to (2.6) and (2.7):

$$NO + C \rightarrow \frac{1}{2}N_2 + CO, \quad (2.6)$$

$$2NO + C \rightarrow N_2 + CO_2. \quad (2.7)$$

Within the $CO_2$ atmosphere, also more CO from gasification reactions is present, which changes the reaction pathways for the two nitric oxides to

$$NO + CO \rightarrow \frac{1}{2}N_2 + CO_2, \quad (2.8)$$

$$NO_2 + CO \rightarrow NO + CO_2. \quad (2.9)$$

ending up in an overall stronger reduction of $NO_x$. By contrast to the other N-species, $NH_3$ tends to show higher yields and an opposite temperature dependency in a $CO_2$ atmosphere.

For the sulphur-containing species depicted in Fig. 2.13, a strong temperature dependency is observable, whereby all three species considered decrease with increasing temperature. For all species, the decrease occurs similarly in $N_2$ and $CO_2$ atmospheres, which makes sense when considering the reaction mechanisms for $SO_2$, $H_2S$, and COS, e.g. given by Chardonneaua et al. [46] or Hildor et al. [47]:

$$2H_2S + SO_2 \rightarrow 3S + 2H_2O, \quad (2.10)$$

$$CaO + COS \rightarrow CaS + CO_2. \quad (2.11)$$

As it can be noticed, the species $N_2$ and $CO_2$ are not involved in the reactions. Both equations shift the equilibrium state for higher temperatures to the product side, where species are formed that are non-detectable with the measurement setup explaining the overall decrease of detected sulphur with increasing temperature.

Hydrogen chloride is shown in Fig. 2.13 as example for a Cl-containing species. The formation of HCl decreases in both atmospheres with increasing temperature, whereby the change in atmosphere from $N_2$ to $CO_2$ slightly favours the decomposition of HCl. In the study by Wei et al. [48], it is shown that with increasing temperature, potassium and chlorine are released from biomass. At high temperature, both react to KCl. But with increased temperature also the formation of calcium dichloride $CaCl_2$ and iron dichloride $FeCl_2$ is increased. This can lead to a fast decrease of HCl formation.

The formation of nitrogenous, sulphur and chloride species under nitrogen and carbon dioxide atmospheres shows two different main dependencies: The oxygenated compounds, like $NO_x$ and $SO_2$, show a high dependency on the temperature and for the nitrogen-based pollutants, also a high dependency on the atmosphere. The hydrogenated compounds, e.g. $NH_3$ and $H_2S$ are less sensitive to temperature and tend to show higher yields in a $CO_2$ atmosphere.

Besides the temperature, the release of minority species depends on the fuel type and especially the present sulphur, nitrous and chlorine contents, whereby large differences between different fuel types have been observed. For torrefied poplar, a biomass with similar S, N, and Cl contents (cf. Table 1.1), very similar yields and trends can be observed for both fuels [49]. However, a comparison with fossil fuels, e.g. a sulphur-rich coal from the US (USHS), reveals significant differences in species yields and trends [50]. While the formation of $H_2S$ is almost not present for biomass, $H_2S$ is the dominant sulphur species in coal pyrolysis [50]. Further on, almost all released S-containing species show an increase with temperature contrary to the observed trends for biomass. This can essentially be explained by the fact that a significant proportion (80–90 %) of the sulphur in coal is bound in the char, while for biomass, the overall char yield and thus also the amount of sulphur in the char are low. The same trends and explanations are valid for Cl, where for coal experiments maximum of 15 % of the chlorine can be found in the gas phase [50].

### 2.4.4 Condensable Tar Products

From the reactor setups described in Sect. 2.2, especially the drop tube reactor (DTR) and the heated strip reactor (HSR) are well-suited for the collection of condensable products. Due to the long gas residence times in the fluidised bed reactor (see Table 2.1), the tar components undergo cracking reactions resulting in high light gas yields and only small amounts of condensable products, as shown in Fig. 2.11. Thanks to the immediate cool-down in the cold gas environment in the HSR, the captured tars can be, in a first approximation, seen as primary tars. At the same time, the product spectrum in the DTR indicates the onset of secondary reactions. From different feedstocks, including the single extracted components and natural biomass,

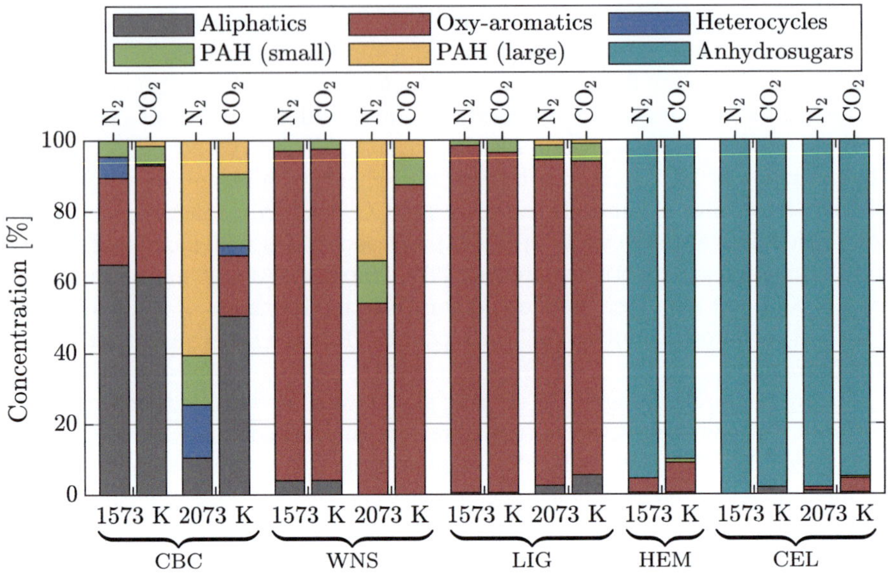


**Fig. 2.14** Lumped composition of tars from heated strip reactor experiments at different temperatures and gas atmospheres for Colombian bituminous coal (CBC), walnut shells (WNS), lignin, hemicellulose, and cellulose (based on the publication by Cerciello et al. [51])

samples of light and heavy tars are collected and analysed according to the procedure described in Sect. 2.2.5. To better compare the results, the species identified in the light tar from the chromatograms are grouped in classes of interest [51].

Figure 2.14 shows species groups detected in HSR experiments for different fuel types and atmospheres. For coal light tar, aliphatic and oxy-aromatic species are—independent of the reaction atmosphere—predominant in the primary tar produced in the HSR at 1573 K. For higher temperatures, the product spectrum shifts to polycyclic aromatic hydrocarbons (PAH), whereby the $N_2$ atmosphere is more affected than the $CO_2$ atmosphere. For comparable experiments in the DTR (data not shown), light aliphatics and PAH are identified as the main products, indicating the ongoing decomposition process.

The primary tar of natural biomass (walnut shells, WNS) strongly differs from that of coal. Instead of aliphatics, oxy-aromatic compounds dominate the product spectrum at lower temperatures. For higher temperatures, also heavy PAH are formed, whereby the same trend as for coal experiments regarding the atmosphere is visible. In the case of the lignin-rich walnut shells, oxo-aromatics were mainly constituted of methoxy-aromatic compounds (syringaldehyde, vanillin, eugenol and coniferyl alcohol) that can be related to the lignin structure [51,52]. However, the results also show that the pyrolysis in natural biomass is much more complex than the individual pyrolysis of the three extracted components.

In the case of biomass pyrolysis, experiments carried out on single components in the HSR allowed to confirm the mechanism of primary pyrolysis of the individual bio-polymers, without mutual interactions. The primary tars produced from cellulose and hemicellulose in fact, reveal that the primary product of pyrolysis of these two biopolymers is mainly levoglucosane (over 95 %) resulting from depolymerisation of constitutive polysaccharides [53,54]. Lignin primary tar shows a predominance of methoxy-aromatic compounds.

## 2.5 Char Morphology

The physicochemical structural changes of the char as porosity, morphology and graphitisation are important parameters for combustion and gasification processes, e.g. as described in Sect. 3.1. These features contribute to the burning mode, conversion rate and efficiency in a boiler. Most of the morphological studies in the literature are based on experiments with low heating rates and long residence times [5,6,55]. For these conditions, particles largely retain their shape [8]. In contrast, during flash pyrolysis, the particles deform plastically, spherodise in many cases, and get highly porous. The structural changes of the char particles concerning metaplasticity, increase of porosity and changes in the cell structure are investigated for coal [31,56] and biomass [19,20,56].

### 2.5.1 Plastic Deformation

The physical softening of particles and the formation of metaplast during pyrolysis are important for the morphological changes of the fuel. For coal particles, Yang et al. [57–59] manage to predict particle swelling during pyrolysis by combining CPD predictions and metaplast formation. Analyses specific to biomass are still missing and the interplay between biomass composition (cellulose, hemicellulose, and lignin) and reaction chemistry, metaplast formation, viscosity, and morphological changes is not understood. Cerciello et al. [19] use predictions from the Bio-CPD model (cf. Sect. 8.2.1) for volatile yields and formed metaplastic species to explain the plastic deformation during the pyrolysis of xylan (a typical hemicellulose) in the DTR (cf. Sect. 2.2.3) and the HSR (cf. Sect. 2.2.4) at temperatures between 1000 and 1573 K. For both test rigs, similar particle phenomena are observable during pyrolysis by comparing SEM images from particles samples collected in the DTR and thermocamera images obtained in the HSR.

The individual images from the IR camera shown in Fig. 2.15 of the HSR experiments allow for a timeline reconstruction of the deformation and reaction process. The melting stage is observed until 0.16 s. Between 0.25 and 1.04 s, corresponding to the strip's temperature range 1160–1450 K, pyrolysis and spheroidisation take place. At the end of the pyrolysis stage, a slight shrinkage is observed.

Figure 2.16 shows SEM images of chars from the DTR for different pyrolysis stages with the associated residence times. It can be observed that the raw sharp-edged particles in Fig. 2.16a) drastically change their morphology during pyrolysis as shown in Figs. 2.16b) to d). Up to a residence time $t = 0.19$ s, which corresponds to a volatile yield $y_{vol} = 0.58$, a huge increase in particle diameter is observed. Particles

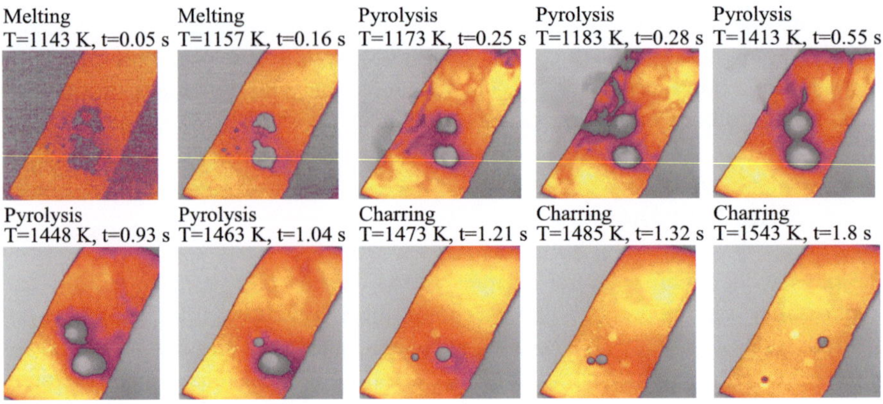


**Fig. 2.15** Thermo camera images from the HSR experiment of xylan at a final strip temperature of 1573 K (reprinted from Cerciello et al. [19], Copyright (2024), with permission from Elsevier and minor editing)

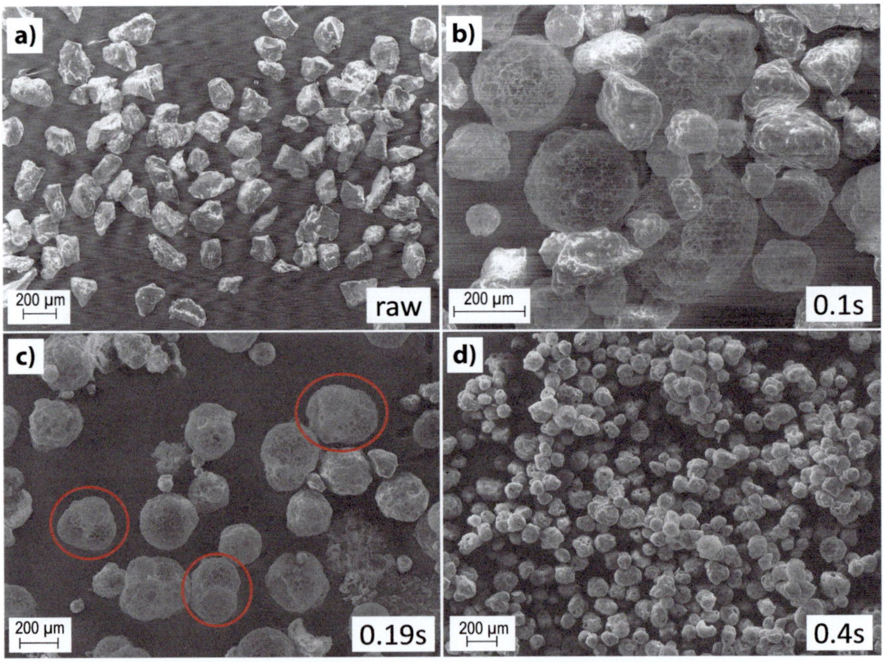


**Fig. 2.16** SEM images of xylan chars at different pyrolysis stages in the DTR [19]: **a)** raw material, **b)** 20 wt. % mass loss, **c)** 58 wt. % mass loss (red circles highlight particles that are stuck), and **d)** 80 wt. % mass loss. (Reprinted from Cerciello et al. [19], Copyright (2024), with permission from Elsevier and minor editing)

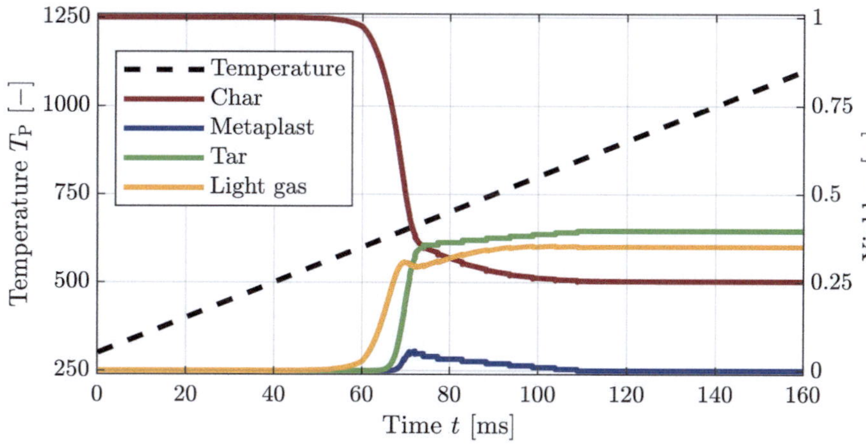


**Fig. 2.17** Bio-CPD predictions for a xylan particle with a fixed particle heating rate of 5000 K s$^{-1}$ approximating DTR reaction conditions

swell and become spherical with a sponge-like morphology. Some particles stick together (see red circles in Fig. 2.16c)) and partially wrap around each other. During the cooling step in the sampling probe, these viscous particles stick together and solidify. At $t = 0.4$ s (Fig. 2.16d), where the devolatilisation stage is completed, it is possible to observe the effects of charring and particle shrinking. Particles are solid and agglomeration phenomena in the probe are no longer relevant suggesting that primary pyrolysis is finished and collected particles are not viscous anymore.

To better understand the observed phenomena, the experimental results are compared with predictions using the Bio-CPD model from Sect. 8.2.1. Figure 2.17 shows Bio-CPD results of xylan pyrolysis using a fixed heating rate of 5000 K s$^{-1}$ to approximate the reaction conditions in the DTR.

The volatile yield (sum of tar and light gas) in the Bio-CPD model is predicted to be $y_{vol} = 0.75$ showing a good agreement with the experimental DTR results, where a volatile yield $y_{vol} = 0.8$ is determined. Bio-CPD simulations predict that after an initial release of gaseous products arising from small molecular fragments, tar and metaplast formation are initiated. The metaplast fraction reaches its maximum well before the tar release is complete. This result is in good agreement with the morphological changes observed in the DTR, and even in the HSR, which show that samples melt and spherodise in the early stage of pyrolysis. In the late stages, metaplast is either cross-linked with the char matrix or released from the particle as soon as the temperature is high enough for the corresponding fragment to vaporise. The phase change from liquid to gas is accompanied by a strong volume expansion, which can lead to a swelling of the particle if the gases cannot escape from the particle quickly enough. This effect can be well seen in the Fig. 2.16b) and c). Finally, the volatiles escape and char formation generates porosity and particle shrinking. However, the Bio-CPD model seems to fail since it underestimates cross-linking reaction (reattachment of the metaplast fragments to the char structure). This discrepancy may arise from the incorrect assignment of the kinetic parameters used in the Bio-CPD simulations. It must be recalled, in this respect, that the kinetic parameters for hemicellulose were evaluated from slow experiments in a TG [60], which might not be the most appropriate for high heating rates and that the kinetics of cross-linking in the Bio-CPD are assumed equal to those of the original coal-based CPD model.

### 2.5.2 Porosity Evolution

A further major morphological parameter is porosity, which changes during pyrolysis. Porosity is a key factor in char conversion models, particularly in cases where reaction takes place under pore-diffusion limitation (also known as regime II) as described in detail in Sect. 3.1.

Several studies have analysed the evolution of porosity of chars during pyrolysis, mainly focusing on low heating rates [5–7,61]. Information on flash pyrolysis under high heating rates in the order of $10^4$ K $s^{-1}$ is provided for coal by Heuer et al. [31] and for biomass by Freisewinkel et al. [20] to analyse the porosity change with progressing pyrolysis. Pyrolysis experiments are conducted in $N_2$ atmosphere but also in $CO_2$ atmosphere to consider effects of oxy-fuel conditions.

The collected char particles from a high volatile bituminous coal are subject to porosity analysis using physisorption measurements ($N_2$ adsorption). Different reactor temperatures (1300 and 1473 K) reveal that porosity increases with increasing temperature. Nevertheless, it should be noted that physisorption measurements with nitrogen are inappropriate to measure pore sizes regimes below 2 nm or closed pores, potentially leading to an underestimation of pore volume. After completion of primary pyrolysis, the influence of the gas atmosphere was negligible. However, with further heat treatment the porosity of the char particles in the $CO_2$ atmosphere increased, which is caused by the onset of gasification reactions. Moreover, when analysing the porosity at different pyrolysis stages, it was noticed that the porosity decreases at the beginning of the pyrolysis before it increases afterwards. This is due to the formation of metaplast, which is confirmed by simulations using the CPD model [31].

As indicated above, characterising the porosity of char is difficult due the wide range of pore size regimes and shapes. Various methods, such as physisorption measurements, mercury intrusion porosimetry, and optical techniques, are applicable for different pore size regimes. However, none of these methods can cover the entire range of the pore sizes. In addition, mercury intrusion can damage the fine wall structures of highly porous chars due to the high pressure of the mercury. Therefore, it is not considered for further consideration.

For the examination of biomass particles [20], an improved physisorption measurement technique was combined with an optical method. Physisorption has been improved by adding a second adsorption isotherm ($CO_2$), to cover a larger spectrum of pore size regimes. With 2-D non-local density functional theory (NLDFT) calculations, for $N_2$ and $CO_2$ adsorption isotherms, it is possible to measure pore regimes in the range of 0.33–50 nm. A more detailed description of the method is described in Sect. 5.2. Due to the limitation of the physisorption measurements, an optical

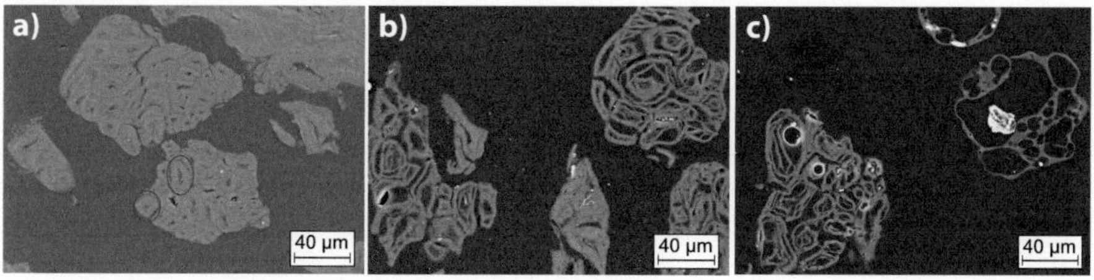


**Fig. 2.18** SEM images of walnut shell particles for porosity analysis: **a)** raw material (red circles indicate the individual sclerenchyma), **b)** ~ 48 wt. % mass loss, and **c)** ~77 wt. % mass loss. (Based on the publication by Freisewinkel et al. [20])

method was added to identify large pores. For the optical method, polished thin film samples are analysed using a scanning electron microscope to examine pores with diameters larger than 1 μm. Exemplary images of raw and two differently pyrolysed walnut shell particles are shown in Fig. 2.18.

First results from the porosity analysis of walnut shells during pyrolysis demonstrate that macropores larger than 1 μm have the most significant impact on overall porosity. The proportions of the different pore size regimes are shown in Table 2.2. The proportion of pores above 1 μm increases significantly with reaction distance, ranging from 8 to 64 %, whereas micropores constitute only a small fraction of the total porosity with 2 to 6 %. Additionally, stronger heat treatment at 1300 K has no particular influence on the overall porosity if the primary pyrolysis is completed. The porosity derived using the methods described above is only accurate for an idealised particle without pores in the range of 50 nm to 1 μm. However, preliminary calculations of density and mass loss based on this porosity suggest that the undetected pore size range has a minor impact on the overall pore volume, particularly for high pyrolysed samples. The influence of the $CO_2$ atmosphere on the porosity has not been investigated yet, but no gasification reactions were observed in a previous study [33].

The images from Fig. 2.18 give further insights into changes in the cell structure and the influence of the tissue structure on plastic deformation. The tissue structure of walnut shells is built of various individual sclerenchyma, which have a thick lignified secondary cell wall providing structural support for the plant [62]. The individual

**Table 2.2** Proportions of different pore size regimes obtained by adsorption ($d_{po} < 50$ nm) and optical measurements ($d_{po} > 1$ μm) for walnut shell particles pyrolysed in the DTR at different reaction distances [20]

| Reactor temp. | | 1125 K | | 1300 K | |
|---|---|---|---|---|---|
| Pore size class | | $d_{po} < 50$ nm | $d_{po} > 1$ μm | $d_{po} < 50$ nm | $d_{po} > 1$ μm |
| Reaction distance | 0 mm | 2.5 % | 8.8 % | 2.5 % | 8.8 % |
| | 50 mm | 3.3 % | 13.7 % | 1.8 % | 36.3 % |
| | 150 mm | 1.8 % | 26.9 % | 3.5 % | 60.0 % |
| | 300 mm | 2.1 % | 57.1 % | 6.0 % | 63.6 % |

sclerenchyma are clearly recognisable in the raw sample (marked with red circles). With progressive pyrolysis, the interior of the sclerenchyma seems to devolatilise and the porosity increases, but the secondary cell wall remains intact and largely retains its shape (cf. Fig. 2.18b)). At a certain stage of pyrolysis, the structure of the individual sclerenchyma changes significantly. The lignified secondary cell wall destabilises and forms large cavities as shown in Fig. 2.18 c). The missing structure of the sclerenchyma leads to the shape change from an edge-like particle to an almost spherical one. The results show the influence of the tissue structure on the morphological change of particles and how the pore characteristics change with progressing pyrolysis. These results presented are the first important data for improving models to predict physical softening, metaplast formation, and porosity evolution. Future work must address how the biomass composition (cellulose, hemicellulose, and lignin) affects the morphology changes. For example, it would be of high interest to close the knowledge gap concerning the viscosity of metaplast and to attempt correlating it with experimental and model results on the composition of primary tars and the size of metaplast clusters, especially for the main biomass components and in $CO_2$ atmospheres.

## 2.6 Conclusion

Pyrolysis describes the thermal decomposition of organic fuels into the three main products light gas, tar, and char. The pyrolysis process is an important substep in the overall conversion process of solid fuels during oxy-fuel combustion. The rate and amount of released volatiles control the volatile combustion and ignition. Tars and minority species are relevant as pollutant precursors, and the char morphology influences the reactivity and heterogeneous char conversion kinetics. The profound development of mathematical models that can be included in CFD simulations of full-scale boilers requires detailed experimental knowledge on the single-particle level for simplified but well-controllable boundary conditions. Various experimental setups—each designed to analyse different aspects and products of the pyrolysis process—cover different thermal boundary conditions regarding temperature and heating rate as well as particle and gas residence times. Targeted modifications of established systems have led, for example, to their use for new biogenic fuels (e.g. new analytical methods in the tar characterisation or the probe adaptation in the drop tube reactor) or to shifting the limits of the application area (e.g. optimised gas transport in the fluidised bed reactor).

Reaction kinetic investigations under flash pyrolysis conditions for extracted single biomass components extend the current database of TG-derived kinetics and build the basis for improving reaction kinetic pyrolysis models. The comparison with natural biomass revealed that for low heating rates, a linear superposition yields satisfactory results, while high heating rates lead to a strong underprediction of the volatile formation, indicating the presence of interactions between the components. The same applies, for example, to the characterisation of the tar fraction. In addition to the fuel type, the temperature and residence time in the gas phase play a decisive role in terms of product distribution. While the change from an $N_2$ to a $CO_2$ atmosphere plays a minor role in the kinetics, the formation of pollutants (especially $NO_x$) is significantly influenced. Detailed investigations on the char morphology gained new insights into plastic deformation and pore formation. These data form an important basis for the validation of char conversion models, where particle-internal transport processes must be considered.

**Acknowledgements** This work has been funded by the Deutsche Forschungsgemeinschaft (DFG, German Research Foundation)—215035359—SFB/TRR 129 'Oxyflame'.

## References

1. Bridgwater, A. V. "Review of Fast Pyrolysis of Biomass and Product Upgrading". *Biomass and Bioenergy* 38 (2012), 68–94. https://doi.org/10.1016/j.biombioe.2011.01.048
2. Fang, Z., Smith, R. L. and Xu, L., eds. *Production of Biofuels and Chemicals*. Singapore: Springer, 2020. https://doi.org/10.1007/978-981-15-2732-6
3. Wafiq, A., Reichel, D. and Hanafy, M. "Pressure Influence on Pyrolysis Product Properties of Raw and Torrefied Miscanthus: Role of Particle Structure". *Fuel* 179 (2016), 156–167. https://doi.org/10.1016/j.fuel.2016.03.092
4. Quicker, P. and Weber, K. *Biokohle: Herstellung, Eigenschaften und Verwendung von Biomassekarbonisaten*. Wiesbaden: Springer Vieweg, 2016. https://doi.org/10.1007/978-3-658-03689-8
5. Lu, S. and Zong, Y. "Pore Structure and Environmental Serves of Biochars Derived from Different Feedstocks and Pyrolysis Conditions". *Environmental Science and Pollution Research International* 25.30 (2018), 30401–30409. https://doi.org/10.1007/s11356-018-3018-7
6. Chen, B., Zhou, D. and Zhu, L. "Transitional Adsorption and Partition of Nonpolar and Polar Aromatic Contaminants by Biochars of Pine Needles with Different Pyrolytic Temperatures". *Environmental Science & Technology* 42.14 (2008), 5137–5143. https://doi.org/10.1021/es8002684
7. Fu, P., Hu, S., Xiang, J., Sun, L., Su, S. and Wang, J. "Evaluation of the Porous Structure Development of Chars from Pyrolysis of Rice Straw: Effects of Pyrolysis Temperature and Heating Rate". *Journal of Analytical and Applied Pyrolysis* 98 (2012), 177–183. https://doi.org/10.1016/j.jaap.2012.08.005
8. Yao, Y., Panahi, A., Schiemann, M. and Levendis, Y. A. "Determination of Size and Porosity of Chars During Combustion of Biomass Particles". *Combustion and Flame* 242 (2022), 112182. https://doi.org/10.1016/j.combustflame.2022.112182
9. Panahi, A., Levendis, Y. A., Vorobiev, N. and Schiemann, M. "Direct Observations on the Combustion Characteristics of Miscanthus and Beechwood Biomass Including Fusion and Spherodization". *Fuel Processing Technology* 166 (2017), 41–49. https://doi.org/10.1016/j.fuproc.2017.05.029
10. Fletcher, T. H. "Review of 30 Years of Research Using the Chemical Percolation Devolatilization Model". *Energy & Fuels* 33.12 (2019), 12123–12153. https://doi.org/10.1021/acs.energyfuels.9b02826
11. Zhu, L. and Zhong, Z. "Effects of Cellulose, Hemicellulose and Lignin on Biomass Pyrolysis Kinetics". *Korean Journal of Chemical Engineering* 37.10 (2020), 1660–1668. https://doi.org/10.1007/s11814-020-0553-y
12. Couhert, C., Commandré, J.-M. and Salvador, S. "Is it Possible to Predict Gas Yields of any Biomass After Rapid Pyrolysis at High Temperature from its Composition in Cellulose, Hemicellulose and Lignin?" *Fuel* 88.3 (2009), 408–417. https://doi.org/10.1016/j.fuel.2008.09.019

13. Zhao, S., Liu, M., Zhao, L. and Zhu, L. "Influence of Interactions Among Three Biomass Components on the Pyrolysis Behavior". *Industrial & Engineering Chemistry Research* 57.15 (2018), 5241–5249. https://doi.org/10.1021/acs.iecr.8b00593
14. Eckhard, T., Pflieger, C., Böttger, J., Telaar, P., Cerciello, F. and Muhler, M. "The Catalytic Effect of Iron and Alkali and Alkaline Earth Metal Sulfates Loading Series on the Conversion of Cellulose-Derived Hydrochars and Chars". *ACS Omega* 8.11 (2023), 10629–10639. https://doi.org/10.1021/acsomega.3c00887
15. Eckhard, T., Pflieger, C., Schmidt, S., Böttger, J., Senneca, O., Schiemann, M., Scherer, V., Muhler, M. and Cerciello, F. "Catalytic Effects for Cellulose-Based Model Fuels Under Low and High Heating Rate in Air and Oxy-Fuel Atmosphere". *Fuel* 324 (2022), 124437. https://doi.org/10.1016/j.fuel.2022.124437
16. Pielsticker, S., Gövert, B., Umeki, K. and Kneer, R. "Flash Pyrolysis Kinetics of Extracted Lignocellulosic Biomass Components". *Frontiers in Energy Research* 9 (2021). https://doi.org/10.3389/fenrg.2021.737011
17. Pielsticker, S., Schlögel, K. U., Kreitzberg, T., Hatzfeld, O. and Kneer, R. "Biomass Pyrolysis Kinetics in a Fluidized Bed Reactor: Measurements and Plausibility Verification for Reaction Conditions". *Fuel* 254 (2019), 115589. https://doi.org/10.1016/j.fuel.2019.05.172
18. Pielsticker, S., Gövert, B., Kreitzberg, T., Habermehl, M., Hatzfeld, O. and Kneer, R. "Simultaneous Investigation into the Yields of 22 Pyrolysis Gases from Coal and Biomass in a Small-Scale Fluidized Bed Reactor". *Fuel* 190 (2017), 420–434. https://doi.org/10.1016/j.fuel.2016.10.085
19. Cerciello, F., Freisewinkel, E., Coppola, A., Ontyd, C., Tarlinski, D., Schiemann, M., Senneca, O., Salatino, P., Allouis, C., Scherer, V. and Fletcher, T. H. "Xylan Fast Pyrolysis: An Experimental and Modelling Study of Particle Changes and Volatiles Release". *Fuel* 357 (2024), 129983. https://doi.org/10.1016/j.fuel.2023.129983
20. Freisewinkel, E., Eisenbach, T., Weise, V., Tarlinski, D., Pörtner, L., Cerciello, F., Schiemann, M., Span, R., Senneca, O. and Scherer, V. "Porosity of Biomass Char: A Novel Determination Method and Changes in Walnut Shell with Progressing Pyrolysis". *48$^{th}$ International Conference on Clean Energy*. Clearwater, USA, 2024
21. Ontyd, C., Pielsticker, S., Yildiz, C., Schiemann, M., Hatzfeld, O., Ströhle, J., Epple, B., Kneer, R. and Scherer, V. "Experimental Determination of Walnut Shell Pyrolysis Kinetics in $N_2$ and $CO_2$ via Thermogravimetric Analysis, Fluidized Bed and Drop Tube Reactors". *Fuel* 287 (2021), 119313. https://doi.org/10.1016/j.fuel.2020.119313
22. Senneca, O., Allouis, C., Chirone, R. and Russo, S. "Set up of an Experimental Apparatus for the Study of Fragmentation of Solid Fuels Upon Severe Heating". *Experimental Thermal and Fluid Science* 34.3 (2010), 366–372. https://doi.org/10.1016/j.expthermflusci.2009.10.015
23. Böttger, J., Eckhard, T., Pflieger, C., Senneca, O., Muhler, M. and Cerciello, F. "Green Coal Substitutes for Boilers Through Hydrothermal Carbonization of Biomass: Pyrolysis and Combustion Behavior". *Fuel* 344 (2023), 128025. https://doi.org/10.1016/j.fuel.2023.128025
24. Pielsticker, S., Gövert, B., Kreitzberg, T., Habermehl, M., Hatzfeld, O. and Kneer, R. "Development of a Rapidly Responding Fluidized Bed Reactor by Theoretical and Experimental Evaluation of Combustion Reactions". *Fuel* 223 (2018), 462–469. https://doi.org/10.1016/j.fuel.2018.02.171
25. Pielsticker, S., Möller, G. H., Gövert, B., Kreitzberg, T., Hatzfeld, O., Yönder, Ö., Angenent, V., Hättig, C., Schmid, R. and Kneer, R. "Influence of Biomass Torrefaction Parameters on Fast Pyrolysis Products Under Flame-Equivalent Conditions". *Biomass and Bioenergy* 119 (2018), 392–410. https://doi.org/10.1016/j.biombioe.2018.08.014
26. Abad, A., Cardona, S. C., Torregrosa, J. I., López, F. and Navarro-Laboulais, J. "Flow Analysis Deconvolution for Kinetic Information Reconstruction". *Journal of Mathematical Chemistry* 38.2 (2005), 271–292
27. Pielsticker, S. and Kneer, R. "Modeling Pyrolysis Kinetics of Extracted Biomass Components with the Bio-Chemical Percolation Devolatilization Model". *Fuel* 360 (2024), 130454. https://doi.org/10.1016/j.fuel.2023.130454

28. Gövert, B. "Char Combustion Kinetics Using a Micro Fluidized Bed Reactor". PhD thesis. RWTH Aachen University, 2018. https://doi.org/10.18154/RWTH-2019-00500
29. Pyle, D. L. and Zaror, C. A. "Heat Transfer and Kinetics in the Low Temperature Pyrolysis of Solids". *Chemical Engineering Science* 39.1 (1984), 147–158. https://doi.org/10.1016/0009-2509(84)80140-2
30. Gunn, D. J. "Transfer of Heat or Mass to Particles in Fixed and Fluidised Beds". *International Journal of Heat and Mass Transfer* 21.4 (1978), 467–476. https://doi.org/10.1016/0017-9310(78)90080-7
31. Heuer, S., Wedler, C., Ontyd, C., Schiemann, M., Span, R., Richter, M. and Scherer, V. "Evolution of Coal Char Porosity from $CO_2$-Pyrolysis Experiments". *Fuel* 253 (2019), 1457–1464. https://doi.org/10.1016/j.fuel.2019.05.071
32. Heuer, S., Senneca, O., Wütscher, A., Düdder, H., Schiemann, M., Muhler, M. and Scherer, V. "Effects of Oxy-Fuel Conditions on the Products of Pyrolysis in a Drop Tube Reactor". *Fuel Processing Technology* 150 (2016), 41–49. https://doi.org/10.1016/j.fuproc.2016.04.034
33. Ontyd, C., Freisewinkel, E., Wedler, C., Eisenbach, T., Span, R., Senneca, O. and Scherer, V. "Flash Pyrolysis of Walnut Shells and Poplar in $N_2$ and $CO_2$: Pyrolysis Kinetics, Pore Evolution and a Model for Evaluation of Pyrolysis Species". *$13^{th}$ European Conference on Industrial Furnaces and Boilers*. Algarve, Portugal, 2022
34. Cerciello, F., Freisewinkel, E., Coppola, A., Ontyd, C., Schiemann, M., Salatino, P., Allouis, C., Scherer, V. and Senneca, O. "Xylan Fast Pyrolysis: An Experimental and Modelling Study of Particle Changes and Volatiles Release". *$12^{th}$ Mediterranean Combustion Symposium*. Luxor, Egypt, 2023
35. Froment, K., Defoort, F., Bertrand, C., Seiler, J. M., Berjonneau, J. and Poirier, J. "Thermodynamic Equilibrium Calculations of the Volatilization and Condensation of Inorganics During Wood Gasification". *Fuel* 107 (2013), 269–281. https://doi.org/10.1016/j.fuel.2012.11.082
36. Murphy, J. J. and Shaddix, C. R. "Combustion Kinetics of Coal Chars in Oxygen-Enriched Environments". *Combustion and Flame* 144.4 (2006), 710–729. https://doi.org/10.1016/j.combustflame.2005.08.039
37. Apicella, B., Senneca, O., Russo, C., Heuer, S., Cortese, L., Cerciello, F., Scherer, V., Schiemann, M. and Ciajolo, A. "Separation and Characterization of Carbonaceous Particulate (Soot and Char) Produced from Fast Pyrolysis of Coal in Inert and $CO_2$ Atmospheres". *Fuel* 201 (2017), 118–123. https://doi.org/10.1016/j.fuel.2016.11.049
38. Senneca, O., Apicella, B., Heuer, S., Schiemann, M., Scherer, V., Stanzione, F., Ciajolo, A. and Russo, C. "Effects of $CO_2$ on Submicronic Carbon Particulate (Soot) Formed During Coal Pyrolysis in a Drop Tube Reactor". *Combustion and Flame* 172 (2016), 302–308. https://doi.org/10.1016/j.combustflame.2016.07.023
39. Orfão, J. J. M., Antunes, F. J. A. and Figueiredo, J. L. "Pyrolysis Kinetics of Lignocellulosic Materials—Three Independent Reactions Model". *Fuel* 78.3 (1999), 349–358. https://doi.org/10.1016/S0016-2361(98)00156-2
40. Yu, J., Paterson, N., Blamey, J. and Millan, M. "Cellulose, Xylan and Lignin Interactions During Pyrolysis of Lignocellulosic Biomass". *Fuel* 191 (2017), 140–149. https://doi.org/10.1016/j.fuel.2016.11.057
41. Xu, L., Jiang Liqun, Zhang, H., Fang, Z. and Smith, R. L. "Introduction to Pyrolysis as a Thermo-Chemical Conversion Technology". *Production of Biofuels and Chemicals*. Ed. by Fang, Z., Smith, R. L. and Xu, L. Singapore: Springer, 2020. https://doi.org/10.1007/978-981-15-2732-6_1
42. Krusch, S., Scherer, V., Solimene, R., Ruoppolo, G., Cerciello, F., Pielsticker, S., Kneer, R., Schlögel, K. U. and Senneca, O. "Comparison of Different Diagnostic Techniques for the Assessment of Coal Pyrolysis Rate". *Joint Meeting the German and Italian Sections of the Combustion Institute*. Sorrento, Italy, 2018
43. Pielsticker, S., Gövert, B., Möller, G. H., Hatzfeld, O. and Kneer, R. "Impact of Torrefaction on Flash Pyrolysis Products of Biomass and Comparison to Fossil Fuels". *$11^{th}$ Asia-Pacific Conference on Combustion*. Sydney, Australia, 2017. https://doi.org/10.18154/RWTH-2017-10920

44. Yildiz, C., Richter, M., Ströhle, J. and Epple, B. "Release of Sulfur and Chlorine Gas Species During Combustion and Pyrolysis of Walnut Shells in an Entrained Flow Reactor". *Energies* 16.15 (2023), 5684. https://doi.org/10.3390/en16155684
45. Wang, C., Du, Y. and Che, D. "Investigation on the NO Reduction with Coal Char and High Concentration CO During Oxy-Fuel Combustion". *Energy & Fuels* 26.12 (2012), 7367–7377. https://doi.org/10.1021/ef301202w
46. Chardonneaua, M., Ibrahim, S., Gupta, A. K. and AlShoaibi, A. "Role of Toluene and Carbon Dioxide on Sulfur Recovery Efficiency in a Claus Process". *Energy Procedia* 75 (2015), 3071–3075. https://doi.org/10.1016/j.egypro.2015.07.630
47. Hildor, F., Leion, H. and Mattisson, T. "Steel Converter Slag as an Oxygen Carrier – Interaction with Sulfur Dioxide". *Energies* 15.16 (2022), 5922. https://doi.org/10.3390/en15165922
48. Wei, X., Schnell, U. and Hein, K. R. G. "Behaviour of Gaseous Chlorine and Alkali Metals During Biomass Thermal Utilisation". *Fuel* 84.7-8 (2005), 841–848. https://doi.org/10.1016/j.fuel.2004.11.022
49. Yildiz, C., Richter, M., Ströhle, J. and Epple, B. "Pollutant Formation Under Nitrogen and Carbon Dioxide Atmosphere of Torrefied Poplar in an Entrained Flow Reactor". *Energy & Fuels* 38.9 (2024), 8157–8167. https://doi.org/10.1021/acs.energyfuels.3c04150
50. Frigge, L., Ströhle, J. and Epple, B. "Release of Sulfur and Chlorine Gas Species During Coal Combustion and Pyrolysis in an Entrained Flow Reactor". *Fuel* 201 (2017), 105–110. https://doi.org/10.1016/j.fuel.2016.11.037
51. Cerciello, F., Senneca, O., Russo, C. and Apicella, B. "Comparison of Primary Volatiles from Coal and Biomass Pyrolysis in $N_2$ and $CO_2$". *Energy & Fuels* 33.12 (2019), 12822–12829. https://doi.org/10.1021/acs.energyfuels.9b03167
52. Cerciello, F., Apicella, B., Russo, C., Cortese, L. and Senneca, O. "Effects of Pressure on Lignocellulosic Biomass Fast Pyrolysis in Nitrogen and Carbon Dioxide". *Fuel* 287 (2021), 119604. https://doi.org/10.1016/j.fuel.2020.119604
53. Senneca, O., Cerciello, F., Russo, C., Wütscher, A., Muhler, M. and Apicella, B. "Thermal Treatment of Lignin, Cellulose and Hemicellulose in Nitrogen and Carbon Dioxide". *Fuel* 271 (2020), 117656. https://doi.org/10.1016/j.fuel.2020.117656
54. Debiagi, P., Piazza, V., Papagni, M., Beretta, A., Frassoldati, A. and Faravelli, T. "Cellulose Pyrolysis Kinetic Model: Detailed Description of Volatile Species". *Proceedings of the Combustion Institute* 40.1-4 (2024), 105651. https://doi.org/10.1016/j.proci.2024.105651
55. Trubetskaya, A., Jensen, P. A., Jensen, A. D., Steibel, M., Spliethoff, H. and Glarborg, P. "Influence of Fast Pyrolysis Conditions on Yield and Structural Transformation of Biomass Chars". *Fuel Processing Technology* 140 (2015), 205–214. https://doi.org/10.1016/j.fuproc.2015.08.034
56. Senneca, O., Ontyd, C., Cerciello, F., Schiemann, M. and Scherer, V. "Extension of the Thermal Annealing Concepts Developed for Coal Combustion to Conversion of Lignocellulosic Biomass". *Energy & Fuels* 34.3 (2020), 3661–3670. https://doi.org/10.1021/acs.energyfuels.9b03172
57. Yang, H., Fletcher, T. H., Li, S., Hu, H., Jin, L. and Li, Y. "Model for the Evolution of Pore Structure in a Lignite Particle During Pyrolysis. 2. Influence of Cross-Linking Reactions, Molten Metaplast, and Molten Ash on Particle Surface Area". *Energy & Fuels* 31.8 (2017), 8036–8044. https://doi.org/10.1021/acs.energyfuels.7b01163
58. Yang, H., Fletcher, T. H., Li, Y., Jin, L., Li, S., Shang, Y. and Hu, H. "Modeling the Influence of Changes in Aliphatic Structure on Char Surface Area During Coal Pyrolysis". *AIChE Journal* 66.2 (2020). https://doi.org/10.1002/aic.16834
59. Yang, H., Li, S., Fletcher, T. H., Dong, M. and Zhou, W. "Simulation of the Evolution of Pressure in a Lignite Particle During Pyrolysis". *Energy & Fuels* 28.5 (2014), 3511–3518. https://doi.org/10.1021/ef500584q
60. Fletcher, T. H., Pond, H. R., Webster, J., Wooters, J. and Baxter, L. L. "Prediction of Tar and Light Gas During Pyrolysis of Black Liquor and Biomass". *Energy & Fuels* 26.6 (2012), 3381–3387. https://doi.org/10.1021/ef300574n

61. Brewer, C. E., Chuang, V. J., Masiello, C. A., Gonnermann, H., Gao, X., Dugan, B., Driver, L. E., Panzacchi, P., Zygourakis, K. and Davies, C. A. "New Approaches to Measuring Biochar Density and Porosity". *Biomass and Bioenergy* 66 (2014), 176–185. https://doi.org/10.1016/j.biombioe.2014.03.059
62. Zhao, S., Niu, J., Yun, L., Liu, K., Wang, S., Wen, J., Wang, H. and Zhang, Z. "The Relationship Among the Structural, Cellular, and Physical Properties of Walnut Shells". *HortScience* 54.2 (2019), 275–281. https://doi.org/10.21273/HORTSCI13381-18

# Char Conversion Kinetics

# 3

David Tarlinski, Stefan Pielsticker, Reinhold Kneer, Martin Schiemann and Viktor Scherer

## Contents



**Abstract**

Char conversion is a highly complex process. Primarily composed of solid carbon, char has a very low vapour pressure, which prevents it from evaporating and undergoing gas-phase oxidation. Instead, char reacts directly at its surface throughout

D. Tarlinski (✉) · M. Schiemann · V. Scherer
Department of Energy Plant Technology, Ruhr-Universität Bochum, Bochum, Germany
e-mail: tarlinski@leat.rub.de

M. Schiemann
e-mail: schiemann@leat.rub.de

V. Scherer
e-mail: scherer@leat.rub.de

S. Pielsticker · R. Kneer
Institute of Heat and Mass Transfer, RWTH Aachen University, Aachen, Germany
e-mail: pielsticker@wsa.rwth-aachen.de

R. Kneer
e-mail: kneer@wsa.rwth-aachen.de


R. Kneer et al. (eds.), *Clean Energy Conversion*,
https://doi.org/10.1007/978-3-031-98956-8_3

oxygen, carbon dioxide, hydrogen, and water vapour. These surface reactions generate carbon monoxide (CO), which then escapes the solid phase and oxidises into $CO_2$ in the surrounding gas. Key factors that affect char conversion include the temperature of the surrounding gas, the composition of the atmosphere near the surface, ambient pressure, and characteristics like fuel composition, particle size, and shape. To capture the entire temperature range relevant in practical applications and to study the full span of char burnout—from the initial phase to the late stage over extended residence times—two experimental setups are employed: a fluidised bed reactor and a laminar plug flow reactor. These two test rigs also allow to study all char conversion regimes: regime I (kinetically controlled), regime II (pore-diffusion controlled) and regime III (film-diffusion controlled). Therefore char conversion is studied across a variety of fossil and biogenic solid fuels, with a broad range of temperatures and oxy-fuel gas atmospheres applied.

## 3.1 Introduction

In solid fuel combustion processes, alongside pyrolysis, char conversion is the second essential step, particularly because char conversion dominates the total conversion time of a particle. Understanding the dynamics of char conversion is paramount but the knowledge is still limited, particularly for the use of renewable solid fuels in connection with oxy-fuel combustion.

The heterogeneous gas-solid reactions occur at the surface and within the pores of the char particle. This process can last up to several seconds under typical conditions of pulverised fuel combustion. The structure and reactivity of the char are shaped and influenced during the pyrolysis phase. Particularly, long residence times at high temperatures lead to thermal deactivation of the char matrix, known as thermal annealing [1–5]. Char particles from coal as well as biomass represent a complex macromolecule after pyrolysis. However, char combustion models typically assume carbon as the primary reactive species. This leads to a total of five reactions, which are categorised into two oxidation reactions (3.1) and (3.2) and three gasification reactions (3.3)–(3.5).

$$C_{(S)} + O_2 \rightarrow CO_2 \tag{3.1}$$

$$C_{(S)} + \frac{1}{2}O_2 \rightarrow CO \tag{3.2}$$

$$C_{(S)} + CO_2 \leftrightarrow 2CO \tag{3.3}$$

$$C_{(S)} + H_2O \leftrightarrow CO + H_2 \tag{3.4}$$

$$C_{(S)} + 2H_2 \leftrightarrow CH_4 \tag{3.5}$$

Under conventional pulverised fuel combustion conditions in an air atmosphere, oxidation dominates the char conversion process. At typical combustion temperatures, over 95% of the carbon is converted to CO [6]. Compared to a conventional atmosphere, gasification reactions gain significance in an oxy-fuel atmosphere, where 20–30% of the solid carbon is typically converted through gasification [7].

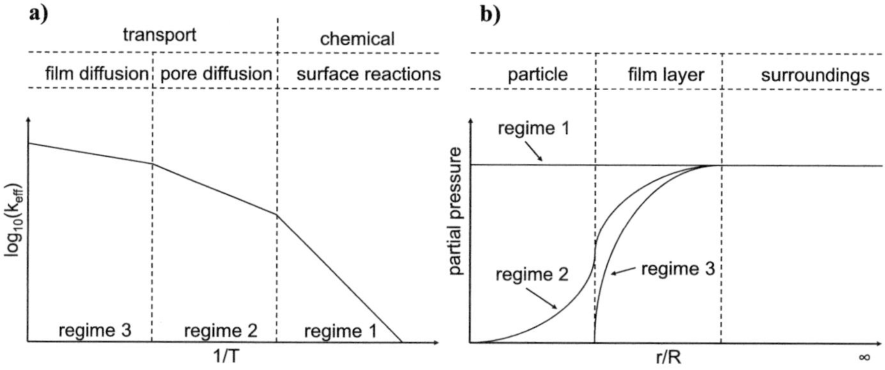


**Fig. 3.1** **a**) Effective reaction rate as a function of the temperature and **b**) the schematic profile of the oxidant partial pressure in dependence of the dimensionless particle radius in accordance with Joos [8]

Char conversion is divided into three reaction regimes, as illustrated in Fig. 3.1. The left diagram shows the effective reaction rate $k_{\text{eff}}$ plotted logarithmically against the reciprocal temperature. On the right side, the variation of the oxygen partial pressure as a function of the dimensionless particle radius for each combustion regime is schematically depicted. The regime I, also known as the kinetic regime, occurs at relatively low temperatures. The rate of reaction is governed by intrinsic kinetics, with an exponential dependency on temperature. The reaction occurs within the internal porosity, and oxygen concentration profiles remain flat across the particle and its surroundings. With further temperature increase, the system enters regime II. In regime II, the chemical reaction rate becomes less sensitive to temperature compared to regime I (activation energy reduces by a factor of 2). The reaction is controlled by both chemical kinetics and internal diffusion. Oxygen concentration decreases within the particle and the surrounding film layer. This regime is also called the mixed regime. As a result, the reaction rate exceeds the transport rate into the pores, leading to a characteristic drop in the oxidant partial pressure within the particle. Upon further temperature increase, the particle gets to regime III. In this regime, the reaction rate is controlled by external diffusion, and the oxygen is completely consumed at the particle surface. The sensitivity of the rate of reaction to temperature is low, according to the temperature dependence of the diffusion coefficient, which is characteristic for this regime. In regime I, the solid carbon reacts within the pores of the particle reducing the particle density. In contrast to regime I, the particle in regime III converts from the outside to the inside reducing the particle size. Regime II exhibits characteristics of both extremes (regime I and regime III). Models are needed to accurately predict the conversion rate of the fuel during the time-dominant step of char combustion.

This necessitates the availability of comprehensive datasets on char conversion that capture the influence of oxy-fuel atmospheres on both biogenic and fossil fuels. Such data sets must cover operation conditions and contain particle parameters that influence the char conversion rate. For oxy-fuel operation, the data sets need to cover a wide range of gas compositions with varying concentrations of $O_2$, $CO_2$, and $H_2O$. Particle temperature, size, and shape as well as porosity development as a function of char conversion degree have to be determined. Furthermore, it must assess the influence of sorption effects and catalytic effects on conversion rates. One of the most widely used datasets for validating char combustion models in an oxy-fuel atmosphere is provided by Geier et al. [9] and Shaddix and Molina [10]. However, the experimental data from *Sandia National Laboratories* (*Combustion Research Facility*, Livermore, US) present several gaps. The most prominent gaps include the exclusive examination of various coals, resulting in the absence of experimental data on biomass. Additionally, Holland [11] and Holland and Fletcher [12] emphasised that comprehensive calibration of measurement data for forthcoming models requires information on the entire particle distribution. The Sandia data only present results for self-illuminating (already ignited) particles. Of particular interest are the particle sizes of those that are not yet ignited or already combusted. Therefore, it is essential to provide new datasets with even more detailed information for further calibration and validation of models concerning pulverised fuel combustion in an oxy-fuel atmosphere. This chapter presents only the results of single particles. The particle-particle interactions, also referred to as group combustion, are discussed in Chap. 10. The following section (Sect. 3.3.2) will present such data sets on three specific fuels: coal, walnut shells, and miscanthus. Both invasive probe-based and non-invasive optical measurement techniques are utilised to examine the char conversion process. Figure 3.2 summarises the previous data from Sandia as well as all newly created data.

Prior to the data sets on char conversion, results on char conversion from the FBR, which operates at lower temperatures, are presented. This aims to investigate char conversion under regime I conditions. The influence of temperature and oxygen content on the reaction rate is examined and discussed.

## 3.2 Char Conversion in the Fluidised Bed Reactor

Char conversion kinetics are investigated in the fluidised bed reactor described in detail in Sect. 2.2.2 for the pyrolysis experiments. Instead of using pure $N_2$ as fluidisation gas, mixtures of $N_2/O_2$, $CO_2/O_2$, and $N_2/CO_2$ have been employed to mimic air combustion, oxy-fuel combustion, and gasification conditions. The partial pressure for all three gas species can be set individually between $p_i = 0$ and 1 bar. Due to the electrical heating of the reactor and the unlimited residence times of the char particles, also very slow reactions at lower temperatures than those provided by the flat flame burner can be analysed. This ensures measuring reaction rates in the kinetically controlled regime I and the pore diffusion controlled regime II. Reaction rates are equally determined than the volatile release rates in (2.2): Instead of balancing the total mass, for char conversion, only the carbon mass is used. To derive kinetic parameters, char conversion models of different complexity (e.g. the uniform reaction model [13, 14], the shrinking core model [14], or the carbon burnout kinetic model [15]) can be fitted to the dimensionless carbon conversion curve.

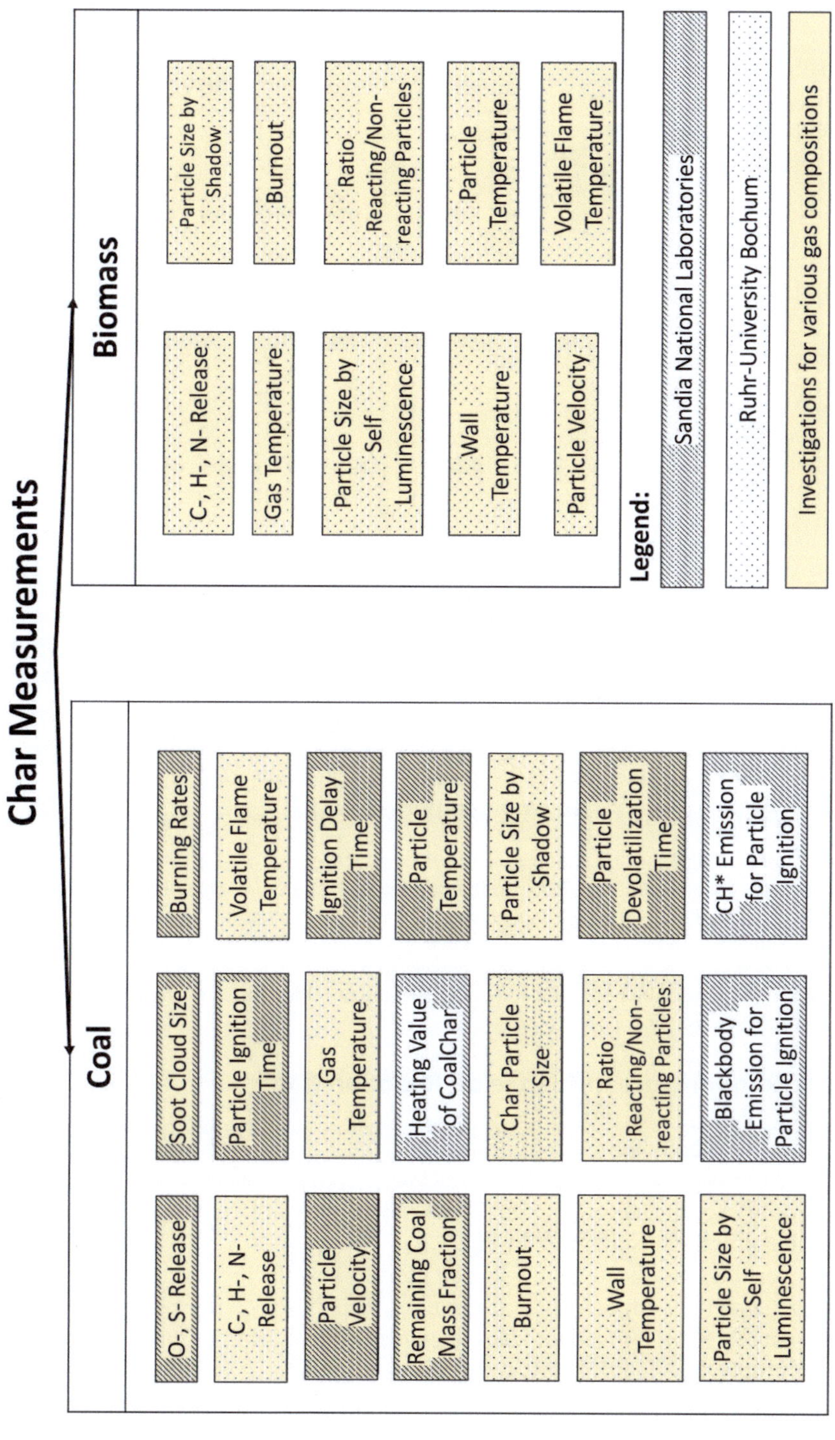


**Fig. 3.2** Overview of previous and newly obtained data sets on char burnout of coal and biomass

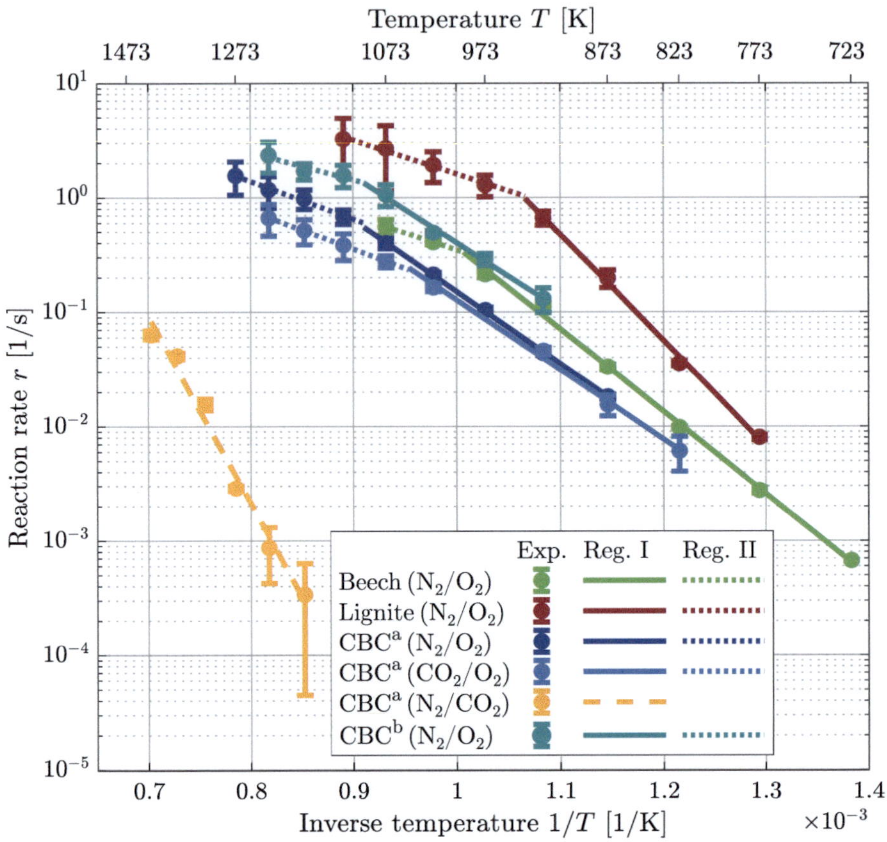


**Fig. 3.3** Arrhenius diagram for combustion ($N_2/O_2$ atmosphere) experiments in the fluidised bed reactor for beech wood char [16], Rhenish lignite char [15], and Colombian bituminous coal char produced in the fluidised bed reactor (superscript "a") [13]. For Colombian bituminous coal (CBC) char, also combustion kinetics in $CO_2/O_2$ and $N_2/CO_2$ atmospheres [13] as well as experiments with char produced in a drop tube reactor (superscript "b", see Fig. 2.4) are given. The kinetically controlled regime I and the pore diffusion controlled regime II are differentiated. Oxidiser partial pressure is always around 0.2 bar

### 3.2.1 Temperature Influence on Reaction Rates

The effect of different temperatures and fuels on the observed reactivity is evaluated using the uniform reaction model. To allow a comparison between the different curves, only experiments with an oxidiser partial pressure around $p_i = 0.2$ bar are shown in the Arrhenius diagram in Fig. 3.3.

Char particles produced from Rhenish lignite show the highest reactivity among all investigated fuels for combustion reactions in an $N_2/O_2$ atmosphere. The reason for this may be the intermediate volatile content (see Table 1.1) that yields a highly porous particle. The intermediate pyrolysis conditions of $T = 1273$ K and $t = 15$ s ensure that functional groups are partly retained. Longer residence times, however,

**Table 3.1** Char production parameters (mean particle diameter $d_{p,50}$, temperature $T$, residence time $t$) and resulting reaction kinetic parameters (activation energy $E_a$, pre-exponential factor $A$ and order of reaction $n$) for regime I and II conditions for reactions in different atmospheres

| | | Beech | Lignite | Bituminous coal (CBC) | | | |
|---|---|---|---|---|---|---|---|
| Char production conditions | | | | | | | |
| Setup | | Oven | FBR | FBR | FBR | FBR | DTR |
| $d_{p,50}$ | [μm] | 132 | 150 | 175 | 175 | 175 | 175 |
| $T$ | [K] | 1073 | 1273 | 1273 | 1273 | 1273 | 1573 |
| $t$ | [s] | 1800 | 15 | 15 | 15 | 15 | 0.12 |
| Reaction kinetics | | | | | | | |
| Atm. | | $N_2/O_2$ | $N_2/O_2$ | $N_2/O_2$ | $CO_2/O_2$ | $N_2/CO_2$ | $N_2/O_2$ |
| $E_{a,I}$ | $[\frac{kJ}{mol}]$ | 138.0 | 177.6 | 120.9 | 116.6 | 333.8 | 112.4 |
| $A_I$ | $[\frac{1}{s\,kPa^n}]$ | $8.3 \times 10^5$ | $4.1 \times 10^8$ | $2.9 \times 10^4$ | $1.8 \times 10^4$ | $5.4 \times 10^9$ | $2.8 \times 10^4$ |
| $T_{shift}$ | [K] | 993 | 938 | 1098 | 1048 | – | 1098 |
| $E_{a,II}$ | $[\frac{kJ}{mol}]$ | 57.5 | 56.6 | 62.9 | 64.3 | – | 45.5 |
| $A_{II}$ | $[\frac{1}{s\,kPa^n}]$ | 49.0 | 78.0 | 54.7 | 42.5 | – | 18.6 |
| $n$ | [–] | 0.6 | 0.89 | 0.72 | 0.66 | 0.49 | – |
| Ref. | | [16,19] | [15] | [13] | [13] | [13] | – |

favour thermal annealing coming along with the loss of functional groups [17] and a transition from an amorphous carbon structure to graphitic or crystalline areas [18]. This effect can be seen in the experiments with beech wood char pyrolysed in a tube furnace for $t = 1800\,s$ at $T = 1073\,K$. Although the volatile content is higher (see Table 1.1) and thus pore formation might be increased, the reactivity is lower. The significantly lower ash content and the associated lower catalytic effect could also reduce the reaction rate.

All other curves show reaction rates obtained with chars produced from Colombian bituminous coal (CBC). The direct comparison of the reaction rates in $N_2/O_2$ atmosphere with the char produced in the FBR reveals a lower reactivity than the other two fuels, potentially evoked by the observed small porosity in SEM images [13]. By contrast, chars produced from the same raw fuel in the drop tube reactor (see Sect. 2.2.3) have significantly more pores [20]. One reason is the higher particle heating rates that favour the formation of metaplast. As a consequence, the reaction rates are higher than for the char particle produced in the FBR. However, activation energies and the shift temperature from regime I to regime II are very similar (see Table 3.1). Focusing back on the conversion of chars produced in the FBR, the two remaining curves show reaction kinetics for different atmospheres. Reaction rates obtained in a $CO_2/O_2$ atmosphere show in regime I an almost identical behaviour than in an $N_2/O_2$ atmosphere. However, regime II starts for the oxy-fuel atmosphere at lower temperatures ($T_{sh} = 1048\,K$ vs. $1098\,K$) and thus reveals a lower reactivity than in the synthetic air atmosphere. The reduced diffusion coefficient of $O_2$ in $CO_2$ might be the reason. The last curve shows the conversion of CBC char in an $N_2/CO_2$ atmosphere reflecting gasification conditions. This reaction is orders of magnitude slower than the combustion reactions in the investigated temperature range. However, the different activation energies indicate that for temperatures above $T = 1500\,\text{K}$, the gasification reaction will significantly contribute to the overall conversion.

### 3.2.2 Oxygen Influence on Reaction Rates

After focusing on the temperature effect on the reaction rate, the following section deals with the impact of oxygen partial pressure on the reactivity. The effect is shown exemplarily for char particles produced from beech wood. Figure 3.4 shows the relation of the characteristic reaction rate $r$ in the uniform reaction model from the oxygen partial pressure $p_{O_2}$. Changes in the oxygen partial pressure are realised by changing the oxygen volume fraction and keeping the absolute pressure constant to $p \approx 1$ bar. The development of a pressurised version of the fluidised bed reactor will enable the extension of the investigated pressure range to $p = 20$ bar [21].

In addition to the experimentally observed reaction rates, different model approaches are evaluated in their capability to describe the oxygen dependency. Kreitzberg [19] summarises the considered effects as follows: One of the most widely used approaches is the $n^{\text{th}}$-order approach, although it is completely detached from any physicochemical modelling of the elementary reactions taking place. The depen-

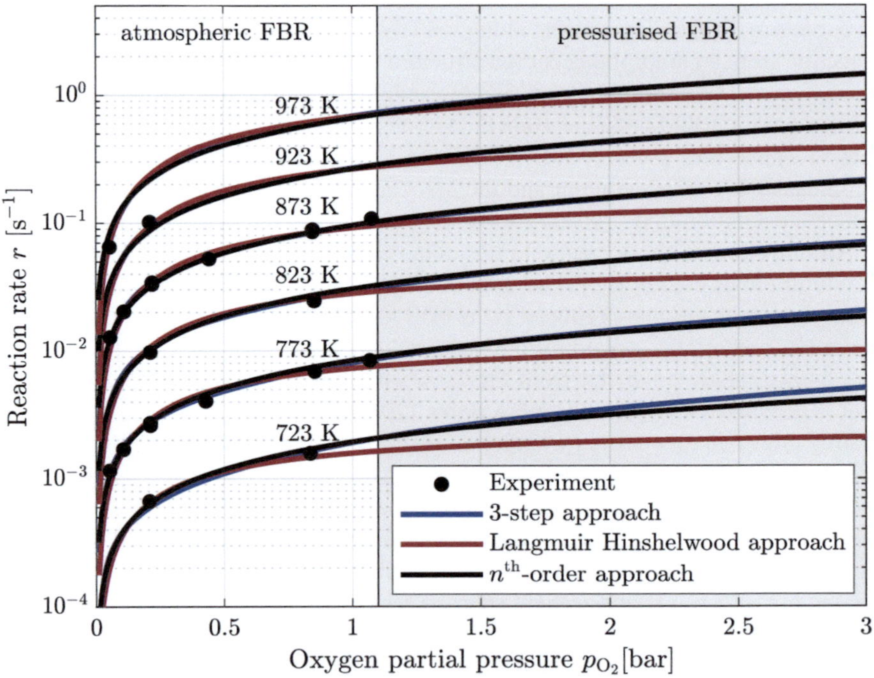


**Fig. 3.4** Characteristic reaction rates for char particles from beech wood reacting in an $N_2/O_2$ atmosphere with different oxygen partial pressures. Comparison of experimental data from the atmospheric FBR [16] and predictions of different models for elevated pressures

dency of the reaction rate from the oxygen partial pressure is given by (3.6):

$$r = k\, p_{O_2}^{n} \tag{3.6}$$

All partial reactions (adsorption, surface reaction, desorption) are lumped into one effective reaction rate. However, typically the different reaction steps have different activation energies and thus limit the overall reaction rate differently. Extrapolation to other boundary conditions than the ones used for calibration is critical. To overcome this shortcoming, the Langmuir Hinshelwood approach was established. It uses different kinetic rates for the adsorption ($k_1$) and desorption ($k_3$) resulting in (3.7) for the reaction rate:

$$r = \frac{k_1\, p_{O_2}}{1 + \frac{k_1}{k_3}\, p_{O_2}} \tag{3.7}$$

The three-step mechanism as given by (3.8) includes additionally the surface reaction ($k_2$):

$$r = \frac{k_1\, k_2\, p_{O_2}^2 + k_1\, k_3\, p_{O_2}}{k_1\, p_{O_2} + 0.5\, k_3} \tag{3.8}$$

The presented approach also forms the basis for the kinetics in the char conversion kinetics model by Hurt and Calo [22]. Figure 3.4 shows the capability of the different model approaches to describe the experimentally observed dependency of the reaction rate from the oxygen partial pressure in the atmospheric fluidised bed reactor. Within the uncertainty of the measurements, all three models are suitable for correctly reflecting the reaction behaviour. However, an extrapolation towards higher oxygen partial pressure reveals differences between the models. While the $n^{\text{th}}$-order approach and the three-step mechanism have a more pronounced dependency on the pressure, the Langmuir Hinshelwood approach predicts a pressure independence of the reaction rate. Current ongoing experimental work will answer the question, which model approach is best suitable to predict reaction rates under high oxygen partial pressures.

## 3.3 Char Conversion in the Laminar Plug Flow Reactor

### 3.3.1 Optically Accessible Reactor for Char Conversion

The experimental setup used in several previous publications [23–26] for char conversion is illustrated in Fig. 3.5. The setup consists of three main components: the laminar flow reactor, the stereoscopic optical camera system, and the solid sampling probe.

The setup allows to assess the reaction progress of pulverised fuel particles under systematic variation of the reaction atmosphere ($O_2$, $H_2O$, $CO_2$), and to collect optical data for single particles in parallel to particle sampling at different residence times. This parallel collection of optical and sampling data delivers complementary information to characterise the details of particle conversion in oxy-fuel atmospheres.

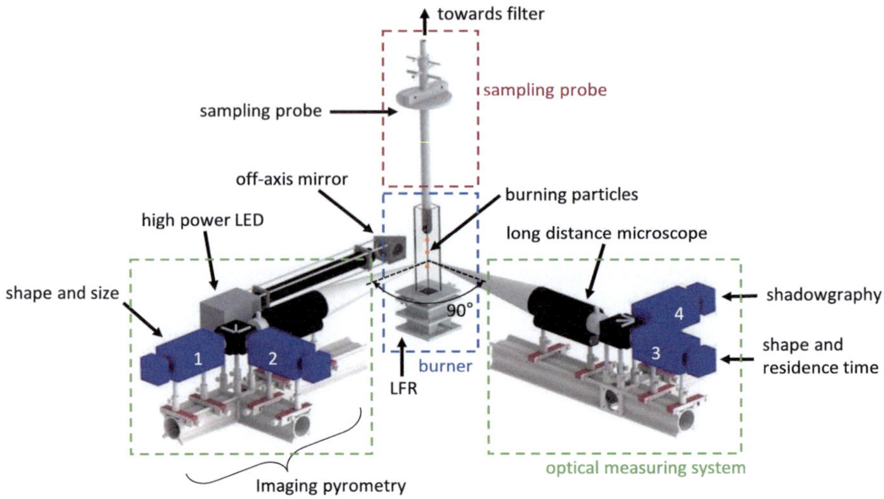


**Fig. 3.5** Experimental setup consisting of the optical measuring system (green box), the burner, (blue box) and the sampling probe (red box)

## Laminar Flow Reactor

A laminar flow reactor, accessible optically, and designed as a flat flame burner, is utilised to examine the behaviour of char conversion [23,25,26]. The heating rates in the reactor exceed $10^4$ K s$^{-1}$, i.e. representing conditions that are typical for pulverised fuel firing scenarios. Figure 3.6 depicts a schematic of the laminar flow reactor.

In the lower section of the reactor, fuel gas ($CH_4/C_2H_2$) and an oxidiser ($O_2/CO_2$) are directed through a ceramic honeycomb structure. Above the ceramic honeycomb, a matrix of uniformly distributed small diffusion flames is formed. The hot exhaust gases from these flames create the reaction atmosphere for the fuel particles, resulting in a laminar flow. Planar quartz glass windows, with square cross-sections measuring $50 \times 50$ mm$^2$, form the walls of the reactor.

The oxy-fuel atmosphere is predominantly composed of three elements: $O_2$, $CO_2$, and $H_2O$. The concentrations of these species affect the reaction rate of the solid fuel, and, hence, must be varied in a sufficient range to cover typical oxy-fuel conditions. Precise supply of $CH_4$, $C_2H_2$, $O_2$, and $CO_2$ to the laminar flow reactor assumes defined reaction conditions in a relevant range.

The particles are injected into the reactor with a low mass flow rate of approximately 1 g h$^{-1}$ using a carrier gas ($CO_2$). To prevent interaction between the particles, it is crucial to maintain a sufficiently large distance between them within the reactor. This distance can be described by the particle number density (PND). Studies have shown that no particle-particle interactions occur below a PND of 0.1 particles/mm$^3$ for particle sizes typical for pulverised fuel combustion (80–200 µm) [24]. The determination of the PND is carried out individually for each image set captured by SCOT, ensuring that single-particle combustion is investigated and group effects do not affect combustion.

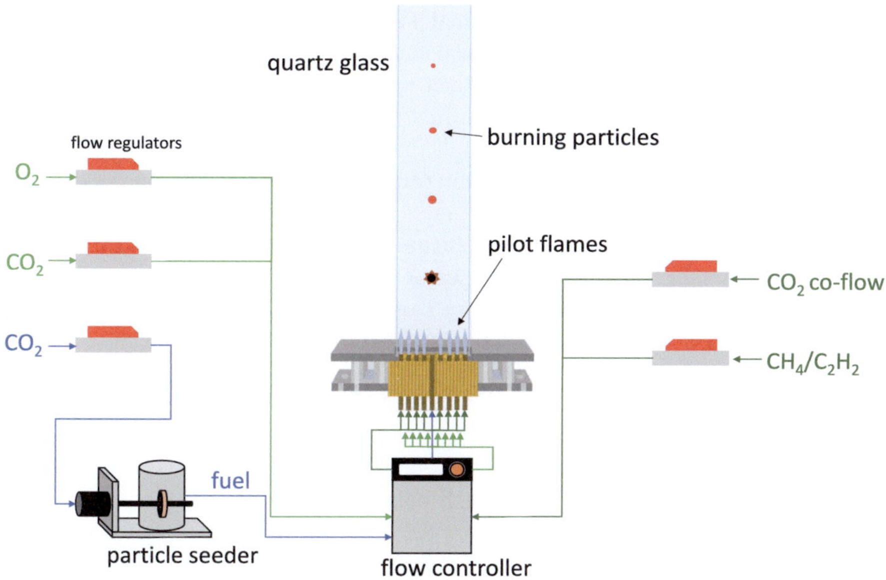


**Fig. 3.6** Schematic chart of the flat flame burner and the peripheral devices

To achieve the above-mentioned PND, a fuel feeding system illustrated in Fig. 3.7 is used. The fuel samples are introduced into a cylindrical reservoir. A perforated disc, submerged within the bed of particles, is rotated by a motor. The upper part of the disc rotates between two tubes. Fuel particles are transported upwards in the pockets of the perforated disc and are pneumatically conveyed into the burner by the carrier gas. Biomass particles are significantly more prone to static charging compared to coal. This static charging of the biomass caused the dosing system to clog. As a countermeasure, a vibrating motor was installed on the tube transporting the particles into the combustion chamber. The vibrating motor aims to prevent statically charged particles from adhering to the tube, thus avoiding blockages in the dosing tube.

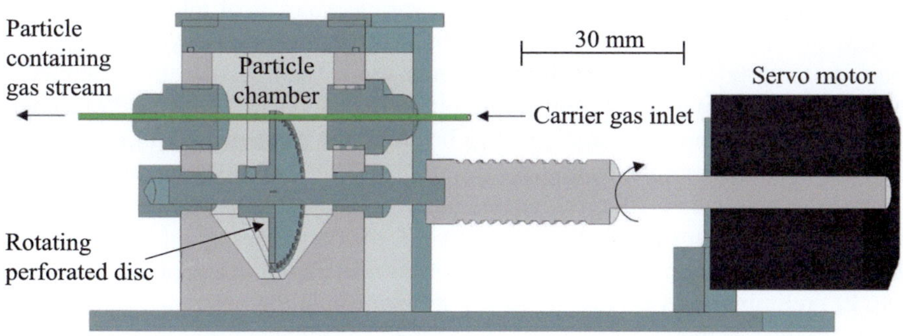


**Fig. 3.7** Schematic figure of the feeding system

**Stereoscopic Camera System for Optical Thermography (SCOT)**
The stereoscopic camera system for optical thermography (SCOT) is an optical camera system, which enables non-invasive measurements of temperature, velocity, size, and shape of single moving micrometre fuel particles. This necessitates high light sensitivity, short exposure times, and high spatial resolution of the optical system. The SCOT system is a development within *Oxyflame*, including some unique design features described below.

The setup comprises four image-intensified CCD cameras, organised into two branches. The first branch, consisting of cameras 1 and 2, measures the temperature via ratio pyrometry. Furthermore, camera 1 of the first branch and camera 3 of the second branch, which are positioned perpendicularly to each other, enable particle shape reconstruction and measurement of particle velocity. Finally, camera 4 of the second branch is used for shadowgraphy to detect non-ignited or already burned out particles.

Each branch is paired with a long-distance microscope *Questar QM1*. The working distance of the microscope can be varied between 559–1676 nm by adjusting the main mirror. The particles are imaged on the detector, which consists of an image intensifier (intensified relay optics, IRO) and a CCD camera Behind the long-distance microscopes, both camera branches utilise filters and beam splitters. Bandpass filters are used for temperature measurement in cameras 1 and 2 by ratio pyrometry. Prior to the bandpass filters, a dichroic long-pass mirror with a 700 nm cut-off wavelength (*Semrock FF700-Di01*) is employed for spectral separation of the particle radiation. Following the mirror, filters with peak wavelengths at centre wavelength (CWL) 650 nm (full width at half maximum (FWHM)= 100 nm, *Semrock FF02-650/100-25*) and 785 nm (FWHM = 62 nm, *Semrock FF01-785/62-25*) are used. The images from the two colour channels are then superimposed and aligned with the calibration curve obtained from a blackbody radiator, assuming the char particles behave as grey radiators. For the second camera pair (camera 3 and 4), an equal beam splitter is used as for the first branch. Camera 1 and 3 are used for shape reconstruction. The equivalent circular diameter (ECD) is calculated using the particle reconstruction method described in Vorobiev et al. [27].

When selecting filters, a compromise must be made between signal yield, spectral detector sensitivity, exposure time, and various sources of interference. In the context of this optical setup, the spectral lines of potassium and sodium represent sources of interference during solid fuel combustion. The filter selection completely bypasses the sodium line. The potassium line is in the area of the channel of camera 1. For measurable emission, sodium must be present in its elemental form and externally excited, for example, by high temperature. These conditions are only achieved during pyrolysis and volatile combustion in the combustion process. However, since the main focus of this part is the char burnout, the above-described drawback was accepted. Examinations were conducted concerning the influence of the potassium line. In preliminary investigations aimed at quantifying the potential impact of the potassium line, cameras 1 and 2 were fitted with specific bandpass filters for potassium and sodium ((CWL 766 nm) and (CWL 589 nm), 10 nm FWHM), instead of the previously mentioned pyrometry filters. In both channels, during both pyrolysis and char burnout, no emissions discernible from the general measurement noise were detected for either sample. For the second camera pair (camera 3 and 4), an equal beam splitter is used as for the first pair. Accordingly, a bandpass filter with CWL of 785 nm (FWHM 63 nm) is employed for camera 3 to ensure that the spectral lines of potassium and sodium do not influence the particle velocity measurement.

As discussed, in cameras 1, 2, and 3, the temperature, velocity, shape and size of the reacting particles are measured through the detection of their self-luminosity. For calibrating models and general understanding of ignition and char conversion, images of particles emitting low and non radiation are necessary. To achieve this, it is necessary to examine these particles using a backlight source in shadowgraphy.

Earlier studies [25,26] were conducted using a controllable *Schott KL1500* halogen lamp. This lamp was positioned at the focus of a parabolic off-axis mirror via a light fibre. The collimated light beam is aligned parallel to the optical axis of the long-distance microscope. This ensures that particles are sharply imaged over a large depth of field, thereby leveraging the advantage of collimated radiation. The light from the halogen lamp is blocked by a short-pass filter above 700 nm. This prevents the signal of camera 3 from being overlaid by background radiation.

With the increasing significance of biogenic fuels, the importance of particle shape has grown. While in fossil fuels like coal particles, a spherical shape of particles has been identified, biomass particles predominantly exhibit an irregular shape. The irregular form of biomass particles significantly impacts the burnout. Consequently, a new backlight system has been installed to achieve more detailed and precise results regarding particle shapes in recent studies [23]. The new backlight system (a high power LED) improves the contrast between the particles and the background illumination. The high power LED (*LPS II, ILA 5150*) is driven in continuous wave mode and operates at a peak wavelength of 523 nm. This allows for a more detailed reconstruction of the biomass particles during the combustion process. This novel system improved the uniformity of background illumination, reducing the intensity variation measured by the camera within a single frame (without particles) from approximately 20% in previous studies [24,25] to about 4% within these studies [23].

**Sampling Probe**

Char particles are extracted at various heights in the flat flame burner to investigate the progression of conversion. Figure 3.8 depicts the diagram of the sampling probe.

The heterogeneous reaction at the particle surface must be immediately stopped upon entering the probe. Therefore, the particles are quenched with nitrogen when they enter the probe. The quench gas is introduced via a thermal mass flow controller. Both, the reaction atmosphere and the added quench gas are evacuated using a membrane pump. The ratio of quench gas flow rate to sucked reaction atmosphere is about 10:1. After the char particles pass through the probe, they are separated from the gas by a coalescence gas filter with a PTFE filter element (pore size $< 2\,\mu m$).

Additionally, the probe is temperature-controlled using an oil thermostat, which is connected to an external cooling circuit. The temperature of the thermal oil is regulated based on the temperature at the tip of the probe. The primary criterion is to prevent water from condensing on the outer tube of the probe and to stop the ongoing

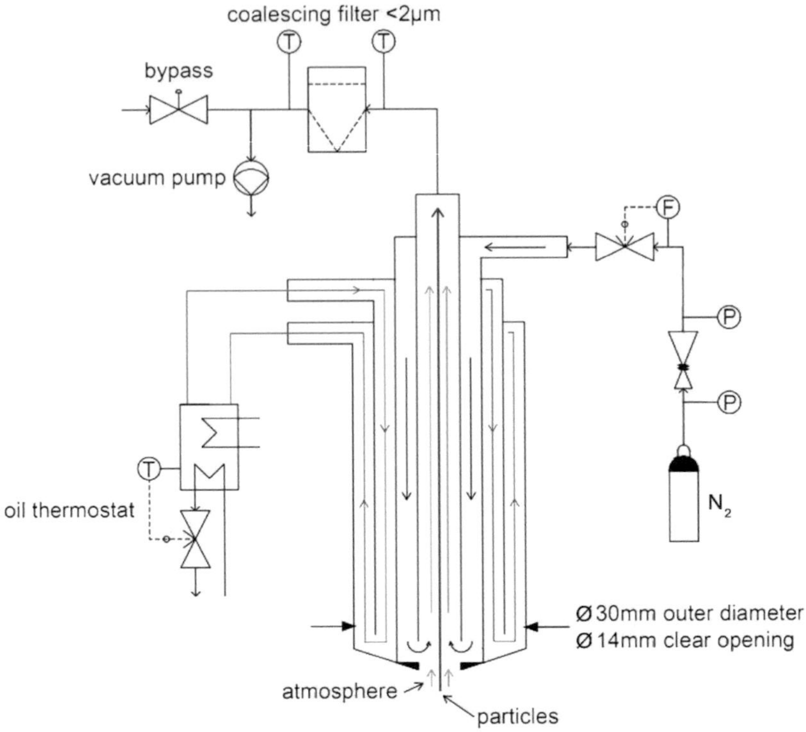


**Fig. 3.8** Flow chart of the sampling probe

reaction of the char particles when they enter the probe. In addition to the temperature measurement point at the probe tip, there are other temperature measurements marked with a "T" in Fig. 3.8. Depending on the atmosphere's characteristics, typical temperatures of 413–443 K are observed at the probe tip, while temperatures before the filter vary between 353–393 K.

Temperature profile measurements within the probe were conducted using a type K thermocouple. It was observed that the temperature gradient for cooling within the probe is approximately 60 K mm$^{-1}$ for the gas temperature.

**Evaluation Procedure**

The creation of a data set for char conversion requires a large number of additional steps from the raw images of the camera system to the results of the data set. Due to the large number and depth of detail of the intermediate steps, only the most important steps are presented in this chapter. For all intermediate steps, the reader is referred to the work of Vorobiev et al. [25,26] and Tarlinski et al. [23,24].

The first step is the calibration and focusing of SCOT. The aim of adjusting and focusing the two camera arms, each with two cameras, is to achieve the optimum overlay of depth of field ranges. A calibration plate is set up at an angle of 45° to both camera arms. The long-distance microscopes must be adjusted so that the

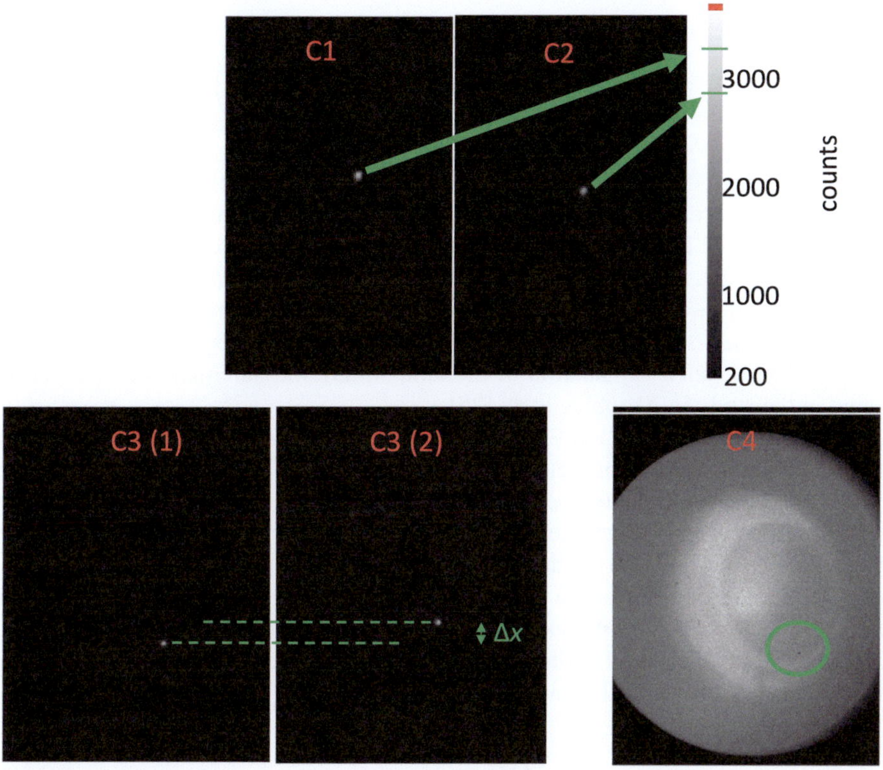


**Fig. 3.9** Raw images from SCOT with the numeration for the respective camera. Camera 3 triggers in double image mode, therefore the number in brackets stands for the respective image of the camera

mark on the calibration plate is in the centre of the two image fields and the depth of field area is in the centre of both images. After the depth of field calibration of the two camera arms in relation to each other, the calibration is carried out with the identical calibration plate for the individual camera arms. For this purpose, the target is positioned orthogonally to the field of view of the cameras. This allows an image scale to be created with several intermediate steps, which is created individually for each camera. The raw images that SCOT records after calibration and focusing are shown as an example for a burning particle in Fig. 3.9.

In Fig. 3.9, the numbers represent the respective images from each camera. As camera 3 runs in double frame mode, the number in the brackets indicates the images at different times. The images from cameras 1 and 2 are used to determine the temperature of the particles. A temperature calibration is carried out for this purpose, which creates the correlation between the ratio of the signal intensity (counts) at certain wavelengths and the temperature of the particle. The calibration function is only valid for the components installed in this experimental setup. The ratio of the measured intensities from the cameras is applied to (3.9) for determining the particle temperature. The physical principle of the temperature calibration is based on Wien's approximation.

$$-\frac{1}{T_\mathrm{p}} = \frac{\lambda_1\,\lambda_2}{c_2\,(\lambda_2-\lambda_1)}\,\frac{\ln\left(\kappa_1\,\frac{\epsilon_2}{\epsilon_1}\left(\frac{\lambda_1}{\lambda_2}\right)^5\,R\right)}{\ln\left(\kappa_2\,R\right)} \tag{3.9}$$

The variables used in (3.9) are defined as follows: The particle temperature to be determined is represented by $T_\mathrm{p}$. The parameters $\lambda_1$ and $\lambda_2$ denote the first and second wavelengths used in two-colour pyrometry, respectively. The calibration constants are given by $\kappa_1$ and $\kappa_2$, while $c_2$ represents the second radiation constant. The intensity ratio of the two wavelengths is expressed as $R$. Under the assumption of a grey body emitter, the ratio of the emission gradients for the two wavelength ranges is set to one. A more detailed explanation of the temperature measurement method can be found in the study by Vorobiev et al. [25].

Images C3(1) and C3(2) in Fig. 3.9 show the particle recorded using camera 3 in double frame mode. The particle in the second image C3(2) is detected in a higher position due to the time difference of 250 µs in comparison to C3(1). The distance $\Delta x$ is measured between the geometric centres of the particle in both frames. From the measured distance $\Delta x$ and the time difference $\Delta t = 250\,\mu\mathrm{s}$ between the two images, an individual velocity $v$ can be determined for each particle. In addition, the exact height $H$ at which the particles are located above the burner entrance, is known. With these three measured variables, the individual residence time of a particle can be calculated using the following equation.

$$t(H) = \int_0^H \frac{1}{v}\,dh \tag{3.10}$$

The image C4 in Fig. 3.9 displays a particle illuminated by a backlight of a high power LED. This information is of particular interest for extending a dataset, as it allows for the detection of both unignited and already fully burned out particles. This enables conclusions to be drawn regarding the entire particle population, not just those that are self-luminous. The ratio of self-luminous detected particles to the total particle population is essential for modelling purposes.

### 3.3.2 Results

#### Measurement Matrix

The achieved datasets for the char burnout of various solid fuels—Colombian bituminous coal (CBC), walnut shells (WNS), torrefied miscanthus giganteus (TMG), beech wood (BW), Rhenish lignite (RL)—include all optical measurement results concerning temperature, equivalent circular diameter (ECD) [27] via pyrometry and shadowgraphy, as well as the ratio between reacting and non-reacting particles. The ratio is calculated from the total number of particles detected by their self-luminosity in cameras 1–3 to the total number of particles that are detected by the backlight illumination. Additionally, experimental results from the probe investigations are available for all mentioned fuels. These results depict the burnout, the release of elements, as well as the pore size distribution along the burnout. Within this chapter, individual results from larger data sets are presented and compared. For a comprehensive overview of the experimental findings, readers are referred to the literature [23,26].

The following settings have been realised: $O_2$ content is varied in steps of 10 mol.% between 10 and 40 mol.%, and for $H_2O$, values of 10 and 20 mol.% are set. The remaining component is $CO_2$, which is typically the dominant species in oxy-fuel atmospheres. These settings lead to eight distinct atmospheres, each represented by a pair of numbers. The first number denotes the $O_2$ concentration, while the second represents the $H_2O$ concentration within the atmosphere (e.g. "40-20" stands for 40 mol.% $O_2$, 20 mol.% $H_2O$, and 40 mol.% $CO_2$ in the atmosphere). The eight atmospheres were specified such that similar adiabatic gas temperatures result. Gas temperatures and wall temperatures were measured using radiation-corrected (see Shaddix [28]) thermocouples of type S and are depicted in Fig. 3.10. Additionally, burner wall temperatures are shown on the right ordinate. As the laminar flow reactor is non-adiabatic, slight differences in the measured gas (and wall temperatures) occur between the different atmospheres, as heat losses depend on the specific composition of the different atmospheres. Hence, completely identical temperatures cannot be achieved, but this is not needed as the calibration procedure to derive kinetic parameters will account for the actual experimental boundary conditions. It just needs to be assured that temperatures are similar to avoid interpolation of kinetic parameters over too large temperature ranges, which might be critical. To further ensure comparability of the data, the gas flow rate to the burner for all experiments was set at 45 L $min^{-1}$, leading to comparable velocities of the gases in the laminar flow reactor.

An overview of the measurements for char burnout in the laminar flow reactor for the fuels investigated is presented in Table 3.2. In total, three highly detailed datasets (Colombian bituminous coal (CBC), walnut shells (WNS), and torrefied miscanthus giganteus (TMG)) were investigated [23,25,26]. This enables further calibration of models using these datasets. In addition to these three highly detailed datasets, beech wood (BW), torrefied beechwood (TBW), and Rhenish lignite (RL) were studied by Vorobiev et al. [27] in just one oxy-fuel atmosphere. Based on these data, kinetic parameters were derived based on an energy balance around single particles [27]. This enables calibration of detailed char conversion models such as the char conversion kinetics (CCK) model decribed in the study by Holland and Fletcher [29]. Such a calibration is shown for the example of walnut shells in Sect. 8.4.

**Comparison of Different Fuels**

In Fig. 3.11, the particle temperatures of the three different samples of miscanthus (particle size 200–250 μm), walnut shells (particle size 125–150 μm), and a high volatile bituminous coal (particle size 90–125 μm) are plotted against residence time for the atmosphere 10–10. The atmosphere 10–10 comprises 10% oxygen, 10% water, and 80% carbon dioxide.

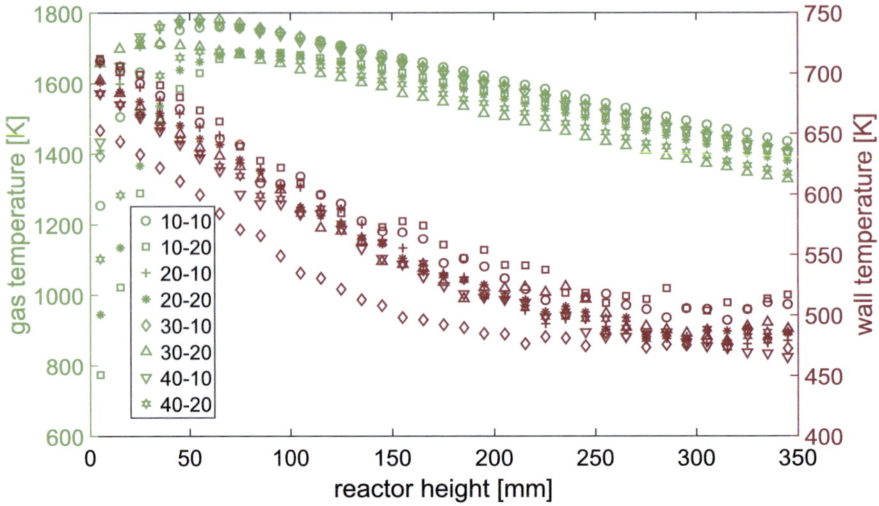


**Fig. 3.10** Average gas temperatures along the centreline of the reactor and wall temperatures of the eight investigated oxy-fuel atmospheres

**Table 3.2** Overview of all produced data sets during the whole time period of *Oxyflame*. All data can be found in the following literature [23,25–27]

| Fuel | Atmosphere composition [mol.%] | | | | | | | | | |
|---|---|---|---|---|---|---|---|---|---|---|
| | 10 | 20 | 30 | 40 | 10 | 20 | 30 | 40 | 30 | $O_2$ |
| | 10 | 10 | 10 | 10 | 20 | 20 | 20 | 20 | 10 | $H_2O$ |
| | 80 | 70 | 60 | 50 | 70 | 60 | 50 | 40 | 10 | $CO_2$ |
| | 0 | 0 | 0 | 0 | 0 | 0 | 0 | 0 | 50 | $N_2$ |
| CBC | Full data[1] | Full data[1] | Full data[1] | Full data[1] | Full data[1] | Red. data[2] | Full data[1] | Red. data[2] | – | |
| TMG | Full data[1] | Full data[1] | Red. data[2] | Full data[1] | Red. data[2] | Red. data[2] | Red. data[2] | Full data[1] | – | |
| WNS | Full data[1] | Full data[1] | Full data[1] | Full data[1] | Full data[1] | Full data[1] | Full data[1] | Full data[1] | – | |
| BW | – | – | – | – | – | – | – | – | Red. data[2] | |
| TBW | – | – | – | – | – | – | – | – | Red. data[2] | |
| RL | – | – | – | – | – | – | – | – | Red. data[2] | |

[1]Full data set includes particle temperature $T_p$, particle size as equivalent circular diameter $d$, aspect ratio (AR) of the reconstructed particle, ratio of reacting (RR) (determined via self-luminescence) to all particles (determined via shadowgraphy), burnout progress $X_{bo}$, elemental release of C, H, N $X_{CHN}$, and pore size distribution (PSD) divided into pore volume and pore surface

[2]Reduced data set includes particle temperature $T$, particle size as equivalent circular diameter $d$, aspect ratio (AR) of the reconstructed particle, ratio of reacting (RR) (determined via self-luminescence) to all particles (determined via shadowgraphy)

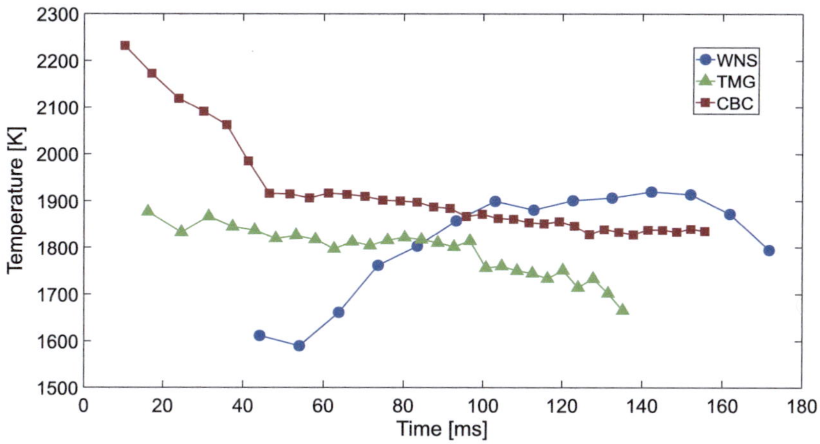


**Fig. 3.11** Mean temperature over the residence time for the torrefied miscanthus (TMG), walnut shells (WNS) and Colombian bituminous coal (CBC) in the 10–10 atmosphere for all detected particles at various residence times. For all further statistical details the reader is referred to the following publications by Tarlinski et al. [23], and Vorobiev et al. [25,26]

For the 10–10 atmosphere, the coal particles exhibit the highest temperature at the beginning. As the residence time increases, the temperature drops and converges to a value of approximately 1900 K. The torrefied miscanthus particles initially show a slightly lower temperature. However, it is observed that with ongoing burnout, the temperature decreases. In comparison to the two previously mentioned fuels, the combustion of walnut shell particles begins at lower temperatures. As the burnout progresses, the temperature rises for these particles, before dropping again towards the end of combustion.

The temperatures of the coal particles can be detected at earlier residence times compared to the two biogenic fuels, as the volatile flames from the coal are detected with SCOT, while no volatile flames are visible for the biogenic fuels. The temperature of the volatile flames surrounding the coal particles is higher than that of the char particles, this is the reason for the higher temperature of the coal particles than that of the biomass particles at earlier residence times. For WNS, a peak is observed shortly before the burnout is complete, as noted in other studies [27]. This peak just before complete burnout appears to be characteristic for raw biomass particles. Another notable aspect is that torrefied biomass exhibits an almost linear temperature curve compared to the other two fuels. A possible reason for this is that the torrefaction process has homogenised the biomass. While the natural components in the chemical composition of the other two fuels (WNS and CBC) may play a role, it is expected that the chemical structures in torrefied miscanthus have already experienced temperatures around 300 °C for a prolonged period due to the torrefaction process.

The ECD of the self-luminosity of particles in a 10–10 atmosphere is initially largest for the CBC particles compared to the other two fuels. After a residence time of approximately 40 ms, the ECD of the self-luminosity of the CBC particles rapidly

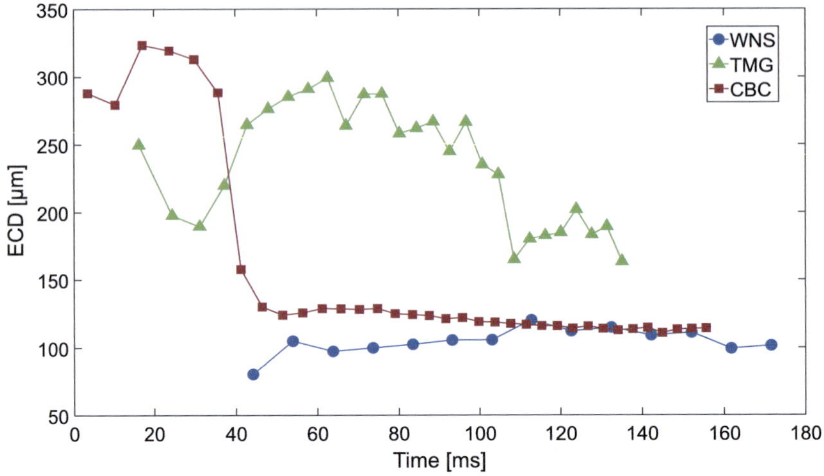


**Fig. 3.12** Equivalent circular diameter (ECD) by pyrometry over the residence time for torrefied miscanthus (TMG), walnut shells (WNS), and Colombian bituminous coal (CBC) in the 10–10 atmosphere

decreases to a third. A visual inspection of the large events allows to identify the majority of these as volatile flames enveloping their parent particles, which are rich in soot and therefore provide a strong signal of self-luminosity. Subsequently, the camera system detects char burnout on the particle surface, leading to a decrease in the ECD. In contrast, experimental investigations did not detect volatile flames with the camera system for walnut shell particles and torrefied miscanthus particles. The ECD of the self-luminosity in the char burning phase is the biggest in the miscanthus samples, as the particles exhibit the highest asphericity (aspect ratio of approximately 2.0–2.2) among the three examined fuel samples (Fig. 3.12).

A possible reason for the presence of volatile flames around coal particles is increased soot formation. While coal particles exhibit higher soot formation, biomass shows little to no soot formation in the flat flame burner. Since the volatile flame consists of numerous very small soot particles that emit thermal radiation, the detection of volatile flames around coal particles can be attributed to soot formation. A common effect for the three fuels is that the ECD measured by pyrometry shows a slight increase after the onset of char combustion. This effect occurs similarly across all fuels, as larger particles ignite at a later stage and are only detected by their self-luminescence in SCOT at a correspondingly later residence time.

The ECD of the fuel particles detected by shadowgraphy is depicted in Fig. 3.13. These results reveal that the ECD of torrefied miscanthus particles is the highest and decreases with increasing residence time. Walnut shell particles exhibit a similar trend, albeit starting at smaller ECD values. This is attributed to the stalk-like nature of torrefied miscanthus particles, reflected in their ECD. CBC particles exhibit a peculiarity in the ECD observed by shadowgraphy. During combustion, there is an initial increase in ECD followed by a decrease. Particle swelling may be a possible

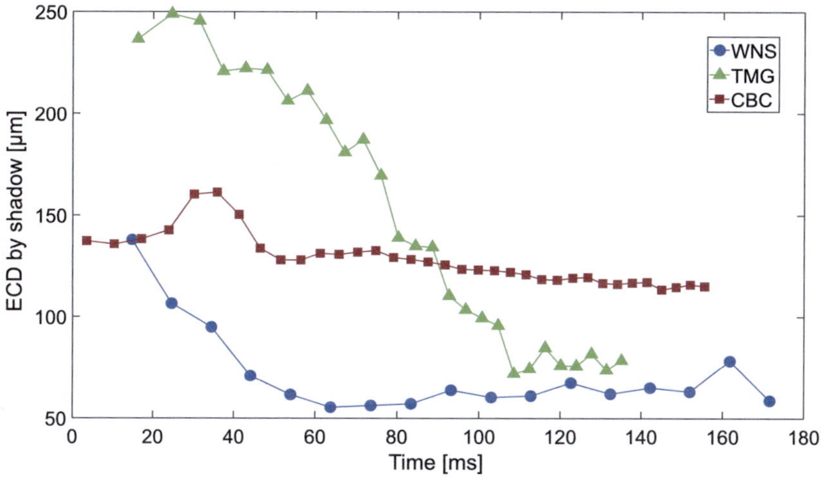


**Fig. 3.13** ECD by shadowgraphy over the residence time for torrefied miscanthus (TMG), walnut shells (WNS), and Colombian bituminous coal (CBC) in the 10–10 atmosphere

cause for this trend [30]. Subsequently, a decreasing trend is observed, similar to the other fuel samples.

Furthermore, both biogenic fuels (WNS and TMG) exhibit a very strong and rapid decrease in ECD as measured by shadowgraphy. The WNS particles shrink to approximately 2.5 times their original size, and TMG particles shrink to about 3 times their initial size over the entire burnout period. In contrast, the high volatile bituminous coal particles only decrease by around 0.2 times compared to their initial value. The approximately tenfold higher ash content in WNS particles compared to CBC particles is associated with a greater reduction in volume during the burnout, and consequently, a more pronounced decrease in diameter. Since only ash remains at the end of combustion, it can be assumed that, relative to the initial state, the CBC contains approximately ten times more mass, which is correspondingly reflected in the volume.

For the validation of individual models and simulations, additional parameters are essential. According to Holland [11] and Holland and Fletcher [12], an important factor for validating the CCK model is the information about ignited (self-luminescence particles) and unignited or already burned particles. While Holland [11] and Holland and Fletcher [12], made assumptions regarding this, results are provided within this dataset. The results of the three fuels from a 10–10 atmosphere are illustrated in Fig. 3.14. The typical characteristics in the results are the same for all three investigated fuels. Initially, it can be observed that all three fuels are only detected by shadowgraphy, as most particles have not yet ignited. Subsequently, the ratio increases for all fuel particles as they ignite and are detected by their self-illumination in the pyrometry cameras. Towards the end, the ratio decreases for all particles as they are already burned out and only a few bigger particles are still detected by the pyrometry cameras.

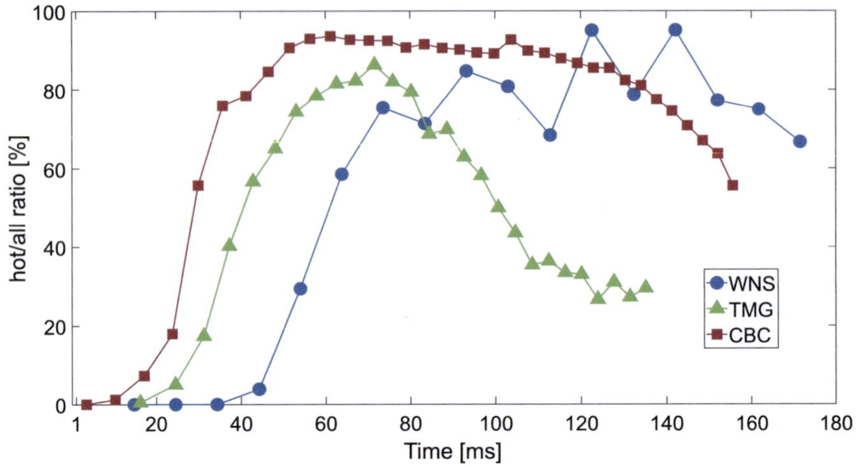


**Fig. 3.14** Ratio of reacting (determined via self-luminescence) to all particles (determined via shadowgraphy) for torrefied miscanthus (TMG), walnut shells (WNS) and Colombian bituminous coal (CBC) samples in the 10–10 atmosphere. Weighted by the number of particles detected in the respective part of the optical system

Figure 3.15 shows the burnout along the vertical position in the burner of the three investigated fuels for the 10–10 atmosphere. The burnout of the biogenic fuels occurs faster than that of the fossil fuels. Raw biomass, in this case walnut shell particles, exhibits a faster burnout than the torrefied miscanthus particles. The slowest progressing burnout is observed for the CBC particles.

In addition to particle burnout, the release of different species (C, H, N) is of further interest for the validation and calibration of models. As an example, within

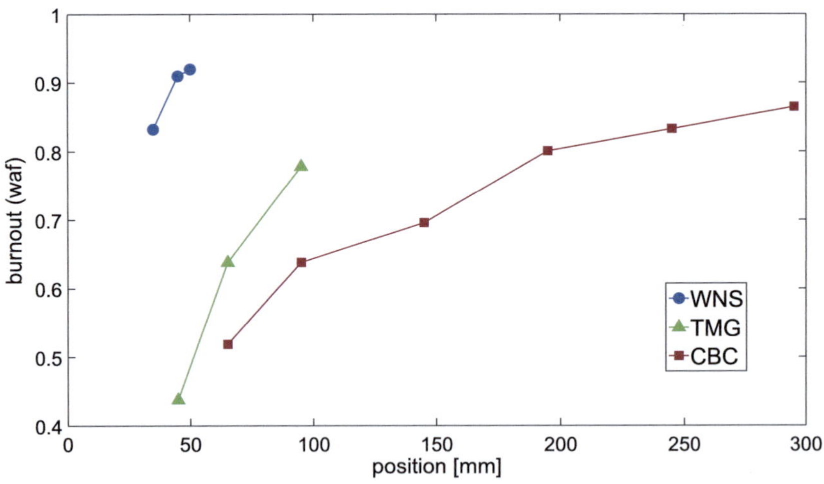


**Fig. 3.15** Burnout progress over the position from the entry in the burner for torrefied miscanthus (TMG), walnut shells (WNS) and Colombian bituminous coal (CBC) in the 10–10 atmosphere

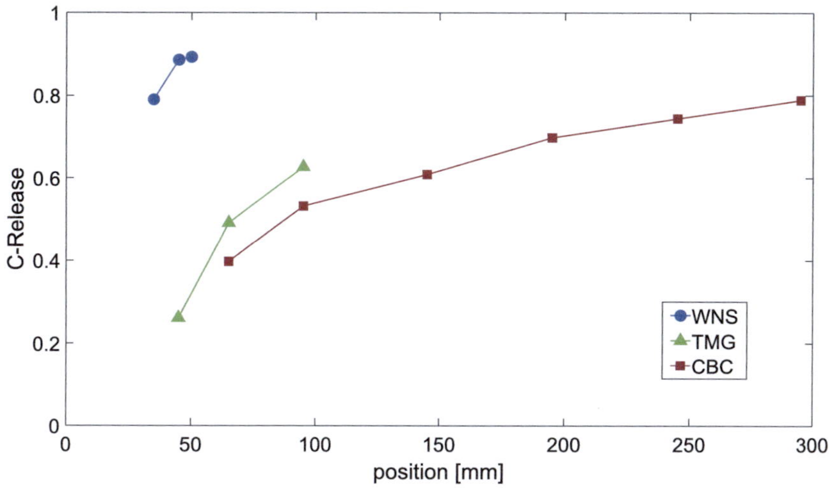


**Fig. 3.16** Carbon release during combustion of torrefied miscanthus (TMG), walnut shells (WNS), and Colombian bituminous coal (CBC) over the position in the burner

this chapter, the release of carbon is presented. Figure 3.16 illustrates the results of carbon release for the three fuels, exemplified for the 10–10 atmosphere. The results of carbon release correlate with the burnout results. It is evident that initially, walnut shell particles exhibit the highest carbon release, followed by torrefied miscanthus particles and CBC particles. Overall, for all three fuels investigated, the trend is observable that the carbon release from the fuel increases with the vertical position in the burner.

In addition to the results presented here, other chemical elements such as Fe, K, Na, Mg, and Ca show catalytic effects on the burnout of various fuels. The catalytic effects of each individual component, as well as mixtures of different components in the respective fuels, can be investigated in the flat flame burner using SCOT. Initial studies on catalytic effects in the flat flame burner were conducted in the work of Eckhard et al. [31]. Details on these catalytic effects are presented in Chap. 4.

**Influence of Oxy-Fuel Atmosphere**

While the previous section presented and discussed the results regarding the differences between individual fuels, this section provides a concise summary of the effects of the eight different oxy-fuel atmospheres studied. In these atmospheres, the oxygen content varies in steps of 10 mol.% from 10 to 40 mol.%. Additionally, for each oxygen concentration, two different water vapour levels of 10 and 20 mol.% were examined, resulting in a total of eight atmospheres.

The influence of water vapour in the atmosphere exhibits a similar effect on particle temperatures for both biogenic and fossil fuels. Higher water vapour (20 mol.%) levels in the atmosphere result in slightly lower particle temperatures compared to those observed under the same conditions with lower water content (10 vol.%). In addition water vapour also affects burnout. Particles in atmospheres with higher water vapour content (20 mol.%) achieve a higher burnout at the same residence times compared to those with lower water vapour content (10 mol.%). A possible reason for this faster burnout at lower particle temperatures may be that, due to reduced temperatures, oxygen can penetrate deeper into the particle, leading to quicker burnout from within. In contrast, at lower water vapour content, oxygen is consumed at the particle surface, limiting its penetration depth and thus reducing the overall burnout achieved.

The level of oxygen concentration in the atmosphere significantly affects char burnout. It is observed that a reduction in oxygen concentration of about 10 mol.% results in a lower particle temperature by approximately 150 K. This effect is evident for both biogenic and fossil fuels. An increase in oxygen concentration in the atmosphere leads to a faster burnout in both types of fuels. The dependence of burnout on different oxy-fuel atmospheres is particularly evident for coal. In biogenic fuels, the difference in burnout is not as clearly identifiable as it is for example for the CBC. Due to the very high volatile content of biomass, it is converted significantly faster than coal. Consequently, especially at shorter residence times, the burnout curves for biogenic fuels are nearly identical.

## 3.4 Conclusion

Char combustion represents the slowest part within the overall combustion sequence. In this regard, it is a crucial subprocess for the design of the entire power plant process in an oxy-fuel atmosphere. Therefore experimental data is required for the validation of models and simulations. The experimental data currently available in the literature has shown several shortcomings. This work aims to fill the gaps in existing datasets by generating new, comprehensive datasets. Two independent methods were employed for this purpose. On the one hand, optical measurement techniques were used to investigate the combustion behaviour of individual pulverised fuels. On the other hand, a solid sample probe was utilised to collect and analyse fuel samples at different stages of combustion. The samples were analysed for elemental composition, pore distribution, and proximate analysis.

The optical measurements provided detailed information on particle temperature, residence times, particle shape and geometry, and the ratio of reacting to non-reacting particles. This resulted in a highly detailed dataset. Experimental data on char combustion was generated for eight different oxy-fuel atmospheres, allowing for the examination of the effects of oxygen, water vapour, and carbon dioxide concentrations. The investigation revealed that the oxygen concentration in the atmosphere exerts the most significant influence on char combustion across all studied fuels, including biomass and various coals. Higher oxygen levels led to increased particle temperatures, faster combustion, and more rapid reduction in particle diameter. A noteworthy finding was the determination of particle shapes using shadowgraphy. The particle geometries obtained via shadowgraphy (using a backlight) differed from those determined by self-luminescence. These findings provide new calibration and validation data for models that were not available in previous datasets.

Consequently, the work on char combustion in oxy-fuel atmospheres establishes a new benchmark for an optical system, a burner for pulverised fuels and a solid sample probe for collecting fuel particles in various residence times. Ultimately, the results underscore the importance of detailed experimental data sets of char burnout of various pulverised fuels for calibration and evaluation of various models and simulations.

**Acknowledgement** This work has been funded by the Deutsche Forschungsgemeinschaft (DFG, German Research Foundation)—215035359—SFB/TRR 129 'Oxyflame'.

## References

1. Wornat, M. J., Hurt, R. H., Yang, N. Y. and Headley, T. J. "Structural and Compositional Transformations of Biomass Chars During Combustion". *Combustion and Flame* 100.1-2 (1995), 131–143. https://doi.org/10.1016/0010-2180(94)00055-W.
2. Senneca, O. and Cortese, L." Thermal Annealing of Coal at High Temperature and High Pressure. Effects on Fragmentation and on Rate of Combustion, Gasification and Oxy-Combustion". *Fuel* 116 (2014), 221–228. https://doi.org/10.1016/j.fuel.2013.07.065.
3. Senneca, O., Ontyd, C., Cerciello, F., Schiemann, M. and Scherer, V. "Extension of the Thermal Annealing Concepts Developed for Coal Combustion to Conversion of Lignocellulosic Biomass". *Energy & Fuels* 34.3 (2020), 3661–3670. https://doi.org/10.1021/acs.energyfuels.9b03172.
4. Holland, T., Fletcher, T. H. and Senneca, O. "Review of Carbonaceous Annealing Effects on $O_2$ and $CO_2$ Coal Reactivity". *Energy & Fuels* 33.11 (2019), 10415–10434. https://doi.org/10.1021/acs.energyfuels.9b02698.
5. Senneca, O., Apicella, B., Russo, C., Cerciello, F., Salatino, P., Heuer, S., Wütscher, A., Schiemann, M., Muhler, M. and Scherer, V. "Pyrolysis and Thermal Annealing of Coal and Biomass in $CO_2$-Rich Atmospheres". *Energy & Fuels* 32.10 (2018), 10701–10708. https://doi.org/10.1021/acs.energyfuels.8b02417.
6. Tognotti, L., Longwell, J. P. and Sarofim, A. F. "The Products of the High Temperature Oxidation of a Single Char Particle in an Electrodynamic Balance". *Symposium (International) on Combustion* 23.1 (1991), 1207–1213. https://doi.org/10.1016/S0082-0784(06)80382-6.
7. Hecht, E. S., Shaddix, C. R., Molina, A. and Haynes, B. S. "Effect of $CO_2$ Gasification Reaction on Oxy-Combustion of Pulverized Coal Char". *Proceedings of the Combustion Institute* 33.2 (2011), 1699–1706. https://doi.org/10.1016/j.proci.2010.07.087.
8. Joos, F. *Technische Verbrennung: Verbrennungstechnik, Verbrennungsmodellierung, Emissionen*. Berlin: Springer, 2006.
9. Geier, M., Shaddix, C. R., Davis, K. A. and Shim, H.-S. "On the Use of Single-Film Models to Describe the Oxy-Fuel Combustion of Pulverized Coal Char". *Applied Energy* 93 (2012), 675–679. https://doi.org/10.1016/j.apenergy.2011.12.097.
10. Shaddix, C. R. and Molina, A. "Particle Imaging of Ignition and Devolatilization of Pulverized Coal During Oxy-Fuel Combustion". *Proceedings of the Combustion Institute* 32.2 (2009), 2091–2098. https://doi.org/10.1016/j.proci.2008.06.157.
11. Holland, T. M. "A Comprehensive Coal Conversion Model Extended to Oxy-Coal Conditions". PhD thesis. Brigham Young University, 2017.
12. Holland, T. M. and Fletcher, T. H. "Global Sensitivity Analysis for a Comprehensive Char Conversion Model in Oxy-Fuel Conditions". *Energy & Fuels* 30.11 (2016), 9339–9350. https://doi.org/10.1021/acs.energyfuels.6b02190.
13. Gövert, B., Pielsticker, S., Kreitzberg, T., Habermehl, M., Hatzfeld, O. and Kneer, R. "Measurement of Reaction Rates for Pulverized Fuel Combustion in Air and Oxyfuel Atmosphere

Using a Novel Fluidized Bed Reactor Setup". *Fuel* 201 (2017), 81–92. https://doi.org/10.1016/j.fuel.2017.03.009.

14. Kreitzberg, T., Haustein, H. D., Gövert, B. and Kneer, R. "Investigation of Gasification Reaction of Pulverized Char Under $N_2/CO_2$ Atmosphere in a Small-Scale Fluidized Bed Reactor". *Journal of Energy Resources Technology* 138.4 (2016), 042207. https://doi.org/10.1115/1.4032791.
15. Gövert, B. "Char Combustion Kinetics Using a Micro Fluidized Bed Reactor". PhD thesis. RWTH Aachen University, 2018. https://doi.org/10.18154/RWTH-2019-00500.
16. Kreitzberg, T., Pielsticker, S. and Kneer, R. "Influence of Oxygen and Water Vapor on the Thermochemical Conversion of Fossil, Biogenic and Surrogate Fuels (Poster)". $3^{rd}$ *International Workshop on Oxy-Fuel Combustion*. Montabaur, Germany, 2020. https://doi.org/10.18154/RWTH-2020-06388.
17. Kreitzberg, T., Eckhard, T., Pielsticker, S., Muhler, M. and Kneer, R. "Änderung von Struktur und Reaktivität durch Pyrolyse fester Brennstoffe". *29. Deutscher Flammentag*. Bochum, Germany, 2019.
18. Kreitzberg, T., Wirch, N., Bormann, C., Pielsticker, S., Hatzfeld, O., Mayer, J. and Kneer, R. "Thermally Induced Changes in Microstructure and Reactivity of Biogenic and Fossil Fuel Particles". *Applied Energy* 254 (2019), 113607. https://doi.org/10.1016/j.apenergy.2019.113607.
19. Kreitzberg, T. "Heterogene Reaktionskinetik alternativer Festbrennstoffe". PhD thesis. RWTH Aachen University, 2021. https://doi.org/10.18154/RWTH-2021-03165.
20. Heuer, S., Wedler, C., Ontyd, C., Schiemann, M., Span, R., Richter, M. and Scherer, V. "Evolution of Coal Char Porosity from $CO_2$-Pyrolysis Experiments". *Fuel* 253 (2019), 1457–1464. https://doi.org/10.1016/j.fuel.2019.05.071.
21. Kreitzberg, T., Pielsticker, S. and Kneer, R. "Einfluss von erhöhtem Druck bei Pyrolyse, Abbrand und Vergasung alternativer Brennstoffe (Presentation)". *Jahrestreffen der ProcessNet-Fachgruppe W ä rme-und Stoff ü bertragung*. Erfurt, Germany, 2020.
22. Hurt, R. H. and Calo, J. M. "Semi-Global Intrinsic Kinetics for Char Combustion Modeling". *Combustion and Flame* 125 (2001), 1138–1149. https://doi.org/10.1016/S0010-2180(01)00234-6.
23. Tarlinski, D., Freisewinkel, E., Eisenbach, T., Span, R., Schiemann, M. and Scherer, V. "Combustion of Single Walnut Shell Particles in a Laminar Flow Reactor Under Oxy-Fuel Conditions: Optical Measurements and Particle Sampling". *Fuel* 369 (2024), 131613. https://doi.org/10.1016/j.fuel.2024.131613.
24. Tarlinski, D., Geschwindner, C., Li, T., Böhm, B. and Schiemann, M. "Particle Temperature and Composition Measurements in the Ignition Phase of Single Coal Particles and Particle Groups Under Conventional and Oxy-Fuel Atmospheres". *Fuel* 332 (2023), 125894. https://doi.org/10.1016/j.fuel.2022.125894.
25. Vorobiev, N., Valentiner, S., Schiemann, M. and Scherer, V. "Comprehensive Data Set of Single Particle Combustion Under Oxy-Fuel Conditions, Part I: Measurement Technique". *Combustion Science and Technology* 56.2 (2020), 1–22. https://doi.org/10.1080/00102202.2020.1743696.
26. Vorobiev, N., Valentiner, S., Schiemann, M. and Scherer, V. "Comprehensive Data Set of Single Particle Combustion Under Oxy-Fuel Conditions, Part II: Data Set". *Combustion Science and Technology* 5.1 (2020), 1–16. https://doi.org/10.1080/00102202.2020.1754207.
27. Vorobiev, N., Becker, A., Kruggel-Emden, H., Panahi, A., Levendis, Y. A. and Schiemann, M. "Particle Shape and Stefan Flow Effects on the Burning Rate of Torrefied Biomass". *Fuel* 210 (2017), 107–120. https://doi.org/10.1016/j.fuel.2017.08.037.
28. Shaddix, C. R. "Correcting Thermocouple Measurements for Radiation Loss: A Critical Review". *1999 National Heat Transfer Conference*. Albuquerque, New Mexico, USA, 1999.
29. Holland, T. M. and Fletcher, T. H. "Comprehensive Model of Single Particle Pulverized Coal Combustion Extended to Oxy-Coal Conditions". *Energy & Fuels* 31.3 (2017), 2722–2739. https://doi.org/10.1021/acs.energyfuels.6b03387.

30. Fletcher, T. H., Kerstein, A. R., Pugmire, R. J., Solum, M. S. and Grant, D. M. "Chemical Percolation Model for Devolatilization. 3. Direct Use of $^{13}C$ NMR Data to Predict Effects of Coal Type". *Energy & Fuels* 6.4 (1992), 414–431. https://doi.org/10.1021/ef00034a011.
31. Eckhard, T., Pflieger, C., Schmidt, S., Böttger, J., Senneca, O., Schiemann, M., Scherer, V., Muhler, M. and Cerciello, F. "Catalytic Effects for Cellulose-Based Model Fuels Under Low and High Heating Rate in Air and Oxy-Fuel Atmosphere". *Fuel* 324 (2022), 124437. https://doi.org/10.1016/j.fuel.2022.124437.

# Catalytic Effects

# 4

Jannik Böttger, Till Eckhard, Christin Pflieger, Katrin Lotz, Viktor Scherer, Francesca Cerciello and Martin Muhler

## Contents




J. Böttger (✉) · T. Eckhard · C. Pflieger · K. Lotz · M. Muhler
Laboratory of Industrial Chemistry, Ruhr-Universität Bochum, Bochum, Germany
e-mail: jannik.boettger@rub.de

T. Eckhard
e-mail: till.eckhard@techem.rub.de

C. Pflieger
e-mail: christin.pflieger@rub.de

K. Lotz
e-mail: katrin.lotz@techem.ruhr-uni-bochum.de

M. Muhler
e-mail: martin.muhler@rub.de

V. Scherer
Department of Energy Plant Technology, Ruhr-Universität Bochum, Bochum, Germany
e-mail: scherer@leat.rub.de

F. Cerciello
Istituto di Scienze e Tecnologie per l'Energia e la Mobilità Sostenibili, Naples, Italy
e-mail: Francesca.cerciello@stems.cnr.it



R. Kneer et al. (eds.), *Clean Energy Conversion*,
https://doi.org/10.1007/978-3-031-98956-8_4

**Abstract**

The conversion of biomass is significantly influenced by the presence and chemical behaviour of catalytically active minerals within the solid fuel. This chapter discusses the role of impregnated minerals during biomass conversion using extensive characterisation techniques. Iron oxides, alkali and alkaline earth metal sulphates undergo complex transformations during heating that significantly affect their catalytic activity. Kinetic analysis showed reduced activation energies for doped fuels in gasification and oxidation following an overall reactivity sequence of K > Na > Fe > Ca > Mg. Fast-heating pyrolysis experiments revealed mineral effects on the formation of all three pyrolysis products influencing tar distribution, light gas formation, and char evolution. Additionally, combustion studies in a flat flame burner demonstrated earlier ignition and varied impacts on the combustion temperature depending on the type of doping. Overall, the results demonstrated a significant influence of minerals on biomass conversion and highlight the importance of mineral characterisation in optimising biomass-based energy production.

## 4.1 Introduction

Biomass as well as coal can contain a significant amount of inorganic elements. Figure 4.1 exemplarily shows the mean metal content for wood, herbaceous biomass, lignite, and bituminous coal from values found in literature [1,2].

The most abundant minerals in biomass are Ca, K, Si, Mg, Al, Fe, and Na as well as Mn, Ti, and others as trace elements [3]. Minerals are essential nutrients for biomass, which have different beneficial effects during the complex growth mechanism of plants including for example an enhanced root water uptake. Furthermore, minerals are a constituent of cell walls and membranes and the activator of dozens of important enzyme-based processes like protein synthesis, N and C metabolism, and sugar transport [4–6]. The type and amount of minerals highly depends on the kind of biomass, location of growth or even the time of harvesting. As shown in Fig. 4.1 herbaceous biomass usually contains a larger amount of minerals than wood. This is due to its different organic structure and faster metabolism which absorbs more nutrients during its growth [7]. Minerals are integrated into the biomass either by natural growth or anthropologically by harvesting or processing procedures and can be separated into four different classes depending on how they are bound in the biomass: dissolved salts, organically bound matter, included mineral matter, and excluded mineral matter [2,8,9].

Also coal contains inorganic elements in significant amounts originating not only from the organic matter, mainly plant material, from which the coal is formed, but also from other sources that infiltrated during the coalification process. Similar to biomass, the mineral type and amount highly depends on the type of coal and its origin. During the combustion process, minerals can lead to the formation of slag and fouling, which reduces the heat transfer efficiency, increases maintenance require-

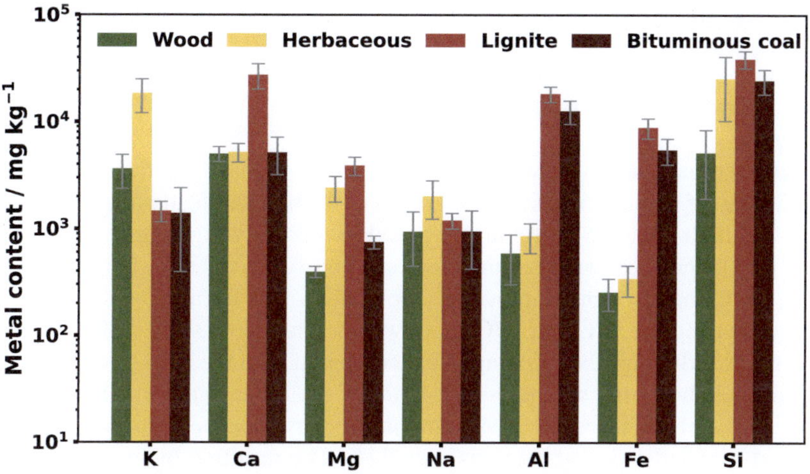


**Fig. 4.1** Averaged inorganic metal content of 26 wood, 17 herbaceous, 9 lignite and 4 bituminous coal samples [1,2]

ments and decreases the overall combustion performance. Furthermore, minerals are known to affect the chemical and physical composition of pyrolysis products and act as catalysts during combustion, promoting or inhibiting certain chemical reactions. In this way, they influence the combustion kinetics. There are several theories explaining the catalytic effect of minerals during combustion, and they do not necessarily contradict each other. One hypothesis suggests that catalytically active sites are formed through the interaction of alkali and alkaline earth metals (AAEM) with the carbon substrate creating new active sites [10,11]. Another hypothesis proposes that minerals catalyse the desorption of gaseous oxidation products. This suggests that the electron-donating properties of the minerals increase the electron density at the carbon surface sites, which promotes oxygen chemisorption and weakens the C–C bonds on the char surface [12]. However, the most widely accepted theory is McKee's oxygen transfer theory. This theory describes the reduction and oxidation cycles of metal oxides, hydroxides or carbonates, which facilitate the transfer of oxygen into the carbon matrix and thus catalyse char conversion [13,14].

Therefore, it is important to investigate the influence of minerals and contribute to a quantification of the individual effect of dopants to enable the implementation into common combustion models. To do so, first a mineral-free model fuel had to be prepared for subsequent selective doping with different minerals. For this purpose, hydrothermal carbonisation (HTC) was chosen as a suitable method.

## 4.2 Methods for Preparation and Characterisation of Model Fuels

### 4.2.1 Model Fuel Preparation

The hydrothermal carbonisation process imitates the natural coalification process in hours instead of several million years. For HTC, the biomass feedstock and water are mixed in a high-pressure vessel and exposed to relatively high temperatures (180–250 °C) [15]. The necessary pressure is autogeneously generated by the water vapour. During HTC mainly dehydration, decarboxylation and decarbonylation reactions occur decreasing the O/C and H/C ratio, which results in a solid called hydrochar with significantly increased heating value compared with the base feedstock.

Figure 4.2 shows the synthesis procedure of the mineral-free model fuels. Cellulose as one of the three biomass components was used as feedstock to produce a mineral-free carbonaceous material via HTC. The resulting hydrochar was compared chemically to leached lignite to validate the synthesis method. Different spectroscopic techniques showed that the cellulose-derived model fuel has comparable properties to leached Rhenish lignite [16]. To exclude the effect of minerals on the devolatilisation stage and to focus only on oxidation/gasification, experiments were also performed with a char obtained by pyrolysing the hydrochar. Therefore, the obtained model hydrochar was pyrolysed at 800 °C with a low heating rate. Shifting the focus of research from coal to biomass, HTC was applied to walnut shells instead of cellulose. Due to the temperature and the acid conditions, HTC is a reliable way to remove the minerals from the biomass and produce a mineral-free model hydrochar.

Different approaches were used to integrate the minerals into the model fuel. First, the mineral was added by in-situ doping during the HTC process ensuring high contact of the mineral with the embedding carbon matrix [17]. However, a controlled and defined concentration of the mineral was not assured, and it was not possible to exclude an effect of the mineral on the carbonisation process. As post-synthetic method physical mixing of the model fuel with the mineral was performed, but the low degree of contact between the mineral particles and the fuel resulted in much less interactions and influence on the combustion process [17]. Wet impregnation using aqueous solutions of mineral salts was found to be the most reliable method to add minerals, since it ensured high contact by the deposition of known and defined amounts of mineral on the fuel surface and in the pores.

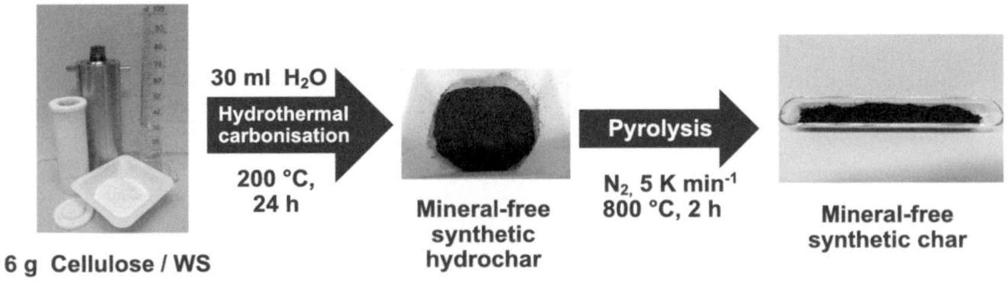


**Fig. 4.2** Synthesis procedure of the mineral-free model fuels

The corresponding model fuel was doped with Fe, K, Na, Ca, and Mg by impregnation. Not only single-doped samples were studied, but also multi-doped ones with more than one mineral to investigate the interactions of the minerals during conversion and the resulting effect on the catalytic activity [18]. During the conducted studies the minerals were doped in different forms with different counterions. One study investigated the influence of the counterion on the catalytic activity by comparing the temperature-programmed oxidation profiles of samples doped with chlorides, nitrates and sulphates. In contrast to the sulphate-doped minerals, Fe and K chlorides and nitrates also seemed to shift devolatilisation to lower temperatures by up to 185 °C. This effect is ambiguous due to the overlap of the chloride/nitrate decomposition with the carbon devolatilisation and partially also oxidation. Similar to the results observed for sulphate-doped samples, Fe and K chlorides and nitrates were found to shift the char oxidation reducing the peak maximum by up to 170 °C. Compared with that, the effect of Mg was only small for the investigated counterions [18].

Chlorides or nitrates are usually present in higher amounts in biomass. However, due to their overlapping decomposition and the resulting more complex conversion profiles, sulphates were chosen for most of the mineral studies. Sulphates are also abundant in biomass and coal, have good solubility in water for the impregnation process, and are thermally more stable.

### 4.2.2 Pyrolysis-GC/MS-GC/TCD

To investigate the broad distribution of small gaseous and larger liquid pyrolysis products, a multi-dimensional gas chromatography system, the pyrolysis-GC/MS-GC/TCD (Pyr-GC/MS), was used [19]. Figure 4.3 shows a schematic drawing of the extended setup.

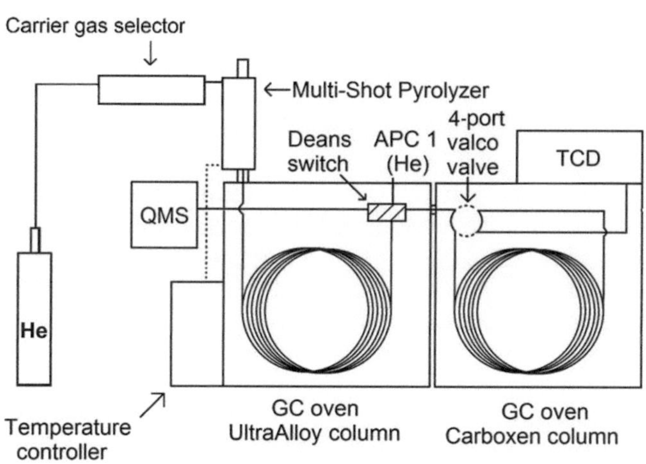


**Fig. 4.3** Schematic drawing of the extended Pyr-GC/MS setup. (Reprinted from Eckhard et al. [19], Copyright (2020), with permission from Elsevier and minor editing)

To avoid air contamination during sample installation, the setup is equipped with a carrier gas selector (*CGS-1050EX, Frontier Laboratories*). A 6-port valve enables the calibration with a sample loop. The multi-shot pyrolyser (*EGA/PY-3030D, Frontier Laboratories*) is heated by a ceramic heater to a temperature of 300–1000 °C and is used for the flash pyrolysis of small sample amounts in the range of 0.2–1 mg. The sample is placed in a crucible that is dropped into the preheated furnace reaching a heating rate in the range of $10^3\,\mathrm{K\,s^{-1}}$. Further, the system consists of two gas chromatographs (*GCMS-QP2010 Ultra, Shimadzu*) connected in series with different columns and detectors. After pyrolysis the evolved gases are transported by the carrier gas (He) to the first GC for on-line analysis. The first column is equipped with an *UltraAlloy* capillary column (30 m, 0.25 mm inner diameter, 0.25 µm average film thickness) separating the larger molecules of condensable gaseous pyrolysis products. For a qualitative analysis, the Deans switch passes the gas flow through a capillary to the QMS (mass range of $m/z = 1.5$–1090 and a mass resolution of 0.5–2.0 u (FWHM)). Light gases evolved during pyrolysis pass the *UltraAlloy* column without interaction and are directed by the Deans switch to the second GC, which is equipped with a fused silica capillary column (*Carboxen 1006 PLOT*, 30 m, 0.53 mm inner diameter, 30 µm average film thickness). The column is connected to a calibrated thermal conductivity detector (TCD) via a 4-port valve for the quantitative analysis of light gases [19].

### 4.2.3 Characterisation Methods

In order to investigate the influence of minerals on fuel conversion, a wide range of characterisation techniques was required that is introduced in the next paragraph, partially supported by results from the actual research. Figure 4.4 provides an overview of the used methods.

**Elemental Analysis**

Element analysis is a basic characterisation method for carbon samples investigating the composition. Proximate analysis determines the moisture, volatile, fixed carbon and also ash content of a sample. Complementary, ultimate analysis provides the C, H, N, S content, and together with the known ash content also the calculated O content. Research has shown that the catalytic influence for most minerals depends highly on the amount of minerals in the fuel [20]. Therefore, its quantitative determination is essential and can be achieved by a variety of techniques, for example inductively coupled plasma mass spectrometry (ICP-MS) or atomic absorption spectroscopy (AAS). The AAS was used to validate the desired and real mineral loadings after impregnation. For doping with Fe and alkali metals, the obtained loadings were in line with the desired values, whereas doping with the alkaline earth metals, especially Ca, resulted in lower loadings than expected due to the lower solubility of $CaSO_4 \cdot 2\,H_2O$ in comparison to the other metal sulphates [18].

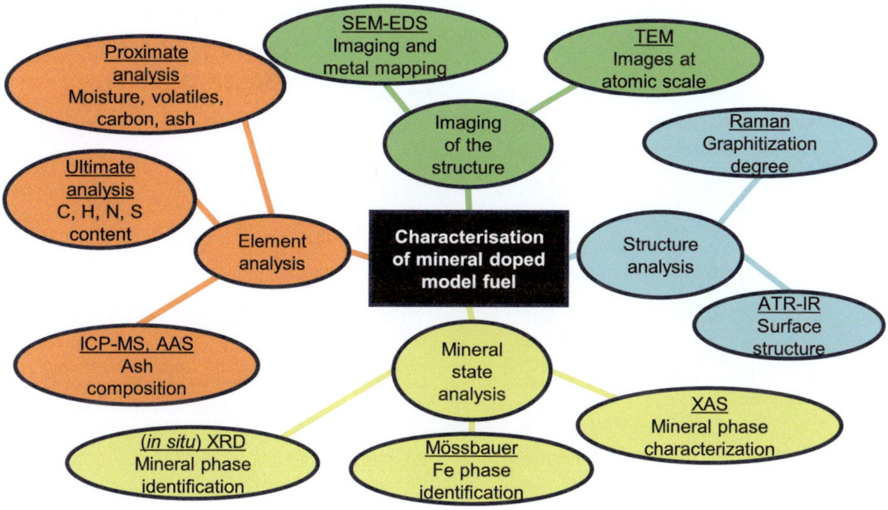


**Fig. 4.4** Overview of the characterisation techniques used for the investigation of the mineral effect on fuel conversion

## Mineral State Analysis

When investigating the influence of minerals on fuel conversion, not only the amount of minerals is important, but also the mineral phases and their phase transformations. They can be determined using different types of characterisation techniques.

X-ray diffraction (XRD) measurements can be used to identify the phase composition of the minerals. The diffraction patterns resulting from X-rays interacting with a material's crystal lattice enable the determination of its atomic arrangement and composition. The phase transformation of minerals can be observed during slow heating and in different atmospheres by in-situ XRD. Due to the carbon background inhibiting the identification of phases and to protect the in-situ cell from released tars, a carbon-free mineral phase present as sulphates was prepared with the same temperature and solubility procedure as the mineral on the impregnated carbonaceous material. Therefore, it has to be considered that no interaction of the minerals with the carbon matrix was taken into account. The in-situ XRD experiments revealed that $K_2SO_4$ and $Na_2SO_4$ were stable up to 800 °C. The alkaline earth metals, on the other hand, were first present in their hydrate form, dehydrated and lost their crystal water (partially over multiple steps) to form the corresponding anhydride. Fe was the only investigated metal which converted from the sulphate into an oxide. In the case of the multi-mineral mixtures, Fe+Mg showed no interaction of the minerals displaying a linear combination of Fe and Mg. For the K+Mg mixture, in-situ XRD (Fig. 4.5) showed that the individual metal sulphate phases disappeared at 200 °C revealing the formation of cubic langbeinite $\left(K_2Mg_2(SO_4)_3\right)$ [18].

For the activity of K its mobility plays an important role [21,22]. This interaction of K and Mg could lead to a decrease of mobility of K resulting in an inhibition of its catalytic activity. The measurements were performed in inert, diluted $O_2$ and diluted $CO_2$ atmosphere, but no influence of the atmosphere was observed during the

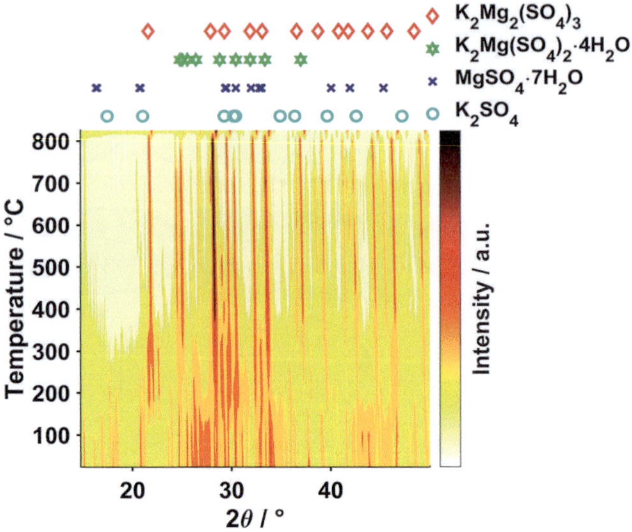


**Fig. 4.5** In-situ XRD pattern of K+Mg sulphates between room temperature and 800 °C in air. (Reprinted from Eckhard et al. [18], Copyright (2022), with permission from Elsevier and minor editing)

in-situ XRD experiments. The investigation of minerals during slow heating XRD experiments highlighted the importance of studying the mineral interactions during conversion processes [18].

The phase transformation of Fe was studied in more detail not just for the pure dopant, but also for the doped carbon material including the interaction of the Fe mineral with the carbon matrix [23]. Due to the small particle size of the Fe-doped char, a thermal treatment at 480 °C in an inert atmosphere was performed to induce sintering and to obtain mineral-related reflections in the XRD measurements. These measurements revealed $\gamma$-$Fe_2O_3$ or $Fe_3O_4$ as possible iron phases.

As these oxides crystallise in the same space group and have similar lattice parameters, they are basically undistinguishable by XRD [24]. So, to differentiate between them, Mössbauer spectroscopy was used, which is a non-destructive highly sensitive method based on the Mössbauer and Doppler effects that are mostly used to study iron phases. The Mössbauer results revealed that iron was present as $Fe^{3+}$ indicating that the phase present after the synthesis procedure was $\gamma$-$Fe_2O_3$ [23].

In addition, the phase transformation during oxidation was investigated. To characterise and quantify the different iron phases, X-ray absorption spectroscopy (XAS) was used. In general, XAS is an element-specific technique that uses synchrotron radiation to provide information about the local structure and composition of a material. In contrast to XRD, no long-range order is required. After an oxidative treatment, a complete transformation of $\gamma$-$Fe_2O_3$ to two other polymorphs was observed depending on the mineral loading: $\epsilon$-$Fe_2O_3$ and $\alpha$-$Fe_2O_3$. Figure 4.6 shows the mineral phase distribution determined by a linear combination fit of the XAS results. Combining the XAS results with XRD and also temperature-programmed (TP) mea-

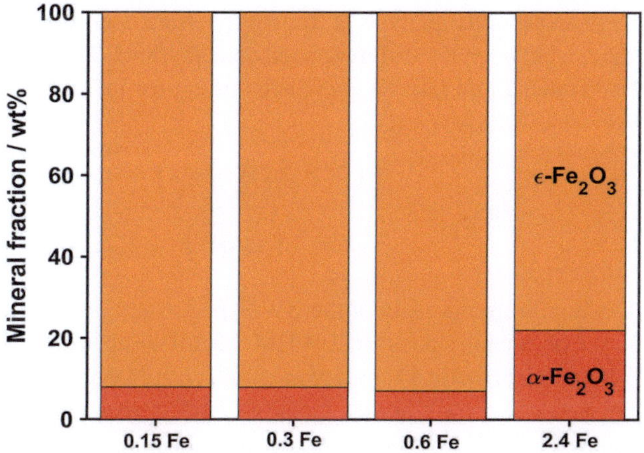


**Fig. 4.6** Mineral phase distributions of the differently loaded chars after an oxidative treatment determined by a linear combination fit of XAS results. (Reprinted from Pflieger et al. [23], Copyright (2024), with permission from Elsevier and minor editing)

surements as an indicator for reactivity revealed that $\epsilon$-$Fe_2O_3$ was the phase mainly present during oxidation. $\epsilon$-$Fe_2O_3$ was highly active lowering the apparent activation energy for char oxidation by up to 20 kJ mol$^{-1}$. With increased loading and progressing char conversion, particle agglomeration is enhanced and, therefore, transformation to the larger $\alpha$-$Fe_2O_3$ particles, which have a much lower catalytic influence on the char oxidation [23].

For gasification the TP measurements indicated the presence of three iron phases with different catalytic influence. Around 650 °C, CO released by carbon conversion induced the reduction of $\gamma$-$Fe_2O_3$ to FeO. FeO lowered the activation energy for gasification up to 42 kJ mol$^{-1}$. At around 900 °C further reduction resulted in the phase transformation to the less active $\gamma$-Fe ($\Delta E_A = 17\,\text{kJ mol}^{-1}$). After the char had been fully consumed, the Fe mineral was partly re-oxidised by $CO_2$, resulting in $Fe_3O_4$ remaining in the ash [23].

The research performed in the field of mineral phase identification during fuel combustion processes using a variety of characterisation methods highlighted the tendency of minerals, especially Fe minerals, to undergo phase transformations. It also showed the dependence of mineral activity on the gas atmosphere and the mineral loading.

**Structure Analysis**

Raman spectroscopy is a non-destructive method to investigate the chemical structure and graphitisation degree of carbonaceous materials. Raman spectroscopy operates on the principle of inelastic scattering, where incident photons interact with molecular vibrations, resulting in a shift in their energy levels, which allows the identification of specific chemical bonds and structures within the sample. In carbon material two bands are usually present: the D and G bands. The G band represents the in-plane vibration of doubly bound carbon atoms, while the D band indicates disorder in a carbon sample. For an accurate analysis of Raman spectra, a deconvolution procedure is used to distinguish between the different band contributions. The G band was fitted by using a Breit-Wigner-Fano (BWF) curve accounting for its asymmetry

$$I_{\mathrm{BWF}} = I_0 \frac{\left(1 + 2\,\frac{\widetilde{\nu}_{\mathrm{s}} - \widetilde{\nu}_{\mathrm{c}}}{\mathrm{FWHM}}\,\frac{1}{\alpha_{\mathrm{BWF}}}\right)^2}{1 + 2\left(\frac{\widetilde{\nu}_{\mathrm{s}} - \widetilde{\nu}_{\mathrm{c}}}{\mathrm{FWHM}}\right)^2} \tag{4.1}$$

with $I_0$ being the maximum intensity of the BWF spectrum, $\widetilde{\nu}_{\mathrm{s}}$ the measured Raman wavenumber, $\widetilde{\nu}_{\mathrm{c}}$ the peak centre position, FWHM the full width at half maximum, and $\frac{1}{\alpha_{\mathrm{BWF}}}$ the BWF asymmetric factor. D bands were fitted with a Lorentzian curve according to (4.1) with $\frac{1}{\alpha_{\mathrm{BWF}}} = 0$ [25]. While for highly ordered samples the $I(\mathrm{D})/I(\mathrm{G})$ ratio increases with a reduced aromatic layer size according to the Tuinstra-Koenig relationship [26], an increase indicates an increase of the atomic layer length for disordered carbon matrices [27]. This increase of graphitisation was observed for chars obtained from fast heating experiments [28]. Further, a significant influence of minerals, here in the form of Fe+Mg and K+Mg multi-doping, on the $I(\mathrm{D})/I(\mathrm{G})$ ratio was found. The incorporation of minerals into the carbon matrix increased the distance between the aromatic layers resulting in a generally decreased $I(\mathrm{D})/I(\mathrm{G})$ ratio. Additionally, minerals seem to enhance the effect of the atmosphere during heat treatment, since doping increased the graphitisation for chars generated in $CO_2$ atmosphere in comparison to chars produced in $N_2$ [28].

Attenuated total reflectance Fourier-transform infrared (ATR-FTIR) spectroscopy is a powerful analytical technique used to characterise biomass-based fuels. ATR-FTIR spectroscopy enables the identification of various functional groups like hydroxyl, carbonyl and aromatic groups that are present in biomass. This helps in determining the chemical reactivity of the biomass, which is crucial for understanding its fuel properties. The technique can also distinguish between the major components of biomass such as cellulose, hemicellulose, and lignin by their characteristic IR bands. This is important for assessing the quality and efficiency of biomass as a fuel source. ATR-FTIR spectroscopy can also offer insights concerning the mineral doping of biomass fuels. Minerals such as silicates, carbonates, sulphates and phosphates exhibit characteristic vibrational bands appearing in specific regions of the IR spectrum, which allows the fast confirmation of mineral doping and the identification of the mineral type.

**Imaging of Structure**

Scanning electron microscopy is a versatile technique based on emitted secondary electrons, which is used to create images of the sample at high resolution to gain information about the surface morphology, composition and properties of the material. The SEM images of the doped model fuel showed no evidence of swelling or shrinking phenomena after the impregnation validating the chosen impregnation method. After fast pyrolysis, a swelling phenomenon was observed by SEM measurements of the collected char particles. Furthermore, the particles were broken

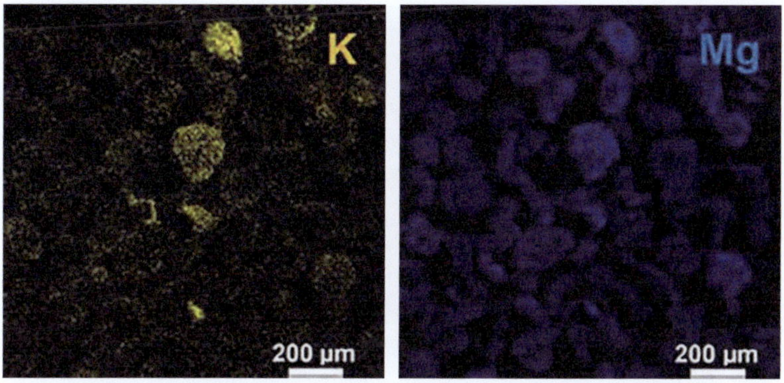


**Fig. 4.7** SEM-EDX mapping of the impregnated material showing K deposition in yellow and Mg deposition in blue. (Reprinted from Eckhard et al. [18], Copyright (2022), with permission from Elsevier)

up due to the forceful volatile release. The open fractured particles were observed independent of the doping, indicating that this behaviour is due to an intrinsic fuel property [18]. Investigating the influence of minerals, the combination of SEM with energy-dispersive X-ray spectroscopy (EDX) is particularly useful. It is possible to investigate the dispersion of the minerals after wet impregnation on the fuel particles by SEM-EDX mapping. An impregnation of 5 wt.% of K and Mg as sulphates resulted in a rather homogeneous distribution of both metals on the fuel particles as shown in Fig. 4.7. Although some particles exhibited a higher concentration of metals compared with others, no pure metal salts without carbon support were detected, and the majority of particles were covered with at least a small amount of both metals [18].

Transmission electron microscopy (TEM) is a microscopy technique to obtain detailed insights into the structure and composition of a material by providing images at the atomic scale. TEM was used to gain additional information on structural ordering caused by iron oxide loading and thermal treatment. The images are shown in Fig. 4.8.

The undoped fuel pyrolysed at 800 °C showed some lattice fringes at the shell of the particles, but was predominantly amorphous with only scarce low-range graphitic ordering. For the iron oxide-loaded fuel, the incorporated iron particles were imaged as dark spheres by TEM. By pyrolysis at 500 °C, a highly amorphous carbon sample was obtained. In contrast, the iron oxide-doped fuel pyrolysed at 800 °C contained both amorphous and ordered fractions. Fringes were observed in multi-layer stacks containing up to 30 layers with a distance of approximately 3.5 Å, which matches with the graphite (002) planes. These graphitic domains, considering their bent shape, had a length of several nanometres. Thus, iron doping catalysed the graphitisation of the carbon matrix at least at high temperatures [29].

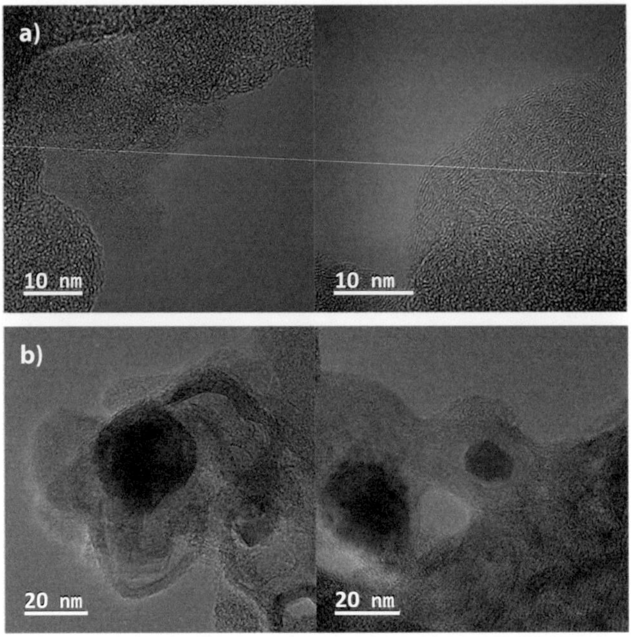


**Fig. 4.8** TEM images of **a**) the pure carbon material and **b**) the iron oxide-loaded sample both pyrolysed at 800 °C. (Reprinted from Lotz et al. [29], Copyright (2019), with permission from ACS Publications and minor editing)

## 4.3 Mineral Influence on Pyrolysis

Char conversion is the more important step in biomass conversion, since most of the energy is released in it. However, the pyrolysis stage has a significant impact on char conversion as it determines the structure of the char and, therefore, also its reactivity. In order to study the relevance of mineral catalytic activity in industrial boilers during pyrolysis, three fast heating setups were used enabling industrial relevant heating rates over $10^3\,K\,s^{-1}$. Figure 4.9 shows an overview of the different setups used for pyrolysis including possible atmospheres, the temperature range and the pyrolysis products that can be analysed/synthesised with the setup. With the three setups it is possible to investigate the mineral influence on all three pyrolysis products: light gas, tar, and char. Detailed descriptions of the HSR and the FFB can be found in Sects. 2.2.4 and 3.3.1, respectively.

The influence of K+Mg and Fe+Mg on the light gas and tar formation was investigated by comparison of the results with the undoped sample [28]. The samples were pyrolysed at 700 °C in a He atmosphere to quantify the permanent gases calibrated for the TCD (CO, $CH_4$, $CO_2$, $C_2H_4$, $C_2H_6$). The undoped sample lost 25 wt.% in form of permanent gases during pyrolysis consisting mostly of CO (14 wt.%) and $CO_2$ (8 wt.%) originating from the decomposition of oxygen-containing functional groups. The remaining 3 wt.% were attributed to small hydrocarbons, with

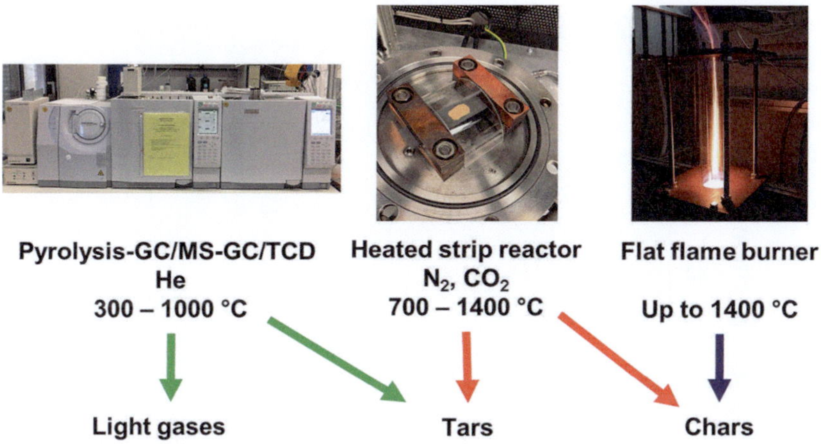


**Fig. 4.9** Overview of the used fast pyrolysis setups including the possible atmospheres, temperature ranges and pyrolysis products that can be analysed to study the catalytic activity of minerals

$CH_4$ clearly being the most abundant. Doping with Fe+Mg significantly increased the amount of detected light gases to 30 wt.%. This was due to an increase of the oxygenated species (CO and $CO_2$), while the amount of formed $CH_4$ slightly decreased. In the K+Mg sample, the overall gas yield did not increase and was slightly lower. However, the ratio of oxygenated to non-oxygenated gaseous products ((CO + $CO_2$)/$CH_4$) increased compared with the undoped sample [28].

Additionally, condensable pyrolysis products were detected by MS, identified using the NIST mass spectral library, integrated, and assigned to product families. Independent of the mineral doping, similar pyrolysis products were identified: aliphatics, aromatic oxygenated compounds, sugar derivates and polyaromatic hydrocarbons (PAH). These pyrolysates are characteristic for primary pyrolysis products. Oxy-aromatics were the primary product family across all samples, representing roughly 30% of the detected pyrolysates. The pyrolysates originating from the undoped sample exhibited a higher proportion of anhydrous sugars (around 12%), linked to the cellulose structure of the precursor. The presence of minerals had a significant catalytic impact, particularly in lowering the levoglucosan fraction formed by the cleavage of 1,4-glycosidic bonds during pyrolysis. At the same time, there was an increase in the yields of benzofurans, phenols, and indenes from overall 21.3 to 25.1% for KMg and 30.3% for FeMg. This suggests an enhanced breakdown of the sugar matrix in the synthetic biomass [28].

The influence of minerals on the tar formation was not only investigated in the Pyr-GC/MS, but also in the heated strip reactor (HSR). Experiments were performed at 700 °C in $N_2$ and $CO_2$ atmosphere. The tars condensed on the *Pyrex* bridge were collected, subsequently dissolved in acetone and analysed in the Pyr-GC/MS at a temperature of 300 °C to avoid further cracking of the molecules. The resulting tar distributions are shown in Fig. 4.10. The different tars showed similar distributions of chemical species independent of doping. When pyrolysing in $N_2$ and $CO_2$, a large contribution of anhydrous sugar was observed, especially in the tar of the

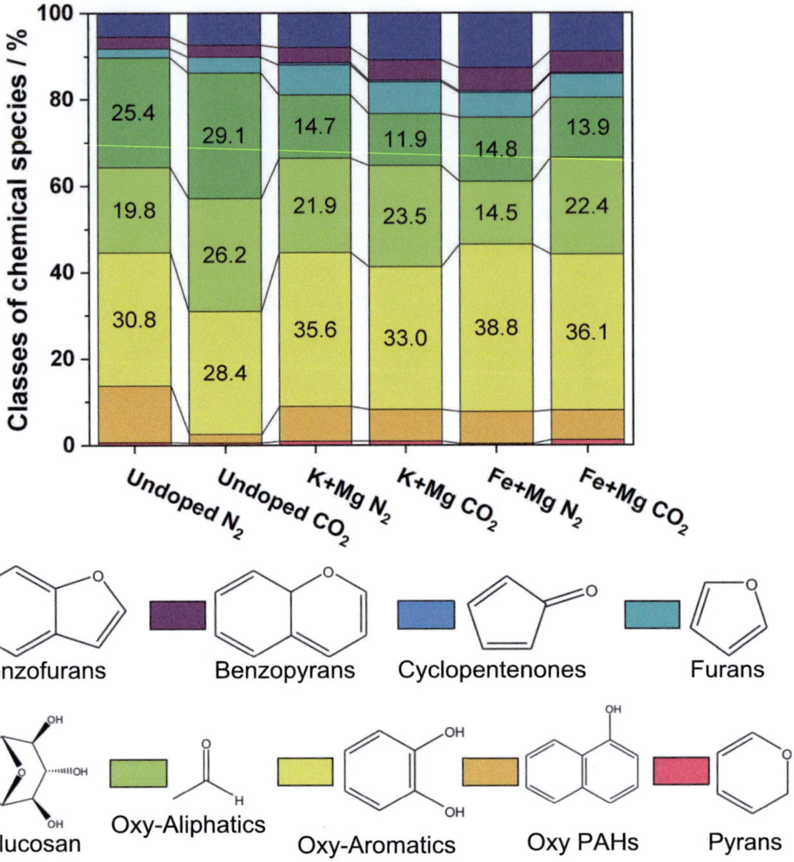


**Fig. 4.10** Distribution of HSR tars collected after pyrolysis of the undoped (MH) and doped fuel at 700 °C in $N_2$ and $CO_2$ volatilised at 300 °C in the Pyr-GC/MS (area %). (Reprinted from Eckhard et al. [28], Copyright (2023), with permission from Elsevier and minor editing)

undoped sample. Oxy-aromatic compounds were the most abundant product family, followed by the oxy-aliphatic compound fraction. No phenol derivates with only the phenolic hydroxyl group as oxygen functional group were found. In contrast, during the online analysis with the Pyr-GC/MS the phenolic compound family made up 15–19% of pyrolysates. Similarly, the amount of cyclopentanone decreased below 1%. The reason for both observations is the suppressed secondary pyrolysis in the HSR due to the cold atmosphere. As already observed for the online analysis of the tars, the anhydrous sugar content decreased due to the presence of minerals. This enhanced the production of furans, (benzo)furan and (benzo)pyran compounds as decomposition products. Additionally, an increase of oxy-aromatics was caused by the minerals, which was more dominant for the Fe+Mg-doped samples than for K+Mg. Notable was also a smaller fraction of oxy-aliphatics for the Fe+Mg-doped sample in $N_2$, which was not observed in $CO_2$ and neither for K+Mg doping [28].

**Table 4.1** Char yields and DTG peak maximum temperatures derived from TG measurements in 20 vol.% $O_2$/He of the HSR chars [28].

| Sample | DTG peak $T_{max}$ [°C] | Char yield [wt.%] |
|---|---|---|
| MH | 300; 470 | – |
| FeMg-MH | 300; 341 | – |
| KMg-MH | 300; 416 | – |
| HSR char | | |
| MH-$N_2$ | 490 | 50.6 |
| MH-$CO_2$ | 490 | 45.5 |
| FeMg-MH-$N_2$ | 356 | 33.0 |
| FeMg-MH-$CO_2$ | 376 | 29.4 |
| KMg-MH-$N_2$ | 431 | 33.8 |
| KMg-MH-$CO_2$ | 414 | 27.8 |

The investigations of the mineral effect on tars formed during pyrolysis in the different pyrolysis setups highlighted the significant catalytic role of minerals in altering tar composition. Both the Pyr-GC/MS and HSR setups show that mineral doping leads to a reduction in anhydrous sugars and to an increase in oxy-aromatics and furans. These findings demonstrate the importance of the mineral content in influencing the distribution and nature of pyrolysis products, providing valuable insights for optimising biomass pyrolysis processes.

In addition to the influence of minerals on tar distribution, the influence on char formation was investigated in the HSR [28]. The different chars were collected from the graphite strip after the pyrolysis experiments and were subsequently analysed. TG measurements, which incorporate a heating to 800 °C with a heating rate of 5 K $min^{-1}$ in synthetic air, were used to investigate their reactivity. The solid remaining after 30 min at 800 °C was recorded as ash content from which the char yield was calculated. Char yields and temperatures correlating with the peak of the DTG curve from (2.1) are summarised in Table 4.1. The char yields indicate that the sample lost the majority of its volatile matter during pyrolysis in $N_2$, with a mass loss of 49.4 wt.% compared with a volatile matter value of 53 wt.% from proximate analysis. In contrast, when pyrolysed in $CO_2$, the undoped sample released all its volatile matter, suggesting that $CO_2$ accelerated the pyrolysis process compared with $N_2$. Mineral doping significantly enhanced devolatilisation, as evidenced by the mass losses in the HSR for the Fe+Mg and K+Mg samples. This effect was independent of the type of mineral, but performing the heating experiments in $CO_2$ atmosphere reduced the char yield even more, although a contribution of heterogeneous gasification cannot be excluded [28].

The DTG profiles of the undoped sample before and after pyrolysis in the HSR are shown in Fig. 4.11. The unconverted samples exhibit two peaks resembling the mass loss of the sample during heating. The first one around 300 °C originates from volatile release of the sample. The second mass loss peak in the range of 341–470 °C is attributed to char oxidation [28]. Doping with minerals did not shift the first peak

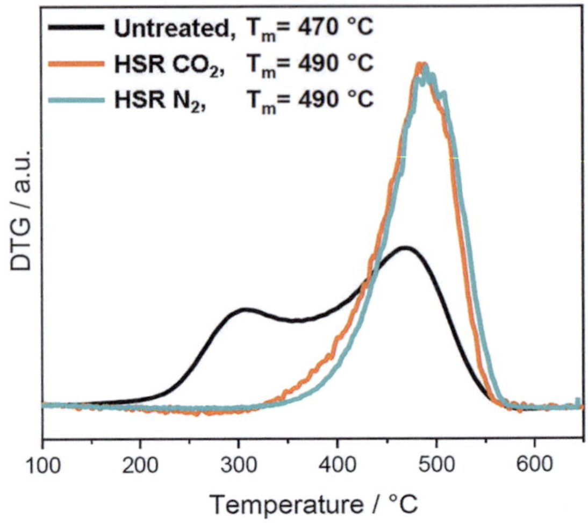


**Fig. 4.11** TPO profiles measured in 20 vol.% $O_2$/He of the undoped sample before and after pyrolysis in the HSR in $N_2$ or $CO_2$ at 700 °C. (Reprinted from Eckhard et al. [28], Copyright (2023), with permission from Elsevier and minor editing)

suggesting that minerals do not affect the volatile release under TGA conditions. On the other hand, the char oxidation peak was significantly shifted to lower temperatures showing the strong catalytic effect of minerals on the combustion stage. After pyrolysis in the HSR at 700 °C, the undoped chars showed a single peak at 490 °C in the DTG profile. Such a high DTG peak temperature indicates a distinctively low reactivity, which is due to the thermal annealing of the carbonaceous structure. The chars produced in the HSR from doped samples were more reactive than the undoped ones, confirming the catalytic activity of minerals also after the fast heat treatment. Notably, the KMg-doped sample treated in $CO_2$ is more reactive than the one treated in $N_2$. As shown in Sect. 4.2.3, the formation of a potassium magnesium sulphate double salt (PMS) is possible, which inhibits the catalytic activity of K. In a $CO_2$-rich atmosphere, this is less pronounced, since K is able to form the mobile and reactive potassium carbonate also under these conditions [22].

## 4.4 Mineral Influence on Char Conversion

The catalytic effect of minerals on the conversion of a biomass-derived model fuel in air- and oxy-fuel-related atmosphere was studied using a TG equipped with a magnetic suspension balance for detecting the weight loss during char conversion. For this investigation, the model fuel was doped with 20 wt.% of minerals by the tight-contact method. Due to their tendency to transform into oxides during combustion, Mg, Ca and Fe were applied as oxide species [30]. For Na and K, carbonates were used, since alkali oxides are not stable against humidity. As char conversion in oxy-fuel atmospheres takes place in a mixture of $O_2$, $CO_2$, and $H_2O$, the individual contributions of the gases were investigated systematically. Isothermal measurements were performed in atmospheres of 20 vol.% $O_2$/He, 20 vol.% $CO_2$/He, 80 vol.% $CO_2$/He, and 5 vol.% $H_2O$/He at 773 K for oxidation and up to 1373 K for gasification.

Overall, a clear increase in reactivity was found for the doped materials. Comparing the different conversion curves a reactivity sequence was derived, which was valid for all the tested conditions: $K_2CO_3 > Na_2CO_3 \gg Fe_2O_3 > CaO > MgO$ [30]. To quantify the catalytic effect of the minerals, first kinetic parameters of the undoped model fuel were determined. Partial pressure variations of the reactive gases at constant temperature were performed to determine the reaction order $n_i$ with respect to the mole fraction $y_g$ of the particular reactive gas $i$, and an overall reaction order $m$ related to the residual char fraction $(1 - X)$ as well as the rate constant $k_i$ using the $n^{th}$-order power-law:

$$\frac{dX}{dt} = k_i \, y_{g,i}^{n_i} \, (1 - X)^m \tag{4.2}$$

Subsequently, the apparent activation energy $E_A$ and pre-exponential factor $A$ were determined based on isothermal char conversion experiments at different reaction temperatures according to the linearised Arrhenius formulation

$$\ln(k_i) = \ln(A) - \left(\frac{E_A}{R}\frac{1}{T}\right) \tag{4.3}$$

with $R$ being the universal gas constant and $T$ the reaction temperature. The determined apparent activation energies for oxidation, $CO_2$ gasification and gasification in $H_2O$ vapour (141 kJ mol$^{-1}$, 240 kJ mol$^{-1}$, and 178 kJ mol$^{-1}$, respectively) of the undoped sample were in the same order of magnitude reported for oxidation and gasification of comparable carbon sources [31–34]. Based on these parameters, the char conversion curves were modeled using the power-law approach to quantify the catalytic effect of the dopants. The reaction order and the pre-exponential factor were fixed to the values determined for the undoped material, while the apparent activation energy was treated as fit parameter. Figure 4.12 shows the difference of the obtained activation energies to the activation energy of the undoped sample in $O_2$, $CO_2$, and $H_2O$ atmosphere [30].

For $O_2$, the reduction of activation energies was relatively low (10 kJ mol$^{-1}$ for the alkali metal-loaded chars) compared with the undoped material. However, these lowered $E_A$ values significantly accelerate char oxidation. The mineral influence resulted in the strongest decrease in the absolute values of apparent activation energies for the $CO_2$ gasification. Unlike oxidation, also for $Fe_2O_3$ and CaO a reduction was observed though not to the same extent as for $K_2CO_3$ and $Na_2CO_3$. This observation also holds true for char gasification in atmospheres containing $H_2O$, but the relative catalytic effects of $Fe_2O_3$ and CaO are less pronounced. Further, the experimental data originating from the isothermal measurements were fitted using the CBK and CRECK-S models. The resulting changes in the apparent activation energy are also shown in Fig. 4.12, which shows that both more detailed models produce identical trends and comparable absolute values for the activation energies, both when compared to each other and to the simple power-law. Minor discrepancies are expected, given the fundamentally different nature of the modelling approaches [30].

In a further study, the influence of the doping amount on the reduction of the apparent activation energy was investigated [20]. The individual effect of different dopants

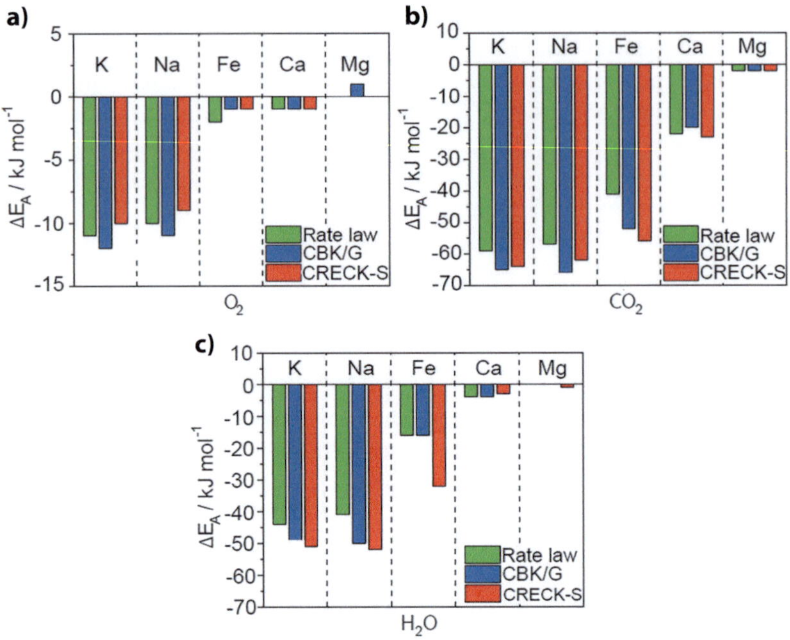


**Fig. 4.12** Activation energy differences due to mineral doping by fitting the experimental data according to the power-law approach, the CBK/G model and the CRECK-S model for char conversion in **a**) $O_2$, **b**) $CO_2$, and **c**) $H_2O$ atmospheres. (Reprinted from Pflieger et al. [30], Copyright (2021), with permission from Elsevier and minor editing)

was quantified using a cellulose-derived biomass fuel by impregnation before and after devolatilisation. This methodology enables a direct assessment of the dopants influence on char conversion, while also indirectly evaluating the effect of catalysed pyrolysis on the resultant char through a comparative analysis of the fuel. The used char was obtained by slow pyrolysis of the model fuel at 800 °C. Both materials were doped by wet impregnation with K, Na, Ca, Mg, and Fe sulphates. To investigate the influence of the minerals on the char conversion, TG measurements were performed in diluted $O_2$ and diluted $CO_2$ atmospheres. The conversion curves obtained from isothermal TG measurements were fitted using the random pore model (RPM) [35]:

$$\frac{\mathrm{d}X}{\mathrm{d}T} = \frac{1}{\beta} A \times \exp\left(-\frac{E_\mathrm{A}}{RT}\right) (1 - X)\, y^n \sqrt{1 - \Psi \ln(1 - X)} \tag{4.4}$$

The RPM accounts for changes of the pore structure during char conversion, which depends on the heating rate $\beta$ as well as on an Arrhenius term. Conversion is modeled to decrease as the amount of char consumed increases and to accelerate with higher availability (mole fraction $y$) and greater apparent reaction order $n$ of the reactive gas. Additionally, the RPM has a pore structure term considering conversion-dependent pore evolution by including the structural parameter $\Psi$.

To simplify the fitting process and minimise uncertainties arising from a large amount of variables, the pre-exponential factor and the reaction order were fixed to values determined for the undoped fuel ($A_{O_2} = 4 \times 10^8$, $n_{O_2} = 0.39$, $A_{CO_2} = 4 \times 10^9$, $n_{CO_2} = 0.31$) or the corresponding char ($A_{O_2} = 9 \times 10^8$, $n_{O_2} = 1.14$, $A_{CO_2} = 9 \times 10^7$, $n_{CO_2} = 0.51$). The modelling of TP curves was conducted using the falling branch of the conversion peak to reduce the influence of overlapping devolatilisation. Least-square fitting was used based on the interior-point algorithm in MATLAB® with maximum $10^6$ iterations, complying to a tolerance of $10^{-10}$. The dependence of loading (mineral amount) on the apparent activation energy was fitted using a Langmuir-type formulation [20]

$$E_A = E_{A,max} \frac{s\,\omega}{1 + s\,\omega} - a\,\omega^b \tag{4.5}$$

with $E_{A,max}$ being the maximum achievable influence approached as a function of the loading $\omega$ with a certain strength of correlation $s$. A power law-type deactivation term with factor $a$ and exponent $b$ was additionally required to model the loading dependence.

Figure 4.13 shows the decrease of the apparent activation energy for the doped samples relative to the corresponding undoped samples as a function of the doped amount of minerals determined by AAS.

For the pre-pyrolysed fuel sample doped with Mg and Ca no results are shown for oxidation, since there was no significant effect on $E_A$ observed for all investigated loadings. The other investigated series showed a decrease of $E_A$ with a nonlinear loading dependence. The changes were more pronounced at lower loadings, resulting in a nearly diminishing effect of further loading and even a reverse effect was observed for the fuel at higher loadings, representing a deactivation by a too high loading of minerals. Independent of the fuel type, the effect on the gasification was approximately twice as strong as for oxidation.

For the doped unpyrolysed fuel upon oxidation, Fe showed the most pronounced effects followed by the alkali metals with a similar maximum effect and strength of loading dependence. For gasification both alkali metals show a high maximum effect, but a low loading dependence so that their catalytic effect on char conversion becomes more dominant at high loadings. In contrast, Fe and Ca are predominant at low loadings, which is typical of the metallic inorganic element content in most natural biomasses. The overall lower maximum catalytic effect compared with alkali metals is due to the higher loading dependence. Similarly, Mg also exhibits a high dependence on the mineral amount, but with a very low maximum effect [20].

For the pre-pyrolysed fuel, similar trends were found compared with the untreated model fuel. Qualitative differences in the two model fuels are mainly related to the absence of a catalytic effect of Ca and Mg during oxidation and the deactivation of alkali metals and Ca upon oxidation. Deactivation was most prominent for K. At higher loadings the mineral particles form agglomerates during the impregnation process limiting the mobility of the mineral and the degree of contact with the carbon matrix. Differences in the extent of loading dependence of both fuels are primarily found during conversion in diluted $O_2$ indicating that the influence of minerals on devolatilisation mainly alters the oxidation behaviour of the char. The stronger dependence for the untreated model fuel may be attributed to the presence

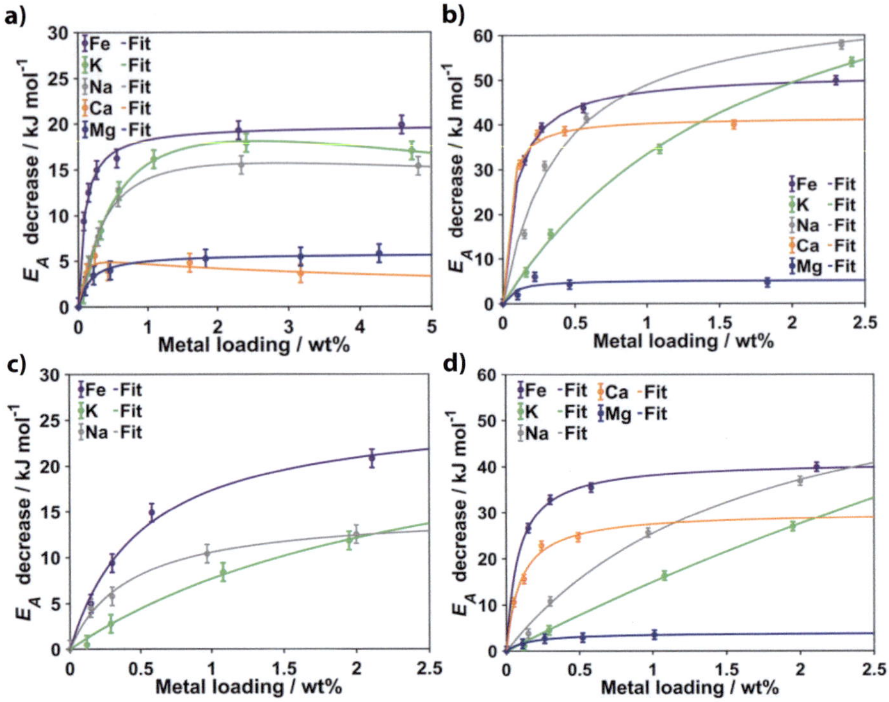


**Fig. 4.13** Decrease of the apparent activation energy dependence on the mineral loading for **a**) fuel during oxidation, **b**) fuel during gasification, **c**) pre-pyrolysed fuel during oxidation, and **d**) pre-pyrolysed fuel during gasification. (Reprinted and fonts adapted from Eckhard et al. [20], licensed under CC BY 4.0)

of minerals that activate the carbon matrix during devolatilisation. The similarity observed when converting both fuels in diluted $CO_2$ suggests that the mineral effects initiated during devolatilisation do not result in a noticeable change of the gasification process. Since gasification occurs at temperatures several hundred degrees higher than oxidation, these catalytic effects are superimposed by the thermal annealing of the carbon matrix [20].

## 4.5 Mineral Influence on Combustion in the Flat Flame Burner

Flat flame burner (FFB) experiments were performed to investigate the influence of different minerals on the combustion behaviour in oxy-fuel atmosphere (20 vol.% $O_2$, 10 vol.%$H_2O$ in $CO_2$) [18]. The model fuel was doped with 5 wt.% of Mg, K+Mg and Fe+Mg by wet impregnation. The analysis focuses on the ignition times, combustion temperatures, and particle behaviour of these samples compared with the undoped fuel.

Figure 4.14 shows the results of the pyrometric measurements displaying the particle combustion temperature with increasing height over the burner. All samples

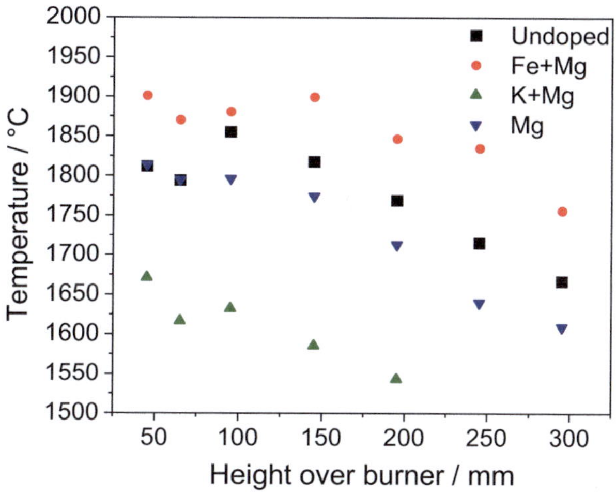


**Fig. 4.14** Particle combustion temperature of the samples at different heights over the burner inlet measured in 20 vol.% $O_2$, 10 vol.% $H_2O$ in $CO_2$. (Reprinted from Eckhard et al. [18], Copyright (2022), with permission from Elsevier and minor editing)

ignited before reaching the first measurement point, which is located 45 mm above the burner. The combustion temperature followed a characteristic pattern: an initial decrease after ignition, followed by an increase, and then a near-linear decline as char burnout progressed. Differences in ignition times and combustion temperatures among the samples indicate varying combustion rates [18].

The Fe+Mg-doped sample ignited earlier and showed higher combustion temperatures compared with the undoped fuel. This observation suggests an enhanced reaction rate due to Fe+Mg-doping, which is in line with the performed TG experiments that also indicate a positive catalytic effect. The single Mg-doped sample also ignited earlier in comparison to the undoped material. Their combustion temperatures were similar to the undoped fuel up to 75 mm above the burner but were 50–70 °C lower at greater distances. This result indicates that Mg doping moderately increases the reaction rate during early pyrolysis and ignition stages, but this effect diminishes during char burnout. The K+Mg-doped samples had moderately earlier ignition times but significantly lower combustion temperatures by 150–240 °C. This observation indicates a reduced combustion rate in the FFB, which differs from the TGA results showing a positive effect on reactivity. During post-ignition, mean particle diameters increased from approximately 100–150 µm with more pronounced swelling in mineral-doped samples. This swelling likely results from mineral deposits hindering the release of volatiles during devolatilisation. This effect was independent of the type of mineral. The observed increase in particle size could explain the discrepancies between the FFB and TGA results for K+Mg doping. Near the burner, catalytic doping may enhance reaction rates, but subsequent particle swelling may inhibit the release of volatiles and introduce mass and heat transfer limitations reducing reactivity. These effects are less pronounced in TGA experiments, which occur under different conditions and do not exhibit significant swelling [18].

## 4.6 Conclusion

Minerals have a significant influence on the biomass conversion that needs to be considered for the development of predictive models of biomass thermal behaviour. This study provided a comprehensive investigation of the influence of minerals on biomass pyrolysis, oxidation as well as gasification. A model fuel synthesis and impregnation procedure was developed allowing systematic doping of minerals. Three different setups were used to enable the analysis of the catalytic impact under fast heating rate conditions on the three pyrolysis products: light gas, tar, and char. The results showed that minerals changed the distribution of tars and emitted light gases as well as the formation of char resulting in a change of graphitisation and reactivity. This highlights the importance of considering mineral influences, since pyrolysis is an important step in the biomass conversion process, which significantly impacts the further thermal process. Characterisation techniques like XRD, Mössbauer spectroscopy and XAS were used to investigate the mineral state during biomass conversion. It was revealed that especially iron was prone to phase transitions resulting in phases with significantly different catalytic activity during char oxidation. Noteworthy was also the formation of a langbeinite phase due to the interaction of potassium and calcium during heating. These studies showed that not only the amount and kind of minerals are important, but also the complex transformations and interactions of minerals have significant influence on the biomass conversion process. The catalytic effects of various minerals at different loading were quantified as reduction of the apparent activation energy for oxidation and gasification. These kinetic parameters can be incorporated into biomass char conversion models enhancing their accuracy and predictive capability. Finally, the experiments in the flat flame burner confirmed that the catalytic activity of the minerals is not only relevant for low heating rate TG experiments, but also for combustion experiments with higher heating rate and short residence times under oxy-fuel conditions.

**Acknowledgement** This work has been funded by the Deutsche Forschungsgemeinschaft (DFG, German Research Foundation)—215035359—SFB/TRR 129 'Oxyflame'.

## References

1. Wang, W., Lemaire, R., Bensakhria, A. and Luart, D. "Review on the Catalytic Effects of Alkali and Alkaline Earth Metals (AAEMs) Including Sodium, Potassium, Calcium and Magnesium on the Pyrolysis of Lignocellulosic Biomass and on the Co-Pyrolysis of Coal with Biomass". *Journal of Analytical and Applied Pyrolysis* 163 (2022), 105479. https://doi.org/10.1016/j.jaap.2022.105479
2. Giudicianni, P., Gargiulo, V., Grottola, C. M., Alfè, M., Ferreiro, A. I., Mendes, M. A. A., Fagnano, M. and Ragucci, R. "Inherent Metal Elements in Biomass Pyrolysis: A Review". *Energy & Fuels* 35.7 (2021), 5407–5478. https://doi.org/10.1021/acs.energyfuels.0c04046
3. Vassilev, S. V., Baxter, D., Andersen, L. K. and Vassileva, C. G. "An Overview of the Composition and Application of Biomass Ash. Part 1. Phase–Mineral and Chemical Composition and Classification". *Fuel* 105 (2013), 40–76. https://doi.org/10.1016/j.fuel.2012.09.041

4. Thor, K. "Calcium-Nutrient and Messenger". *Frontiers in plant science* 10 (2019), 440. https://doi.org/10.3389/fpls.2019.00440
5. Wang, M., Wang, R., Mur, L. A. J., Ruan, J., Shen, Q. and Guo, S. "Functions of Silicon in Plant Drought Stress Responses". *Horticulture research* 8.1 (2021), 254. https://doi.org/10.1038/s41438-021-00681-1
6. Xu, X., Du, X., Wang, F., Sha, J., Chen, Q., Tian, G., Zhu, Z., Ge, S. and Jiang, Y. "Effects of Potassium Levels on Plant Growth, Accumulation and Distribution of Carbon, and Nitrate Metabolism in Apple Dwarf Rootstock Seedlings". *Frontiers in plant science* 11 (2020), 904. https://doi.org/10.3389/fpls.2020.00904
7. Fitria, Liu, J. and Yang, B. "Roles of Mineral Matter in Biomass Processing to Biofuels". *Biofuels, Bioproducts and Biorefining* 17.3 (2023), 696–717. https://doi.org/10.1002/bbb.2468
8. Kleinhans, U., Wieland, C., Frandsen, F. J. and Spliethoff, H. "Ash Formation and Deposition in Coal and Biomass Fired Combustion Systems: Progress and Challenges in the Field of Ash Particle Sticking and Rebound Behavior". *Progress in Energy and Combustion Science* 68 (2018), 65–168. https://doi.org/10.1016/j.pecs.2018.02.001
9. Trubetskaya, A. "Reactivity Effects of Inorganic Content in Biomass Gasification: A Review". *Energies* 15.9 (2022), 3137. https://doi.org/10.3390/en15093137
10. Yu, J., Guo, Q., Gong, Y., Ding, L., Wang, J. and Yu, G. "A Review of the Effects of Alkali and Alkaline Earth Metal Species on Biomass Gasification". *Fuel Processing Technology* 214 (2021), 106723. https://doi.org/10.1016/j.fuproc.2021.106723
11. Chen, S. G. and Yang, R. T. "Unified Mechanism of Alkali and Alkaline Earth Catalyzed Gasification Reactions of Carbon by $CO_2$ and $H_2O$". *Energy & Fuels* 11.2 (1997), 421–427. https://doi.org/10.1021/ef960099o
12. Castoldi, L., Matarrese, R., Lietti, L. and Forzatti, P. "Intrinsic Reactivity of Alkaline and Alkaline-Earth Metal Oxide Catalysts for Oxidation of Soot". *Applied Catalysis B: Environmental* 90.1-2 (2009), 278–285. https://doi.org/10.1016/j.apcatb.2009.03.022
13. McKee, D. W. "Mechanisms of the Alkali Metal Catalysed Gasification of Carbon". *Fuel* 62.2 (1983), 170–175. https://doi.org/10.1016/0016-2361(83)90192-8
14. McKee, D. W. and Chatterji, D. "The Catalytic Behavior of Alkali Metal Carbonates and Oxides in Graphite Oxidation Reactions". *Carbon* 13.5 (1975), 381–390. https://doi.org/10.1016/0008-6223(75)90006-8
15. Pala, M., Kantarli, I. C., Buyukisik, H. B. and Yanik, J. "Hydrothermal Carbonization and Torrefaction of Grape Pomace: A Comparative Evaluation". *Bioresource Technology* 161 (2014), 255–262. https://doi.org/10.1016/j.biortech.2014.03.052
16. Düdder, H., Wütscher, A., Stoll, R. and Muhler, M. "Synthesis and Characterization of Lignite-Like Fuels Obtained by Hydrothermal Carbonization of Cellulose". *Fuel* 171 (2016), 54–58. https://doi.org/10.1016/j.fuel.2015.12.031
17. Lotz, K., Berger, C. M. and Muhler, M. "Catalytic Effect of Iron Phases on the Oxidation of Cellulose-Derived Synthetic Char". *Energy Procedia* 158 (2019), 694–699. https://doi.org/10.1016/j.egypro.2019.01.188
18. Eckhard, T., Pflieger, C., Schmidt, S., Böttger, J., Senneca, O., Schiemann, M., Scherer, V., Muhler, M. and Cerciello, F. "Catalytic Effects for Cellulose-Based Model Fuels Under Low and High Heating Rate in Air and Oxy-Fuel Atmosphere". *Fuel* 324 (2022), 124437. https://doi.org/10.1016/j.fuel.2022.124437
19. Eckhard, T., Wütscher, A. and Muhler, M. "Simultaneous Analysis of Light Gases and Heavy Pyrolyzates Evolved from Lignite and Hard Coal by Pyrolysis–GC/MS–GC/TCD". *Journal of Analytical and Applied Pyrolysis* 149 (2020), 104833. https://doi.org/10.1016/j.jaap.2020.104833
20. Eckhard, T., Pflieger, C., Böttger, J., Telaar, P., Cerciello, F. and Muhler, M. "The Catalytic Effect of Iron and Alkali and Alkaline Earth Metal Sulfates Loading Series on the Conversion of Cellulose-Derived Hydrochars and Chars". *ACS Omega* 8.11 (2023), 10629–10639. https://doi.org/10.1021/acsomega.3c00887

21. Strandberg, A., Holmgren, P., Wagner, D. R., Molinder, R., Wiinikka, H., Umeki, K. and Broström, M. "Effects of Pyrolysis Conditions and Ash Formation on Gasification Rates of Biomass Char". *Energy & Fuels* 31.6 (2017), 6507–6514. https://doi.org/10.1021/acs.energyfuels.7b00688
22. Hüttinger, K. J. and Minges, R. "The Influence of the Catalyst Precursor Anion in Catalysis of Water Vapour Gasification of Carbon by Potassium: 2. Catalytic Activity as Influenced by Activation and Deactivation Reactions". *Fuel* 65.8 (1986), 1122–1128. https://doi.org/10.1016/0016-2361(86)90180-8
23. Pflieger, C., Eckhard, T., Böttger, J., Schulwitz, J., Herrendorf, T., Schmidt, S., Salamon, S., Landers, J., Wende, H., Kleist, W., Muhler, M. and Cerciello, F. "The Catalytic Effect of Iron Oxide Phases on the Conversion of Cellulose-Derived Chars in Diluted $O_2$ and $CO_2$". *Applied Energy* 353 (2024), 122068. https://doi.org/10.1016/j.apenergy.2023.122068
24. Katikaneani, P., Vaddepally, A. K., Reddy Tippana, N., Banavath, R. and Kommu, S. "Phase Transformation of Iron Oxide Nanoparticles from Hematite to Maghemite in Presence of Polyethylene Glycol: Application as Corrosion Resistant Nanoparticle Paints". *Journal of Nanoscience* 2016 (2016), 1–6. https://doi.org/10.1155/2016/1328463
25. Hasdeo, E. H., Nugraha, A. R. T., Dresselhaus, M. S. and Saito, R. "Breit-Wigner-Fano Line Shapes in Raman Spectra of Graphene". *Physical Review B* 90.24 (2014). https://doi.org/10.1103/PhysRevB.90.245140
26. Tuinstra, F. and Koenig, J. L. "Raman Spectrum of Graphite". *The Journal of Chemical Physics* 53.3 (1970), 1126–1130. https://doi.org/10.1063/1.1674108
27. Ferrari, A. C. and Robertson, J. "Interpretation of Raman Spectra of Disordered and Amorphous Carbon". *Physical Review B* 61.20 (2000), 14095–14107. https://doi.org/10.1103/PhysRevB.61.14095
28. Eckhard, T., Pflieger, C., Russo, C., Freisewinkel, E., Eisenbach, T., Böttger, J., Senneca, O., Apicella, B., Schiemann, M., Span, R., Scherer, V., Muhler, M. and Cerciello, F. "Mineral Effects on Chemical and Physical Transformations of Fast Pyrolysis Products of Cellulose-Based Model Fuels in $N_2$ and $CO_2$". *Fuel* 340 (2023), 127477. https://doi.org/10.1016/j.fuel.2023.127477
29. Lotz, K., Wütscher, A., Düdder, H., Berger, C. M., Russo, C., Mukherjee, K., Schwaab, G., Havenith, M. and Muhler, M. "Tuning the Properties of Iron-Doped Porous Graphitic Carbon Synthesized by Hydrothermal Carbonization of Cellulose and Subsequent Pyrolysis". *ACS Omega* 4.2 (2019), 4448–4460. https://doi.org/10.1021/acsomega.8b03369
30. Pflieger, C., Lotz, K., Hilse, N., Berger, C. M., Schiemann, M., Debiagi, P., Hasse, C., Scherer, V. and Muhler, M. "Catalytic Influence of Mineral Compounds on the Reactivity of Cellulose-Derived Char in $O_2$-, $CO_2$-, and $H_2O$-Containing Atmospheres". *Fuel* 287 (2021), 119584. https://doi.org/10.1016/j.fuel.2020.119584
31. Vamvuka, D., Karouki, E., Sfakiotakis, S. and Salatino, P. "Gasification of Waste Biomass Chars by Carbon Dioxide via Thermogravimetry—Effect of Catalysts". *Combustion Science and Technology* 184.1 (2012), 64–77. https://doi.org/10.1080/00102202.2011.618152
32. Di Blasi, C. "Combustion and Gasification Rates of Lignocellulosic Chars". *Progress in Energy and Combustion Science* 35.2 (2009), 121–140. https://doi.org/10.1016/j.pecs.2008.08.001
33. Perander, M., DeMartini, N., Brink, A., Kramb, J., Karlström, O., Hemming, J., Moilanen, A., Konttinen, J. and Hupa, M. "Catalytic Effect of Ca and K on $CO_2$ Gasification of Spruce Wood Char". *Fuel* 150 (2015), 464–472. https://doi.org/10.1016/j.fuel.2015.02.062
34. Klose, W. and Wölki, M. "On the Intrinsic Reaction Rate of Biomass Char Gasification with Carbon Dioxide and Steam". *Fuel* 84.7-8 (2005), 885–892. https://doi.org/10.1016/j.fuel.2004.11.016
35. Prestipino, M., Galvagno, A., Karlström, O. and Brink, A. "Energy Conversion of Agricultural Biomass Char: Steam Gasification Kinetics". *Energy* 161 (2018), 1055–1063. https://doi.org/10.1016/j.energy.2018.07.205

# Sorption Effects

# 5

Tim Eisenbach, Carsten Wedler and Roland Span

## Contents




**Abstract**

This chapter provides a comprehensive analysis of the sorption effects of oxy-fuel-relevant species and the particle pore structure evolution during pyrolysis and char conversion. The methodology considers physisorption measurements and suitable density functional theory models for pore surface area and volume analysis. Furthermore, the mass transport of oxy-fuel-relevant species in the porous fuel particles was investigated by adsorption kinetic measurements. Applying suitable



T. Eisenbach (✉) · R. Span
Chair of Thermodynamics, Ruhr-Universität Bochum, Bochum, Germany
e-mail: t.eisenbach@thermo.rub.de

R. Span
e-mail: roland.span@thermo.rub.de

C. Wedler
Department of Chemical Engineering, Imperial College London, London, UK
e-mail: c.wedler@imperial.ac.uk







R. Kneer et al. (eds.), *Clean Energy Conversion*,
https://doi.org/10.1007/978-3-031-98956-8_5

adsorption kinetic and diffusion models based on the experimental data provides mass transport properties, which are intended to be used in common char conversion and combustion models. A novel pore structure-dependent adsorption kinetic (PSK) model is presented incorporating a physically sound description of the time-dependent intraparticle mass transport process. Moreover, incorporating relative pressure adsorption kinetic data, a relative pressure (RP)-PSK model is developed enabling the prediction of mass transport processes, which depend on a relative pressure difference of oxy-fuel components in the bulk phase and the gas phase in the intraparticle structure.

## 5.1 Introduction

The analysis of porosities of solid fuel chars as a product of pyrolysis and char conversion processes is relevant for the understanding of the conversion process, since the available pore surface areas and volumes in certain pore regimes can determine the accessibility of the intraparticle surface area for oxy-fuel-relevant gas species. Furthermore, char porosities are used as input parameters for extensive char conversion and combustion models, which require a physics-based and reasonable description of pore structure properties. The individual constitution of the pore structure also determines the mass transport of gases involved in the reaction. The diffusive mass transport through the strongly pronounced pore structure of the considered solid fuel particles can be a limiting factor for the subsequent reaction at the surface. To account for an oxy-fuel atmosphere in char conversion and combustion models, it is essential to understand the differing intraparticle mass transport behaviours of $CO_2$ and $O_2$. Literature indicates that $CO_2$ primarily generates small micropores and reduces gasification reaction rates due to its distinct diffusion properties within the microporous structure of the particles. In contrast, $O_2$ predominantly enlarges existing mesopores, with minimal dependence on pore diffusion [1,2]. This chapter focuses on the close investigation of the contrasting mass transport behaviours of these two major oxy-fuel species. It further presents the derivation of suitable mass transport models, providing useful information about diffusion in pores for char conversion and combustion models. The mass transport of oxy-fuel species is investigated with adsorption kinetic measurements on representative chars at different temperatures and pressures, while suitable models should provide a physically sound description of the pore structure-dependent and time-resolved mass transport.

Section 5.2 contains a detailed description of the analysis methods, measurement procedures and model evaluations for the determination of pore structure and mass transport properties. Section 5.3 focuses on representative results for the pore structure evolution using different analysis methods. Furthermore, novel adsorption kinetic models are described, which can be used to predict suitable mass transport properties in extensive char conversion and combustion models.

## 5.2 Experimental Setup and Evaluation Methods

In comparison to coal chars, chars from biomass exhibit a significantly larger internal surface area, which is potentially available for reactions. Especially gasification reactions are allocated at the large intraparticle surface area, which ascribes great importance to the analysis of pore morphology and available pore surface area and volume [2]. The overall reaction kinetics of burnout are influenced by pore structure through the limited mass transport of reactive gases towards the surface where reactions occur. Furthermore, pore properties are important parameters for evaluating reaction products or volatile release during char conversion or pyrolysis, respectively.

The precise investigation of the sorption behaviour of oxy-fuel-relevant gases on solid fuel particles is an important measure for the accessibility of the inner pore surface as a preliminary process to the final reaction of species. In addition to the intraparticle mass transport and diffusion, chemical reaction or film diffusion can determine the rate of the conversion process as well, depending on the temperature and the combustion region of the conversion process [2,3]. The intraparticle mass transport can be accessed by adsorption kinetic measurements, which describe the delay in occupation of the pore surface by gaseous species. The overall affinity and capacity of the pore surface can be analysed by adsorption isotherms, which describe the pressure-dependent equilibrium loading. The overall adsorption capacity depends on the overall available structural pore surface area, on the chemical constitution of the surface, e.g. the presence of functional groups or minerals, and on the overall tendency of a gaseous species to adsorb on a solid surface.

### 5.2.1 Pore Structure Investigation with Physisorption Measurements

The pore structure investigations are based on systematic physisorption measurements, a strict evaluation and a further processing of experimental data with suitable models and calculation methods for pore structure. The chars investigated here are products of pyrolysis and conversion studies, which are conducted either in a drop tube reactor (DTR) for flash pyrolysis (see Sect. 2.2.3), a thermogravimetric analyser (TGA) for slow pyrolysis (see Sect. 2.2.1), or a flat flame burner (FFB) for char conversion (see Sect. 3.3.1). Pore volumes and skeletal densities are used in the calculation of porosities, which are properties required in combustion models [4].

For the assessment of pore structure properties such as pore volume and pore surface area of particles, physisorption measurements are performed, which were supplemented by different adsorption theories. The experimental basis is built upon $N_2$ and $CO_2$ adsorption isotherms for different samples of char particles with different conversion backgrounds. In addition to physisorption measurements, some optical methods are applied to make larger pores accessible (see Sect. 2.5.2).

## Experimental Setups

The adsorption isotherms of $N_2$ and $CO_2$ used for pore structure examination were measured almost exclusively with a volumetric gas sorption device *BELSORP-mini II* by *MicrotracBEL*. A few analyses have been conducted in a *BELSORP-max* by *MicrotracBEL*, which is a comparable adsorption measurement device. Compared to gravimetric devices, volumetric adsorption measurement devices are advantageous with regard to the handling of many samples, resulting in a significantly faster throughput. They do not require time-consuming sample changes and leakage tests and measure three to four different samples at a time. The samples, which are subject to investigation, are weighed into glass cells and attached to the sorption device. The sample cells are connected to a gas dosing system, which is monitored by several pressure sensors. A rotary vane pump establishes a sufficient vacuum inside the sample cells and the gas dosing system. The latter is connected to He, which is used for purging and dead volume measurements, and the measuring gas, which is either $N_2$ or $CO_2$. Figure 5.1 shows the arrangement and appearance of the filled sample containers.

## Measurement Procedure

Before the actual adsorption measurements are carried out, a pretreatment is conducted to remove adsorbed molecules, which are predominantly water molecules from humid air, from the surfaces of the sample. For this purpose, the sample is heated in situ to a temperature between 393.15 and 473.15 K using a temperature-

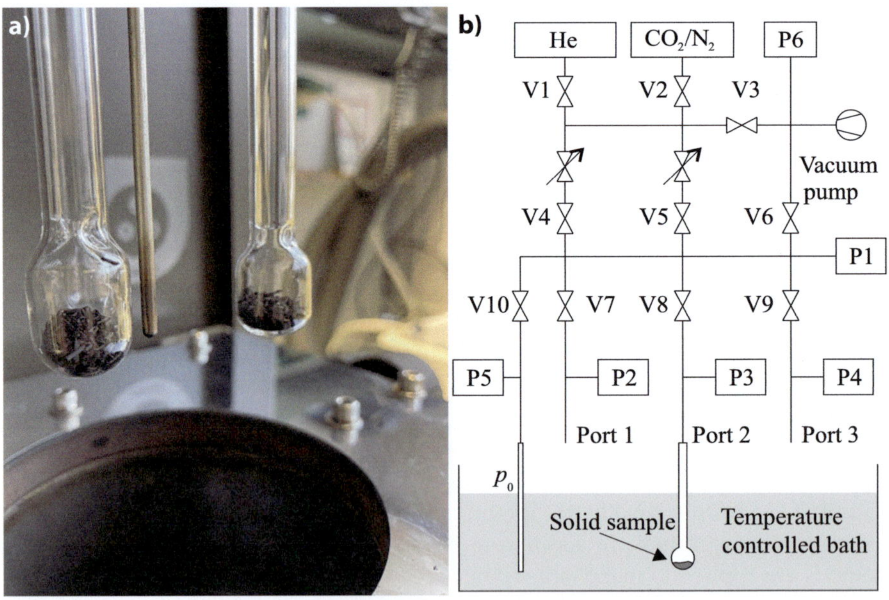


**Fig. 5.1** **a**) Picture of the filled sample containers attached to the gas dosing system of the *BELSORP-mini II* and **b**) a scheme of the gas dosing system including all valves and pressure sensors

controlled jacket. This procedure is also referred to as activation of the sample. The temperature is chosen, depending on the chemical constitution and the tendency for devolatilisation of pyrolysis products from the surface of particles. To maintain the initial state and pore condition of the sample, the temperature of the heating process should not exceed 473.15 K. Ideally, only physisorbed species are desorbing during this pretreatment. Subsequently to the pretreatment, the mass of the remaining sample is determined and the actual measurement procedure is started with the dead volume determination using He in a simple expansion measurement. Helium is expanded from a well-known dosing volume into the unknown dead volume of the sample cell, which is the vacant volume in the sample cell not occupied by the sample while the pressure is monitored simultaneously. Provided that He-adsorption on the surface of the sample is negligible, the accurate determination of the pressure of the gas phase in the dead volume is possible. A precise determination is mandatory since the gas phase volume in the sample cell is important for the calculation of the quantity of adsorbed molecules later on. After a repeated evacuation, the adsorptive ($N_2$ or $CO_2$) is dosed into the sample cells. By measuring the pressure in the dosing volume before the dosing process and in the dosing volume plus the free volume of the measuring cell after the adsorption equilibrium has been established, the adsorbed amount of substance can be determined. Adsorption equilibria at different pressures can be determined by decoupling, filling, and coupling of the dosing volume multiple times. For the required amount of substance or mass balances in the gas phase, virial equations of state can be used. The considered pressure range and adsorption temperature for the individual $N_2$ and $CO_2$ measurements are set by the latest IUPAC report [5] on physisorption experiment protocols for the pore analysis of carbonaceous solids. Accordingly, the $N_2$ adsorption isotherm is measured at 77.36 K and in the pressure range $10^{-3} < p/p_0 < 0.99$, where $p_0 = 0.1\,\text{MPa}$ is the pressure at the normal boiling point at 77.36 K. The $CO_2$ adsorption temperature is set to 273.15 K and the pressure range is set to $10^{-4} < p/p_0 < 0.03$ with the vapour pressure $p_0 = 3.49\,\text{MPa}$ for $CO_2$ at adsorption temperature. The experimental isotherms are used in different evaluation methods to calculate pore properties such as the pore surface area, the pore volume, or the pore size distribution (PSD).

**Evaluation Methods**

A frequently applied method in the analysis of the available pore surface area for adsorption is based on the Brunauer-Emmett-Teller (BET) theory [6]. For an investigated meso-/macroporous solid, a monolayer occupation of pores is assumed for small relative pressures $p/p_0$ [7], depending on a sufficiently well-defined type II or IV(a) $N_2$-isotherm as defined by IUPAC [5]. The characteristic parameter is the monolayer capacity $n_m$, which results from a fit of the BET equation [5,6] to the experimental $N_2$ adsorption data in the pressure range $p/p_0 = 0.05$–$0.3$. From this the BET surface area is calculated with the molecular cross-sectional area for $N_2$, assuming a closely packed monolayer [5]. Providing a well-defined adsorption isotherm and, thus, applicability of the BET theory, the resulting specific surface areas can serve as a useful measure of comparison between different solid fuel particles with different combustion or conversion histories.

The adsorption behaviour of $N_2$ on solid fuels can further be used to determine the specific pore volume. By applying the Gurvich rule [8], it can be assumed that the available specific pore volume is filled by condensing nitrogen at $p/p_0 = 0.99$, where adsorption is predominantly defined by pore condensation close to the saturation pressure of $N_2$. Nonetheless, the evaluation of $N_2$ physisorption for the purpose of pore property analysis has limits when it comes to the analysis of highly microporous solid particles. In particular biomass char particles with high volatile yield as products of pyrolysis or gasification in oxy-fuel atmospheres are considered to have a predominantly microporous structure [2]. Nitrogen has limitations due to diffusion into very narrow micropores due to its quadrupolar nature and the kinetic limitations at cryogenic temperatures. Although not yet standardised and less widely established than $N_2$ adsorption measurements in combination with the BET theory for pore analysis, there are other options to access the microporous structure of chars. According to IUPAC [5], the physisorption of $CO_2$ at 273.15 K in a narrow relative pressure range of $10^{-4} < p/p_0 < 0.03$ is well-suitable for micropore analysis. Based on this, two often-referred methods [9–11] are applied for micropore analysis in the context of pore structure investigations. Both theories, Dubinin-Ashtakov (DA) [12] and Dubinin-Radushkevich (DR), are based on Dubinins theory for volume filling of micropores, where the DR theory is a special case of the more general DA theory. Further theoretical background for the determination of the pore volume $v_{mi}$ for micropores can be found in the study by Wedler and Span [13]. To determine the surface area of the micropores associated with the pore volume $v_{mi}$, assumptions on the average pore diameter and the general shape of the pores need to be made. One approach that is frequently used in the evaluation of surface areas of micropores is the approach by Medek [14]. It considers a micropore distribution with a simplified and homogeneous pore shape.

A more sophisticated approach to pore structure analysis established over the years is the application of density functional theory (DFT) models. In most DFT applications, an external functional is applied, which describes the solid-fluid interactions of gas molecules in confined pore spaces. Based on this functional, density profiles can be calculated, which describe the spatial distribution of molecules in the pores [15]. By assuming and providing simple geometrics for the respective pore structure and solid-fluid interaction parameters, adsorption isotherms can be predicted. The study by Jagiello et al. [16] developed a method for the calculation of a pore size distribution (PSD) based on DFT models and corresponding experimental isotherm data. Those DFT kernels are tailored to the respective adsorptive gas and the adsorbent interaction parameters. The theoretical kernels are fitted to the experimental isotherms data yielding a calculated pore size distribution. The method has undergone several extensions [17, 18], incorporating energetically heterogeneous surfaces, a geometrical corrugation (2-D) to the model pore walls, and a non-local density functional (NLDFT). The most recent evaluations include a dual gas analysis by fitting $N_2$ and $CO_2$ kernels simultaneously to the experimental isotherms [19]. Furthermore, Jagiello et al. [20] showed the applicability for chars from biomass using the above-introduced 2D-NLDFT analysis. All recently published 2D-NLDFT kernels are implemented in the SAIEUS software tool by *Micromeritics*, while the initial method for the solution of the adsorption integral equation (SAIEUS) can be found in the study by Jagiello [21]. The 2D-NLDFT analysis is the preferable method when it comes to a detailed resolution of the PSD in the pore diameter range $d_p = 0.3$–$50\,\text{nm}$, while the limits are set by the underlying 2D-NLDFT kernels and are retained for all studies. As implemented in the SAIEUS software tool, the lower limit of the pore diameter for the $N_2$ kernel is adjustable in order to obtain the best fit of the calculated 2D-NLDFT to the experimental data. According to IUPAC [5], pores can be subdivided in a ultramicropore regime for $d_p < 0.7\,\text{nm}$, a micropore regime for $d_p = 0.7$–$2\,\text{nm}$, a mesopore regime for $d_p = 2$–$50$ nm and in pores with a diameter of $d_p > 50$ nm, which are assigned to macropores. From the PSD across different pore regimes, the cumulative pore surface areas $A_p$ and pore volumes $v_p$ can be calculated. Those are useful parameters in the analysis of the pore evolution during pyrolysis and char conversion processes since the growth or coalescence of pore regimes can give reference to a possible reaction behaviour of gaseous species in an oxy-fuel atmosphere or to a preference of gases to react in specific pore regimes. Furthermore, the pore volume $v_p$ can be used to calculate the porosity:

$$\Phi = \frac{v_p}{\frac{1}{\rho_{\text{skeletal}}} + v_p}. \tag{5.1}$$

The skeletal or also true density $\rho_{\text{skeletal}}$ can be obtained from He-pycnometry [22] or from He-buoyancy measurements. The latter method is explained in Sect. 5.2.2.

### 5.2.2 Analysis of Sorption Effects and Mass Transport with High Resolution Gravimetric Setups

The investigation of sorption effects is based on experimental analysis with gravimetric sorption setups. With these, sorption equilibria as well as the time-resolved sorption kinetics are measurable.

**Experimental Setups**

For the measurement of adsorption equilibria and kinetics of oxy-fuel-relevant species on chars of different origin and constitution, four experimental setups were used for the investigations discussed here. Three of them are based on the same gravimetric measurement principle and core component, which is a magnetic suspension balance (MSB) weighing the adsorptive mass increase on an investigated sample. A major advantage of a MSB is the contactless weighing as the microbalance is separated from the atmosphere in the measuring cell. Detailed information on the measurement principle can be found in the studies by Kleinrahm et al. [23] and Yang et al. [24]. However, the three devices differ in terms of their periphery and intended use. The first gravimetric sorption device shown in Fig. 5.2 is predominantly used for the investigation of the kinetics of sorption processes of supercritical gases ($CO_2$, $O_2$, and CO). It is equipped with a very high-resolution 1 µg-balance by *Mettler Toledo* and a magnetic suspension system by *TA instruments* former *Rubotherm*.

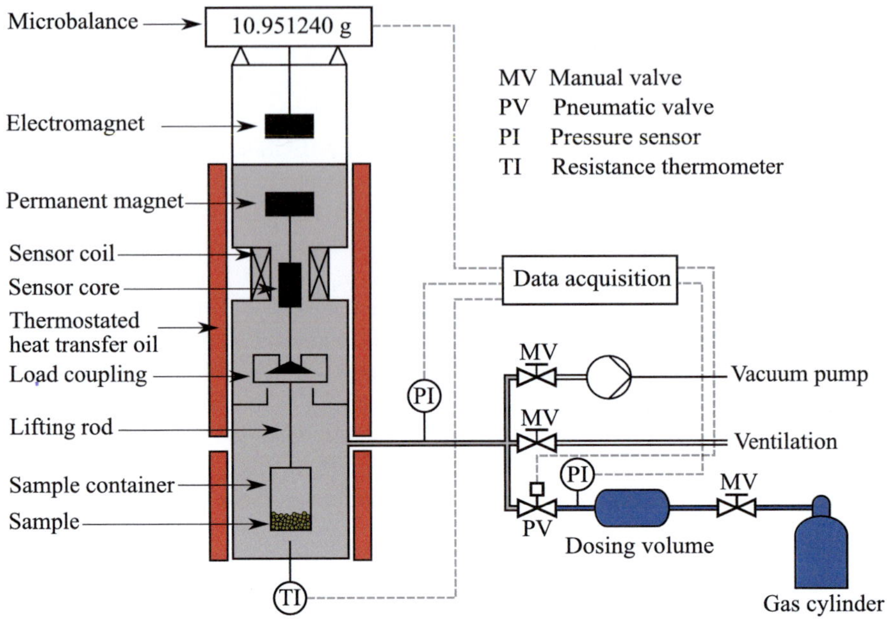


**Fig. 5.2** Experimental setup of the gravimetric sorption device used for measuring adsorption kinetics and equilibria of oxy-fuel-relevant gaseous species. (Based on the publication by Eisenbach et al. [25])

The measuring cell incorporates the sample container filled with the particles (sample) to be analysed. The sample container is attached to the suspended permanent magnet via a load coupling. By changing the electromagnetic field, different parts can be connected to the balance. In the zero-point (ZP) position only the permanent magnet and the sensor core are lifted, while the measuring position (MP1) considers all parts, including the loaded sample container, the lifting rod, and the load coupling. A change in weight is then transduced to the electromagnet, which is attached to the microbalance on top of the apparatus. The temperature of the gas phase in the cell is measured with a PT-100 thermometer connected to the cell and the pressure is monitored with a 10 MPa full-scale pressure sensor. A defined quantity of gas is released by a 300 mL dosing volume into the measuring cell by opening the pneumatic valve. A thermostat unit controls the adsorption temperature in the measuring cell.

For the measurement of condensable fluids, an apparatus was set up by Rösler and Wedler [26] and used for adsorption measurements for the oxy-fuel-component $H_2O$. The major difference in the set up is the incorporation of a vapour-liquid equilibrium (VLE) cell. This cell is able to provide $H_2O$ vapour at a specific pressure, depending on the controlled temperature in the VLE cell. The $H_2O$ vapour is connected to the balance part with a heated pipe to avoid any condensation for vapour pressures at temperatures above ambient temperature. Details on the setup can be found in the studies by Rösler and Wedler [26] and Eisenbach et al. [27]. The $H_2O$ vapour apparatus is equipped with a 10 μg resolution magnetic suspension balance system

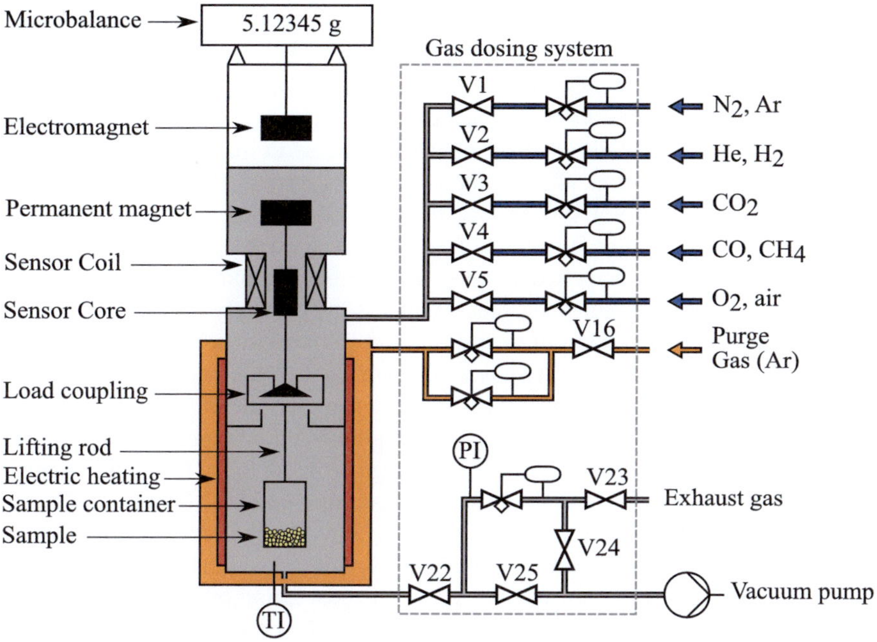


**Fig. 5.3** Schematic picture of the TGA used for the adsorption measurements at elevated temperatures including the core-apparatus and the gas dosing system

by *TA instruments* former *Rubotherm*, which is a sufficient resolution as $H_2O$ vapour shows larger adsorption loadings. Additional adsorption studies were carried out at elevated temperatures using a modified *DynTHERM MP-HT* TGA by *TA instruments* former *Rubotherm*. The apparatus principle is also based on a MSB system with a balance resolution of 10 µg, while the gas to be analysed flows through the measuring cell at a stated pressure and temperature. Theoretically, temperatures of 1573.15 K can be set at a gas pressure up to 3 MPa, but, considering gas adsorption on solid fuel particles, temperatures are significantly more limited due to the onset of conversion processes and associated mass loss. As shown in Fig. 5.3, different atmospheres can be set using a gas dosing and regulation system. The temperature of the measuring cell is controlled with an electric heater which is surrounded by an inert purge gas to prevent high temperature oxidation processes of the outer apparatus parts.

Further studies on the adsorption behaviour of oxy-fuel gases were carried out on an *ASAP 2020* volumetric sorption analyser by *Micromeritics*, which, unlike the *BELSORP-mini II* already mentioned, is capable of recording high-resolution sorption kinetics of various gases. The measurement principle and procedure for a volumetric sorption device were already explained in Sect. 5.2.1, the procedure and evaluation for the gravimetric measurements are explained below.

#### Measurement Procedure

Gravimetric adsorption measurements are based on the detection of the weight of a certain quantity of adsorptive on the porous chars. In contrast to the volumetric principle, the adsorbed mass is measured directly, but a few corrections due to different weight forces in the gravimetric system need to be conducted. All lifted and weighed parts need to be calibrated in terms of their weights and volumes, since they contribute directly to the mass balance of the system or indirectly by causing a buoyancy effect due to the density of the gas phase in the system (see below). After an external calibration, all parts including the investigated sample are attached to the magnetic suspension system (see Fig. 5.2). Before an adsorption measurement, the investigated sample needs to be activated as explained in Sect. 5.2.1. The mass and volume of the activated sample can only be determined in situ at vacuum conditions and with He-buoyancy measurements, respectively. Details on the He-buoyancy measurement procedure can be found in the studies by Eisenbach et al. [25] and Wedler and Span [28]. An adsorption kinetic measurement can use different starting conditions of the measuring cell and the sample. The most common option is to start the measurement from an activated sample and evacuated measuring cell by applying an absolute pressure and a defined gas volume to the cell at the moment of opening the pneumatic valve (see Fig. 5.2). Another option is to impose a higher pressure in relation to an existing cell pressure and an already adsorbed quantity. During the adsorption kinetic measurement, the balance position is kept in MP1 (see above) to detect the change in weight due to the adsorption of gaseous species on the char samples. Reaching the equilibrium state of the time-dependent adsorption process, the measuring position is alternated between ZP and MP1 to cancel out any time-dependent balance shifts. The initial condition of the sample can be restored by evacuating and heating up the measuring cell. Details on the procedure for a gravimetric adsorption (kinetics) measurement can be found in the studies by Eisenbach et al. [25] and Wedler and Span [28].

**Evaluation Methods**

In gravimetric adsorption measurements, the adsorbed molecules are also effected by a buoyancy force. Therefore, the absolute adsorption amount considers only the mass of adsorbed molecules corrected by its buoyancy, while the excess adsorption includes the mass of adsorbed molecules and its reduction due to buoyancy. In all studies, only the excess adsorption is considered, although at moderate pressures the excess adsorption almost corresponds to the absolute quantity. In order to obtain the mass of the excess adsorbed molecules $m_{\mathrm{ads}}$, a mass balance needs to be solved as stated in (5.2):

$$\frac{W_{01}}{\alpha\,\phi} = m_{01} + m_{\mathrm{ads}} + \rho_{\mathrm{fluid}}\,V_{01} \tag{5.2}$$

The left-hand side of (5.2) is the mass difference between the measured weight in MP1 and in ZP and, thus, considers the raw mass of the load coupling, lifting rod, sample container, and sample with adsorbed molecules. During an adsorption kinetic measurement, the mass $W_{01}$ is calculated for each time-step using the time-resolved MP1 values and the ZP value, which is recorded in the finally reached equilibrium state. At adsorption equilibrium conditions, $W_{01}$ is calculated from the MP1 and ZP differences repeatedly recorded at alternating positions. The mass difference $W_{01}$ is corrected by $\alpha$ and $\phi$, which are the balance calibration factor at ambient conditions and the correction factor for the force transmission error [23], respectively. The measured mass difference $W_{01}$ is effected by a buoyancy force $\rho_{\text{fluid}} \times V_{01}$, which is caused by the density of the adsorptive and the volumes of all lifted parts in MP1. The density of the fluid in the system, $\rho_{\text{fluid}}$, is calculated on the basis of measured temperature and pressure and the respective equation of state for the investigated fluid. The contribution of the buoyancy force is considered in the calculation of the mass of the adsorbed molecules (see (5.2)). The mass of all lifted parts $m_{01}$ in MP1 need to be considered in the overall mass balance in (5.2) as well as their respective volumes $V_{01}$, which contribute to the buoyancy effect. Both values can only be determined in situ conducting a He-buoyancy measurement as explained below.

By rearranging (5.2) for $m_{\text{ads}}$ and dividing by the molecular mass of the adsorptive $M_{\text{fluid}}$ and the mass of the solid sample $m_{\text{s}}$, the specific adsorption is determined with

$$q_{\text{ads}} = \frac{m_{\text{ads}}}{M_{\text{fluid}}\, m_{\text{s}}}. \tag{5.3}$$

In the He-buoyancy measurement procedure, different He-densities $\rho_{\text{He}}$ are achieved by different pressure steps along one isotherm. The different densities cause different buoyancy forces, which are measured as $W_{01,\text{He}}$. By assuming that He is not adsorbing, $m_{\text{ads}} = 0$, (5.2) can be reduced to

$$\frac{W_{01,\text{He}}}{\alpha\, \phi} = m_{01} + \rho_{\text{He}}\, V_{01}. \tag{5.4}$$

The values of $V_{01}$ and $m_{01}$ are calculated in a linear regression based on the densities $\rho_{\text{He}}$ and the measured mass differences $W_{01,\text{He}}$. The He-densities are calculated with the equation of state by Ortiz Vega et al. [29]. By substracting all ex-situ calibrated masses and volumes in $m_{01}$ and $V_{01}$, respectively, $m_{\text{s}}$ and $V_{\text{s}}$ of the sample can be derived and used in the calculation of the adsorption loading. Both values can further be used in the calculation of the skeletal density:

$$\rho_{\text{skeletal}} = \frac{m_{\text{s}}}{V_{\text{s}}}. \tag{5.5}$$

The skeletal density $\rho_{\text{skeletal}}$ can be used in the calculation of the porosity $\Phi$ (see (5.1)).

## 5.3 Results

In this section, results of pore structure investigations of various chars from biomass and coal are presented. Furthermore, results for the adsorption behaviour of oxy-fuel components on exemplary chars with different pore structure and chemical constitution are analysed. This section also presents results of adsorption modelling and of the associated derivation of mass transport properties.

### 5.3.1 Pore Structure Evolution

Most of the findings regarding the pore structure evolution analysis are a result of joint studies on chars with different combustion and conversion histories. Intermediate states of chars at different stages of conversion are preserved and analysed accordingly. This section will mainly focus on results for the pore structure of those chars and on the correlations between pore structure evolution and the respective conversion atmosphere, residence time or degree of conversion. Along the discussion of findings for the pore structure evolution of chars from flash pyrolysis (DTR) and char conversion (FFB) the mentioned temperatures always refer to the gas phase temperature. Due to the short residence times, the particle temperatures are lower and are either simulated in case of the DTR or optically measured in the FFB.

**Pore Evolution During Pyrolysis and Char Conversion**

The simplest and most straightforward analysis of pore structure is based on the BET surface area of investigated chars with different conversion histories. Especially during pyrolysis, where solid fuel particles face high temperatures at different heating rates and different residence times, the outbreak of volatiles causes a high degree of porosity. The kinetics of the volatile yield generally correlates with the heating rates assigned to flash and slow pyrolysis. The characteristic release of volatiles can be correlated with the generation and growth of pores along the conversion process. The focus of this study is on chars from biomass with different origins, whereby one exemplary study on coal chars is included as well.

In the following, the pore structure evolution in terms of the BET surface area and the skeletal density for several slow pyrolysis chars originating from Colombian bituminous coal (CBC) is discussed briefly. The pyrolysis chars were produced in a TGA apparatus, where temperatures from 473.15 to 1073.15 K in 100 K steps were applied in an Ar atmosphere. As shown by Wütscher et al. [30], the BET surface area of CBC chars decreases when the raw sample is heated to 573.15 K, which is due to thermal annealing effects. At higher pyrolysis temperatures, the BET surface area increases until 873.15 K is reached, which is a result of volatile and tar release. Reaching 1073.15 K, the highest investigated temperature, the BET surface area decreases due to the coalescence of pores. The skeleton density $\rho_{\text{skeletal}}$ determined by He-buoyancy measurements shows a correlation with the BET surface area. After a initial decrease, densities of the chars increase until 873.15 K reaching a value of $1.81\ \text{kg}\,\text{m}^{-3}$ at 33.6% mass loss. This behaviour can be explained by the densification of the char matrix during pyrolysis. The information on the density of the char is important for estimating thermal and fluid dynamic properties in conversion models. Investigations on the covered pyrolysis chars are reported and analysed by Wütscher et al. [30].

In a flash pyrolysis study [31], experiments with CBC were conducted in a DTR for pyrolysis temperatures of 1300 K and 1475 K in $N_2$ and $CO_2$. The particle heating rate was in the order of $1 \times 10^4\ \text{K}\,\text{s}^{-1}$. Besides a comprehensive analysis of the mass loss kinetics, the pore structure evolution was analysed with porosity calculations on basis of BET pore volumes and skeletal densities obtained from He-buoyancy

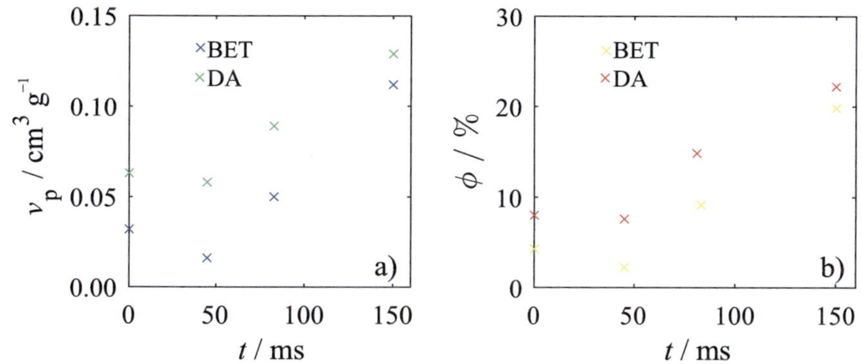


**Fig. 5.4** **a**) BET and DA pore volumes and **b**) porosities based on the respective pore volumes for CBC pyrolysed at $T = 1300\,\text{K}$ in a $CO_2$ atmosphere. (Based on the publication by Wedler et al. [33])

measurements (see (5.1)). As analysed by Heuer et al. [31], the temperature has a clear impact on the porosity as higher porosities are reached along the pyrolysis process. Furthermore, comparing atmospheres consisting of $CO_2$ and $N_2$, porosities are higher for $CO_2$ due to better heat transfer to the particle surface. At high residence times a beginning char reaction is possible.

As already stated in Sect. 5.2.1, the calculation of porosity is derived from physisorption measurements and associated pore volume theories that incorporate some difficulties, since not the entire pore regime spectrum is covered. It therefore makes sense to supplement the physisorption-based pore investigations with optical methods. A novel method for combining both strategies is stated in the study by Freisewinkel et al. [32] and presented in more detail in Sect. 2.5.2. As already proposed in Sect. 5.2.1, BET analyses can also be substituted by other evaluation methods for pore structure. As reported by Wedler et al. [33], the above-mentioned CBC chars from flash pyrolysis can be analysed with $CO_2$ physisorption and the applicable DA theory to improve the possible evaluation of micropores. In Fig. 5.4a), BET and DA pore volumes are compared for pyrolysis temperatures of 1300 K in $CO_2$. It can be seen that the DA micropore volume is generally larger than the BET pore volume, which leads to the conclusion that the DA analysis is capable of detecting pores in the smaller pore regime. This is due to the higher adsorption temperature compared with $N_2$ physisorption at cryogenic temperatures, where $N_2$ is kinetically hindered to diffuse in the smallest pores. Moreover, the differences between both pore volumes become smaller to higher residence times, since micropores develop to larger pores [33].

Considering chars from biomass, a reference feedstock are walnut shells (WNS), for which the pore structure evolution in flash pyrolysis was investigated, accordingly. In comparison to coal, WNS reveal a much faster mass loss kinetic, while releasing a significantly larger amount of volatiles [34]. Comparing the pore structure of two chars from WNS [34] and CBC [33], both pyrolysed in a $CO_2$ atmosphere at 1300 K, the WNS char shows a higher porosity, both in the microporous and in the larger pore regimes as ascribed to BET and DA pore volumes. This can be explained by the much higher volatile yield [34]. In a subsequent study [35], WNS and torrefied poplar chars resulting from flash pyrolysis were analysed with the 2D-NLDFT method in order to investigate the seamless pore evolution of two different biomasses. The original settings for the lower pore diameter limit of the $N_2$ kernel, as implemented in the SAIEUS software [21], have been retained. The WNS and torrefied poplar chars were obtained from a DTR at temperatures up to 1125 and 1300 K in $N_2$ and $CO_2$ atmospheres, covering the complete pyrolysis process. In the following analysis of the results, the PSD obtained from the 2D-NLDFT analysis is used in the calculation of the cumulative pore surface area and volume considering the range of the pore diameter $d_p$ for the three defined pore regimes. For both materials, WNS and torrefied poplar, the meso- to macropore ($2\,\text{nm} \leq d_p < 50\,\text{nm}$) surface area and volume are decreasing with pyrolysis mass loss. Nonetheless, torrefied poplar shows generally larger volume and surface area in this pore regime, an effect that might be associated with the previous torrefaction process. The contributions for pore surface area and volume in the ultramicropore regime ($d_p < 0.7\,\text{nm}$) show an increase for higher pyrolysis mass losses. This suggests a formation of very small pores due to the release of volatiles along the whole pyrolysis process for both materials. Details on the evaluation and results can be found in the study by Ontyd et al. [35].

In a more recent study by Freisewinkel et al. [32], 2D-NLDFT calculations and evaluations for chars from Ontyd et al. [35] are reviewed and adapted. In addition, a new optical porosity evaluation method is applied for the already covered WNS chars, which is intended as a supplemental analysis covering pores with a diameter $d_p > 600\,\text{nm}$. The optical methods were additionally used to cover pores with a size greater than 600 nm to overcome limitations of the 2D-NLDFT analysis. Thereby, SEM images of thin sections of individual particles are analysed with respect to the surface ratio of cavities and the whole particle surface area. From this, porosities can be calculated considering pores greater than 600 nm [32]. Details on the method and evaluation are given in Sect. 2.5. In the following, only improvements in the parameterisation of the 2D-NLDFT calculations are explained and results are shown.

For highly microporous samples like chars from biomass, the dual gas evaluation on the basis of $N_2$ and $CO_2$ physisorption isotherms shows some difficulties regarding the simultaneous 2D-NLDFT fitting of both isotherms. As explained above, the lower boundary for the pore size for measurements with $N_2$ can be adjusted since $N_2$-isotherms might show some weakness and inaccuracy in the lower pore regime due to the limited diffusion of the molecules in small pores at cryogenic adsorption temperatures. The lower pore size limit for $N_2$ for all evaluations of WNS chars was set to about 10 nm. This way, the fit of both interdependent DFT kernels was significantly improved [32]. Figure 5.5 shows the results from the simultaneous dual gas 2D-NLDFT analysis for WNS with different pyrolysis degrees, as assigned with the residence times in the DTR and the respective mass loss in an $N_2$ atmosphere at two temperatures. For all considered WNS chars and with respect to the considered pore size range of $0.3\,\text{nm} < d_p < 50\,\text{nm}$, the ultramicropores have a dominant contribution to the overall pore volume. It is clearly noticeable that the pore volume formation is strongly mass loss-dependent, which is driven by the higher reactor temperature. Up to a shown mass loss of approximately 50%, which is reached in 236 ms residence

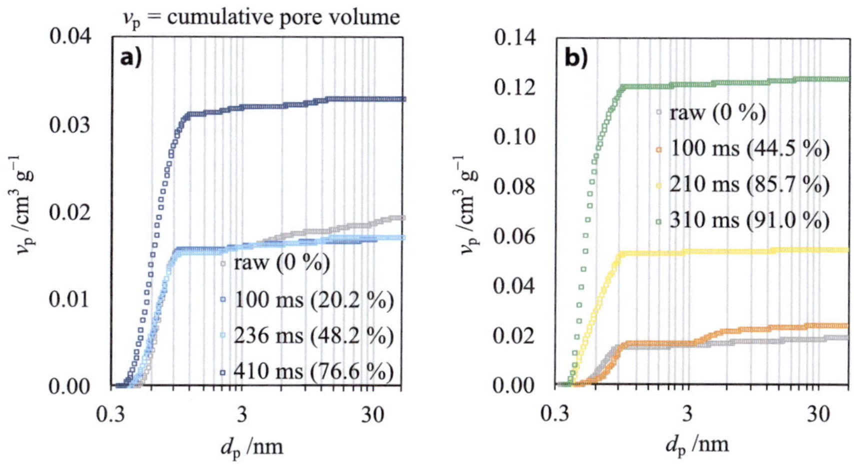


**Fig. 5.5** Evolution of the cumulative pore volume $v_p$ for WNS at different states of the pyrolysis (defined by the residence time and the respective mass loss) in an $N_2$ atmosphere at **a**) $T = 1125$ K and **b**) $T = 1300$ K [32]

time at 1125 K and 100 ms at 1300 K, the pore volume remains almost constant compared to the raw sample. Beyond that pyrolysis degree, the pore volume is increasing significantly from about 0.03 cm$^3$ g$^{-1}$ (raw) considering the whole cumulative pore volume in the range of 0.3 nm to 50 nm to 0.13 cm$^3$ g$^{-1}$ at 1300 K and a residence time of 310 ms (91% mass loss). The volume is predominantly evolved in the ultra- to micropore regime, which leads to the conclusion of a distinct micropore formation due to the release of volatiles, which is dominant over further pore growth in the respective pore regime. However, pores with widths exceeding 50 nm should be considered as well, but fall outside the scope of the 2D-NLDFT analysis. To overcome the limitations of the 2D-NLDFT method, optical methods were additionally used to cover pores with a size greater than 600 nm. Thereby, SEM images of thin sections of individual particles are analysed with respect to the surface ratio of cavities and the whole particle surface area. From this, porosities can be calculated considering pores greater than 600 nm [32]. Details on the method and evaluation are given in Sect. 2.5. The porosities for the 2D-NLDFT pore regime can be calculated with (5.1) in Sect. 5.2.1, using the cumulative pore volumes and the skeletal densities obtained from He-buoyancy measurements or He-pycnometry. The comprehensive results for porosities obtained by 2D-NLDFT calculations and by the optical method as well as their proportion to the overall porosity are reported in the study by Freisewinkel et al. [32]. Furthermore, the impact of FeMg and KMg minerals on the pyrolysis behaviour of a mineral-free hydrochar based on microcrystalline cellulose (MH) was investigated. Therefore, MH was doped with FeMg and KMg to solely consider the impact of individual mineral components [36]. The chars were completely pyrolysed in a DTR in an $N_2$ and $CO_2$ atmosphere at 1300 K without consideration of intermediate states of the process. Subsequently, the chars were analysed using 2D-NLDFT calculations to investigate the resulting pore surface areas and volumes according to the IUPAC [5] defined pore regimes. The pyrolysis of the undoped MH char revealed higher pore surface area and volume for $CO_2$ than for $N_2$ atmospheres, predominantly in the ultramicropore regime, which is reasonable, since $CO_2$ might participate in a reaction creating small pores. Doping with FeMg results in generally stronger pore creation compared to KMg considering the same pyrolysis atmosphere. Compared to the undoped MH char, both mineral species seem to enhance the pore creation predominantly in the ultramicropore regime, while the pore growth in the micropore regime is significantly hindered. However, as for the undoped char, $CO_2$ seems to enhanced pore formation. The detailed analysis for different pore regimes can be found in the study by Eckhard et al. [36]. The studies in the DTR considering a $CO_2$ atmosphere are still treated as pyrolysis as the residence time of the chars in the DTR is too short and the temperature of the particles is still too low for a pronounced gasification. However, considering char conversion, the reaction of oxy-fuel components ($CO_2$, $H_2O$, and $O_2$) with the solid fuel happens in different pore regimes with different kinetic rates [2]. Accordingly, the investigation of pore evolution for char conversion can also provide information about the pore growth in certain regimes due to conversion. The study by Tarlinski et al. [37] considers the char conversion in different oxy-fuel-atmosphere compositions and the influence of individual gaseous species on the properties of biomass-derived chars. Consistent with the procedure used for the pore evolution investigations for pyrolysis processes, different burnout states of char conversion experiments were sampled from the reactor. At the bottom of the used laminar flow reactor (LFR), a methane/ethane flame and mixtures of $CO_2$ and $O_2$ as oxidiser produce oxy-fuel-atmospheres with different compositions. Details on the particle sampling procedure can be found in the study by Tarlinski et al. [37]. Exemplarily, three samples of WNS converted in an atmosphere consisting of 10% $O_2$, 20% $H_2O$, and 70% $CO_2$ are analysed. The three sampled chars correspond to a burnout of 84%, 88%, and 96%. For 84% burnout, the pore structure is dominated by ultramicropores and the cumulative pore volume for all pore regimes is $0.02\,cm^3\,g^{-1}$. With proceeding burnout the proportions of meso- to macropores increase. At 96% burnout, the overall pore volume is $0.13\,cm^3\,g^{-1}$, whereby meso- to macropores dominate. This example shows a typical pore evolution for different burnout states in one specific oxy-fuel atmosphere. In the proceeding char conversion, small pores grow and coalesce into bigger pores. Thus, depending on the reactive gases and solid fuel particles, micropore proportions are decreasing while macropore proportions are increasing. A detailed analysis also considering constant burnout states and changing oxy-fuel atmospheres is shown in the study from Tarlinski et al. [37].

### 5.3.2 Adsorption Behaviour of Oxy-Fuel-Relevant Species

This section shows the experimental results for the adsorption capacities and adsorption kinetics for oxy-fuel-relevant species and provides an analysis of the different mass transport and diffusion properties.

**Adsorption Capacities**

In the study by Seibel et al. [38] adsorption capacities of CBC chars for $CO_2$ and $O_2$ are studied. All measurements were conducted in the TGA, while the adsorption temperatures covered in the experiments are limited by the onset of chemical reactions (gasification in $CO_2$: 723.15 K, oxidation in $O_2$: 423.15 K) at elevated pressures. In Fig. 5.6, the adsorption capacities $q$ for $CO_2$ and $O_2$ are shown along four different isotherms. Overall, the adsorption tendency of $CO_2$ is larger compared to $O_2$. Applying the ideal adsorbed solution theory (IAST), the selectivity of the char surface for one component in a theoretical mixture of both adsorptives can be derived considering information on both isotherms. As stated in the study by Seibel et al. [38], the adsorption on this CBC is strongly selective for $CO_2$ over a wide temperature range.

In the following, the influence of different minerals on the adsorption behaviour of oxy-fuel gases was studied, since their presence at the chars surface can contribute to a chemisorption of molecules or can determine the pore structure formation across conversion processes [39]. The study investigated the adsorption behaviour of $CO_2$, $O_2$, and $CH_4$ on mineral-doped HTC chars ($Fe_2O_3$ and $SiO_2$), that were subsequently pyrolysed. In the adsorption measurements, the adsorption capacities as well as the respective relative adsorption kinetics are considered using the *ASAP 2020* apparatus. The results and findings for the adsorption kinetics are mentioned below in this section. For all three adsorptives, the adsorption capacities are lowest for the $Fe_2O_3$-doped char. This can be explained by the pore structure constitution, which is least pronounced for the respective char [39]. In this context, the Fe content results in a higher structural ordering during char conversion and, thus, the number of available adsorption sites decreases [39]. In contrast, it was found that $O_2$ affinity is highest for the Si-doped chars, which can be explained by chemisorption, while the undoped chars provide the highest adsorption capacities for $CH_4$ and $CO_2$, which can be attributed to the most pronounced pore structure.

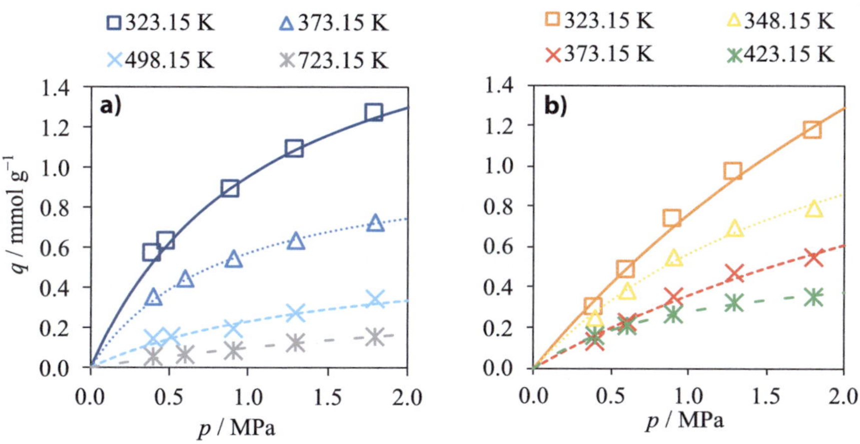


**Fig. 5.6** Adsorption capacity of **a)** $CO_2$ and **b)** $O_2$ on CBC up to 723.15 K and 1.8 MPa along with the corresponding fit of the Langmuir-isotherm model. (Based on the publication by Seibel et al. [38])

Furthermore, the adsorption capacities of $H_2O$ on biomass chars with different pore structures and contents of oxygen functional groups (OFG) are studied [40]. The investigated chars are the hydrothermal-carbonised model char MH800, a torrefied and gasified beech wood (TBW) char and a highly functionalised O-MH800 char, which is based on the MH800 material. Throughout the conversion process, different OFG can be formed as intermediates, which in turn can serve as favoured adsorption sites for $H_2O$ prior to a reaction at the surface. In steam gasification reactions, $H_2O$ reacts preferentially at the intraparticle surface [1], which ascribes great importance to the constitution of the pore structure in subsequent analyses. In the functionalisation process, the MH800 material was exposed to $HNO_3$ vapour at different temperatures and for different duration to control the degree of functionalisation in terms of the molar quantity of synthesised OFG. Details on the functionalisation procedure are given in the study by Pflieger et al. [41]. The respective chars are analysed in order to investigate the correlation between the chemical constitution as defined by the OFG content, the pore structure properties and the adsorption behaviour of $H_2O$. The OFG composition and quantities are investigated by the deconvolution of effluent gas curves generated in temperature-programmed desorption (TPD) experiments. The pore structure of all three chars was analysed by physisorption measurements with $CO_2$ and $N_2$ and analysed via 2D-NLDFT calculations (see Sect. 5.2.1). The analysis of the OFG spectrum showed a comparably low content of OFG for the MH800 and TBW chars, which categorises the chars surfaces as rather hydrophobic [40]. As intended, the O-MH800 char has a significantly higher quantity of all analysed OFG. Considering the entire spectrum of OFG, O-MH800 reaches 10.3 mmol $g^{-1}$ , compared to 2.1 mmol $g^{-1}$ and 1.6 mmol $g^{-1}$ for TBW and MH800, respectively. The analysis of the PSD for MH800 shows a strong (ultra-) microporosity in the pore regime ($d\text{p} = 0.3$–2 nm), which is larger compared to the TBW char, while O-MH800 shows the smallest porosity in the entire pore regime analysed by 2D-NLDFT. The adsorption isotherms of $H_2O$ are measured with the apparatus described in Sect. 5.2.2. The results for 308.15 K for all investigated chars are shown in Fig. 5.7. Both TBW and MH800 show a typical type V isotherm behaviour, which is characteristic of $H_2O$ adsorption on hydrophobic surfaces. The isotherm for O-MH800 can be categorised as type II isotherm according to IUPAC [5], which is representative of highly functionalised material [42]. Despite the comparably small internal pore surface area, the adsorption capacity is found to be much higher for the O-MH800 char. Due to the slightly higher OFG content, TBW shows higher adsorption capacities in the low-pressure region compared to MH800. At low $H_2O$ concentrations, the adsorption on O-MH800 can be mainly attributed to the chemisorption on OFG [42]. The high proportion of micropores in the PSD for the MH800 chars correlates with the pronounced increase of the adsorption capacity in the medium pressure range from $p/p_0 = 0.4$ to 0.6, where the process of micropore-filling dominates. As an extended investigation, the adsorption behaviour of $H_2O$ was modeled in MD simulations considering ideal slit pores with different widths and different quantities of carboxylic acids, which are considered to be the main influence on the adsorption affinity for $H_2O$. The details on the char model, the force field, and the framework of the simulations are given in Sect. 7.2.2.

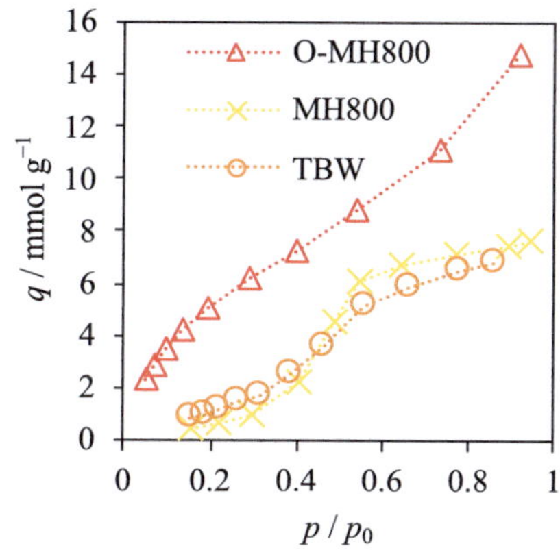


**Fig. 5.7** Adsorption isotherms of $H_2O$ for chars with different degree of functionalisation at $T = 308.15$ K. (Based on the publication by Eisenbach et al. [40])

**Adsorption Kinetics**

The basis for models of mass transport and diffusion of oxy-fuel species in the porous structure of biomass particles is the experimental investigation of adsorption kinetics. In the measurement procedure of adsorption equilibrium loadings at a constant temperature and pressure, time-resolved adsorption kinetics that end in equilibrium can be recorded simultaneously. The corresponding procedure was covered previously in the studies by Eisenbach et al. [25] and Wedler and Span [28].

The study by Seibel et al. [38] investigated adsorption kinetics of $CO_2$ at selected temperatures and pressures for porous CBC char, which were produced in the FFB or in the TGA. By using a gravimetric sorption device comparable to the experimental setup shown in Fig. 5.2, but equipped with a 10 µg-balance, adsorption kinetics were obtained at a constant pressure of 0.2 MPa in a temperature range of 313.15 to 433.15 K [38]. To describe the experimental results, the isothermal solution of Fick's second law of diffusion (IFD) is used to derive the constant effective diffusion coefficient for each adsorption kinetic (constant temperature and pressure). The diffusion coefficients can be compared for different chars to yield a correlation between pore structure and mass transport of gases. Furthermore, diffusion coefficients can be obtained from independent fits to adsorption kinetic data at different temperatures, reflecting the temperature dependency of adsorption. Seibel et al. [38] concluded that the diffusion process is dominated by the outer surface of the particles for the TGA chars, resulting in faster mass transport and, thus, smaller diffusion coefficients. It can be concluded that the process of pore diffusion might be subordinate to that of film diffusion for the TGA chars. In contrast, chars from the FFB reveal a higher pore surface area, as determined by a BET analysis, and, thus, pore diffusion is more dominant and diffusion coefficients are higher. Further details are given in the study by Seibel et al. [38].

Wedler et al. [43] investigated the diffusion behaviour of $CO_2$ and $CH_4$ in the porous structure of the mineral-free MH800 char. A comprehensive adsorption kinetic model was applied to the data, which is also based on Fick's second law of diffusion but considers non-isothermal effects. Generally, it was found that $CH_4$ has a slightly faster diffusion behaviour than $CO_2$. The data basis of the relative adsorption kinetics of $CO_2$ for the MH800 char is further used for comprehensive modelling as presented in Sect. 5.3.3.

Eisenbach et al. [25] and Wedler and Span [28] have each conducted a study on the adsorption kinetic behaviour of $CO_2$ and $O_2$, respectively, on the same representative biomass chars. The first char is the well investigated and covered MH800 char and the second one is a torrefied beech wood (TBW) char, which has been pyrolysed at 1273 K in a $CO_2$ atmosphere. The used adsorption measurement setup was described in Sect. 5.2.2 and details on the measurement procedure and the experimental results are reported in the study by Eisenbach et al. [25] and Wedler and Span [28]. Comparing the results for $O_2$ and $CO_2$, $CO_2$ shows a slower kinetic adsorption process in a comparable temperature and pressure range [25]. As shown by Eisenbach et al. [25] and Wedler and Span [28], all kinetics show a clear temperature and pressure dependency. The experimental results in both studies were also intended to be used for the development and validation of the novel PSK model [25,28]. The use of the experimental adsorption kinetic data for the PSK model is exemplarily shown for $O_2$ and the MH800 char in Sect. 5.3.3.

To investigate the adsorption kinetic behaviour of $H_2O$, the same biomass chars as covered in the study by Eisenbach et al. [25] and Wedler and Span [28] were considered. In comparison to the $O_2$ and $CO_2$ adsorption kinetics, which can be characterised as super-critical gases for the majority of the stated adsorption conditions, $H_2O$ is a condensable fluid and shows significantly different adsorption characteristics. The apparatus setup for vapour adsorption investigation was discussed in Sect. 5.2.2. With respect to adsorption kinetics, the pore-filling process depends on molecule cluster formation, micropore filling and pore condensation [42]. In Fig. 5.8a), the adsorption kinetics taken from a study by Eisenbach et al. [27] are shown for different relative pressures $p/p_0$ at the constant adsorption isotherm temperature of 298.15 K. For rising pressures, the adsorption kinetics get slower, which can be ascribed to the pore-filling process in the micropore regime, where initially formed clusters of molecules diffuse in and condensate [42]. In the same relative pressure regime, the adsorption uptake increases significantly (see Fig. 5.8b), showing a significant increase in the adsorption layer thickness. Interestingly, the adsorption kinetic gets faster again at $p/p_0 = 0.61$, which can be explained by the accelerated formation of the adsorbed phase.

In addition to an analysis of adsorption kinetics starting from a reproducible vacuum state of the sample, the experimental investigation of relative pressure kinetics is crucial, as it more accurately reflects the diffusion process driven by partial pressure differences in a conversion process. An excerpt of the experimental results of an adsorption kinetic study of $CO_2$ on MH800 is presented along with modelling results in Sect. 5.3.3.

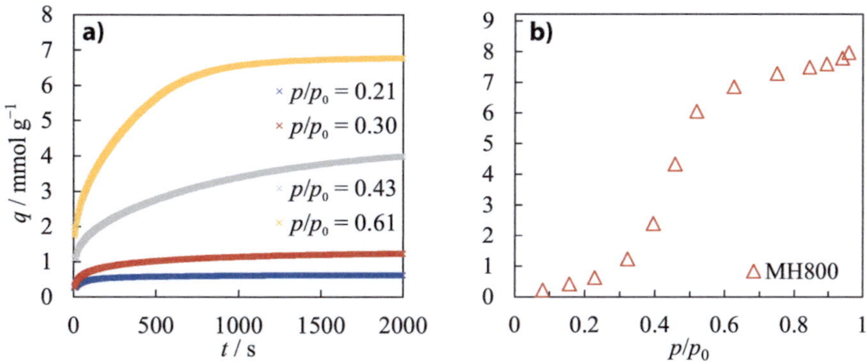


**Fig. 5.8** **a**) Adsorption kinetics of $H_2O$ for MH800 and **b**) corresponding adsorption capacities. (Based on the publication by Eisenbach et al. [27])

### 5.3.3 Adsorption Modelling and the Derivation of Mass Transport Properties

Char conversion and combustion models commonly assume constant pore or diffusion properties, which is not a physically sound representation of the mass transport in highly porous biomass chars [2,25,28]. In this section, the perks and weaknesses of the commonly used simple adsorption kinetic models are discussed and a comprehensive pore structure-dependent adsorption kinetic model is presented, which allows for a sound description of the mass transport process of oxy-fuel-relevant gases in the porous char structure.

#### Common Adsorption Kinetic Models

As shown by Seibel et al. [38] and Wedler et al. [43], common adsorption kinetic models, like the isothermal Fickian diffusion (IFD) model [44] or the non-isothermal model [45,46] are applicable to adsorption kinetics of $CO_2$ on coal char and the mineral-free reference biomass char MH800. As long as the modelled curves show a reasonable agreement with the adsorption kinetic data, their informative value in terms of mass transport can be used as a first approach to the description of mass transport. Their exact term structure and the underlying theory can be found elsewhere [38,43,45,46]. The drawback is that those models incorporate a constant effective diffusion coefficient for the description of the time-dependent mass transport process. As shown by Wedler and Span [28], the above-mentioned adsorption kinetic models do not represent the adsorption kinetic data of $CO_2$ for the TBW and MH800 chars. Furthermore, all mentioned adsorption kinetic models are bound to the temperature and pressure limits of the experiments, and can only model mass transport properties in a limited regime without any extrapolation capabilities. As introduced by Wedler and Span [28], a pore structure-dependent adsorption kinetic (PSK) model improves the modelling and more importantly the prediction of time-resolved mass transport properties, claiming that the mass transport behaviour in highly heterogeneous pore structures can be described in a more rigorous and physically sound way.

**A Pore Structure Dependent Adsorption Kinetic Model**

The PSK model allows a sound temperature- and pressure-dependency by considering more than one adsorption kinetic at a constant temperature and pressure. Based on a fitted independent parameter set (see below) the model is capable of predicting adsorption data for higher temperatures beyond the experimental boundaries and reaching realistic oxy-fuel reaction temperatures. However, the PSK model predicts the time-dependent adsorption capacity and a derivation of suitable mass transport properties like the adsorption flow rate $\dot{q}$ and a time-resolved mass transfer coefficient $k_{\mathrm{LDF}}$ is possible. The PSK model structure as presented in the study by Wedler and Span [28] is shown in (5.6)

$$q_{\mathrm{PSK}}(t, p, T) = \rho_{\mathrm{mol}} \sum_{i=1}^{3} A_i\, c_i \frac{p\, k_i\, \exp\left(\frac{H_i}{RT}\right)}{1 + p\, k_i\, \exp\left(\frac{H_i}{RT}\right)} \left[1 - \exp\left(\frac{-t}{a_i\, \left(\frac{T}{T_{\mathrm{Ref}}}\right)^{-n_i}}\right)\right] \tag{5.6}$$

The mathematical structure is based on the individual contribution of three pore regimes, ultramicropores (3), micropores (2) and meso-to macropores (1), to the overall time-dependent adsorption capacity. Thus, it comprises three independent pore-filling processes for three pore regimes, which are defined by IUPAC [5]. Furthermore, the pore structure information is implemented with the surface area $A_i$, which is a direct result of the pore structure investigations using physisorption measurements and 2D-NLDFT analysis (see Sect. 5.2.1). In general, the PSK model incorporates a time-independent equilibrium part, which is based on the Langmuir model approach for adsorption capacities along isotherms and a time-dependent adsorption kinetic part. The adsorption kinetic part is similar to the linear driving force (LDF) model structure [47], but contains additional parameters. Actually, all remaining independent fitting parameters have a direct physical meaning. The parameter $a_i$ can be considered as the time constant of the individual pore-filling process, which is multiplied by the temperature-dependent part of the adsorption kinetic, which is defined by the temperature exponent $n_i$. The parameter $c_i$ in the equilibrium part can be seen as a constant that is proportional to the thickness of the adsorption layer, while $k_i$ represents the pressure dependence and $H_i$ is the heat of adsorption. As shown by Eisenbach et al. [28] and Wedler and Span [25], the fitting of the 15 parameters to adsorption kinetic data for $CO_2$ and $O_2$ works well and the derivation of mass transport properties is possible. The derivation of the adsorption flow rate $\dot{q}$ and the time-resolved mass transfer coefficient $k_{\mathrm{LDF}}$ from the PSK model is explained in the study by Eisenbach et al. [25]. In Fig. 5.9a), the adsorption flow rate $\dot{q}$ and the mass transfer coefficient $k_{\mathrm{LDF}}$ are shown for temperatures between 400 and 1000 K as calculated from the PSK model by using the parameter set for $O_2$ adsorption on the MH800 char [25].

For both properties, the predicted step-wise decrease is characteristic. This can be attributed to the three distinguished pore-filling processes in three defined pore

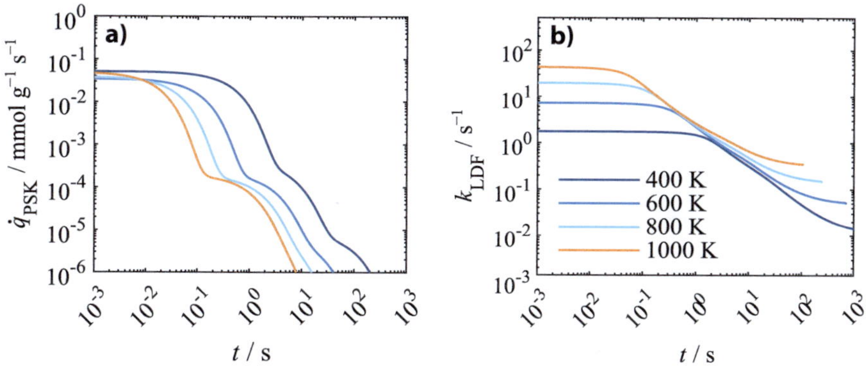


**Fig. 5.9** **a**) Adsorption flow rate and **b**) mass transfer coefficient derived from the data set of Eisenbach et al. [25] for $O_2$ for several temperatures outside of the experimental temperature range

regimes. First, the wide pores in the macropore-regimes are filled with a small time constant $a_1$ in (5.6), describing a fast adsorption process. In parallel, but slightly time-wise shifted, smaller pores are filled with larger respective time constants, $a_2$ and $a_3$, causing a decreasing mass transport and, thus, describing the diffusion limitations due to the microporous structure. The distinct steps in the adsorption flow rate $\dot{q}_{PSK}$ result from the assumption that pores can be assigned to three distinct pore widths. Unless such distinct pore widths are a result of the chemical structure of an adsorbent, the actual pore-width distribution might be a continuous one and the curves for the adsorption flow rate over time will be smoother than predicted. However, to assume more than three characteristic pore sizes seemed inappropriate based on the available data for both PSD and adsorption kinetics. The adsorption flow rate shows higher values for smaller temperatures since it incorporates information on the overall adsorption capacity, which is higher at smaller temperatures. Interestingly, the influence of the high temperature on the adsorption flow rate can be observed. At 1000 K, the predicted values become larger than for 800 K for small times, since the temperature-dependent kinetic behaviour begins to dominate over the overall adsorption capacity. In contrast to this, the mass transfer coefficient is always larger for higher temperatures. The mass transfer coefficient incorporates only kinetic information on the adsorption process, which strongly depends on temperature and less strongly on pressure [25]. Nonetheless, these properties need to be validated by adsorption measurements at elevated temperatures and also by comparison to diffusion properties for molecular diffusion.

The mentioned adsorption kinetics for $CO_2$ and $CH_4$ on the mineral-free MH800 chars, measured with the ASAP 2020 apparatus [43] (see Sect. 5.3.2), are relative pressure adsorption kinetics. This means that after reaching an adsorption equilibrium at a certain pressure $p_1$, the adsorption kinetics towards the next equilibrium point, which is set by increasing the pressure by $\Delta p$, are measured. In this way, the approach of intraparticle diffusion and mass transport processes with a relative pressure description, which is similar to a partial pressure difference between the bulk phase and gas phase in the intraparticle structure during a conversion process is possible. An excerpt of the experimental results of the adsorption kinetic study of $CO_2$ on MH800 is presented along with modelling results in this section. Consistent to the adsorption kinetic data, which are written as a time-dependent increase of the adsorption loading (j + 1) relative to the previous adsorption equilibrium (j), the PSK model formulation can be extended:

$$\Delta q_{\mathrm{PSK}} = q_{\mathrm{PSK,j+1}}(t, p_{\mathrm{j}} + \Delta p, T) - q_{\mathrm{PSK,j}}(t, p_{\mathrm{j}}, T) \tag{5.7}$$

Now, the kinetic part defines the difference between the dynamic adsorption process and the initial equilibrium state (j), which is defined by the pressure $p_{\mathrm{j}}$. The dynamic state tends towards the following equilibrium state at a pressure, which is higher by $\Delta p$ relative to the initial pressure $p_{\mathrm{j}}$. A pressure dependency for the adsorption kinetic behaviour is integrated into the original formulation of the PSK model (see (5.6)) leading to a relative pressure (RP)-PSK model:

$$\begin{aligned} \Delta q_{\mathrm{PSK}} = \rho_{\mathrm{mol}} \sum_{i=1}^{3} A_i\, c_i & \left( \frac{p_{\mathrm{j+1}}\, k_i\, \exp\left(\frac{H_i}{RT}\right)}{1 + p_{\mathrm{j+1}}\, k_i\, \exp\left(\frac{H_i}{RT}\right)} - \frac{p_{\mathrm{j}}\, k_i\, \exp\left(\frac{H_i}{RT}\right)}{1 + p_{\mathrm{j}}\, k_i\, \exp\left(\frac{H_i}{RT}\right)} \right) \\ & \times \left[ 1 - \exp\left( \frac{-t}{a_i \left(\frac{T}{T_{\mathrm{Ref}}}\right)^{-n_i} + b_i \left(1 - \frac{\Delta p}{p_{\mathrm{j}} + \Delta p}\right)^{m_i}} \right) \right] \end{aligned} \tag{5.8}$$

As already mentioned in Sect. 5.3.2, the relative pressure kinetics of $CO_2$ for the MH800 char are measured with the ASAP 2020 and presented here for two exemplary temperatures. The experimental data are evaluated with the RP-PSK model formulation as the additional uptake starting from the equilibrium state with the initial pressure $p_{\mathrm{j}}$:

$$\Delta q(t) = q_{\mathrm{j+1}}(t, p_{\mathrm{j}} + \Delta p, T) - q_{\mathrm{j}}(p_{\mathrm{j}}, T) \tag{5.9}$$

As shown in Fig. 5.10, the additional adsorption loading is highest at low initial pressures $p_{\mathrm{j}}$, which can be explained by the initial gradient of an adsorption isotherm, where a high quantity of molecules adsorb on free adsorption sites. In relation to the overall uptake for the individual kinetics, the adsorption kinetics are slowest for high pressures $p_{\mathrm{j}}$. This indicates that higher initial pressures $p_{\mathrm{j}}$ contribute to diffusion resistance and slow down the adsorption process. Figure 5.10 shows the modelling results for the RP-PSK model on the basis of the presented experimental data. The RP-PSK model represents the data well and is suitable for modelling relative pressure adsorption kinetics. However, a dependency of the kinetic part on the relative pressure increase $\Delta p$ as implemented in (5.8) could not be confirmed by the limited experimental data basis and requires further investigations. Nonetheless, the formulation of the RP-PSK model and the pressure-dependent term of the kinetic part in particular, enables a reduction to the original from of the PSK model for adsorption kinetic process starting from the vacuum state, where $\Delta p = p$ in (5.6)

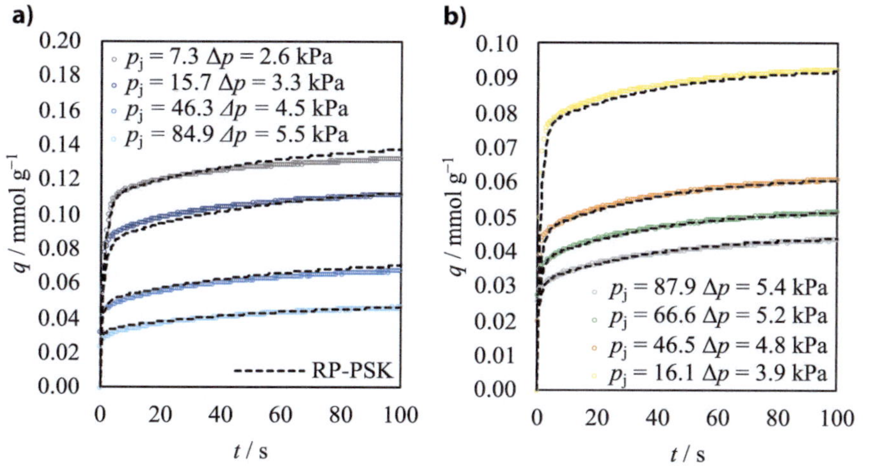


**Fig. 5.10** Relative pressure adsorption kinetics for different initial pressures $p_j$ and relative pressure steps along two adsorption isotherms at **a**) 298.15 K and **b**) 323.15 K for the MH800 char

and $p_1 = 0$. The derivation of mass transport properties can be carried out in the same manner as for the initial PSK model explained above.

## 5.4 Conclusion

This chapter provides an overview for the detailed investigation of pore structure properties for chars across pyrolysis and conversion processes. In combination with well-established sampling methods for chars of different pyrolysis and conversion histories, the investigation of the evolution of pore structure properties in terms of pore surface areas and volumes could be successfully realised. The most comprehensive and extensive analysis method is based on physisorption measurements and 2D-NLDFT kernels, which provide highly-resolved information on the pore evolution in pore regimes of different pore widths. In the analysis of the pore size distribution of different chars, insights could be gained regarding pore formation and growth during pyrolysis and conversion, which is controlled by temperature-dependent volatile yield, residence time, and constitution of the atmosphere. Biomass particles, as exemplified by WNS, exhibit higher porosity in the micropore regime compared to coal chars during pyrolysis. This difference can be attributed to the higher release of volatiles in biomass. During pyrolysis, the pore morphology is predominantly characterised by ultra-micropores. Elevated temperatures and a $CO_2$-enriched atmosphere further promote the formation of a significant proportion of micropores. In char conversion, the pore structure evolves depending on the composition of the oxy-fuel atmosphere. A sufficient proportion of $O_2$ facilitates the growth of small pores into meso- and macropores, while the creation of new small pores is driven by the $CO_2$ content in the atmosphere. All findings and pore structure data are intended for use in the development of a pore structure evolution model, which should be able to predict porosities or comparable pore structure properties for pyrolysis and char conversion models.

The effect of diffusion limitations and mass transport behaviour of oxy-fuel-relevant gases is accessed by comprehensive adsorption kinetic measurements for representative chars. All in all, an extensive data basis of adsorption kinetics for the three oxy-fuel-components $O_2$, $H_2O$, and $CO_2$ was established. The key experimental findings reveal that $CO_2$ exhibits significantly more pronounced micropore-filling, resulting in a slower mass transport in the porous structure of chars compared to $O_2$. This effect is particularly prominent in highly microporous biomass particles, where diffusion plays a critical role in differentiating the intraparticle mass transport of $CO_2$ and $O_2$. Additionally, the affinity of $H_2O$ is strongly influenced by the extent of micropore development and the presence of oxygen-containing functional groups on the surface. In order to derive physically sound and time-resolved mass transport properties, which could ideally be implemented in currently available char conversion and combustion models, a novel PSK model as well as the extended RP-PSK model were developed and tested successfully. In this chapter, two temperature and pressure-dependent mass transport properties, the adsorption flow rate and the mass transfer coefficient are proposed for use in further investigations and modelling approaches. Both can directly be derived from the PSK model. Ideally, the pore structure-dependent and time-resolved mass transport properties can be employed in char conversion and combustion models in order to improve the description of the pore-diffusion limitations. In further studies, sorption models need to be validated for high temperatures and mixtures of the oxy-fuel species $O_2$, $H_2O$, and $CO_2$.

**Acknowledgement** This work has been funded by the Deutsche Forschungsgemeinschaft (DFG, German Research Foundation)—215035359—SFB/TRR 129 'Oxyflame'.

## References

1. Guizani, C., Jeguirim, M., Gadiou, R., Escudero Sanz, F. J. and Salvador, S. "Biomass Char Gasification by $H_2O$, $CO_2$ and Their Mixture: Evolution of Chemical, Textural and Structural Properties of the Chars". *Energy* 112 (2016), 133–145. https://doi.org/10.1016/j.energy.2016.06.065
2. Wu, R., Beutler, J., Price, C. and Baxter, L. L. "Biomass Char Particle Surface Area and Porosity Dynamics During Gasification". *Fuel* 264 (2020), 116833. https://doi.org/10.1016/j.fuel.2019.116833
3. Liu, G.-S. and Niksa, S. "Coal Conversion Submodels for Design Applications at Elevated Pressures. Part II. Char Gasification". *Progress in Energy and Combustion Science* 30.6 (2004), 679–717. https://doi.org/10.1016/j.pecs.2004.08.001
4. Fletcher, T. H. "Review of 30 Years of Research Using the Chemical Percolation Devolatilization Model". *Energy & Fuels* 33.12 (2019), 12123–12153. https://doi.org/10.1021/acs.energyfuels.9b02826
5. Thommes, M., Kaneko, K., Neimark, A. V., Olivier, J. P., Rodriguez-Reinoso, F., Rouquerol, J. and Sing, K. S. W. "Physisorption of Gases, with Special Reference to the Evaluation of Surface Area and Pore Size Distribution (IUPAC Technical Report)". *Pure and Applied Chemistry* 87.9–10 (2015), 1051–1069. https://doi.org/10.1515/pac-2014-1117

6. Brunauer, S., Emmett, P. H. and Teller, E. "Adsorption of Gases in Multimolecular Layers". *Journal of the American Chemical Society* 60.2 (1938), 309–319. https://doi.org/10.1021/ja01269a023
7. Lowell, S., Shields, J. E., Thomas, M. A., Thommes, M. and Scarlett, B. *Characterization of Porous Solids and Powders: Surface Area, Pore Size and Density*. Dordrecht: Springer Netherlands, 2004. https://doi.org/10.1007/978-1-4020-2303-3
8. Gregg, S. J. and Sing, K. S. W. *Adsorption, Surface Area and Porosity*. London: Academic Press, 1982
9. Gale, T. K., Fletcher, T. H. and Bartholomew, C. H. "Effects of Pyrolysis Conditions on Internal Surface Areas and Densities of Coal Chars Prepared at High Heating Rates in Reactive and Nonreactive Atmospheres". *Energy & Fuels* 9.3 (1995), 513–524. https://doi.org/10.1021/ef00051a017
10. Yang, H., Fletcher, T. H., Li, Y., Jin, L., Li, S., Shang, Y. and Hu, H. "Modeling the Influence of Changes in Aliphatic Structure on Char Surface Area During Coal Pyrolysis". *AIChE Journal* 66.2 (2020). https://doi.org/10.1002/aic.16834
11. Carrasco-Marin, F., Lopez-Ramon, M. V. and Moreno-Castilla, C. "Applicability of the Dubinin-Radushkevich Equation to $CO_2$ Adsorption on Activated Carbons". *Langmuir* 9.11 (1993), 2758–2760. https://doi.org/10.1021/la00035a002
12. Dubinin, M. M. and Astakhov, V. A. "Development of the Concepts of Volume Filling of Micropores in the Adsorption of Gases and Vapors by Microporous Adsorbents". *Bulletin of the Academy of Sciences of the USSR Division of Chemical Science* 20.1 (1971), 3–7. https://doi.org/10.1007/BF00849307
13. Wedler, C. and Span, R. "Micropore Analysis of Biomass Chars by $CO_2$ Adsorption: Comparison of Different Analysis Methods". *Energy & Fuels* 35.10 (2021), 8799–8806. https://doi.org/10.1021/acs.energyfuels.1c00280
14. Medek, J. "Possibility of Micropore Analysis of Coal and Coke from the Carbon Dioxide Isotherm". *Fuel* 56.2 (1977), 131–133. https://doi.org/10.1016/0016-2361(77)90131-4
15. Sauer, E. and Gross, J. "Prediction of Adsorption Isotherms and Selectivities: Comparison Between Classical Density Functional Theory Based on the Perturbed-Chain Statistical Associating Fluid Theory Equation of State and Ideal Adsorbed Solution Theory". *Langmuir: the ACS journal of surfaces and colloids* 35.36 (2019), 11690–11701. https://doi.org/10.1021/acs.langmuir.9b02378
16. Jagiello, J., Ania, C. O., Parra, J. B., Jagiello, L. and Pis, J. J. "Using DFT Analysis of Adsorption Data of Multiple Gases Including $H_2$ for the Comprehensive Characterization of Microporous Carbons". *Carbon* 45.5 (2007), 1066–1071. https://doi.org/10.1016/j.carbon.2006.12.011
17. Jagiello, J. and Olivier, J. P. "2D-NLDFT Adsorption Models for Carbon Slit-Shaped Pores with Surface Energetical Heterogeneity and Geometrical Corrugation". *Carbon* 55 (2013), 70–80. https://doi.org/10.1016/j.carbon.2012.12.011
18. Jagiello, J. and Olivier, J. P. "Carbon Slit Pore Model Incorporating Surface Energetical Heterogeneity and Geometrical Corrugation". *Adsorption* 19.2-4 (2013), 777–783. https://doi.org/10.1007/s10450-013-9517-4
19. Jagiello, J., Ania, C., Parra, J. B. and Cook, C. "Dual Gas Analysis of Microporous Carbons Using 2D-NLDFT Heterogeneous Surface Model and Combined Adsorption Data of $N_2$ and $CO_2$". *Carbon* 91 (2015), 330–337. https://doi.org/10.1016/j.carbon.2015.05.004
20. Jagiello, J., Kenvin, J., Celzard, A. and Fierro, V. "Enhanced Resolution of Ultra Micropore Size Determination of Biochars and Activated Carbons by Dual Gas Analysis Using $N_2$ and $CO_2$ with 2D-NLDFT Adsorption Models". *Carbon* 144 (2019), 206–215. https://doi.org/10.1016/j.carbon.2018.12.028
21. Jagiello, J. "Stable Numerical Solution of the Adsorption Integral Equation Using Splines". *Langmuir* 10.8 (1994), 2778–2785. https://doi.org/10.1021/la00020a045
22. Nguyen, H. G. T., Horn, J. C., Bleakney, M., Siderius, D. W. and Espinal, L. "Understanding Material Characteristics Through Signature Traits from Helium Pycnometry". *Langmuir: the ACS journal of surfaces and colloids* 35.6 (2019), 2115–2122. https://doi.org/10.1021/acs.langmuir.8b03731

23. Kleinrahm, R., Yang, X., McLinden, M. O. and Richter, M. "Analysis of the Systematic Force-Transmission Error of the Magnetic-Suspension Coupling in Single-Sinker Densimeters and Commercial Gravimetric Sorption Analyzers". *Adsorption* 25.4 (2019), 717–735. https://doi.org/10.1007/s10450-019-00071-z
24. Yang, X., Kleinrahm, R., McLinden, M. O. and Richter, M. "Uncertainty Analysis of Adsorption Measurements Using Commercial Gravimetric Sorption Analyzers with Simultaneous Density Measurement Based on a Magnetic-Suspension Balance". *Adsorption* 26.4 (2020), 645–659. https://doi.org/10.1007/s10450-020-00236-1
25. Eisenbach, T., Wedler, C. and Span, R. "Experimental Investigation and Modeling of Mass Transport Properties of $O_2$ in Biomass Chars". *Chemical Engineering Science* 287 (2024), 119735. https://doi.org/10.1016/j.ces.2024.119735
26. Rösler, M. and Wedler, C. "Adsorption Kinetics and Equilibria of Two Methanol Samples with Different Water Content on Activated Carbon". *Adsorption* 27.8 (2021), 1175–1190. https://doi.org/10.1007/s10450-021-00341-9
27. Eisenbach, T., Duarte, H. A., Wedler, C., Pflieger, C., Muhler, M. and Span, R. "Experimental Investigations on the Adsorptive Behavior of $H_2O$ Vapor on HTC Char Particles". *14$^{th}$ International Conference on Applied Energy*. Bochum, Germany, 2022. https://doi.org/10.46855/energy-proceedings-10289
28. Wedler, C. and Span, R. "A Pore-Structure Dependent Kinetic Adsorption Model for Consideration in Char Conversion – Adsorption Kinetics of $CO_2$ on Biomass Chars". *Chemical Engineering Science* 231 (2021), 116281. https://doi.org/10.1016/j.ces.2020.116281
29. Ortiz Vega, D. O., Hall, K. R., Holste, J. C., Harvey, A. H. and Lemmon, E. W. *An Equation of State for the Thermodynamic Properties of Helium*. National Institute of Standards and Technology (U.S.) 2023. https://doi.org/10.6028/NIST.IR.8474
30. Wütscher, A., Wedler, C., Seibel, C., Hiltrop, D., Fieback, T., Muhler, M. and Span, R. "On the Alternating Physicochemical Characteristics of Colombian Coal During Pyrolysis". *Journal of Analytical and Applied Pyrolysis* 123 (2017), 12–19. https://doi.org/10.1016/j.jaap.2017.01.007
31. Heuer, S., Wedler, C., Ontyd, C., Schiemann, M., Span, R., Richter, M. and Scherer, V. "Evolution of Coal Char Porosity from $CO_2$-Pyrolysis Experiments". *Fuel* 253 (2019), 1457–1464. https://doi.org/10.1016/j.fuel.2019.05.071
32. Freisewinkel, E., Eisenbach, T., Weise, V., Tarlinski, D., Pörtner, L., Cerciello, F., Schiemann, M., Span, R., Senneca, O. and Scherer, V. "Porosity of Biomass Char: A Novel Determination Method and Changes in Walnut Shell with Progressing Pyrolysis". *48$^{th}$ International Conference on Clean Energy*. Clearwater, USA, 2024
33. Wedler, C., Span, R. and Richter, M. "Comparison of Micro- and Macropore Evolution of Coal Char During Pyrolysis". *Fuel* 275 (2020), 117845. https://doi.org/10.1016/j.fuel.2020.117845
34. Wedler, C., Ontyd, C., Schiemann, M., Scherer, V., Richter, M. and Span, R. "Evolution of Micro- and Mesopore Structure During Pyrolysis of Coal and Biomass in Different Atmospheres (Presentation)". *3$^{rd}$ International Workshop on Oxy-Fuel Combustion*. Montabaur, Germany, 2020
35. Ontyd, C., Freisewinkel, E., Wedler, C., Eisenbach, T., Span, R., Senneca, O. and Scherer, V. "Flash Pyrolysis of Walnut Shells and Poplar in $N_2$ and $CO_2$: Pyrolysis Kinetics, Pore Evolution and a Model for Evaluation of Pyrolysis Species". *13$^{th}$ European Conference on Industrial Furnaces and Boilers*. Algarve, Portugal, 2022
36. Eckhard, T., Pflieger, C., Russo, C., Freisewinkel, E., Eisenbach, T., Böttger, J., Senneca, O., Apicella, B., Schiemann, M., Span, R., Scherer, V., Muhler, M. and Cerciello, F. "Mineral Effects on Chemical and Physical Transformations of Fast Pyrolysis Products of Cellulose-Based Model Fuels in $N_2$ and $CO_2$". *Fuel* 340 (2023), 127477. https://doi.org/10.1016/j.fuel.2023.127477
37. Tarlinski, D., Freisewinkel, E., Eisenbach, T., Span, R., Schiemann, M. and Scherer, V. "Combustion of Single Walnut Shell Particles in a Laminar Flow Reactor Under Oxy-Fuel Condi-

tions: Optical Measurements and Particle Sampling". *Fuel* 369 (2024), 131613. https://doi.org/10.1016/j.fuel.2024.131613
38. Seibel, C., Wedler, C., Vorobiev, N., Schiemann, M., Scherer, V., Span, R. and Fieback, T. "Sorption Measurements for Determining Surface Effects and Structure of Solid Fuels". *Fuel Processing Technology* 153 (2016), 81–86. https://doi.org/10.1016/j.fuproc.2016.08.004
39. Wedler, C., Lotz, K., Arami-Niya, A., Xiao, G., Span, R., Muhler, M., May, E. F. and Richter, M. "Influence of Mineral Composition of Chars Derived by Hydrothermal Carbonization on Sorption Behavior of $CO_2$, $CH_4$, and $O_2$". *ACS Omega* 5.19 (2020), 10704–10714. https://doi.org/10.1021/acsomega.9b04370
40. Eisenbach, T., Jahromi, B. F., Angenent, V., Pflieger, C., Muhler, M., Schmid, R., Wedler, C. and Span, R. "Linking experimental $H_2O$ vapor adsorption on biomass char with physicochemical char properties and MD simulation". Fluid Phase Equilibria 597 (2025), 114460. https://doi.org/10.1016/j.fluid.2025.114460
41. Pflieger, C., Eckhard, T., Schmitz, G., Angenent, V., Göckeler, M., Senneca, O., Schmid, R., Cerciello, F. and Muhler, M. "Thermicity of the Decomposition of Oxygen Functional Groups on Cellulose-Derived". *ACS Omega* 7.51 (2022), 48606–48614. https://doi.org/10.1021/acsomega.2c07429
42. Liu, L., Tan, S. J., Horikawa, T., Do, D. D., Nicholson, D. and Liu, J. "Water Adsorption on Carbon – A Review". *Advances in colloid and interface science* 250 (2017), 64–78. https://doi.org/10.1016/j.cis.2017.10.002
43. Wedler, C., Arami-Niya, A., Xiao, G., Span, R., May, E. F. and Richter, M. "Gas Diffusion and Sorption in Carbon Conversion". *Energy Procedia* 158 (2019), 1792–1797. https://doi.org/10.1016/j.egypro.2019.01.422
44. Crank, J. *The Mathmatics of Diffusion*. Oxford: Oxford University Press, 1975
45. Lee, L.-K. and Ruthven, D. M. "Analysis of Thermal Effects in Adsorption Rate Measurements". *Journal of the Chemical Society, Faraday Transactions 1: Physical Chemistry in Condensed Phases* 75.0 (1979), 2406. https://doi.org/10.1039/F19797502406
46. Kočiřík, M., Struve, P. and Bülow, M. "Analytical Solution of Simultaneous Mass and Heat Transfer in Zeolite Crystals Under Constant-Volume/Variable-Pressure Conditions". *Journal of the Chemical Society, Faraday Transactions 1: Physical Chemistry in Condensed Phases* 80.8 (1984), 2167. https://doi.org/10.1039/F19848002167
47. Glueckauf, E. and Coates, J. I. "241. Theory of Chromatography. Part IV. The Influence of Incomplete Equilibrium on the Front Boundary of Chromatograms and on the Effectiveness of Separation". *Journal of the Chemical Society (Resumed)* (1947), 1315–1321. https://doi.org/10.1039/JR9470001315

# Gas-Phase Chemistry of Solid-Fuel Particle Combustion

6

Francesca Loffredo, Raymond Langer, Maximilian Hellmuth, Sanket Girhe, Anita Meraviglia, Bingjie Chen and Heinz Pitsch

## Contents



**Abstract**

This chapter discusses kinetic modelling of the gas-phase chemistry for biomass combustion under oxy-fuel conditions. Several datasets of species mole frac-

F. Loffredo (✉) · R. Langer · M. Hellmuth · S. Girhe · A. Meraviglia · B. Chen · H. Pitsch
Institute for Combustion Technology, RWTH Aachen University, Aachen, Germany
e-mail: f.loffredo@itv.rwth-aachen.de

R. Langer
e-mail: r.langer@itv.rwth-aachen.de

M. Hellmuth
e-mail: m.hellmuth@itv.rwth-aachen.de

S. Girhe
e-mail: s.girhe@itv.rwth-aachen.de

A. Meraviglia
e-mail: a.meraviglia@itv.rwth-aachen.de

B. Chen
e-mail: b.chen@itv.rwth-aachen.de

H. Pitsch
e-mail: h.pitsch@itv.rwth-aachen.de



R. Kneer et al. (eds.), *Clean Energy Conversion*,
https://doi.org/10.1007/978-3-031-98956-8_6

tion measurements obtained from counterflow flames covering a broad range of conditions and fuels were utilised to assess the progress in detailed kinetic modelling of polycyclic aromatic hydrocarbons and to identify a detailed chemical kinetic model suitable for further investigations of solid fuel combustion under oxy-fuel conditions. The selected model was amended by a $NO_x$ submodel and the oxidation chemistry of oxygenated and nitrogen-containing tar species, i.e. anisole, levoglucosan, propionaldehyde, ammonia, and pyridine. A multistep reduction approach provided a skeletal kinetic model for biomass gas-phase combustion with 104 species (140 species, including the $NO_x$ submodel), which can be used to perform complex simulations at reduced computational cost. The validation of these skeletal models against experimental data representative of atmospheric high-temperature combustion under oxy-fuel conditions and in air demonstrated the models' prediction accuracy and suitability for practically relevant applications.

## 6.1 Introduction

Developing models that accurately predict the properties of solid fuel combustion is a considerable scientific challenge. Combustion of coal or biomass is a multiphase process comprising three main stages: devolatilisation, gas-phase combustion, and char burnout [1]. When solid fuel particles are exposed to heat, their surface temperature rises, initially causing contained water to evaporate. As the temperature increases further, the solid fuel undergoes thermal decomposition, releasing light gases and tars. This volatile matter mixes with oxygen in the surrounding environment, combusting in the gas phase, a stage that is particularly important for biomass as it involves the rapid release of most energy [1,2]. In the final stage, the remaining char is oxidised. This chapter addresses gas-phase combustion (second stage) in the form of chemical kinetic models.

In general, solid fuel combustion is a diffusion-controlled process, where the rate of combustion is dictated by the diffusion of oxygen into the flame zone. However, results from direct numerical simulations [3] show that in practical applications, premixed and non-premixed combustion may be relevant due to local extinction and re-ignition. Chemical kinetic models developed for describing oxy-fuel combustion must be able to predict such phenomena for the broad range of species observed in volatile matter. At the same time, the high computational cost of simulations, the iterative nature of the model development process, and requirements on uncertainty and richness of experimental data for model validation demand simplifications. The counterflow configuration is the simplest configuration that reflects the importance of both non-premixed combustion and Damköhler numbers. Additionally, its relatively simple geometry permits the analysis of kinetic models through numerically obtained one-dimensional similarity solutions [4], reducing the computational effort. Therefore, this work's validation and analysis of chemical kinetic models focuses on measurements in counterflow diffusion flames. These are complemented by experimental data from jet-stirred reactors (JSR) covering the premixed combustion modes.

Besides combustion phenomena, such as heat release, extinction, and re-ignition, solid fuel combustion modelling often requires quantifying pollutants, such as nitrogen oxides ($NO_x$) and soot. Accurate $NO_x$ predictions can be challenging for biomass combustion because of the high content of organically bound nitrogen (N) within the molecular structure of the fuel [5]. The release of nitrogen-containing volatiles leads to potentially high $NO_x$ emissions [6,7], necessitating the investigation of their formation pathways. Further, prediction uncertainties of soot formation and oxidation remain large and can often exceed two orders of magnitude [8]. Although accumulating evidence suggests that polycyclic aromatic hydrocarbons (PAHs) are the gaseous precursors of soot particles [8], several steps in the formation processes of this pollutant are not yet fully understood and remain an active area of research. This work assesses the progress towards quantitatively accurate predictions of these pollutants under oxy-fuel conditions. $NO_x$ and soot formation are relatively slow processes compared to fuel oxidation. Therefore, residence time strongly impacts $NO_x$ and soot formation. The counterflow and JSR configurations facilitate direct access to residence times, which makes it possible to mimic the conditions in technical applications and renders experimental measurements with these setups an excellent choice for investigating pollutant formation.

While predicting combustion phenomena and pollutants requires detailed models containing numerous species, considering all components of light volatiles and tars is neither feasible nor practical. Instead, the composition and properties of the actual fuel are often approximated by simplified surrogate formulations with a small number of representative components [9]. Besides various hydrocarbons, surrogate mixtures typically include oxygenated and nitrogen-containing species. Pyrolysis studies of algae, plant matter, and animal proteins identified the primary nitrogen-containing compounds in light volatiles and tar species [10,11]. Among the light volatiles, ammonia ($NH_3$) and hydrogen cyanide (HCN) are the predominant nitrogen-containing species [5–7]. Chemical analyses of tars revealed pyrrole ($C_4H_5N$), pyridine ($C_5H_5N$), and piperazine ($C_4H_{10}N_2$) as the main nitrogen-containing compounds [12–14]. For oxygenated species, vanillin ($C_8H_8O_3$) and anisole ($A_1OCH_3$) were identified as key surrogate components [13,14].

This chapter first evaluates the recent progress in modelling gas-phase chemistry for solid fuel combustion, focusing on oxy-fuel conditions (Sect. 6.2). Several data sets of species mole fraction measurements obtained within the CRC/TRR 129 and compiled from the literature are utilised to assess the progress in detailed kinetic modelling of PAHs and identify a detailed chemical kinetic model suitable for further investigations. The identified detailed model is amended by nitrogen chemistry and surrogate components relevant to biomass combustion, and subsequently reduced in a series of steps (Sect. 6.3). Concluding, a compact kinetic model is proposed for the investigation of atmospheric, high-temperature combustion under oxy-fuel conditions, which is suitable for simulating complex application cases at reduced computational costs.

## 6.2 Detailed Modelling of PAH Formation Under Oxy-Fuel Conditions

Over the past decade, several studies advanced the understanding of gas-phase kinetics in biomass and coal combustion under oxy-fuel conditions. Besides the $C_0$–$C_4$ core chemistry [15–19], the formation and oxidation of nitrogen-containing and oxygenated tars, such as pyrrole [20], pyridine [21], and anisole [22], were investigated through experimental, theoretical, and modelling work. These works demonstrated that the respective developed or analysed models could describe the oxidation of small hydrocarbons and tars in reasonable agreement with experimental measurements under oxy-fuel conditions. However, predicting pollutants, such as $NO_x$ and PAHs, remains challenging [8], especially for oxygenated fuels [22–24], nitrogen-containing components [20], or complex mixtures of such species as they emerge from biomass devolatilisation [14].

This section evaluates the progress made in predicting PAHs under oxy-fuel conditions. It is well-established that pollutant formation, especially PAH formation [8,25], is sensitively impacted by fuel-specific chemistry, often making it difficult to identify a single cause for deviations between experimental and computed values. For example, the investigation of atmospheric anisole counterflow diffusion flames [22], discussed in more detail below, showed that tentative updates of the PAH chemistry significantly impacted kinetic model predictions. At the same time, deviations between kinetic model predictions and experimental measurements for important PAH precursors, such as ethylene ($C_2H_4$), diacetylene ($C_4H_2$), and vinylacetylene ($C_4H_4$), could exceed a factor of four [22]. Hence, both uncertainties in the oxidation and the PAH chemistry may significantly contribute to the discrepancies observed for PAH mole fractions.

The analysis of Chen et al. [22] illustrates that identifying causes for unsatisfactory PAH predictions can be challenging if simulations involve novel kinetic models with significant prediction uncertainties for small dehydrogenated intermediates. For the complex composition of volatiles, analysis could additionally be hampered by potential interactions of nitrogen-containing and hydrocarbon species [26]. The approach in the present work alleviates this challenge by exploiting the current understanding that molecular growth pathways are primarily initiated by relatively small dehydrogenated species with up to five carbon (C) atoms [8]. Consequently, PAH formation can be investigated by considering the combustion of small hydrocarbons, which reduces relevant uncertainties in PAH precursor predictions. In addition, the effects of fuel-specific chemistry on the evaluation are reduced by considering several fuels.

Burning solid fuels under oxy-fuel conditions can significantly impact combustion. For equal fractions of $O_2$, the maximum flame temperatures are substantially lower in oxy-fuel combustion compared to combustion in air, owing to the higher specific heat of $CO_2$. Excess $CO_2$ can participate in chemical reactions either as a reactant or as a third body for collision, affecting the fuel consumption and pollutant formation pathways. Cai et al. [15] demonstrated that their previously published detailed chemical kinetic model [27] could reasonably reproduce several measurements obtained for methane ($CH_4$) combustion under oxy-fuel conditions, including extinction limits of counterflow flames [15,28–30], ignition delay times [31–33], and laminar flamespeeds [34–38]. The model of Cai et al. [27] also contains detailed PAH chemistry [39] and was extensively validated against various targets relevant to $C_0$–$C_4$ chemistry. Therefore, it constitutes a suitable reference point for evaluating the recent progress in modelling PAH predictions under oxy-fuel conditions.

A review of the analyses performed in the above-mentioned studies [15–18,22] suggests that advancements in recently published kinetic modelling work are relevant to predicting PAH formation from oxy-fuel combustion. Langer et al. [8] and Langer [40] developed a detailed chemical kinetic model for PAHs that builds upon the model of Cai et al. [27]. The most recent revision of the model by Hellmuth et al. [41] introduced updates to the $C_7H_7$ kinetics. This model [41] will be called the "ITV model" in the following. Besides an extensive revision of the PAH chemistry [8,40,41], Langer et al. [8] and Langer [40] introduced several updates to the $C_0$–$C_4$ core chemistry that are expected to impact combustion under oxy-fuel conditions. For example, Langer et al. [8] updated the rate coefficient of the reaction $CO + OH = CO_2 + H$ based on the theoretical work of Senosiain et al. [42]. This reaction may be particularly significant under oxy-fuel conditions because it sensitively influences hydrogen (H) radical availability [15], which may strongly affect PAH formation. Further modifications include the integration of the $C_3$ chemistry from the modelling work of Panigrahy et al. [43]. Baroncelli et al. [16] showed that $C_3$ chemistry plays a crucial role in aromatics formation from the non-premixed combustion of two of the most abundant devolatilisation products [16], methane and acetylene ($C_2H_2$). Updates to the PAH chemistry [8] that are expected to be particularly relevant include radical recombination reactions involving five-membered ring species. Reaction flux analysis of Chen et al. [22] for anisole counterflow diffusion flames suggests that introduced updates of the cyclopentadienyl radical self-recombination reactions ($C_5H_5 + C_5H_5$) are critical, as this was found to be the dominant naphthalene ($A_2$) formation pathway. The ITV model [41] is directly comparable to the model of Cai et al. [27] and was last updated in 2024. Hence, this is another suitable reference point for evaluating recent progress in PAH modelling under oxy-fuel conditions.

Figures 6.1 and 6.2 compare predictions obtained with the model of Cai et al. [27] and the ITV model [41] to measurement data for several fuels. The comparison considers data from time-of-flight mass spectrometry (ToF-MS) [16–18,44–46] and gas chromatography (GC) [41,47,48], including investigations performed within *Oxyflame* [16–18,41]. Figures 6.1 and 6.2 show that deviations between measured and predicted species peak mole fractions decreased for cyclopentadiene ($C_5H_6$), benzene ($A_1$), phenylacetylene ($A_1C_2H$), styrene ($A_1C_2H_3$), ethylbenzene ($A_1C_2H_5$), indene ($C_9H_8$), naphthalene, and acenaphthylene ($C_{12}H_8$) when comparing the ITV model [41] to the model of Cai et al. [27]. The only exception is toluene ($A_1CH_3$), although the predictions remain accurate. The computed mean deviation factors [8] shown in the bottom right of the parity diagrams in Figs. 6.1 and 6.2 highlight a considerable improvement for phenylacetylene and acenaphthylene, whose mean deviation factors reduce by more than a factor of four. This

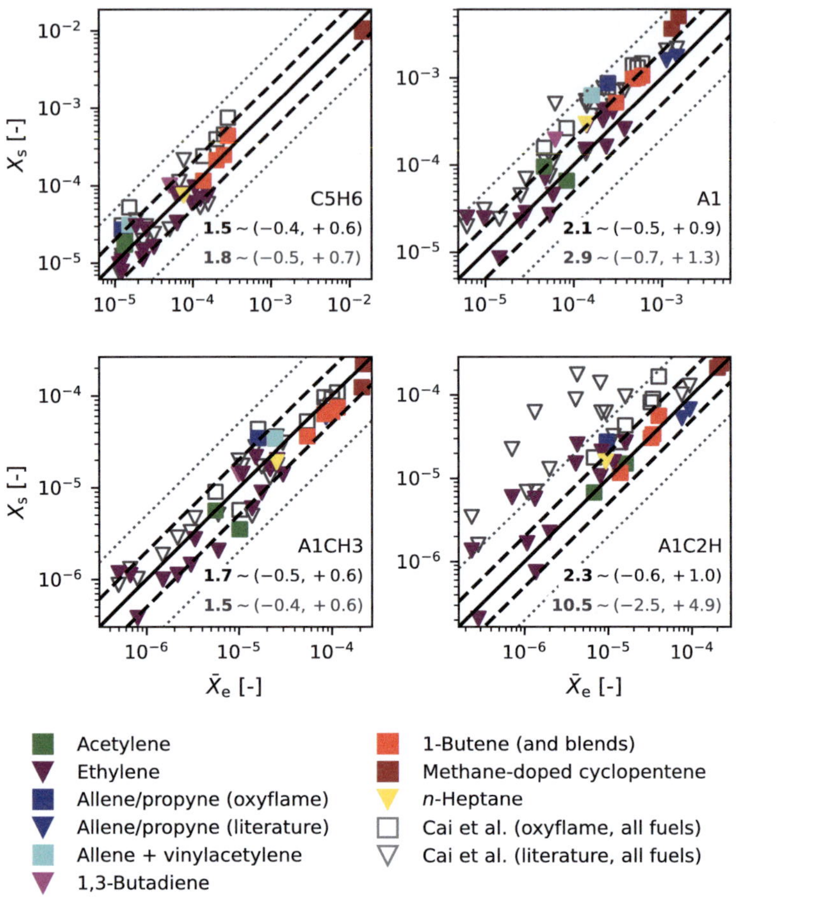


**Fig. 6.1** Parity diagrams comparing peak mole fractions from measurements to simulation results. Single species or isomer mole fractions were selected as appropriate for the respective experiment. Dashed lines indicate a deviation of a factor of two, and dotted lines a factor of five. Results obtained with the ITV model [41] use colours representing the fuel composition (acetylene [16], ethylene [47,48], allene or propyne [41,45], allene/vinylacetylene [41], 1,3-butadiene [46], 1-butene or 1-butene blends [18], methane-doped cyclopentene [17], *n*-heptane [44]). Results obtained with the model of Cai et al. [27] are shown as empty symbols. Experimental data measured as a part of *Oxyflame* are shown as squares, and other literature data are shown as triangles. Values in the bottom right are the average deviation factor and estimates of the associated 90 %-confidence intervals [8] for the ITV model [41] (upper row) and the model of Cai et al. [27] (lower row), respectively

quantitative evaluation illustrates that the ITV model [41] is suitable for investigating oxy-fuel combustion and PAH formation. Therefore, it was chosen as the basis for developing a compact model for biomass combustion in the subsequent sections.

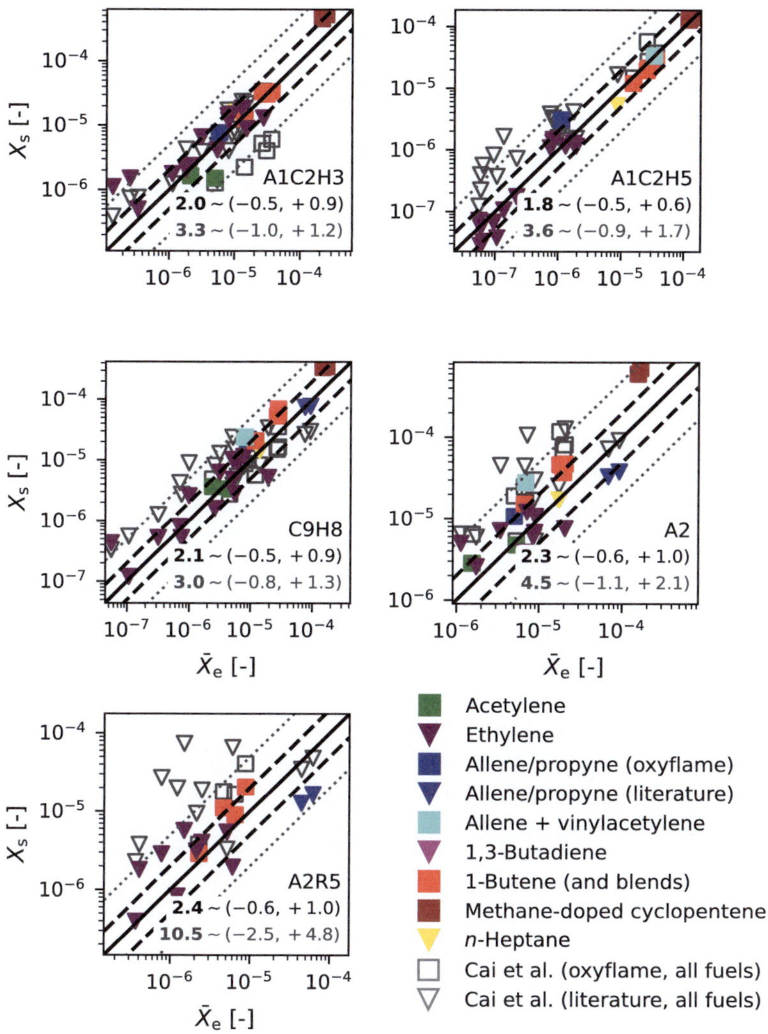


**Fig. 6.2** Parity diagrams comparing peak mole fractions from measurements to simulation results. Results obtained with the ITV model [41] use colours representing the fuel composition. Results obtained with the model of Cai et al. [27] are shown as empty symbols. An explanation of the symbol shapes, the lines, and the numbers at the bottom right, as well as references to the experimental data, can be found in the caption of Fig. 6.1

## 6.3 Skeletal Kinetic Model for Biomass Surrogate Components

Modelling the gas-phase chemistry of biomass fuels requires the description of the underlying chemistry that governs the volatilisation and gas-phase processes. However, a detailed representation of these processes results in complex chemical kinetic models, making their use computationally impractical in large-scale simulations. This section addresses the development of a skeletal, i.e. compact, chemical kinetic model suitable for large-scale simulations, which is based on the ITV model [41] discussed in Sect. 6.2. The skeletal model was specifically designed to predict the gas-phase chemistry of biomass fuels under conditions relevant to practical applications, including high-temperature atmospheric combustion in air or oxy-fuel environments. Figure 6.3 illustrates the development process.

Developing such a model first requires deriving a surrogate formulation to approximate the volatile mixture composition released from the fuels targeted in the numerical investigation. Here, the particle devolatilisation model proposed by Ranzi et al. [9] and further extended by Debiagi et al. [49] was applied to three different biomass particles: raw miscanthus, torrefied miscanthus, and torrefied corn straw. The particle devolatilisation model provides complex mixtures including light hydrocarbons, oxygenated species, aldehydes, tars, aromatics, and several nitrogen-containing species. The latter includes nitrogen tars and light compounds such as ammonia and hydrogen cyanide. Since the ITV model [41] development did not target biomass fuels, the chemistry of missing volatiles found in significant amounts in the mixture was integrated in this step. Anisole was selected as the primary representative of oxygen-containing tars, levoglucosan ($C_6H_{10}O_5$) as the dominant tar component of cellulose devolatilisation, and propionaldehyde ($CH_3CH_2CHO$) as the main representative of

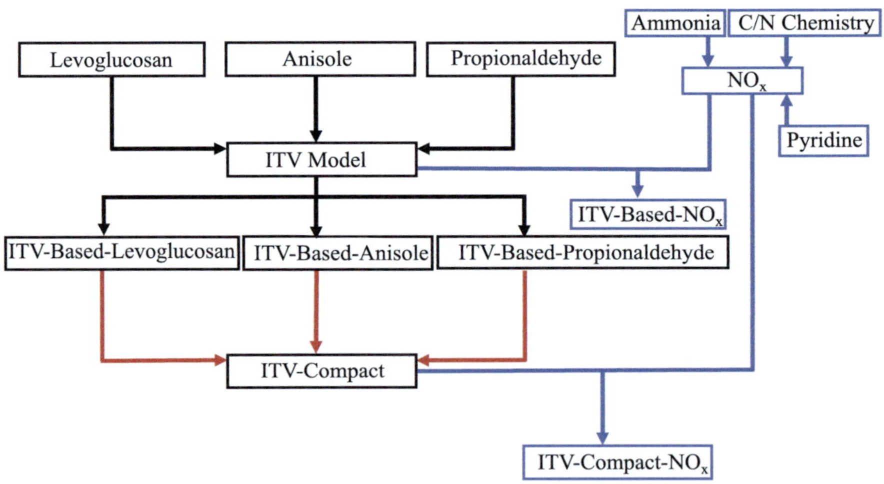


**Fig. 6.3** Development process of the compact surrogate model. Black arrows denote the integration of tars surrogate components with the ITV model, red arrows indicate the reduction and merging steps, and blue arrows denote steps related to the development of the skeletal $NO_x$ submodel and its integration with the other models

aldehydes. The work of Debiagi et al. [13], Wagnon et al. [50] and Pelucchi et al. [51] provided the kinetic models for levoglucosan, anisole, and propionaldehyde, respectively. The subchemistry models of the three species were individually integrated with the ITV model, resulting in three models labeled with the prefix "ITV-Based" followed by the name of the integrated volatile (see Fig. 6.3). The ITV-Based models were reduced to a skeletal level and combined into one compact surrogate model, named "ITV-Compact". To include the nitrogen-containing species from the volatile mixture, an additional skeletal $NO_x$ submodel was developed to accurately describe the primary $NO_x$ formation pathways and the oxidation chemistry of pyridine. The latter was selected as a representative species for nitrogen tars. The following sections provide a more detailed description of the ITV-Compact model development steps.

### 6.3.1 ITV-Compact Model Development

The three detailed ITV-Based models were reduced to a skeletal level using the directed relation graph with error propagation (DRGEP) methodology [52]. The goal of the reduction process was to preserve the accuracy of the detailed models when predicting the oxidation of major volatiles under air and oxy-fuel environments at atmospheric pressure and high temperatures. Ignition delay times were selected as the main combustion characteristic whose accuracy had to be retained during the reduction phase. Experimental data from the literature were used to define model-specific reduction targets, thereby determining the validity range of the skeletal kinetic models. Target species for the reduction procedure included the volatiles of interest (anisole, propionaldehyde, and levoglucosan), major combustion products (carbon monoxide (CO), $CO_2$, and water ($H_2O$)), and smaller intermediate species, such as hydrogen ($H_2$), methane, ethylene, and ethane ($C_2H_6$). The reduced models were then systematically combined to develop the ITV-Compact surrogate model, which comprises 104 species. The merging process was carried out incrementally, starting with the smallest model and progressively integrating larger ones by identifying and incorporating missing species and corresponding reactions. Two key factors facilitated this merging strategy: First, the reduced models were derived from detailed models that shared the same base chemistry, minimising potential conflicts during merging. Second, the reduction targets for each ITV-Based model included a shared set of conditions to represent the oxidation chemistry of small volatiles, differing exclusively in conditions specific to the oxidation of the integrated volatiles. As a result, consistency was ensured among the core chemistry of the reduced models, and differences were primarily confined to the integrated tar-specific chemistry.

The validation of the ITV-Compact model considered zero- and one-dimensional configurations, including measurements of ignition delay times in shock tubes and species mole fractions in counterflow diffusion flames. Predictions from the ITV-Compact model were compared with those from the corresponding detailed model to quantify reduction effects. Furthermore, comparing the ITV-Based models

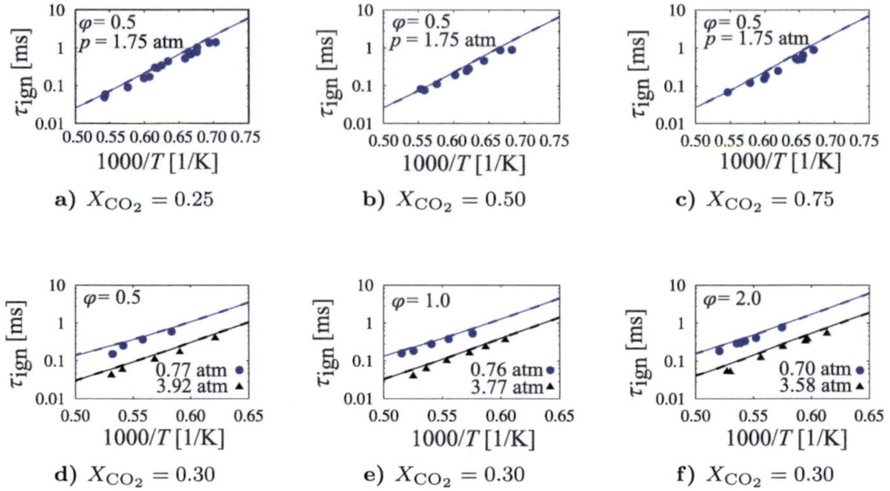


**Fig. 6.4** Comparison of ignition delay times of $CH_4/O_2/CO_2$/argon (Ar) mixtures at various pressures ($p$) and fuel-air equivalence ratios ($\varphi$) in a shock tube. Symbols denote the experimental data [31,32], solid lines denote model predictions with the ITV-Compact model, and dashed lines denote the ITV model [41] predictions

to the reference models from which the chemistry of the integrated volatile species was adopted revealed the impact of different base chemistries.

### Ignition Delay Times for Small Volatiles

Figures 6.4, 6.5 and 6.6 show measurement results [31,32,53,54] of ignition delay times for small volatiles under oxy-fuel conditions in comparison with model predictions using the ITV-Compact and the ITV model [41]. Notably, the ignition delay times predicted with the ITV model [41] are in good agreement with the experimental data. The minor discrepancies between predictions obtained with ITV-Compact and the ITV model [41] indicate that the reduction process introduced no significant errors in the ignition delay time prediction.

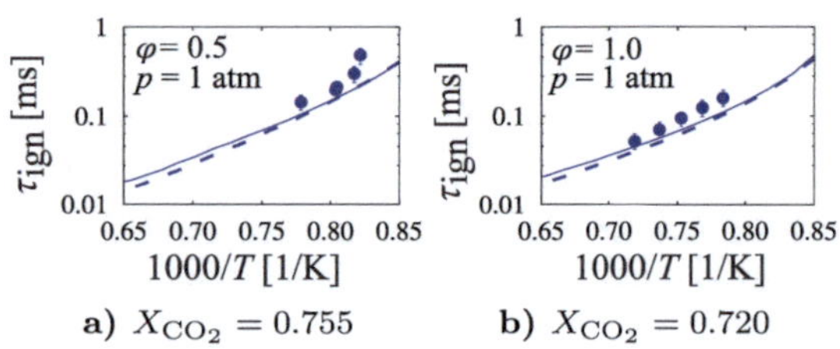


**Fig. 6.5** Comparison of ignition delay times of $C_2H_4/O_2/CO_2$ mixtures at lean and stoichiometric conditions in a shock tube. Symbols denote the measurements [53], solid lines denote model predictions with the ITV-Compact model, and dashed lines denote the ITV model [41] predictions

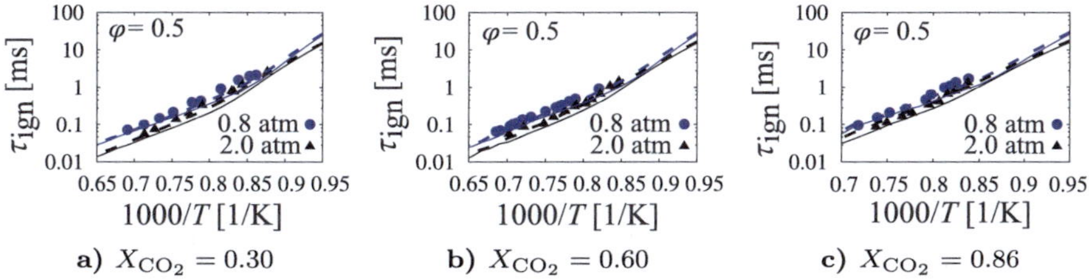


**Fig. 6.6** Comparison of ethane ignition delay times of lean $C_2H_6/O_2/CO_2$ mixtures in a shock tube. Symbols denote the measurements [54], solid lines denote model predictions with the ITV-Compact model, and dashed lines denote the ITV model [41] predictions

**Propionaldehyde Chemistry**

Figure 6.7 displays ignition delay time measurements of Ar/$O_2$/propionaldehyde mixtures from Pelucchi et al. [51] and Akih-Kumgeh and Bergthorson [55] at three different fuel-air equivalence ratios over a broad range of temperatures. Additionally, predictions with the ITV-Based-Propionaldehyde, the ITV-Compact model, and the model of Pelucchi et al. [51] are shown. The latter functions as a reference model to assess the impact of different base chemistries on propionaldehyde ignition delay times. The ITV-Based-Propionaldehyde model and the model from Pelucchi et al. [51] are in good agreement with the experimental data. Furthermore, only minor discrepancies exist between the ITV-Compact and the ITV-Based-Propionaldehyde model for all validation cases considered in Fig. 6.7.

**Anisole Chemistry**

Figures 6.8 and 6.9 show experimental speciation profiles of major and intermediate species from the counterflow diffusion flame of Chen et al. [22] with $CO_2$ as a diluent on the oxidiser side ($CO_2$O-flame). Additionally, prediction results obtained with the ITV-Based-Anisole, the ITV-Compact model, and the model of Wagnon et al. [50] are presented. The latter was chosen since it provided the anisole-specific chemistry for the ITV-Based models. Predictions obtained with the ITV-Compact

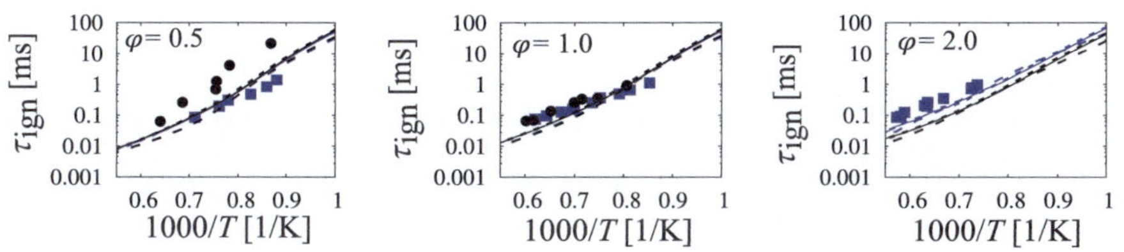


**Fig. 6.7** Comparison of measured ignition delay times of Ar/$O_2$/propionaldehyde mixtures by Pelucchi et al. [51] (blue squares: $p = 1.25$ atm, ($\varphi = 0.5$, $X_{Ar} = 0.91$)/($\varphi = 1.0$, $X_{Ar} = 0.95$)/($\varphi = 2.0$, $X_{Ar} = 0.97$)) and from Akih-Kumgeh and Bergthorson [55] (black circles: atmospheric pressure, ($\varphi = 0.5$, $X_{Ar} = 0.8889$)/($\varphi = 1.0$, $X_{Ar} = 0.939$)/($\varphi = 2.0$, $X_{Ar} = 0.9298$)) to model predictions with the ITV-Compact (solid lines), the ITV-Based-Propionaldehyde model (dotted lines), and the model of Pelucchi et al. [51] (dashed lines). Experimental data from the work of Akih-Kumgeh and Bergthorson [55] were unavailable for the fuel-rich validation case

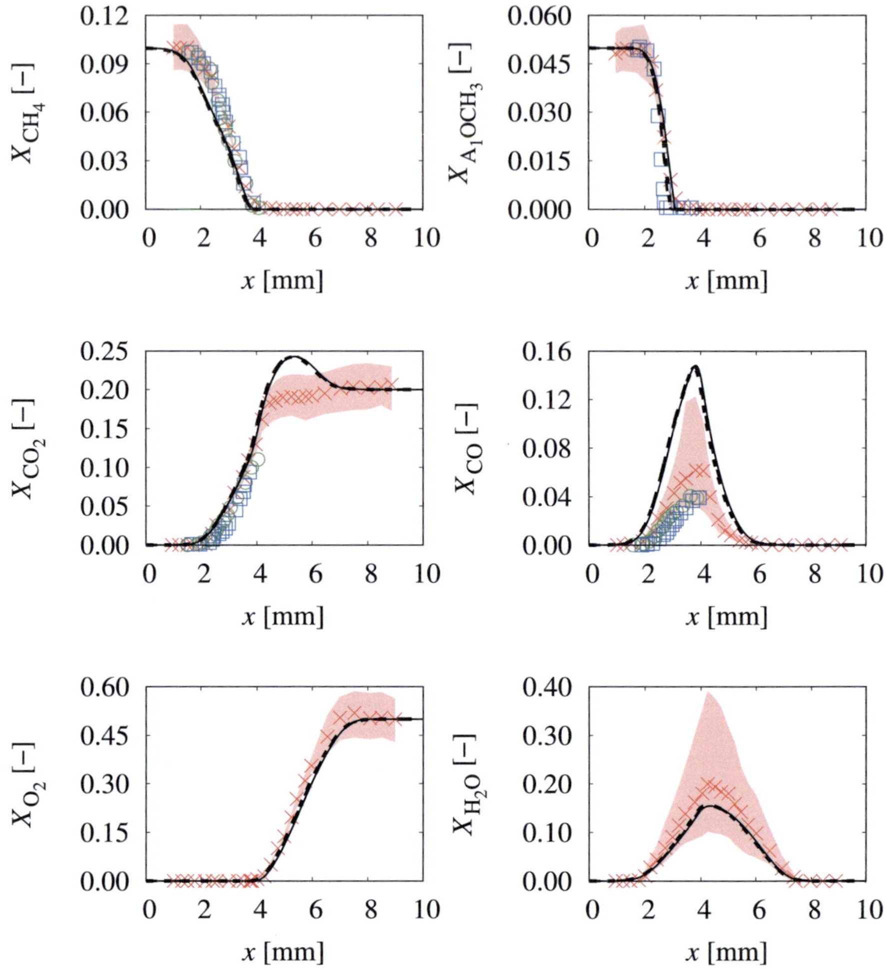


**Fig. 6.8** Comparison of measured and predicted mole fractions of reactants and main products for the $CO_2$O-flame, one of the atmospheric anisole diffusion flames investigated by Chen et al. [22]. Experimental data were obtained using three measurement techniques: ToF-MS (red crosses), GC with a *Restek Rt-Q-Bond* column (blue squares), and GC with an *Agilent DB-Petro* column (green circles). Predictions with the ITV-Compact model, the ITV-Based-Anisole model, and the model of Wagnon et al. [50] are depicted by solid, dotted, and dashed lines. The red-shaded areas indicate the uncertainty in the ToF-MS measurements. The spatial coordinate $x$ refers to the distance from the fuel nozzle

and the ITV-Based-Anisole models show no discernable difference. Moreover, predictions with the ITV-Based-Anisole and the Wagnon et al. [50] models exhibit no discrepancies, indicating that the differences in their base chemistries are negligible.

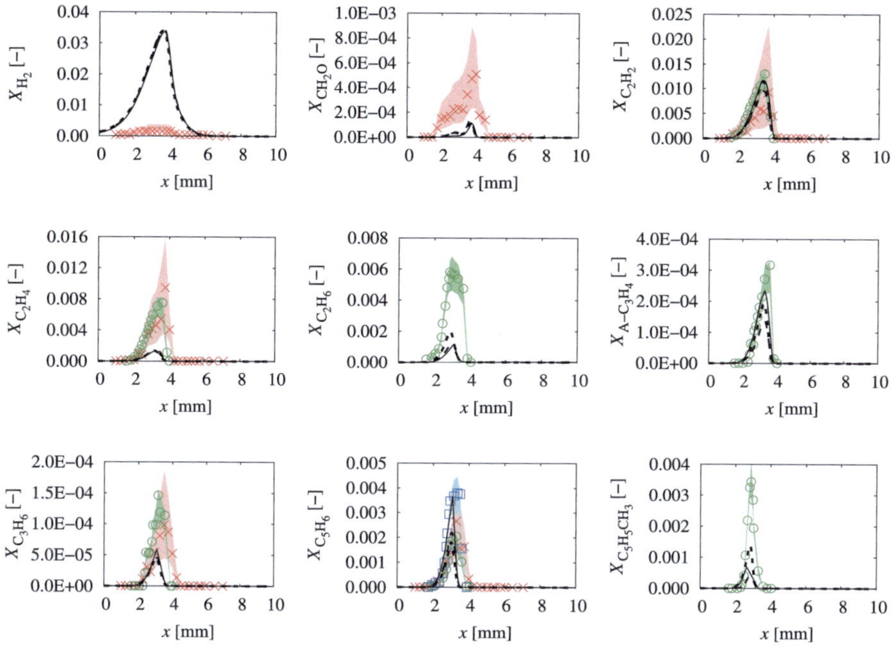

**Fig. 6.9** Comparison of experimentally measured and predicted mole fractions of small intermediates and hydrocarbons for the $CO_2O$-flame, one of the atmospheric anisole diffusion flames investigated by Chen et al. [22]. Experimental data were obtained using three measurement techniques: ToF-MS (red crosses), GC with a *Restek Rt-Q-Bond* column (blue squares), and GC with an *Agilent DB-Petro* column (green circles). Predictions with the ITV-Compact model, the ITV-Based-Anisole model, and the model of Wagnon et al. [50] are depicted by solid, dotted, and dashed lines. Shaded areas with different colours indicate the measurement uncertainty of the respective techniques. The spatial coordinate $x$ refers to the distance from the fuel nozzle

### 6.3.2 $NO_x$ Submodel

Nitrogen oxides are harmful pollutants that pose significant environmental and health risks. In biomass combustion, fuel-$NO_x$ is the predominant nitric oxide (NO) formation mechanism [7], i.e. $NO_x$ primarily originates from the oxidation of light nitrogen-containing species such as hydrogen cyanide, isocyanic acid (HNCO), and ammonia. These species can be directly released during the devolatilisation of solid particles or formed through tar-nitrogen cracking [6,7]. Therefore, simulations of biomass combustion require, first and foremost, an accurate model of fuel-$NO_x$ formation. In addition to fuel-$NO_x$ pathways, nitrous dioxide ($N_2O$) chemistry and prompt-$NO_x$ and thermal-$NO_x$ pathways can contribute to overall $NO_x$ emissions and must be included. This section introduces a newly developed $NO_x$ submodel suitable for biomass combustion. As illustrated in Fig. 6.3, the model consists of three subsets: the C-N, ammonia, and tar-nitrogen (represented by pyridine) chemistry. The subsets are discussed in the following. Since this work aims to develop a $NO_x$ model suitable for large-scale simulations, relevant kinetic mechanisms for the C-N and ammonia chemistry were extracted from the detailed $NO_x$ submodel through systematic reduction using the DRGEP method [52]. The selection of reduction targets for these subsets is also outlined below.

**Subset Model Selection and Adaption**

The developed model for C-N chemistry is based on the comprehensive review of Glarborg et al. [7]. Subsets were adopted from the model of Glarborg et al. [7] because their choice of rate coefficients and thermochemistry parameters was informed by extensive literature surveys that considered both ab initio calculations and experimental measurements. The adopted subsets include the oxidation kinetics of nitrogen-containing species, such as hydrogen cyanide, isocyanic acid, and prompt-$NO_x$ formation pathways. The model also accounts for interactions between small nitrogen-containing and hydrocarbon species, which is particularly relevant in biomass combustion due to the complex and heterogeneous mixtures of volatiles released during devolatilisation. A subset of the validation data compiled by Glarborg et al. [7] was also used to design the reduction targets for the C-N submodel, focusing on experimental conditions covering the oxidation chemistry of hydrogen cyanide, isocyanic acid, hydrocarbon-nitrogen interactions, and prompt-$NO_x$ formation.

The ammonia model provides a detailed description of ammonia combustion behaviour. It includes the kinetics of the thermal DeNOx process, which plays a crucial role in ammonia oxidation, as well as the $N_2O$ and thermal $NO_x$ formation pathways. Ammonia chemistry was incorporated from the model of Zhang et al. [56]. This choice was based on a quantitative evaluation of twelve chemical kinetic models, which demonstrated that the model of Zhang et al. [56] performed best across a wide range of experimental conditions [57]. The model was systematically reduced to a skeletal level, with reduction targets including ignition delay times at pressures between 1 and 5 bar, temperatures between 1200 and 1600 K, and fuel-air equivalence ratios between 0.5 and 1, and laminar burning velocities at ambient conditions for fuel-air ratios between 0.7 and 1.4. The reduction process was conducted while ensuring that the skeletal kinetic model maintained the same level of performance as the detailed model, as indicated by the mean score [58], quantifying the agreement between experimental data and their computed counterparts.

The reduced model developed by Shamooni et al. [59] provided the pyridine chemistry. Modifications to the C-N chemistry [7] helped to improve the agreement with experimentally measured NO speciation profiles from pyridine oxidation, as shown in Fig. 6.10. Specifically, the frequency factor of the reaction $NCO + NO = N_2O + CO$ was decreased by a factor of two, and the rate parameters of the reaction $NCO + O = NO + CO$ were adopted from Shamooni et al. [59].

The result is a $NO_x$ submodule that consists of 36 species. This makes it well-suited for large-scale simulations, which demand accurate $NO_x$ predictions.

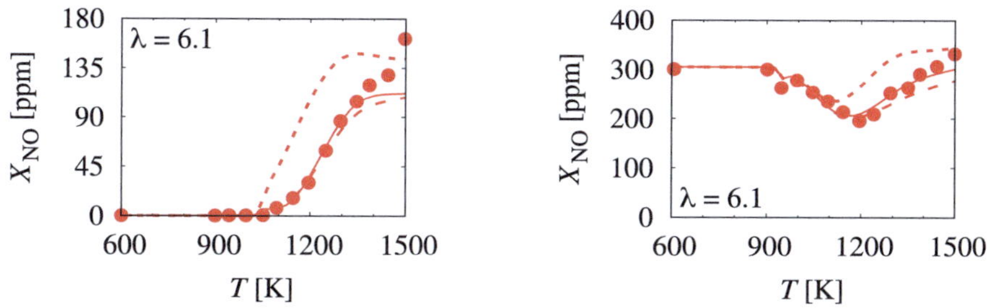


**Fig. 6.10** Mole fractions of NO as a function of temperature for atmospheric isothermal pyridine oxidation in a flow reactor at an air-fuel ratio (λ) of 6.1 (set 7 and 8) [60]. The sets differ in their dilution level and initial share of NO. The $NO_x$ submodel predictions are shown with the modified (solid lines) and unchanged (dotted lines) rate parameters for the reactions $NCO + NO = N_2O + CO$ and $NCO + O = NO + CO$. Dashed lines represent predictions of the model of Shamooni et al. [59], and symbols are experimental data

**$NO_x$ Chemistry Validation for the Oxidation of Light Volatiles and Pyridine** Integrating the $NO_x$ submodel with the ITV-Compact and the ITV model [41] resulted in two different models, referred to as the "ITV-Compact-$NO_x$" and the "ITV-Based-$NO_x$" model (see Fig. 6.3), respectively. Validation starts by evaluating the ITV-Compact-$NO_x$ model's capability to predict $NO_x$ formation from smaller volatile species, i.e. isocyanic acid (Fig. 6.11), hydrogen cyanide (Fig. 6.12), and ammonia (Fig. 6.13). Figures 6.11–6.13 also include predictions with the ITV-Based-$NO_x$ model to assess the impact of differing base chemistries on $NO_x$ formation pathways. This comparison is particularly relevant for cases with potential interactions between nitrogen-containing species and smaller hydrocarbons. Further, the ITV-Based-$NO_x$ model's performance was evaluated against the performance of the model of Glarborg et al. [7], since it is the reference source for C-N chemistry. Flow reactor measurements [7,61,62] served as validation targets for the oxidation of isocyanic acid (Fig. 6.11) and hydrogen cyanide (Fig. 6.12). The predictions with the ITV-Compact-$NO_x$ and ITV-Based-$NO_x$ models closely align and show only minor discrepancies compared to those with the model of Glarborg et al. [7]. Matching the latter highlights the successful preservation of the predictive accuracy of the C-N chemistry. Similarly, Fig. 6.13 illustrates good agreement between predictions with the ITV-Compact-$NO_x$ and ITV-Based-$NO_x$ models for NO formation from $CH_4/NH_3$ oxidation at different fuel-air equivalence ratios. Differences in ammonia consumption compared to the model of Glarborg et al. [7] were attributed to variations in the ammonia chemistry.

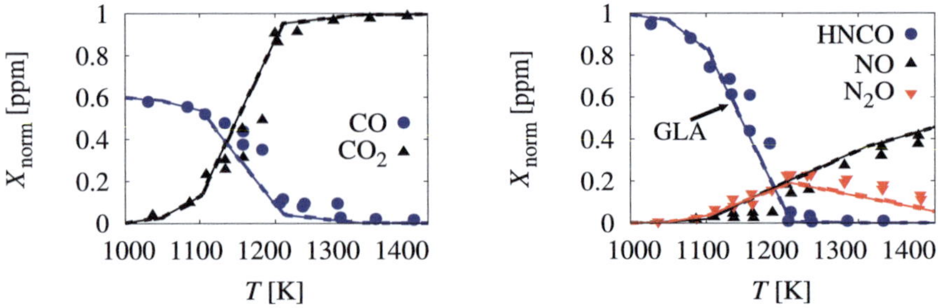


**Fig. 6.11** Comparison of flow reactor measurements (symbols) of Glarborg et al. [63] for HNCO oxidation at a pressure of 1.05 atm to predictions with the ITV-Compact-$NO_x$ (solid lines), the ITV-Based-$NO_x$ (dotted lines) models, and the model of Glarborg et al. [7] (dashed lines, GLA). The inlet HNCO mole fraction normalises the shown mole fractions. Inlet mole fractions were HNCO = 480 ppm, CO = 730 ppm, $O_2$ = 5.5 %, and $H_2O$ = 0.5 %, balanced by nitrogen ($N_2$)

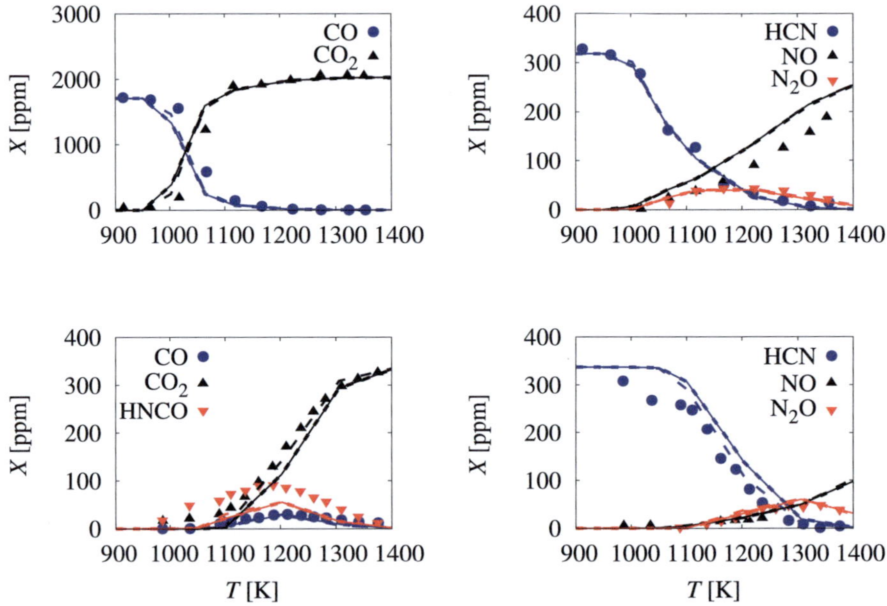


**Fig. 6.12** Comparison of flow reactor experimental data (symbols) from Hulgaard and Dam-Johansen [61] and Glarborg and Miller [62] for HCN oxidation at a pressure of 1.05 atm with (upper row, inlet mole fractions were HCN = 318 ppm, CO = 1710 ppm, $O_2$ = 2.4 %, and $H_2O$ = 2.8 %, balanced by $N_2$) and without CO in the inlet (lower row, inlet mole fractions were HCN = 337 ppm, $O_2$ = 2.6 %, and $H_2O$ = 3.1 %, balanced by $N_2$) against predictions from the ITV-Compact-$NO_x$ (solid lines), the ITV-Based-$NO_x$ (dotted lines) models, and the model of Glarborg et al. [7] (dashed lines)

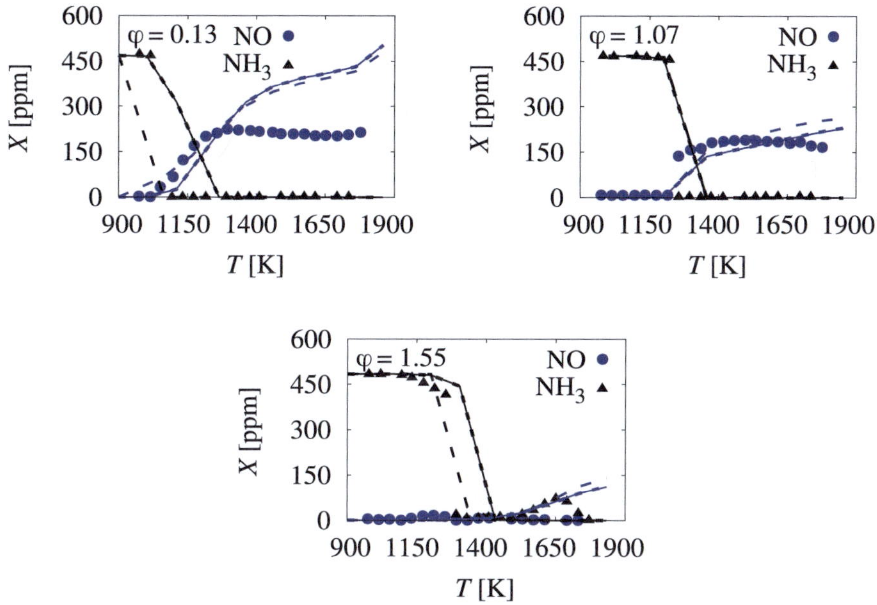


**Fig. 6.13** Comparison of atmospheric flow reactor measurements (symbols) of $CH_4$ and $NH_3$ of Mendiara and Glarborg [64] at different stoichiometry to predictions with the ITV-Compact-$NO_x$ (solid lines), the ITV-Based-$NO_x$ (dotted lines) models, and the model of Glarborg et al. [7] (dashed lines). Inlet mole fractions for $\varphi = 0.13/1.07/1.55$ were $CH_4 = 2508/2513/2515$ ppm, $O_2 = 3480/5040/40100$ ppm, and $NH_3 = 468/468/484$ ppm, balanced by $N_2$

The validation of pyridine oxidation in Fig. 6.14 follows the same scheme as for the smaller volatile species, except that the model of Shamooni et al. [59] was considered since it provided the pyridine chemistry. The predictions with the ITV-Compact-$NO_x$ and ITV-Based-$NO_x$ models agree well and differ little from the ones of the model of Shamooni et al. [59], except for the $N_2O$ profiles, for which they show improved prediction accuracy.

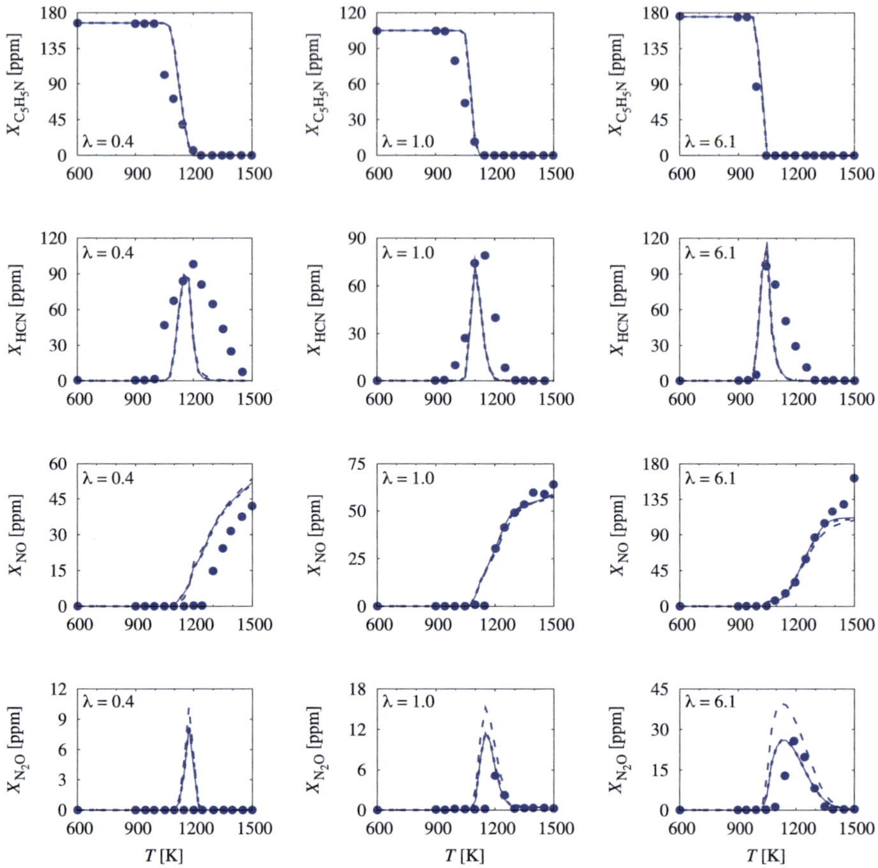

**Fig. 6.14** Comparison of atmospheric flow reactor measurements (symbols) of Alzueta et al. [60] to predictions with the ITV-Compact-$NO_x$ (solid lines), the ITV-Based-$NO_x$ (dotted lines) models, and the model of Shamooni et al. [59] (dashed lines). Initial conditions correspond to set 3 ($\lambda = 0.4$), set 5 ($\lambda = 1.0$), and set 7 ($\lambda = 6.1$) Alzueta et al. [60]

## 6.4 Conclusion

This chapter investigated the gas-phase combustion chemistry of biomass-released volatiles under oxy-fuel conditions and in air. Progress in detailed chemical kinetic modelling of PAH formation from combustion under these conditions was evaluated by comparing predictions of two literature models [27,41]. Both models [27,41] proved suitable for the investigation of combustion under oxy-fuel conditions. However, analyses revealed that the ITV model [41] predicts the mole fractions of aromatic species measured in counterflow flames more accurately than its predecessor [27]. Mean deviations between measured and predicted peak mole fractions of aromatic species obtained with the ITV model [41] are, on average, close to a factor of two for a broad range of conditions and fuels. The enhanced predictive capabilities and revisions of the $C_0$–$C_4$ chemistry make the ITV model [41] the recommended choice for further investigations of oxy-fuel combustion.

Building upon the ITV model [41], a skeletal kinetic model for biomass combustion is developed. First, a particle devolatilisation model [9,49] identified the main volatile species released under conditions representative of practical applications to determine a surrogate formulation. Anisole, levoglucosan, and propionaldehyde emerged as representative biomass surrogate components. The chemistry of these volatiles was incorporated from literature kinetic models into the ITV model [41], yielding three detailed models, one for each component. These models were then reduced to a skeletal level and merged to develop a compact surrogate model with 104 species. To account for nitrogen-containing species released from the volatile mixture, a 36-species skeletal $NO_x$ submodel is developed that combines subset models from different literature works. This submodel accurately captures key $NO_x$ formation pathways and the oxidation of small nitrogen-containing volatiles and tars. All developed models were validated against detailed kinetic models and experimental data, demonstrating their reliability in predicting critical aspects of biomass combustion under conditions relevant to practical applications. The developed $NO_x$ submodel and a further reduced version of the ITV-Compact model were successfully applied in the numerical simulations conducted in Chaps. 19 and 12, demonstrating their prediction accuracy and suitability for large-scale simulations due to their compact size.

**Acknowledgements** This work has been funded by the Deutsche Forschungsgemeinschaft (DFG, German Research Foundation)—215035359—SFB/TRR 129 'Oxyflame'.

## References

1. Williams, A., Jones, J. M., Ma, L. and Pourkashanian, M. "Pollutants from the Combustion of Solid Biomass Fuels". *Progress in Energy and Combustion Science* 38.2 (2012), 113–137. https://doi.org/10.1016/j.pecs.2011.10.001
2. Jenkins, B., Baxter, L. L., Miles Jr., T. and Miles, T. "Combustion Properties of Biomass". *Fuel Processing Technology* 54.1–3 (1998), 17–46. https://doi.org/10.1016/S0378-3820(97)00059-3
3. Hasse, C., Debiagi, P., Wen, X., Hildebrandt, K., Vascellari, M. and Faravelli, T. "Advanced Modeling Approaches for CFD Simulations of Coal Combustion and Gasification". *Progress in Energy and Combustion Science* 86 (2021), 100938. https://doi.org/10.1016/j.pecs.2021.100938
4. Smooke, M. D. "The Computation of Laminar Flames". *Proceedings of the Combustion Institute* 34.1 (2013), 65–98. https://doi.org/10.1016/j.proci.2012.09.005
5. Pielsticker, S., Gövert, B., Kreitzberg, T., Habermehl, M., Hatzfeld, O. and Kneer, R. "Simultaneous Investigation into the Yields of 22 Pyrolysis Gases from Coal and Biomass in a Small-Scale Fluidized Bed Reactor". *Fuel* 190 (2017), 420–434. https://doi.org/10.1016/j.fuel.2016.10.085

6. Glarborg, P., Jensen, A. D. and Johnsson, J. E. "Fuel Nitrogen Conversion in Solid Fuel Fired Systems". *Progress in Energy and Combustion Science* 29.2 (2003), 89–113. https://doi.org/10.1016/S0360-1285(02)00031-X
7. Glarborg, P., Miller, J. A., Ruscic, B. and Klippenstein, S. J. "Modeling Nitrogen Chemistry in Combustion". *Progress in Energy and Combustion Science* 67 (2018), 31–68. https://doi.org/10.1016/j.pecs.2018.01.002
8. Langer, R., Mao, Q. and Pitsch, H. "A Detailed Kinetic Model for Aromatics Formation from Small Hydrocarbon and Gasoline Surrogate Fuel Combustion". *Combustion and Flame* 258 (2023), 112574. https://doi.org/10.1016/j.combustflame.2022.112574
9. Ranzi, E., Cuoci, A., Faravelli, T., Frassoldati, A., Migliavacca, G., Pierucci, S. and Sommariva, S. "Chemical Kinetics of Biomass Pyrolysis". *Energy & Fuels* 22.6 (2008), 4292–4300. https://doi.org/10.1021/ef800551t
10. Hansson, K.-M., Samuelsson, J., Tullin, C. and Åmand, L.-E. "Formation of HNCO, HCN, and $NH_3$ from the Pyrolysis of Bark and Nitrogen-Containing Model Compounds". *Combustion and Flame* 137.3 (2004), 265–277. https://doi.org/10.1016/j.combustflame.2004.01.005
11. Wanjun, T., Cunxin, W. and Donghua, C. "An Investigation of the Pyrolysis Kinetics of Some Aliphatic Amino Acids". *Journal of Analytical and Applied Pyrolysis* 75.1 (2006), 49–53. https://doi.org/10.1016/j.jaap.2005.04.003
12. Apicella, B., Russo, C., Ciajolo, A., Cortese, L., Cerciello, F., Stanzione, F., Wütscher, A., Muhler, M. and Senneca, O. "High Temperature Pyrolysis of Lignite and Synthetic Carbons". *Fuel* 241 (2019), 264–272. https://doi.org/10.1016/j.fuel.2018.12.065
13. Debiagi, P., Gentile, G., Pelucchi, M., Frassoldati, A., Cuoci, A., Faravelli, T. and Ranzi, E. "Detailed Kinetic Mechanism of Gas-Phase Reactions of Volatiles Released from Biomass Pyrolysis". *Biomass and Bioenergy* 93 (2016), 60–71. https://doi.org/10.1016/j.biombioe.2016.06.015
14. Debiagi, P., Trinchera, M., Frassoldati, A., Faravelli, T., Vinu, R. and Ranzi, E. "Algae Characterization and Multistep Pyrolysis Mechanism". *Journal of Analytical and Applied Pyrolysis* 128 (2017), 423–436. https://doi.org/10.1016/j.jaap.2017.08.007
15. Cai, L., Kruse, S., Felsmann, D. and Pitsch, H. "A Methane Mechanism for Oxy-Fuel Combustion: Extinction Experiments, Model Validation, and Kinetic Analysis". *Flow, Turbulence and Combustion* 106.2 (2021), 499–514. https://doi.org/10.1007/s10494-020-00138-w
16. Baroncelli, M., Felsmann, D., Hansen, N. and Pitsch, H. "Investigating the Effect of Oxy-Fuel Combustion and Light Coal Volatiles Interaction: A Mass Spectrometric Study". *Combustion and Flame* 204 (2019), 320–330. https://doi.org/10.1016/j.combustflame.2019.03.017
17. Baroncelli, M., Mao, Q., Galle, S., Hansen, N. and Pitsch, H. "Role of Ring-Enlargement Reactions in the Formation of Aromatic Hydrocarbons". *Physical Chemistry Chemical Physics* 22.8 (2020), 4699–4714. https://doi.org/10.1039/c9cp05854k
18. Baroncelli, M., Mao, Q., Pitsch, H. and Hansen, N. "Effects of $C_1$–$C_3$ Hydrocarbon Blending on Aromatics Formation in 1-Butene Counterflow Flames". *Combustion and Flame* 230 (2021), 111427. https://doi.org/10.1016/j.combustflame.2021.111427
19. Chen, B., Ilies, B. D., Chen, W., Xu, Q., Li, Y., Xing, L., Yang, J., Wei, L., Hansen, N., Pitsch, H., Sarathy, S. M. and Wang, Z. "Exploring Low Temperature Oxidation of 1-Butene in Jet-Stirred Reactors". *Combustion and Flame* 222 (2020), 259–271. https://doi.org/10.1016/j.combustflame.2020.08.051
20. Chen, B., Liu, P., Pelucchi, M., Guidici, C., Maffei, L. P., Faller, S., Xu, Q., Huang, J., Zhang, F., Huang, C., Leonhard, K., Wang, Z., Mehl, M., Roberts, W. L., Faravelli, T. and Pitsch, H. "New Insights into the Oxidation Chemistry of Pyrrole, an N-Containing Biomass Tar Component". *Proceedings of the Combustion Institute* 39.1 (2023), 73–84. https://doi.org/10.1016/j.proci.2022.07.019
21. Chen, B., Lyu, H., Liu, P., Samaras, V. G., Lu, X., Gao, X., Roberts, W. L. and Pitsch, H. "On the Formation of Pyridine, the First Nitrogen Heterocyclic Ring in NPAHs". *Proceedings of the Combustion Institute* 40.1–4 (2024), 105675. https://doi.org/10.1016/j.proci.2024.105675
22. Chen, B., Hellmuth, M., Faller, S., May, L., Liu, P., Cai, L., Roberts, W. L. and Pitsch, H. "Exploring the Combustion Chemistry of Anisole in Laminar Counterflow Diffusion-Flames

Under Oxy-Fuel Conditions". *Combustion and Flame* 243 (2022), 111929. https://doi.org/10.1016/j.combustflame.2021.111929

23. Hellmuth, M., Cameron, F., Faller, S., Schmückert, L., Chen, B., Ren, Y. and Pitsch, H. "Synergistic Effect on PAH and Soot Formation in Ethylene Counterflow Diffusion Flames by the Addition of 1,3-Dioxolane – a Bio-Hybrid Fuel". *Proceedings of the Combustion Institute* 39.1 (2023), 899–908. https://doi.org/10.1016/j.proci.2022.07.164
24. Hellmuth, M., Chen, B., Bariki, C., Cai, L., Cameron, F., Wildenberg, A., Huang, C., Faller, S., Ren, Y., Beeckmann, J., Leonhard, K., Heufer, K. A., Hansen, N. and Pitsch, H. "A Comparative Study on the Combustion Chemistry of Two Bio-Hybrid Fuels: 1,3-Dioxane and 1,3-Dioxolane". *The Journal of Physical Chemistry. A* 127.1 (2023), 286–299. https://doi.org/10.1021/acs.jpca.2c06576
25. Wang, H. "Formation of Nascent Soot and Other Condensed-Phase Materials in Flames". *Proceedings of the Combustion Institute* 33.1 (2011), 41–67. https://doi.org/10.1016/j.proci.2010.09.009
26. Liu, P., Chen, B., Bennett, A., Pitsch, H. and Roberts, W. L. "Probing the Influence of Hydrogen Cyanide on PAH Chemistry". *Proceedings of the Combustion Institute* 39.1 (2023), 1139–1146. https://doi.org/10.1016/j.proci.2022.08.088
27. Cai, L., Ramalingam, A., Minwegen, H., Alexander Heufer, K. and Pitsch, H. "Impact of Exhaust Gas Recirculation on Ignition Delay Times of Gasoline Fuel: An Experimental and Modeling Study". *Proceedings of the Combustion Institute* 37.1 (2019), 639–647. https://doi.org/10.1016/j.proci.2018.05.032
28. Maruta, K., Abe, K., Hasegawa, S., Maruyama, S. and Sato, J. "Extinction Characteristics of $CH_4/CO_2$ Versus $O_2/CO_2$ Counterflow Non-Premixed Flames at Elevated Pressures up to 0.7 MPa". *Proceedings of the Combustion Institute* 31.1 (2007), 1223–1230. https://doi.org/10.1016/j.proci.2006.08.013
29. Li, Q., Wang, X., Xin, Y., Zhang, Z., Zhang, Y., Hao, C., Meng, M., Zheng, L. and Zheng, L. "A Unified Intermediate and Mechanism for Soot Combustion on Potassium-Supported Oxides". *Scientific Reports* 4 (2014), 4725. https://doi.org/10.1038/srep04725
30. Kim, T. H., Park, J. W., Park, H. Y., Park, J., Park, J. H. and Lim, I. G. "Chemical and Radiation Effects on Flame Extinction and NO Formation in Oxy-Methane Combustion Diluted with $CO_2$". *Fuel* 177 (2016), 235–243. https://doi.org/10.1016/j.fuel.2016.03.012
31. Hargis, J. W. and Petersen, E. L. "Methane Ignition in a Shock Tube with High Levels of $CO_2$ Dilution: Consideration of the Reflected-Shock Bifurcation". *Energy & Fuels* 29.11 (2015), 7712–7726. https://doi.org/10.1021/acs.energyfuels.5b01760
32. Koroglu, B., Pryor, O. M., Lopez, J., Nash, L. and Vasu, S. S. "Shock Tube Ignition Delay Times and Methane Time-Histories Measurements During Excess $CO_2$ Diluted Oxy-Methane Combustion". *Combustion and Flame* 164 (2016), 152–163. https://doi.org/10.1016/j.combustflame.2015.11.011
33. Pryor, O., Koroglu, B., Barak, S., Lopez, J., Ninnemann, E., Nash, L. and Vasu, S. S. "Ignition Delay Times of High Pressure Oxy-Methane Combustion with High Levels of $CO_2$ Dilution". *Proceedings of ASME Turbo Expo 2017: Turbomachinery Technical Conference and Exposition GT2017*. Charlotte, North Carolina, USA, 2017.
34. Mazas, A. N., Lacoste, D. A. and Schuller, T. "Experimental and Numerical Investigation on the Laminar Flame Speed of $CH_4/O_2$ Mixtures Diluted with $CO_2$ and $H_2O$". *Proceedings of ASME Turbo Expo 2010: Power for Land, Sea and Air GT2010*. Glasgow, United Kingdom, 2010.
35. Di Benedetto, A., Di Sarli, V., Salzano, E., Cammarota, F. and Russo, G. "Explosion Behavior of $CH_4/O_2/N_2/CO_2$ and $H_2/O_2/N_2/CO_2$ Mixtures". *International Journal of Hydrogen Energy* 34.16 (2009), 6970–6978. https://doi.org/10.1016/j.ijhydene.2009.05.120
36. Ratna Kishore, V., Duhan, N., Ravi, M. R. and Ray, A. "Measurement of Adiabatic Burning Velocity in Natural Gas-Like Mixtures". *Experimental Thermal and Fluid Science* 33.1 (2008), 10–16. https://doi.org/10.1016/j.expthermflusci.2008.06.001

37. Chan, Y. L., Zhu, M. M., Zhang, Z. Z., Liu, P. F. and Zhang, D. K. "The Effect of $CO_2$ Dilution on the Laminar Burning Velocity of Premixed Methane/Air Flames". *Energy Procedia* 75 (2015), 3048–3053. https://doi.org/10.1016/j.egypro.2015.07.621
38. Persis, S. de, Foucher, F., Pillier, L., Osorio, V. and Gökalp, I. "Effects of $O_2$ Enrichment and $CO_2$ Dilution on Laminar Methane Flames". *Energy* 55 (2013), 1055–1066. https://doi.org/10.1016/j.energy.2013.04.041
39. Blanquart, G., Pepiot-Desjardins, P. and Pitsch, H. "Chemical Mechanism for High Temperature Combustion of Engine Relevant Fuels with Emphasis on Soot Precursors". *Combustion and Flame* 156.3 (2009), 588–607. https://doi.org/10.1016/j.combustflame.2008.12.007
40. Langer, R. T. "Adjoint Sensitivity Analysis for Chemical Kinetic Model Development". PhD thesis. RWTH Aachen University, 2023. https://doi.org/10.18154/RWTH-2023-06993
41. Hellmuth, M., Langer, R., Meraviglia, A., Beeckmann, J. and Pitsch, H. "The Role of $C_3$ and $C_4$ Species in Forming Naphthalene in Counterflow Diffusion Flames". *Proceedings of the Combustion Institute* 40.1–4 (2024), 105620. https://doi.org/10.1016/j.proci.2024.105620
42. Senosiain, J. P., Klippenstein, S. J. and Miller, J. A. "A Complete Statistical Analysis of the Reaction Between OH and CO". *Proceedings of the Combustion Institute* 30.1 (2005), 945–953. https://doi.org/10.1016/j.proci.2004.07.009
43. Panigrahy, S., Liang, J., Ghosh, M. K., Wang, Q.-D., Zuo, Z., Nagaraja, S., Mohamed, A. A. E.-S., Kim, G., Vasu, S. S. and Curran, H. J. "An Experimental and Detailed Kinetic Modeling Study of the Pyrolysis and Oxidation of Allene and Propyne Over a Wide Range of Conditions". *Combustion and Flame* 233 (2021), 111578. https://doi.org/10.1016/j.combustflame.2021.111578
44. Wullenkord, J., Graf, I., Baroncelli, M., Felsmann, D., Cai, L., Pitsch, H. and Kohse-Höinghaus, K. "Laminar Premixed and Non-Premixed Flame Investigation on the Influence of Dimethyl Ether Addition on *n*-Heptane Combustion". *Combustion and Flame* 212 (2020), 323–336. https://doi.org/10.1016/j.combustflame.2019.11.012
45. Kukkadapu, G., Wagnon, S. W., Pitz, W. J. and Hansen, N. "Identification of the Molecular-Weight Growth Reaction Network in Counterflow Flames of the $C_3H_4$ Isomers Allene and Propyne". *Proceedings of the Combustion Institute* 38.1 (2021), 1477–1485. https://doi.org/10.1016/j.proci.2020.07.130
46. Moshammer, K., Seidel, L., Wang, Y., Selim, H., Sarathy, S. M., Mauss, F. and Hansen, N. "Aromatic Ring Formation in Opposed-Flow Diffusive 1,3-Butadiene Flames". *Proceedings of the Combustion Institute* 36.1 (2017), 947–955. https://doi.org/10.1016/j.proci.2016.09.010
47. Figura, L. and Gomez, A. "Structure of Incipiently Sooting Ethylene-Nitrogen Counterflow Diffusion Flames at High Pressures". *Combustion and Flame* 161.6 (2014), 1587–1603. https://doi.org/10.1016/j.combustflame.2013.11.023
48. Carbone, F., Gleason, K. and Gomez, A. "Pressure Effects on Incipiently Sooting Partially Premixed Counterflow Flames of Ethylene". *Proceedings of the Combustion Institute* 36.1 (2017), 1395–1402. https://doi.org/10.1016/j.proci.2016.07.041
49. Debiagi, P., Pecchi, C., Gentile, G., Frassoldati, A., Cuoci, A., Faravelli, T. and Ranzi, E. "Extractives Extend the Applicability of Multistep Kinetic Scheme of Biomass Pyrolysis". *Energy & Fuels* 29.10 (2015), 6544–6555. https://doi.org/10.1021/acs.energyfuels.5b01753
50. Wagnon, S. W., Thion, S., Nilsson, E. J. K., Mehl, M., Serinyel, Z., Zhang, K., Dagaut, P., Konnov, A. A., Dayma, G. and Pitz, W. J. "Experimental and Modeling Studies of a Biofuel Surrogate Compound: Laminar Burning Velocities and Jet-Stirred Reactor Measurements of Anisole". *Combustion and Flame* 189 (2018), 325–336. https://doi.org/10.1016/j.combustflame.2017.10.020
51. Pelucchi, M., Somers, K. P., Yasunaga, K., Burke, U., Frassoldati, A., Ranzi, E., Curran, H. J. and Faravelli, T. "An Experimental and Kinetic Modeling Study of the Pyrolysis and Oxidation of *n*-$C_3$–$C_5$–Aldehydes in Shock Tubes". *Combustion and Flame* 162.2 (2015), 265–286. https://doi.org/10.1016/j.combustflame.2014.07.027

52. Pepiot-Desjardins, P. and Pitsch, H. "An Efficient Error-Propagation-Based Reduction Method for Large Chemical Kinetic Mechanisms". *Combustion and Flame* 154.1–2 (2008), 67–81. https://doi.org/10.1016/j.combustflame.2007.10.020
53. Peng, C., Zou, C., Ren, J., Lin, Q., Xia, W., Luo, J. and Xiao, Y. "Ignition Delay Times of $C_2H_4$, $C_2H_4/n\text{-}C_4H_{10}$, and $n\text{-}C_4H_{10}/i\text{-}C_4H_{10}$ Under $O_2/CO_2$ Atmospheres: Shock Tube Experiments and Kinetic Model". *Combustion and Flame* 254 (2023), 112825. https://doi.org/10.1016/j.combustflame.2023.112825
54. Liu, P., Chen, B., Li, Z., Bennett, A., Sioud, S., Sarathy, S. M. and Roberts, W. L. "Evolution of Oxygenated Polycyclic Aromatic Hydrocarbon Chemistry at Flame Temperatures". *Combustion and Flame* 209 (2019), 441–451. https://doi.org/10.1016/j.combustflame.2019.08.018
55. Akih-Kumgeh, B. and Bergthorson, J. M. "Ignition of $C_3$ Oxygenated Hydrocarbons and Chemical Kinetic Modeling of Propanal Oxidation". *Combustion and Flame* 158.10 (2011), 1877–1889. https://doi.org/10.1016/j.combustflame.2011.02.015
56. Zhang, X., Moosakutty, S. P., Rajan, R. P., Younes, M. and Sarathy, S. M. "Combustion Chemistry of Ammonia/Hydrogen Mixtures: Jet-Stirred Reactor Measurements and Comprehensive Kinetic Modeling". *Combustion and Flame* 234 (2021), 111653. https://doi.org/10.1016/j.combustflame.2021.111653
57. Girhe, S., Snackers, A., Lehmann, T., Loffredo, F., Glaznev, R., Beeckmann, J. and Pitsch, H. "Ammonia/Hydrogen Combustion: Comprehensive Quantitative Evaluation of Chemical Kinetic Mechanisms and Improvement". $11^{th}$ *European Combustion Meeting*. Rouen, France, 2023.
58. Olm, C., Zsély, I. G., Pálvölgyi, R., Varga, T., Nagy, T., Curran, H. J. and Turányi, T. "Comparison of the Performance of Several Recent Hydrogen Combustion Mechanisms". *Combustion and Flame* 161.9 (2014), 2219–2234. https://doi.org/10.1016/j.combustflame.2014.03.006
59. Shamooni, A., Debiagi, P., Wang, B., Luu, T. D., Stein, O. T., Kronenburg, A., Bagheri, G., Stagni, A., Frassoldati, A., Faravelli, T., Kempf, A. M., Wen, X. and Hasse, C. "Carrier-Phase DNS of Detailed $NO_x$ Formation in Early-Stage Pulverized Coal Combustion with Fuel-Bound Nitrogen". *Fuel* 291 (2021). https://doi.org/10.1016/j.fuel.2020.119998
60. Alzueta, M. U., Tena, A. and Bilbao, R. "Pyridine Conversion in a Flow Reactor and its Interaction with Nitric Oxide". *Combustion Science and Technology* 174.10 (2002), 151–169. https://doi.org/10.1080/00102200290021470
61. Hulgaard, T. and Dam–Johansen, K. "Homogeneous Nitrous Oxide Formation and Destruction Under Combustion Conditions". *AIChE Journal* 39.8 (1993), 1342–1354. https://doi.org/10.1002/aic.690390811
62. Glarborg, P. and Miller, J. A. "Mechanism and Modeling of Hydrogen Cyanide Oxidation in a Flow Reactor". *Combustion and Flame* 99.3–4 (1994), 475–483. https://doi.org/10.1016/0010-2180(94)90039-6
63. Glarborg, P., Kristensen, P. G., Jensen, S. H. and Dam-Johansen, K. "A Flow Reactor Study of HNCO Oxidation Chemistry". *Combustion and Flame* 98.3 (1994), 241–258. https://doi.org/10.1016/0010-2180(94)90239-9
64. Mendiara, T. and Glarborg, P. "Ammonia Chemistry in Oxy-Fuel Combustion of Methane". *Combustion and Flame* 156.10 (2009), 1937–1949.https://doi.org/10.1016/j.combustflame.2009.07.006

# Reaction Kinetics Modelling

# Atomistic Simulations

# 7

Bastian Schnieder, Babak Farhadi Jahromi, Gunnar Schmitz, Christof Hättig and Rochus Schmid

## Contents




**Abstract**

Efficient solid fuel combustion models use semi-global mechanisms with effective parameters, traditionally derived from a small set of reference fuels and condi-



B. Schnieder (✉) · C. Hättig
Chair for Theoretical Chemistry, Ruhr-Universität Bochum, Bochum, Germany
e-mail: bastian.schnieder@ruhr-uni-bochum.de

C. Hättig
e-mail: christof.haettig@rub.de

B. F. Jahromi · G. Schmitz · R. Schmid
Chair for Inorganic Chemistry II, Ruhr-Universität Bochum, Bochum, Germany
e-mail: babak.farhadi-Jahromi@rub.de

G. Schmitz
e-mail: gunnar.schmitz@rub.de

R. Schmid
e-mail: rochus.schmid@rub.de







R. Kneer et al. (eds.), *Clean Energy Conversion*,
https://doi.org/10.1007/978-3-031-98956-8_7

tions. This limits their transferability. Atomistic simulations can compute thermodynamic properties such as reaction enthalpies and adsorption energies. This enables the determination of rate coefficients for elementary reaction steps and diffusion constants across a broad range of temperatures and pressures, including those inaccessible to experiments. Unlike empirical data tied to specific fuels, kinetic parameters from simulations depend on molecular composition and structure. This allows for seamless kinetic models applicable to different solid fuels. Computational results on diffusion and adsorption also complement experiments, aiding the development of improved sorption models. To achieve this, various computational chemistry tools are employed. Char models are built using reactive and non-reactive molecular dynamics. Small molecule diffusivity is isolated via non-reactive simulations, while reactive dynamics combined with molecular graph analysis are used to identify possible reaction pathways. Accurate electronic structure methods are used to compute rate coefficients.

## 7.1 Introduction

Due to its heterogeneous structure, the combustion of (bio)char is an extremely complex process in which various volatile and non-volatile intermediates occur. Heterogeneous and homogeneous reactions are important and, at the elevated temperatures of the combustion, mass transport through the porous char matrix with pores on different length scales plays a central role. Experimentally, reaction parameters of individual reactions are difficult to access. Widely-used simplified reaction models are based on the formulation of net reactions taking into account boundary layer and pore diffusion, and simplified heterogeneous reactions [1,2]. Such models are ultimately required for multi-scale simulations of the whole combustion process. A comprehensive overview of models can be found in a review by Hasse [1]. It is important in this context that the network of homogeneous gas-phase reactions for the combustion of small volatile molecules is known to some extent and rate parameters are often already available [3]. Also, those reactions are mostly fast and can be regarded as approximately being in equilibrium [2].

In contrast, much less data is available at the atomistic level on the slower, heterogeneous reactions for the combustion of (bio)char matrices. It is commonly modelled with a small set of lumped net reactions with Arrhenius-type expressions for temperature dependence. The parameters of such models are usually fitted to experimental data or empirically adjusted to reproduce the product spectrum and overall net reactions rates. However, such models subsume a large number of elementary reactions for chemically different intermediates. Heteroatoms contained in char, such as nitrogen, sulphur, potassium, etc. are usually not accounted for in such models because of missing experimental data.

Similarly, as shown by Hurt and Calo [4], an experimental determination of adsorption and desorption rates is, especially at high temperatures, often not possible due to transport limitations. This makes it very difficult to disentangle experimentally the contributions from different processes to determine the actual reaction rates for individual elementary steps. In addition, effects such as change in pore structure, thermal deactivation of the char by annealing, or reduction of the apparent reaction rate with increasing ash content during combustion can only be taken into account by very simple approaches with empirically adjusted parameters.

Theoretical methods with atomistic resolution can provide results for elementary steps that are not easily accessible by experiments but needed to separate contributions from different reaction channels. So far, there are only a few studies that have investigated individual elementary reactions of char combustion in this way [5]. First with the advances in theoretical models and the rapid development of computer architectures in recent years such complex reaction systems became accessible for a theoretical modelling at the atomistic and elementary reaction level.

Pioneering in this field has been the joint work of the developers of the reactive force field ReaxFF [6] around van Duin and the group of Mathews, who used molecular dynamics (MD) simulations to model the combustion of an atomistic char model [7]. By combining different sizes of condensed aromatic compounds with stochastically distributed functional groups, they constructed an atomistic model with almost 10,000 atoms for which they subsequently simulated the combustion in a pure oxygen atmosphere. This study demonstrated the possibilities of theoretical calculations and the mechanistic insights that could be obtained from simulations at that time. However, the accuracy of the reactive force field was quite limited. To make the MD simulation feasible, very high temperatures of 3000–4000 K were used to accelerate the computations. The conclusions that can be drawn from such a study based on a single theoretical method and brute force MD, are therefore limited. Nevertheless, this work represented an important feasibility study for the use of reactive force fields.

As a result of the development mentioned above, the treatment of larger systems is now possible at the same accuracy that before was only achievable for small gas-phase species. The determination of reaction energies is now possible with chemical accuracy for systems with 30 or more atoms, provided the molecular structure is well-defined. This is commonly used for homogeneous gas phase reactions of small molecules. However, due to the complex amorphous and hierarchical structure of (bio)char and the coupling of heterogeneous combustion and gasification reactions to mass transport through pores on different length scales, a direct brute force simulation at the atomistic level is still out of reach. There is no question that theoretical modelling of char combustion aiming at predictive accuracy required also at the atomistic level a multi-scale approach. Only by combining different specialised methods with varying degrees of accuracy for different intra- and intermolecular interactions and, thus, varying computational demands with length- and time-scale bridging approaches, the complex processes during char burn out can be captured.

Besides such a multi-scale and multimethod approach, a heuristic methodology to generate realistic structure models with atomic resolution is needed as starting point to simulate transport processes of small molecules and as a stage to investigate reactive processes during combustion. Generating such models presents several significant challenges due to the complex heterogeneous nature of the material. Char exhibits regions of high graphitisation, but also contains numerous defects that disrupt these ordered domains. Additionally, char is interspersed with various functional groups containing elements such as oxygen, nitrogen, sulphur, etc., adding to its chemical complexity. The pyrolysis process creates voids of varying sizes within the char, as the volatile gases are released, further contributing to the irregular and multifaceted nature of its structure. Up to now, char models have mainly been generated ad hoc or by semi-automatic assembling of small graphitic units (flakes) with a moderate amount of non-planar defects [8]. In the present work, an entirely new strategy has been explored to arrive at realistic char models, which is rooted on exploiting ReaxFF to anneal and relax initial structures towards chemically meaningful molecular systems. The approach allows to incorporate experimental information like density, flake size and tortuosity, or concentration of functional groups as boundary conditions into the generation process.

A computational workflow has been developed to produce different stacks of defect-rich graphitic carbon with a predefined size and corrugation. Different functional groups can be distributed in an automated fashion. From these graphitic stacks, or from fragments of these, larger structure models are generated using periodic boundary conditions. Here, a geometric analysis of the pore size distribution allows again comparison with experimental data.

With an increasing size of these structure models in terms of the number of atoms, even the use of rather efficient methods like the reactive force field ReaxFF becomes computationally intractable. In addition, they are not specifically calibrated to reproduce non-bonded interactions. Thus, at these larger length and timescales, where a realistic and chemically meaningful atomistic structure has already been realised, non-reactive generic force fields are employed to investigate transport properties of small gas molecules like oxygen, nitrogen, carbon dioxide, or water. In addition, as a benchmark, also idealised slit pores of differently functionalised graphitic surfaces have been used to study adsorption and diffusion of gaseous molecules.

Due to its high computational performance, ReaxFF is also used extensively in the current study to discover possible reaction pathways. Because of the approximations that are made to achieve the computational performance that is needed for MD simulations, the accuracy of ReaxFF is not sufficient to obtain quantitative results for reaction energies or realistic reaction rates. The latter require accurate electronic structure methods based on a quantum mechanics that are computationally demanding. To bridge the gap between accurate electronic structure methods and the reaction discovery that is done with MD simulations combined with a qualitative method for the potential energy surface, a tool for the analysis of MD trajectories is implemented that automatically detects reaction events and stores them in a database (see below). They serve then as starting points for step-wise refinements of reaction pathways, molecular structures, energies, and entropies that employ electronic structure methods of increasing accuracy and computational demands. It starts with density functional theory (DFT) at the meta generalised gradient approximation (meta-GGA) level followed by a further refinement with a hybrid functional, and explicitly correlated coupled-cluster calculations for the final energies. This combination of methods allows to compute reaction and activation energies with an accuracy of a few kilojoules per mole. The vibrational energy levels are used in combination with a quasi rigid rotor harmonic oscillator (qRRHO) approximation [9, 10] to compute the partition functions needed for reaction rates and entropy contributions to free energies.

Although, as a force field approach, ReaxFF stands out by its high performance for reactive MD simulations it has some shortcomings that limits its applicability for char combustion modelling. One of them is that parameterisations suitable for combustion conditions are not yet available for some atom types that are relevant for char combustion like phosphorus, Ca, K, and Na. Also, as a force field approach, ReaxFF is not aware of the electronic structure and thus of information like the number of electrons or the electronic spin. Therefore, extended tight binding methods from the GFN-xTB family [11] are explored as semi-empirical electronic structure methods as alternative approaches for reactive MD simulations for reaction discovery.

As mentioned above, (bio)char consists of interlinked polyaromatic hydrocarbons (PAHs), functionalised in particular with oxygen, but also with nitrogen and sulphur containing functional groups. The functional groups occur mostly at the edge of the aromatic regions and are the most important points of attack during oxidation. But also hydrogen atoms at the edge of aromatic ring systems can be abstracted by highly reactive radicals like H and O atoms that are present at the elevated temperatures in the combustion atmosphere. The organic radicals that are generated by these initial steps can react with molecular oxygen—which is otherwise not very reactive—and form organic peroxo radicals that play a central role in radical chain branching.

The atomistic simulations are separated into reactive and non-reactive processes in the ensuing sections.

## 7.2 Non-Reactive Processes

For the char-burnout mechanism, the heterogeneous combustion and gasification reactions are crucial. However, due to the entanglement of the chemical reactions with transport processes, also this aspect needs to be considered in atomistic simulations. In particular the transport of small molecules like oxygen, carbon dioxide or water as well as the transport of heat (or the heat capacity) can be determined by molecular dynamics simulations. From an experimental point of view, transport properties like diffusion constants can be determined, however, not for the conditions during char burnout. It is an advantage here that theoretical simulations allow on one side to extrapolate such properties to, e.g. high temperature conditions, but at the same time to constrain the systems not to undergo any chemical reactions, which allows in the end to disentangle transport and reactive contribution to effective rates. However, as pointed out above, a required prerequisite for such simulations are appropriate atomistic structure models for char. From a methodological point of view, also more efficient non-reactive force field methods need to be implemented, capable to describe the complex chemical environments found in the amorphous char.

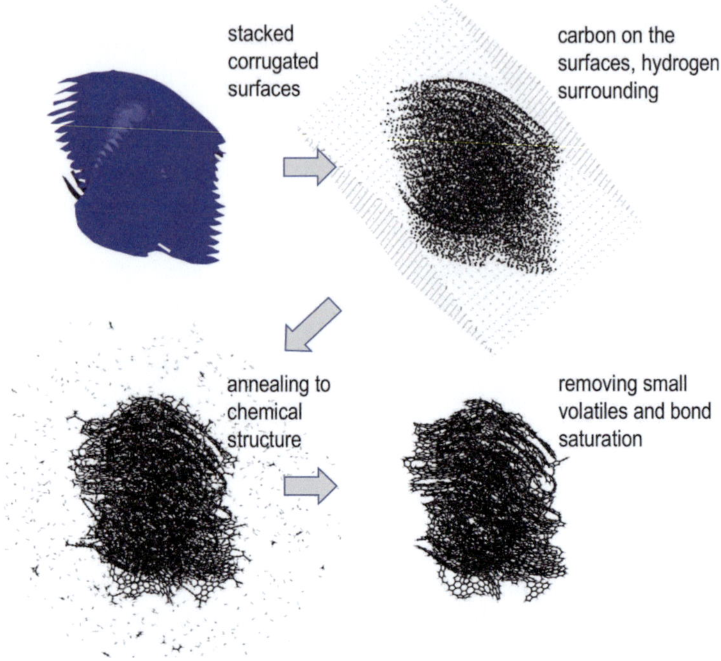


**Fig. 7.1** Schematic visualisation of the top-down method for the construction of stacked char models using curved meshes as blueprints

### 7.2.1 Construction of Atomistic Char Models

The basic workflow of the newly developed char construction methodology [12] is visualised in Fig. 7.1. It begins with a curved mesh as primary input, for which parameters like size and corrugation should in practice be derived from experimental HRTEM data. Stacked meshes are then formed by extrapolating mesh points from the initial mesh via its normal vectors with certain geometric constraints, most importantly a 3.5 Å distance of the new points to the initial mesh to reproduce the experimentally observed distance of stacked graphene sheets. This procedure is continued until the desired stack size is obtained.

Then, carbon atoms are randomly placed on each of these templating surfaces in the stack with the number of carbon atoms slightly below the atomic surface density of graphene. An adequately sized simulation box is then spanned around the stacked carbon atoms and the empty space is populated with hydrogen atoms at least 3 Å apart from each carbon atom in the stack.

The transformation of the resulting system into a chemically reasonable char model is initialised by a restrained energy minimisation of the system using ReaxFF, where the atoms are allowed to structurally relax under the restraint of harmonic spring potentials that force the carbon atoms to their respective mesh planes. Next, the system is dynamically restructured in a MD simulation employing ReaxFF that consists of a 25 ns long heating stage towards 2500 K followed by a 25 ns long

**Table 7.1** Parameters of the three char single planes formed on meshes generated from Gaussian hill functions

| Name | GH15 | GH20 | GH25 |
|---|---|---|---|
| Gaussian height/Å | 15 | 20 | 25 |
| Empirical formula | $C_{2038}H_{313}$ | $C_{2102}H_{376}$ | $C_{2104}H_{360}$ |

annealing stage at 2500 K and a final 2.5 ns long quenching stage towards 100 K during which the system freezes into a chemically reasonable structure.

The frozen structure is relaxed in an energy minimisation and refined with an algorithm which identifies and modifies energetically unfavorable structural features. First, regardless of the actual element, atoms bonded to only a single other atom are replaced by hydrogen, the remaining atoms are considered carbon. Next, carbon atom pairs with interatomic distances below 1 Å are fused into a single carbon atom with an averaged position. Next, 3-membered carbon rings are opened by removing single members and saturating with hydrogen atoms, where appropriate. The remaining undercoordinated carbon atoms are also saturated with hydrogen atoms if the surrounding steric bulk allows it, else they are kept as radicals. Molecules with less than 20 atoms are considered to be volatile species and are removed. Finally, the system is again allowed to relaxed via energy minimisation using ReaxFF to yield a chemically stable char model. This refinement process can be reiterated multiple times if required.

The presented workflow is robust to arbitrary mesh curvatures, which is demonstrated by the application on input meshes generated from two-dimensional Gaussian functions with heights given in Table 7.1. Truncating the process after the formation of the initial layer yields the char monosheets displayed in Fig. 7.2. These achieve the desired curvatures in a chemically plausible fashion by formation of several types of defects with respect to pristine graphene sheets. On the one hand, certain local curvatures are achieved by the incorporation of 4- to 8-membered carbon rings, on the other hand the formation of holes in the monosheet increases local flexibility around the holes allowing for certain curvatures.

Based on the foundational char models obtained by this workflow, more realistic char models can be derived by taking experimental data into account, specifically the fringe lengths of real char flakes computed from HRTEM images and the amount of functional groups from TPD measurements. The distribution of such measured fringe lengths serve as the experimental anchor data for the developed second generation of char models. These are generated by first cutting out stacks of three to five carbon layers from the initial model with fringe lengths in the range of the experimentally measured values. These cutouts then undergo a series of refinement steps before they are finally recombined to a realistic char model. In the first algorithmic refinement step, oxygen-carrying functional groups are hierarchically introduced by replacing carbon and hydrogen atoms at suitable sites identified by the algorithm until a functional group distribution close to experiment is reached. A functionalised char stack obtained form this procedure is displayed in Fig. 7.3.

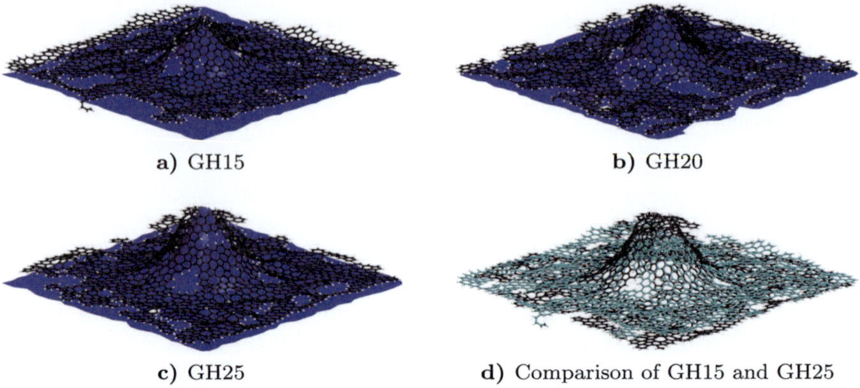


**Fig. 7.2** Char single planes with meshes generated from Gaussian hills of three different heights (reprinted from Angenent et al. [12], licensed under CC BY-SA 4.0)

Following the oxygen functionalisation, the geometries of the char stacks are optimised using GFN2-xTB [11] to remove structural artifacts introduced by all preceding construction steps, for example by quenching radicalic sites through interlayer bond formation or by resolving overcoordination via bond breaking, which may result in elimination of small molecules that are subsequently purged from the stack. A realistic char model is then obtained by positioning a large number of optimised stacks of varying size and shape in a periodic simulation box as displayed in Fig. 7.3. Here, the PACKMOL [13] package was used for the positioning.

Due to the large system size that is needed for realistic char models constructed in this fashion, any further processing and, more importantly, MD simulation of these at time scales that are sufficiently long to cover relevant dynamical properties requires a potential energy model that is computationally more efficient than ReaxFF. For sampling phenomena that do not involve the breaking of chemical bonds, such as adsorption and diffusion of small molecules in char, the reactivity of ReaxFF can be traded for efficiency by employing a non-reactive force field. Furthermore, ReaxFF is not calibrated for the non-bonded interactions important for diffusion and adsorption

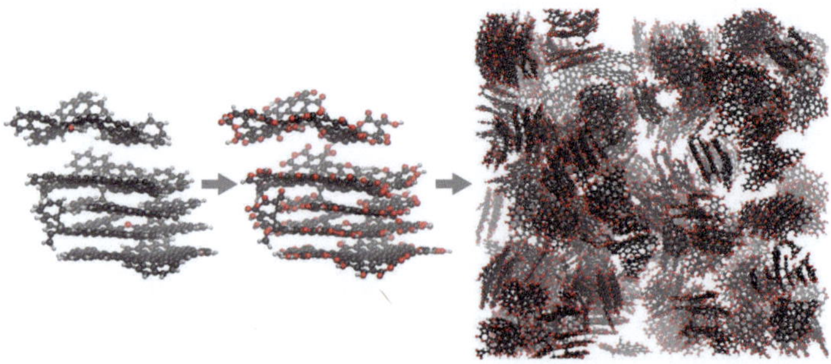

**Fig. 7.3** Examples for oxygen functionalisation of a stack of small char flakes and for a porous char model consisting of 92 unique functionalised stacks with in total system size of 27,907 atoms and a box length of 109.1 Å

in the char matrix. To this end, the generic UFF [14] was employed, which uses a rule-based system to generate force field parameters for the entire periodic table from a set of purely atomic parameters. However, due to the chemical complexity of char it could not be used unmodified because of particular chemical environments present in char models that UFF was not designed for, such as certain three and four-membered rings, allenes, and ketenes, some of which are to be expected in functionalised char and some being artificially favoured by ReaxFF which was used for the initial char construction steps. With respect to the latter, some rather reactive sites contained in the constructed char models can be traced back to insufficient saturation with hydrogen in the initial construction steps, which could not be fully quenched by the reoptimisation with GFN2-xTB. To remedy this, certain modifications to UFF had to be made, such as extending the generic rules for the setup of the UFF potential terms when such chemical environments are identified, as well as entirely eliminating certain potential energy contributions that turn numerically instable in edge cases that are frequently encountered in the generated char models. The atom types that dictate the setup rules for UFF are reliably identified using the *Open Babel* [15] package. For the electrostatic contributions to the potential energy, rather than using a fluctuating charge model based on charge equilibration [16] as in the standard UFF, the modified UFF uses a simple static point charge model, that enters the potential energy via Coulombic interaction terms between these, located at the respective atomic sites. The atomic partial charges are initialised by a topological charge equilibration model (QEq) developed by some of the authors of this chapter, which allows the use of realistic, well-balanced charges independent of the geometry to reduce the numerical effort. Such an electrostatic model is of prime importance to capture, e.g. the non-bonded quadrupole interaction of $CO_2$ with the char matrix. The parametrisation for the topological QEq model has been generated from a large set of reference data for porous hybrid materials with a wide variety of chemical environments.

The above workflow allows now to construct realistic char models with periodic boundary conditions, using corrugated, yet stacked graphitic planes with the ability to systematically vary stack sizes, functional groups and defect concentrations. Densities and pore size distributions can be adjusted either by choosing differently sized stacks or by further compressing the models alongside with an annealing sequence. The implementation of an extended UFF setup allows efficient non-reactive MD simulations to target (mixture) adsorption, and diffusion also at experimentally inaccessible conditions as, e.g. the higher temperatures of char combustion.

### 7.2.2 Simulation of Diffusion in Char

In Chap. 5, the impact of oxygen functionalisation of char on the adsorption and diffusion of water is investigated experimentally [17], indicating a stronger interaction of water with the functionalised hydrophilic interface as compared to the more hydrophobic native char. To shed light on this behaviour on a molecular level, an idealised scenario has been investigated by non-reactive force field simulations. Since the major part of carbon in char appears as graphitic stacks, it was chosen to use a

**Fig. 7.4** Idealised graphite slit pore. The COOH-terminated 2-D graphite slab with a pore width of 14 Å is shown for a water concentration of around 2.0 mol $L^{-1}$ of water, viewed in $x$ direction

highly idealised surface, namely a hydrogen terminated 2-D slab of ideal crystalline graphite. By slicing the crystal structure of graphite in different hkl-planes, either zig-zag or armchair type surfaces are formed. Note, that the entirely hydrophobic graphene-like interface was intentionally omitted, since it can not be functionalised. The dangling bonds resulting from this slicing have been saturated with hydrogen atoms. For the functionalised variant of the 2-D slab, a number of these bonds have instead been saturated with carboxylic acid (COOH) groups (54 groups in the simulation box). This is roughly equivalent to the estimated number of functional groups per surface area determined from the experimental measurements on the high-functionalised model char O-MH800 which is based on the native MH800 [18] char. In this setup, the $z$ direction is the surface normal, whereas $x$ and $y$ are defined to be parallel and perpendicular to the graphene sheets, respectively (see Fig. 7.4).

Figure 7.4 shows the simulation setup for a pore width of 14 Å and a low water concentration. In addition, a narrow 5 Å, and a very wide 100 Å split pore have been considered. The amount of water was varied from a low concentration of 0.1 mol $L^{-1}$ up to a near-bulk concentration (see Fig. 7.4 for the example of 2 mol $L^{-1}$).

Note, that the actual pore volume needed to determine the concentrations are approximated using the pore width, which is measured between the saturating hydrogen atoms as shown in Fig. 7.4. The generic force field UFF [14] was used for the graphite slabs together with the SPC/Fw flexible water potential [19]. The slit pores have been loaded with water molecules and, using MD simulations in the NVT ensemble, the molecular trajectories of the water molecules were determined. From Einstein's diffusion equation, the water self-diffusion constants $D_{H_2O}$ can be derived via the slope of the mean square displacement $MSD$

$$MSD(t) = \frac{1}{N} \sum_{i=1}^{N} [\mathbf{r}_i(t) - \mathbf{r}_i(0)]^2 , \tag{7.1}$$

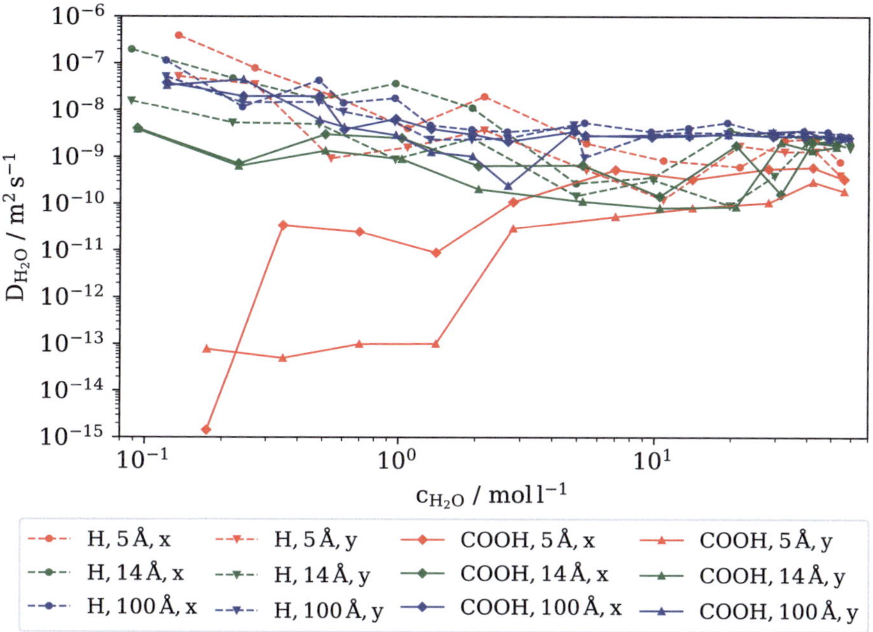


**Fig. 7.5** Computed self-diffusion constants of water in the slit pores in the directions parallel to the graphite slab surface with unfunctionalised (C–H) and functionalised (additional –COOH groups) termination

where $\mathbf{r}_i(0)$ and $\mathbf{r}_i(t)$ are the particle positions at the time origin and at time $t$, respectively, plotted over $t$. Multiple time origins are used to improve statistics. Self-diffusion constants in the two directions parallel to the slab surface ($x$ and $y$) were determined for the different slab geometries and water loadings (see Fig. 7.5). An anisotropy in the diffusivity can be observed especially for lower loading, due to the different corrugation of the surface in the two surface orientations, which can be traced back to the layered nature of graphite. As expected, the speed of diffusion is reduced for smaller pore sizes and generally decays towards high water concentrations, except for the COOH-functionalised 5 Å pore, where the diffusivity at lower concentrations is reduced by several orders of magnitude. Here, the strong interaction of the guest molecules with the pore walls significantly reduces their mobility. This effect is also observable in the 14 Å pore, albeit to a much lesser degree. Thus, in smaller pore size regimes of functionalised char, host-guest interactions increasingly dominate the diffusive behaviour of adsorbed water.

## 7.3 Reactive Processes

Previous investigations of elementary reaction steps for the combustion of solid fuels have mostly been limited to pyrolysis and primarily concerned with the unimolecular decomposition of small functionalised aromatic hydrocarbons [20–22].

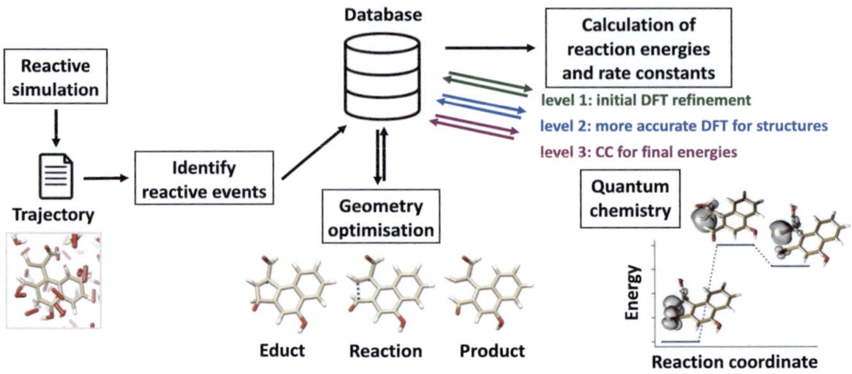


**Fig. 7.6** Workflow for the automatised reaction discovery and refinement

In difference to pyrolytic conversion, combustion comprises a large variety of reactions with oxygen-containing radical species. Accurate detailed mechanisms at the level of elementary reactions are only available for a number of liquid and gaseous fuels [23–26].

In the current study, elementary reaction steps for the oxidation of molecular surrogates for char were investigated with a workflow that makes use of several methods, each specialised for a particular task.

### 7.3.1 Multi-Method Approach for Reaction Networks and Rate Coefficients

The chemical space of combustion of even small char surrogate models is vastly huge and automatic tools for reaction discovery are indispensable for the exploration of the reaction network. There are various options for reaction sampling techniques. An approach was chosen which utilises reactive MD simulations with cost-effective methods like ReaxFF and GFN-xTB. This allows to sample the potential energy surface and to discover key reactions.

These methods alone are, however, not sufficient: The amount of reaction events makes it infeasible to analyse them by hand. Furthermore, the limited accuracy of ReaxFF or GFN-xTB does not allow for quantitative predictions. Both issues are addressed by introducing an automated workflow from MD simulations to accurate quantum chemical calculations [27] sketched in Fig. 7.6. In a first step, relevant reactions are identified and filtered to remove redundancies during the reactive simulation by comparing to previously detected reactions already stored in a database.

In the second step these reactions are post-processed on higher levels of theory, which allows to target quantitative predictions. The treatment on various levels of theory allows to employ accurate, but computationally costly electronic structure methods only for those reactions, which before on a lower level of theory were found to be most likely relevant.

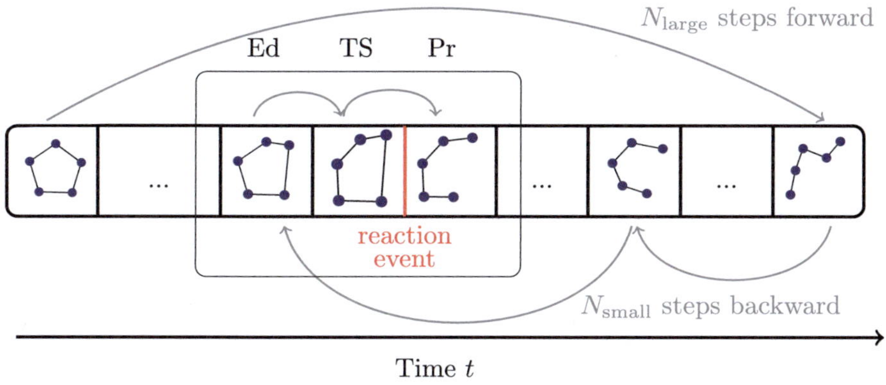


**Fig. 7.7** Illustration of the trajectory scan. If a reaction event is found three frames for the educt (Ed), transition state (TS) and product (Pr) are stored for post-processing

For each start system, usually 5 MD simulations with different random start velocities are run and the trajectories stored in a portable HDF5 file. After completion of the simulation, the reactions are identified as bond breaking or formation events and extracted from the trajectory in the HDF5 file. For this, the structures for each time step of the MD simulation are mapped onto an (undirected) molecular graph. The connectivity for the graph stems from bond orders. A bond order larger 0.5 is used as criterium for a bond between two atoms to determine the connectivity of the molecular structure and thus the graph. The graph representation is used to compare the frames within a trajectory and thereby locate reactive events. For efficiency reasons this search is not done frame by frame but, as illustrated in Fig. 7.7, the frame index is incremented by a user-defined step size $N_{\text{large}}$ (default 200). If a change in a molecular graph is detected, the search direction is reversed employing a smaller step size $N_{\text{small}}$ (default 10), which will be reduced down to 1 to locate the frames, where the reaction event took place. If a reaction event was found, three frames representing the educt(S), product(S), and the transition state, are stored for post-processing. It is distinguished between unimolecular rearrangements and non-unimolecular reactions involving either on the educt or product side more than one species.

**Table 7.2** Statistics of filtered reactions [27]. Listed are the total number of located reactions $n_r$ and the number of unique reactions $n_{\bar{r}}$. Additionally the number of unique reactions in which the molecular C-graph changes $n_{\bar{r}}^{(C)}$ and the screening ratio $Screen = (n_r - n_{\bar{r}})/n_r$ are given

| | $C_{16}H_9$ | | | | | $C_{20}H_{11}$ | | | | |
|---|---|---|---|---|---|---|---|---|---|---|
| run | 1 | 2 | 3 | 4 | 5 | 1 | 2 | 3 | 4 | 5 |
| $n_r$ | 1643 | 2494 | 1166 | 809 | 978 | 3829 | 3733 | 50 | 498 | 3933 |
| $n_{\bar{r}}$ | 103 | 32 | 11 | 34 | 14 | 11 | 8 | 9 | 41 | 7 |
| $n_{\bar{r}}^{(C)}$ | 38 | 10 | 3 | 8 | 4 | 2 | 2 | 20 | 1 | 0 |
| $Screen/\%$ | 94 | 99 | 99 | 96 | 99 | 99.9 | 99.9 | 82 | 92 | 99.9 |

For an efficient analysis, it is vital to filter the found reactions to extract a set of relevant unique reaction events. This filtering process is also done on the basis of the graph representation. In addition Cartesian coordinates are stored to distinguish stereo isomers. During the comparison, it is tracked if the molecular species belong to the educt or product side for the classification as forward or backward reaction. If a reaction $r$ is not found in the database it is added as new unique reaction. The first reaction of its kind is always classified as forward reaction. Table 7.2 highlights the importance of the filtering process, as screening ratios up to 90 % have been observed.

**Reaction and Activation Energies**

After the reaction discovery, the structures of educts, products, and transition states of the newly identified unique reactions are in a first step refined at the DFT level using a computationally cheap meta-GGA functional and a small basis set, TPSS-D3/SVP. Depending on the method used in the reaction discovery step, only information on the molecular geometries and possibly charges are available from the latter. However, additional information is needed to unambigously specify the electronic states of the reactants when the electronic structure is taken into account. This includes in addition to the charges the information on the spin state: at the DFT level the $Z$ component of the electronic spin, $M_\mathrm{S}$, which sets a lower bound for the spin quantum number $S \geq |M_\mathrm{S}|$ and the number of unpaired electrons. To determine $M_\mathrm{S}$, a single-point DFT calculation with Fermi smearing [27,28] is done. After the Fermi smearing calculation, fractional occupation numbers are rounded to the closest integer and an initial value for $M_S$ is determined as $M_\mathrm{S}^\mathrm{init} = (n_\alpha - n_\beta)/2$. The DFT energy without Fermi smearing is then evaluated for $M_\mathrm{S}^\mathrm{init}$, and, where possible, for $M_\mathrm{S}^\mathrm{init} \pm 1$. The spin orbital occupations resulting in the lowest DFT energies are then used for the subsequent calculations.

Next, the geometries of the reactants are optimised at the DFT/TPSS-D3/SVP level and the molecular graphs of the optimised structures are compared with the initial graphs from the reaction discovery. If the molecular graphs changed, the further refinement is skipped because the respective reaction has then no counterpart on the DFT potential energy surface.

To ensure that along a reaction the spin states of the educts, the transition state, and the products are compatible also for non-unimolecular cases, e.g. $\mathrm{A} + \mathrm{B} \longrightarrow [\mathrm{TS}] \longrightarrow \mathrm{C} + \mathrm{D}$, for the educts and for the products the sets of possible total $M_\mathrm{S}$ values, $M_\mathrm{S}^\mathrm{educts}$ and $M_\mathrm{S}^\mathrm{products}$, respectively, is determined that can be reached by coupling of the individual spins. Only values contained in both sets and, thus, compatible with a spin-conserving reaction, are considered. In the workflow for the location of transition states on the DFT potential energy surface (PES) there is the possibility to either refine the structure from the MD trajectory using an algorithm for the optimisation for saddle points or, if this fails, to use a double ended reaction pathway optimisation. If the sets of possible values of $M_S$ for the educts and products do not contain a common value, the workflow tries to optimise the transition state structure starting from the initial guess found during the reaction discovery with all values for $M_S$ possible for the assumed products or the assumed educts. The final educts and products of the reaction on the DFT PES are then determined from the DFT transition state by following the reaction coordinate (gradient extremal) pathways downwards in forward and backward direction.

As mentioned above, the workflow uses two alternative algorithms to locate transition states. The first one is partially adopted from the studies by Döntgen et al. [29,30] and Kröger et al. [31]. It starts from the structures identified during the reaction discovery step as those where a bond breaking or formation event is happening (see above). These structures are first preoptimised at the DFT level with fixed bond distances for the „reacting“ bonds to bring all other coordinates close to equilibrium. Then the Hessian for the nuclear coordinates is calculated for this pre-optimised structure and a regular saddle point optimisation with a trust-region image minimisation algorithm is carried out.

In any case, it is verified by a final Hessian calculation that transition states are first-order saddle points. Forward and backward intrinsic reaction coordinate calculations are used to verify the educts and products that are connected by the located transition state. For non-unimolecular reactions where the IRCs end at structures with (weakly interacting) separate reactants, the spin orbitals are localised and assigned to the individual fragments which allows to identify the individual $M_S$ spin quantum numbers. Then, the geometries of the separate fragments are optimised. The molecular graphs of the optimised IRC end points and the respective values for $M_S$ are compared to those of the assumed reference educts and products that were obtained from the refinement of the structures identified in the reaction discovery. If they do not agree, the results are assigned to a new reaction and a second algorithm to locate the correct transition state for the searched-for reaction is used which starts with a double-ended reaction path optimisation between the DFT-optimised educt and product structures.

For those reactions that could successfully refined with the low-accuracy DFT method, the geometries of the reactants and the transition states are further refined with the more accurate meta-GGA hybrid functional TPSSh-D3 and the more accurate, larger TZVP basis. The same level of theory, DFT/TZVP-D3/TZVP is also used to compute within the harmonic approximation the vibrational frequencies for the zero-point vibrational energy contributions and for the calculation of the partition functions.

While DFT/TZVP-D3/TZVP gives structures and harmonic vibrational energies that are sufficiently accurate for the purposes of this work, the electronic energies obtained at this level of theory are not sufficient for the requirements on reaction energies and rate coefficients. Therefore, in the last step of the refinement workflow accurate electronic energies at the DFT-optimised structures are computed for the reactants and transition states. For this, the explicitly correlated coupled-cluster singles-and-doubles method is employed with a scaled perturbative triples corrections, CCSD(F12*)(T*), with optimised basis set for this level, cc-pVDZ-F12. For open-shell systems, the CCSD(F12*)(T*) calculations are done based on restricted open-shell Hartree-Fock (ROHF) reference wavefunctions to avoid that the accuracy of the results is impaired by spin-contamination.

**Rate Coefficients**

The rate coefficients are calculated within the framework of transition state theory with transmission coefficients evaluated in the high temperature limit of the Eckart tunneling correction. The partition functions needed for this are calculated within a modified qRRHO approximation [9, 10]. The rate coefficients are temperature dependent, most importantly through an exponential factor depending on the activation barrier. For a compact representation of the rate coefficients $k(T)$ over a wide temperature range, parameters $A, n$, and $E$ for a modified Arrhenius expression

$$k(T) = A\,T^n \exp\left(-\frac{E}{T}\right) \tag{7.2}$$

are fitted with a non-linear least-square fit to explicitly computed rate coefficients on a grid for the temperature range 300–3000 K.

### 7.3.2 Reaction Discovery with Reactive MD

The performance of reaction MD simulations for polyaromatic hydrocarbons in an oxy-fuel atmosphere in terms of the required simulation length and the outcome is mostly determined by the concentration of radicals that can initiate combustion reactions and the simulated temperature. In reactive MD simulations at physical combustion temperatures of $\approx$ 1000 K, $O_2$ reacts only very rarely with hydrocarbons without radical centres. Usually, aryl and oxo (RO·) radicals are formed in the first steps of hydrocarbon combustion. These radicals are much more reactive and, thus, are efficient starting points for reactive MD simulations.

In early ReaxFF studies on char combustion [7], large coal models were oxidised at high temperatures of 3000–4000 K for 250 ps to resolve the bottleneck for the initial reaction steps. However, applying simulation temperatures higher than 3000 K causes an unphysical bias towards pyrolytic cracking reactions. The oxidation of perylene ($C_{20}H_{12}$) is investigated in two distinct combustion atmospheres at 2000 and 4000 K (Table 7.3). At 4000 K and in the absence of radical species, perylene undergoes rapid fragmentation into different $C_2$–$C_7$ gas-phase species before these fragments are oxidised by molecular oxygen. For example, after 0.1 ns two acetylene fragments were eliminated which were subsequently oxidised by $O_2$ to carbon dioxide. In fact, the unimolecular fragmentation reactions are similar to pyrolytic cracking and do not involve radicals or molecular oxygen in the combustion atmosphere. At lower temperatures (2000 K) perylene is nonreactive and no oxidation event was detected even after 10 ns. Only if in addition hydroxyl radicals are added to the atmosphere, perylene is oxidised rapidly at both simulation temperatures. But even with HO· present fragmentation reactions involving dehydrogenation and acetylene or ethene elimination reactions are at 4000 K the main initial reaction channels that lead to the decomposition of the aromatic system. Hence, radical species are crucial for an efficient but realistic simulation of combustion reactions of PAHs and open up different reaction channels involving molecular oxygen. These open-shell gas-phase

**Table 7.3** The influence of radical species and temperature in ReaxFF simulations using perylene as a typical PAH system

| Temperature | Radicals | Oxygen | Observation |
|---|---|---|---|
| 2000 K | None | 40 $O_2$ | Nonreactive |
| 4000 K | None | 40 $O_2$ | Fragmentation (after 0.1 ns) |
| 2000 K | 10 OH radicals | 40 $O_2$ | Edge carbon oxidation |
| 4000 K | 10 OH radicals | 40 $O_2$ | Fragmentation (after 0.1 ns) |

molecules trigger oxidation reactions, e.g. by abstracting hydrogen atoms from the edge of a PAH system, thereby forming intermediate aryl radicals.

To minimise a potential bias of simulations caused by unreasonably high radical concentrations in the simulation box, a preprocessing tool for generating transition states for the radical initiation reactions is implemented. The hydrogen abstraction transition state explorer (HyATraX) [32] can be used as a tool to generate reactive intermediates that are generated via common initial reaction channels involving hydrogen atoms or hydroxyl radicals. The formation of these aryl radicals is time consuming during the MD simulations at physically reasonable temperatures ($\lesssim$ 2000 K). Also PAHs with heteroatomic functional groups also tend to be nonreactive in the absence of gas-phase radical species in the atmosphere and do not solve this problem. The application of HyATraX facilitates the quick generation of open-shell intermediates and the automatised calculations of the respective activation and reaction energies which are depicted schematically in Fig. 7.8 for two selected nitrogen-containing PAHs. The generated radical intermediates can be used as an input for MD simulations, thereby reducing the computational time by lifting the bottleneck of the slow initial abstraction reactions. The analysis of multiple trajectories with different randomly chosen starting velocities leads to a reaction network that can be further expanded by sampling additional trajectories for stable intermediate structures. This allows different combustion products to be linked and more complete reaction networks to be created. Also in general it is important for the completeness of the reaction network that enough trajectories with different start conditions are generated. While for oxygen- and sulphur-containing PAHs it was found that five individual runs are sufficient to explore the reaction space of these systems, this changes when nitrogen is included in the carbon framework. Nitrogen stays carbon-bound during the oxidation reactions and forms $NO_x$ precursors, whereas oxygen and sulphur functionalities are eliminated in the initial reaction steps. It turned out that typically about 15 MD runs are needed to cover the reaction network for nitrogen-containing PAHs.

As alternative to reactive force fields, semi-empirical xTB methods can be applied for the reaction discovery. Specifically for char models containing potassium and other alkali metal atoms for which currently no ReaxFF parametrisations suitable for combustion is available, GFN2-xTB is currently the only feasible option to sample the reaction space for combustion. GFN2-xTB is parameterised for almost the whole

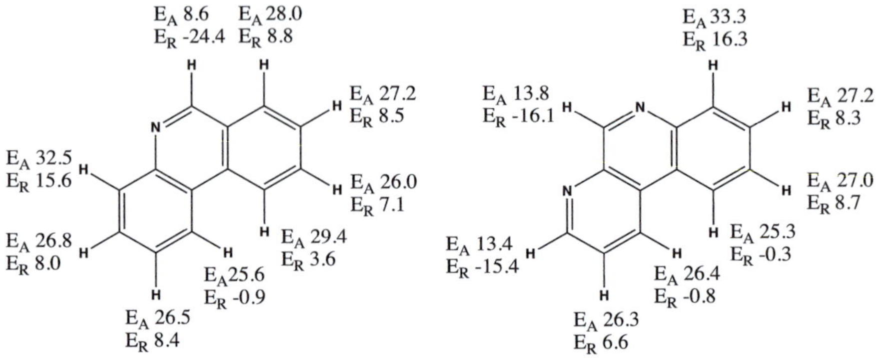


**Fig. 7.8** Reaction energies ($E_R$) and activation barriers ($E_A$) in kJ mol$^{-1}$ of initial hydrogen abstraction reactions by H atoms in two nitrogen containing PAHs. Transition states and products were generated with the HyATraX tool at the DFT/TPSS/TZVP level

periodic table and, in contrast to ReaxFF, does not require different parameterisations for different reaction conditions. In addition, it uses an electronic wavefunction which has some advantages for the transfer of the MD results to the DFT refinement. The price to be paid for this are the larger computational costs of GFN2-xTB simulations that are at least one order of magnitude higher than for ReaxFF.

For the MD simulations with GFN2-xTB, a nanoreactor setup is employed, where the reactants are placed in a spherical confinement (Fig. 7.9). Furthermore, an artificial central force is applied which partially compensates for the increased computational complexity of GFN2-xTB by increasing the number of collisions and thus the reactivity of the system:

$$\mathbf{F} = \text{sign}(\mathbf{r}_i - \mathbf{o})\, k_F\, m_i\, (\mathbf{r}_i - \mathbf{o})^2 \,. \tag{7.3}$$

In (7.3), $m_i$ denotes the mass of atom $i$, $\mathbf{r}_i$ its position vector and $\mathbf{o}$ the centre of the simulation box. Moreover, the force is defined by a predetermined force constant $k_F$. For the construction of the molecular graph, approximate bond orders are calculated

**Fig. 7.9** Spherical confinement (blue) used in the GFN2-xTB MD simulations

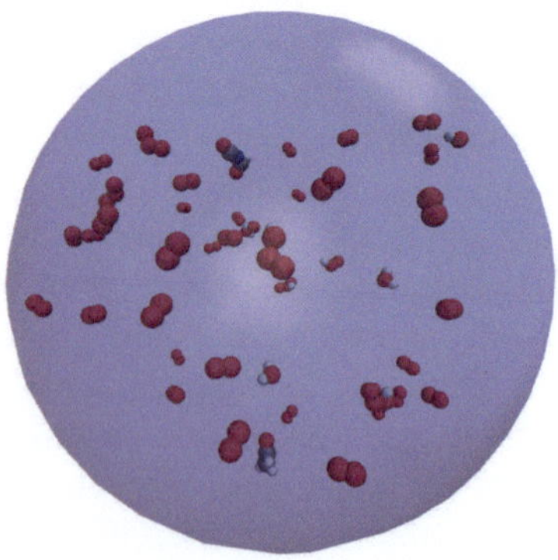

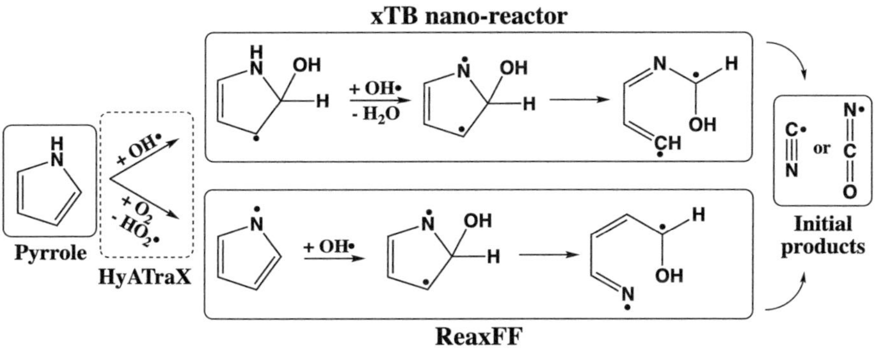


**Fig. 7.10** Initial oxidation events of pyrrole in the xTB nanoreactor and ReaxFF simulations. The MD simulation of the hydrogen abstraction reaction can be skipped by applying HyATraX

as [33]:

$$\xi(r_{i,j}) = \frac{1}{1 + \exp\left[-k_1\left(k_2\left(R_{i,\text{cov}} + R_{j,\text{cov}}\right)/r_{ij}\right) - 1\right]}, \tag{7.4}$$

where $R_{i,\text{cov}}$ are pre-tabulated covalent radii and $k_1 = 16$ and $k_2 = 4/3$.

The agreement between the initial reaction products and reaction mechanisms in both approaches is confirmed by a direct comparison of the reaction events observed in the ReaxFF simulations and GFN2-xTB nanoreactors. A suitable example of the initial reactions of pyrrole in an oxygen atmosphere containing hydroxyl radicals in both setups is provided in Fig. 7.10.

Pyrrole reacts in both simulations to cyanides and isocyanates. The first step is the abstraction of hydrogen atoms from the aromatic system that can be directly generated with HyATraX. Thereafter, hydroxyl radicals add to the carbon framework before the heterocycle breaks apart and reacts with additional radicals in the combustion atmosphere. Besides, the nanoreactor setup leads to the same initial products as observed in the ReaxFF simulations.

### 7.3.3 Initial Reactions and Influence of Functional Groups

With increasing temperature during combustion, radical species open up new reaction channels leading to reactive intermediates that increase the reactivity of char subsystems. Next to the initially generated radicals, small volatile reactive species are eliminated from the carbon framework in the later stages of the combustion process, e.g. formic acid, formaldehyde, their analogue radicals, carbon monoxide and carbon dioxide. At the beginning of char combustion, the abstraction of hydrogen atoms from aromatic C–H bonds is one of the most important initiation reaction channels during combustion. The abstraction of individual hydrogen atoms from the edge of PAHs leads to the formation of aryl radicals as highly reactive species that can react with species in the combustion atmosphere including $O_2$. The molecular composition and quantity of highly structured and amorphous PAH-areas in (bio)chars is

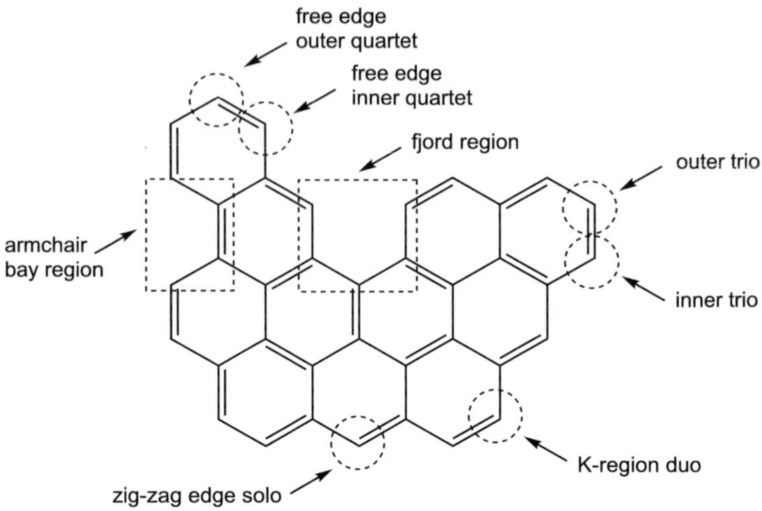


**Fig. 7.11** Nomenclature of reaction sites in a generalised PAH

a non-uniquely defined and feedstock-dependent property. Their reactivity can not be modelled at the atomistic level with just a single molecular surrogate. In general, they contain polyaromatic substructures with different shapes and sizes offering individual reaction sites of which the most important ones are schematically depicted in Fig. 7.11.

To account for the diversity in reaction sites for the hydrogen abstraction by atomic hydrogen, Yönder et al. [9] calculated temperature-depended reaction rate coefficients for PAHs with different sizes and reported generalised site-specific kinetic parameters for hydrogen abstraction reactions at the CCSD(F12*)(T*) level (Table 7.4) that allow to extrapolate the kinetic data to differently sized and shaped PAHs.

Beside interlinked aromatic and aliphatic hydrocarbons, char contains mostly oxygen, nitrogen, and sulphur functional groups that can enhance its reactivity by allowing for reaction channels with lower activation barriers. During combustion,

**Table 7.4** Non-linear least-square fit of the calculated rate coefficients $k$ to a modified Arrhenius equation (see (7.2)) for studied reaction sites [9]

| Site type | $A$/cm$^3$ mol$^{-1}$ s$^{-1}$K$^{-n}$ | $n$ /- | $E$/K |
|---|---|---|---|
| Zigzag edge solo | $1.25 \times 10^8$ | 1.76 | 7585 |
| K-region duo | $1.30 \times 10^8$ | 1.78 | 7437 |
| Inner trio | $1.82 \times 10^8$ | 1.76 | 7469 |
| Outer trio | $1.27 \times 10^8$ | 1.80 | 7333 |
| Free edge inner quartet | $1.17 \times 10^8$ | 1.81 | 7361 |
| Free edge outer quartet | $1.20 \times 10^8$ | 1.82 | 7319 |
| Armchair bay region | $9.26 \times 10^7$ | 1.75 | 7468 |

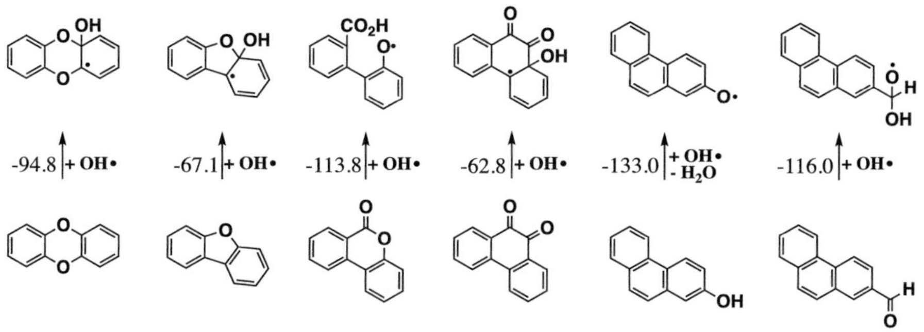


**Fig. 7.12** Barrierless initial reactions of OFGs with hydroxyl radicals leading to reactive open-shell oxo-intermediates shown with the corresponding reaction energies (kJ mol$^{-1}$) at CCSD(F12*)(T*)/cc-pVDZ-F12 level. From left to right: dioxine, furan, lactone, dicarbonyls, aromatic alcohols, and aldehydes [34]

aldehydes, carboxylic acids, aromatic alcohols, quinones, furans, and dioxins can react exothermically with hydroxyl radicals (OH·) without significant barriers, forming reactive open-shell intermediates.

Figure 7.12 shows an overview of initial hydroxyl radical addition reactions in phenanthrene-sized systems with oxygen-containing functional groups (OFGs) together with the reaction energies. The generated radical species can react further with molecular oxygen ($O_2$) or peroxyl radicals ($HO_2$·) to create highly oxidised intermediates initiating the break-down of the carbon framework.

Carboxylic acids (COOH), for example, tend to undergo decarboxylation after the hydrogen atom from the acid functionality has been abstracted by a hydroxyl radical. The generated aryl radical reacts with molecular oxygen, opening reaction

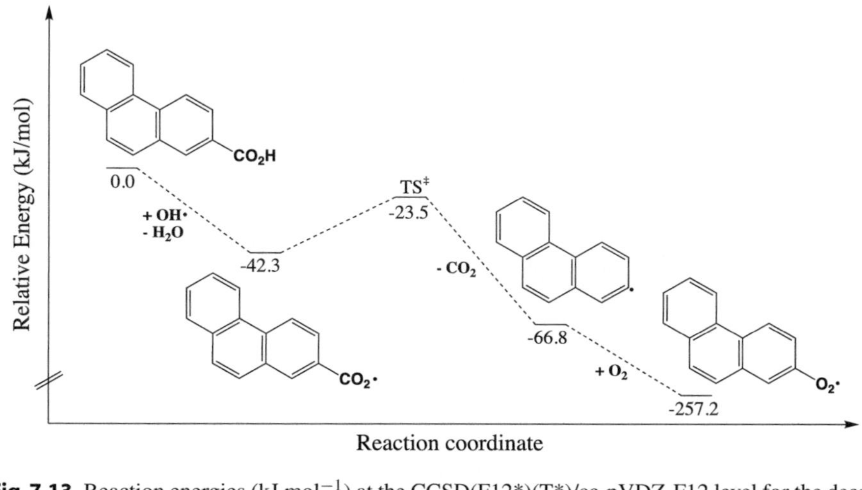


**Fig. 7.13** Reaction energies (kJ mol$^{-1}$) at the CCSD(F12*)(T*)/cc-pVDZ-F12 level for the decarboxylation phenanthrene-2-carboxylic acid and the subsequent addition of $O_2$. Energies are relative to the reactants [34]

channels for radical chain branching reactions. This reaction type was observed, e.g. for phenanthrene-2-carboxylic acid (Fig. 7.13). The decarboxylation has a comparatively small activation barrier of 18.8 kJ mol$^{-1}$ with a reaction energy of $-24.5$ kJ mol$^{-1}$. The addition of dioxygen ($O_2$) to the resulting aryl radical is barrierless and highly exothermic releasing an energy of $-190.4$ kJ mol$^{-1}$ (cf. Fig. 7.13).

The most abundant sulphur functional groups (SFGs) in char are thiophenes, thiols (R–SH), sulphonic acids (R–$SO_3$H), thioketones ($R_2C = S$), and cyclic sulphones ($R - SO_2$-R). Terminal SFGs are oxidised during combustion, e.g. by hydroxyl radicals step-wise to sulphenic, sulphoxylic, and sulphonic acids before they are eliminated as the $H_ySO_x$ gas-phase species (compare Fig. 7.14) [35]. In thiophenes the sulphur atom is part of the aromatic ring system and, thus, reacts differently in the initial reaction steps compared to terminal SFGs. Thiophene itself reacts e.g. with hydroxyl radicals that add to the aromatic ring system before a C–S bond is broken. The generated thiyl radical can then recombine with different radicals present in the combustion atmosphere, e.g. a second hydroxyl radical. At this step, sulphur can eliminate as a thioperoxyl radical (SOH). Embedded thiophenes react slightly differently: For these species, sulphur tends to leave the aromatic framework via elimination of carbonyl sulphide (COS), the mono-sulphur analogue of carbon dioxide.

The most common nitrogen functional groups (NFGs) in char are pyrroles, pyridines, and diketopiperazine groups. Different initial radical sites generated with the HyATraX [32] tool revealed large variations in reactivities. While, e.g. pyrrole 1-yl did not react further during the course of the ReaxFF simulations, the 2-yl and 3-yl radicals exhibit high reactivity and are oxidised rapidly by molecular oxygen or hydroxyl radicals. Typically nitrogen tends to stay carbon-bonded during the initial oxidation reactions until the NFGs are converted into $NO_x$ precursors like hydrogen cyanide (HCN), methanimine ($H_2$CNH) or (iso)cyanates (NCO, CNO). Heterocyclic N-compounds either undergo hydrogen abstraction reactions or direct

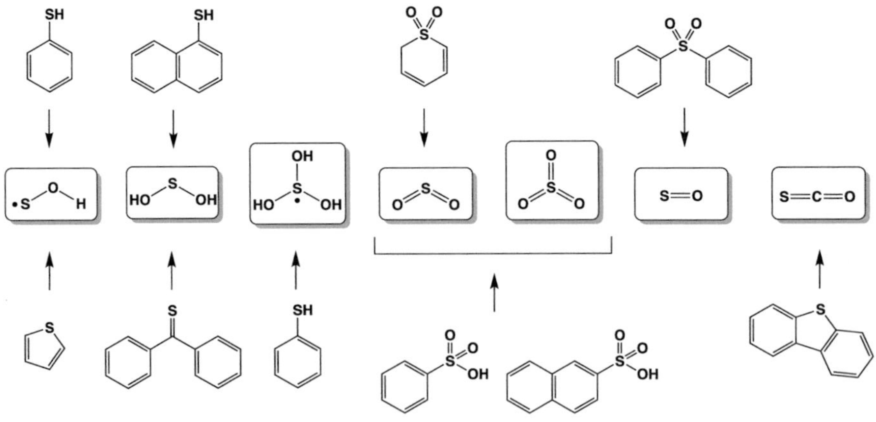

**Fig. 7.14** Initial $H_ySO_x$ elimination products of different sulphur functional groups (SFGs) in varying system sizes observed in ReaxFF

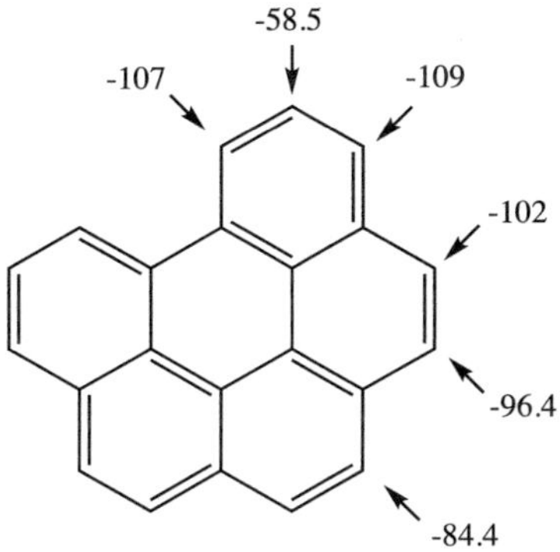


**Fig. 7.15** Reaction energies ($\mathrm{kJ\ mol^{-1}}$) for the addition of atomic oxygen at different reaction sites in benzo[ghi]perylene [36]

oxidation, e.g. by hydroxyl radicals from the combustion environment. Additionally, C–C or C–N bond dissociations lead to fragmentation and further elimination reactions, e.g. the formation of formaldehyde and different $NO_x$ precursors. Non-aromatic diketopiperazines react similarly forming cyanides and primarily ketenes as initial elimination products. The $NO_x$ precursors then react with atomic and molecular oxygen to nitrogen monoxide (NO) and dioxide ($NO_2$).

### 7.3.4 Break-Down of the Carbon Framework

After terminal functional groups have eliminated or induced ring-opening reactions the carbon framework is gradually decomposed by the oxygen species present in the combustion atmosphere. One important reaction channel that initiates the break-down of aromatic carbon frameworks involves the addition of molecular (Eq. (7.5)) or atomic oxygen (Eq. (7.6)) to the hydrocarbon system, forming highly reactive oxyradicals. Aryl radicals ($C_xH_{y-1}$) reacts with molecular oxygen forming oxyl radicals ($C_xH_{y-1}O$) and highly reactive atomic oxygen:

$$C_xH_{y-1} + O_2 \rightarrow C_xH_{y\text{-}1}O_2 \rightarrow C_xH_{y\text{-}1}O + O(^3P) \tag{7.5}$$

Atomic oxygen ($O(^3P)$) can add aromatic rings which then rearrange, e.g. by eliminating atomic hydrogen to oxyl radicals:

$$C_xH_y + O \rightarrow C_xH_yO \rightarrow C_xH_{y\text{-}1}O + H\cdot \tag{7.6}$$

Figure 7.15 shows for example the reaction energies for the $O(^3P)$ addition at different sites in benzo[ghi]perylene.

The aryl oxyl radicals can subsequently eliminate CO [36] leaving behind a reactive radical that can react further, e.g. with molecular oxygen. The CO elimination can occur in a single-step, as observed in pentacene, or through a multi-step pathway via additional intermediates as observed for anthanthrene. The mechanistic details for the elimination reaction are influenced by geometric strain and steric effects at the various reaction sites in specific PAHs. The geometrical effects significantly influence the activation energies of the CO elimination. Due to geometrical strain,

**Table 7.5** Activation energy ranges in kJ mol$^{-1}$ for CO elimination at different site types in PAHs [36]

| Site type | $\Delta E^{\ddagger}$/kJ mol$^{-1}$ |
|---|---|
| Outer trio | 140–162 |
| Outer quartet | 170–194 |
| Duo & 2 neighbours | 182–220 |
| Inner quartet | 189–242 |
| Inner trio | 240–257 |
| Solo | 304–380 |

so-called solo sites located in the centre of edges generally exhibit the highest activation barriers for CO elimination reactions. Table 7.5 lists an overview of activation energies for the time-determining step of the CO elimination.

Oxygen functional groups (OFGs) are contributing to the decomposition of the carbon backbone by increasing the local reactivity. The attack of different radicals onto the carbon framework can lead to a distortion of the aromaticity. Commonly, the oxidation of two adjacent carbon atoms initiates a C–C bond dissociation which then either opens a ring or causes the elimination of a volatile fragment. Figure 7.16 shows typical ring opening and contraction reactions for aromatic system with oxygen functionalities. These reactions generate functional groups which react easily further with molecular oxygen or different radicals and eliminate CO or $CO_2$. Also, nitrogen functional groups (NFGs) facilitate the decomposition of aromatic carbon frameworks. Nitrogen tends to stay carbon-bound throughout the char burnout process and leads to the formation of $NO_x$ precursors that still contain carbon atoms as, e.g. $CH_2NH$, HCN, or CNO.

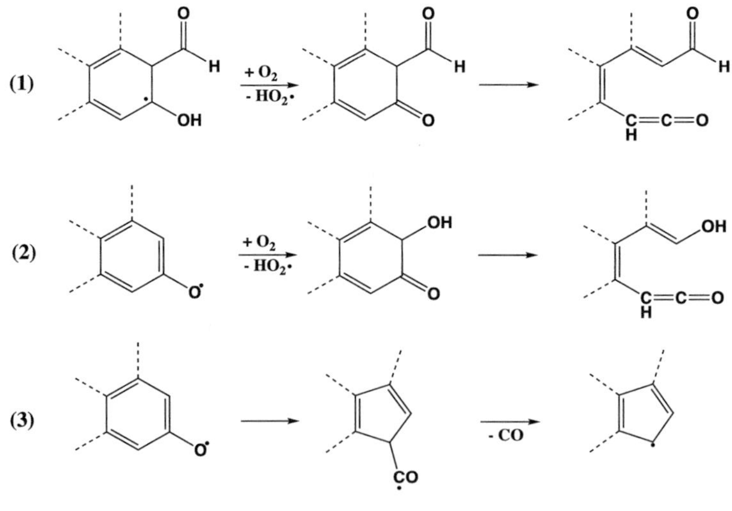


**Fig. 7.16** Ring opening (1,2) and contraction (3) reactions during char burnout involving radicals from the combustion atmosphere

## 7.4 Conclusion

During the last two decades, the theoretical modelling of char combustion at the atomistic level has made enormous progress on all scales from the electronic structure up to the dynamics of diffusion. The combination of tools specialised for different tasks on the various scales allows to study the processes involved in combustion, not only with less assumptions and less harsh approximations, but also more systematically and with higher accuracy.

Advanced computational tools have been used to establish an improved workflow to generate realistic char models. Since char consists mainly of graphitic carbon with a certain stacking of planes, but at the same time having a high defect ratio, a computational workflow is needed. This protocol has to adhere to the experimental boundary conditions like flake size or stack height, but should on the other hand not be restricted to artificial combinations of ideal graphene-like flakes. This was achieved by combining different methods and annealing steps. In addition, atomistic simulations for idealised model environments, like functionalised graphene slit pores, give access to transport properties of volatiles in pressure and temperature ranges inaccessible by experiment.

For an efficient, automatised and as much as possible unbiased reaction discovery, reactive MD simulations with ReaxFF and GFN-xTB were combined with an automatised generation of initial radical sites. This allowed to use more realistic temperatures and combustion atmospheres for the MD simulations than in previous studies and to avoid unphysical biases, e.g. towards fragmentation reactions because of unphysical high temperatures.

In general, the oxidation of the char model compounds can be divided into initiation reactions that generate reactive radical sites aromatic ring structures, and subsequent reactions that break-down the ring structures and eventually oxidise the fragments to CO, $CO_2$, and $H_2O$. The initiation reactions, such as hydrogen abstraction from the edge of polyaromatic regions, have high barriers or require highly reactive gas-phase radical species like HO·, H, or O. The so activated structures can react with molecular oxygen and other gas species in ring-opening and contraction reactions releasing CO and $CO_2$. Heteroatoms and functional groups increase the reactivity of char by reducing the activation barriers for the initiation reactions. Nitrogen atoms tend to remain carbon-bound during the initial steps, forming $NO_x$ precursors like $CH_2NH_2$ or HCN. In contrast, sulphur is predominantly eliminated as $H_ySO_x$ species. Besides, it was found that carbonyl sulphide (COS) can be released from embedded thiophenes. Oxygen-containing groups primarily decompose under the release of carbon oxides.

The decomposition of the carbon backbone begins with the oxidation of edge carbon atoms and continues as gradual decomposition from the outer edge to the inner core of the hydrocarbon. At the atomistic level, a large variety of reaction channels, reaction sites, and a number of volatile reactive radical species is observed in the combustion atmosphere that contribute to the char conversion.

Overall, the electronic structure calculations and molecular dynamics simulations provide a detailed atomistic insight into elementary reaction steps and individual contributions to diffusion and sorption processes. This helps to improve the understanding of the physical and chemical processes important for the combustion of solid fuels.

With MD simulations it has been possible to complement experimental measurements (see Sect. 5.3.2) for diffusion and sorption processes in char to temperature and pressure regimes that are difficult to access experimentally. The detailed atomistic insight from the simulations can be used to validate for example the assumptions made in development of a pore structure dependent adsorption kinetic model (see Sect. 5.3.3).

Similarly, the DFT calculations for the oxidation reactions at char surrogate models provide insight into the reaction mechanisms that are often experimentally not accessible for complex materials like (bio)char. The level of detail achievable with such calculations for elementary reaction steps, implies that for complex solid fuels it is not possible to cover all potential reaction channels and predict 1:1 the outcome of experimental combustion studies. However, they can complement experimental studies by providing atomistic insight into the reaction mechanisms that are difficult to obtain or even inaccessible. Most importantly, they provide kinetic data that is based on defined molecular compositions and, thus, transferable between different solid fuels independent of their origins. Such data is necessary for implementing and parameterising comprehensive and seamless char conversion models like the CRECK-S models, discussed in Sect. 8.2.3.

**Acknowledgements** This work has been funded by the Deutsche Forschungsgemeinschaft (DFG, German Research Foundation)—215035359—SFB/TRR 129 'Oxyflame'. The authors are indebted to Vanessa Angenent, Özlem Yönder, and LisaWarczinski for their contributions to this work. The authors are also grateful to Kai Leonhard and the authors of ChemTraYzer for helpful discussions and access to their code.

## References

1. Hasse, C. "Models for Coal Kinetics". *Reference Module in Chemistry, Molecular Sciences and Chemical Engineering*. Elsevier, 2015. https://doi.org/10.1016/B978-0-12-409547-2.11520-3
2. Tremel, A. and Spliethoff, H. "Gasification Kinetics During Entrained Flow Gasification – Part II: Intrinsic Char Reaction Rate and Surface Area Development". *Fuel* 107 (2013), 653–661. https://doi.org/10.1016/j.fuel.2012.10.053
3. Miller, J. A., Pilling, M. J. and Troe, J. "Unravelling Combustion Mechanisms Through a Quantitative Understanding of Elementary Reactions". *Proceedings of the Combustion Institute* 30.1 (2005), 43–88. https://doi.org/10.1016/j.proci.2004.08.281
4. Hurt, R. H. and Calo, J. M. "Semi-Global Intrinsic Kinetics for Char Combustion Modeling". *Combustion and Flame* 125 (2001), 1138–1149. https://doi.org/10.1016/S0010-2180(01)00234-6
5. Wang, D.-m., Xin, H.-h., Qi, X.-y., Dou, G.-l., Qi, G.-s. and Ma, L.-y. "Reaction Pathway of Coal Oxidation at Low Temperatures: A Model of Cyclic Chain Reactions and Kinetic Characteristics". *Combustion and Flame* 163 (2016), 447–460. https://doi.org/10.1016/j.combustflame.2015.10.019

6. van Duin, A. C. T., Dasgupta, S., Lorant, F. and Goddard, W. A. "ReaxFF: A Reactive Force Field for Hydrocarbons". *The Journal of Physical Chemistry A* 105.41 (2001), 9396–9409. https://doi.org/10.1021/jp004368u
7. Castro-Marcano, F., Kamat, A. M., Russo, M. F., van Duin, A. C. T. and Mathews, J. P. "Combustion of an Illinois No. 6 Coal Char Simulated Using an Atomistic Char Representation and the ReaxFF Reactive Force Field". *Combustion and Flame* 159.3 (2012), 1272–1285. https://doi.org/10.1016/j.combustflame.2011.10.022
8. Huang, Y., Cannon, F. S., Watson, J. K., Reznik, B. and Mathews, J. P. "Activated Carbon Efficient Atomistic Model Construction that Depicts Experimentally-Determined Characteristics". *Carbon* 83 (2015), 1–14. https://doi.org/10.1016/j.carbon.2014.11.012
9. Yönder, Ö., Schmitz, G., Hättig, C., Schmid, R., Debiagi, P., Hasse, C., Locaspi, A. and Faravelli, T. "Can Small Polyaromatics Describe Their Larger Counterparts for Local Reactions? A Computational Study on the H-Abstraction Reaction by an H-Atom from Polyaromatics". *The Journal of Physical Chemistry A* 124.46 (2020), 9626–9637. https://doi.org/10.1021/acs.jpca.0c07133
10. Grimme, S. "Supramolecular Binding Thermodynamics by Dispersion-Corrected Density Functional Theory". *Chemistry* 18.32 (2012), 9955–9964. https://doi.org/10.1002/chem.201200497
11. Bannwarth, C., Ehlert, S. and Grimme, S. "GFN2-xTB-an Accurate and Broadly Parametrized Self-Consistent Tight-Binding Quantum Chemical Method with Multipole Electrostatics and Density-Dependent Dispersion Contributions". *Journal of Chemical Theory and Computation* 15.3 (2019), 1652–1671. https://doi.org/10.1021/acs.jctc.8b01176
12. Angenent, V., Oelert, D. and Schmid, R. "Towards Realistic Structural Char Models: Generation of Stacked Graphene-Like Layers Using Constrained Reactive Molecular Dynamics Simulations". *Fuel* 372 (2024), 132064. https://doi.org/10.1016/j.fuel.2024.132064
13. Martínez, L., Andrade, R., Birgin, E. G. and Martínez, J. M. "PACKMOL: A Package for Building Initial Configurations for Molecular Dynamics Simulations". *Journal of Computational Chemistry* 30.13 (2009), 2157–2164. https://doi.org/10.1002/jcc.21224
14. Rappe, A. K., Casewit, C. J., Colwell, K. S., Goddard, W. A. and Skiff, W. M. "UFF, a Full Periodic Table Force Field for Molecular Mechanics and Molecular Dynamics Simulations". *Journal of the American Chemical Society* 114.25 (1992), 10024–10035. https://doi.org/10.1021/ja00051a040
15. O'Boyle, N. M., Banck, M., James, C. A., Morley, C., Vandermeersch, T. and Hutchison, G. R. "Open Babel: An Open Chemical Toolbox". *Journal of cheminformatics* 3 (2011), 33. https://doi.org/10.1186/1758-2946-3-33
16. Rappe, A. K. and Goddard, W. A. "Charge Equilibration for Molecular Dynamics Simulations". *The Journal of Physical Chemistry* 95.8 (1991), 3358–3363. https://doi.org/10.1021/j100161a070
17. Eisenbach, T., Duarte, H. A., Wedler, C., Pflieger, C., Muhler, M. and Span, R. "Experimental Investigations on the Adsorptive Behavior of $H_2O$ Vapor on HTC Char Particles". *$14^{th}$ International Conference on Applied Energy*. Bochum, Germany, 2022. https://doi.org/10.46855/energy-proceedings-10289
18. Pflieger, C., Eckhard, T., Schmitz, G., Angenent, V., Göckeler, M., Senneca, O., Schmid, R., Cerciello, F. and Muhler, M. "Thermicity of the Decomposition of Oxygen Functional Groups on Cellulose-Derived Chars". *ACS Omega* 7.51 (2022), 48606–48614. https://doi.org/10.1021/acsomega.2c07429
19. Wu, Y., Tepper, H. L. and Voth, G. A. "Flexible Simple Point-Charge Water Model with Improved Liquid-State Properties". *The Journal of Chemical Physics* 124.2 (2006), 024503. https://doi.org/10.1063/1.2136877
20. Pielsticker, S., Möller, G. H., Gövert, B., Kreitzberg, T., Hatzfeld, O., Yönder, Ö., Angenent, V., Hättig, C., Schmid, R. and Kneer, R. "Influence of Biomass Torrefaction Parameters on Fast Pyrolysis Products Under Flame-Equivalent Conditions". *Biomass and Bioenergy* 119 (2018), 392–410. https://doi.org/10.1016/j.biombioe.2018.08.014

21. Ling, L.-X., Zhang, R.-G., Wang, B.-J. and Xie, K.-C. "Density Functional Theory Study on the Pyrolysis Mechanism of Thiophene in Coal". *Journal of Molecular Structure: THEOCHEM* 905.1-3 (2009), 8–12. https://doi.org/10.1016/j.theochem.2009.02.040
22. Jiang, B., Yu, C.-F., Yuan, L., Lu, K., Tao, W., Lin, H. and Zhou, Y. "Investigation on Oxidative Pyrolysis Characteristics of Bituminous Coal Through Thermal Analysis and Density Functional Theory". *Applied Energy* 349 (2023), 121680. https://doi.org/10.1016/j.apenergy.2023.121680
23. Stagni, A., Cavallotti, C., Arunthanayothin, S., Song, Y., Herbinet, O., Battin-Leclerc, F. and Faravelli, T. "An Experimental, Theoretical and Kinetic-Modeling Study of the Gas-Phase Oxidation of Ammonia". *Reaction Chemistry & Engineering* 5.4 (2020), 696–711. https://doi.org/10.1039/C9RE00429G
24. Huang, C., Zhao, Y., Roy, I. S., Chen, B., Hansen, N., Pitsch, H. and Leonhard, K. "Pathway Exploration in Low-Temperature Oxidation of a New-Generation Bio-Hybrid Fuel 1,3-Dioxane". *Proceedings of the Combustion Institute* 39.1 (2023), 385–393. https://doi.org/10.1016/j.proci.2022.09.057
25. Kopp, W. A., Huang, C., Zhao, Y., Yu, P., Schmalz, F., Krep, L. and Leonhard, K. "Automatic Potential Energy Surface Exploration by Accelerated Reactive Molecular Dynamics Simulations: From Pyrolysis to Oxidation Chemistry". *The Journal of Physical Chemistry A* 127.50 (2023), 10681–10692. https://doi.org/10.1021/acs.jpca.3c05253
26. Jacobs, S., Döntgen, M., Alquaity, A. B. S., Hesse, R., Kruse, S., Beeckmann, J., Kröger, L. C., Morsch, P., Leonhard, K., Pitsch, H. and Heufer, K. A. "A Comprehensive Experimental and Kinetic Modeling Study of the Combustion Chemistry of Diethoxymethane". *Energy & Fuels* 35.19 (2021), 16086–16100. https://doi.org/10.1021/acs.energyfuels.1c01988
27. Schmitz, G., Yönder, Ö., Schnieder, B., Schmid, R. and Hättig, C. "An Automatized Workflow from Molecular Dynamic Simulation to Quantum Chemical Methods to Identify Elementary Reactions and Compute Reaction Constants". *Journal of Computational Chemistry* 42.32 (2021), 2264–2282. https://doi.org/10.1002/jcc.26757
28. Kresse, G. and Furthmüller, J. "Efficiency of Ab-Initio Total Energy Calculations for Metals and Semiconductors Using a Plane-Wave Basis Set". *Computational Materials Science* 6.1 (1996), 15–50. https://doi.org/10.1016/0927-0256(96)00008-0
29. Döntgen, M., Przybylski-Freund, M.-D., Kröger, L. C., Kopp, W. A., Ismail, A. E. and Leonhard, K. "Automated Discovery of Reaction Pathways, Rate Constants, and Transition States Using Reactive Molecular Dynamics Simulations". *Journal of Chemical Theory and Computation* 11.6 (2015), 2517–2524. https://doi.org/10.1021/acs.jctc.5b00201
30. Döntgen, M., Schmalz, F., Kopp, W. A., Kröger, L. C. and Leonhard, K. "Automated Chemical Kinetic Modeling via Hybrid Reactive Molecular Dynamics and Quantum Chemistry Simulations". *Journal of Chemical Information and Modeling* 58.7 (2018), 1343–1355. https://doi.org/10.1021/acs.jcim.8b00078
31. Kröger, L. C., Kopp, W. A., Döntgen, M. and Leonhard, K. "Assessing Statistical Uncertainties of Rare Events in Reactive Molecular Dynamics Simulations". *Journal of Chemical Theory and Computation* 13.9 (2017), 3955–3960. https://doi.org/10.1021/acs.jctc.7b00524
32. Schnieder, B., Schmid, R. and Hättig, C. *HyATraX (Data set)*. 2024. https://doi.org/10.5281/zenodo.12526535
33. Grimme, S., Antony, J., Ehrlich, S. and Krieg, H. "A Consistent and Accurate Ab Initio Parametrization of Density Functional Dispersion Correction (DFT-D) for the 94 Elements H-Pu". *The Journal of Chemical Physics* 132.15 (2010), 154104. https://doi.org/10.1063/1.3382344
34. Yönder, Ö., Schmitz, G., Schmid, R., Schnieder, B. and Hättig, C. "A Computational Chemistry Study on the Evolution of Oxygen Functional Groups During Char Burnout". *Fuel* 365 (2024), 131217. https://doi.org/10.1016/j.fuel.2024.131217
35. Schnieder, B., Schmid, R. and Hättig, C. "Quantum Chemical Study on the Evolution of Sulfur Functional Groups During Char Burnout". *The Journal of Physical Chemistry A* (2025). https://doi.org/10.1021/acs.jpca.4c07973

36. Yönder, Ö. and Hättig, C. "Density Functional Theory Study of CO Formation Through Reactions of Polycyclic Aromatic Hydrocarbons with Atomic Oxygen ($O(^3P)$)". *Fuel* 241 (2019), 851–860. https://doi.org/10.1016/j.fuel.2018.12.047

# Kinetic Modelling of Thermochemical Conversion of Solid Fuels

8

Leon Loni Berkel, Paulo Debiagi, Stefan Pielsticker, David Tarlinski, Osvalda Senneca, Tiziano Faravelli, Viktor Scherer, Reinhold Kneer and Christian Hasse

## Contents



L. L. Berkel (✉) · P. Debiagi · C. Hasse
Institute for Simulation of Reactive Thermo-Fluid Systems, Technische Universität Darmstadt, Darmstadt, Germany
e-mail: berkel@stfs.tu-darmstadt.de

P. Debiagi
e-mail: Paulo.Debiagi@nottingham.edu.cn

C. Hasse
e-mail: hasse@stfs.tu-darmstadt.de

P. Debiagi
Nottingham Ningbo China Beacons of Excellence Research and Innovation Institute, University of Nottingham Ningbo China, Ningbo, China

S. Pielsticker · R. Kneer
Institute of Heat and Mass Transfer, RWTH Aachen University, Aachen, Germany
e-mail: pielsticker@wsa.rwth-aachen.de

R. Kneer
e-mail: kneer@wsa.rwth-aachen.de

D. Tarlinski · V. Scherer
Department of Energy Plant Technology, Ruhr-Universität Bochum, Bochum, Germany
e-mail: tarlinski@leat.rub.de

V. Scherer
e-mail: scherer@leat.rub.de

O. Senneca
Istituto di Scienze e Tecnologie per l'Energia e la Mobilità Sostenibili, University of Naples, Naples, Italy
e-mail: osvalda.senneca@stems.cnr.it

T. Faravelli
Technische Universität Darmstadt, STFS, Darmstadt, Germany
e-mail: tiziano.faravelli@polimi.it



R. Kneer et al. (eds.), *Clean Energy Conversion*,
https://doi.org/10.1007/978-3-031-98956-8_8

**Abstract**

This chapter presents the development of models for the thermochemical conversion of coal and biomass, focusing on advances in biomass modelling. Chapter 8 introduces the characteristics of solid fuels, compares pyrolysis behaviours, and outlines the molecular structure of biomass, providing a foundation for kinetic modelling. Key models discussed include the chemical percolation devolatilisation (CPD) model, char conversion kinetics (CCK) model, and the CRECK-S model. CPD and CCK models carry an increased level of phenomenological description, ideal for benchmarking fuel conversion predictions. In contrast, CRECK-S model offers balanced complexity, predictivity and seamless coupling of pyrolysis and char reactions. The CRECK-S model is chosen to be the basis of solid fuel conversion modelling in the *OxySim-129* framework, and hence is the main focus of this chapter. The outcome is a comprehensive model capable of handling diverse operating conditions that can be coupled directly or indirectly to CFD simulations. Major developments include extensions for oxy-fuel char conversion, increasing details of released products, validation for biomass feedstock samples of *Oxyflame* (i.e. walnut shells).

# 8.1 Introduction

## 8.1.1 General Decomposition Behaviour

The overall thermochemical conversion process of solid fuels starts with the pyrolysis (devolatilisation) step [1–3], which is very sensitive to the specific characteristics of the feedstock, showing strong differences between coal and biomass, and also between different ranks of coal, types of biomass, and even among different samples of the same feedstock. During heating, moisture evaporates and numerous chemical bonds break, producing light gases and tars. The release of heavier tars is often slower, which potentially leads to depolymerisation to the solid matrix. The remaining solid material after the pyrolysis step is called char. Because of the fuel complexity and the multi-step nature of this process, involving both chemistry and transport phenomena, solid fuel pyrolysis is very complex, and requires a comprehensive understanding [4]. The wide variety of volatile species formed and released during the pyrolysis step are subject to secondary gas-phase reactions.

The nascent char resulting from the devolatilisation step continues to release entrapped species. Weaker chemical bonds are broken to form and release additional volatile species, resulting in further solid mass loss and an increase in carbon content. At high temperatures, char undergoes physico-chemical reordering (thermal annealing), gradually increasing the graphitisation degree of the structure and reducing its reactivity to oxidation. Different from conversion of volatiles, the conversion of char in combustion is a heterogeneous process and is directly dependent on the structural and chemical properties of the char, as well as on the transport and availability of oxygen to the external and internal (pores) surface of the solid particles.

The numerous characteristic complexities of the overall process are a challenge to achieving satisfactory predictive kinetic models for the thermochemical conversion of solid fuels. Accurate and reliable pyrolysis models must be able to describe the rate of pyrolysis reactions, the yield and composition of pyrolysis products, and the evolution of the solid morphology during the formation of char. Gas-phase kinetic models must describe the reactivity and products of volatiles in diverse atmospheric composition, temperature and pressure. Heterogeneous char conversion models must be able to account for the changing structure and composition of char during the process, while being able to describe the reactivity and resulting products.

Besides having similar steps for overall conversion, coal and biomass strongly differ in rate of conversion and product distribution. Compared to biomass, coal is typically lower in volatile matter, higher in fixed carbon, and higher in calorific value. These differences significantly impact the overall combustion process, as the role of char conversion step for coal, and the role of conversion of volatiles for biomass are largely more important in each case. These differences must be carefully taken care of in the design of burners.

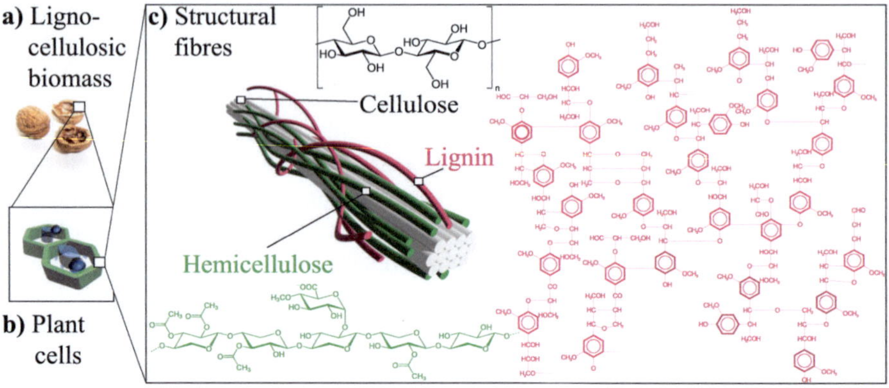


**Fig. 8.1** Structure of lignocellulosic biomass on **a)** macroscopic, **b)** microscopic, and **c)** molecular level. Molecular structures are adapted from Faravelli et al. [7] and Ranzi et al. [8]

### 8.1.2 Molecular Structure of Biomass

Whereas coal consists of aromatic clusters with varying amounts of bridges, side chains, and functional groups, biomass primarily contains organic polymers—cellulose (25–60 wt. %), hemicellulose (15–40 wt. %), and lignin (15–45 wt. %)—along with minor amounts of extractives, proteins and inorganic ash [5]. Cellulose, hemicellulose and lignin are polymeric chains that create an intricate chain of microfibrils (Fig. 8.1c)). These fibres compose the walls of the plant cells and have a key role in the plant structure (Fig. 8.1b)), constituting most of the dry weight in adult plants. At a macroscopic level, cells form tissues and organs that are macroscopically observable structures (Fig. 8.1a)). The non-structural extractive species are involved in plant metabolism. During thermochemical conversion, inorganic materials form metal oxides that create the ash residue. Finally, water can be found in various forms (hygroscopic, capillary, vapour) [6].

The individual amounts of each component in biomass vary significantly depending on the type and source of biomass. Three categories can be identified based on their structural similarities: softwoods, hardwoods and grass/cereals. The softwood category includes gymnosperms with most common types pine, fir, larch and spruce. Hardwoods are angiosperms with woody flowering and major examples are beech, birch, olive and oak. The grass/cereals category covers all the non-woody plants. Examples include agricultural products/residues like wheat straw and rice husk while others are cultivated for the energy sector, such as miscanthus.

As described previously, cellulose is the most abundant component in biomass in general. It is a polymer of glucose monomers, a 6-carbon sugar molecule, connected by β-1,4 glycosidic bonds. Several of these long polymer chains can combine with hydrogen bonds, creating crystalline structures called microfibrils. The second common polymer, hemicellulose, contains either 6-carbon (hexose) or 5-carbon (pentose) sugars. In hemicellulose, common hexoses are glucose and mannose while common pentoses are galactose and arabinose. Hemicellulose chains are shorter and more amorphous compared to cellulose due to branches on the chains and a less regular composition. Depending on the biomass type, the monosaccharides forming hemicellulose chains vary significantly. The microfibrils formed with hemicellulose are connected to the cellulose chains. The total polysaccharide content is commonly referred to as holocellulose, which is the combination of cellulose and hemicellulose. Lignins are intricate racemic polymers formed by three primary hydroxycinnamyl alcohol monomers: p-coumaryl, coniferyl, and sinapyl alcohols, all derived from the aminoacid phenylalanine. These monomers, known as monolignols, form p-hydroxyphenyl, guaiacyl, and syringyl phenylpropanoid units, which constitute the fundamental building blocks of the lignin polymer. The extractive species can be categorised into resins and phenolics. Resins are hydrophobic, soluble in alcohols, ethers, and chloroform. They are composed of terpenes and fatty acids. Some softwood plants can contain up to 15 wt. % oil resins, mostly containing terpenes. Hardwood plants usually contain smaller amounts of wood resins, whose main component is triglycerides of fatty acids. The other category of extractives, the phenolics, is common in evergreen plants, especially in their bark. Phenolics can be further grouped into flavonoids or non-flavonoids. Flavonoids have two aromatic rings and are usually connected to sugar molecules. Non-flavonoids are mostly of tannin-type molecules, which are formed by polymerisation of flavonoids (condensed tannins).

The amounts of structural and non-structural compounds, including cellulose, hemicellulose, lignins, extractives and proteins can be measured with biochemical analysis methods. Analytical techniques required for biochemical analysis methods are usually more complex than elemental analysis methods [9]. These complex techniques involve thermal, chemical and/or enzymatic treatments of biomass samples, which have the potential to modify the original structure of biomass. As a result, open-access literature describing both elemental and biochemical data is not very common. This leads to difficulties in characterising the fuel and modelling its conversion.

Debiagi et al. [5] created a database summarising a large set of measurements from literature which include ultimate, proximate and biochemical analysis for a variety of common biomass samples.

### 8.1.3 General Modelling Approaches

Many models describing the thermochemical conversion of solid fuels have been proposed over the past decades. They can be sorted into three main categories: empirical, semi-empirical and theoretical (phenomenological).

The single-first order reaction (SFOR), the competing 2-step (C2SM), and the distributed activation energy (DAEM) models are the most typical examples of empirical pyrolysis models, widely used in academia and industry for simulation. However, for lacking a theoretical background, they are only able to reproduce observable behaviour through experiments, which reflects on the limited predictive capabilities of these models when operating conditions are different, including fuel composition, heating rate, peak temperature and residence times. Nevertheless, their simplicity and low computation costs can be beneficial given that they are calibrated reliably for the targeted conditions. Computationally demanding tasks like CFD simulations can favourably use correctly calibrated empirical models. A detailed description of this approach can be found in Chap. 19.

On the other end of the spectrum, the theoretical models have a much higher detail and complexity. The most well-known types include chemical percolation and devolatilisation (CPD), the FLASHCHAIN®, and FG-DVC, all of which are network pyrolysis models. They consider an accurate characterisation of the fuels, which directly correlates with a complex reaction mechanism, reflecting the predictive capabilities in terms of devolatilisation rate and distribution of products across a wide range of operating conditions and fuel compositions. Their complexity, however, limits their applicability in complex particle- and reactor-scale CFD simulations. These limitations arise due to the challenges of numerically implementing and interconnecting the information generated by these models, as well as the exponential increase in computational cost arising from such complex models.

For char conversion, the carbon conversion kinetics (CCK) model is a semi-empirical approach to accurately represent char conversion processes in oxy-fuel atmospheres. The CCK model incorporates detailed physical and chemical mechanisms within its submodels to account for char conversion through gasification by carbon dioxide, water vapour, and oxidation through oxygen. These submodels capture the essential chemical reactions and transport phenomena involved in char conversion. The CCK model employs an 8-step mechanism for surface reactions including oxidation and gasification reactions.

The CRECK-S is an example of a semi-empirical model which uses theoretical reaction pathways to describe and determine kinetic parameters for both the pyrolysis and char conversion processes. It is later simplified and calibrated into a reduced number of representative species and reactions that can describe the process in a comprehensive and predictive manner. An important feature of this model is the detailed description of the volatile composition, and the use of the CHEMKIN format, which enables a seamless coupling of char conversion and pyrolysis reactions as well as the coupling of the overall model with detailed gas-phase kinetic mechanisms. On the other hand, the semi-empirical nature of the model presents decreased accuracy on poorly characterised fuels, for atypical fuel feedstocks and for ranges of operating conditions not previously validated. Due to its balanced features in terms of complexity and predictivity, this model has been chosen by several research groups for coupling with CFD simulations.

Figure 8.2 qualitatively shows a selection of models based on their applicability, fidelity, and scope. In general, higher fidelity models result in higher computation costs, which limits their applicability to large-scale CFD simulations. CRECK-S was ultimately chosen as the foundation for *OxySim-129* (described in Chap. 19) due to its ability to seamlessly couple pyrolysis and char conversion within a unified computational framework, addressing a key limitation of the stepwise CPD and CCK models. Unlike CPD and CCK, which operate independently, CRECK-S integrates these processes, enabling predictive analysis of how pyrolysis conditions influence the initial reactivity of char. This capability, combined with a balanced trade-off

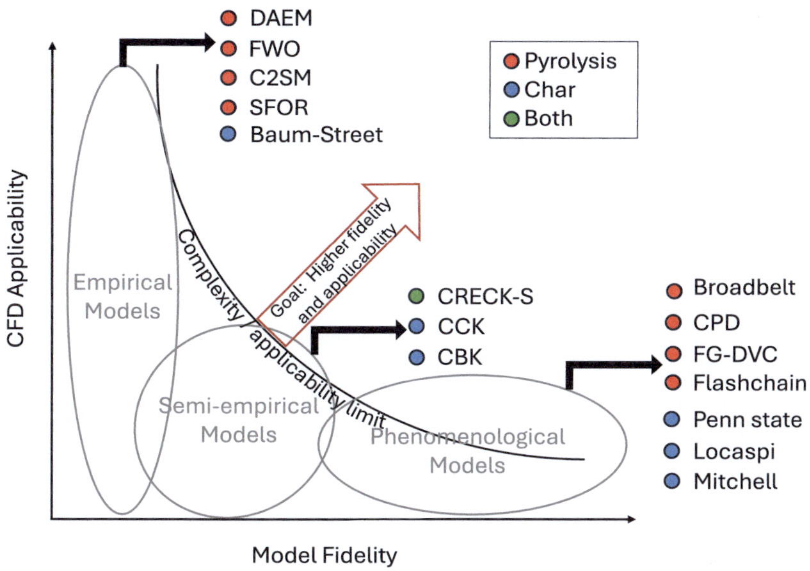


**Fig. 8.2** Qualitative comparison of models in terms of their scope, applicability (depending also on model complexity) and fidelity. Models are the FWO [10], C2SM [11], SFOR [1], Baum-Street [12], CRECK-S [13], CCK [14], CBK [15], Broadbelt [16], CPD [17], FG-DVC [18], FLASHCHAIN® [19], Penn State [20], and Locaspi [21]

between computational cost and model detail, makes CRECK-S particularly suitable for applications like *OxySim-129*. A more detailed review on the theoretical background and CFD applications of various solid fuel conversion models can be found in the study by Hasse et al. [3].

The following sections of this chapter focus on three key modelling approaches used, analysed, and improved for oxy-fuel combustion of coal and biomass: (Bio-)CPD, CCK, and CRECK-S. First, their main features and working principles are described, followed by individual sections for results of each approach, reflecting recent developments. CPD and CCK contain a fundamentally higher degree of fidelity to the phenomenological description of pyrolysis and char conversion, which render them comprehensive and accurate. This characteristic—together with the extensive validation already performed and available in the literature—make them ideal for benchmarking the predictions of reaction rates and characteristic conversion times, which support the developments proposed for CRECK-S model. While all three models are integral to *Oxyflame*, special emphasis is placed on CRECK-S, as it forms the foundation of *OxySim-129*. Moreover, significant efforts within *Oxyflame* have advanced the modelling of biomass fuels, which previously lagged behind coal in terms of maturity and predictive capability. These improvements in biochemical characterisation and kinetic modelling of biomass aim to address this gap, reflecting the commitment to developing a robust and comprehensive framework for biomass conversion in oxy-fuel conditions.

## 8.2 Description of Modelling Approaches

### 8.2.1 Bio-CPD Model for Pyrolysis

The (bio-)chemical percolation devolatilisation (Bio-CPD) model is a phenomenologically based model that attempts to take into account the chemical-physical processes during pyrolysis. In the CPD model, the solid is assumed to be a network of aromatic macromolecules (clusters) that are linked together by various chemical bonds (bridges), characterised by different stabilities. Although the model idea with cross-linked aromatic clusters can only be directly applied to the chemical structure of lignin, the approach is also used for the polymers cellulose and hemicellulose, whereby the monomer building block is used as the cluster unit. Figure 8.3 illustrates the reaction pathways within the (Bio-)CPD model.

The reaction mechanism, originally proposed by Grant et al. [17] for coal, begins with the conversion of labile bridges into an activated intermediate state. These activated bridges are immediately decomposed and can either break into two side chains or be converted into a stable bridge. The side chains formed during the bridge separation are subsequently split off and form light gas. Additional light gas is also formed during the stabilisation reaction. The reaction rates for all steps follow an Arrhenius approach with distributed activation energies depending on the progress of the respective reaction. For coal, these kinetic parameters are postulated to be fuel-independent and the different reaction behaviours can be solely controlled by the structural parameters (coordination number, initial fraction of bridges, fragment and bridge molecular weights). However, for biomass, multiple

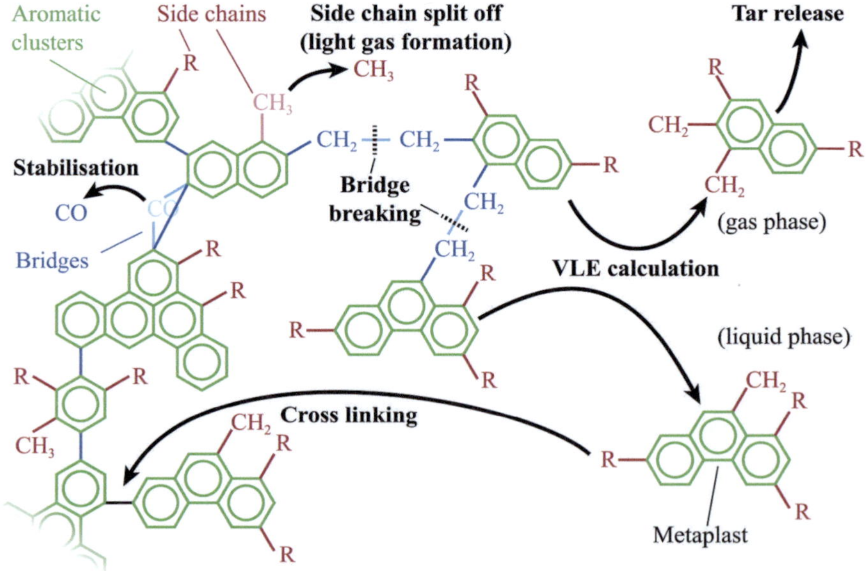


**Fig. 8.3** Reaction pathways in the (Bio-)CPD model

sets for kinetic and structural parameters have been proposed in the literature for each component [22–25].

During the pyrolysis process, more and more bridges in the network are successively decomposed. This can result in finite fragments (mergers of several clusters) that are no longer connected to the infinite lattice. By using the fraction of intact bridges and the number of possible connection points per cluster (coordination number), the probabilities of occurrence for finite fragments with different numbers of clusters are calculated via the percolation theory. Although the finite fragments no longer have a direct connection to the infinite lattice, they first remain in the liquid phase (metaplast) in the particle. To leave the particle, the fragments formed must be in the gaseous phase. The ratio of liquid metaplastic fragments and fragments in the gas phase is determined by a vapour-liquid equilibrium [26]. The position of the equilibrium depends on the respective fragment size as well as the particle temperature and pressure. In the original model [26], all gaseous fragments are removed from the particle in each time step. The adapted version [27] used here, instead, uses a differential balance of the fragments and provides for an outgassing term in which gaseous fragments can leave the particle via a first-order approach. This allows a more flexible coupling with other submodels, e.g. for particle heat-up or gas phase modelling. Instead of changing from the liquid phase to the gas phase, metaplast fragments can also recombine with the infinite lattice in crosslinking reactions.

While the coal-based version is extensively validated, the biomass version is only used in a handful of studies. All have in common that the natural biomass is modelled as a linear superposition without any interactions between the components. Typically, validation is carried out using integrally released pyrolysis products after the process has been completed. This means that the evaluation of reaction rates and kinetic parameters can only be undertaken to a limited extent, as the integral amount of product released essentially depends on the molecular fuel structure. The use of natural biomass instead of extracted components carries the risk that interactions between the components are not correctly reflected or are included as fuel bias in the kinetic parameters. To minimise this risk, the study by Pielsticker and Kneer [28] uses the time-dependent pyrolysis behaviour of extracted structural components to evaluate the predictive accuracy of existing parameter sets and find optimal new parameters. In addition to pure superposition, the newly developed model structure also allows parallel simulation of the three structural components with interactions at different model levels.

### 8.2.2 CCK Model for Char Conversion

The carbon conversion kinetics (CCK) model combined with computational fluid dynamics (CFD) allows for the simulation of pulverised coal fired combustion systems.

The CCK model can also serve to train low computational cost, reduced-order models, enabling the description of particular scenarios. Previous models such as the

CBK/G and CBK/E model were not able to reproduce the influence of an oxy-fuel atmosphere. CBK/E involves a 3-step char oxidation process with oxygen, while CBK/G utilises a 5-step gasification model involving carbon dioxide, water vapour, and hydrogen. Therefore, CBK/E and CBK/G were combined into the CCK model with eight surface reactions. The eight reactions of the CCK model are listed in (8.1) to (8.8). Holland [29] calibrated the CCK model based on the reactions for coal in an oxy-fuel atmosphere.

$$2\mathrm{C} + \mathrm{O_2} \rightarrow \mathrm{C(O)}_\alpha + \mathrm{CO} \tag{8.1}$$

$$\mathrm{C} + \mathrm{O_2} + \mathrm{C(O)}_\alpha \rightarrow \mathrm{C(O)}_\alpha + \mathrm{CO_2} \tag{8.2}$$

$$\mathrm{C(O)}_\alpha \rightarrow \mathrm{CO} \tag{8.3}$$

$$\mathrm{CO_2} + \mathrm{C} \leftrightarrow \mathrm{C(O)}_\beta + \mathrm{CO} \tag{8.4}$$

$$\mathrm{C(O)}_\beta \rightarrow \mathrm{CO} \tag{8.5}$$

$$\mathrm{C} + \mathrm{HO_2} \leftrightarrow \mathrm{C(O)}_\gamma + \mathrm{H_2} \tag{8.6}$$

$$\mathrm{C(O)}_\gamma \rightarrow \mathrm{CO} \tag{8.7}$$

$$\mathrm{C} + 2\mathrm{H_2} \rightarrow \mathrm{CH_4} \tag{8.8}$$

In addition to the eight surface reactions, the CCK model contains several other submodels. The most substantial ones are: the CPD model [17,26] to capture the influence of pyrolysis on the subsequent char conversion, the random pore model [29] to account for pore size evolution during the conversion, the mode of burning model [29] to discriminate between a shrinking core behaviour during char conversion or the development of an excavated char particle, the model of Shurtz [30] that describes potential swelling of particles and a thermal annealing model.

**Thermal Annealing in the CCK Model**

Thermal annealing describes the loss of char reactivity under thermal treatment. Despite the complexity and variety of phenomena involved in thermal annealing, within a first-order approximation, the overall effect has been regarded as a linear loss of active sites. Under this assumption, thermal annealing can be easily included into global kinetic expressions of combustion and gasification by introducing an appropriate deactivation factor, which reduces the value of the pre-exponential factor.

In CCK, the basis of the description of the effect of thermal annealing, is the well-known model proposed by Hurt et al. [15]. In this model, a lognormal distribution of active sites is assumed to be initially present in the coal. The annealing model calculates the ratio of the number of active sites to the initial number of active sites as a function of particle temperature and time of heat treatment $\sum N_i/N_0 = f(T_\mathrm{p}(t))$ and ultimately evaluates an annealing factor $\sum f_{i,\mathrm{A}} = \sum N_i/N_0$ which is a simple multiplier to the char oxidation pre-exponential factor $\mathrm{A_0}$. However, when complex reaction schemes are considered, annealing expressions with the capacity to adjust multiple pre-exponential factor could be potentially more accurate.

**Thermal Annealing for Biomass**

Such a thermal annealing model was proposed for coal by Senneca and Salatino [31]

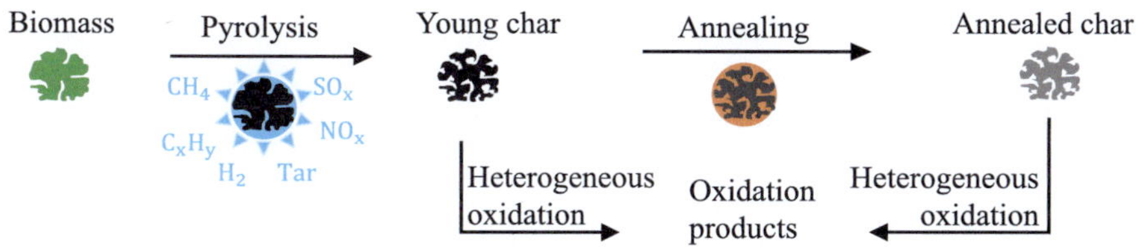


**Fig. 8.4** Mechanism for annealing model [33]

to describe post pyrolysis thermal annealing in parallel with gasification reactions. Application of the model to the reaction with $O_2$ in combination with kinetics of Niksa et al. [32] is also documented by Senneca and Salatino [33]. The model assumes that pyrolysis is complete and a young char $C_Y$ is produced. Along further heat, treatment the young char is progressively transformed into annealed carbon $C_A$, and the reactivity shifts from that of the young char, to that of the fully annealed char (see Fig. 8.4). The annealing coordinate $\xi$ is defined on the basis of the char reactivity. The following (8.9)–(8.11) hold:

$$\mathrm{C}(t) = \mathrm{C_Y} + \mathrm{C_A} = (1 - \xi)\,\mathrm{C_{Y0}} + \xi\,\mathrm{C_{Y0}} \tag{8.9}$$

$$R = r_0\,C_{Y0}\,(1 - \xi) + r_\mathrm{A}\,\mathrm{C_{Y0}}\,\xi \tag{8.10}$$

$$\xi = \frac{R_0 - R}{R_0 - R_\mathrm{A}} \tag{8.11}$$

where $R$ is the reactivity per unit mass at the given time, $R_0$ is the reactivity of the young non-annealed char, and $R_\mathrm{A}$ the reactivity of the fully annealed carbon per unit sample mass. The rate of annealing is assumed to follow a power law of the type:

$$\frac{\mathrm{d}\xi}{\mathrm{d}t} = k_\mathrm{A}\,(1 - \xi)^n \tag{8.12}$$

The kinetic constant $k_\mathrm{A}$, increases with temperature according to an Arrhenius law with activation energy $E_\mathrm{A}$.

This phenomenological approach has been extended to biomass, which is not straightforward due to the co-existence of several components, in particular cellulose, hemicellulose and lignin. An annealing model for biomass should, therefore, take into account the possibility that in a heat-treated sample undevolatised matter from one component can coexist with young or even annealed char originated from other components. Under the simplifying assumption that interactions among components are negligible (which is not always true), the overall reactivity $R$ of a heat-treated biomass sample should in principle be calculated summing up the contributions of all the $k$ components. Each component multiplied by its relative weight fraction $w_k$ leads to

$$R = \sum_{k=1}^{n} \left[ r_{\mathrm{V},k}\,(1 - x_k)\,\mathrm{C}_{v0} + r_{0,k}\,\mathrm{C_{Y0}}\,(1 - \zeta_k) + r_{\mathrm{A},k}\,\mathrm{C_{Y0}}\,\zeta_k) \right] w_k, \tag{8.13}$$

where $x_k$ and $\zeta_k$ define the degree of pyrolysis and of annealing of the $k^{th}$ component throughout heat treatment, while $r_{V,k}$ $r_{0,k}$ and $r_{A,k}$ are the combustion reactivity of undevolatised, young and annealed carbon of the $k^{th}$ component.

First results on how to determine the overall reactivity $R$, based on TGA measurements of walnut shells are presented in the study by Senneca et al. [34]. Within Sect. 8.4, results of the presented novel biomass annealing approach integrated into the CCK model are shown.

### 8.2.3 CRECK-S for Seamless Pyrolysis and Char Conversion

The CRECK-S kinetic model family is designed to describe thermochemical conversion of carbonaceous solid fuels. The successful format and applicability of this approach allow it to describe the decomposition process of coal, biomass and plastics. Due to the usage of CHEMKIN format, kinetic submodels describing various processes such as pyrolysis of different components, heterogeneous reactions (i.e. char conversion) and detailed gas-phase reactions can be combined seamlessly with each other depending on their relevance.

**CRECK-S-C Model for Coal**

Coal is composed of aromatic clusters with various bridges, side chains, and functional groups. Its composition and structure vary significantly across different types of coal. Low-rank coals have higher oxygen content, longer side chains, and smaller aromatic clusters. As coal matures (i.e. higher rank coals), its oxygen content decreases, side chains shorten, and aromatic clusters grow, eventually resembling graphite. The pyrolysis and devolatilisation processes mimic this ageing, leading to increased carbon content and a higher apparent rank of the char. The characterisation of coal in terms of its initial structure of atoms, bonds, and chemical structure is necessary to describe its pyrolysis behaviour. Sommariva et al. [4] introduced a characterisation procedure based on elemental analysis (C, H, O, and ashes) of coals. Ashes, which vary with coal grade and origin, are inert and remain in the residue. As a first step for the characterisation, elemental analysis is adjusted and normalised to the C, H, and O content on a dry, ash-free basis. Typically, the carbon content of coals exceeds 60–65 wt. %, while hydrogen content is below 7 wt. %. As carbon content increases, the coal rank rises from lignite to bituminous coal, and then to anthracite.

Figure 8.5 shows a van Krevelen diagram for a variety of coal samples as well as the reference species in the CRECK-S model. All the coal samples fall into a triangle with vertices representing highly graphitised pure carbon anthracite CHAR-C, high-oxygen lignite COAL3, and an oxygen-free, hydrogen-rich coal anthracite COAL1. A third reference coal COAL2, positioned near many bituminous coals, is included in the middle of this triangle. These reference coals are depicted by three lumped monomer structures reflecting their elemental C/H/O composition. Figure 8.6 illustrates these monomer structures: COAL1 ($C_{12}H_{11}$), COAL2 ($C_{14}H_{10}O$), and COAL3 ($C_{12}H_{12}O_5$). The subsequent multi-step kinetic model describes the pyrolysis of these reference coals.

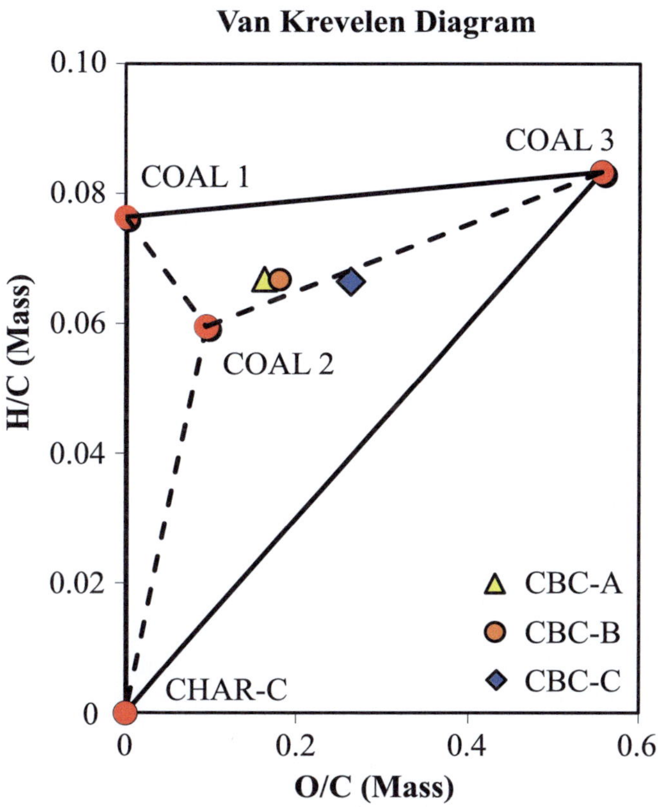


**Fig. 8.5** Van Krevelen diagram for the reference species in CRECK-S and three real exemplary coal samples (CBC-A, CBC-B, CBC-C). (Based on the publication by Debiagi et al. [35])

During pyrolysis, coal initially creates a metaplastic phase which later releases gases and tars with varying mechanisms at low and high temperatures. At lower temperatures, reference coals create gas and tars that are entrapped in the condensed

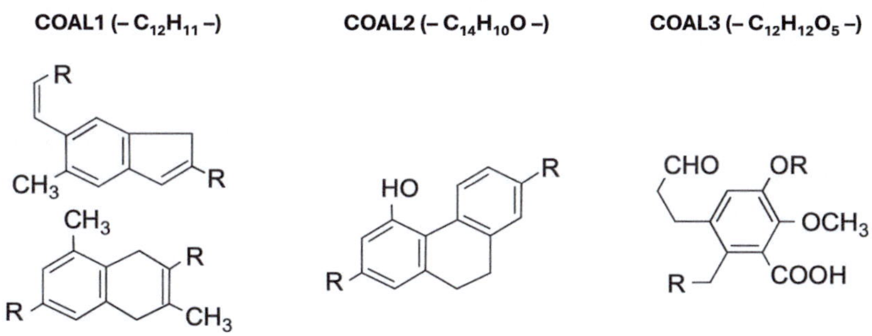


**Fig. 8.6** Structure of reference coals in CRECK-S model. (Based on the publication by Sommariva et al. [4])

phase. These metaplastic tars can either be released to the gas phase or interact with char in cross-linking and reticulation reactions. Higher temperatures result in more direct release of products into the gas phase and more aromatic chars are created in the solid.

The van Krevelen plot in Fig. 8.5 is divided into three triangles by connecting these reference coals. Each coal is treated as a linear combination of the three closest reference coals, with its devolatilisation modelled as a weighted combination of the pyrolysis of these reference coals.

Previously made available as standalone submodels, the approach for sulphur [36] and nitrogen [37] release from coal was fully integrated to the main C/H/O model seamlessly [38]. The sulphur model represents the sulphur-containing species in three main classes: inorganic sulphur, pyritic sulphur, and sulphate sulphur. Analysing a range of coal samples from various experimental measurements, the ratios of these classes of sulphur species are found to be correlated to the carbon fraction of coal samples. Accordingly, correlations are created based on the elemental sulphur and carbon fractions of coal. This procedure is included in the characterisation process previously described. For calculating the sulphur release, the three sulphur-containing classes are represented with reference compounds, and each of their decomposition is modelled with a set of reactions including high and low-temperature decomposition paths. Considering nitrogen species, coals contain three common classes: quaternary, pyridinic and pyrolytic compounds. Unlike sulphur, these compounds are not correlated to the coal rank (or another physical property). The nitrogen-containing volatiles and total volatiles from coal are shown to have similar behaviours. As a result, the nitrogen submodel distributes the total nitrogen content (as obtained from elemental analysis) to another set of reference species that are closely related to the original mechanism: COAL1-N, COAL2-N, COAL3-N, and CHAR-N. These reference species share the previously described pyrolysis characteristics: High and low-temperature reaction paths exist where either volatiles are directly released or more metaplastic species are created respectively.

**CRECK-S-B: Lignocellulosic Biomass Characterisation**

The biochemical molecules including cellulose, hemicellulose and lignin are used as the base for characterising biomass samples with the CRECK-S-B model. The biomass samples are represented as a mixture of reference species representing these molecules. The model also proposes a characterisation method based on elemental analysis of the biomass to obtain the composition of reference species for when the biochemical composition measurements are unavailable [39]. This method is later extended [5,8,40,41] to include a wider range of measurements and additional reference species representing the extractives. In the overall model, the cellulose is represented by a corresponding reference species CELL, resembling the molecular formula of its monomer glucose (with one water molecule extracted, due to polymerisation). The highly variable behaviour of hemicellulose among biomasses is accounted by the reference species XYHW, XYGR, and GMSW characterising hemicellulose structures typically present in hardwood, grass and cereals, and softwood types of biomass, respectively. Lignins present a larger variability and hence,

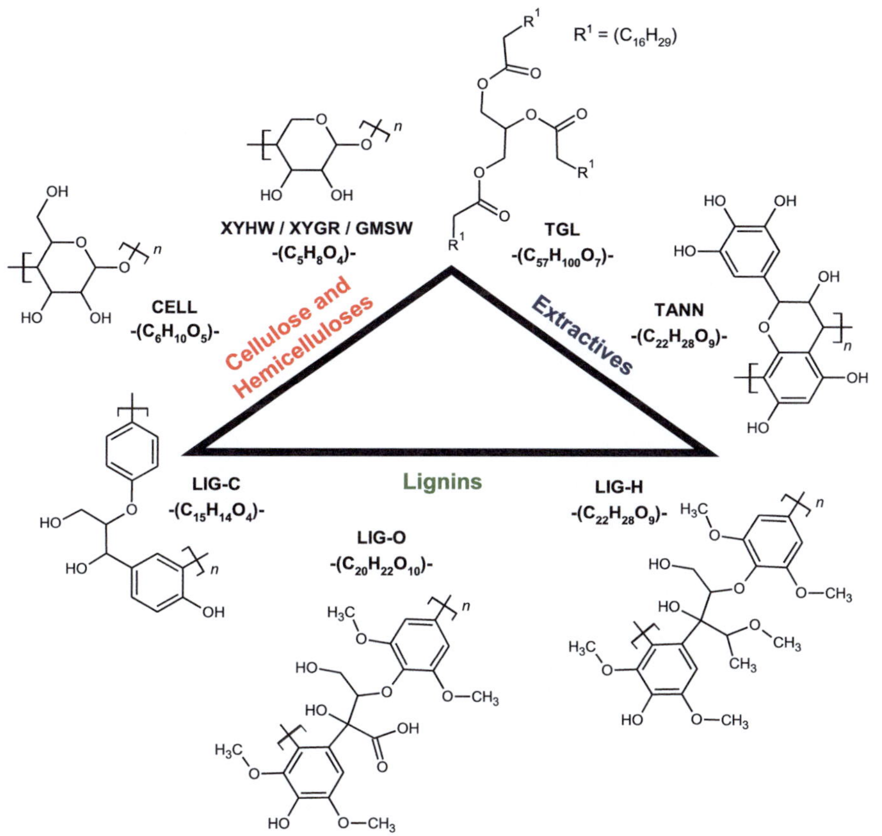


**Fig. 8.7** Structure of reference species in CRECK-S-B model. (Based on the publication by Berkel et al. [42])

are represented by three different structures based on whether they are rich in carbon (LIGC), hydrogen (LIGH), or oxygen (LIGO) [7]. Different mixtures of these three species are able to characterise a wide range of natural lignins. Finally, two lumped extractive species represent hydrophobic (TGL) extractives and hydrophilic (TANN) ones. As a result, biomass is represented as a combination of these seven reference species, whose molecular structures are shown in Fig. 8.7.

For characterising biomass samples with seven reference species based on their elemental C/H/O content, the number of degrees of freedom must be reduced. To this aim, three reference mixtures, RM-1, RM-2, and RM-3, are described as combinations of these seven reference species. Reference mixture RM-1 represents holocellulose, a mixture of cellulose and hemicellulose. Mixture RM-2 is a mixture of hydrophobic extractives and lignins. Lastly, RM-3 represents the mixture of hydrophilic extractives and lignins. The exact ratios of the reference mixtures can be adjusted based on the actual biomass type. The van Krevelen diagram for the reference species, reference mixtures and a collection of biomass samples is shown in Fig. 8.8.

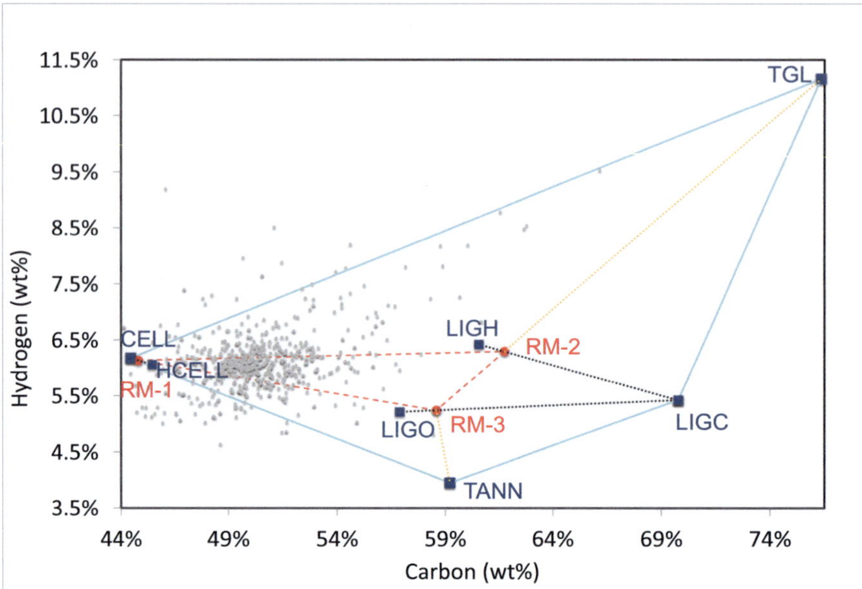


**Fig. 8.8** Van Krevelen diagram for CRECK-S-B reference species (blue squares), reference mixtures (red circles), and several real biomass examples (grey dots) (based on the publication by Debiagi et al. [5])

In addition, the model accounts for the nitrogen content in biomasses in the form of proteins. A similar approach to lignin is employed to characterise proteins, using a combination of three reference components, PROT-C, PROT-H, and PROT-O, which are rich in carbon, hydrogen and oxygen, respectively. Different combinations allow to encompass the large variability of protein composition. In particular, the decomposition of these compounds leads to the formation of N-containing tars and $NO_x$ precursors, supporting the prediction of fuel-$NO_x$ formation from biomass decomposition. This characterisation is done separately from the other biomass components, as this step accounts only for the nitrogen content, and the resulting mixture of PROT species is then subtracted from the remaining biomass content to be characterised.

**CRECK-S-B: Pyrolysis Kinetics**

Similar to coal pyrolysis, biomass pyrolysis is represented as the linear combination of pyrolysis products of reference species. Each reference species has its own decomposition pathway. Several competing reactions with different activation energies allow for representing distinct reaction paths at low and high temperatures. The reactions produce volatiles (gas and tar) as well as char and metaplastics, which are a range of oxygen/hydrogen-containing groups embedded in the char matrix. The subsequent reactions of metaplastic species can also release varying amounts of volatiles and chars. The individual reaction mechanisms for the reference species are summarised next.

**Cellulose**

The cellulose (CELL) pyrolysis in the CRECK model is based on the findings from the study by Shafizadeh and Fu [43] and Bradbury et al. [44]. Cellulose first forms an intermediate species, active cellulose, representing cellulose with a smaller degree of polymerisation. This initial depolymerisation has an apparent activation energy of 47 kcal $mol^{-1}$. The active cellulose then decomposes through two competitive reactions: a tar release reaction and a decomposition reaction. The tars in the former reaction are lumped into a levoglucosan (LVG) species since this is the most abundant sugar. The latter reaction prevails at high temperatures (above 750 K) and forms char and permanent gases. The products of the lumped reactions include hydroxyl-acetaldehyde, glyoxal, acetone, hydroxy-acetone, furfural, 5-hydroxymethyl-furfural, formaldehyde and formic acid as well as $CO_2$ and CO. The lumped product distributions and stoichiometries of reactions are based on previous mechanistic [16,45–47] and theoretical/experimental works [45,48,49]. The mechanism also includes a side charring reaction directly from the CELL, prevailing at lower temperatures.

**Hemicellulose**

The hemicellulose pyrolysis (XYHW or GMSW) follows a similar decomposition behaviour to cellulose [50–53]. First, two intermediate species, HCE1 and HCE2 are formed. These species represent the partially depolymerised hemicellulose for hardwood or softwood. In general, volatile species released from hemicellulose include acetic acid, hydroxyacetaldehyde, hydroxypropanone, formic acid and furfuryl alcohol along with permanent gases. The hardwood hemicellulose (XYHW) has a higher reactivity, yields a higher amount of residual char, and forms more acetic acid in the volatiles. On the other hand, the softwood hemicellulose (GMSW) releases higher amounts of volatiles.

**Lignin**

The lignin pyrolysis occurs in a wide range of temperatures and follows a complex decomposition behaviour, releasing phenolic compounds. The detailed lignin kinetics by Faravelli et al. [7] is simplified in the CRECK model presented here. The lumped volatiles in the mechanism include phenol, anisole (metoxy-benzene), 2,6-dimethoxy-phenol, 4-(3-hydroxy-1-propenyl)phenol, and 3-(4-hydroxy-3, 5-dimethoxyphenyl)-acrylaldehyde. Additionally, one lumped species represents heavy molecular weight lignins, with the molecular formula $C_{24}H_{28}O_4$.

**Extractives**

The hydrophilic condensed tannin (TANN) decomposition is defined by a two-step mechanism in the CRECK model. The decomposition step occurs around $300 - 400$ °C, releasing phenol and catechol as volatiles along with a polymeric intermediate lumped species ITANN. The lumped intermediate polymer undergoes further cross-linking reactions at higher temperatures. This creates char and some permanent gases.

The mechanism for the resins and triglycerides (TGL) is supported by soybean and corn oil TGA. The TGL releases acrolein and individual fatty acid molecules into the gas phase. The pyrolysis happens in a single step, without significant solid residues.

**Proteins**

The primary decomposition reaction of the three reference protein species is assumed as a first degradation and/or depolymerisation step with the release, together with gas species, of a couple of intermediate proteins with a lower molecular weight (PROTCC: $C_{90}H_{70}O_{25}N_{10}$ and PROTOH: $C_{45}H_{70}O_{25}N_{10}$). Successive pyrolysis of these intermediate protein species forms nitrogen-containing tar components (TARN), together with other gas and tar species. The species TARN is assumed as a fixed combination of pyrrole ($C_4H_5N$), pyridine ($C_5H_4ON$), and diketopiperazine (DKP: $C_4H_6O_2N_2$). The details of the protein pyrolysis submechanism are reported in the work by Debiagi et al. [54].

## 8.3 Bio-CPD Model Results

### 8.3.1 Analysis and Optimisation of Parameters for Biomass

Within the literature, four different sets [22–25] for kinetic and structural parameters have been proposed based on different experimental data for the three biomass components cellulose, hemicellulose and lignin. However, the parameters have mostly not been derived from extracted single components but are the result of fitting the approach of three individually reacting species to experiments of natural biomass neglecting any interactions between the components. Consequently, the study by Pielsticker and Kneer [28] uses the detailed experimental data from the extracted and separately pyrolysed samples in the fluidised bed reactor as reference data [55]. Figure 8.9 shows extracts of the analysis of time-dependent volatile release rates [28] and integrally released total volatile yields [56].

Remarkably, all parameter sets show—with some exceptions—only a slight deviation between model prediction and experimental data when determining the integrally released amount of pyrolysis. However, the same parameter sets show strong differences both among themselves and in comparison with the experimental data with regard to the predicted rate of volatiles released. For cellulose, the overall predictivity is the best compared to the other components. The reason for this is probably the comparatively simple molecular structure, which leads to very similar reaction behaviour under different conditions. A particularly good agreement between model and experimental data was achieved for the data set of Vizzini et al. [23], where pure cellulose was also pyrolysed in a fluidised bed reactor. Other parameters tend to overpredict the volatile release rate at low temperatures. For hemicellulose (not shown in Fig. 8.9), also one of the already existing parameter sets (Sheng and Azevedo [22]) could be identified to satisfyingly predict the volatile release rates [28]. By contrast, for lignin, all of the available parameter sets tend to strongly overpredict the rates

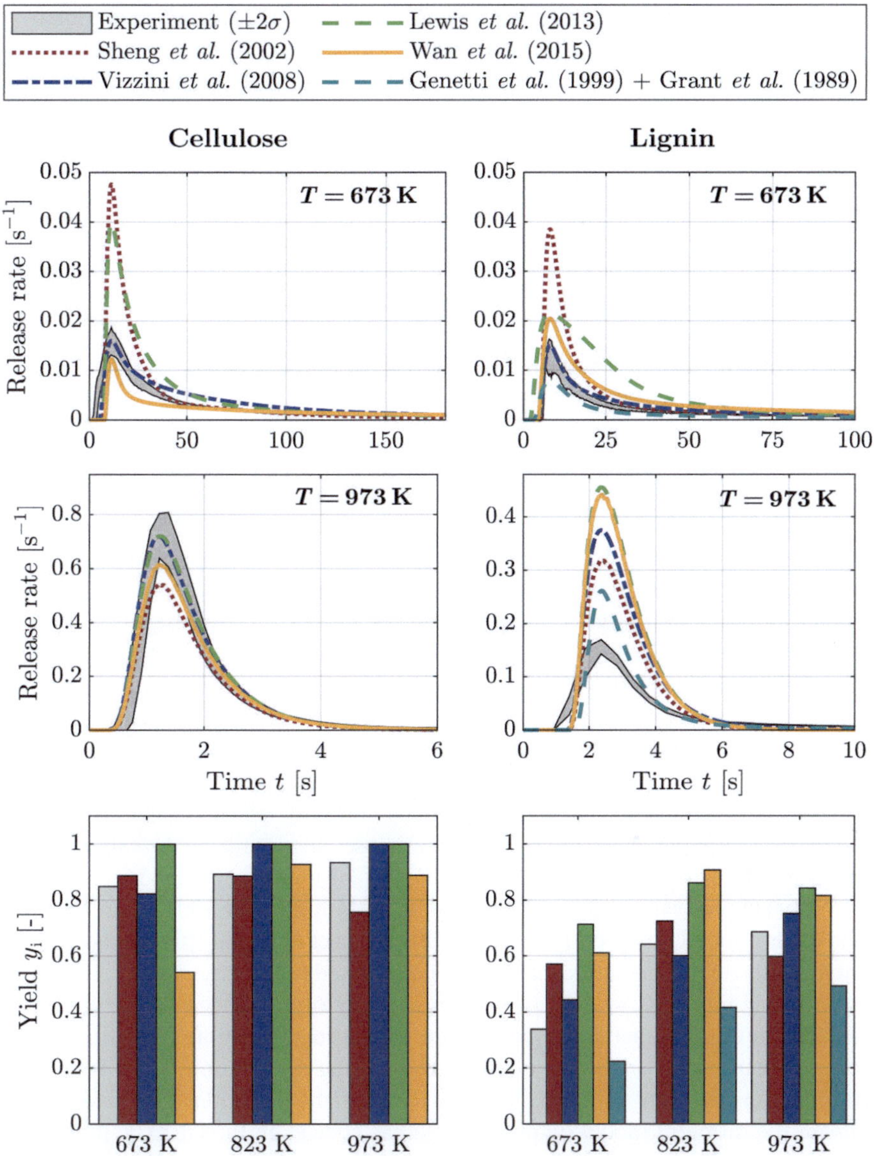


**Fig. 8.9** Predictions of volatile release rates (reprinted from Pielsticker and Kneer [28], Copyright (2024), with permission from Elsevier and minor editing) and integrally released volatile yields (reprinted and slightly adapted from Pielsticker et al. [56], licensed under CC BY 4.0) using different kinetic and structural parameter sets from the literature in comparison with experimental data from the fluidised bed reactor [55]

although the final yields are mostly correctly predicted. This shows the problem of using only integral measurement data for the calibration of kinetic parameters. The alternatively suggested use of the Genetti correlation [57] to estimate the structural parameters in combination with the coal-based kinetics results in a significantly better prediction of the volatile release rates. However, the predicted total yields are 28–35 % lower than the experimental ones. The large variety of different bond types in the lignin molecule leads to the fact that only approximately 50 % of the mass is released during the primary peak and the other part transfers into the gas phase through very slow decomposition reactions [55].

### 8.3.2 Importance of Secondary Gas-Phase Reactions in Experimental Setups

To derive kinetic parameters for primary pyrolysis reactions, multiple reactor setups are available as presented in Sect. 2.2. The setups differ in particle heating rate, maximum reachable temperature, and residence times (see Table 2.1). The gas residence of the volatile products formed during the primary pyrolysis reaction influences the decomposition into secondary pyrolysis products due to gas-phase reactions. From the listed setups, especially the fluidised bed reactor (FBR) described in Sect. 2.2.2 has high gas residence times under hot conditions favouring gas-phase reactions. Thus, this section evaluates the role of secondary gas phase reaction in the FBR setup and highlights the consequences for the determination of kinetic parameters and model development of primary pyrolysis processes described in Sects. 8.2.1 and 8.2.3.

As shown in Sect. 8.3.1, the rates and yields of totally released volatiles from the three biomass components cellulose, hemicellulose, and lignin can be captured with minor limitations by both the CRECK-S-B and Bio-CPD models [56]. However, both primary pyrolysis models predict significantly higher tar yields than observed in the experimental study performed by Pielsticker et al. [55]. This can be seen by comparing the experimental results with a deactivated secondary gas-phase reaction mechanism in Fig. 8.10, where yields of cellulose pyrolysis in the FBR are shown for three different bed temperatures [56]. Figure 8.10 further illustrates the effect of different gas-phase reaction models on the light gas to tar distribution. The division into the lumped species light gas and tar is made based on the molecular weight to directly compare the different model outputs (single gas species in the CRECK model and fragments of different sizes in the Bio-CPD model). The different model outputs also limit the possibility of coupling the respective primary pyrolysis model with different secondary gas-phase models. While the CRECK-S-B can be directly connected to the detailed CRECK-G-2003 [58] gas-phase model, the Bio-CPD model allows only coupling with less complex single first-order tar cracking models, e.g. proposed by Fagbemi et al. [59].

While the simulations with the active CRECK-G-2003 gas-phase mechanism only have a significant effect at the highest investigated temperature, the simple tar cracking mechanism reveals a decomposition effect at all temperatures. However,

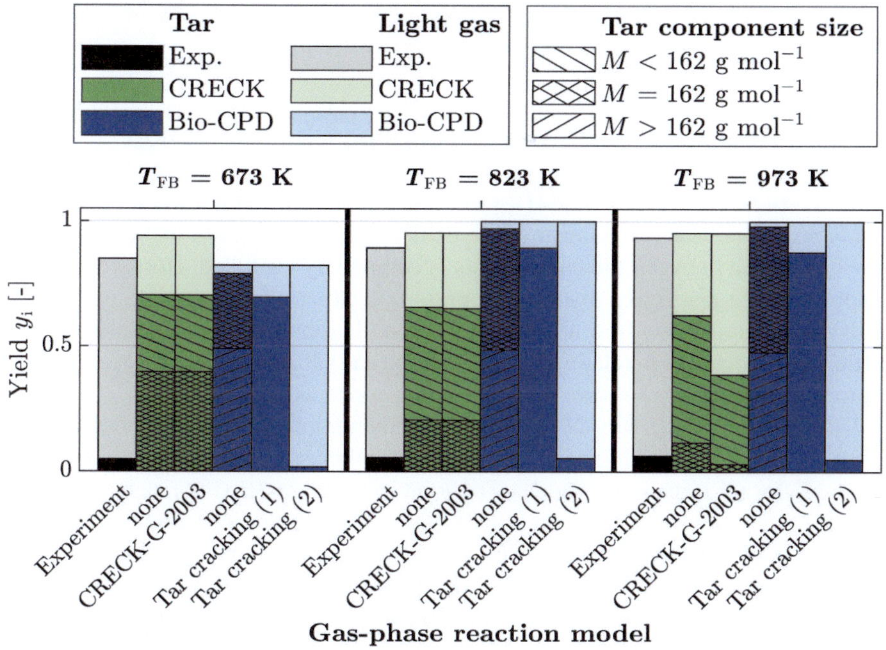


**Fig. 8.10** Comparison of experimental light gas and tar yields obtained from cellulose pyrolysis in the FBR at three bed temperatures with model predictions using different secondary gas-phase reaction models (CRECK-G-2003 [58], first-order tar cracking model with either the parameter set from (1) Fagbemi [59] or (2) Pielsticker [28]). The hatched areas characterise the subdivision of the tar fraction into different molecular sizes. (Reprinted from Pielsticker et al. [56], licensed under CC BY 4.0)

the predicted tar fraction is still much higher than the experimentally observed one. To match the experimental results, kinetic parameters proposed by Pielsticker and Kneer [28] are required. While the activation energies in the two sets are almost identical, the pre-exponential factor is more than two orders of magnitude higher. The contact between gas and hot solid particles in the bed and the specific linear structure of the cellulose molecule are suspected as possible reasons [28,60].

However, the subdivision of the tar fraction by the molecular weight (marked by hatched areas) reveals further differences in the pyrolysis models and identifies development potential: In the CRECK model, increasing temperature and progressive reaction in the gas phase lead to a decrease of the cellulose monomer ($C_6H_{10}O_5$) fraction at the expense of smaller tar molecules and light gases. By contrast to the CRECK model, the Bio-CPD model in its current implementation does not allow the primary release of fragments that are smaller than the monomer unit. The implementation of an additional fragment decomposition mechanism inside the particle resulting in smaller subfragments ($M < 162$ g mol$^{-1}$) could lead to an improvement of the model prediction. As already shown in studies by Ranzi et al. [61] and Hameed et al. [62], it is also advisable to consider the effect of catalytically active minerals in the model. In addition to model adaptation, improvements can also be targeted in the experimental analysis, for instance, a systematic condensation.

### 8.3.3 Linear Superposition and Coupling with Interaction

Previous studies [22,24,25,63,64] on the application of the Bio-CPD model to natural biomass are based on a linear superposition of the three individual components. Any interaction between the components is completely excluded. However, since a coupling of the respective reaction progresses is expected in natural biomass, the following section shows how a simultaneous simulation of all three components in a particle (mix) behaves in relation to the separated and subsequent linear superposition approach. Figure 8.11 a) shows both modelling concepts.

The structure of the Bio-CPD model implementation allows interactions in different submodels: In the mixed version, energy and mass balances are solved simultaneously for all three components. The joint particle model (PM) ensures that all reactions are analysed at the same time with the same particle temperature. Further on, all formed fragments are fed to a joint vapour-liquid equilibrium. In the other submodels—bridge breaking (BBM), percolation theory (PT) and mass release (RM)—no interactions between the components are considered. For simplicity, an artificial biomass with equivalent proportions of the three components and a mean particle diameter $d_P = 110\ \mu m$ reacting in the fluidised bed reactor at a bed temperature $T_b = 973$ K is considered for the comparison here. At this temperature, the time scales of heat-up and reaction are in the same order and thus, the heat-up history influences the pyrolysis process significantly.

Figure 8.11 b)–d) show the differences between linear superposition and simultaneous simulation (mix) with regard to the particle temperature, the volatile release rate at the particle level and the observed volatile release rate in the FTIR gas cell. Figure 8.11 b) shows that the history of the particle temperature $T_p$ of the mix differs significantly from the histories of the individual components. Responsible for this is, on the one hand, the different effective heat capacity $c_{p,eff}$ of the particle (calculated based on the correlation derived by Merrick [65]) and on the other hand the different mass loss $\frac{dm_p}{dt}$, which in the energy balance influences both the remaining mass $m_p$ to be heated and the enthalpy flow $\dot{H}$ by the (endothermic) chemical reaction as illustrated by (8.14):

$$m_p \underbrace{\left[\sum_i^3 w_i\, c_{p,i}\right]}_{=c_{p,eff}} \frac{dT_p}{dt} = \dot{Q}_{con} + \dot{Q}_{rad} + \underbrace{\left[\sum_i^3 \frac{dm_i}{dt} h_i\right]}_{=\dot{H}} \tag{8.14}$$

Convective and radiative heat flows are equally modelled in both cases but may differ due to the different temperature histories. This means that the mixed particle heats up faster than the pure components in the relevant temperature range of the incipient reaction for hemicellulose and cellulose, as the endothermic effect of the reaction

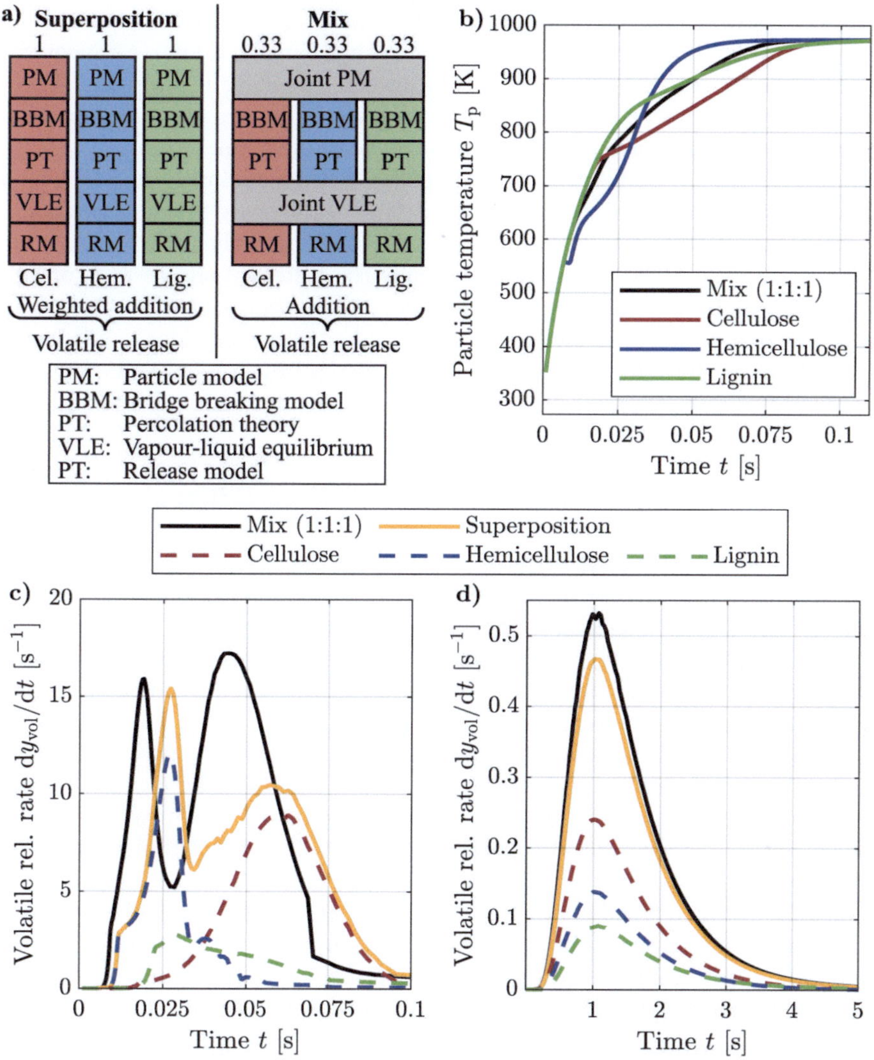


**Fig. 8.11** Differences between the linear superposition and the mixed approaches **a)** modelling concept with interacting submodels **b)** particle temperature for reaction in the fluidised bed reactor **c)** obtained volatile release rate at the particle **d)** observed volatile release rate in the FTIR gas cell

is lower. Consequently, the peaks of the mass release for those two components are shifted to earlier times in the mixed approach as shown in Fig. 8.11 c). As the overall contribution of lignin is small, the effect of a lower temperature in the mixed case and thus reduced reactivity is not that significant. However, the allowed interaction with the other components nevertheless leads to an increase of volatiles from $y_{\mathrm{Lig}} = 0.139$ to $y_{\mathrm{Lig}} = 0.175$. This symbiosis effect is highest for hemicellulose (+26.5 %) but not present for cellulose as the entire material is fully decomposed in both cases here.

Figure 8.11 d) shows the effects of both modelling approaches on the release rate of the volatiles obtained in the FTIR gas cell. The increased release rates of light gas and tar in the simultaneous simulation lead to an approximately 14 % higher maximum release rate compared to the superposition of the individual components. Also, the integrally released volatile yield increases from $y_{\mathrm{vol}} = 0.665$ to $y_{\mathrm{vol}} = 0.752$ (+13.2 %) by using the mixed approach instead of the sole superposition.

### 8.3.4 Prediction of Kinetics and Yields of Natural Biomass

After showing the principal effects of linear superposition and interactive mixtures in the previous section, both model approaches are now applied to natural biomass. First, the time-dependent reactive behaviour is analysed for beech wood pyrolysed in the fluidised bed reactor [66]. Afterwards, total volatile yields and (Bio-)CPD model predictions are compared for various fuels including different raw biomass, torrefied beech wood and fossil fuels of different ranks.

Figure 8.12 shows the total volatile release rate obtained for pyrolysis experiments at two different bed temperatures in comparison with the Bio-CPD model predictions using either the linear superposition approach or the mixed simulation with interactions amongst the three basic components. The distribution of cellulose, hemicellulose, and lignin is chosen based on the study by Jahirul et al. [67] to 45, 20, and 25 %. The remaining fraction is attributed to non-reactives and extractives.

Two main aspects are visible in the plots: By comparing the two different model approaches, significant differences are only present at the higher temperature. At lower temperatures, the time scale of the pyrolysis reaction is much higher than the one of particle heat-up. Thus, the interaction effect in the particle model becomes negligible. This, however, indicates that the coupling of mass and energy balance in the joint particle model is of higher importance than the VLE coupling. By comparing the model predictions with the experimental results, a satisfying agreement is only achieved for the higher temperature. At lower temperatures, the Bio-CPD strongly overpredicts the volatile release rate and yield. Similar behaviour is also observed by Couhert et al. [68], where differences up to 60 % between the superposition and the natural biomass experiment are reported. The authors name interactions between the components and catalytically active minerals as the most relevant factors. Especially the extraction process with solvents yields an enrichment of catalytically active sodium carbonate $Na_2CO_3$ and calcium carbonate $CaCO_3$, which favours a stronger conversion in the single components than in the natural biomass [69]. The strong catalytic effect of $Na_2CO_3$ on the reactivity of different conversion processes

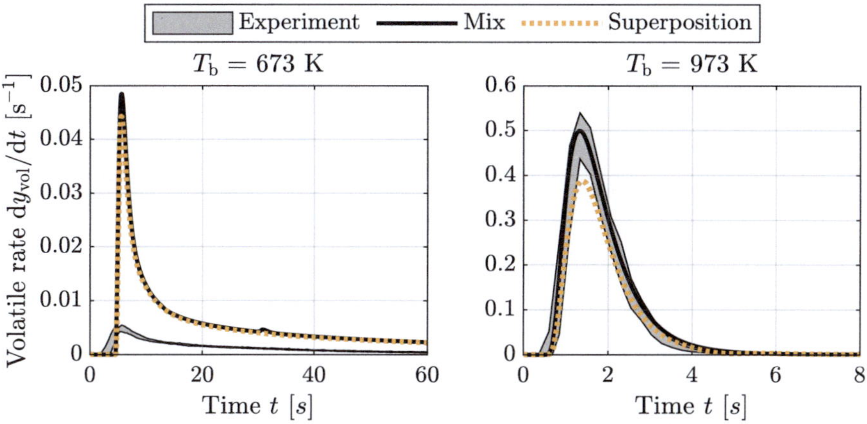


**Fig. 8.12** Experimentally obtained total volatile release rates from beech wood pyrolysis in the fluidised bed reactor [66] in comparison with predictions from the Bio-CPD model using the mixed and linear superposition approaches

is also outlined in Sect. 4.4. Furthermore, the co-existence of the three components in natural biomass can lead to interlinking reactions taking place between the individual components at low temperatures, thereby changing the global reaction behaviour. This feedback loop from the reaction process to the structural parameters of the CPD model has not yet been adequately modelled. A first attempt was made in a study by Wang et al. [70], where the cross-linking mechanism was adopted for low temperatures. Likewise, an adjustment of the vapour pressure correlations in the VLE for the fragments of the biomass structure components as well as partially exothermic reaction enthalpies in this temperature range should be reconsidered.

Figure 8.13 shows the predictivity of the Bio-CPD model for biomass with variuos fractions of cellulose, hemicellulose and lignin. As the previously mentioned limited accuracy in the low temperature range is observable for all biomass, the fuel effect is only evaluated for the higher temperature $T_b = 973$ K here.

Using the literature data from Jahirul et al. [67], the trends for the three raw biomass can be well reproduced. Thereby, the mixed approach yields approximately 13 % higher yields and thus comes closer to experimentally obtained yields. The Bio-CPD model is also well-suited to predict the pyrolysis yields of the pretreated biomass samples via torrefaction [71]. Since the determination of the structural components for the torrefied biomass is difficult, the initial mass fractions are adopted instead: Based on the composition of the raw beech wood, parts of the structural components are decomposed and the undecomposed components remain in larger relative proportions. At a torrefaction temperature of 553 K, a complete hemicellulose decomposition and a 50 % reduction of the cellulose component is assumed. At the higher torrefaction temperature of 573 K, cellulose and hemicellulose are both completely decomposed. As torrefied biomass cis closer to fossil fuels, the predictivity using the coal-based CPD version is also tested. The predicted yields are only

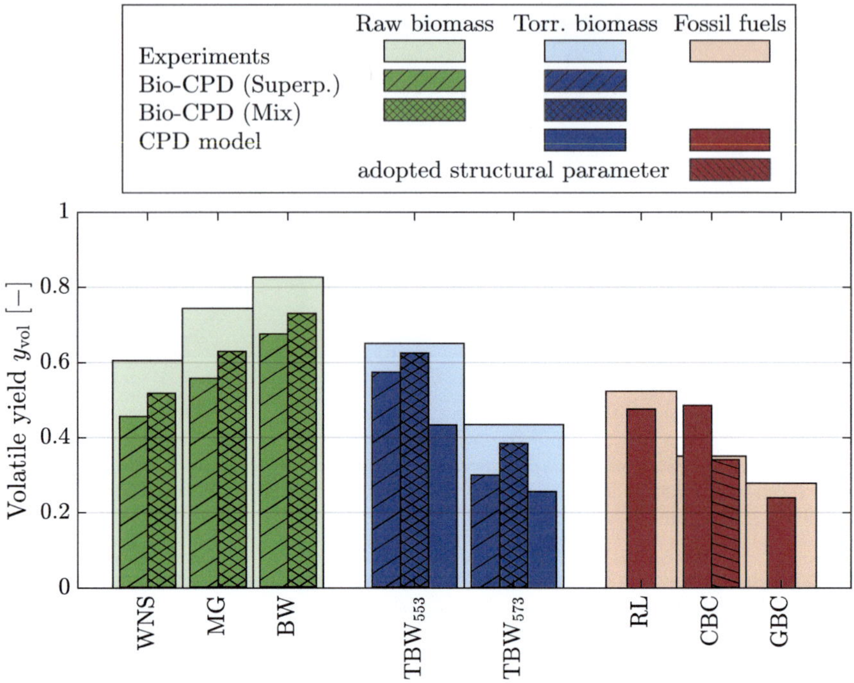


**Fig. 8.13** Experimental volatile yields (see details in Fig. 2.12) obtained in the fluidised bed reactor at $T_b = 973$ K in comparison with (Bio-)CPD model predictions using different approaches (superpositioned and the mixed Bio-CPD model as well as the coal-based CPD model

slightly lower than those using the superposition method, but significantly lower than the experimental data or the yields from the mixed approach.

As torrefied biomass may serve as an alternative to fossil fuels, the performance of the adapted CPD version [27] is evaluated for different fuel types. While a very good performance—also for the kinetics in the low-temperature range [27]—is achieved for the Rhenish lignite with the Genetti approximation [57], the Colombian bituminous coal required the adjustment of the structural parameters.

Generally, it is recognisable that for all fuels, the experimentally determined volatile fractions are systematically larger than the predicted ones. This systematic deviation can result from a measurement error in the reactor and analysis system (e.g. FTIR calibration) or an inadequate model assumption. Overall, it can be seen that the integral pyrolysis products can be predicted much more consistently with the Bio-CPD model than the time-dependent release behaviour. A possible reason for this could be the validation/calibration data of the Bio-CPD model, which usually only takes integral products into account.

## 8.4 CCK Model Results

The original CCK model code developed by Holland [29] was designed and calibrated for oxy-fuel coal combustion. To extend the CCK model for the prediction of pulverised biomass combustion, mainly three extensions and modifications have been made to the original CCK model:

1. The standard CPD model has been replaced by the Bio-CPD model developed by Lewis and Fletcher [26] and Fletcher et al. [24].
2. The original thermal annealing model was substituted with the thermal annealing model for biomass as proposed by Senneca et al. [34], described in more detail in Sect. 8.2.2.
3. Following the suggestions by Holland [29], the kinetic parameters of the surface reactions listed in (8.1) to (8.8) were calibrated based on the measurement of particle temperature, size and conversion degree in different oxy-fuel atmospheres [72]. These measurements are described in Sect. 5.

For the integration of biomass pyrolysis in the CCK model, the original CPD model had to be extended to consider the three main components of biomass (hemicellulose, cellulose and lignin). The respective kinetics of the biomass components have been taken from the study by Lewis and Fletcher [24]. For further information on the Bio-CPD model, see Sect. 8.2.1.

The parameters used for the novel biomass thermal annealing model in the CCK model are summarised in Table 8.1, and are the parameters needed for solving the annealing model in (8.9) to (8.13). The kinetic parameters are derived by using a nonlinear multi fitting procedure described below. The main difference between the original thermal annealing model and the extended annealing model for lignocellulosic biomass is the temperature dependent boundary conditions. Holland [29] classify annealing effects into two regimes: below 1500 K and above 1500 K. In contrast, Senneca et al. [34] defined three zones of annealing effects: The first zone encompasses temperatures up to 1100 K, the second spans the range from 1100 to 1600 K, and the third applies to temperatures exceeding 1600 K. The extended annealing model for biomass demonstrates for temperatures below 1100 K negligible annealing effects and much more noticeable effects for temperatures above 1600 K.

For the surface reactions (8.1)–(8.8), the kinetic constants were derived based on the optical measurements for walnut shells presented in Sect. 5. The procedure to derive the constants is the same as used by Holland and Fletcher [73] for the calibration of the CCK model for oxy-fuel coal combustion. It is important to note that Holland and Fletcher [73] has reduced the number of free variables (reaction rate constants $k$) from eight to just two, namely $k_3$ and $k_7$ from reactions (8.3) and (8.7). All other reactions rate constants depend on these two reaction rate constants. Calibration of these constants is carried out following the integration of Bio-CPD and thermal annealing model into the CCK model. This calibration was carried out based on a nonlinear multivariable fitting using the MATLAB® function `fmincon`.

**Table 8.1** Parameters of the thermal annealing model given in (8.9) to (8.12) taken from the study by Senneca et al. [34] for walnut shells. The ratio $R/R_0$ represents the ratio of the reactivity at the given time to the reactivity of the young non-annealed char

| | | | |
|---|---|---|---|
| Activation energy | $E_{av}$ | [kJ mol$^{-1}$] | −288.7 |
| Pre-exponential factor | $A_v$ | [s$^{-1}$] | $9.76 \times 10^{-3}$ |
| Activation energy | $E_{a0}$ | [kJ mol$^{-1}$] | 75.31 |
| Pre-exponential factor | $A_0$ | [s$^{-1}$] | $5.7 \times 10^{-4}$ |
| Activation energy | $E_{aA}$ | [kJ mol$^{-1}$] | 750.31 |
| Pre-exponential factor | $A_A$ | [s$^{-1}$] | $4.8 \times 10^{-4}$ |
| Order of reaction | $n$ | [–] | 2 |
| Reactivity ratio | $R/R_0$ | [–] | 340–610 |

This function identifies the optimal parameters while adhering to all constraints defined. The same procedure has been used by Holland and Fletcher [73]. The newly parameters derived for reaction kinetics are summarised in Table 8.2. For comparison, the kinetic parameters from Holland and Fletcher [73] for the original coal-based CCK model are also provided.

Figure 8.14 gives a comparison of measured and simulated particle temperatures as a function of residence time for walnut shells as an example for a lignin rich biomass. Obviously, the original CCK model (dashed line) completely fails to predict the particle temperature, while the burnout time is much too short. The CCK model predicts the end of combustion after around 40 ms, whereas experimental investigations capture particle temperatures indicating reaction starting at approximately 70 ms. Additionally, the duration of the total combustion time in the CCK model is notably shorter than indicated by the experimental data. In comparison, the modified CCK model presents the measurement data with good accuracy and also burnout times are in reasonable agreement (CCK model modified 140 ms, measurement 160 ms).

In summary, the extension of the CCK model with the Bio-CPD model, a novel thermal annealing model and adapted kinetic reaction rates improved predictions significantly. This approach can easily be extended to other biomass, as long as detailed single particle measurements for temperature, size and burnout as a function of residence time are available. Such data are presented in Sect. 5 for walnut shells,

**Table 8.2** Values for the kinetic parameters from Holland and Fletcher [73] for coal and newly calibrated constants for walnut shells

| Parameters | | Black Thunder | North Antelope | Pitts-burgh 8 | Utah Skyline | Walnut shells |
|---|---|---|---|---|---|---|
| $k_3$ | [$10^{10}$ s$^{-1}$] | 2.18 | 1.39 | 0.0213 | 2.39 | 1.87 |
| $k_7$ | [$9^{10}$ s$^{-1}$] | 1.21 | 1.60 | 0.363 | 7.28 | 3.23 |
| $E_{a,3}$ | [kJ/mol] | 117 | 117 | 180 | 130 | 134 |
| $E_{a,7}$ | [kJ/mol] | 210 | 240 | 267 | 272 | 418 |

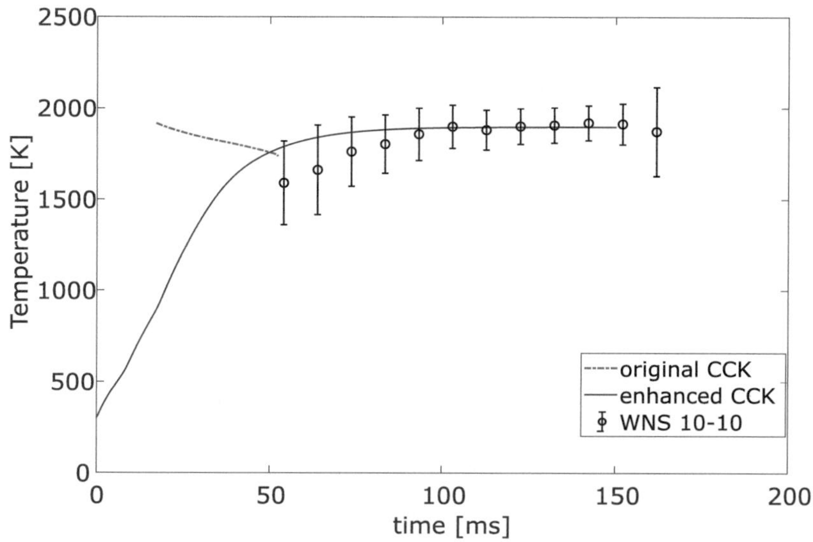


**Fig. 8.14** Comparison of measurement data (black line with error bar) with the results of the original (orange line) and the modified (purple line) CCK model for an oxy-fuel atmosphere of 10 vol.% oxygen, 10 vol.% water and 80 vol.% carbon dioxide

miscanthus and beech wood. When further data are needed, they can be achieved by the experimental approach presented in Sect. 3.3.1.

## 8.5 CRECK-S Results

### 8.5.1 Extension of Coal Model to Oxy-Fuel Conditions

With the work by Debiagi et al. [35], the CRECK-S model is calibrated for improved accuracy in oxy-fuel conditions in a wide range of heating rates. The selected fuel is a high volatile Colombian bituminous coal (CBC, cf. Table 1.1). TGA experiments and drop tube reactor measurements (DTR) are used for calibrating the model. Afterwards, the calibrated model is validated using measurements from a flat flame burner (FFB) and a fluidised bed reactor (FBR).

For higher temperatures, effectiveness factors are used in simulations, based on the study by Senneca et al. [74] to account for the oxygen diffusion effects on to char surface. Five main changes are made to the original CRECK-S mechanism based on TGA and DTR results:

1. The CHAR-H (lumped char species containing hydrogen) content predicted by the model is decreased by increasing its consumption in combustion reactions, and decreasing its formation rate through calibrated stoichiometric coefficients.

2. Metaplastic content is increased for low temperatures, and decreased for high temperatures. This results in progressive release of volatiles at low temperatures through metaplast decomposition and increased char at high temperatures.
3. The frequency factors of char combustion reactions are adjusted. This was necessary due to the volumetric description of reactions in CRECK-S model which can only capture the surface area effects by the frequency factors.
4. A new, less reactive pure carbon char (CHAR-$CO_2$) is introduced, representing the increased char annealing effects in the presence of $CO_2$ containing atmospheres.
5. Combustion reactions for metaplasts are included in the mechanism to match the elemental composition of the solid residues with the experimental observations.

Fig. 8.15 shows the experimental data used for calibration along with the calibrated model results for TGA and DTR measurements. When air is used for the TGA experiments, the char can be burned, yielding smaller amounts of solid residue. In the TGA-air measurements of chars obtained from DTR experiments, the reactivities vary depending on whether $CO_2$ or $N_2$ is used during the DTR. Observing the DTR

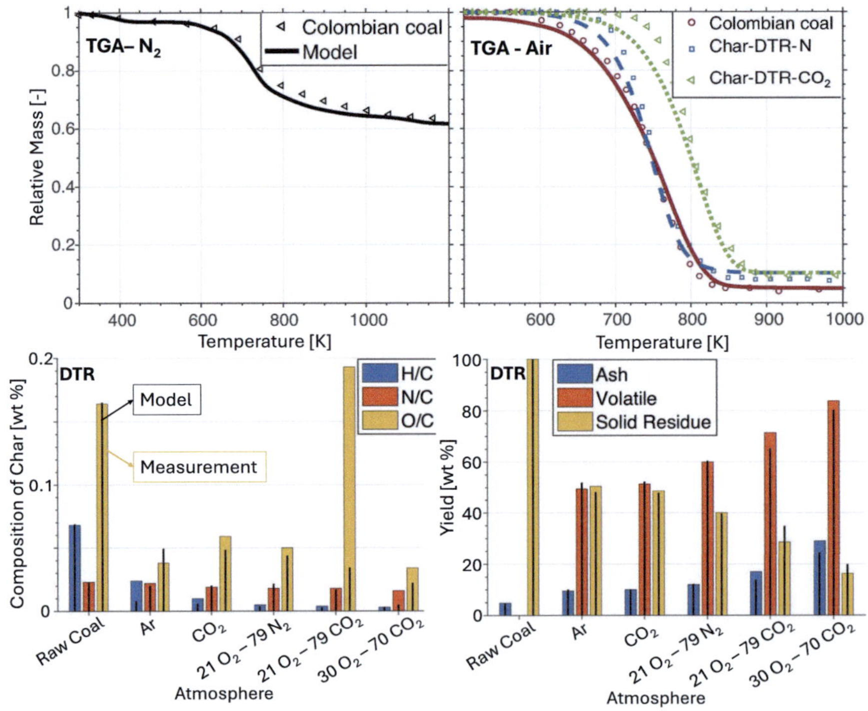


**Fig. 8.15** Comparison of TGA (top) and DTR (bottom) measurements with calibrated CRECK-S model results. Chars obtained from the DTR are also used for TGA experiments with air (based on the publication by Debiagi et al. [35])

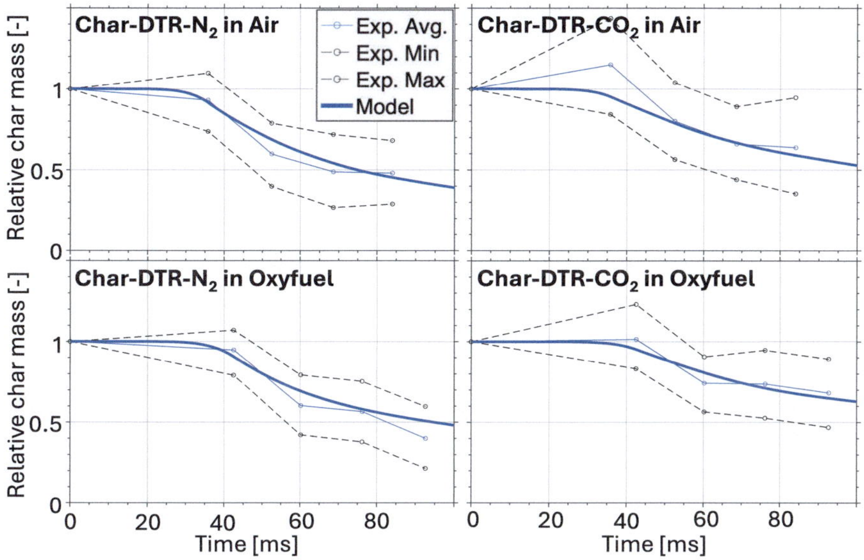


**Fig. 8.16** Comparison of experiments (symbols) and model predictions (lines) of char mass loss (Char-DTR-$N_2$ and Char-DTR-$CO_2$) in the flat flame burner. Red marks indicate the average experimental values and black marks the experimental deviations. Model predictions are reported in the solid lines. (Based on the publication by Debiagi et al. [35])

data, the calibrated model successfully captures the composition and yields of the solid residue.

The validation of the calibrated CRECK-S model with FFB is shown in Fig. 8.16. Chars obtained from DTR experiments either with $CO_2$ or $N_2$ (DTR-$CO_2$ or DTR-$N_2$, respectively) are burned with either air or oxy-fuel atmosphere. DTR-$CO_2$ chars are found to be more reactive, similar to the TGA cases. Considering the combustion atmosphere, it should be noted that oxy-fuel cases exhibit lower flame temperatures than the corresponding air ones. As a result, for the same type of char, the air combustion achieves a higher conversion compared to the corresponding oxy-fuel case, especially in the simulations.

The validation of the calibrated CRECK-S model with the char compositions obtained from FBR is shown in Table 8.3. A good agreement can be observed between the experimental measurements and model predictions. A more detailed comparison of the volatiles from FBR experiments and the CRECK-S model can be found in the study by Debiagi et al. [35], including additional discussions on potential measurement errors/uncertainties.

Overall, the work from Debiagi et al. [35] successfully extends the CRECK-S model for oxy-fuel conditions and validates the model predictions using a wide range of test rigs and operating conditions.

A similar calibration procedure was also performed for lignite coals in the work by Özer et al. [75]. Furthermore, the calibrated kinetics were used for simulating a wide range of particle trajectories within the Darmstadt burner. Detailed information

**Table 8.3** Comparison of char elemental fractions obtained by fluidised bed reactor measurements and CRECK-S model (based on the publication by Debiagi et al. [35])

| Char composition | Measurement [wt %] | Model [wt %] |
|---|---|---|
| C | 95.22 | 95.18 |
| H | 0.68 | 0.26 |
| O | 1.58 | 1.75 |
| N | 1.87 | 2.44 |
| S | 0.65 | 0.43 |

on particle trajectories was obtained from high-fidelity LES. The simulation of these cases generated an extensive dataset that was employed in the training and validation of a machine learning-driven procedure, being able to showcase a reduced-order HDMR model [76]

In the following sections, the studies and results from the CRECK-S-B model will be presented for biomass.

### 8.5.2 Improvement of Volatile Yields from Cellulose

Cellulose decomposition in the CRECK mechanism has been improved [77] in the light of new experimental data [78]. Both the predicted devolatilisation process and composition of volatiles are adjusted in the model, accounting for additional reacting paths and improved characterisation of the volatile species. For the devolatilisation, the activation energy of the initial cellulose depolymerisation step is decreased, and a slow-releasing metaplastic LVG species is introduced with higher activation energy.

Several changes are made for the product distribution focusing on the second step, where depolymerised cellulose decomposes with two competing reactions into either sugar or light oxygenate and permanent gases. First, the kinetic parameters of the two competing reactions are adjusted in favour of the sugar-yielding reaction, in the expense of light-oxygenates. Second, the lumped LVG product of the sugar-yielding reaction is delumped by introducing another lumped $C_6H_{10}O_5$ sugar, along with the actual LVG molecule (which is partially trapped into metaplastic species as previously mentioned). Lastly, furan and furfural species are included to represent $C_4$ and $C_5$ products in the oxygenate-releasing reaction products.

The thermogravimetric (i.e. mass loss) curves obtained with previous (v20) and updated (v23) models as well as experimental data are shown in Fig. 8.17. The decreased activation energy of the initial activation step in v23 results in an earlier onset of mass loss at lower temperatures, in agreement with the experimental data. Further, the metaplastic LVG introduced in v23 results in slower and longer mass loss at higher temperatures, also improving the agreement with experimental data.

The effects of the modifications on the product distribution are compared in Fig. 8.18. The updated model (v23) releases less light gas and light oxygenates but yields more sugar compared to the previous model (v20) as a result of changed kinetic parameters for sugar-yielding and oxygenate-yielding reactions. A better agreement

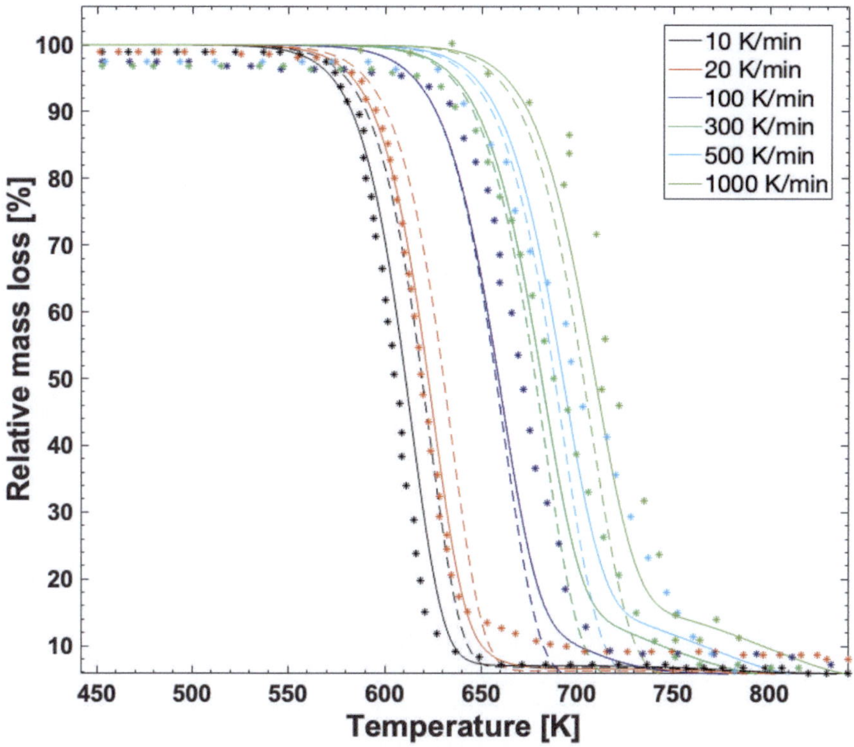


**Fig. 8.17** Comparison of mass loss curves obtained from the experiments, the original v20 model and the updated v23 model. Stars: Experimental data. Solid lines: v23. Dashed lines: v20. (Based on the publication by Debiagi et al. [77])

with the experiments is obtained as a result. Additionally, the previously lacking (in model v20) $C_4$ and $C_5$ species are included in v23, yielding satisfactory agreement with the experimental results.

### 8.5.3 Char Formation During Biomass Pyrolysis

The composition of the chars obtained by the pyrolysis of a variety of biomasses Zou et al. [79] was further investigated. The previous model (baseline CRECK model) is found to predict significantly less oxygen inside the formed char compared to experiments, when the pyrolysis temperatures are above 1073 K. Analysing the model predictions, most of the oxygen content is found in a metaplastic species in the model, G{$COH_2$}. With further heating, this species further decomposes and releases gas-phase species with the following reaction (R30 in the previous CRECK model):

$$G\{\mathrm{CH_2O}\} \rightarrow 0.8\,\mathrm{CO} + 0.8\,\mathrm{H_2} + 0.2\,\mathrm{H_2O} + 0.2\,\mathrm{CHAR} \qquad (8.15)$$

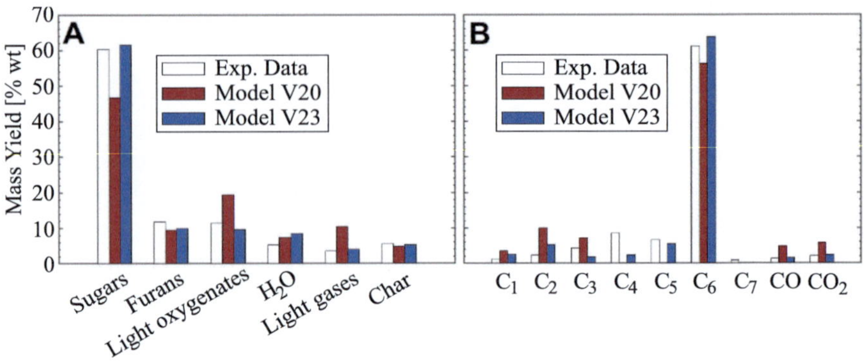


**Fig. 8.18** Comparison of CRECK v20 and v23 models on product distribution. (Based on the publication by Debiagi et al. [77])

This reaction releases all the elemental oxygen into gas-phase and creates only a single solid-phase species CHAR (pure carbon). The improvements proposed and validated by Zou et al. [79] are to introduce a new reaction with a similar form to R30, but yielding a new metaplastic species G{$CO_{stiff}$} (R32), and its subsequent decomposition (R33). Half of the previously released CO is then attributed to this new species. Additionally, G{$CO_{stiff}$} can release CO and form a new permanent char species CHAR-O (equimolar composed by C and O) at higher temperatures. The new reactions R32 and R33 (here expressed as (8.16) and (8.17)) are given as

$$G\{\mathrm{CH_2O}\} \rightarrow 0.4\,\mathrm{CO} + 0.4\,G\{\mathrm{CO_{stiff}}\} + 0.8\,\mathrm{H_2} + 0.2\,\mathrm{H_2O} + 0.2\,\mathrm{CHAR} \quad (8.16)$$

$$G\{\mathrm{CO_{stiff}}\} \rightarrow 0.8\,\mathrm{CO} + \mathrm{CHAR\text{-}O} \quad (8.17)$$

The introduction of these reacting paths hinders the release of elemental oxygen into the gas phase, and keeps a permanent portion of it in the char structure with species CHAR-O. The results of the new (with R32 and R33) and previous models are first demonstrated on rice husk (RH) and corn straw (CS) biomasses in Fig. 8.19, in comparison to experimental data. The model can reproduce the mass loss curves successfully. In addition, the elemental composition of the char is significantly improved with the new model, especially at higher temperatures.

To confirm the validity of the changes over a broad range of biomass feedstocks, a collection of 376 experiments was analysed with the previous and new models. For each data point, the success criterion is defined as 10 % relative error. The percentage of successful simulations is used to obtain the final accuracy for previous and new models, over the whole dataset collection. The results are shown in Fig. 8.20. For pyrolysis temperatures below 773 K, both models have the same accuracy between 80 and 90 % for elemental fractions in the char residue. However, at higher temperatures, the accuracy of the previous model drops drastically, while the new model shows significant improvements. This is a particularly important improvement that is reflected in the subsequent reactivity of the char conversion step. Moreover, such

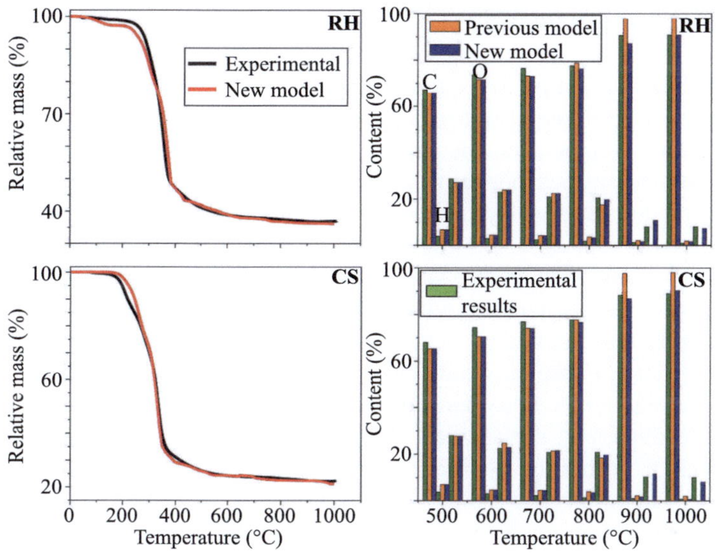


**Fig. 8.19** TG curves (left) and elemental content of the solid residue (right) for rice husk (top) and corn straw (right) (based on the publication by Zou et al. [79])

a detailed and accurate prediction of char composition from a pyrolysis process is a unique feature of this model.

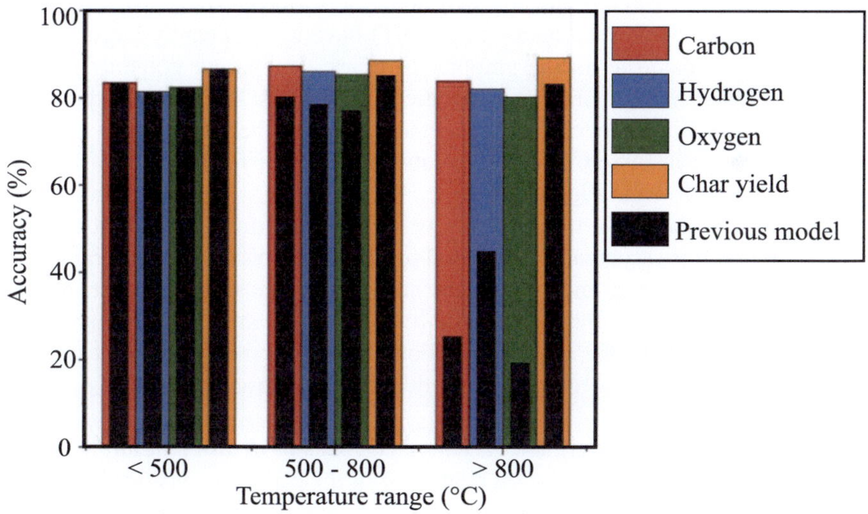


**Fig. 8.20** Statistical accuracy of previous and new models based on a collection of 376 experiments with a range of biomasses. (Based on the publication by Zou et al. [79])

### 8.5.4 Biochar Conversion

An accurate description of char combustion involves higher complexity due to the surface reactions. For combustion to happen, oxygen molecules in the gas phase have to be in contact with the available surface area of the char inside the solid particle. Consequently, the physical phenomena like gas adsorption on the surface and surface area evolution should also be addressed along with the chemistry of surface reactions. The work by Locaspi et al. [21] developed a predictive physico-chemical model addressing the mentioned issues. The developed surface reaction model represents the carbonaceous solid material with a collection of bulk species that are not in contact with the surrounding gas, and surface species upon which ambient oxygen can adsorb. In total, 17 pseudo-species characterise different chemical functionalities including various types of carbon bonds, carbonyl groups and inorganic species (ash). The evolution of surface and bulk species—including a range of reaction types such as gas adsorption to surface, surface isomerisation, surface decomposition (desorption) and volumetric (bulk) isomerisation—are described by 205 reactions. Certain decomposition reactions also expose some of the underlying bulk species to the surrounding gas environment, effectively consuming bulk species to create surface species. The material balances for surface, bulk and gas phases are solved using the proposed mechanism.

The modelling approach is validated with a variety of experiments. Figure 8.21 shows the model predictions and measurements for a variety of operating conditions during thermogravimetric analysis. In all cases, the model accurately predicts the trends from the experiments and has satisfactory agreement.

In Fig. 8.21a), the carbon nanotubes (CNT) have a significantly larger specific surface area compared to all the graphite particles. Comparing graphite particles, smaller particle size results in larger specific surface areas. Both in the experiments and in the model, higher specific surface areas result in higher reactivity as expected. Hence, the char conversion onset temperatures are lower for particles with larger surface areas as shown in Fig. 8.21a).

Figure 8.21b) shows the mass loss curves of hydrothermal carbonised cellulose (HTC) at 800 K, for varying oxygen partial pressures. In this temperature range, surface adsorption of oxygen and consequent surface decomposition dominate over the pyrolytic pathway, hence, the high sensitivity to oxygen partial pressure is observed both in experiments and the model prediction. The higher oxygen content results in faster adsorption and overall char consumption rates, which are successfully captured by the model.

In both Figs. 8.21c) and d), the varying feedstocks have different initial compositions, which affect the combustion behaviour of the carbonaceous materials.

This detailed approach is still a challenge to seamlessly couple to the rest of the CRECK-S-B model. One possibility is to reduce the detailed char model into a volumetric description for the specific fuel and conditions, which can be seamlessly coupled to the pyrolysis model. In particular, this reduced volumetric char conversion model has been calibrated for char oxidation and gasification of Colombian coal, and similar calibration can be performed for any different feedstock. Moreover, the

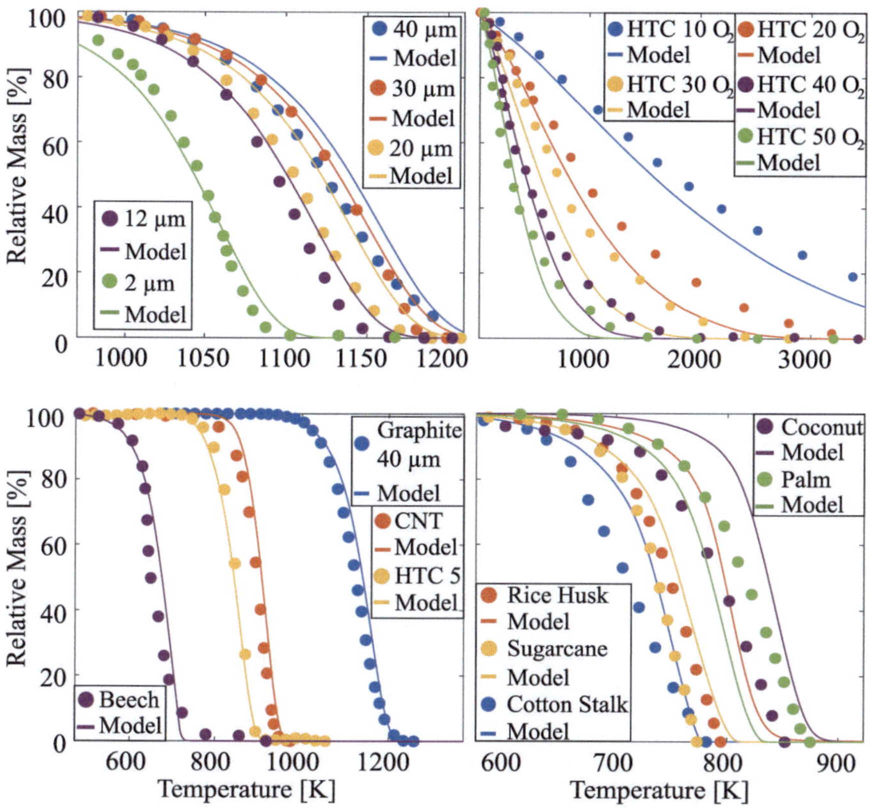


**Fig. 8.21** Validation of the char model with varying conditions: **a)** varying graphite particle sizes, **b)** varying oxygen partial pressures, **c)** varying carbonaceous materials and **d)** varying biochars from different biomass sources. (Based on the publication by Locaspi et al. [21])

catalytic effects of most important metals present in the ash were evaluated and introduced in this version of the kinetic mechanism Pflieger et al. [80], which are described in Chap. 4.

## 8.5.5 CRECK-S-B Model for Walnut Shells

As one of the core biomass feedstocks investigated in *Oxyflame*, the improvements to the CRECK-S-B models described in the previous sections are assessed for reproducing the pyrolysis behaviour of walnut shells as observed in different experimental setups. TGA experiments and fluidised bed reactor (FBR) measurements are used for assessing the model performance, covering both low and high heating rates. The details on the experimental setups can be found in Sect. 2.2. For the simulations, the standard characterisation approach proposed for the CRECK-S-B model is employed, which is based on the ultimate analysis results of the biomass feedstock [5,54] and includes proteins to account for the presence of nitrogen. The results of ultimate analysis measurements and the corresponding reference species distribution for walnut shells are presented in Table 8.4. Walnut shells are rich in lignin compared to many other biomass feedstocks. Preliminary simulations of walnut shells as well as other lignin-rich biomass samples showed some inaccuracies of the CRECK-S-B model. Until recently, the CRECK-S-B model has neglected the component interactions which can become especially important when there is a significant amount of lignin in the biomass. It is known that lignins can be attached to carboydrates (cellulose/hemicellulose) to form the so called lignin-carbohydrate-complex (LCC) in native biomasses [81].

As an initial step to account for the presence of LCC, a single interaction reaction is added to the CRECK-S-B model where lignin decomposes faster in the presence of hemicellulose. The default and modified model are compared with respect to the thermogravimetric analysis (TGA) and its differential form (DTG) under low heating rates and inert atmospheres ($N_2$). Results are shown in Fig. 8.22. For all the considered heating rates between 5 and 20 K $min^{-1}$, the modified model has significantly higher accuracy. The DTG curves of the default model shows three distinct peaks at approximately 280, 360, and 420 °C, corresponding to hemicellulose, cellulose and lignin decomposition respectively. Both in the measurement data and the modified model, the first peak at 280 °C is more pronounced due to the LCC breakdown where lignin is consumed together with hemicellulose. Given enough time for pyrolysis, the final solid yields are not effected with the modification and are captured well with both the default and the modified model. Recently, the parameters of the new LCC reaction have been refined and the approach has been validated across different heating rates and biomass feedstocks with varying compositions [82].

Walnut shell pyrolysis experiments under fast heating rate are conducted in a fluidised bed reactor Ontyd et al. [84] in inert ($N_2$) atmosphere. Different reactor temperatures are set between 500 and 800 °C range. Figure 8.23 compares the measurements and modified model predictions for the rate of volatiles release (cor-

**Table 8.4** Ultimate analysis and CRECK-S-B reference species weight fractions for walnut shells

| Elemental composition [–] | | Reference species [–] | |
|---|---|---|---|
| C | 0.522 | CELL | 0.3158 |
| H | 0.0577 | XYHW | 0.1715 |
| O | 0.4178 | LIGO | 0.3052 |
| N | 0.0028 | LIGH | 0.0372 |
| | | LIGC | 0.0521 |
| | | TANN | 0.0687 |
| | | TGL | 0.0239 |
| | | PROT-C | 0.0072 |
| | | PROT-H | 0.0071 |
| | | PROT-O | 0.0071 |
| | | ASH | 0.0042 |

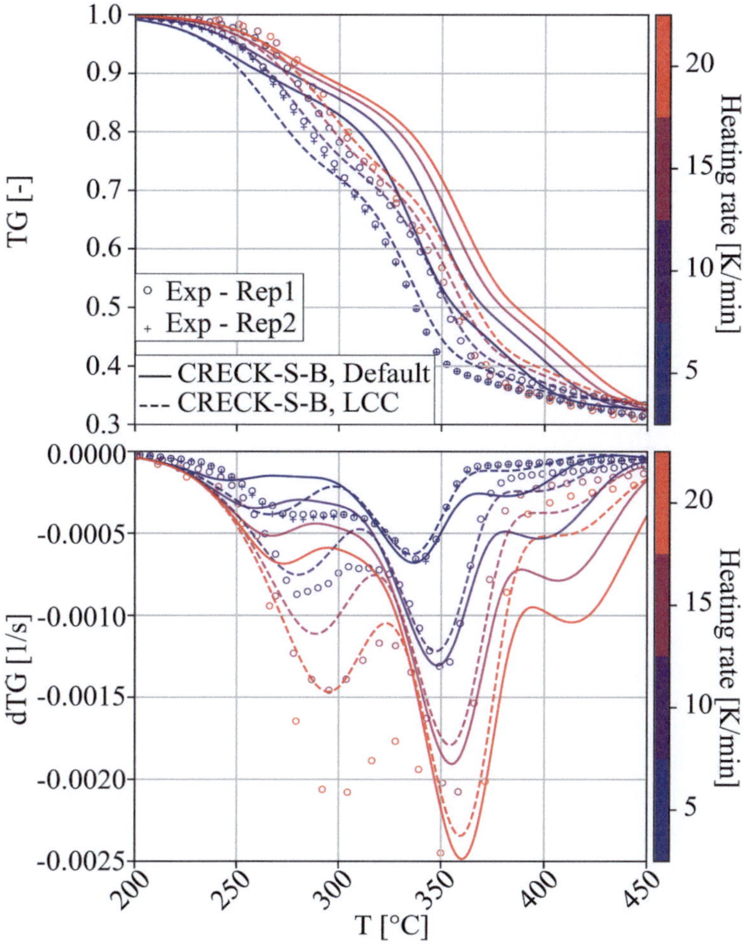


**Fig. 8.22** The TGA and DTG curves from experiments by Senneca et al. [83] and Ontyd et al. [84], and model predictions with default and modified CRECK-S-B model. The modified model includes hemicellulose-lignin interaction to account for the presence of LCC in the walnut shells

responding to the mass loss rate), revealing an increasing rate as reactor temperature increases. Onset of volatile release are also reached faster with the increase in reactor temperature, which is explained by an increased heat transfer (perceived heating rate) to the biomass particles injected in the reactor. The overall trends for decomposition rate are well reproduced in the model predictions, in particular for the higher temperatures, while the decomposition rate is overpredicted for lower temperatures below 650 °C. Potential reasons for the deviations include a different LCC reaction path under high heating rates that is currently missing or uncertainties in the calculation of residence times of the gas in the reactor. The latter has an effect since the released gas species are compared (not the solid residue) and residence time also depends

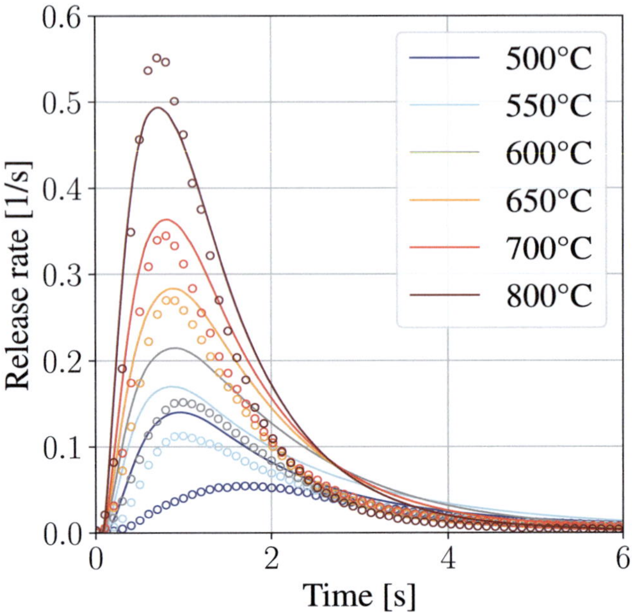


**Fig. 8.23** Comparison of volatile release rate measurements (based on the publication by Ontyd et al. [84]) (circles) and CRECK-S-B model predictions including LCC (lines)

on the amount of released volatiles, becoming an unsteady quantity based on the injection profile of the biomass to the fluidised bed.

Figure 8.24 compares the normalised yield of volatiles species obtained from experiments and model predictions, revealing good agreement with the observed relative yields and distribution of species. The preliminary results are obtained by assuming the gas yields would be the same in steady and unsteady operation. The automatic network generation algorithm described in the study by Berkel et al. [42] is used. The trends are also clearly reproduced: a decreasing relative yield of $H_2O$ and increase of CO with the increase of temperature. The progressive increase in the biomass conversion degree with the increase of reactor temperature leads to the dominating yields of CO.

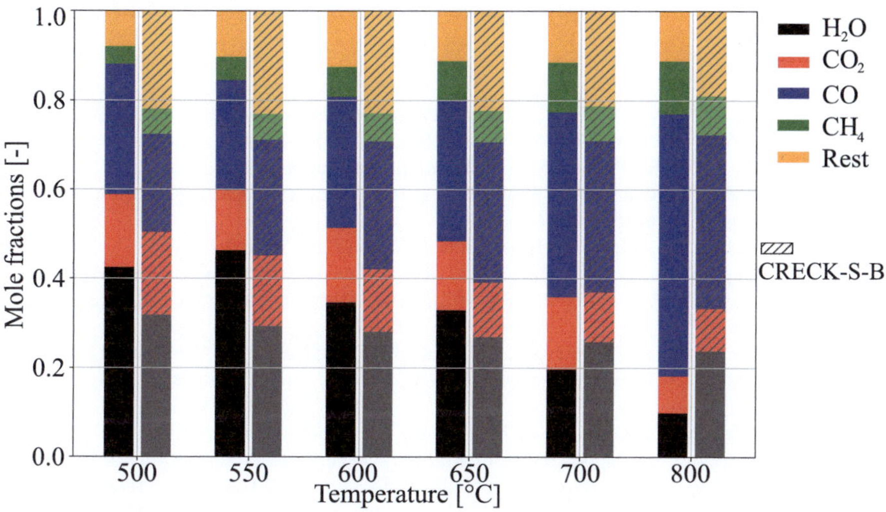


**Fig. 8.24** Comparison of normalised volatile compositions (based on the publication by Ontyd et al. [84]) and the CRECK-S-B model (hatched bars) for the FBR

## 8.6 Conclusion

This chapter presented a detailed description of lignocellulosic biomass biochemistry and the modelling approaches describing its thermochemical conversion. In particular CCK, Bio-CPD, and CRECK-S models are described and selected results from each are presented.

For most modelling approaches, knowledge from coal serves as a good basis due to the large amounts of available data and literature. The adaptations to biomass are based on the molecular structure of the lignocellulosic biomass, which mainly contains cellulose, hemicellulose and lignins. The individual decomposition behaviour of these components creates the basis for CRECK-S and Bio-CPD models. Several different experimental setups with varying heating rates are captured succesfully with both CRECK-S-B and Bio-CPD models.

Development of detailed kinetic models with predictive capabilities is an indispensable task for in-depth understanding and forms the basis for more complex tasks. The knowledge gained from these models can be used for optimising processes and simulating reactors or burners. One example is the CFD simulations where Bio-CPD and CRECK-S-B models can be used either directly or indirectly. These applications can be found in Chap. 19.

**Acknowledgements** This work has been funded by the Deutsche Forschungsgemeinschaft (DFG, German Research Foundation)—215035359—SFB/TRR 129 'Oxyflame'.

## References


1. Anthony, D. B. and Howard, J. B. "Coal Devolatilization and Hydrogasification". *AIChE Journal* 22.4 (1976), 625–656. https://doi.org/10.1002/aic.690220403
2. Vascellari, M., Arora, R., Pollack, M. and Hasse, C. "Simulation of Entrained Flow Gasification with Advanced Coal Conversion Submodels. Part 1: Pyrolysis". *Fuel* 113 (2013), 654–669. https://doi.org/10.1016/j.fuel.2013.06.014
3. Hasse, C., Debiagi, P., Wen, X., Hildebrandt, K., Vascellari, M. and Faravelli, T. "Advanced Modeling Approaches for CFD Simulations of Coal Combustion and Gasification". *Progress in Energy and Combustion Science* 86 (2021), 100938. https://doi.org/10.1016/j.pecs.2021.100938
4. Sommariva, S., Maffei, T., Migliavacca, G., Faravelli, T. and Ranzi, E. "A Predictive Multi-Step Kinetic Model of Coal Devolatilization". *Fuel* 89.2 (2010), 318–328. https://doi.org/10.1016/j.fuel.2009.07.023
5. Debiagi, P., Pecchi, C., Gentile, G., Frassoldati, A., Cuoci, A., Faravelli, T. and Ranzi, E. "Extractives Extend the Applicability of Multistep Kinetic Scheme of Biomass Pyrolysis". *Energy & Fuels* 29.10 (2015), 6544–6555. https://doi.org/10.1021/acs.energyfuels.5b01753
6. Wang, Z., Popolan-Vaida, D. M., Chen, B., Moshammer, K., Mohamed, S. Y., Wang, H., Sioud, S., Raji, M. A., Kohse-Höinghaus, K., Hansen, N., Dagaut, P., Leone, S. R. and Sarathy, S. M. "Unraveling the Structure and Chemical Mechanisms of Highly Oxygenated Intermediates in Oxidation of Organic Compounds". *Proceedings of the National Academy of Sciences of the United States of America* 114.50 (2017), 13102–13107. https://doi.org/10.1073/pnas.1707564114
7. Faravelli, T., Frassoldati, A., Migliavacca, G. and Ranzi, E. "Detailed Kinetic Modeling of the Thermal Degradation of Lignins". *Biomass and Bioenergy* 34.3 (2010), 290–301. https://doi.org/10.1016/j.biombioe.2009.10.018
8. Ranzi, E., Debiagi, P. E. A. and Frassoldati, A. "Mathematical Modeling of Fast Biomass Pyrolysis and Bio-Oil Formation. Note I: Kinetic Mechanism of Biomass Pyrolysis". *ACS Sustainable Chemistry & Engineering* 5.4 (2017), 2867–2881. https://doi.org/10.1021/acssuschemeng.6b03096
9. Sluiter, J. B., Ruiz, R. O., Scarlata, C. J., Sluiter, A. D. and Templeton, D. W. "Compositional Analysis of Lignocellulosic Feedstocks. 1. Review and Description of Methods". *Journal of agricultural and food chemistry* 58.16 (2010), 9043–9053. https://doi.org/10.1021/jf1008023
10. Flynn, J. H. and Wall, L. A. "A Quick, Direct Method for the Determination of Activation Energy from Thermogravimetric Data". *Journal of Polymer Science Part B: Polymer Letters* 4.5 (1966), 323–328. https://doi.org/10.1002/pol.1966.110040504
11. Kobayashi, H., Howard, J. B. and Sarofim, A. F. "Coal Devolatilization at High Temperatures". *Symposium (International) on Combustion* 16.1 (1977), 411–425. https://doi.org/10.1016/S0082-0784(77)80341-X
12. Baum, M. M. and Street, P. J. "Predicting the Combustion Behaviour of Coal Particles". *Combustion Science and Technology* 3.5 (1971), 231–243. https://doi.org/10.1080/00102207108952290
13. Migliavacca, G., Parodi, E., Bonfanti, L., Faravelli, T., Pierucci, S. and Ranzi, E. "A General Mathematical Model of Solid Fuels Pyrolysis". *Energy* 30.8 (2005), 1453–1468. https://doi.org/10.1016/j.energy.2004.02.011
14. Shurtz, R. C. and Fletcher, T. H. "Coal Char-$CO_2$ Gasification Measurements and Modeling in a Pressurized Flat-Flame Burner". *Energy & Fuels* 27.6 (2013), 3022–3038. https://doi.org/10.1021/ef400253c
15. Hurt, R. H., Sun, J.-K. and Lunden, M. "A Kinetic Model of Carbon Burnout in Pulverized Coal Combustion". *Combustion and Flame* 113.1–2 (1998), 181–197. https://doi.org/10.1016/S0010-2180(97)00240-X
16. Mayes, H. B. and Broadbelt, L. J. "Unraveling the Reactions that Unravel Cellulose". *The Journal of Physical Chemistry. A* 116.26 (2012), 7098–7106. https://doi.org/10.1021/jp300405x

17. Grant, D. M., Pugmire, R. J., Fletcher, T. H. and Kerstein, A. R. "Chemical Model of Coal Devolatilization Using Percolation Lattice Statistics". *Energy & Fuels* 3.2 (1989), 175–186. https://doi.org/10.1021/ef00014a011
18. Solomon, P. R., Hamblen, D. G., Carangelo, R. M., Serio, M. A. and Deshpande, G. V. "General Model of Coal Devolatilization". *Energy & Fuels* 2.4 (1988), 405–422. https://doi.org/10.1021/ef00010a006
19. Niksa, S. and Kerstein, A. R. "FLASHCHAIN Theory for Rapid Coal Devolatilization Kinetics. 1. Formulation". *Energy & Fuels* 5.5 (1991), 647–665. https://doi.org/10.1021/ef00029a006
20. Wang, C., Watson, J. K., Louw, E. and Mathews, J. P. "Construction Strategy for Atomistic Models of Coal Chars Capturing Stacking Diversity and Pore Size Distribution". *Energy & Fuels* 29.8 (2015), 4814–4826. https://doi.org/10.1021/acs.energyfuels.5b00816
21. Locaspi, A., Debiagi, P., Pelucchi, M., Hasse, C. and Faravelli, T. "A Predictive Physico-Chemical Model of Biochar Oxidation". *Energy & Fuels* 35.18 (2021), 14894–14912. https://doi.org/10.1021/acs.energyfuels.1c01559
22. Sheng, C. and Azevedo, J. L. T. "Modeling Biomass Devolatilization Using the Chemical Percolation Devolatilization Model for the Main Components". *Proceedings of the Combustion Institute* 29.1 (2002), 407–414. https://doi.org/10.1016/S1540-7489(02)80054-2
23. Vizzini, G., Bardi, A., Biagini, E., Falcitelli, M. and Tognotti, L. "Prediction of Rapid Biomass Devolatilization Yields with an Upgraded Version of the Bio-CPD Model". *31st Meeting of the Italian Section of the Combustion Institute*. Turin, Italy, 2008
24. Lewis, A. D. and Fletcher, T. H. "Prediction of Sawdust Pyrolysis Yields from a Flat-Flame Burner Using the CPD Model". *Energy & Fuels* 27.2 (2013), 942–953. https://doi.org/10.1021/ef3018783
25. Wan, K., Wang, Z., He, Y., Xia, J., Zhou, Z., Zhou, J. and Cen, K. "Experimental and Modeling Study of Pyrolysis of Coal, Biomass and Blended Coal–Biomass Particles". *Fuel* 139 (2015), 356–364. https://doi.org/10.1016/j.fuel.2014.08.069
26. Fletcher, T. H., Kerstein, A. R., Pugmire, R. J., Solum, M. S. and Grant, D. M. "Chemical Percolation Model for Devolatilization. 3. Direct Use of $^{13}C$ NMR Data to Predict Effects of Coal Type". *Energy & Fuels* 6.4 (1992), 414–431. https://doi.org/10.1021/ef00034a011
27. Pielsticker, S., Ontyd, C., Kreitzberg, T., Hatzfeld, O., Schiemann, M., Scherer, V. and Kneer, R. "Adaptation of the Chemical Percolation Devolatilization Model for Low Temperature Pyrolysis in a Fluidized Bed Reactor". *Combustion Science and Technology* 52 (2019), 1–18. https://doi.org/10.1080/00102202.2019.1682433
28. Pielsticker, S. and Kneer, R. "Modeling Pyrolysis Kinetics of Extracted Biomass Components with the Bio-Chemical Percolation Devolatilization Model". *Fuel* 360 (2024), 130454. https://doi.org/10.1016/j.fuel.2023.130454
29. Holland, T. M. "A Comprehensive Coal Conversion Model Extended to Oxy-Coal Conditions". PhD thesis. Brigham Young University, 2017
30. Shurtz, R. C. "Effects of Pressure on the Properties of Coal Char Under Gasification Conditions at High Initial Heating Rates". PhD thesis. Brigham Young University, 2011
31. Senneca, O. and Salatino, P. "Loss of Gasification Reactivity Toward $O_2$ and $CO_2$ Upon Heat Treatment of Carbons". *Proceedings of the Combustion Institute* 29.1 (2002), 485–493. https://doi.org/10.1016/S1540-7489(02)80063-3
32. Niksa, S., Liu, G.-S. and Hurt, R. H. "Coal Conversion Submodels for Design Applications at Elevated Pressures. Part I. Devolatilization and Char Oxidation". *Progress in Energy and Combustion Science* 29.5 (2003), 425–477. https://doi.org/10.1016/S0360-1285(03)00033-9
33. Senneca, O. and Salatino, P. "A Semi-Detailed Kinetic Model of Char Combustion with Consideration of Thermal Annealing". *Proceedings of the Combustion Institute* 33.2 (2011), 1763–1770. https://doi.org/10.1016/j.proci.2010.08.011
34. Senneca, O., Ontyd, C., Cerciello, F., Schiemann, M. and Scherer, V. "Extension of the Thermal Annealing Concepts Developed for Coal Combustion to Conversion of Lignocellulosic Biomass". *Energy & Fuels* 34.3 (2020), 3661–3670. https://doi.org/10.1021/acs.energyfuels.9b03172

35. Debiagi, P., Ontyd, C., Pielsticker, S., Schiemann, M., Faravelli, T., Kneer, R., Hasse, C. and Scherer, V. "Calibration and Validation of a Comprehensive Kinetic Model of Coal Conversion in Inert, Air and Oxy-Fuel Conditions Using Data from Multiple Test Rigs". *Fuel* 290 (2021). https://doi.org/10.1016/j.fuel.2020.119682
36. Maffei, T., Sommariva, S., Ranzi, E. and Faravelli, T. "A Predictive Kinetic Model of Sulfur Release from Coal". *Fuel* 91.1 (2012), 213–223. https://doi.org/10.1016/j.fuel.2011.08.017
37. Maffei, T., Frassoldati, A., Cuoci, A., Ranzi, E. and Faravelli, T. "Predictive One Step Kinetic Model of Coal Pyrolysis for CFD Applications". *Proceedings of the Combustion Institute* 34.2 (2013), 2401–2410. https://doi.org/10.1016/j.proci.2012.08.006
38. Debiagi, P., Ranzi, E., Faravelli, T. and Hasse, C. "Coal Combustion: Integrating Nitrogen and Sulfur Fraction in a Single Kinetic Mechanism". *1st International Conference on Smart Energy Carriers*. Naples, Italy, 2019
39. Ranzi, E., Cuoci, A., Faravelli, T., Frassoldati, A., Migliavacca, G., Pierucci, S. and Sommariva, S. "Chemical Kinetics of Biomass Pyrolysis". *Energy & Fuels* 22.6 (2008), 4292–4300. https://doi.org/10.1021/ef800551t
40. Debiagi, P. "A Kinetic Model of Thermochemical Conversion of Biomass". PhD thesis. Politecnico Milano, 2018
41. Debiagi, P., Gentile, G., Cuoci, A., Frassoldati, A., Ranzi, E. and Faravelli, T. "A Predictive Model of Biochar Formation and Characterization". *Journal of Analytical and Applied Pyrolysis* 134 (2018), 326–335. https://doi.org/10.1016/j.jaap.2018.06.022
42. Berkel, L., Debiagi, P., Nicolai, H., Amjed, M. A., Stagni, A., Hasse, C. and Faravelli, T. "Development of a Multiphase Chemical Reactor Network Method as a Tool for Simulating Biomass Gasification in Fluidized Beds". *Fuel* 357 (2024), 129731. https://doi.org/10.1016/j.fuel.2023.129731
43. Shafizadeh, F. and Fu, Y. L. "Pyrolysis of Cellulose". *Carbohydrate research* 29.1 (1973), 113–122. https://doi.org/10.1016/S0008-6215(00)82074-1
44. Bradbury, A. G. W., Sakai, Y. and Shafizadeh, F. "A Kinetic Model for Pyrolysis of Cellulose". *Journal of Applied Polymer Science* 23.11 (1979), 3271–3280. https://doi.org/10.1002/app.1979.070231112
45. Zhou, X., Nolte, M. W., Mayes, H. B., Shanks, B. H. and Broadbelt, L. J. "Experimental and Mechanistic Modeling of Fast Pyrolysis of Neat Glucose-Based Carbohydrates. 1. Experiments and Development of a Detailed Mechanistic Model". *Industrial & Engineering Chemistry Research* 53.34 (2014), 13274–13289. https://doi.org/10.1021/ie502259w
46. Seshadri, V. and Westmoreland, P. R. "Concerted Reactions and Mechanism of Glucose Pyrolysis and Implications for Cellulose Kinetics". *The Journal of Physical Chemistry. A* 116.49 (2012), 11997–12013. https://doi.org/10.1021/jp3085099
47. Seshadri, V. and Westmoreland, P. R. "Roles of Hydroxyls in the Noncatalytic and Catalyzed Formation of Levoglucosan from Glucose". *Catalysis Today* 269 (2016), 110–121. https://doi.org/10.1016/j.cattod.2015.10.033
48. Paulsen, A. D., Mettler, M. S. and Dauenhauer, P. J. "The Role of Sample Dimension and Temperature in Cellulose Pyrolysis". *Energy & Fuels* 27.4 (2013), 2126–2134. https://doi.org/10.1021/ef302117j
49. Shen, D., Jin, W., Hu, J., Xiao, R. and Luo, K. "An Overview on Fast Pyrolysis of the Main Constituents in Lignocellulosic Biomass to Valued-Added Chemicals: Structures, Pathways and Interactions". *Renewable and Sustainable Energy Reviews* 51 (2015), 761–774. https://doi.org/10.1016/j.rser.2015.06.054
50. Anca-Couce, A. and Obernberger, I. "Application of a Detailed Biomass Pyrolysis Kinetic Scheme to Hardwood and Softwood Torrefaction". *Fuel* 167 (2016), 158–167. https://doi.org/10.1016/j.fuel.2015.11.062
51. Prins, M. J., Ptasinski, K. J. and Janssen, Frans J. J. G. "Torrefaction of Wood: Part 1. Weight Loss Kinetics". *Journal of Analytical and Applied Pyrolysis* 77.1 (2006), 35–40. https://doi.org/10.1016/j.jaap.2006.01.001

52. Branca, C., Di Blasi, C., Mango, C. and Hrablay, I. "Products and Kinetics of Glucomannan Pyrolysis". *Industrial & Engineering Chemistry Research* 52.14 (2013), 5030–5039. https://doi.org/10.1021/ie400155x
53. Grønli, M. G., Várhegyi, G. and Di Blasi, C. "Thermogravimetric Analysis and Devolatilization Kinetics of Wood". *Industrial & Engineering Chemistry Research* 41.17 (2002), 4201–4208. https://doi.org/10.1021/ie0201157
54. Debiagi, P., Trinchera, M., Frassoldati, A., Faravelli, T., Vinu, R. and Ranzi, E. "Algae Characterization and Multistep Pyrolysis Mechanism". *Journal of Analytical and Applied Pyrolysis* 128 (2017), 423–436. https://doi.org/10.1016/j.jaap.2017.08.007
55. Pielsticker, S., Gövert, B., Umeki, K. and Kneer, R. "Flash Pyrolysis Kinetics of Extracted Lignocellulosic Biomass Components". *Frontiers in Energy Research* 9 (2021). https://doi.org/10.3389/fenrg.2021.737011
56. Pielsticker, S., Debiagi, P., Cerciello, F., Hasse, C. and Kneer, R. "Comparative Analysis of Pyrolysis Models Including SFOR, CRECK, and Bio-CPD to Predict Reaction Kinetics and Products from Extracted Biomass Components". *Fuel* 371 (2024), 131867. https://doi.org/10.1016/j.fuel.2024.131867
57. Genetti, D., Fletcher, T. H. and Pugmire, R. J. "Development and Application of a Correlation of $^{13}$C NMR Chemical Structural Analyses of Coal Based on Elemental Composition and Volatile Matter Content". *Energy & Fuels* 13.1 (1999), 60–68. https://doi.org/10.1021/ef980074k
58. Ranzi, E., Frassoldati, A., Stagni, A., Pelucchi, M., Cuoci, A. and Faravelli, T. "Reduced Kinetic Schemes of Complex Reaction Systems: Fossil and Biomass-Derived Transportation Fuels". *International Journal of Chemical Kinetics* 46.9 (2014), 512–542. https://doi.org/10.1002/kin.20867
59. Fagbemi, L., Khezami, L. and Capart, R. "Pyrolysis Products from Different Biomasses: Application to the Thermal Cracking of Tar". *Applied Energy* 69.4 (2001), 293–306. https://doi.org/10.1016/S0306-2619(01)00013-7
60. Liden, A. G., Berruti, F. and Scott, D. S. "A Kinetic Model for the Production of Liquids from the Flash Pyrolysis of Biomass". *Chemical Engineering Communications* 65.1 (1988), 207–221. https://doi.org/10.1080/00986448808940254
61. Ranzi, E., Debiagi, P. and Frassoldati, A. "Mathematical Modeling of Fast Biomass Pyrolysis and Bio-Oil Formation. Note II: Secondary Gas-Phase Reactions and Bio-Oil Formation". *ACS Sustainable Chemistry & Engineering* 5.4 (2017), 2882–2896. https://doi.org/10.1021/acssuschemeng.6b03098
62. Hameed, S., Wagh, A. S., Sharma, A., Pareek, V., Yu, Y. and Joshi, J. B. "Kinetic Modelling of Pyrolysis of Cellulose Using CPD Model: Effect of Salt". *Journal of Thermal Analysis and Calorimetry* 147.17 (2022), 9763–9777. https://doi.org/10.1007/s10973-021-11192-5
63. Rabaçal, M., Costa, M., Vascellari, M. and Hasse, C. "Kinetic Modelling of Sawdust and Beech Wood Pyrolysis in Drop Tube Reactors Using Advanced Predictive Models". *Chemical Engineering Transactions* 37 (2014), 79–84. https://doi.org/10.3303/CET1437014
64. Wang, Z., Wan, K., Xia, J., He, Y., Liu, Y. and Liu, J. "Pyrolysis Characteristics of Coal, Biomass, and Coal–Biomass Blends Under High Heating Rate Conditions: Effects of Particle Diameter, Fuel Type, and Mixing Conditions". *Energy & Fuels* 29.8 (2015), 5036–5046. https://doi.org/10.1021/acs.energyfuels.5b00646
65. Merrick, D. "Mathematical Models of the Thermal Decomposition of Coal". *Fuel* 62.5 (1983), 534–539. https://doi.org/10.1016/0016-2361(83)90222-3
66. Pielsticker, S., Schlögel, K. U., Kreitzberg, T., Hatzfeld, O. and Kneer, R. "Biomass Pyrolysis Kinetics in a Fluidized Bed Reactor: Measurements and Plausibility Verification for Reaction Conditions". *Fuel* 254 (2019), 115589. https://doi.org/10.1016/j.fuel.2019.05.172
67. Jahirul, M., Rasul, M., Chowdhury, A. and Ashwath, N. "Biofuels Production Through Biomass Pyrolysis—A Technological Review". *Energies* 5.12 (2012), 4952–5001. https://doi.org/10.3390/en5124952
68. Couhert, C., Commandré, J.-M. and Salvador, S. "Is it Possible to Predict Gas Yields of any Biomass After Rapid Pyrolysis at High Temperature from its Composition in Cellulose, Hemicellulose and Lignin?" *Fuel* 88.3 (2009), 408–417. https://doi.org/10.1016/j.fuel.2008.09.019

69. Niksa, S. "On the Primary Devolatilization of Hemicellulose". *Journal of Analytical and Applied Pyrolysis* 164 (2022), 105515. https://doi.org/10.1016/j.jaap.2022.105515
70. Wang, D., Fletcher, T. H., Mohanty, S., Hu, H. and Eddings, E. G. "Modified CPD Model for Coal Devolatilization at Underground Coal Thermal Treatment Conditions". *Energy & Fuels* 33.4 (2019), 2981–2993. https://doi.org/10.1021/acs.energyfuels.8b04425
71. Pielsticker, S., Möller, G. H., Gövert, B., Kreitzberg, T., Hatzfeld, O., Yönder, Ö., Angenent, V., Hättig, C., Schmid, R. and Kneer, R. "Influence of Biomass Torrefaction Parameters on Fast Pyrolysis Products Under Flame-Equivalent Conditions". *Biomass and Bioenergy* 119 (2018), 392–410. https://doi.org/10.1016/j.biombioe.2018.08.014
72. Tarlinski, D., Freisewinkel, E., Eisenbach, T., Span, R., Schiemann, M. and Scherer, V. "Combustion of Single Walnut Shell Particles in a Laminar Flow Reactor Under Oxy-Fuel Conditions: Optical Measurements and Particle Sampling". *Fuel* 369 (2024), 131613. https://doi.org/10.1016/j.fuel.2024.131613
73. Holland, T. M. and Fletcher, T. H. "Comprehensive Model of Single Particle Pulverized Coal Combustion Extended to Oxy-Coal Conditions". *Energy & Fuels* 31.3 (2017), 2722–2739. https://doi.org/10.1021/acs.energyfuels.6b03387
74. Senneca, O., Vorobiev, N., Wütscher, A., Cerciello, F., Heuer, S., Wedler, C., Span, R., Schiemann, M., Muhler, M. and Scherer, V. "Assessment of Combustion Rates of Coal Chars for Oxy-Combustion Applications". *Fuel* 238 (2019), 173–185. https://doi.org/10.1016/j.fuel.2018.10.093
75. Özer, B., Debiagi, P., Hasse, C., Faravelli, T. and Kazanç, F. "An Experimental and Numerical Study on the Combustion of Lignites from Different Geographic Origins". *Fuel* 278 (2020), 118320. https://doi.org/10.1016/j.fuel.2020.118320
76. Debiagi, P., Nicolai, H., Han, W., Janicka, J. and Hasse, C. "Machine Learning for Predictive Coal Combustion CFD Simulations—from Detailed Kinetics to HDMR Reduced-Order Models". *Fuel* 274 (2020), 117720. https://doi.org/10.1016/j.fuel.2020.117720
77. Debiagi, P., Piazza, V., Papagni, M., Beretta, A., Frassoldati, A. and Faravelli, T. "Cellulose Pyrolysis Kinetic Model: Detailed Description of Volatile Species". *Proceedings of the Combustion Institute* 40.1-4 (2024), 105651. https://doi.org/10.1016/j.proci.2024.105651
78. Piazza, V., Da Silva Junior, R. B., Frassoldati, A., Lietti, L., Chiaberge, S., Gambaro, C., Siviero, A., Faravelli, T. and Beretta, A. "Detailed Speciation of Biomass Pyrolysis Products with a Novel TGA-Based Methodology: The Case-Study of Cellulose". *Journal of Analytical and Applied Pyrolysis* 178 (2024), 106413. https://doi.org/10.1016/j.jaap.2024.106413
79. Zou, X., Debiagi, P., Amjed, M. A., Zhai, M. and Faravelli, T. "Impact of High-Temperature Biomass Pyrolysis on Biochar Formation and Composition". *Journal of Analytical and Applied Pyrolysis* 179 (2024), 106463. https://doi.org/10.1016/j.jaap.2024.106463
80. Pflieger, C., Eckhard, T., Böttger, J., Cerciello, F. and Muhler, M. "Catalytic Influence of Minerals on Cellulose-Based Model Fuels and Model Chars in Oxidation and Gasification". *30. Deutscher Flammentag*. Hannover, Germany, 2021.
81. Giummarella, N., Pu, Y., Ragauskas, A. J. and Lawoko, M. "A Critical Review on the Analysis of Lignin Carbohydrate Bonds". *Green Chemistry* 21.7 (2019), 1573–1595. https://doi.org/10.1039/C8GC03606C
82. Berkel, L. L., Debiagi, P., Nicolai, H., Cerciello, F., Suleiman, M. Y., Senneca, O., Hasse, C. and Faravelli, T. "Biomass Pyrolysis Kinetics Considering Lignin-Hemicellulose Interaction". *Fuel*, 409 (2026), 137793. https://doi.org/10.1016/j.fuel.2025.137793
83. Senneca, O., Cerciello, F., Heuer, S. and Ammendola, P. "Slow Pyrolysis of Walnut Shells in Nitrogen and Carbon Dioxide". *Fuel* 225 (2018), 419–425. https://doi.org/10.1016/j.fuel.2018.03.094
84. Ontyd, C., Pielsticker, S., Yildiz, C., Schiemann, M., Hatzfeld, O., Ströhle, J., Epple, B., Kneer, R. and Scherer, V. "Experimental Determination of Walnut Shell Pyrolysis Kinetics in $N_2$ and $CO_2$ via Thermogravimetric Analysis, Fluidized Bed and Drop Tube Reactors". *Fuel* 287 (2021), 119313. https://doi.org/10.1016/j.fuel.2020.119313

# Sensitivity Analysis and Uncertainty Quantification of a Single Combusting Particle

9

Ahmed Hassan and Taraneh Sayadi

## Contents



**Abstract**

This study explores the complex phenomena involved in simulations of unsteady solid fuel particle combustion. These simulations use models that depend on parameters from experiments or small-scale simulations, making it important to quantify the sensitivities and uncertainties of predicted quantities. The study

A. Hassan (✉) · T. Sayadi
Institute for Combustion Technology, RWTH Aachen University, Aachen, Germany
e-mail: a.hassan@itv.rwth-aachen.de

T. Sayadi
e-mail: taraneh.sayadi@lecnam.net





R. Kneer et al. (eds.), *Clean Energy Conversion*,
https://doi.org/10.1007/978-3-031-98956-8_9

focuses on combustion under different freestream compositions, such as air and oxy-fuel atmospheres. The first step involves identifying the most sensitive model parameters, a process developed in a previous study. The framework is extended to study the evolution of sensitivities over time for a single unsteady particle undergoing devolatilisation. The dominant model parameters affecting particle temperature and burning rate are identified for both air and oxy-fuel atmospheres. The impact of different devolatilisation models on sensitivities is also examined. Additionally, an adjoint-based active subspace method (AASM) is employed to quantify uncertainties in predictions. The results show that uncertainties in particle parameters are much greater than those in gas-phase reaction rates. Furthermore, uncertainties are larger in oxy-fuel atmospheres than in air, emphasising the effect of freestream composition on model predictions.

## 9.1 Introduction

As demonstrated in this book, the new solid fuel sources require careful modelling of the combustion process as it occurs in the solid particle as well as its interaction with the surrounding gas. To this end, experimental campaigns have been undertaken in order to characterise and model the existing heterogeneous and homogeneous reactions. Numerical investigations using the developed models, such as char burnout, devolatilisation and gas-phase reactions, are then used to predict the overall efficiency of the underlying system. As the models used in these simulations are developed through experiments, they are reported with a confidence range (i.e. probabilistic variables) and therefore the predictions of the models need to be assessed in terms of the existing uncertainties. This information would be an indication of the robustness of the predictions of the simulations. In this chapter, dedicated methods to extract such sensitivities and uncertainties in the predictions with respect to the combustion model parameters are presented, both in the solid particle and in the gas phase.

The most common method for calculating the gradient of various objective functions with respect to model parameters is forward sensitivity estimation. However, as can be seen from the work presented in this book, the size of the parameter space (i.e. the number of parameters representing the kinetic models) makes the forward approach prohibitively expensive. An alternative approach is the adjoint-based method. This method is an efficient alternative to extract the sensitivities of relevant quantities of interest (QoI): The performance of the adjoint-state method does not depend on the number of parameters and is easily applicable to unsteady configurations. The application of the adjoint-state method in fluid mechanics dates back to the pioneering work by Pironneau [1], followed by Jameson et al. [2] with application to aeronautical shape optimisation. Since then adjoint-based techniques have been used for optimal control of non-linear problems [3], and in particular, in the combustion community, to study ignition delay time of zero-dimensional reactors [4], steady-state flames [5], and more recently, to high-fidelity turbulent reacting flow [6]. For multiphase combustion (solid/gas), the adjoint-state method to steady and unsteady char combustion is applied [7,8].

Based on this methodology, the analysis of more sophisticated combustion models is presented in this chapter. More importantly, the variations in the sensitivities of the most dominant parameters over time are also studied. The results of this investigation will be important for transient problems such as that of single particle combustion, which can go through different stages of combustion from degassing, ignition and char burnout. This strategy will allow to determine whether any of the existing physical phenomena can cause a significant change in the evolution of sensitivities.

While informative in identifying the sensitive parameters in the model, sensitivity analysis provides only a local estimate of the variation in the quantity of interest with respect to each parameter. To gain a global perspective, it is necessary to perform uncertainty quantification. Traditional methods for performing uncertainty quantification (UQ), such as Monte-Carlo, become infeasible when dealing with systems with a large parameter space (curse of dimensionality). Therefore, several stochastic models have been proposed, such as polynomial chaos expansion (PCE) [9], as well as other techniques that consider only a small part of the parameter space to make the analysis affordable. Alternative methods to reduce computational time are high dimensional model representations (HDMR) [10] and artificial neural networks (ANN) [11]. Despite ongoing efforts to overcome the curse of dimensionality, these methods become inefficient as the number of parameters increases. As an alternative, the active subspace method (ASM) has been proposed [12] to identify the key directions in the function space of the QoI, allowing for a more efficient analysis. ASM has been applied to several studies in the field of combustion, proving to be a successful framework for quantifying uncertainties in various reactive configurations with a large number of parameters [13,14]. The ASM method has been successfully combined with the adjoint state method to further reduce the cost of the algorithm and improve its efficiency by increasing the accuracy of the extracted gradients [15,16]. The technique is called the adjoint active subspace method (AASM).

In addition, an alternative method using only the adjoint information, called the linear approximation adjoint-based method (LAAM) is proposed, to approximate the resulting uncertainties. A comparison of this strategy with the state-of-the-art in the context of a zero-dimensional reactor (Table 9.1) shows that the cost of LAAM is generally about two orders of magnitude lower than that of ASM. The computational cost of ASM compared to LAAM increases significantly as the size of the considered chemical kinetic mechanism increases. This remarkable cost reduction highlights the advantage of LAAM, especially as a first estimation of the uncertainties of large mechanisms.

The resulting algorithm is also implemented in the open-source package `Apophis.jl`. The package is coded in the Julia language. It is important to note that Julia is a new programming language designed to facilitate the computation of scientific applications. Julia combines the execution speed of statically typed languages with the active performance of dynamic languages. This package provides the community with a framework for investigating the uncertainties of chemical kinetic models in an isochoric adiabatic reactor configuration.

**Table 9.1** Comparison of LAAM to ASM for various gas-phase mechanisms. For more information regarding the mechanisms the reader is referred to Hassan et al. [15].

| Mechanism | h2vb | GRI 1.2 | GRI 2.11 | GRI 3.0 | ITV |
|---|---|---|---|---|---|
| # Species | 10 | 32 | 49 | 53 | 490 |
| # Reaction | 21 | 177 | 297 | 325 | 2072 |
| Time (LAAM) [s] | $160 \times 10^{-3}$ | 2.6 | 6 | 15 | 2268 |
| Time (ASM) [s] | 50 | 1203 | 3716 | 11571 | $2.5 \times 10^6$ |

Despite the successful use of AASM and LAAM to extract model parameter uncertainties, the above-mentioned contribution only studied simple configurations in unsteady zero-dimensional or steady one-dimensional settings. Additionally, previous studies on sensitivities of solid fuel combustion also considered char models only, neglecting important phenomena such as devolatilisation. Thus, in what follows: (i) the evaluation of the effectiveness of adjoint-based methods and AASM in extracting sensitivities and uncertainties in more complex unsteady systems, including devolatilisation, (ii) the identification of the most significant gas-phase reactions by incorporating detailed gas-phase combustion mechanisms, and (iii) the examination of the effect of freestream composition on the extracted sensitivities and uncertainties.

This chapter is structured as follows: Sect. 9.2 briefly introduces the underlying configuration and the models used to represent the reaction in the particle and gas phases, as well as the respective freestream compositions. The simulations of unsteady axisymmetric particle are then presented and discussed in Sect. 9.3. Section 9.4 discusses the methodology used to extract the sensitivities and discusses the extracted sensitivities for the different quantities of interest with respect to the existing modelling parameters. The evolution of the important sensitivities over time is also presented and discussed in Sect. 9.5. Section 9.6 presents the AASM algorithms and the results of the uncertainty quantification for different model parameters. The temporal evolution of the extracted uncertainties is also discussed. Finally, Sect. 9.7 discusses the summary and conclusion of this work.

## 9.2 Models and Configuration

The configuration considered here is a reactive axisymmetric particle in a fluid, as shown in Fig. 9.1.

### 9.2.1 Governing Equations

The governing equations that describe the flow around a reactive axisymmetric particle in a fluid are presented in a similar way as in the study by Hassan et al. [8]:

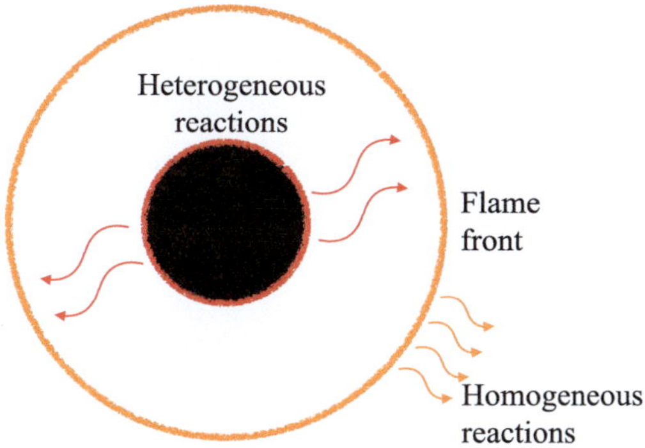


**Fig. 9.1** Schematic of an axisymmetric configuration used in this chapter

$$\frac{\partial \rho}{\partial t} + \frac{1}{r^\alpha}\frac{\partial}{\partial r}\left(\rho\, r^\alpha\, u\right) = 0, \tag{9.1}$$

$$\frac{\partial(\rho\, Y_i)}{\partial t} + \frac{1}{r^\alpha}\frac{\partial}{\partial r}\left[\rho\, r^\alpha\left(u\, Y_i - D_i\frac{\partial Y_i}{\partial r}\right)\right] - \omega_i = 0, \tag{9.2}$$

$$c_\mathrm{p}\frac{\partial(\rho\, T)}{\partial t} + \frac{c_p}{r^\alpha}\frac{\partial}{\partial r}\left[\rho r^\alpha\left(u\, T - \frac{\lambda}{c_p}\frac{\partial T}{\partial r}\right)\right] + \sum_{i=1}^{N_\mathrm{s}} h_i\, \omega_i = 0, \tag{9.3}$$

where $t$ and $r$ are the independent variables, $i = 1, \ldots, N_\mathrm{s}$ denotes the species and $\alpha = 2$ for spherical coordinates. The diffusion model is based on Fick's law, and Fourier's law has been used to model heat transfer. Also, $c_p$ denotes the thermal capacity of the mixture at constant pressure, $\lambda$ its heat conductivity, and $D_i$ is the diffusion coefficient of species $i$. The mixture density is denoted by $\rho$, and the radial mass-centred velocity is $u$. The mass fractions and temperature are denoted as $Y_i$ and $T$, respectively. In (9.3), the term $\sum h_i\, \omega_i$ represents the heat release due to chemical reactions, where $h_i$ denotes the specific enthalpy and $\omega_i$ denotes the production rate of the $i^{\text{th}}$ species.

**Mass-Based Coordinates**

For ease of implementation, and without loss of generality, the governing equations can be mapped from the physical coordinates $(r, t)$ to the mass-based coordinate system $(\psi, \tau)$. This transformation satisfies the continuity equation by introducing a new algebraic equation in which the discrete form of the mapping equation has a closed-form solution. Moreover, the direct dependency on the velocity variable is eliminated from the governing equations. The transformation is carried out using (9.4) and (9.5):

$$\frac{\partial}{\partial r} = \rho\, r^2\, \frac{\partial}{\partial \psi}, \tag{9.4}$$

$$\frac{\partial}{\partial t} = \frac{\partial}{\partial \tau} - \rho\, r^2\, u_r\frac{\partial}{\partial \psi} + \dot{m}_0^\alpha\frac{\partial}{\partial \psi} \tag{9.5}$$

where $\dot{m}_0^\alpha$ represents the mass flux per unit surface area. Applying chain rule to (9.4) and (9.5), under the assumption of unity Lewis number, yields to (9.6) and (9.7) that govern the evolution of species and energy in the transformed coordinates:

$$\frac{\partial Y_i}{\partial \tau} + \dot{m}_0^\alpha \frac{\partial Y_i}{\partial \psi} - \frac{\partial}{\partial \psi}\left(\rho r^{2\alpha} \frac{\lambda}{c_\mathrm{p}} \frac{\partial Y_i}{\partial \psi}\right) - \frac{\omega_i}{\rho} = 0, \tag{9.6}$$

$$\frac{\partial T}{\partial \tau} + \dot{m}_0^\alpha \frac{\partial T}{\partial \psi} - \frac{\lambda}{c_p} \frac{\partial}{\partial \psi}\left(\rho\, r^{2\alpha} \frac{\partial T}{\partial \psi}\right) + \frac{1}{\rho\, c_p} \sum_{i=1}^{N_\mathrm{s}} h_i\, \omega_i = 0, \tag{9.7}$$

where $r$ is computed from the following equation, which in turn enforces mass conservation:

$$\frac{\partial r}{\partial \psi} = \frac{1}{\rho\, r^2}. \tag{9.8}$$

## 9.2.2 Boundary Conditions and Freestream Compositions

The first set of boundary conditions at the particle surface balances the mass transfers from the solid phase and the gas phase, which are due to phase change from solid to gas, resulting from devolatilisation and surface reactions. They take the form of coupled Robin boundary conditions and, once recast in the transformed coordinate system, read as

$$\dot{M}_\mathrm{s}\, Y_{i,\mathrm{s}} - \mathrm{C}_\alpha\, \rho\, r_0^{2\alpha} \frac{\lambda}{c_\mathrm{p}} \left.\frac{\partial Y_i}{\partial \psi}\right|_\mathrm{s} - \omega_{i,\mathrm{s}} = 0 \tag{9.9}$$

where $\mathrm{C}_\alpha = 4\pi$. The variables associated with the particle surface are annotated with the subscript “s”. The burning rate $\dot{M}_\mathrm{s}$ is the sum of the char reaction rate and the volatile release rate, which can be expressed as:

$$\dot{M}_\mathrm{s} = \dot{m}_\mathrm{c} + \sum_{i=1}^{N_\mathrm{v}} \dot{m}_{\mathrm{v},i} \tag{9.10}$$

where $\dot{m}_\mathrm{c}$ is the char burning rate, $\dot{m}_{\mathrm{v},i}$ is the volatile mass release rate of species $i$, and $\omega_{i,\mathrm{s}}$ is the total chemical source of species $i$ at the particle surface. Assuming thermal equilibrium between the particle surface and the adjacent gas-phase layer, the gas-phase temperature at the particle surface ($T_\mathrm{s}^\mathrm{g}$) matches the particle temperature ($T_\mathrm{p}$),

$$T_\mathrm{s}^\mathrm{g} = T_\mathrm{p}. \tag{9.11}$$

The particle temperature is in turn evaluated by stating the energy conservation in a spherical control volume delimited by the particle surface

$$M_\mathrm{p}\, C_\mathrm{s} \frac{\partial T_\mathrm{p}}{\partial \tau} - \mathrm{C}_\alpha\, \rho\, r_0^{2\alpha}\, \lambda_\mathrm{s} \frac{\partial T}{\partial \psi} - \left(\sum_{i=1}^{\mathrm{N}} \dot{m}_{\mathrm{v},i}\, h_{\mathrm{v},i} + \dot{m}_\mathrm{c}\, h_\mathrm{c}\right) = 0, \tag{9.12}$$

**Table 9.2** Compositions of oxygen-rich air and oxy-fuel atmospheres.

| Test case | Diluent | $Y_{O_2}^{\infty}$ [%] | $Y_{CO}^{\infty}$ [%] | $Y_{CO_2}^{\infty}$ [%] | $Y_{N_2}^{\infty}$ [%] | $T^{\infty}$ [K] |
|---|---|---|---|---|---|---|
| Air | $N_2$ | 32.6 | 0 | 10.5 | 56.9 | 1673 |
| Oxy-fuel | $CO_2$ | 26.2 | 0 | 73.8 | 0 | 1673 |

where $M_p$, $C_s$, and $\lambda_s$ are the particle mass, heat capacity and heat conductivity, respectively. The enthalpy of the char reaction $h_c$ and the devolatilisation $h_{v,i}$ will be discussed in Sect. 9.2.3 together with the different source terms. The thermodynamic properties of the particle are taken from Hassan et al. [8]. The variation in the particle mass due to combustion and devolatilisation can be computed as

$$\frac{dM_p}{dt} = -\dot{M}_s \tag{9.13}$$

Finally, the boundary conditions in the far-field are set to the following Dirichlet conditions:

$$Y_i\,(\tau, \psi \to \infty) = Y_i^{\infty}, \tag{9.14}$$

$$T\,(\tau, \psi \to \infty) = T^{\infty}. \tag{9.15}$$

The compositions considered in this study are presented in Table 9.2, and are similar to those of Farmand et al. [17] and Farazi et al. [18].

### 9.2.3 Chemical and Thermodynamic Models

Several terms introduced in (9.9)–(9.12) require modelling. In this section, the chemical and thermodynamic models used to closed the above system are discussed.

**Char Combustion**

To model the char burning rate, a single-step reaction is considered with the reaction

$$C(s) + O_2(g) \to CO_2(g). \tag{9.16}$$

The char burning rate, denoted as $\dot{m}_c$, is modelled as the rate of oxygen consumption at the particle surface. This can be expressed as:

$$\dot{m}_c = \omega_{O_2,s} = \rho_c\, K_c\, Y_{O_2,s}\, \pi\, d_p^2, \tag{9.17}$$

where $\rho_c$ is the particle density, $K_c$ is the char reaction rate, $Y_{O_2,s}$ is the oxygen mass fraction at the particle surface, and $d_p = 120\ \mu m$ is the particle diameter. The char reaction rate is computed using Arrhenius law with $E_c = 100.9$ kJ mol$^{-1}$, $\beta_c = 0$ and $A_c = 3020.75\ s^{-1}$. Finally, the heat of reaction is $h_c = 3 \times 10^4$ kJ kg$^{-1}$, following the model described in the study by Du and Annamalai [19].

**Table 9.3** Volatile species composition.

| Species | H | CO | $CO_2$ | $CH_4$ |
|---|---|---|---|---|
| Composition [%] | 32.6 | 0 | 10.5 | 56.9 |

**Devolatilisation**

A two-step reaction model [17, 19] with first-order kinetics is used to represent the mass release of devolatilised species, given by

$$\dot{m}_{\mathrm{v},i} = \zeta_i \, K_{\mathrm{v},i} \, M_{\mathrm{p}}, \tag{9.18}$$

where $\zeta_1 = \zeta_2 = 0.5$ are the release constants. The volatile release rate $K_{\mathrm{v},i}$ follows the Arrhenius law, with the rate constant values for the multiple steps given by $E_{\mathrm{v},1} = 74.1$ kJ mol$^{-1}$, $A_{\mathrm{v},1} = 1.34 \times 10^5$ s$^{-1}$, $\beta_{\mathrm{v},1} = 0$, $E_{\mathrm{v},2} = 251$ kJ mol$^{-1}$, $A_{\mathrm{v},2} = 1.46 \times 10^5$ s$^{-1}$ and $\beta_{\mathrm{v},2} = 0$. The heat release associated with devolatilisation is $h_{\mathrm{v},1} = h_{\mathrm{v},2} = 1 \times 10^3$ kJ kg$^{-1}$ [19]. The species composition resulting from devolatilisation processes is listed in Table 9.3.

**Gas-Phase Kinetics**

The gas-phase chemistry is presented by the GRI 1.2 mechanism which includes 177 reactions between 32 species. The different thermodynamic properties for each species are computed using NASA polynomial correlations [20] and averaged over the composition as:

$$\Phi = \frac{\phi_i \, Y_i}{\mathrm{W_i}}, \tag{9.19}$$

where $\phi_i \in \left\{c_{p,i}, h_i, s_i\right\}$ represents any thermodynamic specific quantity of the $i^{\mathrm{th}}$ species. The corresponding mixture average quantity is denoted as $\Phi$. Finally, the closure is completed using the ideal gas law.

## 9.3 Simulations of an Axisymmetric Solid Particle

The combination of the models and governing equations results in the solutions demonstrated in Fig. 9.2 for the dominant species. As shown in Fig. 9.2a, carbon dioxide peaks away from the particle surface. This can be attributed to the high reactivity of the reaction involving $\mathrm{OH} + \mathrm{CO} \rightleftharpoons \mathrm{H} + \mathrm{CO_2}$, which leads to the consumption of carbon monoxide and the production of carbon dioxide. It should be emphasised that during this phase, char combustion, also known as the heterogeneous ignition, is negligible due to the unavailability of oxygen at the particle surface. This is due to the fact that oxygen is consumed in gas-phase reactions during the devolatilisation process.

As quantities of interest, burning rate and particle temperature are considered and shown in Fig. 9.3. It is essential to identify different stages of temporal evolution of quantities of interest such as burning rate and particle temperature [8].

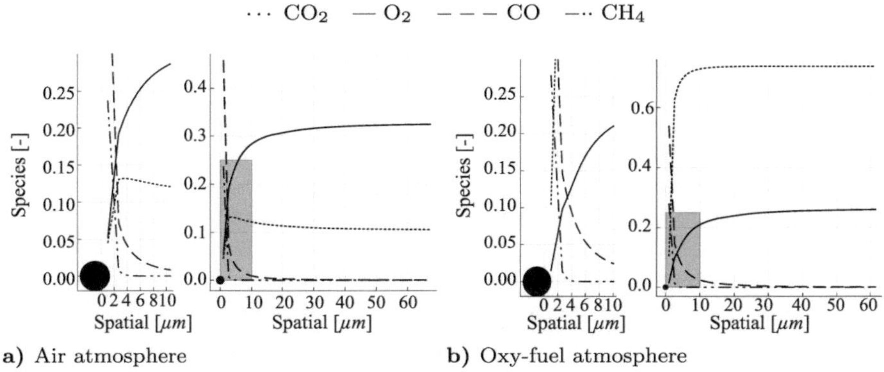


**Fig. 9.2** Distribution of species in air and oxy-fuel atmospheres at $t = 4$ ms (reprinted from Hassan et al. [16], Copyright (2024), with permission from Elsevier and minor editing)

These quantities will later be used to characterise parametric sensitivities. As these QoI indicate the different stages of particle combustion (preheating, devolatilisation, ignition, etc.), they help to study the mutual effects and transitions between different phenomena. Additionally, particle temperature is a major parameter measured in different experiments, and can be used to compare and validate numerical results against experimental data [21]. In both the air and oxy-fuel environments, the particle temperature $T_\text{p}$ increases due to the temperature gradient between the hot medium and the colder particle. During the initial heating stage ($t < 0.5$ ms), both cases experience similar heating rates due to the significant temperature difference between the gas and solid phases, as depicted in Fig. 9.3a. However, the heating rate for the air atmosphere is lower than that of the oxy-fuel atmosphere due to the more rapid increase in gas-phase temperature for oxy-fuel relative to air. An inflection point can be observed in the $T_\text{p}$ evolution curve in Fig. 9.3a around $t \approx 6$ μs,

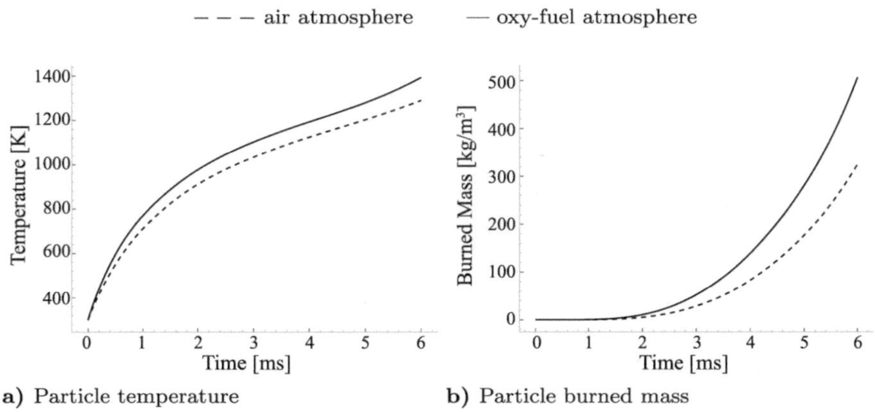


**Fig. 9.3** Temporal evolution of particle temperature and burning rate (reprinted from Hassan et al. [16], Copyright (2024), with permission from Elsevier and minor editing)

which is related to the initiation of char ignition. Notably, the inflection point is not observed for the air atmosphere due to the lower particle heating rate.

## 9.4 Sensitivity Analysis

This section aims to compute the sensitivities of various quantities of interest (QoI) with respect to different model parameters. The gradients will be computed for both the air and oxy-fuel atmospheres, whose compositions are presented in Table 9.2. Additionally, the evolution of sensitivities in time will be examined. All the sensitivities are normalised, helping to distinguish the dominant parameters.

Sensitivity analysis is a method that enables the determination of the rate of change of a particular output measure, such as burning rate or particle temperature, with respect to various model parameters (thermal conductivity, reaction rates, etc.) or initial conditions. By examining the parameter sensitivities obtained, one can identify the most influential parameters for the accurate evaluation of the quantity of interest. Due to the advantages of the adjoint-based algorithms, as highlighted in the introduction of this chapter, they provide an economical and accurate alternative to existing approaches such as finite difference or forward estimation method. Using adjoint-based strategies, the computational cost scales with the number of quantities of interests as opposed to the number of system variables. In other words, the gradient with respect to all systems parameters can be evaluated in a single (forward then backward) simulation, making this method very attractive in cases with a large parameter space, as is the case of interest in this study [8, 15].

Adjoint equations can be derived in various ways: continuous, discrete, or semi-discrete. In this study, the focus will be on the semi-discrete approach, where the adjoint operators are directly extracted for the spatial operators, whereas the time variable is kept continuous. Taking a general form of the system equation in temporally continuous form gives

$$\begin{cases} \frac{\mathrm{d}\mathbf{u}^d}{\mathrm{d}\tau}(\tau) = \mathbf{F}^d\left[\tau, \mathbf{u}^a(\tau), \mathbf{u}^d(\tau), \mathbf{p}\right], \\ \mathbf{0} = \mathbf{F}^a\left[\tau, \mathbf{u}^a(\tau), \mathbf{u}^d(\tau), \mathbf{p}\right] \end{cases} \tag{9.20}$$

where $\mathbf{u} = \left(\mathbf{u}^a, \mathbf{u}^d\right)$ denotes the state vector. This system of equations is in the form of a differential algebraic equation (DAE) system of index 1, thus the state vector consists of differential $\mathbf{u}^d$ (species mass fraction, gas-phase temperature, particle temperature, and particle mass) and algebraic $\mathbf{u}^a$ (mass fractions at the surface) variables. The vector of the chosen parameters is denoted $\mathbf{p}$. The operator $\mathbf{F}^a$ contains the discrete form of the algebraic equations (i.e. Robin boundary and Dirichlet boundary conditions). Similarly, $\mathbf{F}^d$ presents the semi-discrete differential equations and particle temperature equations. It should be stressed that the DAE of index-1 behaves differently from ordinary differential equations (ODE) in that no rate of change of the algebraic variable $\mathbf{u}^a$ appears in the system. After discretising the spatial operators of the forward problem, the first step in deriving the adjoint equations is to form the following Lagrangian functional

$$L = \mathrm{QoI}\left(\mathbf{u}^a, \mathbf{u}^d, \mathbf{p}\right) - \int_0^{\mathcal{T}} \eta^d \cdot \left\{ \frac{\mathrm{d}\mathbf{u}^d}{\mathrm{d}\tau} - \mathbf{F}^d \left[\tau, \mathbf{u}^a(\tau), \mathbf{u}^d(\tau), \mathbf{p}\right] \right\} \mathrm{d}\tau + \int_0^{\mathcal{T}} \eta^a \cdot \mathbf{F}^a \left[\tau, \mathbf{u}^a(\tau), \mathbf{u}^d(\tau), \mathbf{p}\right] \mathrm{d}\tau, \tag{9.21}$$

where $\mathcal{T}$ is the time horizon under consideration, and $\eta^a$ and $\eta^d$ are the algebraic and differential Lagrangian (adjoint) variables, respectively. The derivation of the adjoint equations changes based on the type of QoI (integral or local [8]). In this study, the interest lies in an average of the QoI in time, written in integral form as

$$\mathrm{QoI} = \int_0^{\mathcal{T}} g\left[\tau, \mathbf{u}^a(\tau), \mathbf{u}^d(\tau), \mathbf{p}\right] \mathrm{d}\tau. \tag{9.22}$$

Here $g$ represents the time varying quantity of interest such as particle temperature or burning rate. Substituting this QoI in the Lagrangian functional from (9.21) leads to the following adjoint equations

$$\frac{\mathrm{d}\eta^d}{\mathrm{d}\tau} + \left(\frac{\partial \mathbf{F}^d}{\partial \mathbf{u}^d}\right)^\dagger \eta^d + \left(\frac{\partial \mathbf{F}^a}{\partial \mathbf{u}^d}\right)^\dagger \eta^a = -\frac{\partial g}{\partial \mathbf{u}^d}, \tag{9.23}$$

$$\left(\frac{\partial \mathbf{F}^d}{\partial \mathbf{u}^a}\right)^\dagger \eta^d + \left(\frac{\partial \mathbf{F}^a}{\partial \mathbf{u}^a}\right)^\dagger \eta^a = -\frac{\partial g}{\partial \mathbf{u}^a}, \tag{9.24}$$

where $\dagger$ denotes the transpose operator. In this case, the initial condition of the adjoint equation is

$$\eta^d(\mathcal{T}) = \mathbf{0}. \tag{9.25}$$

The sensitivities can then be evaluated by extracting the gradient of the Lagrangian with respect to the system parameters $\mathbf{p}$. This leads to

$$\frac{\mathrm{dQoI}}{\mathrm{d}\mathbf{p}} = \int_0^{\mathcal{T}} \left(\frac{\partial g}{\partial \mathbf{p}} + \eta^d \cdot \frac{\partial \mathbf{F}^d}{\partial \mathbf{p}} + \eta^a \cdot \frac{\partial \mathbf{F}^a}{\partial \mathbf{p}}\right) \mathrm{d}\tau. \tag{9.26}$$

For a detailed derivation of (9.26), the reader is referred to the study by Hassan et al. [8].

### 9.4.1 Sensitivities of Particle Temperature

To evaluate sensitivities over all combustion phases and to avoid the complexities of particle ignition, the time-cumulative particle temperature $T_\mathrm{p}$ is examined rather than its instantaneous value, resulting in

$$\mathrm{QoI} = \frac{1}{\mathcal{T} - \tau_0} \int_{\tau_0}^{\mathcal{T}} T_\mathrm{p}\, d\tau. \tag{9.27}$$

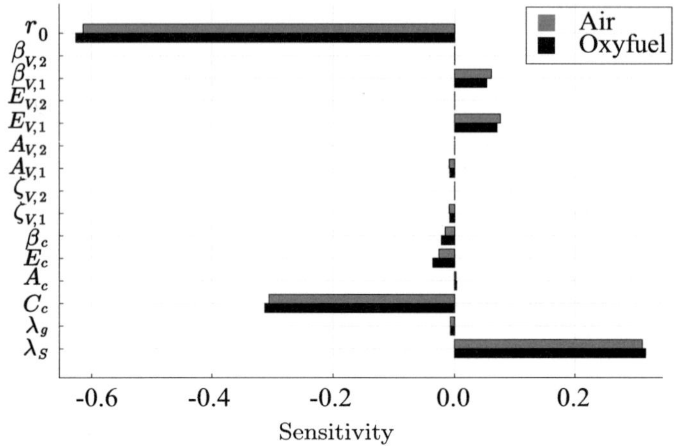


**Fig. 9.4** Sensitivity of the average particle surface temperature with respect to the particle parameters, for air and oxy-fuel atmospheres (reprinted from Hassan et al. [16], Copyright (2024), with permission from Elsevier and minor editing)

The integration limits in (9.27) are $\tau_0 = 0$ and $\mathcal{T} = 5$ ms chosen to ensure that gas-phase ignition occurs in both air and oxy-fuel atmospheres, while avoiding char combustion.

The sensitivities of particle temperature with respect to different particle (surface) parameters are shown in Fig. 9.4. It shows that particle size is the most influential parameter, where increasing particle size results in lower temperatures. Sensitivities to particle thermodynamic parameters ($C_c$ and $\lambda_s$) are the most prominent during the heat-up phase. Sensitivities with respect to char combustion parameters $E_c$, $\beta_c$ and $A_c$ are insignificant due to the absence of heterogeneous ignition. Oxygen is mainly consumed by gas-phase combustion. Particle temperature also shows low sensitivities with respect to various devolatilisation parameters. These parameters are however important when considering particle burning rate, as demonstrated in the study by Farazi et al. [18] and subsequently gas-phase ignition, and will be highlighted in the following section. Comparing sensitivities in air and oxy-fuel atmospheres reveals a similar pattern for both. Notably, the sensitivity to particle parameters $r_0$, $C_c$ and $\lambda_s$ are slightly lower in air atmosphere due to weak flammability, where the temperature gradient between gas and particle decreases.

To assess the effect of the mechanism used to model the gas-phase chemistry on the extracted sensitivities, a simpler reduced kinetics mechanism (DRM22), consisting of 104 reactions and 22 species, is implemented in the simulation. The sensitivities of the particle temperature with respect to the different gas-phase reaction rates are shown in Fig. 9.5. A comparison with results from the detailed mechanism shows that the top ten most sensitive reactions are indeed the same in both cases. The first two dominant reactions in air and oxy-fuel atmosphere remain unchanged, regardless of the mechanism used.

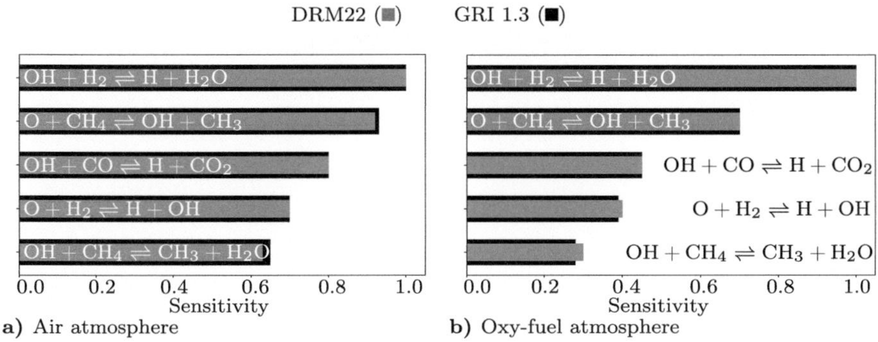


**Fig. 9.5** Sensitivity of the average particle surface temperature with respect to the rates of gas-phase reactions of different mechanisms for air and oxy-fuel atmospheres

### 9.4.2 Sensitivities of Burning Rate

The burning rate, denoted by $\dot{m}_t$, is a crucial QoI in particle combustion analysis. Similar to $T_p$, the burning rate is consistently increasing over time. Therefore, it is reasonable to define the time-averaged value of the burning rate as the QoI. This value can be defined as follows:

$$\mathrm{QoI} = \int_{\tau_0}^{\mathcal{T}} \dot{m}_t \, d\tau. \tag{9.28}$$

Rather than analysing the separate effects of char combustion and devolatilisation ($\dot{m}_t = \dot{m}_c + \dot{m}_v$), it is possible to construct a QoI based on the variation of the total particle mass as shown in (9.13),

$$\mathrm{QoI} = \int_{\tau_0}^{\mathcal{T}} \dot{m}_t \, d\tau = -\int_{\tau_0}^{\mathcal{T}} \frac{\partial M_p}{d\tau} \, d\tau, \tag{9.29}$$

$$\mathrm{QoI} = -\int_{\tau_0}^{\mathcal{T}} \frac{\partial M_p}{d\tau} \, d\tau = M_p|_{\tau_0} - M_p|_{\mathcal{T}}, \tag{9.30}$$

This approach allows for a more comprehensive evaluation of the overall behaviour of the combustion process. The values of $\tau_0$ and $\mathcal{T}$ are chosen to be similar to $T_p$ in order to facilitate their comparison. Figure 9.6 displays the sensitivity analysis results with respect to the various system parameters. The sensitivity values for $r_0$ are significant, similar to that of the particle temperature, owing to the strong correlation between $T_p$ and $\dot{m}_t$. However, in contrast, the devolatilisation parameters have the most dominant effect when the char combustion rate is low, as $\dot{m}_t$ is approximately equal to $\dot{m}_v$, as there is no heterogeneous ignition. In comparison to the extracted sensitivities of particle temperature (Fig. 9.4), Fig. 9.6 indicates lower sensitivities for the thermodynamic parameters.

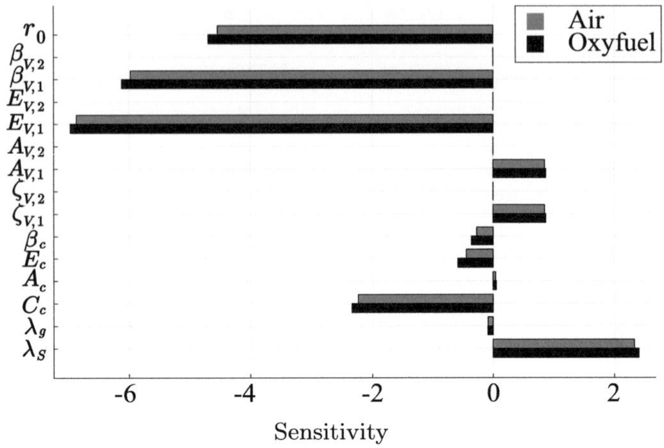


**Fig. 9.6** Sensitivity of the average particle burning rate with respect to the particle parameters, for air and oxy-fuel atmospheres (reprinted from Hassan et al. [16], Copyright (2024), with permission from Elsevier and minor editing)

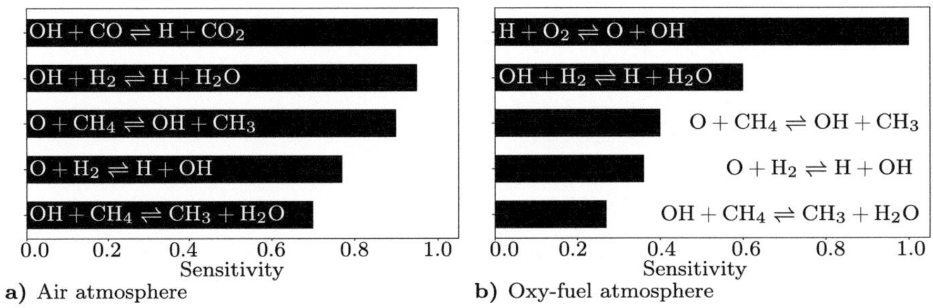


**Fig. 9.7** Sensitivity of the average particle burning rate with respect to the rates of gas-phase reactions

Figure 9.7 highlights the sensitivities of the particle burning rate with respect to the gas-phase reaction rates. These results are consistent with the time-cumulative particle temperature results depicted in Fig. 9.5. The agreement is expected as the ignition process, particle temperature, and the burning rate are highly correlated in this setting. The only notable difference between Figs. 9.7a and 9.7b is that the reaction $OH + CO \rightleftharpoons H + CO_2$ is the most dominant in air atmosphere compared to oxy-fuel atmosphere. This is due to the high release rate of carbon dioxide in the air atmosphere, as shown in Fig. 9.2a.

## 9.5 Evolution of Sensitivities in Time

Given the unsteady nature of the problem, it is important to investigate the temporal evolution of sensitivities with respect to the most influential system parameters. In other words, instead of reporting sensitivities at a discrete time, as in the previous sections, the evolution of these sensitivities for different time horizons needs to be reported. This information will allow to assess whether the predicted sensitivities depend on the time at which they are evaluated and, by definition, on the state of the time-varying system. This is achieved by adjusting the limits of the integral, for example in (9.27), to encompass the entire time domain of the QoI. The time domain is then discretised into several sequential segments, and the sensitivities are computed individually for each segment and then aggregated to provide the complete time evolution. It should also be noted that the sensitivities are in fact the rate of change of the quantity of interest with respect to the parameter. This rate could be either negative, meaning an inverse relationship, or positive. In any case, what is important for this study is the value of this change, as it implies a strong dependency of the quantity of interest on the respective parameter.

### 9.5.1 Particle Temperature as QoI

The results of the evolution of particle parameter sensitivities, when considering particle temperature as the quantity of interest, are shown in Fig. 9.8. These results demonstrate that the sensitivity of particle temperature ($T_\mathrm{p}$) with respect to surface heat conductivity ($\lambda_\mathrm{s}$) for both air and oxy-fuel atmosphere increases until $t \approx 1.5$ ms. This is because the gradient between the adjacent gas layer and particle temperatures decreases significantly at this stage, minimising the effect of $\lambda_\mathrm{s}$. Gas conductivity $\lambda_\mathrm{g}$ follows a similar trend but with a more rapid decay rate due to the propagation of the flame towards the particle surface (not shown here).

Both volatiles activation energy and temperature exponent, ($E_{\mathrm{v},1}$ and $\beta_{\mathrm{v},1}$), show a similar behaviour independent of the atmosphere (air/oxy-fuel). They begin to have an impact after the early non-linear heating phase, close to the gas-phase ignition onset, as demonstrated in Fig. 9.2a. The sensitivities to particle radius and thermal capacity ($\mathrm{r}_0$ and $C_\mathrm{c}$, respectively) increase in both atmospheres as the heating rate of the particle $\frac{\mathrm{d}T_\mathrm{p}}{\mathrm{d}t}$ reaches its maximum. Following the ignition onset, their sensitivities decrease slightly as the heating rate of the particle diminishes and the flame spreads towards the particle surface.

### 9.5.2 Burning Rate as QoI

Evolution of particle burning rate sensitivities in time with respect to the top sensitive parameters is depicted in Fig. 9.9. Similar to particle temperature the sensitives of the dominant parameters reach their maximum when ignition occurs ($t = 1$ ms). These results are expected due to the strong impact of gas-phase ignition on devolatilisation rate and temperature gradient between the gas and particle surface. Comparison of Figs. 9.9 and 9.8 shows that the sensitivity of the temperature exponent $\beta_{\mathrm{v},1}$ in the devolatilisation model with respect to burning rate remains nearly constant regardless of the atmosphere. However, it exhibits a higher sensitivity value when compared to particle temperature.

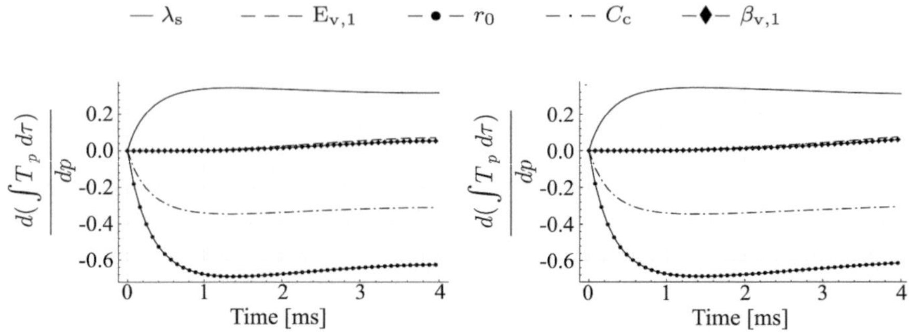


a) Air atmosphere b) Oxy-fuel atmosphere

**Fig. 9.8** Temporal evolution of the average particle temperature sensitivities with respect to the system parameters (reprinted from Hassan et al. [16], Copyright (2024), with permission from Elsevier and minor editing)

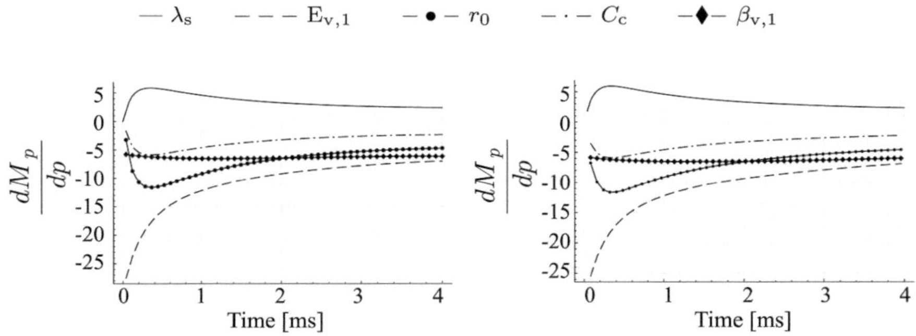


a) Air atmosphere b) Oxy-fuel atmosphere

**Fig. 9.9** Temporal evolution of the average particle burning rate sensitivities with respect to the system parameters (reprinted from Hassan et al. [16], Copyright (2024), with permission from Elsevier and minor editing)

In both atmospheres, the particle burning rate is highly sensitive to the volatile activation energy $E_{v,1}$, especially at an early stage, where $\dot{m}_t \approx \dot{m}_v$ and char combustion is negligible. After gas-phase ignition, the particle temperature $T_p$ becomes high enough and the sensitivity of the volatile activation energy dwindles with a similar rate independent of the atmosphere. It is worth noting that the signs of the sensitivities remain consistent and align with the findings presented in Fig. 9.6.

Comparison of Figs. 9.9 and 9.8 shows that while the atmosphere does not seem to influence the value and evolution of sensitivities with respect to different particle parameters, the selected QoI could have a large influence. This suggests that depending on the quantity to be predicted in such systems, different particle parameters could be influential, and how influential depends on the time instance that the analysis is performed.

## 9.6 Uncertainty Quantification

This section provides a brief description of the adjoint-based active subspace method (AASM), used to identify the dominant directions in the parameter space and to construct the resulting low-dimensional response surface.

The active subspace methodology aims to construct an $M$-dimensional subspace of the $n$-dimensional parameter space ($n \ll M$) that describes most of the variation of quantity of interest (QoI). The low-dimensional approximation of QoI is

$$\mathrm{QoI(x)} \approx G(\mathrm{y}), \quad \mathrm{y} = \mathrm{S}^\top \mathrm{x} \tag{9.31}$$

where $G(\mathrm{y})$ is the response surface, $\mathrm{y} \in R^M, \mathbf{x} \in R^n$, and S is an orthogonal matrix. The active subspace is then defined as (S). One way to identify the active subspace is to perform an eigenvalue decomposition of the expectation matrix M defined as

$$\mathrm{M} = \frac{1}{N} \sum_{i=1}^{M} \left( \frac{\mathrm{dQoI}}{\mathrm{dx}_i} \right) \left( \frac{\mathrm{dQoI}}{\mathrm{dx}_i} \right)^\top = \mathrm{W} \boldsymbol{\Lambda} \mathrm{W}^\top, \tag{9.32}$$

where $N$ denotes the number of samples, and W and $\boldsymbol{\Lambda}$ the eigenvalues and eigenvectors of the expectation matrix. Selecting the first $R$ eigenvectors of this decomposition then constructs a reduced-order response surface $G\,(\mathrm{y})$ that spans over the active dimensions. The gradients in (9.32) in turn are computed using the gradient extraction technique introduced in the previous section, leading to the adjoint-based active subspace method (AASM). The access to adjoint capabilities allows a cost effective method of extracting the required derivatives and has been shown in a previous study [15] to reduce the overall cost of the active subspace methodology considerably when applied to a zero-dimensional reactor. In this study, where a higher-dimensional configuration (i.e. axisymmetric particle) is considered, the gains are expected to be even more substantial.

To construct the response surface of a QoI with respect to the existing system parameters, the following steps are taken:

1. Choose the appropriate distribution for input variables
2. Assemble $M$ random points of the input space, where $M = \eta \ln{(D)}$
3. Use adjoint-based method to compute the expectation matrix C from (9.32)
4. Apply eigenvalue decomposition to matrix $C$ and choose the dominant eigenvalues and eigenvectors
5. Estimate upper and lower bounds for the eigenvalues with bootstrapping (vary $M$ and repeat 3 and 4)
6. Construct the mapping matrix S from the dominant eigenvectors and compute the new variable $\mathrm{y} = \mathrm{S}^\top \mathrm{x}$
7. Construct the response surface $G\,(\mathrm{y}) \approx f\,(\mathrm{x})$ based on the number of active dimensions

For more details, regarding the algorithm see the study by Hassan et al. [15].

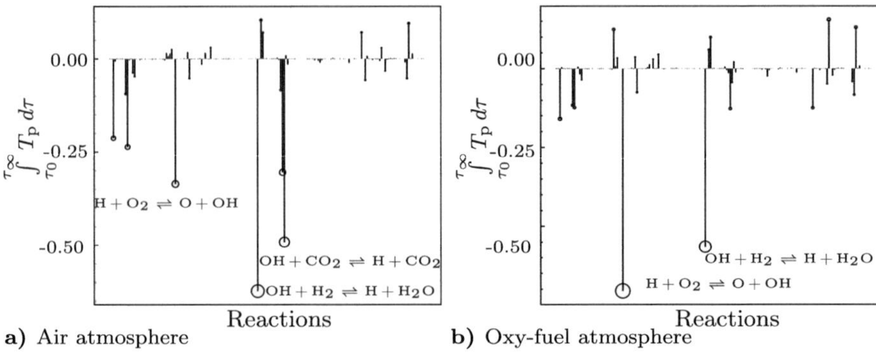


**Fig. 9.10** Uncertainties of gas-phase reaction pathways for cumulative particle temperature (reprinted from Hassan et al. [16], Copyright (2024), with permission from Elsevier and minor editing)

### 9.6.1 Uncertainties in the Prediction of Particle Temperature and Burning Rate

As shown in the previous section and due to their impact on both particle temperature ($T_p$) and burning rate ($\dot{m}_t$), it is crucial to quantify the impact of uncertainties in gas-phase kinetics and particle parameters on the prediction of these quantities of interest. The uncertainties will be extracted as explained earlier. Due to the cost associated with the forward problem and gradient extraction, as well as the high accuracy of AASM, a total of 100 samples ($N = 100$) are utilised to construct the covariance matrix. A single dominant direction (the eigenvector related to the first eigenvalue) is picked to build the response surface of particle temperature $T_p$ and the particle burning rate $\dot{m}_t$.

First the uncertainties with respect to the gas-phase kinetic models will be investigated. In this context, the dominant eigenvector is used as a representative of the dominant reaction pathway, shown in Fig. 9.10, using the average particle temperature as the QoI. These results can be directly compared to the results of the sensitivity analysis shown in Fig. 9.5. Figure 9.10 shows that $OH + H_2 \rightleftharpoons H + H_2O$ (R1) and $H + O_2 \rightleftharpoons O + OH$ (R2) are the most dominant reactions in air and oxy-fuel atmospheres, respectively, contributing significantly to the overall uncertainty of predictions.

It is worth noting that R2 also strongly contributes (within the top three reactions) to the predicted uncertainties of the air atmosphere. This is in contrast with the sensitivity analysis shown in Fig. 9.5a, where R2 ranks lower in the hierarchy of sensitive reactions. This highlights the limitation of a local method such as sensitivity analysis to analyse a complex system such as the one presented here. The reaction $OH + CO \rightleftharpoons H + CO_2$ remains important in the air atmosphere due to the elevated production of $CO_2$. This result is in alignment with the findings of the local sensitivity analysis in Fig. 9.5.

Finally, in order to examine the uncertainties of particle parameters on the overall predictions of the burning rate and particle temperature, a separate analysis is per-

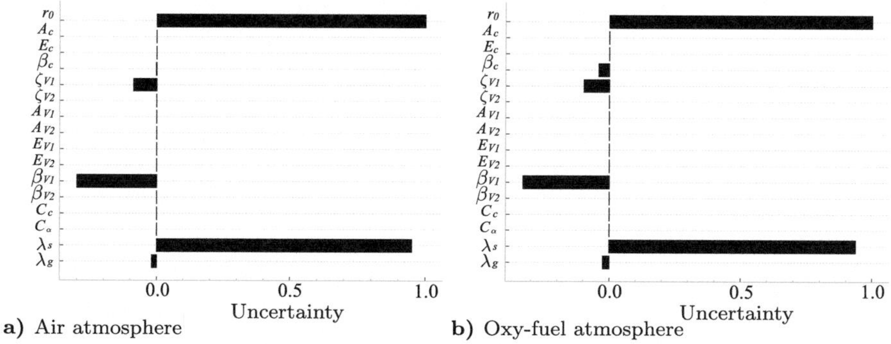


**Fig. 9.11** Uncertainty of particle temperature with respect to particle parameters (reprinted from Hassan et al. [16], Copyright (2024), with permission from Elsevier and minor editing)

formed, where the kinetic parameters of gas-phase reactions are fixed and the AASM method is applied only on the samples extracted from the particle parameters shown in Figs. 9.4 and 9.6.

Since the number of selected particle parameters is smaller than the parameters of the gas-phase kinetic model, only 70 samples are used to build the response surface. The dominant eigenvectors, extracted using the AASM approach, highlighting the global sensitivity of particle temperature with respect to system parameters for each atmosphere are shown in Fig. 9.11 and can be compared directly to Fig. 9.4. When comparing global and local sensitivities, depicted in Figs. 9.11 and 9.4, respectively, it becomes evident that the particle radius $r_0$ emerges as the most sensitive parameter. Furthermore, the global sensitivity of particle thermal conductivity $\lambda_s$, as illustrated in Fig. 9.11, exceeds the sensitivity observed in the local analysis presented in Fig. 9.4. In contrast to the local sensitivity, Fig. 9.11 demonstrates minimal global sensitivity for particle thermal capacity $C_c$. Comparing Figs. 9.11a and 9.11b reveals that the global sensitivity remains consistent, irrespective of atmospheric conditions. These results are consistent with the findings from the local sensitivity analysis.

### 9.6.2 Evolution of Uncertainties in Time

However, as stressed previously, it is essential to assess the evolution of uncertainties over time in unsteady configurations. To accomplish this, the integration limits in (9.28) and (9.27) are sequentially changed from 0 to 5 ms. Figure 9.12 depicts the evolution of the predicted mean value and standard deviation, amplified 50 times for better visibility, superimposed. It reveals that the uncertainties are negligible during the heating phase regardless of the atmosphere and the QoI, and it becomes relatively more significant after the gas-phase ignition. This is because the gas-phase chemistry has minimal influence before the ignition onset. In addition, the uncertainties in the oxy-fuel atmosphere is consistently higher than that in the air atmosphere for all quantities of interest. Moreover, the uncertainties in the burning rate exceeds that of

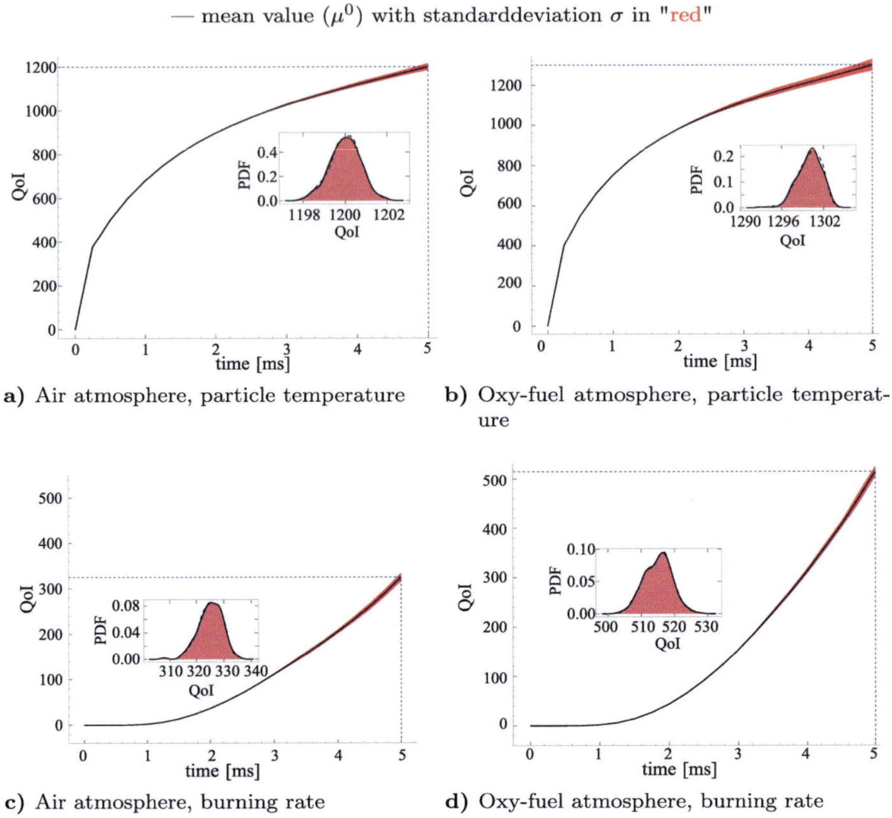


**Fig. 9.12** Evolution of uncertainties in particle temperature and burning rate with respect to gas-phase kinetic parameters (reprinted from Hassan et al. [16], Copyright (2024), with permission from Elsevier and minor editing)

the particle temperature. This is attributed to the fact that the burned mass is more sensitive to gas-phase kinetics compared to the particle temperature.

Figure 9.13 displays the evolution of uncertainties in particle temperature and burning rate with respect to system parameters. By comparing Figs. 9.13 and 9.12, it is observable that system parameters have a much more substantial influence on uncertainties than the reactions rates. This can be attributed to the pre-ignition stage's primary role in controlling the particle's heating rate. Although the atmosphere has a negligible impact on sensitivities (Figs. 9.4 and 9.6), Fig. 9.13 demonstrates that the uncertainties in an oxy-fuel atmosphere are higher than those in an air atmosphere. Moreover, the uncertainties in the oxy-fuel atmosphere increase as ignition progresses, in line with the trends seen in reaction rate Fig. 9.12.

Uncertainties in the burning rate before ignition are almost negligible. However, these uncertainties become significantly more pronounced in both air and oxy-fuel atmospheres after ignition. Comparing the results of Fig. 9.13 reveals that the uncertainties in burning rate are much higher than those in particle temperature. This

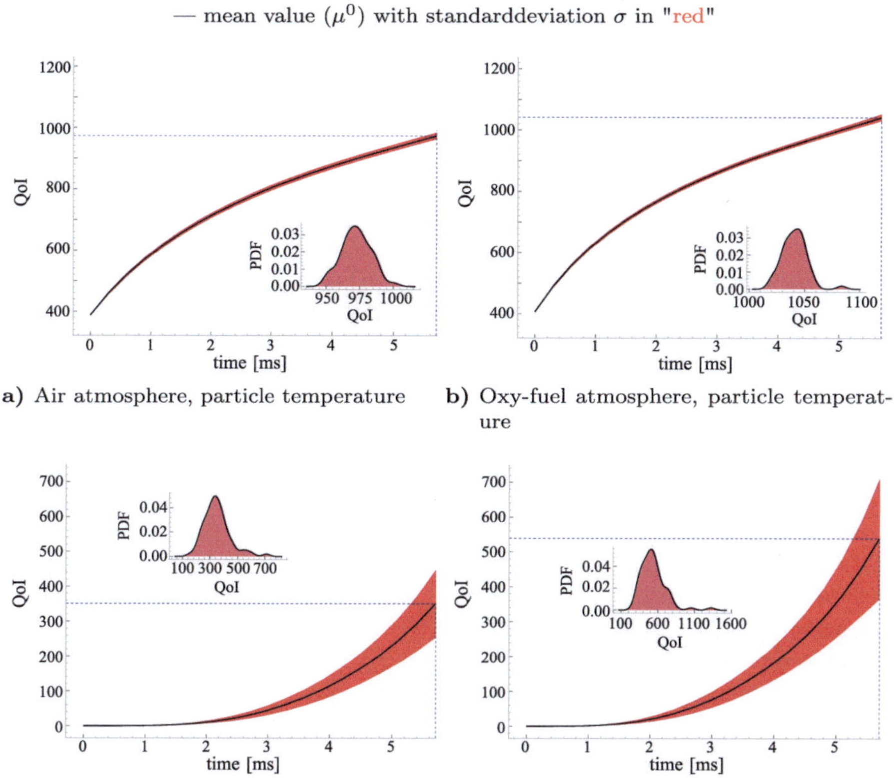


a) Air atmosphere, particle temperature

b) Oxy-fuel atmosphere, particle temperature

c) Air atmosphere, burning rate

d) Oxy-fuel atmosphere, burning rate

**Fig. 9.13** Evolution of uncertainties in particle temperature and burning rate with respect to particle parameters (reprinted from Hassan et al. [16], Copyright (2024), with permission from Elsevier and minor editing)

suggests that particle burning rate is more globally sensitive to system parameters. Moreover, it is important to note that the results of the uncertainties are highly dependent on the uncertainty factor (UF). Therefore, even a small variation in the value of UF can significantly affect the evolution of uncertainties. Due to the shortage of reliable information regarding the value of the uncertainty factor for particle parameters, the chosen value is (UF $= 1.1$), as a further increase in this value of UF could lead to unrealistic results (not show here). Comparing Figs. 9.13 and 9.8 reveals that while the sensitivities of system parameters remain unaffected by the atmosphere, the uncertainties are significantly influenced by atmospheric conditions. This underscores the importance of performing uncertainty quantification in determining the influence of system parameters on prediction of different quantities of interest.

Comparing the different probability density functions (PDF) in Fig. 9.13 for particle temperature $T_\mathrm{p}$ and the average burning rate $\dot{m}_\mathrm{t}$ reveals a higher skewness value for the $\dot{m}_\mathrm{t}$ PDF compared to the $T_\mathrm{p}$ PDF. Furthermore, similar to PDF in Fig. 9.12, the PDF for an oxy-fuel atmosphere exhibit higher skewness than those for an air atmosphere.

## 9.7 Conclusion

This chapter presents an overview of the sensitivity analysis and the uncertainty quantification of several models for solid fuel combustion, including char burnout, devolatilisation and detailed gas-phase kinetics. These models are applied in the simulation of an unsteady axisymmetric reacting particle. To determine the influence of the atmosphere, all the results were compared in both air and oxy-fuel atmospheres. The evolution of species shows that carbon dioxide levels increase in the air atmosphere compared to the oxy-fuel atmosphere due to the high rate of reaction $OH + CO \rightleftharpoons H + CO_2$. Analysing the particle burning rate reveals exponential growth after gas-phase ignition as the devolatilisation process becomes highly active, while the char combustion is negligible due to the limitation of oxygen accessibility to the particle surface. This result is valid for both air and oxy-fuel atmospheres.

To determine the most critical parameters, a semi-discrete adjoint approach is utilised. The resulting sensitivities showed that the particle size is the most influential parameter, regardless of the atmosphere (air/oxy-fuel), when considering the average particle temperature as the QoI. These results are consistent with the findings of [22]. However, when considering the average burning rate as the QoI, devolatilisation parameters become more significant due to the insignificant char combustion during the initial stage of devolatilisation. Additionally, analysis of the sensitivity of different quantities of interest with respect to gas-phase reaction rates indicated that volatile species play a vital role. Comparing the dominant reactions in air and oxy-fuel atmospheres shows that the reactions involving hydroxyl and hydrogen radicals are the most sensitive reactions.

By analysing the evolution of the sensitivities of the most dominant parameters over time, it is revealed that the maximum values occur during the early heating phase of the particle. Moreover, the sensitivity values for most parameters evolve until the onset of gas-phase ignition, at which point the sensitivity trend changes. However, the local gradient is not indicative of the actual reaction path, so the AASM is used to assess global sensitivities, as discussed in the study by Hassan et al. [15]. The reaction pathways were extracted to gain a global insight into the most effective reactions. High values were extracted for the $OH + H_2 \rightleftharpoons H + H_2O$ reaction in air atmosphere and the $H + O_2 \rightleftharpoons O + OH$ reaction in the oxy-fuel atmosphere.

Following the same analogies, the evolution of uncertainties for different quantities of interest (QoI) was extracted and compared. It is observed that there are negligible uncertainties during the heating up phase, but uncertainties gradually increase after the onset of gas-phase combustion. In the case of an oxy-fuel atmosphere, the uncertainties are larger compared to an air atmosphere due to higher gas-phase temperatures and earlier ignition. To validate these extracted uncertainties, probability density functions (PDF) of various quantities of interest related to gas-phase reaction rates at specific times were extracted.

Applying AASM in order to extract the global sensitivities of the particle temperature with respect to particle parameters reveals that the most dominant parameter is particle radius $r_0$ independent of the atmosphere. Also, the evolution of uncertainties in system parameters is analysed over time and compare them with the uncertainties in reaction rates. The comparison demonstrates that the uncertainties in reaction rates are negligible compared to those of system parameters. Additionally, similar to the uncertainties in reaction rates, the oxy-fuel atmosphere exhibits higher uncertainties than the air atmosphere. These findings highlight the higher influence of system parameters on various quantities of interest (QoI) in comparison to gas-phase chemical rates.

When comparing air and oxy-fuel atmospheres, the PDF reveal that the oxy-fuel atmosphere displays pronounced skewness and kurtosis owing to its heightened flammability and extended ignition delay time in gas-phase ignition, in contrast to the air atmosphere. This underscores the significance of investigating gas-phase ignition time and location.

Finally, it should be noted that among the parameters that could be influential in such analysis, initial temperature and pressure are also of great importance. While the impact of temperature on the burning rate has been previously demonstrated in the study by Hassan et al. [8], the analysis of the pressure effects has been left as part of future work, since the models used in this study are all calibrated for atmospheric pressure.

**Acknowledgements** This work has been funded by the Deutsche Forschungsgemeinschaft (DFG, German Research Foundation)—215035359—SFB/TRR 129 'Oxyflame'.

## References

1. Pironneau, O. "On Optimum Design in Fluid Mechanics". *Journal of Fluid Mechanics* 64.1 (1974), 97–110. https://doi.org/10.1017/S0022112074002023
2. Jameson, A., Martinelli, L. and Pierce, N. A. "Optimum Aerodynamic Design Using the Navier-Stokes Equations". *Theoretical and Computational Fluid Dynamics* 10.1–4 (1998), 213–237. https://doi.org/10.1007/s001620050060
3. Fikl, A., Le Chenadec, V. and Sayadi, T. "Control and Optimization of Interfacial Flows Using Adjoint-Based Techniques". *Fluids* 5.3 (2020), 156. https://doi.org/10.3390/fluids5030156
4. Lemke, M., Cai, L., Reiss, J., Pitsch, H. and Sesterhenn, J. "Adjoint-Based Sensitivity Analysis of Quantities of Interest of Complex Combustion Models". *Combustion Theory and Modelling* 23.1 (2019), 180–196. https://doi.org/10.1080/13647830.2018.1495845
5. Braman, K., Oliver, T. A. and Raman, V. "Adjoint-Based Sensitivity Analysis of Flames". *Combustion Theory and Modelling* 19.1 (2015), 29–56. https://doi.org/10.1080/13647830.2014.976274
6. Kord, A. and Capecelatro, J. "A Discrete-Adjoint Framework for Optimizing High-Fidelity Simulations of Turbulent Reacting Flows". *Proceedings of the Combustion Institute* 39.4 (2023), 5375–5384. https://doi.org/10.1016/j.proci.2022.06.021
7. Hassan, A., Sayadi, T., Le Chenadec, V., Pitsch, H. and Attili, A. "Adjoint-Based Sensitivity Analysis of Steady Char Burnout". *Combustion Theory and Modelling* 25.1 (2021), 96–120. https://doi.org/10.1080/13647830.2020.1838614
8. Hassan, A., Sayadi, T., Le Chenadec, V. and Attili, A. "Sensitivity Analysis of an Unsteady Char Particle Combustion". *Fuel* 287 (2021). https://doi.org/10.1016/j.fuel.2020.119738

9. Eldred, M. and Burkardt, J. "Comparison of Non-Intrusive Polynomial Chaos and Stochastic Collocation Methods for Uncertainty Quantification". *47th AIAA Aerospace Sciences Meeting*. Orlando, Florida, USA, 2009. https://doi.org/10.2514/6.2009-976
10. Xiu, D. and Hesthaven, J. S. "High-Order Collocation Methods for Differential Equations with Random Inputs". *SIAM Journal on Scientific Computing* 27.3 (2005), 1118–1139. https://doi.org/10.1137/040615201
11. Li, H., Tang, Y., Ren, X., Xu, L. and Xu, X. "The Way to Develop Software Towards Exascale Computing". *Proceedings of the 2016 International Conference on High Performance Computing & Simulation*. Innsbruck, Austria, 2016, pp. 334–341.
12. Constantine, P. G., Dow, E. and Wang, Q. "Active Subspace Methods in Theory and Practice: Applications to Kriging Surfaces". *SIAM Journal on Scientific Computing* 36.4 (2014), A1500–A1524. https://doi.org/10.1137/130916138
13. Ji, W., Wang, J., Zahm, O., Marzouk, Y. M., Yang, B., Ren, Z. and Law, C. K. "Shared Low-Dimensional Subspaces for Propagating Kinetic Uncertainty to Multiple Outputs". *Combustion and Flame* 190 (2018), 146–157. https://doi.org/10.1016/j.combustflame.2017.11.021
14. Ji, W., Ren, Z., Marzouk, Y. and Law, C. K. "Quantifying Kinetic Uncertainty in Turbulent Combustion Simulations Using Active Subspaces". *Proceedings of the Combustion Institute* 37.2 (2019), 2175–2182. https://doi.org/10.1016/j.proci.2018.06.206
15. Hassan, A., Sabry, M., Le Chenadec, V. and Sayadi, T. "Uncertainty Quantification of Kinetic Models Using Adjoint-Driven Active Subspace Algorithms". *Proceedings of the Combustion Institute* 39.4 (2023), 5209–5218. https://doi.org/10.1016/j.proci.2022.07.177
16. Hassan, A., Schumacher, L., Sabry, M., Le Chenadec, V. and Sayadi, T. "Uncertainty Quantification and Sensitivity Analysis of an Unsteady Axisymmetric Reacting Solid Particle". *Combustion and Flame* 262 (2024), 113365. https://doi.org/10.1016/j.combustflame.2024.113365
17. Farmand, P., Nicolai, H., Schumann, C., Attili, A., Berger, L., Li, T., Geschwindner, C., di Mare, F., Hasse, C., Böhm, B., Janicka, J. and Pitsch, H. "Numerical Investigation and Assessment of Flamelet-Based Models for the Prediction of Pulverized Solid Fuel Homogeneous Ignition and Combustion". *Combustion and Flame* 235 (2022), 111693. https://doi.org/10.1016/j.combustflame.2021.111693
18. Farazi, S., Sadr, M., Kang, S., Schiemann, M., Vorobiev, N., Scherer, V. and Pitsch, H. "Resolved Simulations of Single Char Particle Combustion in a Laminar Flow Field". *Fuel* 201 (2017), 15–28. https://doi.org/10.1016/j.fuel.2016.11.011
19. Du, X. and Annamalai, K. "The Transient Ignition of Isolated Coal Particle". *Combustion and Flame* 97.3–4 (1994), 339–354. https://doi.org/10.1016/0010-2180(94)90025-6
20. McBride, B. J., Gordon, S. and Reno, M. A. *Coefficients for Calculating Thermodynamic and Transport Properties of Individual Species*. 1993.
21. Hassan, A., Sayadi, T., Schiemann, M. and Scherer, V. "Adjoint-Based Sensitivity Analysis of Char Combustion Surface Reaction Kinetics". *Fuel* 287 (2021). https://doi.org/10.1016/j.fuel.2020.119503
22. Gonzalo-Tirado, C., Jiménez, S., Johansson, R. and Ballester, J. "Comparative Study of Four Alternative Models for CO Oxidation Around a Burning Coal Char Particle". *Combustion and Flame* 161.4 (2014), 1085–1095. https://doi.org/10.1016/j.combustflame.2013.09.025

# Particle Flow Interaction

# 10 Experimental Investigation of Particle-Flow-Chemistry Interactions

Christopher Geschwindner, David Tarlinski, Tao Li, Jan Köser, Martin Schiemann, Andreas Dreizler and Benjamin Böhm

## Contents



---

C. Geschwindner (✉) · T. Li · J. Köser · A. Dreizler · B. Böhm
Institute of Reactive Flows and Diagnostics, Technische Universität Darmstadt, Darmstadt, Germany
e-mail: geschwindner@rsm.tu-darmstadt.de

T. Li
e-mail: tao.li@rsm.tu-darmstadt.de

J. Köser
e-mail: koeser@rsm.tu-darmstadt.de

A. Dreizler
e-mail: dreizler@rsm.tu-darmstadt.de

B. Böhm
e-mail: boehm@rsm.tu-darmstadt.de

D. Tarlinski · M. Schiemann
Department of Energy Plant Technology, Ruhr-Universität Bochum, Bochum, Germany
e-mail: tarlinski@leat.rub.de

M. Schiemann
e-mail: schiemann@leat.rub.de



R. Kneer et al. (eds.), *Clean Energy Conversion*,
https://doi.org/10.1007/978-3-031-98956-8_10

**Abstract**

The combustion of solid particulate fuels involves the interplay of particle phase, fluid flow, and thermochemistry, where investigating and quantifying the influence of each aspect of this interaction triangle is essential to modelling and understanding the combustion processes. This chapter presents an in-depth experimental investigation of particle-flow-chemistry interactions in solid fuels, with a focus on high-volatile bituminous coal and biomass. The fundamental physico-chemical processes of solid fuel combustion are outlined, illustrated by the behaviour of a single solid fuel particle in a hot gaseous environment. Two experimental configurations are described: a laminar flat flame burner for controlled particle-particle interaction studies, and a turbulent hot gas test rig for the investigation of the interplay of turbulent carrier flows and particulate fuels. Various optical diagnostic techniques were advanced and applied to obtain detailed data on particle and gas-phase interactions. The experimental investigations provide valuable insights and unique data that are essential for validating and improving combustion models, thus contributing significantly to the understanding and modelling of solid fuel combustion.

## 10.1 Introduction

### 10.1.1 Background

The ignition of fuel particles ($<150\,\mu m$) in hot gas flows ($\approx 1500$ K) under high heating rates ($> 1 \times 10^4\ K\,s^{-1}$) occurs either in the region of the outgassing volatiles surrounding the particle (homogeneous ignition), on the particle surface (heterogeneous ignition) or during a combination of both [1]. When simulating dust flames, ignition is modelled as a sequence of drying, pyrolysis, and char burnout [1], although in the hetero-homogeneous case the two exothermic processes may occur simultaneously. The outgassing process plays an important role in ignition [1,2] and flame stabilisation [3], while the transport processes at the particle surface are of crucial importance for char combustion [4].

To investigate the fundamental underlying processes in solid fuel combustion, laminar flow reactors are typically used. Individual solid fuel particles are introduced into a hot, oxygen-rich flue gas by means of a central carrier flow [5,6]. In addition to classical intrusive measurement techniques such as thermocouples to determine gas temperatures or sampling to characterise the particle composition, non-invasive optical methods are preferred and used where possible [7]. Ignition delay times as well as the duration of volatile combustion can be determined by high-speed imaging of the flame's luminosity [5,6,8] or by laser-induced fluorescence (LIF) of the OH radical [7]. Particle temperatures can be measured using multicolour pyrometry [9], simultaneous particle sizes and temperatures using extended stereoscopic pyrometers [10], or particle velocities using high-speed imaging [8]. Such detailed experimental investigations of individual solid fuel particles represent an important building block for modelling [1].

In real pulverised solid fuel flames, however, ignition delay times are also a function of particle number densities [11,12] and turbulence [13]. As the particle load increases, ignition shifts from the close vicinity of individual particles to enveloping flames surrounding entire particle clouds, changing the ignition delay times [11]. An increase in the Reynolds number of the carrier flow, switching from laminar to turbulent conditions, increases particle dispersion. The increased convective transport leads to better mixing of the particle stream with the hot gas atmosphere, which shortens the ignition delay time and burnout [13]. In turbulent flows, the mutual interaction of the particle and gas phase can also become relevant, which has been observed previously in non-reacting flows [14]. Investigations of particle clouds in turbulent environments have been carried out, for example, in a gas-assisted jet [15] or swirling flames [16], but these have not investigated the phenomena at the level of individual particles.

### 10.1.2 Research Objectives and Approach

The research programme within this work pursued three primary goals:

- To improve the understanding of the ignition and combustion processes of fuel particles as a function of particle number density (PND), gas composition, and flow field.
- To develop and implement advanced optical diagnostic methods for simultaneous multi-parameter measurements that capture the crucial aspects of the physico-chemical interactions.
- To build an experimental data set with well-characterised boundary conditions for modelling and validation of numerical simulations.

To achieve the above-mentioned objectives, the simplest possible geometries were used from a fluid mechanics point of view, in which the chemical, thermodynamic, and fluid mechanical parameters can be varied under reproducible and defined boundary conditions. The complexity of the measurement sequence was successively increased from a single particle in a laminar environment to group combustion in a laminar environment, individual particle combustion in a turbulent environment and finally group combustion in a turbulent environment. The relevant parameters were recorded with high spatial and temporal resolution. Laser-optical and imaging measurement methods were used to record the gas phase properties. Imaging pyrometric methods and sampling techniques were used to characterise the particle phase. These experimental results formed a unique data set for the validation and refinement of ignition and combustion models for *OxySim-129* (Chap. 19) to describe coal and biomass particle combustion under air and oxy-fuel conditions.

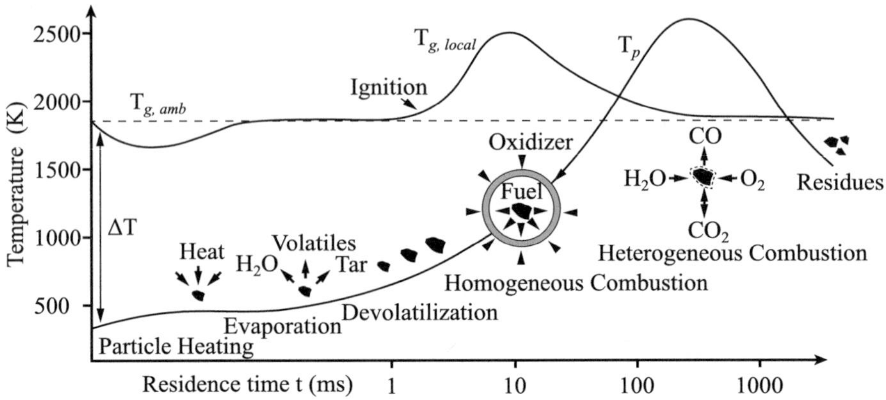


**Fig. 10.1** Schematic of combustion stages of single CBC particles in high-temperature oxidising atmospheres (reprinted from Li et al. [7], Copyright (2023), with permission from Elsevier and minor editing)

### 10.1.3 Fundamental Physico-Chemical Processes in Solid Fuel Combustion

This section outlines the fundamental physico-chemical processes involved in the combustion of carbonaceous solid fuels, illustrated in Fig. 10.1 by a single high volatile Colombian bituminous coal (CBC) particle in a hot gaseous environment. Figure 10.1 shows typical particle and gas temperatures over the particle residence time, with actual values dependent on factors such as coal rank, particle size, and oxygen concentration.

A representative CBC particle of diameter $d_\mathrm{p}$ and initial temperature $T_\mathrm{p} \approx 300$ K is injected into a high temperature oxidising environment. The particle undergoes rapid heating by convective and radiative heat transfer, followed by pyrolysis, homogeneous and heterogeneous combustion phases. The duration and temperature rise of each phase depends on particle size, flow conditions, and fuel type. Assuming a homogeneous ambient gas temperature of 1800 K and constant gas velocities, Fig. 10.1 illustrates the temporal behaviour of particle and gas temperature rise.

Initially, heat transfer from the gas to the particle causes moisture to evaporate, which cools the surrounding gas slightly. As the particle temperature rises, devolatilisation releases fuel gases such as $CH_4$, $H_2$, and tars. The heating rate affects the duration of moisture evaporation, the rate of volatile release, and the fuel gas composition. The particle temperature changes with time, balancing internal energy, and heat transfer [17]:

$$\frac{\pi}{6}\, d_\mathrm{p}^3\, \rho_\mathrm{p}\, c_\mathrm{p} \frac{dT_\mathrm{p}}{dt} = \pi\, d_\mathrm{p}^2\, h\, (T_\mathrm{g} - T_\mathrm{p}) - \pi\, d_\mathrm{p}^2\, \sigma\, (\epsilon_\mathrm{g}\, T_\mathrm{g}^4 - \epsilon_\mathrm{p}\, T_\mathrm{p}^4) + \dot{Q}_\mathrm{evap} + \dot{Q}_\mathrm{devol}, \tag{10.1}$$

where $h$ is the heat transfer coefficient, $T_\mathrm{g}$ and $T_\mathrm{p}$ the gas and particle temperatures, $\rho_\mathrm{p}$ the particle density, $c_\mathrm{p}$ the heat capacity of the particle, $\epsilon_\mathrm{g}$ and $\epsilon_\mathrm{p}$ the gas phase and particle emissivities, and $\sigma$ the Stefan-Boltzmann constant. The heat required for evaporation and released from devolatilisation reactions are denoted by $\dot{Q}_{\text{evap}}$ and $\dot{Q}_{\text{devol}}$, respectively.

For CBC or biomass particles, two combustion modes occur: homogeneous (gas phase) and heterogeneous (surface reactions). During homogeneous combustion, a volatile-oxidiser mixture is ignited, while heterogeneous combustion involves surface reactions at elevated temperatures. During devolatilisation, a flammable gas mixture forms, leading to auto-ignition when the mixture reaches ignition temperature. The ignition delay time $\tau_{\text{i}}$ for auto-ignition is given by

$$\tau_{\text{i}} = \frac{c_v T_0^2 / T_{\text{a}}}{q_{\text{c}}\, Y_{\text{F},0}\, A\, \exp\left(-\frac{E_{\text{a}}}{RT}\right)}, \tag{10.2}$$

where $c_v$ is the specific heat capacity, $q_{\text{c}}$ the chemical reaction heat release rate, $T_0$ the initial temperature, $T_{\text{a}}$ the activation temperature, $E_{\text{a}}$ the activation energy, $Y_{\text{F},0}$ the fuel mass fraction, and $A$ the pre-exponential factor [18]. Homogeneous combustion involves complex reactions of volatile matter, forming $CO_2$, $H_2O$, and pollutants like $SO_x$ and $NO_x$. At the single-particle level, an enveloping diffusion flame forms, directly visible through thermal radiation and chemiluminescence. Heterogeneous combustion includes char oxidation, gasification, and secondary reactions. The char combustion rate depends on temperature, oxygen concentration, and gas diffusivity. Increasing oxygen mole fraction decreases burnout times and increases temperatures for both volatile and char combustion [7].

Understanding the fundamental processes of solid fuel combustion is essential for model development and exploration of the underlying phenomena. Keeping these fundamentals in mind, the following sections detail the measurements of key properties of solid fuel combustion. Custom-designed burner systems and minimally intrusive optical diagnostics have been used to investigate the details of these processes. These experimental methods provide valuable insights and unique data that are essential for validating and improving combustion models.

## 10.2 Methods

### 10.2.1 Experimental Configurations and Burners

**Laminar Flat Flame Burner**

In the literature, different burner setups are used to study the combustion of solid fuels under laminar flow conditions, each offering different advantages. Common configurations include drop tube furnaces (DTF), flat flame burners (FFB), jet-in-crossflow reactors, and counterflow burners [7]. DTF, typically electrically heated, offer flexibility in gas composition and can reach wall temperatures of up to 1600 K. Flame-assisted reactors, such as FFB, generally achieve higher gas temperatures in excess of 2000 K due to premixed or diffusion flames. Flat flame burners come in a variety of designs, including the Hencken burner, McKenna burner, and laminar flow

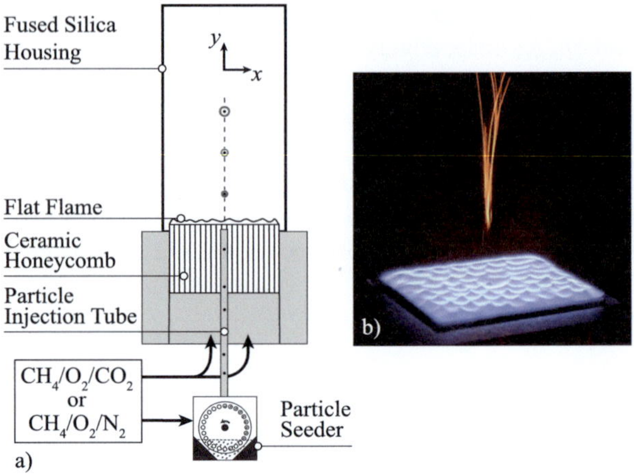


**Fig. 10.2** Laminar flat flame burner used at *TU Darmstadt* for optical diagnostics of single particle and particle group combustion: **a**) sketch of the reactor and **b**) long-exposure colour photograph of individual coal particles burning in the flue gas of the fully premixed $CH_4$ flame of the flat flame burner (reprinted from Li et al. [20], Copyright (2021), with permission from Elsevier and minor editing)

reactor (LFR), each offering different degrees of control over gas temperature and composition. In this work, a flat flame burner was chosen for its ability to provide a controlled and stable environment, which is essential for studying fundamental processes such as particle heating, ignition, and volatile combustion.

The laminar flat flame burner used in this work and conceptualised at *TU Darmstadt* [19] is designed to study single particle combustion and particle-particle interactions in laminar flows. This setup provides stable boundary conditions allowing for precise control over temperature and oxygen concentration, essential for investigating fundamental processes such as particle heating, ignition, and volatile combustion under well-defined conditions. The burner setup is displayed in Fig. 10.2.

The burner consists of a ceramic honeycomb with a cross-sectional area of approximately $80 \times 80\ mm^2$ and a height of 40 mm. Premixed mixtures of $CH_4/O_2/N_2$ or $CH_4/O_2/CO_2$ are issued from 3600 rectangular holes, each $1 \times 1\ mm^2$ in size. The flat flame is stabilised approximately 1.5 mm above the honeycomb surface by optimising the flow bulk velocities. The flow reactor is enclosed by a quartz-glass chimney of 300 mm in length, providing optical access and preventing secondary air entrainment.

Coal particles are transported by a carrier gas through an exchangeable central tube with variable inner diameters (0.8 mm for single and group combustion experiments [21] and 2.9 mm for group combustion experiments [22]). The particle density can be adjusted to ensure the combustion of individual coal particles up to dense clouds of solid fuel particles. The operational conditions can be varied by the gas composition used for the premixed flame and the composition of the particle carrier gas. Typical operating conditions include a base case with a $CH_4/O_2/N_2$ mixture resulting in 20.5 vol% $O_2$ in the exhaust gas, and variations with pure $N_2$ as the carrier gas, reduced $O_2$ content, or oxy-fuel conditions where $N_2$ is replaced by $CO_2$.

The laminar flat flame burner can achieve high particle heating rates, with steep gas temperature rises on the order of $1 \times 10^5\ K\,s^{-1}$, allowing detailed investigations of particle heat-up and combustion processes. The maximum gas phase temperature of 1800 K was measured using nanosecond rotational-vibrational coherent anti-Stokes

Raman spectroscopy (CARS) [23] and quantitative laser-induced fluorescence of the OH radical [20]. The fused silica enclosure ensures optical access from all sides, enabling the application of various optical diagnostic techniques.

All the studies discussed in this chapter used sieved fractions of solid fuels, including high volatile bituminous coal (CBC), walnut shells (WNS), and miscanthus particles (MG). These fuel samples were meticulously prepared, with particles sieved into multiple distinct ranges with mesh sizes ranging from 60 μm to 200 μm. Comprehensive proximate and ultimate analyses of the raw fuel compositions prior to combustion were carried out and are detailed in Table 1.1. These analyses provide information on the elemental composition, moisture content, volatile matter, and ash content of the fuels, ensuring that the experimental investigations are supported by well-characterised fuel properties. These fuel properties are essential for accurate interpretation of the combustion behaviour observed in the flat flame burner studies and for kinetic modelling.

**Turbulent Hot Gas Test Rig**

The transition from controlled laminar experiments to turbulent conditions presents challenges in investigating the influence of individual parameters. While laminar reactors provide well-defined boundary conditions for detailed investigations of specific factors like gas atmosphere and temperature, turbulent burners, such as swirled lab-scale systems examined in Chap. 13, involve numerous overlapping effects that complicate the analysis of individual influences. To bridge this gap, a turbulent hot gas test rig initially developed for auto-ignition studies of gaseous fuels [24] was adapted at *TU Darmstadt*. This setup operates between the highly controlled environment of laminar reactors and the large-scale industrial burners, facilitating detailed studies of particle combustion in turbulent flows.

The turbulent hot gas test rig utilises a microwave plasma heater to achieve the high gas temperatures necessary for initiating self-ignition and subsequent combustion of solid fuels. The test rig is displayed in Fig. 10.3. The microwave plasma heats the oxidiser stream using a microwave source (magnetron) operating at 915 MHz with a maximum power of 75 kW. Air is injected tangentially and axially into the plasma region, with controlled flow ratios to maintain stable plasma operation. This setup achieves oxidiser temperatures up to 1400 K with bulk exit velocities of up to 35 $m\,s^{-1}$.

After plasma heating, the oxidiser stream passes through a ceramic-insulated tube and a static mixer in the flow conditioning unit to homogenise the temperature profile, ensuring a top-hat temperature distribution essential for controlled experiments. The burner head features a Morel-type nozzle with a contraction ratio of 4 and turbulence-generating grids. Fuel particles are introduced through a central water-cooled injection lance, made of concentric stainless steel tubes with ceramic insulation, which prevents premature pyrolysis of solid fuels. As solid fuel particles are injected into the heated oxidiser stream through the central lance, they heat up through mixing of the hot co-flow and the cold jet flow and finally ignite and burn. Despite the capabilities of the burner system, in this chapter, only experiments under non-reactive conditions are summarised [26,27], highlighting the interaction of par-

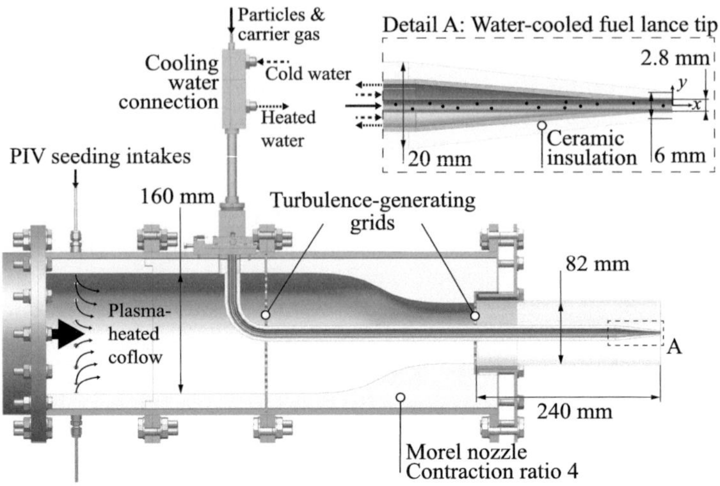


**Fig. 10.3** Schematic of the turbulent hot gas test rig at *TU Darmstadt* including a detail view of the water-cooled fuel lance tip

ticles and their surrounding carrier-gas turbulence. Ongoing studies will also include detailed reactive measurements.

As demonstrated by both burner setups in this work, the experimental environments were specifically designed to accommodate optical and laser-based measurement methods. These techniques, extensively used in gas-phase combustion studies, have been adapted and applied to solid fuel combustion in the experiments described in this chapter. The following sections detail the experimental methods employed and the key parameters measured, providing valuable insights and data essential for validating and improving combustion models.

### 10.2.2 Measurement Methods

Optical measurement techniques offer several advantages in the study of solid fuel combustion processes. These methods are minimally intrusive, ensuring that the combustion environment remains largely undisturbed. They also provide high spatial and temporal resolution, allowing detailed characterisation of rapid and localised phenomena at the single particle scale. In addition, optical techniques can be combined simultaneously, allowing multi-parameter measurements and the generation of conditioned statistics. This capability is critical for investigating the effects of specific variables on combustion processes, rather than relying on uncorrelated statistics, thereby improving understanding of the complex interactions in solid fuel combustion. A recent review by Li et al. [7], which is a product of the work presented in this chapter, summarises the current state of research in optical diagnostics for solid fuel combustion at the particle level.

The studies presented in this chapter have made use of a variety of advanced optical diagnostic techniques (and intrusive techniques) to obtain detailed data on particle-flow-chemistry interactions. These methods have been integrated to provide comprehensive multi-parameter measurements that provide insight into the complex interactions during combustion. The individual measurement techniques are presented as follows.

**Diffuse Back-Illumination (DBI)**
Particle-resolved DBI is utilised to measure particle properties such as size, shape, and particle number density (PND). This technique provides high-resolution imaging, enabling detailed analysis of particle behaviour and interactions within the combustion environment. DBI involves illuminating particles with a high-power pulsed LED, shaped into a Lambertian source using diffusers and capturing their shadows with a CMOS camera at high repetition rates. A band-pass filter centred at the LED emission wavelength suppresses thermal radiation from the particle surface, ensuring undisturbed particle images. This method allows precise measurements of relevant particle properties, crucial for understanding particle phenomena such as swelling [28] and PND measurements in clouds [22,29].

**Laser-Induced Fluorescence of the OH Radical (OH-LIF)**
The OH-LIF method can be used to map reaction zones and visualise flame structures [30]. It detects the fluorescence of hydroxyl radicals, key intermediates in volatile combustion reactions of carbonaceous solid fuel particles [19,31]. Combined with a tomographic or laser scanning approach, this technique is able to provide three-dimensional flame topologies essential for understanding the spatial distribution of reaction zones around burning particles. The basic measurement principle of OH-LIF is to excite OH radicals at specific wavelengths in the ultraviolet to induce electronic transitions, and to capture the emitted fluorescence as the radicals transition back to their temperature-dependent ground state using intensified cameras. Typically, this is done using a laser sheet leading to planar laser-induced fluorescence of OH radicals termed OH-PLIF. This method is particularly useful for identifying ignition delay times of single particles and complex flame structures in group combustion environments.

**Luminosity Imaging (LU)**
Imaging of combustion luminosity is a valuable and straightforward technique for visualising and analysing the combustion process of solid fuel particles. This method captures the light emitted from burning particles using (high-speed) cameras, providing a direct visual representation of combustion stages such as ignition, volatile combustion, and char combustion. It is especially useful for studying flame structures and burn times without the need for complex laser excitation. While effective for visualising flame structures and determining the end of volatile combustion, luminosity imaging is not as precise as OH-PLIF for detecting ignition delay times [20]. The OH-PLIF method provides higher signal intensities for tracking ignition by specifically visualising OH radicals, offering more accurate ignition delay times

compared to LU [23]. Combining luminosity imaging with OH-PLIF enhances the overall understanding of combustion dynamics by leveraging the strengths of both techniques.

**Particle Tracking Velocimetry (PTV) and Particle Image Velocimetry (PIV)** Both techniques, PTV and PIV, are essential for measuring flow fields and particle dynamics in two-phase flow systems, including combustion studies [32]. The PTV method involves tracking individual particles over time using high-speed imaging, Mie scattering, or DBI to provide detailed information about their trajectories. This method is particularly useful for studying the specific motion and behaviour of individual particles, allowing precise analysis of particle dynamics within the flow [7]. In contrast, PIV is more commonly used to characterise the flow field of the fluid phase [32,33]. In its simplest form, it involves illuminating tracer particles in the flow with a laser sheet and capturing their motion using Mie scattering images with cameras. By cross-correlating sequential images of these tracer particles within interrogation windows, PIV provides statistically robust estimates of local fluid velocities. Together, PTV and PIV provide complementary insights: PTV provides granular data on individual particle motion, while PIV provides a comprehensive view of the flow field, both of which are critical to a thorough understanding of combustion processes.

By systematically combining the presented advanced diagnostic techniques, comprehensive insights into the physico-chemical interactions during solid fuel combustion were achieved in this work. The simultaneous application of these methods enabled the generation of conditioned statistics, allowing detailed analysis of how specific particle characteristics interacted with the surrounding flow field and combustion environment at any given instant. This integrated approach provided a unique data set for comparison with numerical simulations, significantly advancing the understanding of reacting flows involving carbonaceous solid fuels as outlined in the following sections.

## 10.3 Results and Discussion

Following the research objectives detailed in Sect. 10.1.2, the results presented below highlight the main achievements in high-resolution diagnostics of solid fuel combustion. To lay the foundation for understanding such a complex subject with multiple interactions, starting with single particle combustion in laminar flows is helpful. This first part focuses on the development of diagnostic techniques and key findings related to single particle combustion of coal and biomass. The second part examines the effects of PND in laminar environments, detailing the transition from single particle to particle group combustion. Finally, the interaction of particles with a turbulent carrier flow is explored, with an emphasis on particle-resolved high-resolution diagnostics.

### 10.3.1 Single Particle Combustion in Laminar Flows

The simplest configuration of solid fuel combustion involves a single particle in a laminar hot flow environment. This scenario is particularly relevant for ignition and volatile combustion model development and validation due to its simplicity. For carbonaceous solid fuel particles with high volatile content, such as bituminous coal or biomass, combustion typically starts homogeneously, as shown in Fig. 10.1. In this case, the pyrolysis of the volatiles and their mixing with the surrounding oxidising environment determines the ignition delay time according to Eq. (10.2). Accurate measurements of the ignition delay requires precise knowledge of the duration of exposure of an initially cold particle to a high temperature oxidising environment under controlled conditions. The laminar flat flame burner with central particle injection shown in Fig. 10.2 was developed for this purpose. The burner provides a sharp temperature rise across the premixed methane flame front and maintains a constant temperature level downstream of the flame. An application of high-speed laser diagnostics is essential to follow the evolution of the homogeneous ignition over time, providing accurate data for individual particles. This approach contrasts with statistical methods using low-speed cameras, which require particle uniformity and high experimental repeatability, both of which are difficult to achieve in real systems. In the following, the development of multi-parameter high-speed measurement systems for single particle combustion diagnostics is presented.

In this work, high-speed optical diagnostics were applied to single particle combustion to capture the entire temporal history of each particle. High-speed OH-PLIF was specifically developed to detect the homogeneous volatile ignition of single particles. This technique was first applied to coal particles in a Hencken burner, as described by Köser et al. [31], and later in the laminar flat flame burner detailed earlier to measure ignition delay times [19]. Figure 10.4 shows the temporal evolution of the volatile flame around a single CBC particle. The laser sheet for LIF excitation, directed from the left, casts a shadow on the opposite side of the particle, ensuring a central laser sheet impact. This setup allows accurate determination of the flame stand-off distance from the OH-PLIF images, facilitating the extraction of radial OH-LIF profiles as described in the study by Köser et al. [19]. The normalised LIF intensity indicates that the hot flue gas containing water vapour decomposes to some extent into OH radicals, which can be used as a background to correct the OH-PLIF images. This correction takes into account the inhomogeneous beam profiles of the UV laser and the pulse-to-pulse fluctuations of the high-speed dye laser, allowing the calculation of a normalised OH-LIF intensity. This normalisation ensures accurate determination of the ignition delay time by comparing the LIF intensity with the background signal of the hot flue gas.

As explained in Sect. 10.2.2, high-speed OH-PLIF is highly effective for ignition detection. However, the field of view (FOV) required to monitor the entire volatile combustion phase from ignition to the transition to char combustion is limited by the available laser pulse energy, as the laser is formed into a sheet of a certain thickness and height. As accurate measurements of the full volatile combustion phase are essential, an additional suitable measurement technique operating simultaneously is

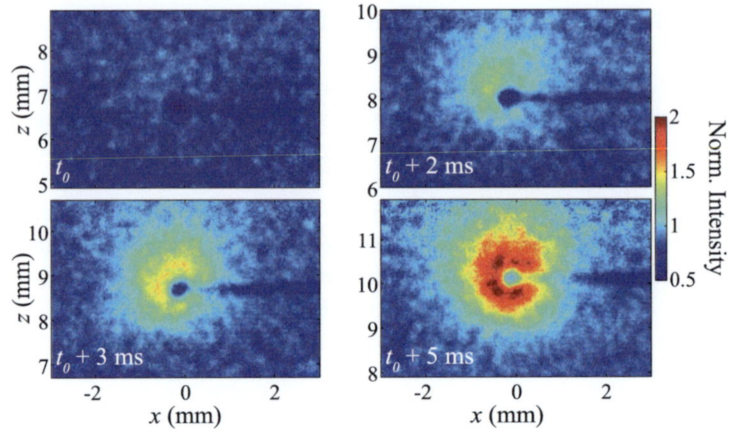


**Fig. 10.4** Sequential OH-PLIF images showing the volatile flame around a single bituminous coal particle injected into the laminar flat flame burner. The initial particle diameter is in a sieved range of 90–125 μm. Time $t_0$ marks the instant of ignition, occurring 7 ms post-injection for the shown particle (reprinted from Attili et al. [34], Copyright (2020), with permission from Elsevier and minor editing)

required. To this end, luminosity imaging (LU) has been combined with OH-PLIF to capture the full volatile combustion duration.

Figure 10.5 shows an exemplary time evolution of the luminescence intensity area of a single particle, highlighting the ignition, volatile combustion, and char combustion phases. The area of the particle is determined by binarising the LU images using a predefined threshold. It is observed that the LU signal increases several milliseconds after ignition, which is first detected by OH-PLIF. This delay is due to the higher sensitivity and signal intensity of OH-PLIF compared to LU imaging [20, 23]. The limited FOV of OH-PLIF precludes observation of the entire volatile combustion process, whereas LU can effectively monitor this phase but with lower spatial resolution. The volatile combustion duration is determined by combining the ignition delay time from OH-PLIF measurements and the end time of volatile combustion from luminescence measurements. The combination of these diagnostic techniques ensures that both the initiation and duration of volatile combustion are captured with high accuracy, providing data essential for the validation and development of ignition and volatile combustion models.

**Fig. 10.5** Temporal variation of the luminescence signal area of a single particle, showcasing ignition, volatile combustion, and char combustion phases (reprinted from Li et al. [20], Copyright (2021), with permission from Elsevier and minor editing)

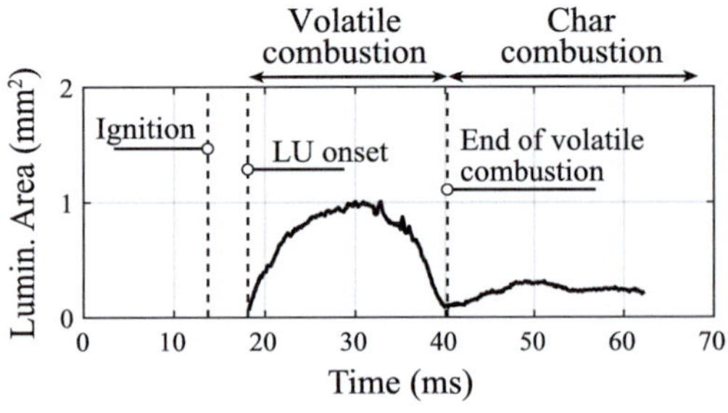

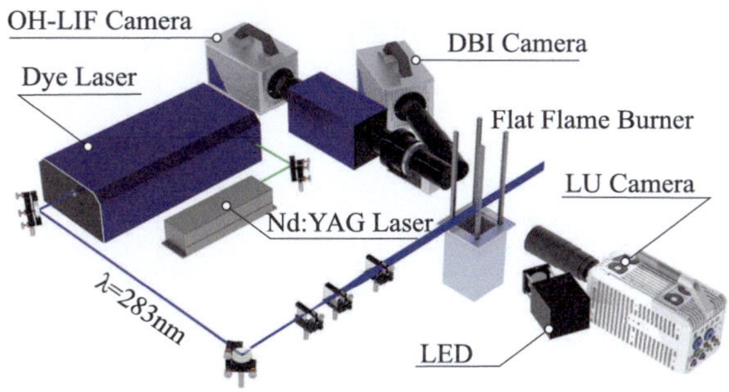


**Fig. 10.6** Experimental setup of simultaneous multi-parameter measurements using OH-PLIF, LU, and DBI (reprinted from Li et al. [20], Copyright (2021), with permission from Elsevier and minor editing)

Although sieved particles have been used in all of the investigations included in this chapter, particles vary in size and shape, which affects several combustion characteristics. Therefore, in addition to OH-PLIF and LU for combustion imaging, it is important to investigate how particle size affects combustion behaviour. This requires another simultaneous system to determine particle size and shape in situ along with the combustion characteristics derived from LU and OH-PLIF. Diffuse back-illumination was used for this purpose. The combination of LU, OH-PLIF, and DBI is a highly effective combination of diagnostic techniques for the generation of conditioned statistics by measuring particle and combustion parameters simultaneously. This multi-parameter approach was first demonstrated by Köser et al. [23] and has become the backbone of subsequent experimental investigations.

Based on the demonstration of multi-parameter single particle combustion measurements by Köser et al. [23], Li et al. [20] adapted the experimental setup to the configuration shown in Fig. 10.6. Using this measurement system, a comprehensive investigation of the influence of various factors on the homogeneous ignition and volatile combustion of single solid fuel particles was performed [20] and further extended diagnostically and in terms of the investigated parameters in the study of Li et al. [35]. These studies combine detailed experimental data with high resolution simulations to provide an in-depth understanding of the combustion behaviour of CBC and walnut shell particles.

One of the key results of these studies regarding the ignition of single particles is shown in Fig. 10.7 [35]. Here the ignition delay time is plotted as a function of particle diameter for various size sieves of CBC and walnut shell particles in air and oxy-fuel atmospheres with a volumetric oxygen content of 20 % (termed AIR20 and OXY20). In Fig. 10.7, each data point represents a single particle measurement with ignition delay time and particle diameter measured simultaneously using OH-PLIF and DBI, allowing a one-to-one correlation. These data points are coloured according to the sieve ranges. For each sieve range, mean values and standard deviations for both ignition delay time and particle diameter are displayed by error bars. In addition, the data points are linearly fitted, as indicated by the dashed lines, to provide a clearer dependence of ignition time on particle diameter.

The data shown in Fig. 10.7 illustrates the effects of particle size, surrounding atmosphere, and particle composition on the ignition delay time. It is evident that the ignition delay increases with particle diameter for both CBC and walnut shell

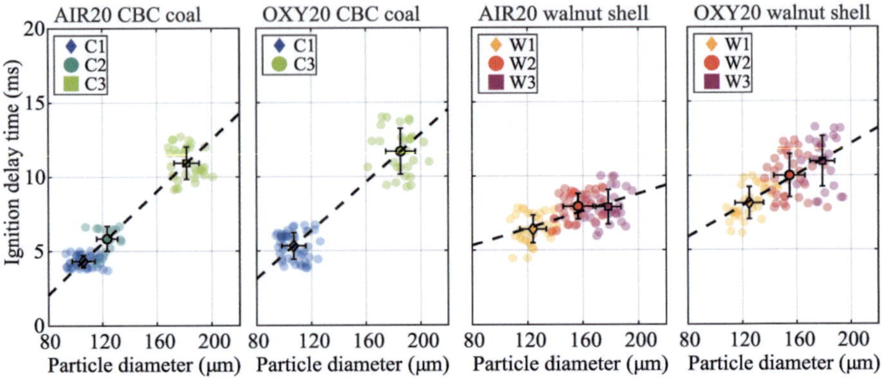


**Fig. 10.7** Ignition delay time as a function of particle diameter for Colombian bituminous coal (CBC) particles (C1: 90–106 μm, C2: 106–125 μm, C3: 160–200 μm) and walnut shell (WNS) particles (W1: 90–125 μm, W2: 125–150 μm, W3: >150 μm) in both AIR20 and OXY20 atmospheres (reprinted from Li et al. [35], Copyright (2024), with permission from Elsevier and minor editing)

particles. This trend is consistent for both AIR20 and OXY20 atmospheres, indicating that larger particles require a longer time period to reach the conditions necessary for ignition due to slower heating rates and delayed volatile release compared to smaller particles.

The influence of the ambient atmosphere on ignition varies between CBC and walnut shell particles, with greater impact on WNS than CBC ignition. The effect of $CO_2$-enriched gas atmospheres on ignition depends on the boundary conditions. The $CO_2$ atmosphere can result in a delayed gas-phase ignition due to its lower oxygen diffusion rate compared to $N_2$. However, at elevated gas temperatures, such as those in the reactor, $CO_2$-enriched gas mixtures can maintain higher local gas temperatures around the particle, compensating for its retarding effect. For walnut shell particles, a change from $N_2$ to $CO_2$ as the main gas component surrounding fuel particles causes an increase in the ignition delay time. Several factors contribute to this phenomenon. Particle heating depends on gas temperature, particle slip velocity, and particle diameter. Walnut shell particles release more volatiles than CBC particles which can lead to faster ignition. Additionally, a higher slip velocity present in AIR20 conditions can increase the heating rate of the particles, leading to earlier ignition. The higher moisture content of walnut shells initially delays ignition due to the energy required to evaporate the moisture. This may explain why small CBC and walnut shell particles have similar ignition delay times. For larger particles, the effect of higher volatile content becomes more significant than the moisture content, which is typically released prior to the pyrolysis of ignitable gases.

In summary, the ignition delay of carbonaceous solid fuel particles is influenced by particle size, the surrounding atmosphere, and the inherent properties of the fuel. Larger particles exhibit longer ignition delays due to slower heating rates. The presence of $CO_2$ in the atmosphere can either retard or enhance ignition depending on the conditions. The higher volatile content and reactivity of walnut shells results in different ignition behaviour compared to CBC, highlighting the importance of considering particle size and composition in combustion studies. For the studies shown, the detailed simulations illustrated in Sect. 12.3 providing data on particle heating and pyrolysis rates are an effective tool to combine experimental and numerical research.

Another key study combining numerical simulations with experimental data is the investigation of particle motion and swelling during early pyrolysis. During this phase, particles release moisture and pyrolysis products, leading to swelling, a process that can be effectively observed using high-resolution DBI, as demonstrated by Li et al. [28]. Understanding the swelling behaviour of coal particles prior to the onset of volatile combustion is crucial as it has a significant impact on the overall combustion process and ignition delay. Figure 10.8 illustrates the results from Li et al. [28], which include the distribution of initial particle diameters, time-resolved swelling ratio, numerically derived particle temperature, and heating rate evolution, and a phase diagram linking swelling ratio to particle heating rate. The data provided highlight the time evolution of single particles undergoing swelling in the reactor and provide valuable insights into the interplay between particle size, heating rates, and combustion behaviour.

The results indicate that swelling is synchronised with the pyrolysis process and is strongly dependent on particle size. Larger particles exhibit more pronounced swelling due to the higher internal pressure generated by the rapid release of volatiles. The swelling ratio, defined as the ratio of the time resolved particle diameter to the initial particle diameter, decreases as the particle heating rate increases. This inverse relationship suggests that faster heating rates result in more controlled and limited swelling.

The phase diagram in Fig. 10.8d shows that the maximum swelling ratio decreases from 1.22 to 1.07 as the heating rate increases from approximately $3 \times 10^4\ \mathrm{K\,s^{-1}}$ to $8 \times 10^4\ \mathrm{K\,s^{-1}}$. This behaviour is attributed to the balance between the rate of volatile release and the mechanical strength of the particle structure, which can only expand so far before fragmenting or contracting. As can be seen, the combined experimental and numerical approach provides valuable insights into the complex interactions between particle size, heating rate, and swelling behaviour during single particle combustion.

### 10.3.2 Combustion of Particle Groups in Laminar Flows

Building on the detailed understanding of single particle combustion in laminar flows provided by high resolution multi-parameter diagnostics, the research focus was extended to investigate the effects of particle number density (PND) in group combustion scenarios. In practical combustion systems, such as those found in industrial boilers, particles rarely burn in isolation but they interact with each other and significantly influence the overall combustion process. It is therefore essential to systematically investigate these interactions under similarly controlled laminar conditions to study the effect of increasing PND on the combustion behaviour.

The shift from single particles to particle groups introduces significant complexity as the physical and chemical phenomena are increasingly influenced by particle-particle interactions. These interactions add a layer of complexity as they not only

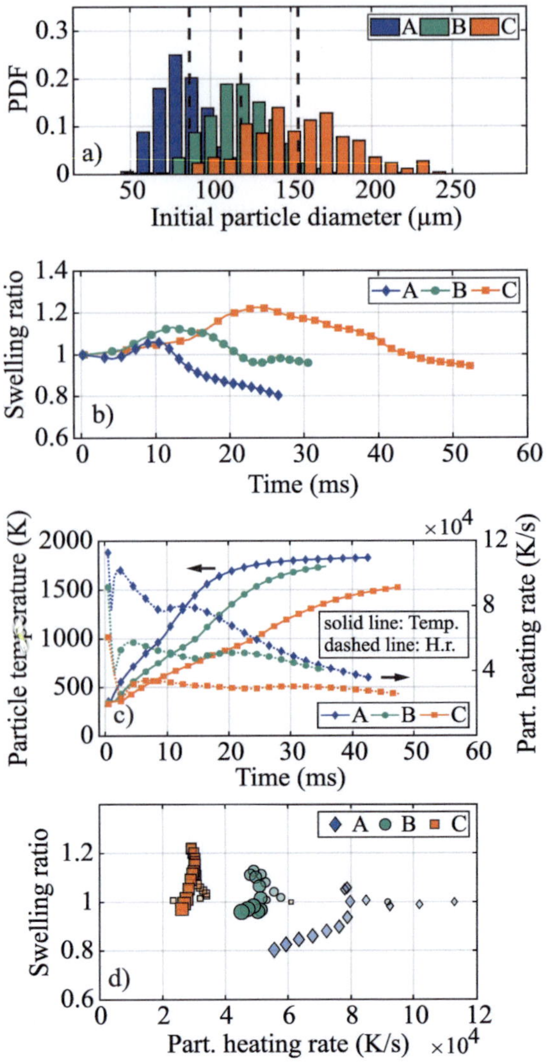


**Fig. 10.8** **a**) Distribution of initial particle diameters, **b**) temporal evolution of the swelling ratio for three representative particle sizes, **c**) particle temperature and heating rate histories for these particles and **d**) phase diagram showing the relationship between swelling ratio and particle heating rate (reprinted from Li et al. [28], Copyright (2023), with permission from Elsevier and minor editing)

affect combustion behaviour, but also require the adaptation of diagnostic methods to capture the resulting three-dimensional nature of the problem. In group combustion, particle competition for heat and oxygen becomes a critical factor that significantly alters combustion dynamics. This competition, together with the variability of ambient conditions, such as air vs. oxy-fuel environments, further complicates the process. To address these challenges, it has been necessary to move beyond traditional planar imaging techniques, which are inadequate for capturing the three-dimensional reaction zones that develop within particle clouds. Instead, volumetric measurement techniques have been developed and applied to accurately measure these complex interactions. The following section presents a step-by-step adaptation of these advanced diagnostic techniques to capture these phenomena adequately. As discussed in Sect. 10.3.1, laser-induced fluorescence of the OH radical (OH-LIF) has proven to be an effective technique for visualising reaction zones in the combustion of volatile carbonaceous solid fuels. However, in group combustion scenarios where the local PND increases, it becomes essential to measure the volatile flame structures volumetrically to account for the altered flame topologies.

Two primary methods have been explored to extend OH-LIF to volumetric imaging: tomographic LIF (Tomo-LIF) and LIF based on laser scanning (Scanning-LIF). Tomographic LIF involves volumetric rather than planar illumination, with the UV fluorescence excitation laser illuminating the measurement volume. Multiple cameras simultaneously capture the OH fluorescence from different angles and reconstruct the flame shape using tomographic reconstruction algorithms such as the multiplicative algebraic reconstruction technique (MART) [36]. While this method allows for high temporal resolution and the acquisition of a full volume per image, it has certain drawbacks. Volumetric illumination results in a lower signal-to-noise ratio (SNR) than planar techniques because the laser fluence—and therefore the fluorescence signal—decreases when the laser is spread over a volume rather than a sheet. In addition, tomographic reconstruction requires significant computational effort, making it a complex and resource-intensive method. Despite these challenges, Tomo-LIF has been successfully demonstrated in lifted jet flames of hydrocarbon fuels [37].

Given the limitations of Tomo-LIF, in particular the need to use a high power UV laser to achieve sufficient SNR, an alternative method known as Scanning-LIF has been developed for volumetric flame imaging. In this technique, a laser sheet is rapidly scanned in depth using a suitable scanning device. Initially, laser scanning techniques used rotating polygon mirrors, which are constrained by the mechanical inertia of moving parts, limiting the maximum scanning frequency and raising safety concerns [38]. A more advanced approach uses an acousto-optic deflector (AOD), which deflects the laser beam by inducing a high-frequency acoustic wave through a crystal, periodically changing its refractive index. By varying the frequency of the acoustic wave, the deflection angle of the laser beam can be precisely controlled, providing improved SNR and planar resolution compared to planar-LIF [39]. In addition, Scanning-LIF requires only a single camera with sufficient depth of field, as opposed to the multiple camera setup required for Tomo-LIF, further simplifying the experimental effort.

Despite these advantages, Scanning-LIF also involves certain compromises. While it improves SNR and planar resolution, it does so at the expense of depth and temporal resolution. In particular, temporal resolution is affected by the number of planes scanned per sweep. Scanning more planes reduces the temporal resolution between full volumetric sweeps. This trade-off between measurement volume size, depth resolution, and temporal resolution must be carefully balanced according to the specific requirements of the study. The first successful demonstration of laser scanning in combustion diagnostics, focusing on Mie scattering imaging of evaporating droplets, was performed by Li et al. [39].

Based on the knowledge gained from the development of volumetric flame imaging in lifted turbulent gas flames, the next step was to apply these methods to the study of solid fuel combustion, in particular the combustion of particle groups. The first applications of these methods are documented in the study by Li et al. [21] for the diagnostic development of Scanning-LIF based on the fluorescence of the OH radical in solid fuel flames, leading to the study by Li et al. [29]. In this work, the transition from single particle combustion to particle group combustion was investigated using the laminar flow reactor described in Sect. 10.2.1, which had previously been used for single particle combustion studies. Here, Scanning-LIF was combined with a simultaneously recording DBI system, which was used to obtain local PND and particle velocity fields derived from particle tracking velocimetry.

For single particles and small slip to the gas phase, a spherical flame can be expected to surround the particle, with a topology that is easily captured using planar techniques. However, as the system transitions to a group of particles with interacting flames from individual particles, a three-dimensional flame shape emerges, requiring volumetric imaging. Scanning-LIF was applied in such a way that high-speed cameras equipped with image intensifiers recorded at a repetition rate of 10 kHz in synchronisation with the UV LIF excitation laser, capturing 10 scan planes. These planes were then combined into volumetric flames using linear interpolation, with corrections made for particle motion during the acquisition of individual scan planes. This approach effectively results in a quasi-volumetric flame image repetition rate of 1 kHz. Figure 10.9 shows the transition in flame topology as local PND increase, with four selected snapshots labelled A, B, C, and D.

At low PND, the flame forms distinct spherical shapes around individual particles, which then merge into a larger flame volume with the highest LIF intensities concentrated in the core, as seen in snapshot A. As the PND increases slightly, the flame expands radially, accompanied by a noticeable reduction in reactivity at the core, as seen in snapshot B. This progression continues with rising values of PND, leading to the formation of a non-flammable core in snapshot C. Further increases in PND result in a flame topology similar to a lifted, non-premixed jet flame typical of gaseous fuels, where the flame base is disrupted, as shown in snapshot D. Scanning-LIF provides remarkable visualisations of this transition, capturing details of the emerging non-flammable central zone that planar techniques can only partially reveal.

The observed flame transition from isolated spherical flames around individual particles to a lifted conical flame with an inner non-flammable zone can be explained by revisiting the fundamental processes of solid fuel particles in the laminar flat flame burner. As particles pass through the premixed flat flame at the bottom of the reactor, they undergo rapid heating, resulting in swelling and the release of volatiles, as shown in Fig. 10.1. These volatiles must then mix with the ambient oxygen. If the oxygen concentration is sufficiently high to form a combustible mixture and the local gas temperature is high enough, ignition and subsequent combustion will occur. As the PND increases, so does the heat required per unit volume to raise the temperature of these particles. Consequently, for a given laminar flow rate of hot oxidant gas, the local gas temperature around the particles decreases as the PND increases. This phenomenon is evident in the OH-LIF intensity line profiles taken below the reaction zones, as shown in Figs. 10.10a and b, which correspond to the snapshots in Fig. 10.9. Here, a decrease in normalised OH-LIF intensity relative to an

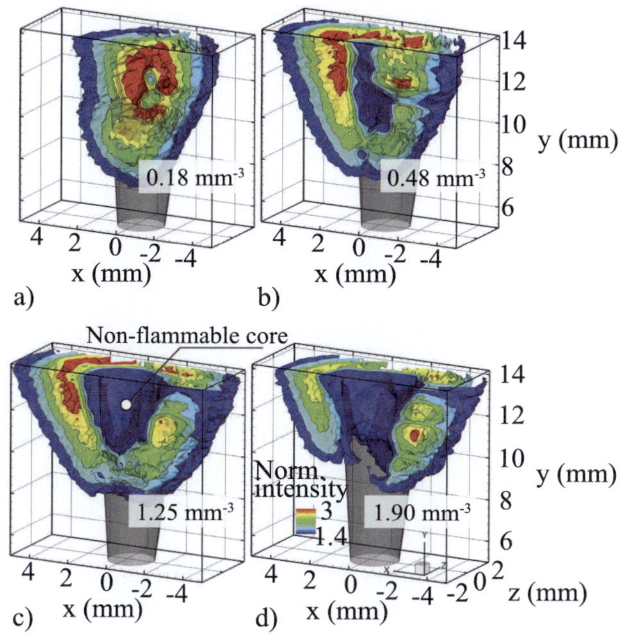


**Fig. 10.9** Flame topology transition from single to particle group combustion illustrated by instantaneous OH-LIF reconstruction of individual realisations A, B, C, and D with global PND of **a**) 0.18 mm$^{-3}$, **b**) 0.48 mm$^{-3}$, **c**) 1.25 mm$^{-3}$ and **d**) 1.9 mm$^{-3}$ (reprinted from Li et al. [29], Copyright (2021), with permission from Elsevier and minor editing)

undisturbed region in the flat flame exhaust gas is evident, leading to the assumption that the local temperature in the jet centre decreases with increasing PND values. In addition, as more particles undergo pyrolysis and release volatiles, a fuel-rich core may develop. In this core, the oxygen concentration may become too low to sustain ignition. These coupled effects can ultimately lead to the opening of the flame base as the PND increases, as shown schematically in Fig. 10.10c.

To determine which of these effects is more dominant, detailed simulations were performed as described in the study by Nicolai et al. [40] using the framework described in Chap. 12. The simulations showed that at high PND, the gas temperature at the tip of the flame drops below the threshold required for local ignition, even though the mixture fraction remains within flammable limits. The reduced heating rate at elevated PND results in a slower release of volatiles, which further disrupts flame formation and stabilisation. Tracking the particles from their initial heating in the flat flame to the volatile flame zone reveals the following sequence: As the particles pass through the flat flame, the local gas phase temperature rises sharply due to heat release from the flat flame, leading to rapid particle heating. This temperature then stabilises until volatile ignition occurs, at which point a smaller increase in gas-phase temperature at the particle positions is observed, driven by the exothermic reactions of volatile combustion. After this volatile combustion phase, the gas temperature experiences a slight decrease as the particles continue to heat. Meanwhile,

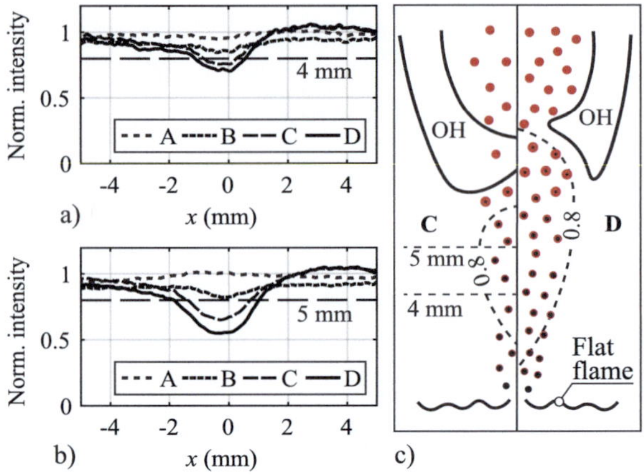


**Fig. 10.10** Horizontal OH-LIF intensity profiles of the background exhaust gas at **a**) 4 mm and **b**) 5 mm, and **c**) scheme of the flame transition of group combustion from individual visualisations C and D (reprinted from Li et al. [29], Copyright (2021), with permission from Elsevier and minor editing)

particle temperatures continue to rise steadily, with significant differences between low and high PND. These simulations confirm that the heat loss in the gas phase due to particle heating has a significant impact on flame formation and stabilisation, explaining the observed flame transitions in high PND scenarios.

Having established an explanation for the flame topology behaviour observed during the transition from single particle to group combustion, the validation of models used in the simulation requires the robust calculation of parameters that further quantify this transition. Similar to the single particle combustion studies, the ignition delay time is of particular interest to determine. However, for large PND, it becomes almost impossible to track each individual particle in the group and its individual ignition delay time with the optical techniques used. Due to particle-particle interactions, the pyrolysis gases released from the particles mix, and it becomes challenging to track the movement of individual particles in the cloud by DBI while simultaneously observing the flames formed from these particles, as particle interactions dominate flame stabilisation at higher PND. To overcome this challenge, a different approach was used to determine the PND-dependent ignition delay time. First, the ignition height, which represents the lowest position of the flame tip, was extracted from the volumetric OH-PLIF reconstructions. As shown in Fig. 10.11, the ignition height first decreases slightly with PND and then increases as the PND continues to increase. As described in Sect. 10.1.3, the residence time of particles and their released volatiles in the hot oxidising gases determines the ignition delay. Therefore, the PND-dependent velocity of the particles must be considered.

Due to particle-particle interactions in the tube, the exit velocity of particles entering the flat flame decreases with increasing PND. In addition, the lower local gas temperatures around the particles and therefore reduced thermal expansion of the hot gases in the flat flame exit at the particle positions at high PND result in a reduced acceleration of the particles. As a result, particles ascend the reactor more slowly when ejected as dense clouds than when ejected as individual particles. This information can be used to calculate a PND-dependent ignition delay time as shown in Fig. 10.11. The results clearly show that the ignition delay increases steadily with PND, leading to the conclusion that the competition for heat and oxygen gradually

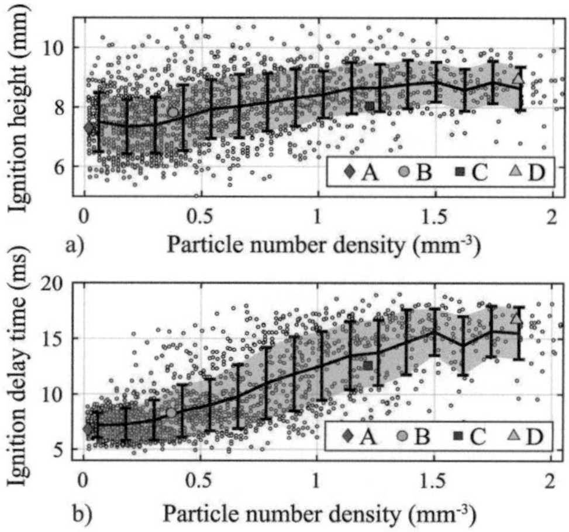


**Fig. 10.11** **a**) Ignition height over local PND in the flame tip region and **b**) ignition delay time over PND. Individual realisations A, B, C and D are highlighted (reprinted from Li et al. [29], Copyright (2021), with permission from Elsevier and minor editing)

causes a longer ignition delay. This conclusion was also confirmed by the numerical study of Nicolai et al. [40]. It should be noted again that such studies are only possible due to the simultaneous multi-parameter measurement of PND and volatile flame structures by DBI and Scanning-OH-LIF.

Based on the knowledge gained from the first study of the transition effects in the combustion of solid fuel particle groups, an adapted experiment was carried out using the same burner but with a larger jet orifice diameter of 2.9 mm instead of the previous 0.8 mm particle tube [22]. In this study, the experimental setup of Scanning-OH-LIF and DBI was extended to include Mie scattering measurements using a high-speed Nd:YAG laser operating at a wavelength of 532 nm to visualise soot formation during solid fuel combustion. In addition, the investigation of different parameters such as atmospheric oxygen content and the differences between air and oxy-fuel environments on the formation of the non-flammable core observed at high PND were investigated.

A key observation of this study is the evaluation of the ratio between the volume of the non-flammable core and the total volume of the flame, including both reaction zones and extinguished zones. This ratio, termed the non-flammability ratio, provides a measure of how much of the flame volume is occupied by regions where combustion does not occur. Figure 10.12 shows this ratio for varying oxygen concentrations relative to the PND for air and oxy-fuel atmospheres.

The general shape of all the curves in Fig. 10.12 is similar, showing an S-shaped trend where the non-flammability ratio increases steadily with PND. As the oxygen concentration increases, the curves are shifted towards higher PND values, indicating that higher oxygen availability makes the occurrence of non-flammability less likely. This observation suggests that increased $O_2$ concentrations improve flame stability and reduce the volume of non-flammable zones in the jet core. This effect is attributed to the increased reactivity and higher flame temperatures achievable with higher oxygen levels, which promote more efficient combustion of the volatiles released from the solid fuel particles.

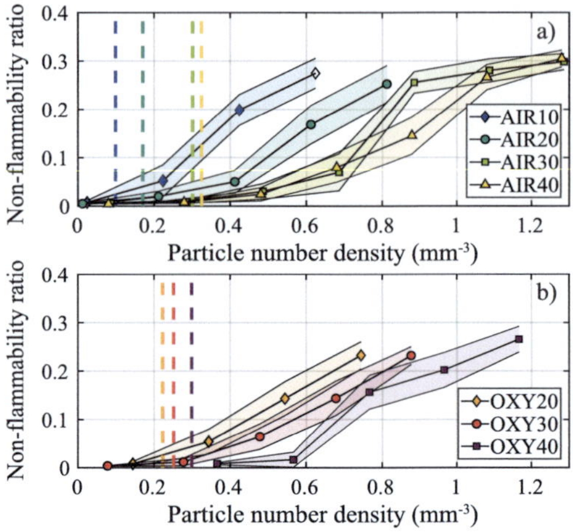


**Fig. 10.12** Comparison of the non-flammability ratio over the PND for varying $O_2$ concentrations in **a**) air and **b**) oxy-fuel atmospheres. Dashed lines represent the average PND at which no non-flammable regions are present, representing a transition limit from weak to strong particle-flame interaction leading to local flame extinction (reprinted from Li et al. [22], Copyright (2023), with permission from Elsevier and minor editing)

When comparing air and oxy-fuel environments, the overall effect on the non-flammability ratio is found to be minimal. The oxygen diffusivity in $CO_2$ is approximately 0.8 times that in $N_2$ at 1123 °C [41], meaning that oxygen does not mix as rapidly with the fuel in oxy-fuel conditions, which could theoretically increase ignition delay and promote non-flammability. On the other hand, $CO_2$ has a higher heat capacity than $N_2$, approximately 1.7 times larger at 1123 °C [41]. This higher heat capacity means that $CO_2$ can store more heat per unit volume, which helps to moderate temperature drops during particle heating.

The experiments in the laminar flat flame burner were carried out at constant ambient temperatures in the flue gas, similar to previous single particle combustion studies [20]. The replacement of $N_2$ by $CO_2$ introduces two counteracting effects: increased heat capacity, which stabilises gas-phase temperatures, and lower oxygen diffusivity, which could hinder combustion. Looking at the flammability curves, there are only small differences between the two diluents, suggesting that these effects largely cancel each other out. However, it is important to note that in practical applications the relative importance of these effects may vary depending on the specific combustion system. In self-sustained solid fuel flames using burners that do not rely on an external hot co-flow, but instead produce heat autonomously through the heat release by solid fuel conversion, the balance between these effects could change, leading to different results in terms of flame stability and non-flammability.

As presented, the introduction of particle interactions in solid fuel flames within laminar environments introduces complex behaviours, which have been captured experimentally using multi-parameter optical diagnostics. Factors such as PND, oxygen concentration and diluent composition have been systematically investigated and quantified, with accompanying numerical simulations providing detailed accompanying explanations of the observed behaviour. However, in realistic burner systems, the transition from laminar to turbulent flow has a significant impact on mixing, particle dynamics, and ultimately flame stabilisation. The following section explores the first step in this additional complexity, focusing on how particles interact with turbulence and how these dynamics influence key parameters in particle-laden flows.

### 10.3.3 Solid Fuel Particle Dynamics in Turbulent Jets

In real burner systems, the high mass flow rates of both air and solid fuels inherently introduce turbulence, raising the question of how turbulent flows affect the complex interactions between particles, chemical reactions, and the surrounding flow field. The accurate prediction and modelling of particle-laden turbulent flows remains a significant challenge, even for non-reactive cases [42]. Modern computational approaches typically use a Lagrangian point-particle framework coupled with Eulerian fluid simulation, known as Euler-Lagrange methods. However, these methods rely on empirical correlations for drag and torque, as the direct fluid-particle interactions are not fully resolved, but modelled. The complexity increases further when moving from mostly spherical coal particles to aspherical, fibrous biomass particles such as miscanthus. These aspherical particles affect the surrounding turbulence not only through their translational motion but also through rotational dynamics, introducing orientation-dependent drag characteristics [43]. Traditionally, correlations for these effects are derived from fully resolved direct particle-fluid simulations (DPFS), as detailed in Sect. 11.4 [44]. While such simulations have been successfully applied to decaying homogeneous isotropic turbulence involving particle clouds, they require extremely high computational resources [45], making them impractical for simulating the high Reynolds number flows found in actual solid fuel boiler systems. This necessitates the development and application of high-resolution experimental techniques capable of capturing the intricate particle-turbulence interactions of complex particle shapes within technically relevant turbulent flows.

Since turbulent flows are highly unsteady, characterised by temporal and spatial fluctuations over a wide range of scales, the interaction between dispersed particles and the surrounding carrier phase must be captured instantaneously, i.e. both the fluid and particle phases must be measured simultaneously. To achieve this, two main classes of experimental methods are commonly used. The simplest method uses a single camera to image both phases, which are then separated on the basis of particle size and scattered light intensity [46]. A more sophisticated method involves the use of multiple cameras to separate the two phases using different wavelengths of light, achieved by applying colour filters based on phosphorescence, fluorescence, or incandescence [7]. After phase separation, the motion of the dispersed phase is tracked using particle tracking techniques, while the velocity field of the carrier phase is typically calculated using particle image velocimetry (PIV). In the present study, single-camera methods were used to resolve the complex interactions of biomass particles within the turbulent flow.

When attempting to resolve solid fuel particles in turbulent flows, which involve significantly higher velocities than the laminar environments studied in Sect. 10.3.1, it becomes evident that the high-speed lasers previously used at repetition rates of 10 kHz are inadequate. In PIV, velocity measurements are based on the displace-

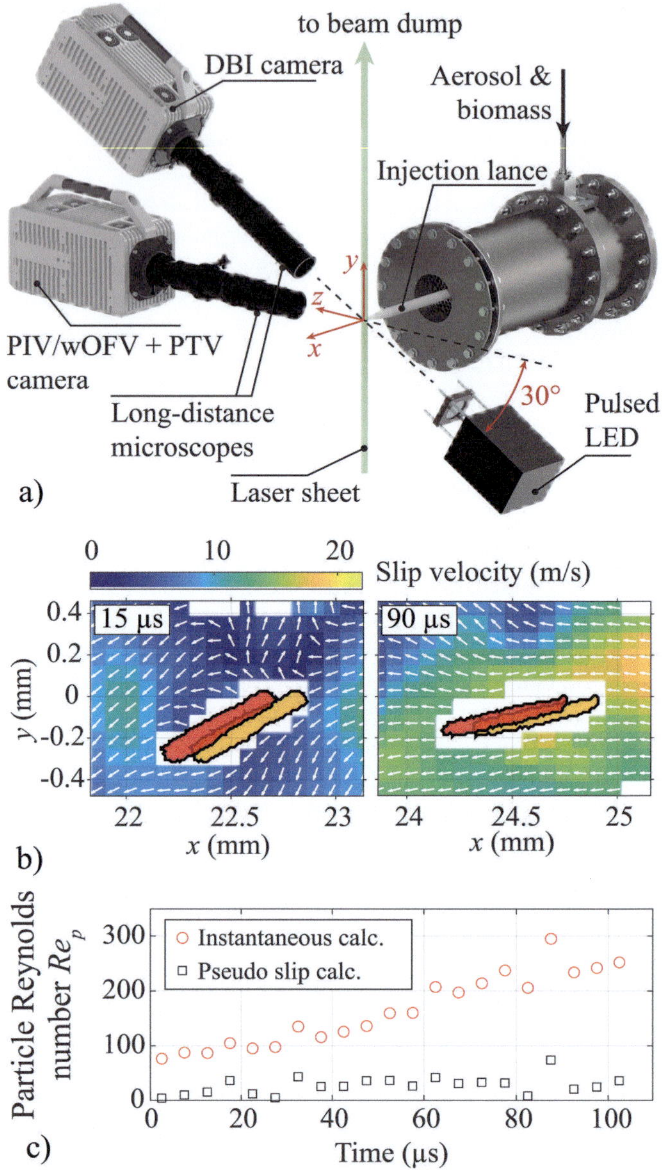


**Fig. 10.13** **a**) Experimental setup for simultaneous ultra-high-speed two-phase velocimetry and DBI at the turbulent hot gas test rig, **b**) slip velocity fields of miscanthus particle moving through a region of high turbulence at two selected instants and **c**) temporal dynamics of the Reynolds particle numbers $\mathrm{Re}_{\mathrm{p},\langle \underline{u} \rangle}$ and $\mathrm{Re}_\mathrm{p}$ (reprinted from Geschwindner et al. [26], Copyright (2023), with permission from Elsevier and minor editing)

ment between two successive frames, which dictates the size of the interrogation window—typically three to four times larger than the displacement in one dimension. To achieve the required spatial resolution for solid fuel particles of around 100 μm at flow velocities exceeding 30 $\mathrm{m\,s^{-1}}$, ultra-high repetition rates beyond 100 kHz are required. To meet this challenge, a novel Yb-doped fibre laser has been used, initially characterised for time-resolved PIV (TR-PIV) at repetition rates of 400 kHz [47], to capture the non-reactive single-phase turbulent jet in the turbulent hot gas test rig described in Sect. 10.2.1. This fibre laser, using a master oscillator power amplifier (MOPA) architecture, amplifies a Gaussian beam distributed over an active multi-core fibre. Due to the high surface to volume ratio inherent in fibre lasers, effective air cooling is sufficient to maintain consistent beam quality even at high optical output powers of 50 W. This efficient cooling mechanism mitigates thermal stress and avoids the thermal lensing effects, allowing the laser to operate at ultra-high repetition rates without compromising performance.

The successful application of this laser system was subsequently extended to ultra-high-speed two-phase measurements of biomass particles. The experimental setup used in this study, as shown in Fig. 10.13a, involved two synchronised high-speed cameras that simultaneously captured light scattering of oil droplet tracers and biomass particles from the fibre laser, as well as diffuse back-illumination images, at a repetition rate of 200 kHz and a pixel resolution of 10 $\mathrm{\mu m\,px^{-1}}$ [26,48].

This configuration allowed the spatio-temporal resolution of aspherical miscanthus particles moving through the turbulent flow field in the centreline region of the turbulent jet. With simultaneous measurements of the biomass particle and fluid velocity fields, a slip velocity field could be calculated based on the in-plane velocity components, as shown for two different times in Fig. 10.13b. With both particle size and slip velocity known, the particle Reynolds number $\mathrm{Re_p}$, defined in (10.3), could be measured instantaneously.

$$\mathrm{Re_p} = \frac{|\underline{u} - \underline{u}_\mathrm{p}|\, d_\mathrm{p}}{\nu} \tag{10.3}$$

In (10.3), $\underline{u}$ denotes the fluid velocity vector and $\underline{u}_\mathrm{p}$ denotes the particle velocity vector. The quantity $d_\mathrm{p}$ is the circle area equivalent particle diameter, while $\nu$ is the kinematic viscosity of the carrier phase. Figure 10.13c shows the time evolution of $\mathrm{Re_p}$ for the single representative miscanthus particle shown in Figs. 10.13b. As the particle moves from a region of high to low fluid velocity, it maintains its velocity due to its high relaxation time, resulting in an increase in slip velocity and hence $\mathrm{Re_p}$ over time to values indicating flow detachment based on the highly resolved simulations described in Chap. 11 [44]. If the instantaneous fluid velocity had not been resolved and the mean velocity field $\langle \underline{u} \rangle$ derived from single-phase statistical measurements had been used, the particle Reynolds number

$$\mathrm{Re}_{\mathrm{p},\langle \underline{u} \rangle} = \frac{|\langle \underline{u} \rangle - \underline{u}_\mathrm{p}|\, d_\mathrm{p}}{\nu} \tag{10.4}$$

would have been significantly underestimated using this pseudo-slip approach. This underestimation occurs because the instantaneous slip velocity is not captured by the statistical approach, which also fails to account for the modulation of the carrier phase turbulent flow by the presence of particles. This highlights the critical importance of simultaneous multi-parameter diagnostics of both the fluid and particle phases in turbulent flows, an aspect that has not been as critical in the context of slip velocity measurements in laminar flows.

A limitation of the previous approach was that, due to the high frame rates, the high-speed camera sensor had to be cropped, reducing the number of active pixels. This cropping, combined with the fact that PIV resolution is inherently limited by the size of the interrogation window [49], in this case 32 pixels, resulted in a significant loss of spatial resolution. To overcome this limitation and increase the spatial resolution of flow field measurements, a wavelet-based optical flow velocimetry (wOFV) method was employed. Originally developed by Dérian et al. [50] and implemented in this work by Nicolas et al. [51], this method allows for more detailed resolution of the velocity field in turbulent flows.

Optical flow methods estimate velocity fields by assuming the conservation of some quantity, typically brightness intensity $I(\underline{x}, t)$, across frames. This principle leads to the displaced frame difference equation:

$$I_0(\underline{x}) - I_1(\underline{x} + \underline{v}(\underline{x})) = 0, \tag{10.5}$$

where $I_0$ and $I_1$ represent the intensities at different frames, and $\underline{v}(\underline{x})$ is the displacement field to be estimated. However, this problem is inherently ill-posed as it relates two unknown velocity components to a single observable intensity variable. To address this, variational approaches are used that minimise a penalty function combining a data term $J_\mathrm{D}$ and a regularisation term $J_\mathrm{R}$ that depends on the velocity field derivatives [52]:

$$\hat{\underline{v}} = \arg\min_{\underline{v}} \left[ J_\mathrm{D}(I_0, I_1, \underline{v}) + \lambda J_\mathrm{R}(\underline{v}) \right]. \tag{10.6}$$

The scalar parameter $\lambda$ balances the weighting between the data term and the regularisation term, with higher values forcing greater smoothness in the estimated displacement field. In this study, a Laplacian regularisation is applied over the image domain, making it well suited for resolving turbulent flow and gradient quantities [51,53].

Wavelet-based optical flow velocimetry differs from traditional optical flow methods by performing this minimisation in the wavelet domain rather than the spatial domain. This approach introduces a natural multi-resolution scheme that is particularly effective in capturing the wide range of scales present in turbulent flows [50]. The wOFV method has been shown to outperform PIV over a range of regularisation parameters [51]. For the present study, a regularisation parameter of 0.1 was chosen based on the physical properties of the turbulence spectrum, ensuring that no artificial kinetic energy is introduced at smaller scales, consistent with turbulence cascade theory [54].

Figure 10.14a compares the slip velocity fields obtained using PIV and wOFV. The analysis shows significant differences, particularly close to the particle, where the velocity gradient is more pronounced in wOFV, especially in the upper left

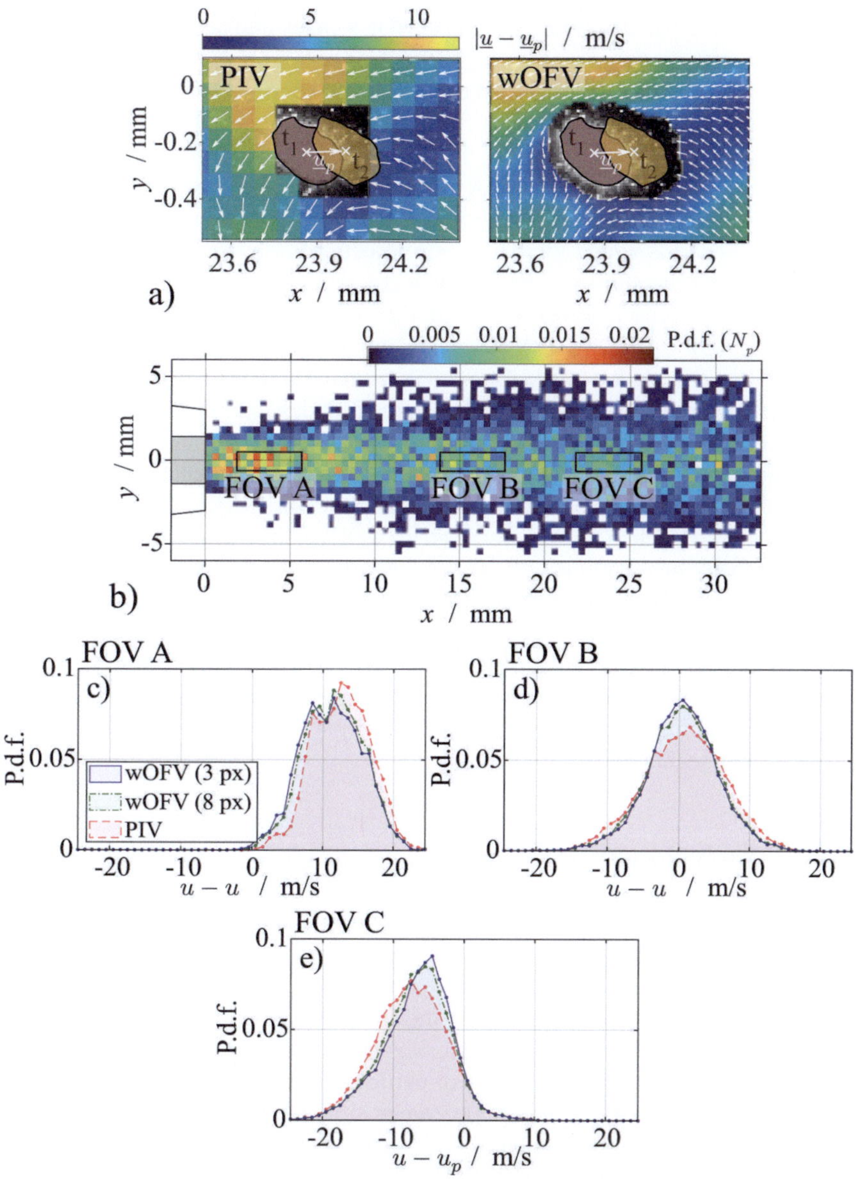


**Fig. 10.14** **a**) Slip velocity field around a single walnut shell particle, derived from both PIV and wOFV, with every fourth vector shown in the latter for clarity, **b**) probability distribution of particle presence indicating the fields of view (FOV) A, B, and C and **c**)–**e**) axial slip velocity (reprinted from Geschwindner et al. [48], Copyright (2024), with permission from Elsevier and minor editing)

region close to the particle. This is because wOFV takes into account the movement of structures close to the particle, which PIV does not.

The subsequent analysis within the three fields of view (FOV) highlighted in Fig. 10.14b shows the local axial slip velocity for each particle across all FOV. A clear pattern emerges: Particles initially lag behind the flow as they exit the tube in FOV A, forming a wake in the axial direction. A wake inversion is observed in FOV B, shifting the wake to the left of the particles, which could have significant implications for combustion scenarios where local transport affects volatile combustion. Consistent with the superior resolution of wOFV near the particle, the absolute value of the local axial slip velocity is lower with wOFV compared to PIV. Aligning the 8 pixel ring edge of wOFV with the PIV sampling region confirms this observation. The larger sampling ring shows minor differences from the smaller 3-pixel ring of wOFV and consistently reports lower slip velocity magnitudes than PIV, indicating better resolution of velocity gradients near the particle.

As the research presented in this section demonstrates, the integration of advanced instrumentation, such as the fibre laser, and novel processing techniques, such as wavelet-based optical flow velocimetry (wOFV), has enabled unprecedented particle-resolved observations of particle-turbulence interactions for individual biomass particles. These innovative approaches not only provide insights into the complex dynamics at play in solid fuel combustion systems, but also open up opportunities to address a wide range of challenges in dispersed multi-phase flows. The techniques developed and applied here provide a powerful set of tools for researchers to study and optimise multi-phase flow systems, paving the way for more efficient and sustainable technologies in the future.

## 10.4 Conclusion and Outlook

This chapter has presented a comprehensive experimental investigation of the interactions between particles, fluid flow, and chemical reactions in the combustion of solid fuels, focusing on both highly volatile bituminous coal and biomass. A stepwise approach has been taken, moving from the study of single particle combustion in controlled laminar environments, to exploring the complexities of group combustion, and finally to the dynamic interactions of particles with turbulent flows.

The detailed study of single particle combustion in laminar flows provided new insights into the ignition and combustion processes, revealing how particle size, composition, and the surrounding atmosphere influence ignition delay times and volatile combustion. The multi-parameter diagnostic techniques developed, including high-speed OH-PLIF, DBI, and LU, allowed the full temporal history of individual particles to be captured, providing a unique dataset for model validation.

In group combustion scenarios, the shift from single to multiple particles presented new challenges, particularly in measuring the three-dimensional flame topologies that arise as particle number densities increase. The adaptation of volumetric imaging techniques, such as Scanning-LIF combined with DBI, allowed the detailed study of particle interactions, highlighting how competition for heat and oxygen affects combustion dynamics. The systematic investigation of factors such as oxygen concentration and diluent composition provided a deeper understanding of flame stability and the formation of non-flammable zones in particle clouds.

The study of solid fuel particle dynamics in turbulent jet flows marked a significant advance in methodology for the study of particle-turbulence interactions. The combination of state-of-the-art instrumentation, such as an ultra-high-speed Yb-doped fibre laser, and novel data processing techniques, such as wavelet-based optical flow velocimetry, enabled unprecedented particle-resolved observations. These innovative approaches have provided unprecedented insight into slip velocity fields, particle Reynolds numbers, and turbulence modulation, providing a powerful set of tools for future multi-phase flow research.

Looking ahead, the methods and results presented in this chapter provide the basis for significant advances in the understanding and modelling of solid fuel combustion in complex environments at the particle scale. The integration of high-resolution experimental data with detailed numerical simulations is essential for the development of more accurate and predictive combustion models. The techniques developed here extend far beyond solid fuel combustion and offer potential solutions to a wide range of challenges in dispersed multi-phase flows.

**Acknowledgements** This work has been funded by the Deutsche Forschungsgemeinschaft (DFG, German Research Foundation)—215035359—SFB/TRR 129 'Oxyflame'.

## References

1. Goshayeshi, B. and Sutherland, J. C. "A Comparison of Various Models in Predicting Ignition Delay in Single-Particle Coal Combustion". *Combustion and Flame* 161.7 (2014), 1900–1910. https://doi.org/10.1016/j.combustflame.2014.01.010
2. Vascellari, M., Xu, H. and Hasse, C. "Flamelet Modeling of Coal Particle Ignition". *Proceedings of the Combustion Institute* 34.2 (2013), 2445–2452. https://doi.org/10.1016/j.proci.2012.06.152
3. Moroń, W. and Rybak, W. "Ignition Behaviour and Flame Stability of Different Ranks Coals in Oxy Fuel Atmosphere". *Fuel* 161 (2015), 174–181. https://doi.org/10.1016/j.fuel.2015.08.065
4. Hecht, E. S., Shaddix, C. R. and Lighty, J. S. "Analysis of the Errors Associated with Typical Pulverized Coal Char Combustion Modeling Assumptions for Oxy-Fuel Combustion". *Combustion and Flame* 160.8 (2013), 1499–1509. https://doi.org/10.1016/j.combustflame.2013.02.015
5. Shaddix, C. R. and Molina, A. "Particle Imaging of Ignition and Devolatilization of Pulverized Coal During Oxy-Fuel Combustion". *Proceedings of the Combustion Institute* 32.2 (2009), 2091–2098. https://doi.org/10.1016/j.proci.2008.06.157
6. Simões, G., Magalhães, D., Rabaçal, M. and Costa, M. "Effect of Gas Temperature and Oxygen Concentration on Single Particle Ignition Behavior of Biomass Fuels". *Proceedings of the Combustion Institute* 36.2 (2017), 2235–2242. https://doi.org/10.1016/j.proci.2016.06.102
7. Li, T., Geschwindner, C., Dreizler, A. and Böhm, B. "Particle-Resolved Optical Diagnostics of Solid Fuel Combustion for Clean Power Generation: A Review". *Measurement Science and Technology* 34.12 (2023), 122001. https://doi.org/10.1088/1361-6501/acef49

8. Zhang, L., Binner, E., Qiao, Y. and Li, C.-Z. "In Situ Diagnostics of Victorian Brown Coal Combustion in $O_2/N_2$ and $O_2/CO_2$ Mixtures in Drop-Tube Furnace". *Fuel* 89.10 (2010), 2703–2712. https://doi.org/10.1016/j.fuel.2010.04.020
9. Bejarano, P. A. and Levendis, Y. A. "Single-Coal-Particle Combustion in $O_2/N_2$ and $O_2/CO_2$ Environments". *Combustion and Flame* 153.1–2 (2008), 270–287. https://doi.org/10.1016/j.combustflame.2007.10.022
10. Schiemann, M., Vorobiev, N. and Scherer, V. "Stereoscopic Pyrometer for Char Combustion Characterization". *Applied Optics* 54.5 (2015), 1097–1108. https://doi.org/10.1364/AO.54.001097
11. Du, X., Gopalakrishnan, C. and Annamalai, K. "Ignition and Combustion of Coal Particle Streams". *Fuel* 74.4 (1995), 487–494. https://doi.org/10.1016/0016-2361(95)98350-N
12. Liu, Y., Geier, M., Molina, A. and Shaddix, C. R. "Pulverized Coal Stream Ignition Delay Under Conventional and Oxy-Fuel Combustion Conditions". *International Journal of Greenhouse Gas Control* 5 (2011), S36–S46. https://doi.org/10.1016/j.ijggc.2011.05.028
13. Xu, M., Li, S., Wu, Y., Jia, L. and Lu, Q. "Effects of $CO_2$ on the Fuel Nitrogen Conversion During Coal Rapid Pyrolysis". *Fuel* 184 (2016), 430–439. https://doi.org/10.1016/j.fuel.2016.06.130
14. Cheng, Y., Pothos, S. and Diez, F. J. "Phase Discrimination Method for Simultaneous Two-Phase Separation in Time-Resolved Stereo PIV Measurements". *Experiments in Fluids* 49.6 (2010), 1375–1391. https://doi.org/10.1007/s00348-010-0878-0
15. Hwang, S. m., Kurose, R., Akamatsu, F., Tsuji, H., Makino, H. and Katsuki, M. "Application of Optical Diagnostics Techniques to a Laboratory-Scale Turbulent Pulverized Coal Flame". *Energy & Fuels* 19.2 (2005), 382–392. https://doi.org/10.1021/ef049867z
16. Balusamy, S., Schmidt, A. and Hochgreb, S. "Flow Field Measurements of Pulverized Coal Combustion Using Optical Diagnostic Techniques". *Experiments in Fluids* 54.5 (2013). https://doi.org/10.1007/s00348-013-1534-2
17. Fletcher, T. H. "Time-Resolved Temperature Measurements of Individual Coal Particles During Devolatilization". *Combustion Science and Technology* 63.1 (1989), 89–105. https://doi.org/10.1080/00102208908947120
18. Law, C. K. *Combustion Physics*. Cambridge and New York: Cambridge University Press, 2006. https://doi.org/10.1017/CBO9780511754517
19. Köser, J., Becker, L., Goßmann, A.-K., Böhm, B. and Dreizler, A. "Investigation of Ignition and Volatile Combustion of Single Coal Particles within Oxygen-Enriched Atmospheres Using High-Speed OH-PLIF". *Proceedings of the Combustion Institute* 36.2 (2017), 2103–2111. https://doi.org/10.1016/j.proci.2016.07.083
20. Li, T., Farmand, P., Geschwindner, C., Greifenstein, M., Köser, J., Schumann, C., Attili, A., Pitsch, H., Dreizler, A. and Böhm, B. "Homogeneous Ignition and Volatile Combustion of Single Solid Fuel Particles in Air and Oxy-Fuel Conditions". *Fuel* 291 (2021). https://doi.org/10.1016/j.fuel.2020.120101
21. Li, T., Schiemann, M., Köser, J., Dreizler, A. and Böhm, B. "Experimental Investigations of Single Particle and Particle Group Combustion in a Laminar Flow Reactor Using Simultaneous Volumetric OH-LIF Imaging and Diffuse Backlight-Illumination". *Renewable and Sustainable Energy Reviews* 136 (2021), 110377. https://doi.org/10.1016/j.rser.2020.110377
22. Li, T., Geschwindner, C., Dreizler, A. and Böhm, B. "An Experimental Study of Coal Particle Group Combustion in Conventional and Oxy-Fuel Atmospheres Using Multi-Parameter Optical Diagnostics". *Proceedings of the Combustion Institute* 39 (2023), 3259–3269. https://doi.org/10.1016/j.proci.2022.07.081
23. Köser, J., Li, T., Vorobiev, N., Dreizler, A., Schiemann, M. and Böhm, B. "Multi-Parameter Diagnostics for High-Resolution In-Situ Measurements of Single Coal Particle Combustion". *Proceedings of the Combustion Institute* 37.3 (2019), 2893–2900. https://doi.org/10.1016/j.proci.2018.05.116
24. Eitel, F., Pareja, J., Geyer, D., Johchi, A., Michel, F., Elsäßer, W. and Dreizler, A. "A Novel Plasma Heater for Auto-Ignition Studies of Turbulent Non-Premixed Flows". *Experiments in Fluids* 56.10 (2015). https://doi.org/10.1007/s00348-015-2059-7

25. Geschwindner, C., Westrup, K., Li, T., Dreizler, A. and Böhm, B. "Experimentelle Untersuchung der Interaktion von Biomasse mit einem turbulenten plasmabeheizten Oxidatorgasstrom". *30. Deutscher Flammentag*. Hannover, Germany, 2021.
26. Geschwindner, C., Westrup, K., Dreizler, A. and Böhm, B. "Pulse Picking of a Fiber Laser Enables Velocimetry of Biomass-Laden Jets at Low and Ultra-High Repetition Rates". *Proceedings of the Combustion Institute* 39.1 (2023), 1325–1335. https://doi.org/10.1016/j.proci.2022.07.138
27. Geschwindner, C., Werner, M., André, L., Nicolas, A., Peterson, B., Meinke, M., Schröder, W., Dreizler, A. and Böhm, B. "High-Resolution Two-Phase Velocimetry of Aspherical Particles Using Wavelet-Based Optical Flow Velocimetry (wofv) Benchmarked with Fully Resolved Direct Numerical Simulations". *21st International Symposium on Applications of Laser and Imaging Techniques to Fluid Mechanics*. Lisbon, Portugal, 2024.
28. Li, T., Li, B., Farmand, P., Dreizler, A., Pitsch, H. and Böhm, B. "Motion and Swelling of Single Coal Particles During Volatile Combustion in a Laminar Flow Reactor". *Proceedings of the Combustion Institute* 39.3 (2023), 3333–3341. https://doi.org/10.1016/j.proci.2022.07.111
29. Li, T., Geschwindner, C., Köser, J., Schiemann, M., Dreizler, A. and Böhm, B. "Investigation of the Transition from Single to Group Coal Particle Combustion Using High-Speed Scanning OH-LIF and Diffuse Backlight-Illumination". *Proceedings of the Combustion Institute* 38 (2021), 4101–4109. https://doi.org/10.1016/j.proci.2020.06.314
30. Kohse-Höinghaus, K. "Laser Techniques for the Quantitative Detection of Reactive Intermediates in Combustion Systems". *Progress in Energy and Combustion Science* 20.3 (1994), 203–279. https://doi.org/10.1016/0360-1285(94)90015-9
31. Köser, J., Becker, L., Vorobiev, N., Schiemann, M., Scherer, V., Böhm, B. and Dreizler, A. "Characterization of Single Coal Particle Combustion within Oxygen-Enriched Environments Using High-Speed OH-PLIF". *Applied Physics B* 121.4 (2015), 459–464. https://doi.org/10.1007/s00340-015-6253-3
32. Raffel, M., Willert, C. E., Scarano, F., Kähler, C. J., Wereley, S. T. and Kompenhans, J. *Particle Image Velocimetry*. Cham: Springer International Publishing, 2018. https://doi.org/10.1007/978-3-319-68852-7
33. Scharnowski, S. and Kähler, C. J. "Particle Image Velocimetry – Classical Operating Rules from Today's Perspective". *Optics and Lasers in Engineering* 135 (2020), 106185. https://doi.org/10.1016/j.optlaseng.2020.106185
34. Attili, A., Farmand, P., Schumann, C., Farazi, S., Böhm, B., Li, T., Geschwindner, C., Köser, J., Dreizler, A. and Pitsch, H. "Numerical Simulations and Experiments of Ignition of Solid Particles in a Laminar Burner: Effects of Slip Velocity and Particle Swelling". *Flow, Turbulence and Combustion* 106 (2020), 515–531. https://doi.org/10.1007/s10494-020-00222-1
35. Li, T., Farmand, P., Chen, H., Boehme, C., Nicolai, H., Hasse, C., Pitsch, H. and Böhm, B. "Homogeneous Ignition and Volatile Flame Structure of Single Bituminous Coal and Walnut Shell Particles: Effects of Particle Size and Gas Atmosphere". *Fuel* 371 (2024), 131955. https://doi.org/10.1016/j.fuel.2024.131955
36. Debasish Mishra, K. M. P. "A Robust MART Algorithm for Tomographic Applications". *Numerical Heat Transfer, Part B: Fundamentals* 35.4 (1999), 485–506. https://doi.org/10.1080/104077999275857
37. Li, T., Pareja, J., Fuest, F., Schütte, M., Zhou, Y., Dreizler, A. and Böhm, B. "Tomographic Imaging of OH Laser-Induced Fluorescence in Laminar and Turbulent Jet Flames". *Measurement Science and Technology* 29.1 (2018), 015206. https://doi.org/10.1088/1361-6501/aa938a
38. Weinkauff, J., Greifenstein, M., Dreizler, A. and Böhm, B. "Time Resolved Three-Dimensional Flamebase Imaging of a Lifted Jet Flame by Laser Scanning". *Measurement Science and Technology* 26.10 (2015), 105201. https://doi.org/10.1088/0957-0233/26/10/105201
39. Li, T., Pareja, J., Becker, L., Heddrich, W., Dreizler, A. and Böhm, B. "Quasi-4D Laser Diagnostics Using an Acousto-Optic Deflector Scanning System". *Applied Physics B* 123.3 (2017), 1243. https://doi.org/10.1007/s00340-017-6663-5

40. Nicolai, H., Li, T., Geschwindner, C., di Mare, F., Hasse, C., Böhm, B. and Janicka, J. "Numerical Investigation of Pulverized Coal Particle Group Combustion Using Tabulated Chemistry". *Proceedings of the Combustion Institute* 38.3 (2021), 4033–4041. https://doi.org/10.1016/j.proci.2020.06.081
41. Toftegaard, M. B., Brix, J., Jensen, P. A., Glarborg, P. and Jensen, A. D. "Oxy-Fuel Combustion of Solid Fuels". *Progress in Energy and Combustion Science* 36.5 (2010), 581–625. https://doi.org/10.1016/j.pecs.2010.02.001
42. Balachandar, S. and Eaton, J. K. "Turbulent Dispersed Multiphase Flow". *Annual Review of Fluid Mechanics* 42.1 (2010), 111–133. https://doi.org/10.1146/annurev.fluid.010908.165243
43. Voth, G. A. and Soldati, A. "Anisotropic Particles in Turbulence". *Annual Review of Fluid Mechanics* 49.1 (2017), 249–276. https://doi.org/10.1146/annurev-fluid-010816-060135
44. Fröhlich, K., Meinke, M. and Schröder, W. "Correlations for Inclined Prolates Based on Highly Resolved Simulations". *Journal of Fluid Mechanics* 901 (2020), A5. https://doi.org/10.1017/jfm.2020.482
45. Schneiders, L., Meinke, M. and Schröder, W. "Direct Particle-Fluid Simulation of Kolmogorov-Length-Scale Size Particles in Decaying Isotropic Turbulence". *Journal of Fluid Mechanics* 819 (2017), 188–227. https://doi.org/10.1017/jfm.2017.171
46. Schmidt, B. E. and Sutton, J. A. "Evaluation of Gas- and Particle-Phase Separation Methods for Velocity Measurements in Turbulent Multiphase Flows". *Experiments in Fluids* 61.12 (2020). https://doi.org/10.1007/s00348-020-03082-7
47. Geschwindner, C., Westrup, K., Dreizler, A. and Böhm, B. "Ultra-High-Speed Time-Resolved PIV of Turbulent Flows Using a Continuously Pulsing Fiber Laser". *Experiments in Fluids* 63.4 (2022). https://doi.org/10.1007/s00348-022-03424-7
48. Geschwindner, C., Nicolas, A., Westrup, K., Dreizler, A., Peterson, B. and Böhm, B. "Resolving Biomass-Turbulence Interactions at the Particle Scale Using Ultra-High-Speed Wavelet-Based Optical Flow Velocimetry (wofv)". *Proceedings of the Combustion Institute* 40.1–4 (2024), 105461. https://doi.org/10.1016/j.proci.2024.105461
49. Kähler, C. J., Scharnowski, S. and Cierpka, C. "On the Resolution Limit of Digital Particle Image Velocimetry". *Experiments in Fluids* 52.6 (2012), 1629–1639. https://doi.org/10.1007/s00348-012-1280-x
50. Dérian, P., Héas, P., Herzet, C. and Mémin, E. "Wavelets and Optical Flow Motion Estimation". *Numerical Mathematics: Theory, Methods and Applications* 6.1 (2013), 116–137. https://doi.org/10.4208/nmtma.2013.mssvm07
51. Nicolas, A., Zentgraf, F., Linne, M., Dreizler, A. and Peterson, B. "Assessment and Application of Wavelet-Based Optical Flow Velocimetry (wofv) to Wall-Bounded Turbulent Flows". *Experiments in Fluids* 64.3 (2023), 50. https://doi.org/10.1007/s00348-023-03594-y
52. Horn, B. K. and Schunck, B. G. "Determining Optical Flow". *Artificial Intelligence* 17.1–3 (1981), 185–203. https://doi.org/10.1016/0004-3702(81)90024-2
53. Kadri-Harouna, S., Dérian, P., Héas, P. and Mémin, E. "Divergence-Free Wavelets and High Order Regularization". *International Journal of Computer Vision* 103.1 (2013), 80–99. https://doi.org/10.1007/s11263-012-0595-7
54. Schmidt, B. E. and Sutton, J. A. "High-Resolution Velocimetry from Tracer Particle Fields Using a Wavelet-Based Optical Flow Method". *Experiments in Fluids* 60.3 (2019).https://doi.org/10.1007/s00348-019-2685-6

# Modelling and Analysis of Fluid-Particle Interactions

# 11

Laurent André, Konstantin Fröhlich, Matthias Meinke and Wolfgang Schröder

## Contents




L. André (✉) · K. Fröhlich · M. Meinke · W. Schröder
Chair of Fluid Mechanics and Institute of Aerodynamics, RWTH Aachen University, Aachen, Germany
e-mail: l.andre@aia.rwth-aachen.de

K. Fröhlich
e-mail: k.froehlich@aia.rwth-aachen.de

M. Meinke
e-mail: m.meinke@aia.rwth-aachen.de

W. Schröder
e-mail: office@aia.rwth-aachen.de







R. Kneer et al. (eds.), *Clean Energy Conversion*,
https://doi.org/10.1007/978-3-031-98956-8_11

**Abstract**

The particle-fluid interaction of non-reactive ellipsoidal particles is studied in turbulent particle-laden flows. Highly resolved direct particle-fluid simulations, capturing all turbulent scales, provide necessary reference solutions for the development of Lagrangian point-particle models. Studies in decaying isotropic turbulence highlight the effect of particle shape on turbulence modulation and reveal the limits of existing point-particle models for non-spherical particles, unable to accurately predict the dynamics of Kolmogorov-length scale sized ellipsoidal particles. To address this, correlations that account for finite particle Reynolds numbers, particle shape, and orientation have been developed for drag, lift, torque, and heat transfer. Simulations of isotropic and anisotropic turbulent flows reveal that conventional spherical and ellipsoidal point-particle models without the novel correlations tend to overestimate the particle velocity. In addition, the non-spherical particles tend to align their major axes along the flow direction if finite particle Reynolds number effects are not considered by the model. In contrast, applying the new correlations results in a preferred orientation perpendicular to the flow direction.

## 11.1 Introduction

The interaction of non-reactive ellipsoidal particles with turbulent flows, in particular the influence of the non-spherical shape of biomass particles, is investigated. The particle shape and its orientation can have a significant impact on the macroscopic features of turbulent particle-laden flows. Therefore, highly-resolved direct particle-fluid simulations (DPFS) are conducted where all turbulent scales, including the particle boundary layers and wakes are resolved. Due to the immense computational effort required for DPFS of Kolmogorov length-scale sized particles, particle resolved simulations are not feasible, e.g. for design studies of a full combustion chamber. Therefore, accurate point-particle models for non-spherical particles that can be used in large-eddy simulations (LES) are needed. Point-particle models are not only important to correctly predict the particle dynamics and turbulence modulation, but also the particle temperature and the conjugate heat transfer between the particles and the fluid. Herein, reference solutions generated by DPFS generate important input for the development and validation.

At the time, when the CRC/TRR 129 was started, fully resolved particle laden flows for thousands of particles at a length scale in the range of the Kolmogorov scale could hardly be simulated even on high performance computing systems. The development of a cut-cell method based on an unstructured Cartesian formulation for arbitrary moving geometries in viscous flows was a must to tackle such challenging problems. The smooth formulation for the coupling of the fluid and the structural motion was complemented by a new predictor-corrector Runge-Kutta method (PC-RK). The coupling to solvers for structural deformation, Lagrangian particle transport, level-set methods, or heat conduction was substantially improved since no interfacial states had to be resolved at intermediate time levels. To couple the flow solver and the level-set solver, a hierarchical dual-mesh framework was developed. The grids of both solvers are adapted independently from each other such that solution adaptive mesh refinement for the flow solver and the level-set solver is straightforward. The hierarchical octree data structure allows an efficient data exchange between the Lagrangian point-particle solver and the Navier-Stokes solver. In brief, the development of this simulation framework [1–3] made it possible to perform direct particle-laden flow simulations with a number of particles on the order of $10^4$ to $10^5$.

Since then, extensive studies have been conducted in decaying isotropic turbulence for various particle shapes. One such study investigated the turbulence modulation caused by 45,000 heavy spherical particles of a size that is on the order of the Kolmogorov length scale [3]. These highly resolved particle simulations highlighted that heavy particles of a similar size as the smallest turbulent scales impact the decay of the kinetic energy in the fluid across all length scales by induced local velocity disturbances, suggesting severe implications on the accuracy of common Lagrangian point-particle models, where the local flow field around the particles is not resolved. The accuracy of a commonly used spherical and ellipsoidal point-particle model was assessed in isotropic turbulence by comparison with DPFS results [4]. In this regard, the interaction of 45,000 heavy spherical and prolate particles of a size that is again on the order the smallest turbulent scales was investigated. Compared to spherical particles with an equivalent volume, prolate particles exhibited significantly reduced velocities, a phenomenon that could not be captured by either point-particle model. On the one hand, the spherical model does not consider the particle shape and thus cannot predict a preferred alignment of the particle's major axis. On the other hand, the ellipsoidal model does not account for finite particle Reynolds numbers. Both models also significantly underpredict the overall viscous dissipation due to the particle turbulence interaction. In a rigorous follow-up study, the DPFS performed by Schneiders et al. [3] were extended to non-spherical particles by Schneiders et al. [5]. The turbulence modulation of $\mathcal{O}(10^4)$ non-spherical particles was investigated in isotropic turbulence. It was shown, that the linear part of the two-way coupling rate depends on the particle shape as well as the orientation and as such indirectly on the rotational motion of the particle. The analysis highlights the importance of an orientation-dependent model for drag, lift, and torque as well as corrections for finite particle Reynolds numbers in Lagrangian point-particle models. Unless the rotational contribution is negligible compared to the linear contribution, the study emphasises the necessity of a torque coupling scheme to ensure the conservation of energy. The development of such orientation-dependent correlations is subject of this contribution.

In Sect. 11.2, an overview of the relevant mathematical models is given. Followed by a description of the numerical implementation details in Sect. 11.3, including the twofold modelling of the particle phase, i.e. the particle-resolved and the point-particle approaches. Section 11.4 describes the development of novel correlations for prolate particles at higher particle Reynolds numbers, considering both the particle aspect ratio and orientation. The extended point-particle model is compared to existing models and validated by DPFS in Sect. 11.5. Finally, the analysis of a particle-laden turbulent round jet is presented in Sect. 11.6.

## 11.2 Mathematical Model

### 11.2.1 Governing Equations for Viscous Fluid Flow

The motion of a compressible and viscous fluid is governed by the general conservation equations for mass, momentum and energy in their integral form. The non-dimensional integral form of the governing equations reads

$$\int_V \frac{\partial \mathbf{Q}}{\partial t}\,\mathrm{d}V + \oint_{\partial V} \left(\underline{\mathbf{H}}^{\mathrm{inv}} - \underline{\mathbf{H}}^{\mathrm{vis}}\right)\cdot\mathbf{n}\,\mathrm{d}A = \int_V \mathbf{F}\,\mathrm{d}V\,, \tag{11.1}$$

where $V$ refers to the control volume, $\partial V$ its surface, and $\mathbf{n}$ the normal vector. The vector $\mathbf{Q} = [\rho, \rho\mathbf{u}, \rho E]^T$ includes the conservative variables with the density $\rho$, the velocity vector $\mathbf{u}$, the total specific energy $E = e + 0.5\,\mathbf{u}^2$, and the specific internal energy $e$. The vector $\mathbf{F}$ denotes external forces exerted on the volume, such as gravity. The flux vector is split up in its inviscid $\underline{\mathbf{H}}^{\mathrm{inv}}$ and viscous part $\underline{\mathbf{H}}^{\mathrm{vis}}$. They are given by

$$\underline{\mathbf{H}}^{\mathrm{inv}} - \underline{\mathbf{H}}^{\mathrm{vis}} = \begin{bmatrix} \rho\,(\mathbf{u} - \mathbf{u}_{\partial V}) \\ \rho\,\mathbf{u}\,(\mathbf{u} - \mathbf{u}_{\partial V}) + p\,\underline{\mathbf{I}} \\ \rho\,E\,(\mathbf{u} - \mathbf{u}_{\partial V}) + p\,\mathbf{u} \end{bmatrix} - \frac{1}{\mathrm{Re}_0}\begin{bmatrix} 0 \\ \underline{\boldsymbol{\tau}} \\ \underline{\boldsymbol{\tau}}\,\mathbf{u} - \mathbf{q} \end{bmatrix}, \tag{11.2}$$

with the velocity vector of the control volume surface $\mathbf{u}_{\partial V}$, the pressure $p$, the identity tensor $\underline{\mathbf{I}}$, the stress tensor $\underline{\boldsymbol{\tau}}$ and the heat conduction vector $\mathbf{q}$. The Reynolds number $\mathrm{Re}_0$ is defined at the stagnation point as $\mathrm{Re}_0 = \rho_0\,a_0\,L/\mu_0$, where $\rho_0$ is the reference density, $\mu_0$ the dynamic viscosity, $L$ the reference length, and $a_0 = \sqrt{\gamma\,p_o/\rho_0}$ the speed of sound with the heat capacity ratio $\gamma$. For a Newtonian fluid with zero bulk viscosity, the stress tensor $\underline{\boldsymbol{\tau}}$ can be formulated as

$$\underline{\boldsymbol{\tau}} = \mu\left(\nabla\mathbf{u} + (\nabla\mathbf{u})^T\right) - \frac{2}{3}\,\mu\,(\nabla\cdot\mathbf{u})\,\underline{\mathbf{I}}\,, \tag{11.3}$$

where the non-dimensional dynamic viscosity is calculated from the temperature $T$ using the Sutherland's law

$$\mu = T^{3/2}\frac{1+S}{T+S}\,, \tag{11.4}$$

with the constant $S = 110.4/273.15$ for air. The heat conduction vector $\mathbf{q}$ is expressed by Fourier's law for a constant Prandtl number $\mathrm{Pr}_0 = \mu_0\,c_{p,0}\,/k_0$, where $c_{p,0}$ is the specific heat at constant pressure and $k_0$ the thermal conductivity as

$$\mathbf{q} = -\frac{\mu}{\mathrm{Pr}_0\,(\gamma - 1)}\nabla T\,. \tag{11.5}$$

Closure is achieved through the ideal gas law in its non-dimensional form

$$\gamma p = \rho T . \tag{11.6}$$

### 11.2.2 Governing Equations for Rigid Particle Motion

The translational motion of a rigid particle of mass $m$ is governed by Newton's second law of motion and can be described in an inertial frame of reference, yielding for a particle with a centre of mass $\mathbf{r}_\mathrm{p}$ and a velocity $\mathbf{v}_\mathrm{p}$

$$\frac{\mathrm{d}\mathbf{r}_\mathrm{p}}{\mathrm{d}t} = \mathbf{v}_\mathrm{p} , \tag{11.7}$$

$$m \frac{\mathrm{d}\mathbf{v}_\mathrm{p}}{\mathrm{d}t} = \mathbf{F}_\mathrm{p} . \tag{11.8}$$

The quantity $\mathbf{F}_\mathrm{p}$ denotes the total force exerted on the particle. The rotational motion of a rigid particle, where a torque $\mathbf{T}_\mathrm{p}$ is applied, is described by Euler's rotation equations. For a particle-fixed frame that is aligned with the particle's principal axes of inertia, the governing equation reads

$$\underline{\hat{\mathbf{J}}} \frac{\mathrm{d}\hat{\boldsymbol{\omega}}}{\mathrm{d}t} + \hat{\boldsymbol{\omega}} \times \left( \underline{\hat{\mathbf{J}}} \, \hat{\boldsymbol{\omega}} \right) = \hat{\mathbf{T}}_\mathrm{p} , \tag{11.9}$$

where $\underline{\mathbf{J}}$ represents the diagonal tensor of the principal moments of inertia and $\hat{\boldsymbol{\omega}}$ is the angular velocity of the particle. Quantities denoted by a hat $(\hat{\cdot})$ are defined in the particle-fixed frame and are computed from the inertial-frame quantities through an orthogonal rotation matrix

$$\underline{\mathbf{R}}(\boldsymbol{\varepsilon}) = \begin{bmatrix} \varepsilon_0^2 + \varepsilon_1^2 - \varepsilon_2^2 - \varepsilon_3^2 & 2(\varepsilon_1 \, \varepsilon_2 - \varepsilon_0 \, \varepsilon_3) & 2(\varepsilon_1 \, \varepsilon_3 + \varepsilon_0 \, \varepsilon_2) \\ 2(\varepsilon_1 \, \varepsilon_2 + \varepsilon_0 \, \varepsilon_3) & \varepsilon_0^2 - \varepsilon_1^2 + \varepsilon_2^2 - \varepsilon_3^2 & 2(\varepsilon_2 \, \varepsilon_3 - \varepsilon_0 \, \varepsilon_1) \\ 2(\varepsilon_1 \, \varepsilon_3 - \varepsilon_0 \, \varepsilon_2) & 2(\varepsilon_2 \, \varepsilon_3 + \varepsilon_0 \, \varepsilon_1) & \varepsilon_0^2 - \varepsilon_1^2 - \varepsilon_2^2 + \varepsilon_3^2 \end{bmatrix} , \tag{11.10}$$

where $\varepsilon_i$ refers to the component of the quaternion $\boldsymbol{\varepsilon} = [\varepsilon_0, \varepsilon_1, \varepsilon_2, \varepsilon_3]^T$ used to describe the particle orientation in a three-dimensional space. The time evolution of the particle orientation is expressed by

$$\frac{\mathrm{d}\boldsymbol{\varepsilon}}{\mathrm{d}t} = \underline{\boldsymbol{\Omega}}(\hat{\boldsymbol{\omega}}) \, \boldsymbol{\varepsilon} , \tag{11.11}$$

$$\underline{\boldsymbol{\Omega}}(\boldsymbol{\omega}) = \frac{1}{2} \begin{bmatrix} 0 & -\omega_1 & -\omega_2 & -\omega_3 \\ \omega_1 & 0 & \omega_3 & -\omega_2 \\ \omega_2 & -\omega_3 & 0 & \omega_1 \\ \omega_3 & \omega_2 & -\omega_1 & 0 \end{bmatrix} . \tag{11.12}$$

The hydrodynamic force and torque exerted by the fluid on the particle's surface $\Gamma_\mathrm{p}$ is given by the surface integrals

$$\mathbf{F}_\mathrm{p} = \oint_{\Gamma_\mathrm{p}} (-p\,\mathbf{n} + \underline{\tau}\cdot\mathbf{n})\,\mathrm{d}A\,, \tag{11.13}$$

$$\mathbf{T}_\mathrm{p} = \oint_{\Gamma_\mathrm{p}} (\mathbf{x}_\Gamma - \mathbf{r}_\mathrm{p}) \times (-p\,\mathbf{n} + \underline{\tau}\cdot\mathbf{n})\,\mathrm{d}A\,. \tag{11.14}$$

Depending on the modelling approach, $\mathbf{F}_\mathrm{p}$ and $\mathbf{T}_\mathrm{p}$ can be explicitly computed (Sect. 11.3.1) or determined by analytical or semi-empirical relations (Sect. 11.3.2).

## 11.3 Numerical Methods

A hierarchical Cartesian mesh is used to discretise the system of equations from (11.1). For the inviscid fluxes, an upwind-biased scheme is used, while for the viscous fluxes, a central scheme is applied. The time integration is performed by a second-order-accurate five-stage Runge-Kutta scheme. The Cartesian mesh is generated by a fully automated parallel mesh generator, and solution-adaptive mesh refinement is used to ensure a sufficiently high resolution during the simulation [6]. Available adaptation sensors range from a distance sensor to solution-dependent sensors such as the vorticity sensor [2,6,7]. A dynamic load balancing approach [8] is used to ensure an efficient usage of the parallel computational resources, accounting for varying loads due to the adaptive mesh-refinement or point-particle movement. As indicated in Sect. 11.2, the modelling approaches for the particle phase differ in how the hydrodynamic forces and torques acting on the particle surfaces are computed. If the mesh is locally sufficiently refined, i.e. the particle surface and boundary layer are resolved, the surface integrals in (11.13) and (11.14) can be numerically integrated. Details of these accurate but computationally expensive direct particle-fluid simulations are described in Sect. 11.3.1. Considering the computational effort, the more commonly used approach for sufficiently small particles is the Lagrangian point-particle method. In this approach, particles are considered as point masses, and the acting forces are estimated through analytical or semi-empirical relations. Section 11.3.2 introduces two models: the spherical and the ellipsoidal point-particle model.

### 11.3.1 Particle Resolved Simulation

When the spatial step of the computational mesh is smaller than the particle diameter, the particle surface $\Gamma_\mathrm{p}$ is resolved using a cut-cell approach. The location of the particle surface is defined by a signed distance, i.e. a level-set function [2]. When particles approach each other, multi-cut cells can be formed, which are defined by using multiple level-sets, allowing the resolution of particle-particle collisions. Due to the immense resolution requirement for DPFS with particles that are on the order of the Kolmogorov length scale, dynamic mesh refinement around the individual particles is crucial. Using the previously mentioned adaptation sensors, the mesh around the particles is continuously adapted as the particles move through the domain, greatly reducing the overall number of cells compared to a uniformly refined mesh. Since the particles are fully resolved, the surface integrals for the hydrodynamic force and torque can be computed as

$$\mathbf{F}_{\mathrm{p}} = \sum_{i \in \Gamma_{\mathrm{p}}} [-p\,\mathbf{n} + \underline{\boldsymbol{\tau}} \cdot \mathbf{n}]_i \, A_i \ , \tag{11.15}$$

$$\mathbf{T}_{\mathrm{p}} = \sum_{i \in \Gamma_{\mathrm{p}}} [(\mathbf{x}_{\Gamma} - \mathbf{r}_{\mathrm{p}}) \times (-p\,\mathbf{n} + \underline{\boldsymbol{\tau}} \cdot \mathbf{n})]_i \, A_i \ . \tag{11.16}$$

On the surface of the particles a no-slip boundary condition is imposed [1]

$$\mathbf{u}(\mathbf{x}) = \mathbf{u}_{\Gamma}(\mathbf{x}), \quad \mathbf{x} \in \Gamma(t) \ , \tag{11.17}$$

with respect to heat transfer, the imposed boundary condition can be of adiabatic type

$$\frac{\partial T}{\partial n} = 0, \quad \mathbf{x} \in \Gamma^{\mathrm{ad}}(t) \ , \tag{11.18}$$

or isothermal

$$T(\mathbf{x}) = T_{\Gamma}, \quad \mathbf{x} \in \Gamma^{\mathrm{iso}}(t) \ . \tag{11.19}$$

### 11.3.2 Point-Particle Models

In the following, two Lagrangian point-particle models are presented, the spherical Lagrangian model, which does not consider the particle orientation and only solves the translational motion, and the ellipsoidal Lagrangian model, which also solves the rotational motion equation and considers the particle orientation.

**Spherical Lagrangian Model (SLM)**

For a sufficiently small solid particle of spherical shape, the particle motion can be modelled by the equation of Maxey-Riley [9]. Assuming Stokes flow conditions, i.e. vanishing particle Reynolds numbers such that $\mathrm{Re}_{\mathrm{p}} \ll 1$, and a particle size that is significantly smaller than the characteristic length scale of the flow, the equation of Maxey-Riley reads

$$m \frac{\mathrm{d}\mathbf{v}_\mathrm{p}}{\mathrm{d}t} = m_\mathrm{f} \frac{\mathrm{D}\mathbf{u}}{\mathrm{D}t} + \frac{1}{2} m_\mathrm{f} \left( \frac{\mathrm{D}\mathbf{u}}{\mathrm{D}t} - \frac{\mathrm{d}\mathbf{v}_\mathrm{p}}{\mathrm{d}t} \right) + (m - m_\mathrm{f})\, \mathbf{g}$$
$$+ 3\pi\, \mu\, d_\mathrm{p}\, (\mathbf{u} - \mathbf{v}_\mathrm{p}) + 3\pi\, \mu\, d_\mathrm{p} \int_0^t K(t-\tau) \frac{\mathrm{d}(\mathbf{u} - \mathbf{v}_\mathrm{p})}{\mathrm{d}\tau}\, \mathrm{d}\tau\,. \qquad (11.20)$$

Here, $\mathbf{u}$ refers to the undisturbed fluid velocity vector at the particle position, $m_\mathrm{f}$ the mass of the displaced fluid, while $\frac{D}{Dt}$ and $\frac{d}{dt}$ refer to the material derivative and the derivative along the particle trajectory respectively. The right-hand side of (11.20) consists of contributions from the force exerted by the undisturbed flow, the added-mass force, buoyancy, the quasi-steady Stokes drag, as well as the Basset-Boussinesq history force with its kernel function $K(t)$. Several extensions exist to expand (11.20) to finite particle Reynolds numbers through empirical correlations. For instance, the drag force can be approximated as

$$\mathbf{F} = 3\pi\, \mu\, f_\mathrm{d}\, d_\mathrm{p} (\mathbf{u} - \mathbf{v}_\mathrm{p})\,, \qquad (11.21)$$

where $f_\mathrm{d}$ is correction term such as the popular Schiller-Naumann drag function $f_\mathrm{d} = 1 + 0.15\, \mathrm{Re}_\mathrm{p}^{0.687}$. Equation (11.20) further reduces under the assumption of heavy particles to only the contributions of the drag and gravity [10] and can be reformulated after introduction of the particle relaxation time $\tau_\mathrm{p} = (\rho_\mathrm{p}\, d_\mathrm{eq}^2)/(18\, \mu)$ as

$$\frac{\mathrm{d}\mathbf{v}_\mathrm{p}}{\mathrm{d}t} = \frac{f_\mathrm{d}}{\tau_\mathrm{p}} (\mathbf{u} - \mathbf{v}_\mathrm{p}) + \mathbf{g}\,. \qquad (11.22)$$

The reduced equation is solved by a predictor-corrector scheme [4] where the flow velocity is interpolated through a least-squares interpolation routine. The rotational motion of spherical particles is generally considered of minor importance [11], which is why it is neglected in this model. For non-spherical particles, an equivalent volume approach can be used. However, upon selecting this approach, the particle orientation is neglected.

**Ellipsoidal Lagrangian Model (ELM)**

Under the assumption of creeping flow, the expression for the hydrodynamic force for an ellipsoidal particle with semi-axes $a$, $b$, and $c$ can be described as

$$\mathbf{F} = \mu\, \underline{\mathbf{R}}^T\, \underline{\hat{\mathbf{K}}}\, \underline{\mathbf{R}}\, (\mathbf{u} - \mathbf{v}_\mathrm{p})\,, \qquad (11.23)$$

where $\underline{\mathbf{R}}$ is the rotation matrix introduced in Sect. 11.2, and $\underline{\hat{\mathbf{K}}}$ [12] is a resistance tensor defined as

$$\underline{\hat{\mathbf{K}}} = 16\pi\, a\, b\, c\, \mathrm{diag} \begin{pmatrix} \frac{1}{\chi_0 + a^2\, \alpha_0} \\ \frac{1}{\chi_0 + b^2\, \beta_0} \\ \frac{1}{\chi_0 + c^2\, \gamma_0} \end{pmatrix}, \qquad (11.24)$$

and depending on the shape parameters, that for a general ellipsoid read

$$\chi_0 = a\,b\,c \int_0^\infty \frac{\mathrm{d}\lambda}{\Delta}\,, \tag{11.25}$$

$$\alpha_0 = a\,b\,c \int_0^\infty \frac{\mathrm{d}\lambda}{(a^2+\lambda)\Delta}\,, \tag{11.26}$$

$$\beta_0 = a\,b\,c \int_0^\infty \frac{\mathrm{d}\lambda}{(b^2+\lambda)\Delta}\,, \tag{11.27}$$

$$\gamma_0 = a\,b\,c \int_0^\infty \frac{\mathrm{d}\lambda}{(c^2+\lambda)\Delta}\,, \tag{11.28}$$

$$\Delta = \sqrt{(a^2+\lambda)(b^2+\lambda)(c^2+\lambda)}\,. \tag{11.29}$$

For prolate particles, analytical solutions for the integrals (11.25)–(11.28) can be derived [4]. The torque governing the rotational motion of the particle can be approximated in creeping shear flow [13] as

$$\hat{\mathbf{T}}_\mathrm{p} = \mu\,\underline{\hat{\mathbf{K}}}_\mathrm{s}\,\hat{\mathbf{s}} + \mu\,\underline{\hat{\mathbf{K}}}_\zeta\,(\hat{\boldsymbol{\zeta}} - \hat{\boldsymbol{\omega}})\,, \tag{11.30}$$

with the resistance tensors $\underline{\hat{\mathbf{K}}}_\mathrm{s}$ and $\underline{\hat{\mathbf{K}}}_\zeta$ defined as

$$\underline{\hat{\mathbf{K}}}_\mathrm{s} = \frac{16}{3}\pi\,a\,b\,c\,\mathrm{diag}\begin{pmatrix} \frac{b^2-c^2}{b^2\,\beta_0+c^2\,\gamma_0} \\ \frac{c^2-a^2}{c^2\,\gamma_0+a^2\,\alpha_0} \\ \frac{a^2-b^2}{a^2\,\alpha_0+b^2\,\beta_0} \end{pmatrix}, \tag{11.31}$$

$$\underline{\hat{\mathbf{K}}}_\zeta = \frac{16}{3}\pi\,a\,b\,c\,\mathrm{diag}\begin{pmatrix} \frac{b^2+c^2}{b^2\,\beta_0+c^2\,\gamma_0} \\ \frac{c^2+a^2}{c^2\,\gamma_0+a^2\,\alpha_0} \\ \frac{a^2+b^2}{a^2\,\alpha_0+b^2\,\beta_0} \end{pmatrix}, \tag{11.32}$$

and the strain and fluid vorticity

$$\hat{\mathbf{s}} = \frac{1}{2}\left(\frac{\partial u_{\hat{z}}}{\partial\hat{y}} + \frac{\partial u_{\hat{y}}}{\partial\hat{z}},\ \frac{\partial u_{\hat{x}}}{\partial\hat{z}} + \frac{\partial u_{\hat{z}}}{\partial\hat{x}},\ \frac{\partial u_{\hat{y}}}{\partial\hat{x}} + \frac{\partial u_{\hat{x}}}{\partial\hat{y}}\right)^T, \tag{11.33}$$

$$\hat{\boldsymbol{\zeta}} = \frac{1}{2}\left(\frac{\partial u_{\hat{z}}}{\partial\hat{y}} - \frac{\partial u_{\hat{y}}}{\partial\hat{z}},\ \frac{\partial u_{\hat{x}}}{\partial\hat{z}} - \frac{\partial u_{\hat{z}}}{\partial\hat{x}},\ \frac{\partial u_{\hat{y}}}{\partial\hat{x}} - \frac{\partial u_{\hat{x}}}{\partial\hat{y}}\right)^T. \tag{11.34}$$

**Extended Ellipsoidal Lagrangian Model (EELM)**

In an analogous procedure to the extension of the spherical point-particle model towards finite particle Reynolds numbers, the ellipsoidal point-particle model is extended through newly derived correlations for the drag, lift, and torque of prolate particles. The derivation process, the resulting correlations, and the reformulated model are described in the following section.

## 11.4 Novel Correlations for Lagrangian Particle Models

This section presents the derivation process of novel correlations to be used in existing point-particle models through individual particle simulations. In the first section, the focus is on the force and moment correlations of prolate particles (Sect. 11.4.1). In the subsequent section, the focus is shifted to heat transfer with the derivation of a new correlation for the Nusselt number (Sect. 11.4.2). Simulations of the uniform flow past a single, fixed prolate particle were conducted. Input parameters for the simulations are the particle Reynolds number $\mathrm{Re_p}$, the particle orientation defined as the inclination angle $\phi$ between the particle major axis and the freestream velocity, as well as the particle aspect ratio $\beta$. The coordinate system of the global frame $(x, y, z)$ is fixed at the centre of gravity of the particle, the particle-fixed frame $(\hat{x}, \hat{y}, \hat{z})$ is aligned with the principal axes of the prolate as shown in Fig. 11.1a. In all of the conducted simulations, the $y$ and $\hat{y}$ axes are aligned. A vast parameter space is covered with $1 \leq \mathrm{Re_p} \leq 100$, $1 \leq \beta \leq 8$, and $0° \leq \phi \leq 90°$. This parameter space is extended in the second study by $0.35 \leq T_\mathrm{R} \leq 1.65$, where $T_\mathrm{R}$ is the ratio between surface temperature $T_\mathrm{S}$ and the freestream fluid temperature $T_\infty$. The particle's centre of gravity is fixed at $(0.25\,L_x, 0.5\,L_y, 0.5\,L_z)$ within a domain size of $L_x \times L_y \times L_z = 96 \times 48 \times 48$ scaled by the spherical volume equivalent-diameter $d_\mathrm{eq}$. The finest resolution is selected such that $\Delta_\mathrm{min}/d_\mathrm{eq} = 1/48$. This resolution ensures the accurate resolution of the sharp edges of elongated particles with aspect ratios in the higher end of the selected parameter space as depicted in Fig. 11.1b. The domain size as well as the resolution were selected after convergence studies conducted by Fröhlich et al. [14] and Kiwitt et al. [15]. The use of adaptive mesh refinement (Sect. 11.3) significantly reduces the total number of required grid cells compared to a uniformly refined mesh. At the inlet and lateral boundaries, freestream conditions are prescribed for velocity and density, while the pressure is extrapolated from the field. The freestream velocity $u_\infty$ is determined by the Mach number $\mathrm{Ma} = 0.1$. For

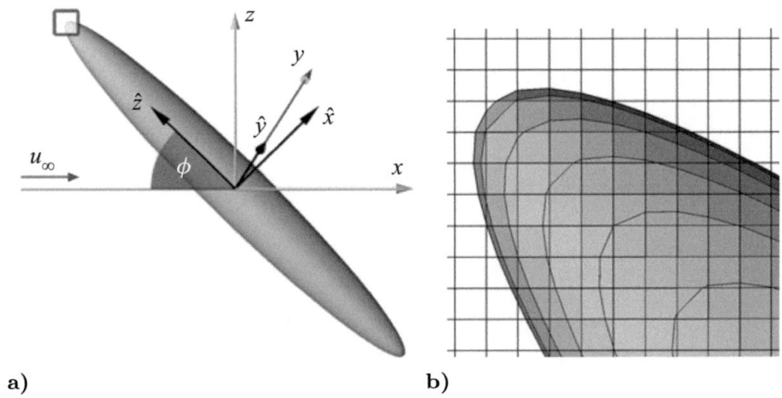


**Fig. 11.1** **a**) Global coordinate system $(x, y, z)$ and the particle-fixed frame $(\hat{x}, \hat{y}, \hat{z})$ and **b**) the sharp discretisation of the elongated particle tip (Fröhlich et al. [14], 2020 © Cambridge University Press, reproduced with permission)

the outflow, boundary conditions of von Neumann type are imposed. At the particle surfaces, a no-slip, isothermal boundary condition is chosen (Sect. 11.3.1).

### 11.4.1 Drag, Lift, and Torque Correlations

Motivated by the limitation of the ELM to creeping flow conditions, and its shortcomings in accurately predicting the rotational motion in decaying isotropic turbulence, new correlations are derived in the study by Fröhlich et al. [14] to extend the ELM for $\mathrm{Re_p} > 0$. To this extent, a large database with over 4440 single DPFS is generated and analysed. A dense coverage of the investigated parameter space is necessary, especially since the simulations revealed a high sensitivity with regards to the particle's aspect ratio. The analysis of the flow within the chosen parameter space reveals that the critical particle Reynolds number, at which the flow starts to separate, depends on both the shape (aspect ratio $\beta$) of the prolate and its orientation (inclination angle $\phi$). In general, the likelihood that flow separation occurs is higher for bluff bodies than for bodies that have a streamlined shape. For a sufficiently high particle Reynolds number, there is an angle $\phi_{\mathrm{sep}}$ where flow separation occurs for all angles $\phi > \phi_{\mathrm{sep}}$. The analysis of the simulation results further reveals that, while separation is observed for a spherical particle ($\beta = 1$) at a Reynolds number of $\mathrm{Re_p} \geq 20$, prolate particles may act as bluff bodies if the inclination angle is sufficiently large such that the separation occurs at much lower $\mathrm{Re_p}$. However, if the particle is not inclined ($\phi = 0$) the particle shape is streamlined and for larger aspect ratios ($\beta > 3$) no separation is observed. In general the topology of the flow and separation onset for particles with aspect ratios $\beta > 3$ is similar, while the geometries with $1 < \beta \leq 3$ are in a transitional regime between the spherical and elongated shape. The vastly different properties of the flow field around the non-spherical particles show again the necessity for accurate correlations that consider both, the shape and orientation of non-spherical particles. The differences in the flow field are visualised in Fig. 11.2 for prolate particles with aspect ratios $\beta = 2$ and 4, at inclination angles of 0° and 75°.

Within the selected parameter space, the flow is steady and mirror symmetric, such that two orientation-dependent coefficients $C_{\mathrm{D},\phi}$ and $C_{\mathrm{L},\phi}$ of the form

$$C_{i,\phi} = \frac{F_i}{\frac{1}{2}\,\rho\,u_\infty^2\,\frac{\pi\,d_{\mathrm{eq}}^2}{4}}\,, \tag{11.35}$$

are sufficient to represent the hydrodynamic forces acting on the particle. Here $F_{\mathrm{D}}$ and $F_{\mathrm{L}}$ refer to drag and lift forces. For the analytical solutions under the assumption of creeping flow conditions presented in the previous section, a correction through an additional torque component was proposed by Dabade et al. [16] for low but finite particle Reynolds numbers. Suggesting that for a particle whose major axis is neither aligned nor perpendicular to the freestream velocity, i.e. for $0° < \phi < 90°$, the particle experiences a pitching torque that rotates the particle towards the stable orientation of $\phi \to 90°$. The orientation-dependent pitching torque coefficient is

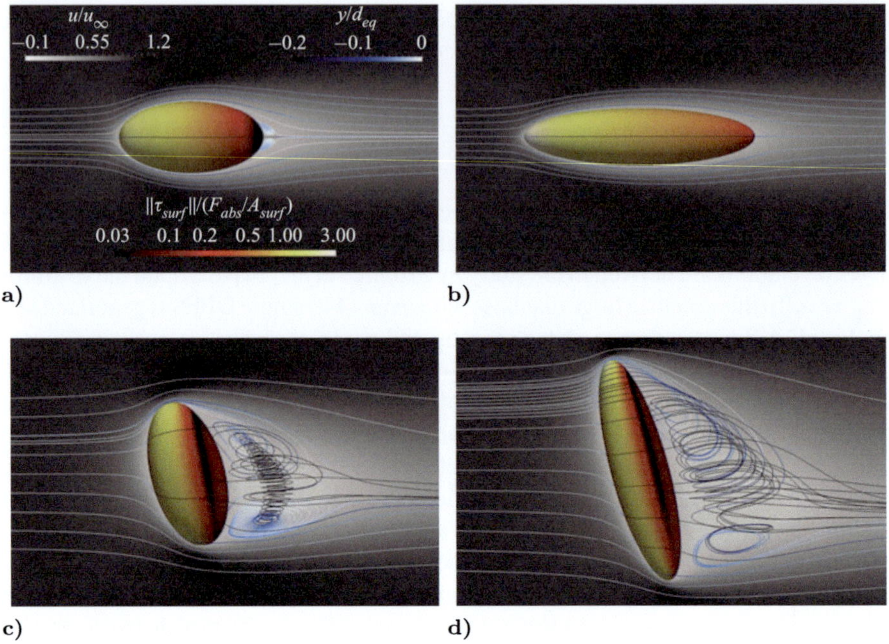


**Fig. 11.2** Visualisations of the flow around prolate particles at $\mathrm{Re_p} = 100$ with different aspect ratios $\beta$ and inclination angles $\phi$: **a**) $\beta = 2$ and $\phi = 0°$, **b**) $\beta = 4$ and $\phi = 0°$, **c**) $\beta = 2$ and $\phi = 75°$, and **d**) $\beta = 4$ and $\phi = 75°$ (Fröhlich et al. [14], 2020 © Cambridge University Press, reproduced with permission)

defined as

$$C_{\mathrm{T},\phi} = \frac{T_\mathrm{p}}{\frac{1}{2}\,\rho\, u_\infty^2 \,\frac{\pi\, d_\mathrm{eq}^3}{8}} \,. \tag{11.36}$$

Based on the database generated through the DPFS, correlations for the drag, lift, and pitching torque are derived using a Levenberg-Marquardt optimisation process [17]. The detailed development process and rationale are provided in the following. The specific values of the fitted constants are summarised in Table 11.1.

Inspired by the drag models for spherical particles, an orientation-dependent correction function $f_{\mathrm{d},\phi}$ of the Stokes flow solution is introduced

$$f_{\mathrm{d},\phi}(\mathrm{Re_p}, \beta) = \frac{C_{\mathrm{D},\phi}}{C_{\mathrm{D,Stokes},\phi}} \,. \tag{11.37}$$

As base for this correction term, the Schiller-Naumann drag correlation for spherical particles introduced in Sect. 11.3.2 is used with an additional term to account for the effects of the non-sphericity. The parameter space of the orientation-dependent drag correction term is reduced through the relationship [18]

$$C_{\mathrm{D},\phi} = C_{\mathrm{D},0} + (C_{\mathrm{D},90} - C_{\mathrm{D},0}) \sin^2(\phi) \,, \tag{11.38}$$

**Table 11.1** Empirical parameters for the drag, lift, and torque correlations fitted using a Levenberg-Marquardt optimisation

| $i$ | 1 | 2 | 3 | 4 | 5 | 6 | 7 | 8 |
|---|---|---|---|---|---|---|---|---|
| $c_{d,i}$ | −0.007 | 1.0 | 1.17 | −0.07 | 0.047 | 1.14 | 0.7 | −0.008 |
| $c_{l,i}$ | 0.01 | 0.86 | 1.77 | 0.34 | 0.88 | −0.05 | – | – |
| $c_{t,i}$ | 0.931 | 0.675 | 0.162 | 0.657 | 2.77 | 0.178 | 0.177 | – |

where

$$C_{D,i}(\mathrm{Re_p}, \beta) = C_{D,\mathrm{Stokes},i}(\mathrm{Re_p}, \beta)\, f_{d,i}(\mathrm{Re_p}, \beta) \quad \forall i \in \{0, 90\}\,. \tag{11.39}$$

The two resulting shape- and orientation-dependent drag correction functions $f_{d,0}$ and $f_{d,90}$ then read

$$f_{d,0}(\mathrm{Re_p}, \beta) = 1 + 0.15\,\mathrm{Re_p^{0.687}} + c_{d,1}(\log(\beta))^{c_{d,2}}\,\mathrm{Re_p}^{c_{d,3}+c_{d,4}\log(\beta)}\,, \tag{11.40}$$

$$f_{d,90}(\mathrm{Re_p}, \beta) = 1 + 0.15\,\mathrm{Re_p^{0.687}} + c_{d,5}(\log(\beta))^{c_{d,6}}\,\mathrm{Re_p}^{c_{d,7}+c_{d,8}\log(\beta)}\,. \tag{11.41}$$

The additional term has a form such that for a particle Reynolds number approaching zero, the total drag converges to the analytical solution for creeping flow conditions. The particle shape is modeled using a logarithmic approach based on the earlier investigated behaviour of the separation onset, which revealed a higher sensitivity in the transitional regime ($1 < \beta \leq 3$) and a reduced sensitivity at higher aspect ratios ($\beta > 3$). For particles orientated such that their major axes are aligned with the freestream direction, a lower correction value is obtained, decreasing further as the aspect ratio increases and the shape becomes more streamlined (i.e. $c_{d,1} < 0$). Conversely, particles orientated with the major axis perpendicular to the freestream velocity, have shown a reduction of the critical Reynolds number for separation, resulting in an increased drag as $\beta$ becomes more important, which is accounted for by $c_{d,5} > 0$. The drag correction function is depicted in Fig. 11.3a and a comparison of the simulation results with the drag correlation is given in Fig. 11.3b. A good agreement of the drag correlation and simulation results is obtained with a total deviation below 5% using the definition of

$$\Delta_{\mathrm{tot}}(\mathrm{Re_p}, \beta) = \max_{\forall \phi \in \phi_{\mathrm{sim}}} \left| \frac{C_{D,\phi}(\mathrm{Re_p}, \beta) - C_{D,\mathrm{sim}}}{C_{D,\mathrm{sim}}} \right|\,. \tag{11.42}$$

For the lift correlation, the parameter space is again reduced by the use of the relation

$$C_{L,\phi} = 2\sin(\phi)\,\cos(\phi)\,C_{L,\max}(\mathrm{Re_p}, \beta)\,. \tag{11.43}$$

This relationship is, however, only accurate for lower particle Reynolds numbers and becomes less accurate as $\mathrm{Re_p}$ is increased. For a prolate with $\beta = 8$, the inclination angle with the highest lift deviates up to 6° as the Reynolds number $\mathrm{Re_p}$ increases

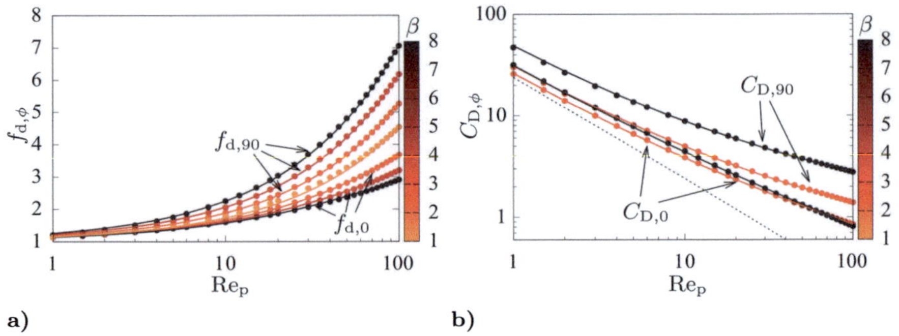


**Fig. 11.3** **a**) Drag correction function $f_{\mathrm{d},\phi}$ for aspect ratios $\beta = 1, 2, 4$ and 8, and **b**) the corresponding drag coefficient $C_{\mathrm{D},\phi}$ for $\beta = 2$ and 8, where $\phi \in \{0°, 90°\}$. DPFS results are shown as points and the correlation function as lines, the dotted line is the drag coefficient $24/\mathrm{Re}_\mathrm{p}$ for spheres at creeping flow conditions (Fröhlich et al. [14], 2020 © Cambridge University Press, reproduced with permission)

from 1 to 100. Therefore, a correction term in form of a coordinate transformation is added such that

$$C_{\mathrm{L},\phi} = 2\sin(\Psi_\phi(\mathrm{Re}_\mathrm{p}, \beta))\cos(\Psi_\phi(\mathrm{Re}_\mathrm{p}, \beta))\, C_{\mathrm{L,max}}(\mathrm{Re}_\mathrm{p}, \beta) \;, \tag{11.44}$$

with

$$\Psi_\phi(\mathrm{Re}_\mathrm{p}, \beta) = \frac{\pi}{2}\left(\frac{\phi}{\pi/2}\right)^{f_{\mathrm{l,shift}}(\mathrm{Re}_\mathrm{p}, \beta)} \;. \tag{11.45}$$

A curve fit is used to fit the exponent for higher particle Reynolds numbers resulting in

$$f_{\mathrm{l,shift}}(\mathrm{Re}_\mathrm{p}, \beta) = \begin{cases} 1 + c_{\mathrm{l},1}(\log(\beta))^{c_{\mathrm{l},2}}(\log(\mathrm{Re}_\mathrm{p}))^{c_{\mathrm{l},3}} & \mathrm{Re}_\mathrm{p} > 1 \;, \\ 1 & \text{otherwise} \;. \end{cases} \tag{11.46}$$

The lift coefficient $C_\mathrm{L}$, normalised by the maximum lift coefficient $C_{\mathrm{L,max}}$, is shown in Fig. 11.4. The term $C_{\mathrm{L,max}}$ in (11.44) is defined by the Stokes drag corrected by $f_{\mathrm{l,max}}$ and reads

$$C_{\mathrm{L,max}}(\mathrm{Re}_\mathrm{p}, \beta) = f_{\mathrm{l,max}}(\mathrm{Re}_\mathrm{p}, \beta)\frac{C_{\mathrm{D,Stokes},90} - C_{\mathrm{D,Stokes},0}}{2} \;, \tag{11.47}$$

where

$$f_{\mathrm{l,max}}(\mathrm{Re}_\mathrm{p}, \beta) = 1 + c_{\mathrm{l},4}\,\mathrm{Re}_\mathrm{p}^{c_{\mathrm{l},5} + c_{\mathrm{l},6}\log(\beta)} \;. \tag{11.48}$$

To avoid vanishing reference values, the definition of the total deviation is revised

$$\bar{\Delta}_{\mathrm{tot}}(\mathrm{Re}_\mathrm{p}, \beta) = \max_{\forall\phi\in\phi_{\mathrm{sim}}}\left|\frac{C_{\mathrm{L},\phi}(\mathrm{Re}_\mathrm{p}, \beta) - C_{\mathrm{L,sim}}}{\max(C_{\mathrm{L,sim}}, 2/\pi\, C_{\mathrm{L,sim,max}})}\right| \;. \tag{11.49}$$

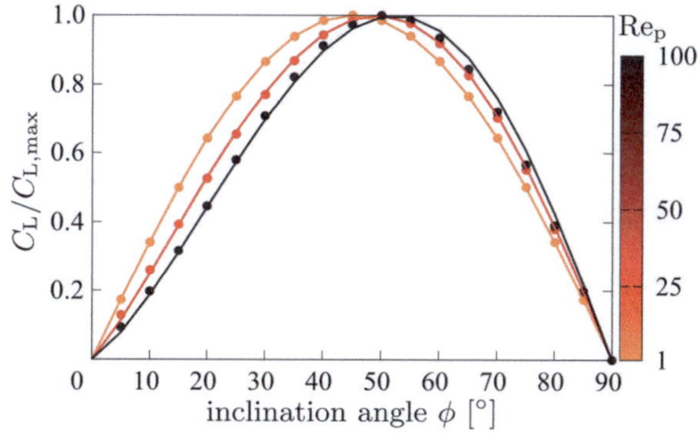


**Fig. 11.4** Lift coefficient $C_L$ normalised by the maximum lift coefficient $C_{L,max}$ for aspect ratio $\beta = 8$ and $Re_p = 1, 25$, and 100. DPFS results are shown as points and the correlation function as lines (Fröhlich et al. [14], 2020 © Cambridge University Press, reproduced with permission)

For the new lift correlation the total error is generally less than 5% reaching a maximum of 7% at the end of the parameter space with $\beta = 8$ and $Re_p = 100$.

Given that the pitching torque vanishes for vanishing particle Reynolds numbers, the orientation-dependent torque coefficient $C_{T,\phi}$ has to be directly modelled and the analytical solutions for creeping flow cannot be used as a base. For this, the terms that dominate the correlation proposed by Sanjeevi et al. [19] are identified for $Re_p < 100$ and generalised for $1 \leq \beta \leq 8$ such that

$$C_{T,\phi}(Re_p, \beta) = 2\sin(\phi)\cos(\phi)\, C_{T,max}(Re_p, \beta)\,, \tag{11.50}$$

with

$$C_{T,max}(Re_p, \beta) = \frac{c_{t,1}(\log(\beta))^{c_{t,2}}}{Re_p^{c_{t,3}}} + \frac{c_{t,4}(\log(\beta))^{c_{t,5}}}{Re_p^{c_{t,6}+c_{t,7}\log(\beta)}}\,. \tag{11.51}$$

A comparison of the DPFS and the torque correlations is depicted in Fig. 11.5 with a comparison of $C_{T,max}$ for various aspect ratios as a function of $Re_p$ and $C_T/C_{T,max}$ as function of $\phi$ for various $Re_p$. For the maximum torque a qualitatively good agreement was found when two exponential functions are used for the modelling. However, for an elongated prolate with $\beta = 8$ larger deviations of $C_{T,\phi}$ are observed at higher Reynolds numbers. Given that the deviations are not systematic, a correction as for the lift coefficient, using a coordinate transformation is not possible. A more complex base model would be necessary to accommodate these deviations, which would increase the complexity of the correlation. The comparison of the correlation with the simulation results revealed that the maximum deviation, using (11.49) as definition, is less than 15%.

The ELM presented in Sect. 11.3.2 is reformulated to incorporate the derived correlations, resulting in the following expressions:

$$\hat{\mathbf{F}}_D = \rho\,\frac{\pi}{8}\,d_{eq}^2\,\left|\hat{\mathbf{u}} - \hat{\mathbf{v}}_p\right|^2 C_{D,\phi}(Re_p, \beta)\,\hat{\mathbf{d}}_D\,, \tag{11.52}$$

$$\hat{\mathbf{F}}_L = \rho\,\frac{\pi}{8}\,d_{eq}^2\,\left|\hat{\mathbf{u}} - \hat{\mathbf{v}}_p\right|^2 C_{L,\phi}(Re_p, \beta)\,\hat{\mathbf{d}}_L\,, \tag{11.53}$$

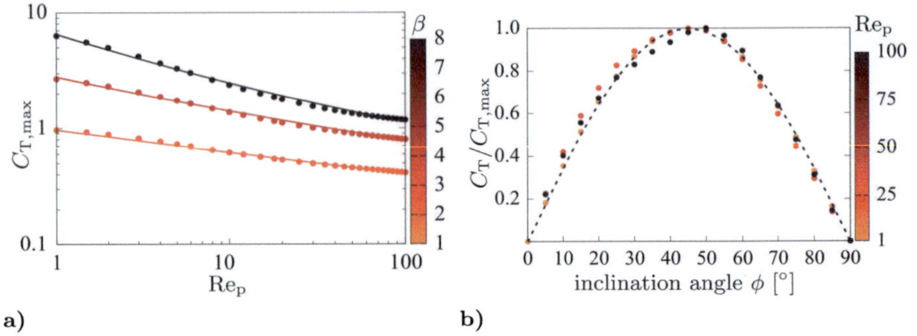


**Fig. 11.5** **a**) Maximum torque coefficient $C_{T,max}$ for aspect ratios $\beta = 2$, 4 and 8 and **b**) the torque coefficient $C_T$ normalised by the maximum torque coefficient $C_{T,max}$ for $\beta = 8$ and $Re_p = 1$, 25 and 100. DPFS results are shown as points and the correlation functions as lines (Fröhlich et al. [14], 2020 © Cambridge University Press, reproduced with permission)

$$\hat{T}_p = \mu\,\underline{\hat{\mathbf{K}}}_s\,\hat{s} + \mu\,\underline{\hat{\mathbf{K}}}_\zeta(\hat{\boldsymbol{\zeta}} - \hat{\boldsymbol{\omega}}) + \rho\,\frac{\pi}{16}\,d_{eq}^3\,\left|\hat{\mathbf{u}} - \hat{\mathbf{v}}_p\right|^2 C_{T,\phi}(Re_p, \beta)\,\hat{\mathbf{d}}_T\ . \tag{11.54}$$

Under the conditions of creeping flow, the directions of the forces were given by the resistance tensor. When the newly derived correlations are used, the directions $\hat{\mathbf{d}}_D$ and $\hat{\mathbf{d}}_L$ are given by

$$\hat{\mathbf{d}}_D = \frac{\hat{\mathbf{u}} - \hat{\mathbf{v}}_p}{\left|\hat{\mathbf{u}} - \hat{\mathbf{v}}_p\right|}\ , \tag{11.55}$$

$$\hat{\mathbf{d}}_L = \hat{\mathbf{d}}_D \times \hat{\mathbf{d}}_T\ . \tag{11.56}$$

The direction of the torque is given as

$$\hat{\mathbf{d}}_T = \frac{\hat{\mathbf{d}}_D \times \hat{\mathbf{e}}_3}{\left|\hat{\mathbf{d}}_D \times \hat{\mathbf{e}}_3\right|}\ \mathrm{sign}(\hat{\mathbf{d}}_D \cdot \hat{\mathbf{e}}_3)\ , \tag{11.57}$$

where $\hat{\mathbf{e}}_3$ denotes the direction of the prolate's major axis.

### 11.4.2 Nusselt Number Correlation

In a subsequent study [15], the focus is on the heat transfer and the derivation of a new correlation for the Nusselt number. The Nusselt number $Nu = \frac{h\,d_{eq}}{k}$, where $h$ is the convective heat transfer coefficient and $k$ the thermal heat conductivity, is a dimensionless number quantifying the heat transfer between a solid boundary and the surrounding fluid. The novel correlation is derived based on the simulation results of more than 6600 single DPFS. Over the full parameter space, introduced in Sect. 11.4, the Prandtl number $Pr = 0.72$ is considered constant, which is representative of a wide range of gases used in combustion processes of biomass. The numerical method of the DPFS with respect to heat transfer is first validated by comparison of the numerical results of heated and cooled spheres with literature results [15]. The good agreement with the literature results shows the capabilities of the numerical method, such that in the next step an analysis of the flow over ellipsoidal particles for various $\mathrm{Re_p}$, $\phi$, $\beta$ and $T_\mathrm{R}$ is presented.

While keeping the temperature ratio constant, i.e. $T_\mathrm{R} = 1.05 = \text{const.}$, the impact of the other parameters is investigated. For a prolate particle with $\beta = 2$, orientated in the streamwise direction ($\phi = 0°$) it can be noted that as the Reynolds number is increased, a compression of the heat transfer on the windward side and conversely, a stretch of the heat transfer is seen on the leeward side. This trend is linked to the local velocity field. In general, an increase of the heat transfer, especially at higher particle Reynolds numbers and larger inclination angles is observed. For a particle orientated with its major axis perpendicular to the freestream velocity and an aspect ratio of $\beta = 3$, the increase of the Reynolds number from $\mathrm{Re_p} = 10$ to 90 results in significant flow alterations due to the separation onset discussed in the previous section. The removal of the hot fluid on the leeward side is promoted by the recirculating flow resulting in a higher heat transfer and as such of the Nusselt number. Maintaining the particle shape with $\beta = 3$ and $\mathrm{Re_p} = 90$, the effect of the inclination angle on the Nusselt number is analysed. As $\phi$ is increased, the mean Nusselt number increases. This is on the one hand due to the mentioned flow behaviour on the leeward side, on the other hand attributed to an increase of the exposed surface at the windward side. The configurations for 30° and 60° are visualised in Fig. 11.6. Notably, the highest local Nusselt number, calculated in slices and circumferentially averaged, occurs independently of $\phi$ at the tip of the particle, however, its value decreases as $\phi$ is increased. The highest local Nusselt number is observed at $\phi = 0°$. In contrast to the flow topology, which remains rather stable for inclined prolates with $\beta \geq 3$, as discussed in the previous section, the mean Nusselt number keeps increasing as the aspect ratio is further increased. Grounded on the fact that the maximal local Nusselt number remains similar as the particles become more elongated, the higher heat transfer is related to the increase of the exposed surface. In the investigated

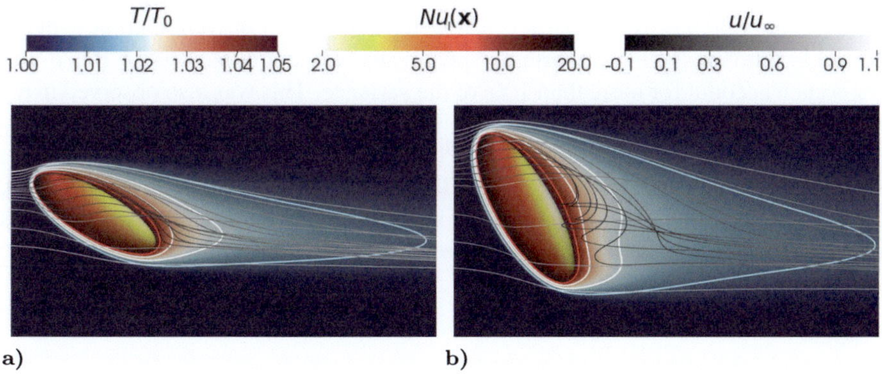


**Fig. 11.6** Streamlines, temperature, and Nusselt number contours for $\mathrm{Re_p} = 90$ and aspect ratio $\beta = 3$ for inclination angles **a)** $\phi = 30°$ and **b)** $\phi = 60°$ (reprinted from Kiwitt et al. [15], Copyright (2022), with permission from Elsevier)

Reynolds parameter space, the temperature ratio $T_R$ has minimal influence on the separation on the leeward side, thus resulting in only slight changes of the temperature field. This is attributed to the strong coupling of flow and temperature field. However, in the lower Reynolds number range, as the inertial forces become less important and the heat transfer is dominated by conduction, the influence of the temperature ratio $T_R$ increases, while the influence of the particle orientation becomes less important. For $\mathrm{Re_p} = 1$, the Nusselt number is strongly influenced by $T_R$. The smallest observed Nusselt number for a cooled particle is $\mathrm{Nu} > 2$, whereas for a heated particle it is $\mathrm{Nu} < 2$.

A genetic programming approach is applied to accurately determine a new Nusselt number correlation based on the DPFS results using symbolic regression grounded on a tree structure. The obtained correlation with slightly rounded coefficients reads

$$\mathrm{Nu} = 1.5 + \frac{0.5 - 0.05\sqrt{\beta}^{-1}}{T_R} + 0.37\frac{\sqrt{\mathrm{Re_p}}}{T_R^{1.1}} + 0.1\frac{\sqrt{\mathrm{Re_p}\,\beta^{\sin(\phi)}\beta}}{T_R}. \tag{11.58}$$

The behaviour of the new correlation is compared with respect to the conclusions drawn from the earlier analysed DPFS. For vanishing Reynolds numbers, the Nusselt number depends on the particle aspect ratio $\beta$ and the temperature ratio $T_R$. As noted before, the minimum Nusselt number is $\mathrm{Nu} > 2$ for $T_R < 1$ and $\mathrm{Nu} < 2$ for $T_R > 1$. The second term on the right-hand side of the new correlation models the impact of the surrounding fluid on the particle's surface. The third and final term establishes a link between the fluid ($\mathrm{Re_p}$ and $T_R$) and the particle shape as well as its orientation through $\beta$ and $\phi$. This formulation ensures that the correlation is still valid for spheres with $\beta = 1$. A sensitivity analysis is conducted to estimate the influence of the input parameters on the heat transfer. The highest impact is due to the Reynolds number and the temperature ratio. Over half of the output variance is linked to the combination $\mathrm{Re_p}\,T_R$. The influence of the particle shape and orientation in combination with the Reynolds number is also important. Specifically, the combinations of $\mathrm{Re_p}\,\beta$ and $\mathrm{Re_p}\,\phi$ each account for more than 10% of the variance. This was also observed in the analysis of the flow and temperature field. The accuracy of the correlation is assessed through the deviation to the reference DPFS results. Over the full parameter space, the mean deviation is $\left|\overline{\Delta \mathrm{Nu}}\right| < 4\%$ and the maximum deviations are below 18%. The highest deviations occur at the limits of the parameter space. Figure 11.7 depicts the correlation function compared to the reference DPFS results.

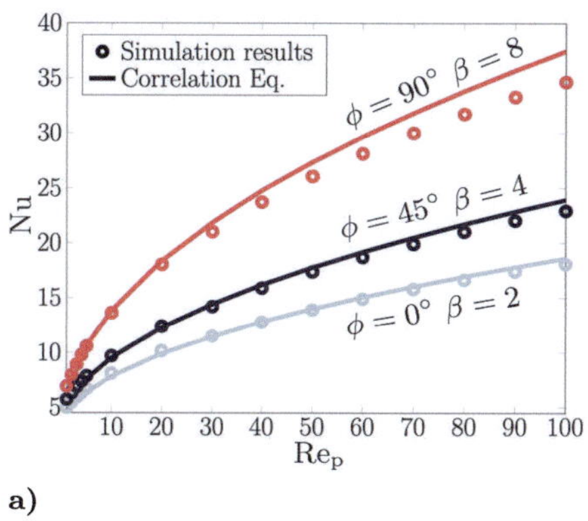


**Fig. 11.7** Nusselt number Nu as a function of the particle Reynolds number $\mathrm{Re_p}$ for varying inclination angles $\phi$ and aspect ratios $\beta$ at **a**) $T_R = 0.35$ and **b**) $T_R = 1.65$. DPFS results are shown as points and the correlation function as lines (reprinted from Kiwitt et al. [15], Copyright (2022), with permission from Elsevier)

## 11.5 Validation of Point-Particle Models

### 11.5.1 Isotropic Turbulence

The settling behaviour of prolate particles using the ellipsoidal Lagrangian model without the novel correlations (ELM) and the model using the newly derived correlations in Sect. 11.4.1 (EELM) is compared in decaying isotropic turbulence [20]. The same setup as in the studies by Schneiders et al. [3,5] is used. After an initial decay, the microscale Reynolds number $\mathrm{Re}_\lambda$ remains stable in the range $31.5 \leq \mathrm{Re}_\lambda \leq 34.5$ during the time frame $2 \leq t/t_{\mathrm{ref}} \leq 4$, a period during which the mean statistics, denoted by the tilde symbol, are sampled. 100,000 particles are placed randomly within the domain. Particle velocity and angular velocity are initialised using the local fluid velocity and angular velocity, while the particle orientation is randomly defined. In the sampling time frame, a Stokes number based on the Kolmogorov time scale $\tau_k$ of $\tau_p/\tilde{\tau}_k = 1.23$ is chosen, a value where phenomena such as preferential concentration and sweeping are observed. The behaviour of heavy particles is analysed with a particle-to-fluid density ratio of $\rho_p/\rho_f = 400$ and an equivalent particle diameter of $d_{eq}/\tilde{\eta} = 0.24$, where $\tilde{\eta}$ refers to the mean Kolmogorov length-scale during the sampling period. The terminal velocity $v_t$ is defined by the particle relaxation time and the gravity in the $z$-direction $v_t = \tau_p \|\mathbf{g}\|$. To ensure an identical flow field between the results with and without the force and moment correlations, one-way coupled simulations are conducted, i.e. the feedback on the flow field is not included. A similar trend for all aspect ratios within the parameter space of the correlations is observed, therefore the following discussion focuses, unless otherwise stated, on the results for prolates with $\beta = 4$. First, the influence of the terminal velocity is inves-

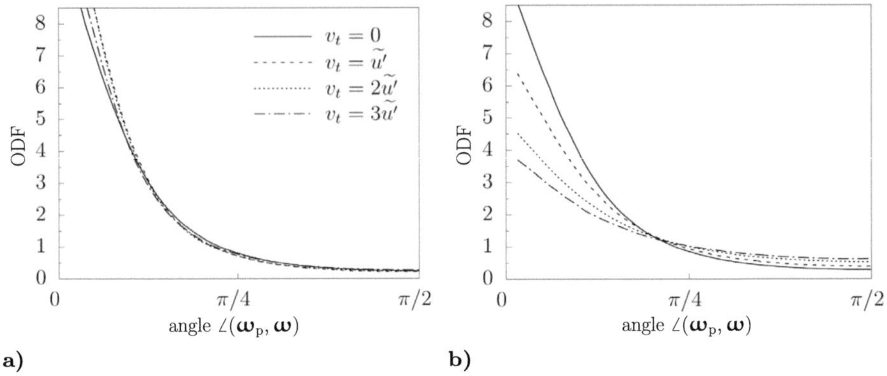


**Fig. 11.8** Orientation distribution function of the angle between the particle angular velocity $\boldsymbol{\omega}_\mathrm{p}$ and the fluid angular velocity $\boldsymbol{\omega}$ **a**) without and **b**) with the novel correlations for different terminal velocities $v_\mathrm{t}$ (reprinted from Fröhlich et al. [20], Copyright (2021), with permission from Elsevier)

tigated. For vanishing $v_\mathrm{t}$, the results for the ELM and EELM are almost identical, which is attributed to the low mean particle Reynolds number that is $\langle \mathrm{Re_p} \rangle \leq 1$. For higher terminal velocities, a significant influence of the model formulations can be observed. Simulations using the EELM show that the initially randomly oriented particles quickly rotate to a stable orientation, achieving a balance between the pitching torque and the torque resulting from the fluid velocity gradients. Hence, the pitching torque term added in the EELM tends to align the prolates with their major axes perpendicular to the settling velocity vector. The simulations using the ELM show the opposite behaviour, i.e. neglecting the finite particle Reynolds number results in an alignment of the major particle axis with the settling velocity vector. In addition, the rotation of the particles aligns with the local angular fluid velocity independently of the terminal velocity, when the ELM is used. For the EELM this remains valid only for $v_\mathrm{t} = 0$, while for an increasing terminal velocity, a reduction of the alignment is observed. The resulting orientation distribution functions (ODF) with respect to the alignment of the angular velocities are summarised in Fig. 11.8.

Next, the influence of the particle Reynolds number is investigated by varying the equivalent particle diameter and density ratio, while maintaining a constant particle relaxation time. For the ELM, no influence of $\mathrm{Re_p}$ on the settling statistics is observed. Given the nature of the model assumption of $\mathrm{Re_p} \ll 1$ this is to be expected. However, a strong impact of $\mathrm{Re_p}$ on the settling velocity is observed when the EELM is used. In fact, a reduction of the settling velocity of up to 46% compared to the case without the novel correlations is seen. Furthermore, preferential sweeping becomes more important, attributed to the fact that as the settling velocities decrease, the particles have more time to respond to the turbulent fluctuations. The alignment of the particle and fluid velocity is depicted for prolates with $\beta = 8$ and $d_\mathrm{eq}/\tilde{\eta} = 0.48$ in Fig. 11.9.

Overall, the new correlations have been shown to significantly influence the results, affecting the preferential orientation, settling velocity as well as preferential sweeping. In addition, the importance of the particle Reynolds number has been highlighted, which is not negligible even for relatively low values of $\mathrm{Re_p}$.

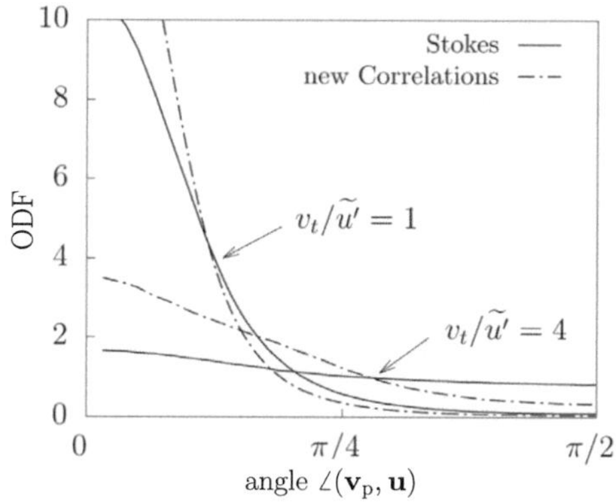


**Fig. 11.9** Orientation distribution function of the angle between the particle velocity $\mathbf{v}_\mathrm{p}$ and the fluid velocity $\mathbf{u}$ at the particle position for prolates with aspect ratio $\beta = 8$ and $d_\mathrm{eq}/\tilde{\eta} = 0.48$ for different terminal velocities (reprinted from Fröhlich et al. [20], Copyright (2021), with permission from Elsevier)

## 11.5.2 Single-Particle Simulations

The settling dynamics of a single, heavy ellipsoidal particle with an aspect ratio of $\beta = 3$ and a density ratio of $\rho_\mathrm{p}/\rho_\mathrm{f} = 600$ is investigated. The particle Reynolds number $\mathrm{Re}_\mathrm{p}$, based on the terminal velocity $v_\mathrm{t}$ and the equivalent spherical diameter $d_\mathrm{eq}$, lies within the validity range of the novel correlations used in the EELM model, that is $0 \leq \mathrm{Re}_\mathrm{p} \leq 100$. Figure 11.10 illustrates the results for the translational and rotational motion in the case of $\mathrm{Re}_\mathrm{p} = 24$. The DPFS reference in Fig. 11.10a shows that the ellipsoidal particle reaches approximately 80% of the reference terminal velocity $v_\mathrm{ref}$ for a spherical particle with an equivalent volume. This result is closely matched by the EELM. In contrast, the SLM and the ELM, which is derived for creeping flow conditions, significantly overestimate the settling velocity. When the particle settles under the influence of the gravitational force, an oscillatory motion is observed. Figure 11.10b shows the angle between the settling direction $\mathbf{x}$ and the particle's major axis $\hat{z}$. The EELM model captures the amplitude of the rotational motion well compared to the reference simulation, though it slightly overpredicts the oscillation frequency. The ELM model fails to predict the oscillatory particle's rotation, while the spherical model predicts no rotation at all. The superior predictive capability of the newly developed EELM model, compared to existing point-particle models, is demonstrated by its ability to deliver accurate results for both the settling velocity and the rotational dynamics for moderate particle Reynolds numbers.

## 11.5.3 Particle-Laden Turbulent Pipe Flow

The properties of the EELM are further analysed by performing simulations of a canonical anisotropic flow, i.e. a turbulent, particle-laden flow in a pipe with circular cross-section and a diameter $D = 2R$ (see Fig. 11.11). The pipe is laden with 1830 heavy particles with an equivalent spherical diameter of $d_\mathrm{eq} = 0.016D$. The particles have a density ratio of $\rho_\mathrm{p}/\rho_\mathrm{f} = 800$ and an aspect ratio of $\beta = 4$. The pipe length is chosen to be $L = 5D$, where periodic boundary conditions are applied in the

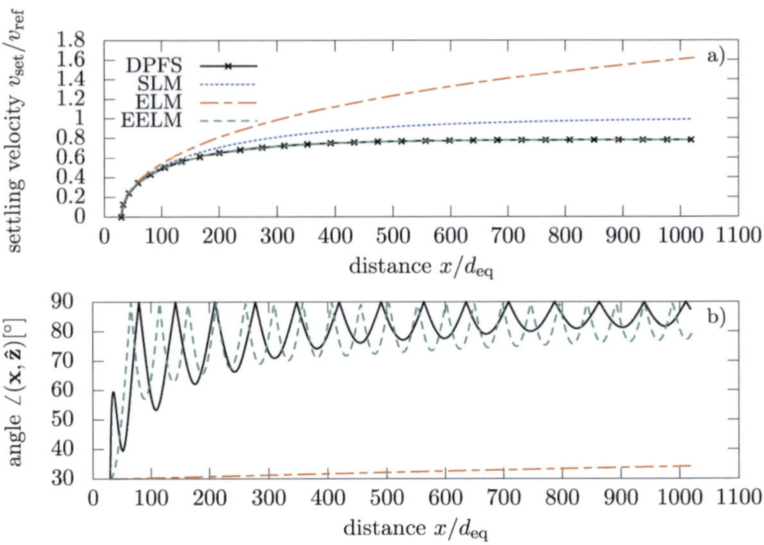


**Fig. 11.10** Settling dynamics of a heavy ellipsoidal particle (aspect ratio $\beta = 3$, $\mathrm{Re}_p = 24$) for the DPFS and different point-particle models. **a**) Shows the non-dimensional settling velocity $v_{set}/v_{ref}$ and **b**) depicts the angle between the settling direction **x** and the particle's major axis $\hat{z}$ as functions of the distance $x/d_{eq}$ from the initial particle position

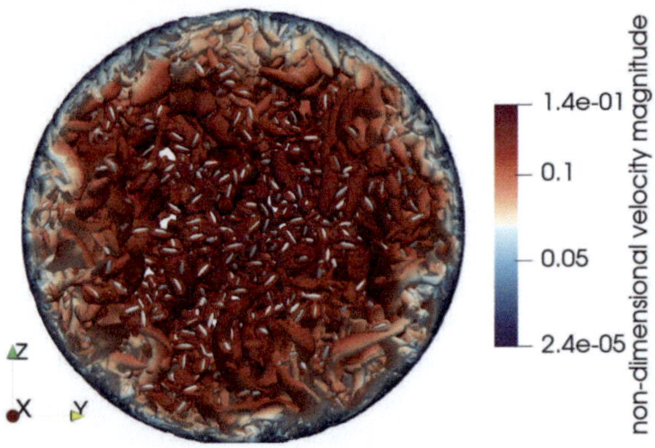


**Fig. 11.11** Q-criterion contours in the particle-laden pipe flow, coloured by the non-dimensional fluid velocity magnitude

streamwise direction and a no-slip boundary condition is used at the pipe wall. The Reynolds number based on the pipe diameter is $\mathrm{Re}_D = 7500$.

Statistical results covering the particle distribution and preferential orientation are compared for the SLM, ELM, and EELM and are validated against reference results from DPFS. After achieving statistical convergence, radial mean quantities are computed over a span of 15 flow-through cycles. Figure 11.12 shows the particle distribution and preferential orientation of prolate particles with an aspect ratio of $\beta = 4$, comparing different point-particle models alongside the DPFS reference results. Similar trends were observed for $\beta = 3$ and 5. The ELM and EELM results show good agreement regarding the particle distribution in the radial direction, with a

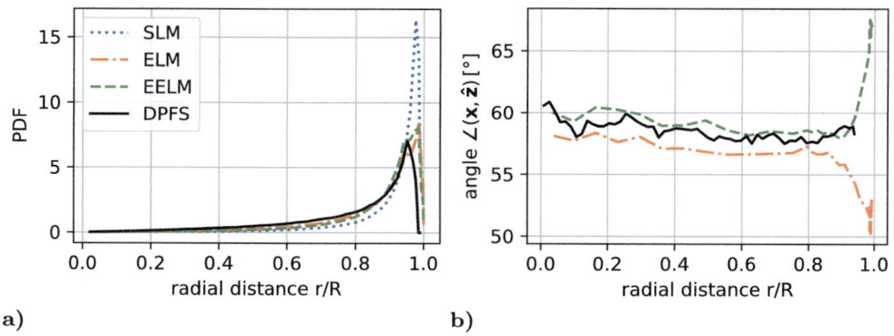


**Fig. 11.12** Turbulent pipe flow laden with prolate particles of aspect ratio $\beta = 4$ for the different point-particle models and DPFS. **a**) Probability density function (PDF) of the particle distribution and **b**) the angle between the particle's major axis $\hat{z}$ and the streamwise direction $\mathbf{x}$ as functions of the normalised radial distance $r/R$ from the pipe centre

pronounced peak near the wall. This is consistent with the mechanics of turbophoresis according to which particles within a wall bounded turbulent flow tend to accumulate near the wall. Compared to the reference solution, the SLM results indicate a significantly higher concentration maximum which is also located closer to the pipe wall. The ellipsoidal point particle models ELM and EELM exhibit a maximum concentration value in reasonable agreement with the DPFS results. All point particle models, however, predict the concentration peak at a similar distance, $\Delta r/R = 4\%$ closer to the wall than the DPFS reference. More substantial differences in the predictions are observed concerning the rotational motion and preferential orientation. The EELM consistently predicts slightly larger inclination angles of about $1 - 2°$ than the DPFS, independently from the radial position, while the ELM underpredicts the inclination angle by $2 - 3°$. This behaviour is consistent with prior observations in isotropic turbulence discussed in Sect. 11.5. It can be attributed to the additional pitching torque term incorporated in the EELM, which tends to align the non-spherical particles with their major axes perpendicular to the streamwise flow direction. Approaching the wall, larger differences in the inclination angle can be observed for the point particle models in comparison to the results of the fully resolved simulation. Note, however, that the currently implemented numerical method for particle-wall collisions does not yet consider the non-spherical particle shape.

## 11.6 Particle-Laden Turbulent Round Jet

A turbulent round jet is investigated in a setup that more closely resembles parts of the combustion chamber of an industrial burner. The numerical setup shown in Fig. 11.13 mimics the experimental configuration described in Sect. 10.2.1 and also allows the addition of a co-flow to the jet. The simulation setup includes two separate domains. The first corresponds to a particle-laden turbulent pipe flow, as analysed in the previous section, where periodic boundary conditions are used in the streamwise direction. The second domain comprises the round jet, where a slicing technique is

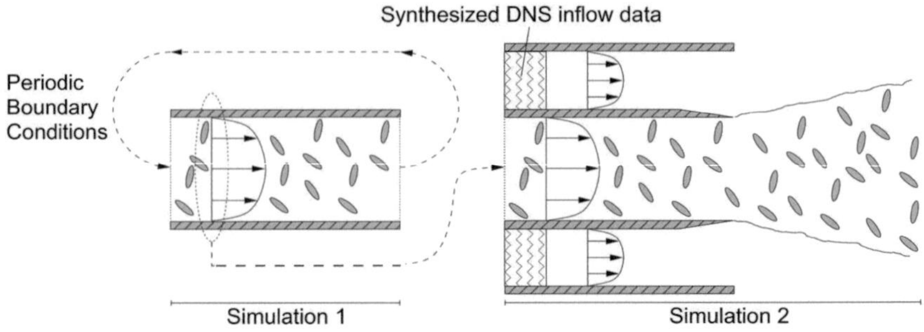


**Fig. 11.13** Numerical setup for the simulation of the turbulent particle-laden round jet

used to transfer the statistically converged pipe flow solution to the inflow boundary of the jet domain. This slicing technique is essential to ensure correct inflow conditions for the resolved particle simulations, in particular for the turbulent flow field and the particle distribution. In the following sections, the slicing technique is validated by focusing on a configuration without a co-flow.

### 11.6.1 Single Phase Jet

Results of a large-eddy simulation for a single-phase flow serve as a reference to evaluate the influence of the particulate phase on the turbulent flow field. The Reynolds number based on the pipe/jet diameter $D = 2\,R$ and the bulk velocity $u_0$ is chosen to be $\mathrm{Re}_D = 7500$. The length of the pipe domain is chosen to be 10 $D$. Figure 11.14 presents a comparison between numerical and experimental data. Excellent agreement is observed in the decay of the non-dimensional centreline velocity of the jet.

### 11.6.2 Particle-Laden Jet

The same setup as for the single phase jet is simulated using heavy prolate particles with an aspect ratio of $\beta = 3$, an equivalent spherical diameter of $d_{\mathrm{eq}}/D = 0.036$ and a density ratio of $\rho_{\mathrm{p}}/\rho_{\mathrm{f}} = 600$. The volume fraction is $\phi_{\mathrm{v}} = 10^{-3}$. Particles from the pipe domain are injected into the jet domain using a similar slicing technique as for the flow field. In the following, results for the different point-particle models are discussed. The results using the SLM, the ELM and the EELM are compared at various axial locations downstream of the jet nozzle ($x/D = 0$, 10, 20, and 30).

First, the evolution of the radial particle distribution is analysed. At both axial positions in Fig. 11.15, the ELM exhibits a higher probability towards the centre of the jet compared to the SLM and the EELM. Overall, the ELM and EELM show similar results for $r/R > 0.5$ at $x/D = 10$, while, in the near field, it is observed that the particles tend to stay more outwards with the spherical model. At the more

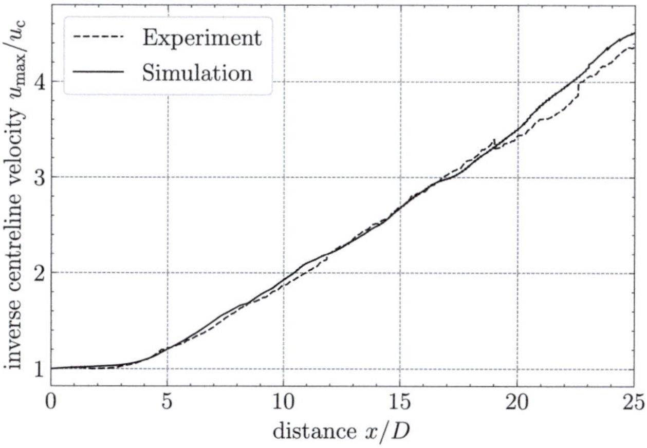


**Fig. 11.14** Comparison of the inverse of the non-dimensional time-averaged centreline velocity $u_c$ between numerical and experimental results (Sect. 10.2.1). The centreline velocity is normalised by the maximum centreline velocity $u_{max}$ at the pipe exit

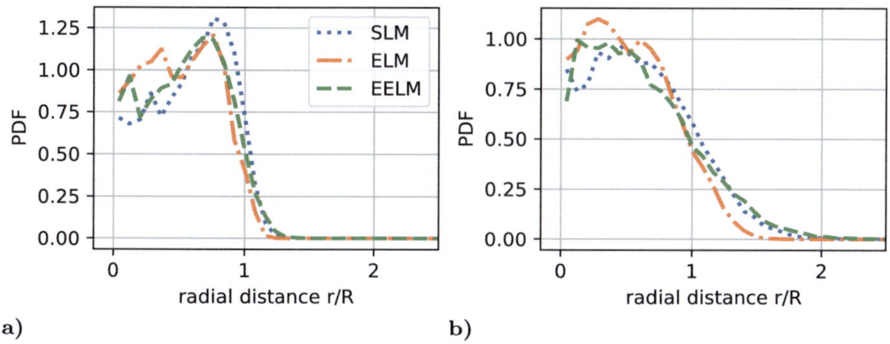


**Fig. 11.15** Probability density function of the particle distribution as a function of the non-dimensional radial distance $r/R$ from the centreline at locations **a)** $x/D = 10$ and **b)** $x/D = 30$ for different point-particle models

downstream location $x/D = 30$, however, the ELM shows a considerably smaller dispersion of the particles off the jet core compared to the SLM and EELM.

More significant differences are seen when the statistics of the particle velocity are analysed. In Fig. 11.16, the ELM overestimates the particle velocity compared to the SLM and EELM, similar to the results in decaying isotropic turbulence. This is especially true further downstream, where the particle Reynolds number increases, since the flow velocity decays faster than the particle velocity due to the inertia of the particles. The differences between the SLM and EELM are mainly visible in a less pronounced maximum of the probability in the velocity distribution. Further downstream at $x/D = 30$, the EELM predicts the maximum at a slightly lower velocity value compared to the SLM, as illustrated in Fig. 11.16b.

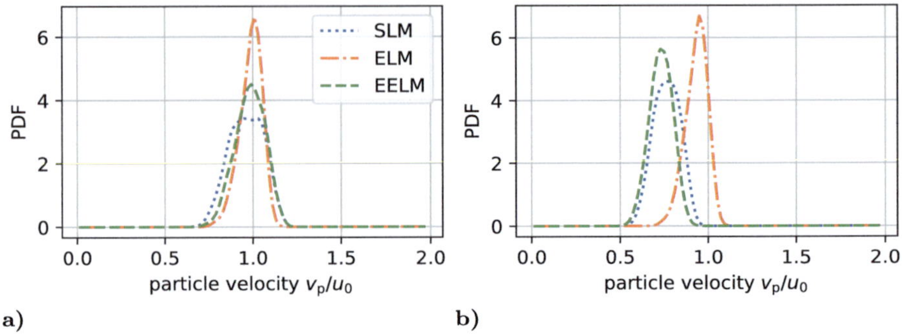


**Fig. 11.16** Probability density function of the non-dimensional particle velocity $v_\mathrm{p}$ at **a**) the nozzle exit $x/D = 0$ and at **b**) $x/D = 30$ for different point-particle models, encompassing all particles passing through these positions. The particle velocity is normalised by the bulk jet velocity $u_0$

**Fig. 11.17** Orientation distribution function of the angle between the particle's major axis $\hat{z}$ and the streamwise direction $\mathbf{x}$ at 10 $D$ downstream of the jet outlet

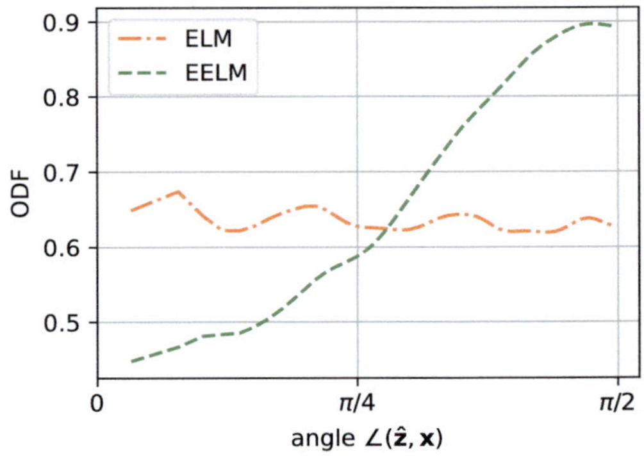


Finally, the preferential orientation of the non-spherical particles is investigated. Since the SLM has no information about the particle orientation, only the results for the ELM and EELM models are compared. Figure 11.17 visualises the orientation distribution function at $x/D = 10$. Similar results are observed at the other investigated axial positions. The results confirm the previously discussed findings. That is, the EELM tends to align the particles with a higher angle between their major axis and the streamwise direction. This holds true independently from the investigated axial position. The ELM exhibits no such preference. In contrast, it tends to show a higher probability for lower angles, i.e. for the particles to align with their major axis almost parallel to the primary flow direction.

These results will be extended by reference DPFS to assess the prediction accuracy of the particle dynamics for the different point-particle models in a particle-laden jet.

## 11.7 Conclusion

Motivated by the investigations in decaying isotropic turbulence, the need for accurate Lagrangian point-particle models for non-spherical particles was highlighted and the derivation process of new correlations for prolate particles was presented in this chapter.

Correlations for drag, lift, and torque that consider the particle orientation and shape were derived after analysing fully resolved simulations of single prolate particles within the parameter space of $1 \leq \mathrm{Re_p} \leq 100$, $1 \leq \beta \leq 8$, and $0° \leq \phi \leq 90°$. The influence of particle shape, orientation, and particle Reynolds number on phenomena such as the flow separation on the particle leeward side were characterised. For spherical particles, i.e. $\beta = 1$, the proposed drag correlation reduces to the popular Schiller-Naumann drag and the lift coefficient vanishes. For finite particle Reynolds numbers, the added pitching torque rotates the particle towards its most stable orientation and as $\mathrm{Re_p} \to 0$ the dynamic equation falls back to the analytically derived solution for Stokes flow. The derived correlation functions show good agreement for the drag and lift coefficients, with total deviations from the detailed simulations below 5% and 7%, respectively, while slightly higher deviations were observed for the pitching torque coefficient with a maximum deviation below 15%. However, without increasing the complexity of the underlying correlation functions, the deviations could not be reduced further. A follow-up study extended the results by deriving a new correlation for the Nusselt number, considering the same parameter space and for a temperature ratio in the range of $0.35 \leq T_\mathrm{R} \leq 1.65$. The largest influence on the Nusselt number was found to be the particle Reynolds number, which significantly influences the flow field and heat transfer. Besides the influence on the separation, parameters such as the particle aspect ratio and orientation also result in geometrical changes, such as an increase of the exposed surface. In the lower Reynolds number range, a strong influence of the temperature ratio $T_\mathrm{R}$ was observed. Overall, the new Nusselt correlation, derived using a genetic programming approach, shows good agreement with the reference simulations.

In a first step towards a full validation of the new correlations, the settling behaviour of prolate particles was compared through simulations using a common ellipsoidal Lagrangian model without novel correlations and a reformulated version, which includes the new correlations for drag, lift, and torque in decaying isotropic turbulence. The influence of the new correlations is seen, among others, in the preferential particle orientation, as the particles tend to orientate themselves with their major axis perpendicular to the settling direction. This is opposite to the simulations without the new correlations, where the particle major axis aligns with the settling direction. Apart from the orientation, the prediction accuracy of the settling velocity is also improved by the new correlations. A deviation of up to 46% was seen for the cases without the novel correlations. The limitations of the ellipsoidal point-particle model without the new correlations were also evident in simulations of a turbulent particle-laden round jet. Using this model resulted in significantly overestimated particle velocities and a reduced particle dispersion downstream of the nozzle compared to both the spherical model and the ellipsoidal model with the new force and moment correlations. Similar to its behaviour in decaying isotropic turbulence, the model with the new correlations showed a higher likelihood for particles to align their major axes at larger angles relative to the primary flow direction than the model without these correlations. To fully assess the new model, further validations against DPFS reference data will be conducted.

**Acknowledgements** This work has been funded by the Deutsche Forschungsgemeinschaft (DFG, German Research Foundation)—215035359—SFB/TRR 129 'Oxyflame'. Computing resources were provided by the *High Performance Computing Centre Stuttgart* (HLRS) and by the *Jülich Supercomputing Centre* (JSC) within a large-scale project of the *Gauss Centre for Supercomputing* (GCS).

## References

1. Schneiders, L., Günther, C., Meinke, M. and Schröder, W. "An Efficient Conservative Cut-Cell Method for Rigid Bodies Interacting with Viscous Compressible Flows". *Journal of Computational Physics* 311 (2016), 62–86. https://doi.org/10.1016/j.jcp.2016.01.026
2. Schneiders, L., Hartmann, D., Meinke, M. and Schröder, W. "An Accurate Moving Boundary Formulation in Cut-Cell Methods". *Journal of Computational Physics* 235 (2013), 786–809. https://doi.org/10.1016/j.jcp.2012.09.038
3. Schneiders, L., Meinke, M. and Schröder, W. "Direct Particle-Fluid Simulation of Kolmogorov-Length-Scale Size Particles in Decaying Isotropic Turbulence". *Journal of Fluid Mechanics* 819 (2017), 188–227. https://doi.org/10.1017/jfm.2017.171
4. Schneiders, L., Meinke, M. and Schröder, W. "On the Accuracy of Lagrangian Point-Mass Models for Heavy Non-Spherical Particles in Isotropic Turbulence". *Fuel* 201 (2017), 2–14. https://doi.org/10.1016/j.fuel.2016.11.096
5. Schneiders, L., Fröhlich, K., Meinke, M. and Schröder, W. "The Decay of Isotropic Turbulence Carrying Non-Spherical Finite-Size Particles". *Journal of Fluid Mechanics* 875 (2019), 520–542. https://doi.org/10.1017/jfm.2019.516
6. Schneiders, L., Grimmen, J. H., Meinke, M. and Schröder, W. "An Efficient Numerical Method for Fully-Resolved Particle Simulations on High-Performance Computers". *PAMM* 15.1 (2015), 495–496. https://doi.org/10.1002/pamm.201510238
7. Hartmann, D., Meinke, M. and Schröder, W. "An Adaptive Multilevel Multigrid Formulation for Cartesian Hierarchical Grid Methods". *Computers & Fluids* 37.9 (2008), 1103–1125. https://doi.org/10.1016/j.compfluid.2007.06.007
8. Niemöller, A., Schlottke-Lakemper, M., Meinke, M. and Schröder, W. "Dynamic Load Balancing for Direct-Coupled Multiphysics Simulations". *Computers & Fluids* 199 (2020), 104437. https://doi.org/10.1016/j.compfluid.2020.104437
9. Maxey, M. R. and Riley, J. J. "Equation of Motion for a Small Rigid Sphere in a Nonuniform Flow". *The Physics of Fluids* 26.4 (1983), 883–889. https://doi.org/10.1063/1.864230
10. Kuerten, J. G. M. "Point-Particle DNS and LES of Particle-Laden Turbulent Flow – a State-of-the-Art Review". *Flow, Turbulence and Combustion* 97.3 (2016), 689–713. https://doi.org/10.1007/s10494-016-9765-y
11. Lucci, F., Ferrante, A. and Elghobashi, S. "Modulation of Isotropic Turbulence by Particles of Taylor Length-Scale Size". *Journal of Fluid Mechanics* 650 (2010), 5–55. https://doi.org/10.1017/S0022112009994022
12. Brenner, H. and Cox, R. G. "The Resistance to a Particle of Arbitrary Shape in Translational Motion at Small Reynolds Numbers". *Journal of Fluid Mechanics* 17.4 (1963), 561–595. https://doi.org/10.1017/S002211206300152X

13. Jeffery, G. B. "The Motion of Ellipsoidal Particles Immersed in a Viscous Fluid". *Proceedings of the Royal Society of London. Series A, Containing Papers of a Mathematical and Physical Character* 102.715 (1922), 161–179. https://doi.org/10.1098/rspa.1922.0078
14. Fröhlich, K., Meinke, M. and Schröder, W. "Correlations for Inclined Prolates Based on Highly Resolved Simulations". *Journal of Fluid Mechanics* 901 (2020), A5. https://doi.org/10.1017/jfm.2020.482
15. Kiwitt, T., Fröhlich, K., Meinke, M. and Schröder, W. "Nusselt Correlation for Ellipsoidal Particles". *International Journal of Multiphase Flow* 149 (2022), 103941. https://doi.org/10.1016/j.ijmultiphaseflow.2021.103941
16. Dabade, V., Marath, N. K. and Subramanian, G. "Effects of Inertia and Viscoelasticity on Sedimenting Anisotropic Particles". *Journal of Fluid Mechanics* 778 (2015), 133–188. https://doi.org/10.1017/jfm.2015.360
17. Press, W. H., Teukolsky, S. A., Vetterling, W. T. and Flannery, B. P. *Numerical Recipes in C: The Art of Scientific Computing*. Cambridge: Cambridge University Press, 1992.
18. Sanjeevi, S. K. P. and Padding, J. T. "On the Orientational Dependence of Drag Experienced by Spheroids". *Journal of Fluid Mechanics* 820 (2017). https://doi.org/10.1017/jfm.2017.239
19. Sanjeevi, S. K., Kuipers, J. and Padding, J. T. "Drag, Lift and Torque Correlations for Non-Spherical Particles from Stokes Limit to High Reynolds Numbers". *International Journal of Multiphase Flow* 106 (2018), 325–337. https://doi.org/10.1016/j.ijmultiphaseflow.2018.05.011
20. Fröhlich, K., Farmand, P., Pitsch, H., Meinke, M. and Schröder, W. "Particle Reynolds Number Effects on Settling Ellipsoids in Isotropic Turbulence". *International Journal of Multiphase Flow* 139 (2021). https://doi.org/10.1016/j.ijmultiphaseflow.2021.103566

# Modelling Ignition, Combustion, and Pollutant Emissions

# 12

Pooria Farmand and Heinz Pitsch

## Contents



**Abstract**

In this chapter, the ignition, combustion, and pollutant formation of solid fuel particles under different operating conditions are investigated using detailed numerical

P. Farmand (✉) · H. Pitsch
Institute for Combustion Technology, RWTH Aachen University, Aachen, Germany
e-mail: p.farmand@itv.rwth-aachen.de

H. Pitsch
e-mail: h.pitsch@itv.rwth-aachen.de





R. Kneer et al. (eds.), *Clean Energy Conversion*,
https://doi.org/10.1007/978-3-031-98956-8_12

simulations. First, coal and biomass combustion in a laminar flow reactor configuration are compared, showcasing similar combustion characteristics between coal and biomass. Then, particle/particle interaction effects in the particle group configuration are investigated, indicating later ignition time and flame opening behaviour in higher particle number densities. Particle/chemistry/turbulence interactions are also studied in particle cloud combustion under isotropic turbulence, where ignition is found to occur most likely outside of particle clusters. To assess the reduced-order flamelet models for solid fuel combustion, the optimal estimator concept is employed using the generated dataset, revealing that modelling errors increase near particles. Finally, $NO_x$ formation during solid fuel combustion is analysed. The highest $NO_x$ formation is found to occur near the particle groups around stoichiometric mixtures. Important pathways for $NO_x$ formation under different operating conditions are studied, revealing the dominance of the fuel-$NO_x$ pathways.

## 12.1 Introduction

The advantages of numerical methods for improving solid fuel combustion systems are still not used to their full potential. Direct numerical simulation (DNS) is an established tool to study fundamental aspects of combustion. However, due to the complexities associated with biomass combustion, DNS requires various submodels for the biomass's thermo-chemical conversion. Solid fuel combustion can be simulated via particle-resolved or point-particle approaches. Due to several challenges and limitations of reactive simulations with a particle-resolved approach, solid fuel combustion has been investigated mostly by means of point-particle approaches in the Euler-Lagrange framework. Since such simulations cannot resolve the immediate surroundings of the particle, the interactions of flow and particle are typically described by so-called film models. In these models, all relevant features, such as the typical geometrical characteristics and the internal structure of the solid particles, should be considered. Appropriate model formulations can be derived from experimental results or simplified resolved simulations.

The overarching goal of developing this type of numerical framework is to enable the investigation of ignition, combustion, and pollutant formation of solid fuels under various conditions. The correct prediction of the combustion characteristics of solid fuels involves many factors. Fuel type, particle geometry and internal structure, and the operating conditions of the burner significantly affect the combustion characteristics and pollutant formation pathways in solid fuels. The inclusion of detailed submodels in the point-particle framework is crucial for predicting the correct physical behaviour for ignition and combustion, which can be validated based on experimental data. By resolving the Eulerian field and using detailed kinetics, point-particle DNS can provide data that are otherwise inaccessible in experiments. These data can serve as a reference guideline for the development of the reduced order models required for large-scale simulations, highlighting the value and innovation of this research.

In this chapter, a point-particle direct numerical simulation framework is employed to study the ignition, combustion, and $NO_x$ formation of single particles and groups of coal and biomass particles under laminar and turbulent conditions. In Sect. 12.2, the relevant models and methods for the point-particle DNS in the fully coupled Euler-Lagrange framework are described. Then, in Sect. 12.3, the ignition and combustion of single particle and particle groups are investigated. To investigate the particle/chemistry/turbulence interactions, ignition and combustion modelling for clustering particle clouds in turbulent conditions is presented in Sect. 12.4, in which the PP-DNS dataset is also used for a priori model assessment of the flamelet progress variable (FPV) models. Finally, in Sect. 12.5, the $NO_x$ formation pathways are studied during ignition and combustion of solid fuels.

## 12.2 Numerical Framework and Modelling

As mentioned in Sect. 12.1, mass, momentum, and energy transfers between a solid particle and its surrounding gas can be either particle-resolved or modelled by a point-particle approximation or a so-called Euler-Lagrange approach, as shown schematically in Fig. 12.1. In the particle-resolved approach, the particle-gas interface and the boundary layer are chemically and spatially resolved. The particle-resolved approach requires computational cells with a size $\Delta x$ much smaller than the particle diameter $D_p$. On the other hand, in the point-particle approach, particles are tracked as Lagrangian points, in which $\Delta x \geq D_p$, and the mass, momentum, and energy exchange between the gas phase and particles corresponds to source terms in the conservation equations.

Since the combustion of solid fuels is a multiphase process, gas phase, solid phase, and the interface between the solid and gas phases should be modelled to capture the correct physical behaviour of the particles in the reactive environment. In the following sections, the modelling framework for each of these parts will be presented using the point-particle approach, which is the most used modelling approach for solid fuel combustion.

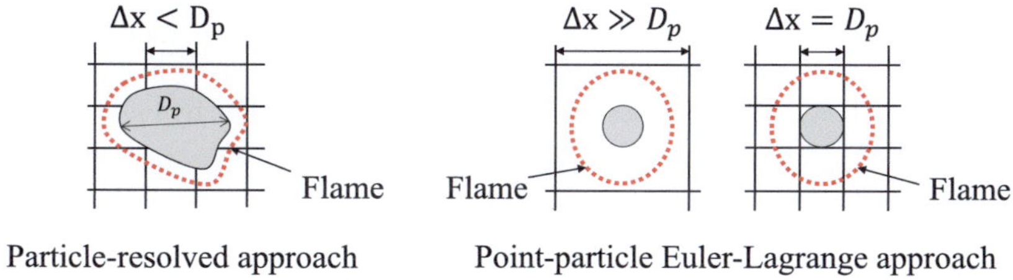


**Fig. 12.1** Schematics of the particle-resolved and Euler-Lagrange approaches

### 12.2.1 Gas-Phase Modelling

Conservation of mass, species, momentum, and energy in the gas phase within the low-Mach formulation leads to the governing equations similar to those applied by Attili et al. [1] as

$$\frac{\partial \rho_g}{\partial t} + \frac{\partial}{\partial x_\beta}\left(\rho_g\, U_{g,\beta}\right) = \dot{S}_m, \tag{12.1}$$

$$\frac{\partial (\rho_g\, Y_i)}{\partial t} + \frac{\partial}{\partial x_\beta}\left(\rho_g \left(U_{g,\beta} + V_{\beta,i}\right) Y_i\right) = \omega_i + \dot{S}_{Y_i}, \tag{12.2}$$

$$\frac{\partial (\rho_g U_{g,\alpha})}{\partial t} + \frac{\partial}{\partial x_\beta}\left(\rho_g\, U_{g,\alpha}\, U_{g,\beta}\right) = -\frac{\partial \Pi}{\partial x_\alpha} + \frac{\partial \tau_{\alpha\beta}}{\partial x_\beta} + \dot{S}_{U,\alpha}, \tag{12.3}$$

$$\begin{aligned} c_{p,g}\frac{\partial (\rho_g\, T_g)}{\partial t} + c_{p,g}\frac{\partial}{\partial x_\beta}\left(\rho_g\, U_{g,\beta}\, T_g\right) &= -\rho_g \frac{\partial T_g}{\partial x_\beta}\sum_i^{N_s} c_{p,g,i}\, Y_i\, V_{\beta,i} \\ &\quad + \frac{\partial}{\partial x_\beta}\left(\lambda_g \frac{\partial T_g}{\partial x_\beta}\right) + \dot{Q}_g + \dot{S}_T. \end{aligned} \tag{12.4}$$

Detailed information on (12.1)–(12.4) can be found the study by Farazi et al. [2]. It should be noted that in (12.1)–(12.4), $\dot{S}_i$ are the source terms from the dispersed phase per unit volume, for which the physical models are described in Sect. 12.2.3. Also, $\dot{Q}_g$ accounts for the rate of heat released per unit volume from the homogeneous reactions in the gas phase. Gas-phase chemistry is modelled using finite-rate chemistry, adopting a specially designed mechanism validated for the oxidation of coal and biomass in both air and oxy-fuel atmospheres. The details of the gas-phase chemical kinetic models can be found in Cai et al. [3] for coal and Chap. 6 for biomass combustion.

### 12.2.2 Solid-Phase Modelling

In the point-particle approximation, the solid fuel combustion is modelled with a collection of point-particles within the Lagrangian framework by solving (12.5)–(12.8) for trajectory, velocity, mass, and temperature as described by Farazi et al. [2].

The Lagrangian equations are given as

$$\frac{dx^k_{\mathrm{p},\alpha}}{dt} = U^k_{\mathrm{p},\alpha}, \tag{12.5}$$

$$\frac{dU^k_{\mathrm{p},\alpha}}{dt} = f^k_{\mathrm{d}} \frac{18\,\mu_{\mathrm{g}} \left(U_{\mathrm{g},\alpha} - U^k_{\mathrm{p},\alpha}\right)}{\rho_{\mathrm{g}} \left(D^k_{\mathrm{p}}\right)^2}, \tag{12.6}$$

$$\frac{dm^k_{\mathrm{p}}}{dt} = \dot{m}^k_{\mathrm{p,evap}} + \dot{m}^k_{\mathrm{p,dev}}, \tag{12.7}$$

$$c^k_{p,\mathrm{p}}\, m^k_{\mathrm{p}} \frac{dT^k_p}{dt} = \dot{m}^k_{\mathrm{p,evap}} \Delta H_{\mathrm{e}} + \dot{m}^k_{\mathrm{p,dev}} \Delta H_{\mathrm{d}} + \Psi^k_{\mathrm{p,c}} + \Psi^k_{\mathrm{p,r}}, \tag{12.8}$$

where for the $k^{\mathrm{th}}$ particle, $m^k_{\mathrm{p}}$ is the total particle mass and $D^k_{\mathrm{p}}$ its diameter. The term $x^k_{\mathrm{p},\alpha}$ refers to the particle position and $U^k_{\mathrm{p},\alpha}$ to the velocity along the $\alpha$ direction. In (12.6), $f^k_{\mathrm{d}}$ is the correction coefficient which accounts for the effect of the Stefan flow on the drag force, which can be computed based on the empirical correlation by Miller and Bellan [4]. Equation (12.8) describes the change of particle temperature assuming a homogeneous temperature inside the particle, which is derived from balancing the heat transfer on the particle surface and the change of thermal energy. Here, the first and second terms on the right hand side refer to the required energy for the evaporation and devolatilisation processes. The heat of evaporation is expressed as $\Delta H_e$ and $\Delta H_d$ is the heat of devolatilisation. The term $\psi^k_{\mathrm{p,c}}$, accounts for the energy transfer by convection and conduction between the particle and the surrounding gas, while the term $\psi^k_{\mathrm{p,r}}$, accounts for the energy transfer due to radiation. Detailed information on these (12.5)–(12.8) can be found in the studies by Farazi et al. [2,5]. Additionally, various submodels are required to capture the full conversion process of the solid particles. The solid kinetic model, which describes the solid fuel's thermal decomposition, provides information on the mass release rate from the particle, alongside the composition of the released gaseous matter, such as light gases and heavy tars, and the thermal effect of this process on the solid particle during the thermal decomposition process. For the reactive simulations of solid particles in the following sections, different solid kinetic models such as single first-order reaction (SFOR), competing two-step model (C2SM), chemical percolation devolatilisation (CPD), and CRECK-S are used, and the detailed description of these models can be found in Chap. 8.

### 12.2.3 Solid-Gas Interface Coupling

In order to predict the correct physical behaviour of solid fuel during ignition and combustion, the particle and gas phases should be fully coupled. Therefore, the exchange of species, mass, and energy between the solid phase and the surrounding gaseous region at the solid-gas interface should be considered using a coupling method. In the point-particle approach, the gas phase and the solid phase are fully coupled using the two-way coupling approach through the source terms appearing in the governing equations at each computational cell with volume $\Omega_{\mathrm{g}}$. In the Eulerian-Lagrangian framework, (12.5)–(12.8) are derived according to the film model, assuming a uniform gas field around the particle. To capture the gas-phase ignition, the domain size is discretised with cubic cells of length $dx$, which is typically equal to the diameter of the mono-disperse particles [2,5]. Since using the gas phase quantities from grid cells with the same size as the particles is not consistent with the film model assumption, a filter should be applied to provide a smoother field in the gas phase to evaluate the state of the gas surrounding the particle consistently with the film model. This approach has been proposed and validated by Farazi et al. [2]. Also, to avoid large particle source terms caused by the small cell-to-particle ratios and to provide grid-independent values, the distribution coefficient $\phi^k$ for each particle $k$ is adopted [2,5]. The distribution coefficient is computed by a Gauss function with a characteristic width $L_{\mathrm{d}}$, centred at the $k^{\mathrm{th}}$ particle position. Then, the source terms $\dot{S}_{\mathrm{i}}$ in the Eulerian equations (12.1)–(12.4) can be written as

$$\dot{S}_m = -\frac{1}{\Omega_{\mathrm{g}}} \sum_{k=1}^{n_{\mathrm{p}}} \phi^k \left( \frac{dm_{\mathrm{p}}^k}{dt} \right) \tag{12.9}$$

$$\dot{S}_{U,\beta} = -\frac{1}{\Omega_{\mathrm{g}}} \sum_{k=1}^{n_{\mathrm{p}}} \phi^k \left( \frac{dm_{\mathrm{p}}^k}{dt} U_{\mathrm{p},\beta}^k + m_{\mathrm{p}}^k \frac{dU_{\mathrm{p},\beta}^k}{dt} \right) \tag{12.10}$$

$$\dot{S}_{Y_i} = -\frac{1}{\Omega_{\mathrm{g}}} \sum_{k=1}^{n_{\mathrm{p}}} \phi^k \left( \frac{dm_{\mathrm{p,dev},i}^k}{dt} \right) \tag{12.11}$$

$$\dot{S}_T = -\frac{1}{\Omega_{\mathrm{g}}} \sum_{k=1}^{n_{\mathrm{p}}} \phi^k \left( \frac{dm_{\mathrm{p}}^k}{dt} c_{p,\mathrm{p}}^k T_{\mathrm{p}}^k + \dot{\Psi}_{\mathrm{p,c}}^k \right) \tag{12.12}$$

## 12.3 Modelling Ignition and Combustion of Solid Fuels Under Laminar Conditions

Based on the Euler-Lagrange formulation, the source terms for the Eulerian equations are obtained, which then can be solved to compute the reactive gas-phase behaviour during ignition and combustion. This facilitates the prediction of the reactive behaviour of the solid fuel particles in more complex configurations using direct numerical simulations, which can provide data inaccessible in the experiments. In this section, the application of the point-particle approximation to capture the underlying physics behind the ignition and combustion of single particles and particle groups in laminar conditions is presented.

**Modelling Single Particle Combustion**

The first step in predicting the correct behaviour of solid fuel particles in a reactive environment is predicting the ignition time. Particle geometry, internal structure, elemental composition, and operating conditions of the burner can significantly impact the ignition and combustion processes. Therefore, it is essential first to investigate the effect of these parameters on the ignition and combustion of single particles in a laminar flow reactor configuration, in which multi-parameter optical measurements provide experimental data for validation of the numerical framework. This configuration is described in detail in Chap. 10. Solid fuel type (coal or biomass) is one of the most important parameters affecting ignition. Due to the different characteristics of biomass compared to coal, the combustion behaviour of biomass particles is expected to be different than that of coal. Therefore, in this section, the reactive behaviour of different solid fuel types is compared. In this context, single high-volatile bituminous Colombian coal (CBC) and walnut shell biomass (WNS) particles have been chosen for this comparison. The respective solid fuel composition for these fuels can be found in Table 1.1.

Since modelling ignition and combustion of solid particles requires a significant number of physical models and simplifying assumptions, a validation of ignition delay times of single particles with available experimental measurements [6] is presented. The comparison with the experiments displayed in Fig. 12.2 shows good agreement with measured ignition delay times for variations in particle diameter $D_\mathrm{p}$. In the simulations, ignition delay time $\tau_\mathrm{ign}$ is defined as the point when the mass fraction of OH radicals $Y_\mathrm{OH}$ exceeds 5% of its maximum. This definition is consis-

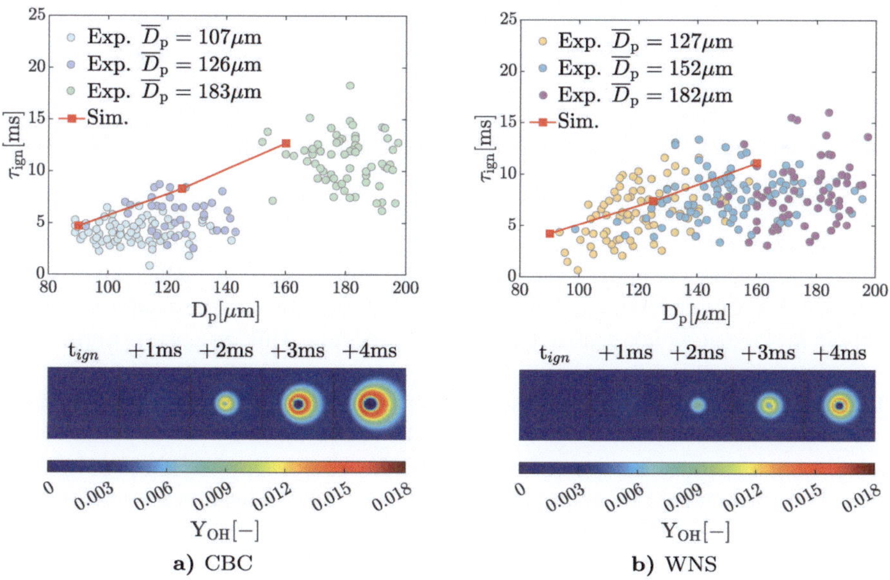


**Fig. 12.2** Ignition delay times $\tau_\mathrm{ign}$ for various particle sizes compared with experimental data reported by Li et al. [6] (top row) and evolution of $Y_\mathrm{OH}$ after ignition for a single particle with $D_\mathrm{p} = 125\,\mu$m (bottom row) for **a)** CBC particles and **b)** walnut shell particles

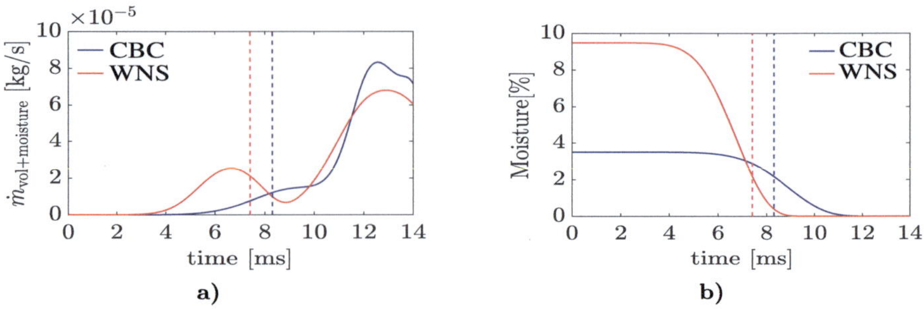


**Fig. 12.3** **a)** Evolutions of total devolatilisation and evaporation rate ($\dot{m}_{\text{vol+moisture}}$), and **b)** remaining moisture content of CBC in comparison with walnut shell during ignition and combustion of single particles with $D_p = 125\,\mu m$. Vertical dashed lines show the ignition times respectively

tent with the experimental definition of the ignition event based on the imaged flame intensities (by OH-LIF) exceeding the background level.

The comparison between the coal and walnut shell ignition shows a slightly faster ignition for walnut shells compared to coal particles. Additionally, as shown in Fig. 12.2, walnut shell particles have a weaker flame with lower OH intensity around the particle compared to coal. This is related to the lower amount of released mass during devolatilisation for walnut shells compared to coal due to the lower apparent density in walnut shells [6].

The lower apparent density of the walnut shells compared to coal alongside with the higher volatile content based on the proximate analysis of these fuels reported by Li et al. [6] is expected to significantly increase the particle heating rate and reduce the ignition delay time. Although, as shown in Fig. 12.3, the higher amount of moisture for biomass particles, which must be evaporated before the release of fuel species, and is known to delay the ignition process are the primary reasons for the similarity in ignition behaviour between coal and walnut shells because of a cancellation of competing effects between drying and volatile release processes.

After studying the ignition characteristics, the flame structure is further investigated by analysing the flame stand-off distance from the particle surface as shown in Fig. 12.4a). In this context, the stand-off distance $r_f$ is defined as the distance from the particle surface at which peak activity $Y_{OH,max}$ occurs. This distance is normalised by the particle radius $r_p$ to yield a dimensionless measure. When considering volatile combustion in a diffusion regime, the chemical reaction is relatively faster than mass and thermal diffusion. The flame location is then determined by two key factors: the volatile release and the oxygen diffusion [6]. Increasing the particle size leads to lower particle heating rate and consequently lower devolatilisation per unit area, as shown in Fig. 12.4b), which results in the flame burning closer to the particle surface. In the comparison of the flame distance between coal and biomass, the main reason is the higher released mass per unit area from the coal particles compared to walnut shells.

In order to investigate the effect of oxygen diffusion on the flame location with respect to particle surface, the flame stand-off distance at different oxygen content of 10, 20, and 40% in air atmosphere are compared for walnut shells with the same particle size. Figure 12.5a) depicts a smaller stand-off distance for the flame for

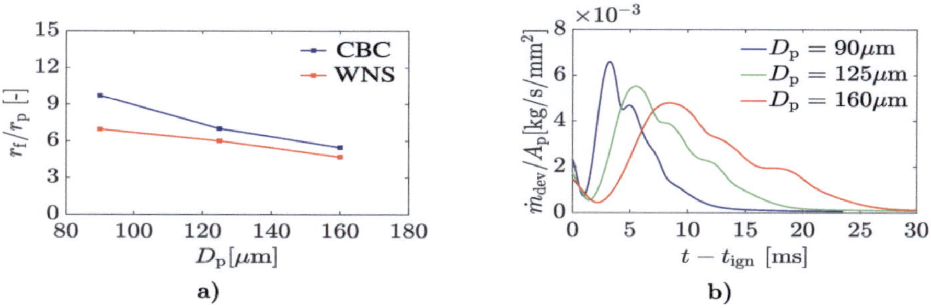


**Fig. 12.4** **a)** Effect of particle size $D_p$ in normalised flame distance $r_f/r_p$ of coal and walnut shells in air and **b)** volatile release rate per particle surface area for walnut shell particles with increasing particle diameter

higher oxygen contents for both coal and biomass. Higher oxygen contents increase the $O_2$ diffusion towards the particle surface. Also, this will increase the reactivity of the mixture by producing more radicals based on $O_2 + H \rightarrow OH + O$, which is the most important chain branching reaction for combustion. This leads to increasing the gas-phase temperature in the vicinity of the particle and, as a result, a higher devolatilisation rate per unit area as shown in Fig. 12.5b). Therefore, it can be concluded that the $O_2$ diffusion rate effect is stronger than the volatile release rate and can significantly impact the flame structure.

More detailed information regarding the differences between the ignition and combustion characteristics of single coal and biomass particles can be found in the study by Li et al. [6].

#### Modelling Particle Group Combustion

In addition to the study of single particles, the ignition process of particle streams in air and oxy-fuel atmospheres needs to be studied. To investigate the effect of interacting volatile flames, the laminar hot co-flow particle stream configuration based on the laminar flat flame burner experiment [7], is presented. This configuration is a 3-D channel with inlet and outlet boundaries in the stream-wise direction, while

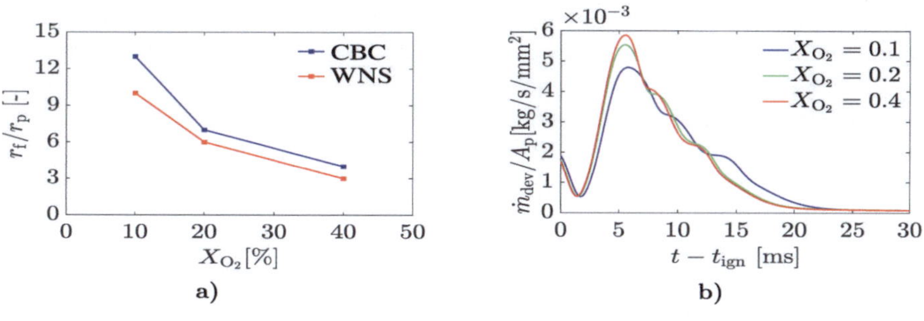


**Fig. 12.5** Effect of oxygen content on **a)** $r_f$ of coal and walnut shells in air and **b)** volatile release rate per particle surface area for walnut shell particles

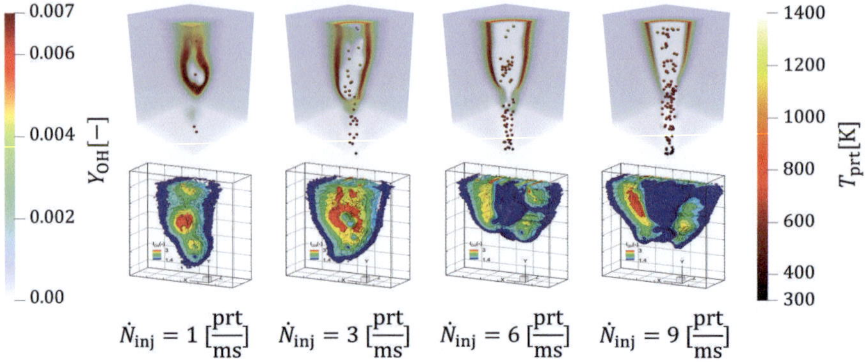


**Fig. 12.6** Flame shapes visualised by OH mass fraction and particle temperature at statistically steady-state conditions in the co-flow jet configuration at different particle injection rates (top row) compared with experimental 3-D OH-LIF signals for the corresponding particle injection rates [8] (bottom row) (reprinted from Farmand et al. [9], Copyright (2022), with permission from Elsevier and minor editing)

periodicity is imposed in the two span-wise directions. Solid fuel particles are seeded directly into a hot oxidiser stream for the channel configuration inlet. This configuration is representative of a typical industrial burner, where the mixing time is faster than the particle heating time.

After injection, due to the heat transfer between particles and the gas phase, particles heat up, and their temperatures gradually increase. Moving further downstream, at a certain distance from the inlet for each case, volatile ignition occurs, which can be identified with the increase in OH radical in the domain. The ignition delay time can be determined after reaching a statistically steady state by means of the ignition position and the averaged particle velocity for different cases of varying particle number density, gas temperature, and injection velocity. Figure 12.6 shows the OH radical fields visualising the flame for different particle injection rates in comparison with the experimental 3-D OH-LIF measurements by Li et al. [8]. Analysing the numerical and experimental OH radical fields demonstrated a transition in ignition location and flame topology from single particles towards dense particle streams. It was observed that increasing the particle injection rate, which leads to higher local particle number density in the domain, shifts the ignition location downstream and increases the ignition delay time because of the increased amount of thermal energy needed for particle heating. The in-depth analysis of the physical processes occurring during this transition highlights that increasing the particle injection rates affects the volatile flame shape, devolatilisation rate, velocity profile, and particle distribution. At low particle injection rates, the volatiles mainly burn in diffusion flames formed around individual particles similar to the single-particle behaviour studied by Farazi et al. [5]. By increasing the particle injection rate, due to oxygen deficiency and relatively lower gas temperatures around the centreline, a delay in the ignition and a more continuous flame region around the particle clouds were observed. This behaviour is more pronounced for higher particle injection rates, in which the flame is suppressed

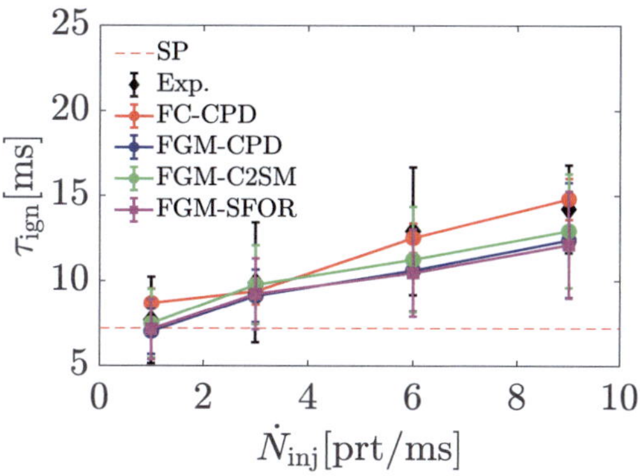


**Fig. 12.7** Comparison of computed ignition delay times in detailed simulations (FC-CPD) with experiments and FGM model predictions with different devolatilisation models (FGM-CPD, FGM-C2SM, FGM-SFOR) for different injection rates. The error bars corresponds to different thresholds of $Y_{\mathrm{OH}}$. The horizontal dashed red line correspond to detailed single particle (SP) simulation reprinted from Farmand et al. [9], Copyright (2022), with permission from Elsevier and minor editing)

in the centre and consequently limited to the area in the boundary region, where the ignitable mixtures are formed due to the mixing of volatiles with the oxidiser [2,9].

In addition to the investigation of physical processes, the complete thermo-chemical state can be extracted from detailed simulations, which allows for the detailed assessment of the reduced-order models for particle group ignition and combustion. Therefore, different modelling assumptions for devolatilisation, ignition, and combustion were also assessed. To this end, the performance of reduced order gas-phase chemistry utilising flamelet generated manifolds (FGM) model and reduced-order particle devolatilisation using CPD, C2SM, and SFOR models in comparison with detailed simulations (finite rate chemistry and CPD (FC-CPD)) was evaluated. As shown in Fig. 12.7, the detailed model simulations (FC-CPD) for both gas and solid phases were successfully validated with the experimentally measured ignition times by Li et al. [8].

It was also observed that the predicted ignition by the simulations using different simplified modelling assumptions for gas-phase and particle devolatilisation (FGM-CPD, FGM, C2SM, FGM-SFOR) agrees well with the experimental data. However, due to the simplified assumptions and model inaccuracies, these reduced-order models under-predict ignition times compared to the detailed model (FC-CPD). To investigate the most reduced-order model (FGM-SFOR) inaccuracies compared with the most detailed model (FC-CPD) and to evaluate the assumptions involved in pulverised fuel FGM modelling, a comprehensive investigation of each reduction step using a systematic model simplification approach was employed. In the detailed simulations, the dynamic volatile composition CPD model is used [5]. In contrast, the typically applied FGM formulation is limited to using a fixed volatile composition to create the flamelet tables. It was found that using this simplifying fixed volatile composition assumption leads to a shift in the release of different volatiles. However, it has a minor effect on particle group ignition, which shows the validity of this assumption for this case [9]. In contrast, the comparison of radial gas-phase

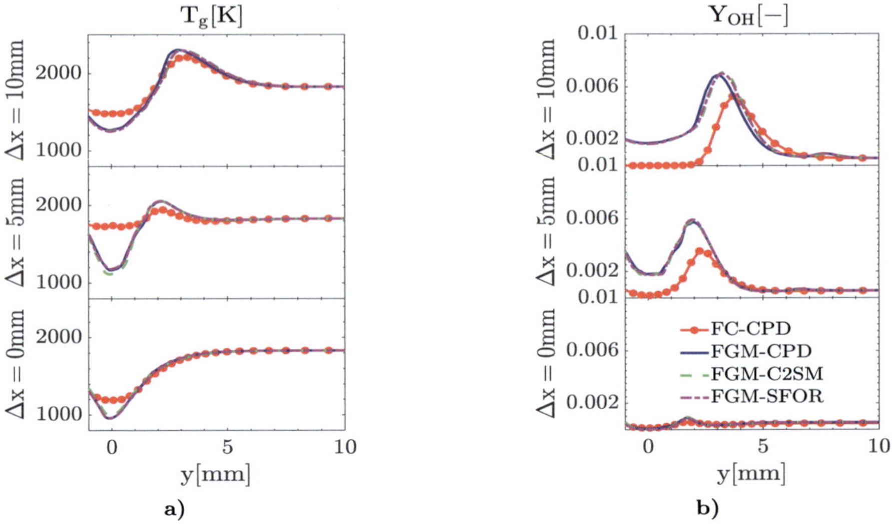


**Fig. 12.8** Effect of different reduced order solid- and gas-phase models (FGM-CPD, FGM-C2SM, FGM-SFOR) in prediction of radial mean profiles of **a)** gas-phase temperature and **b)** OH mass fraction at different axial distances from the ignition height ($\Delta x$) for particle group configuration with $\dot{N}_{\text{inj}} = 6\text{ms}^{-1}$ compared to detailed simulations (FC-CPD) (reprinted from Farmand et al. [9], Copyright (2022), with permission from Elsevier and minor editing).

temperature and OH mass fraction profiles at different heights above the ignition location, as shown in Fig. 12.8 demonstrated that tabulated chemistry has deficits, especially in the central region, where the fuel-rich mixture exceeds the flammability limits and extrapolation in the flamelet table is used.

It should be noted that simplified yet well-fitted devolatilisation models (C2SM and SFOR) can perform similarly compared to detailed devolatilisation model (CPD). However, as the parameters are quite sensitive to the heating rate, the quantitative agreement is very dependent on the heating rates used for fitting the parameters. Detailed analyses are presented in the study by Farmand et al. [9].

## 12.4 Modelling Ignition and Combustion of Solid Fuels Under Turbulent Conditions

Different studies of single-particle and particle group ignition in laminar conditions show that the slip velocity significantly impacts ignition [2, 10, 11]. When particles are exposed to turbulent conditions, they face different slip velocities due to the velocity fluctuations in a turbulent flow, which significantly impact the particle ignition behaviour. In the following sections, first, the physical aspects of the particle cloud ignition under turbulent conditions are investigated. Then, the available models and their applicability for predicting the ignition process in the particle clouds are discussed. It should be mentioned that the general conclusion of this chapter is valid for both high-volatile coal and biomass fuels since both fuels exhibit similar combustion characteristics, as shown in Sect. 12.3.

To understand the effects of turbulent velocity fluctuations on particle ignition and combustion, the homogeneous isotropic turbulent (HIT) configuration is employed to remove the complexities caused by turbulent shear forces.

Due to the absence of mean gradients, HIT has no turbulence production, and as a result, turbulence decays in time. Therefore, the effect of turbulence interactions on the ignition and the combustion process was investigated utilising a fully developed HIT field with a linear forcing in physical space [12] to keep turbulence in a statistically steady state during ignition and combustion [13]. The motion of particles in turbulent flows depends on the time scales of turbulence and particle inertia. Since solid fuel particles have a much higher density than the surrounding gas, they do not fully follow the flow. This behaviour leads to preferential particles concentrations in turbulent flows. This phenomenon typically occurs when the characteristic particle time is in the same order as the flow time scale and is defined by the Stokes number. When the Stokes number is around unity, particles tend to form clusters, resulting in regions almost void of particles [14]. Such preferential concentration can occur in practical applications and has a significant effect on particle ignition process, which is investigated in this section.

### 12.4.1 Particle/Turbulence Interaction Effects on Ignition

In this analysis, turbulence characteristics and particle size were chosen so that particle preferential concentration could be observed. In addition, these conditions can represent the smaller-scale turbulent fluctuations under practically relevant conditions [15, 16]. To study the preferential concentration of the particles, 10,000 CBC particles (see Table 1.1) with $D_\mathrm{p} = 20\,\mu\mathrm{m}$ and $T_{\mathrm{p},0} = 300\,\mathrm{K}$ and with an apparent density of $700\,\mathrm{kg\,m^{-3}}$ are randomly initialised in the forced HIT environment with $\eta \simeq 100\,\mu\mathrm{m}$ and the physical size of $12.8 \times 12.8 \times 12.8\,\mathrm{mm}^3$ with periodic boundary conditions in all directions. Air with 20% oxygen and an initial gas temperature of $T_{\mathrm{g},0} = 1500\,\mathrm{K}$ was chosen as an oxidiser, which is relevant to solid fuel combustion practical applications. To quantify the ignition process, an ignition onset needs to be defined. Typically, a certain OH radical threshold is utilised to identify the ignition onset in joint numerical-experimental studies based on OH-LIF measurements [7, 9].

In the current complex turbulent configuration, due to non-existing experimental measurements, a definition for the ignition similar as in the previous section was used [9]. The flame characteristics are not significantly affected by the choice of OH mass fraction thresholds used to define ignition. This definition was also employed in previously validated studies [9, 17] to quantify the ignition onset in comparison with experimentally measured ignition delay time, was used to ensure having a realistic benchmark for the analysis.

The clustering of the particles leads to increased local particle number density at specific locations and, as a result, reduced gas temperature at those locations due to the

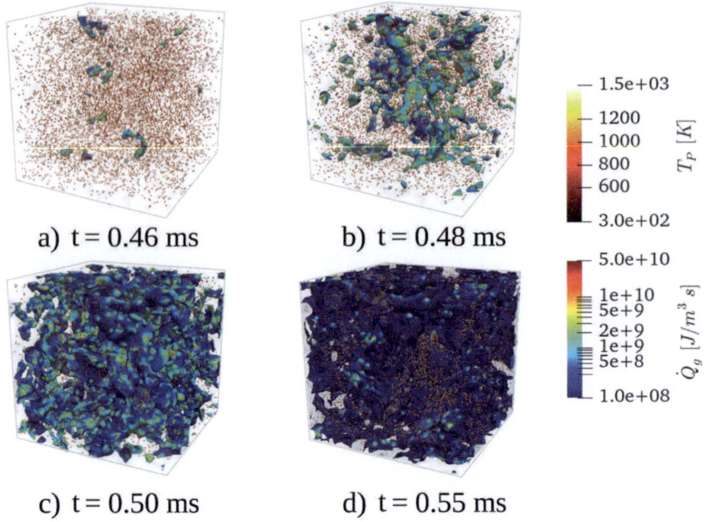


**Fig. 12.9** Evolution of the ignited regions illustrated by the iso-surfaces of the 10% of the maximum $Y_{\mathrm{OH}}$ during the entire combustion process coloured by heat release rate. Particles are marked with spherical points coloured by particle temperature (reprinted and slightly adapted from Farmand et al. [13], licensed under CC BY 4.0)

energy sink required for heating up the particles. Also, higher oxygen availability in these regions results in more favourable thermo-chemical conditions for ignition. The ignited regions can then spread similarly to the laminar flame described in Sect. 12.3 through the entire domain by the growth and merging of individual ignited kernels. The evolution of the ignited region in HIT configuration is depicted in Fig. 12.9.

The particle preferential concentration can be quantified using the Voronoi algorithm [13, 18]. In the Voronoi algorithm, the domain volume is discretised by the Voronoi cells, which are created based on the distance between each two adjacent particles [18]. After discretising the domain by the Voronoi cells, the clustering volume threshold $V_{\mathrm{C}}$ can be obtained by comparing the normalised Voronoi volume distribution in the clustered field $P(V/\overline{V})$ with a randomly distributed particle field $P(V_{\mathrm{R}}/\overline{V})$, in which $\overline{V}$ in the average Voronoi cell volume, calculated based on the Voronoi volume around each particle [18]. Analysing the first ignited regions in the Voronoi space as illustrated in Fig. 12.10a) quantitatively shows that the first ignition kernels predominantly occur outside of particle clusters, where a higher number of ignited points can be observed at volumes larger than clustering limit $V_{\mathrm{C}}$. It was found that the higher gas temperatures outside of particle clusters are the main reason for this ignition behaviour [13]. A detailed description of calculating the clustering limit can be found in the study by Farmand et al. [13].

Calculating the number and the volume of ignited regions during ignition as shown in Fig. 12.10b) can quantify the evolution of the ignition process from auto-ignition to interaction ignition [13]. At interaction ignition, no new individual ignition kernels

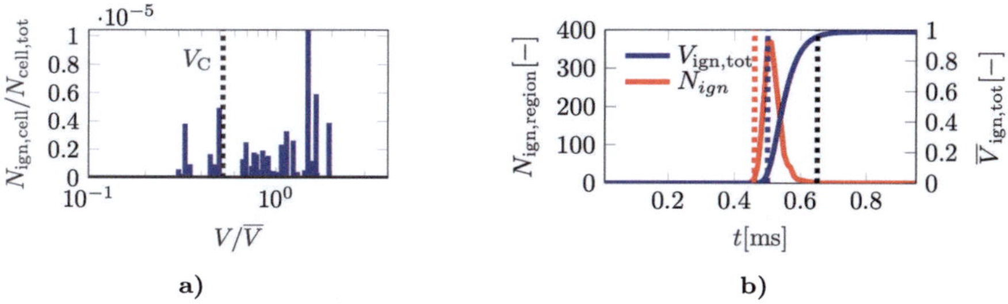


**Fig. 12.10** **a)** Normalised number of the ignited cells in normalised Voronoi volumes at $\tau_{auto.ign} \simeq 0.46$ms. Vertical dashed line shows the clustering threshold $V_C$. **b)** Evolution of the number of separated ignited regions ($N_{ign}$) and the total normalised volume $\sum V_{ign}$ of the ignited regions. The vertical red, blue, and black dotted lines show the auto-ignition time, interaction ignition time, and end of the ignition process (reprinted and slightly adapted from Farmand et al. [13], licensed under CC BY 4.0)

are formed, and flame propagation occurs due to the rapid merging and growth of the ignited regions, resulting from heat release rate from the ignited regions [19].

### 12.4.2 A Priori Model Assessment and Flamelet Modelling of Ignition

The particle cloud ignition process is inherently transient, and the presence of particle preferential concentration further affects the reactivity [20]. Therefore, it is essential to carefully evaluate the assumptions made in reduced-order models to correctly predict ignition and flame characteristics during solid fuel combustion. In this section, a systematic model assessment based on error decomposition method described in the work from Farmand et al. [13] is used to evaluate the model's overall performance from different perspectives. To this end, the flamelet/progress variable (FPV) model is assessed compared to DNS to quantify the suitability of the tabulated thermochemical state. The main focus of this study was placed on modelling the progress variable source term ($\dot{\omega}_C$) as the quantity of interest to predict particle cloud ignition and combustion. The reference progress variable is defined by Farmand et al. [13] and Nicolai et al. [15] as

$$\mathcal{C}_{ref} = Y_{CO_2}/\mathcal{M}_{CO_2} + Y_{CO}/\mathcal{M}_{CO} - Y_{O_2}/\mathcal{M}_{O_2}, \tag{12.13}$$

where $\mathcal{M}_i$ is the molar mass of species $i$.

The progress variable source term in the FPV model shows a shift compared to the DNS, which indicates over-prediction of ignition delay time by the FPV model [13]. To understand the origin of this ignition deviation in the FPV model from the DNS results, the model's performance inside and outside the particle clusters during ignition needs to be studied in more detail. To this end, $t = 0.5$ ms with the maximum number of ignited regions during the ignition process, is shown in Fig. 12.10b).

The optimal estimator concept is used to quantify the errors inside and outside of particle clusters [13]. The reduced-order modelling of unclosed terms in partial differential equations typically involves two steps: First, a set of input parameters should be specified for the model, and second, a specific functional form needs to be defined to model the unclosed terms by the input parameters [21]. Each of these steps introduces errors that need to be carefully investigated in the model assessment process. The error corresponding to the first step is known as the irreducible error, while the error corresponding to the second step is referred to as the functional error. The total modelling errors, computed by the overall difference between the quantities of interest obtained by the model and the DNS, can be systematically decomposed using the optimal estimator method into irreducible and functional errors. The general idea of the optimal estimator concept is to determine for a given parameter set $\Pi$, the mean of a quantity of interest $\phi$ conditioned on the set of parameters $\langle \phi \mid \Pi \rangle$ and obtain a measure for the scatter of the quantity of interest with respect to $\Pi$. Formally, the irreducible error is given by Berger et al. [21] as

$$\epsilon_{\mathrm{irr}}^2 = \left\langle (Q_{\mathrm{int}} - \langle Q_{\mathrm{int}} \mid \Pi \rangle)^2 \right\rangle . \tag{12.14}$$

In the following, irreducible error $\epsilon_{\mathrm{irr}}$ and the error caused by the functional form of the model $\epsilon_{\mathrm{func}}$, are discussed for identifying the potential for model improvements. First, the model parameters chosen to parameterise the flamelet table are assessed using the optimal estimator method. This enables the calculation of irreducible errors for different sets of input parameters to predict the quantity of interest.

Figure 12.11a) demonstrates that the inclusion of the progress variable and enthalpy alongside the mixture fraction strongly decreases the overall irreducible error, especially inside the clusters, where the errors are the highest. Finally, it can be concluded that the combination of mixture fraction $Z$, progress variable $C_{\mathrm{n}}$, and enthalpy $h_{\mathrm{n}}$ leads to the smallest irreducible error. Using these parameters as the best set of input parameters, a non-premixed flamelet table can be generated as described in Farmand et al. [13]. The progress variable source term predicted from the FPV model can be compared with the corresponding value from the DNS datasets to cal-

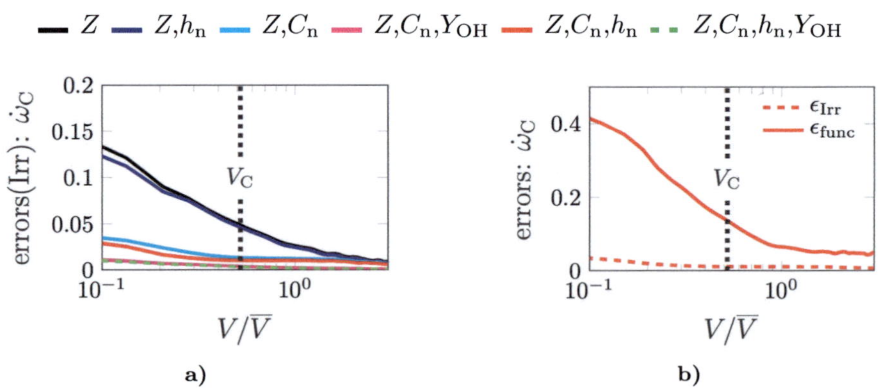


**Fig. 12.11** **a)** Irreducible errors of different input parameters and **b)** decomposed errors involved in the reference FPV model in comparison with the DNS results in the prediction of progress variable source term (reprinted and slightly adapted from Farmand et al. [13], licensed under CC BY 4.0).

**Table 12.1** Different flamelet tables input combinations

| Input parameter combinations | Definition of $C$ |
|---|---|
| $Z, C_n, h_n$ | $C_{ref}$ |
| $Z, C_n$ | $C_{ref}$ |
| $Z, C_{n,1}, h_n$ | $Y_{CO_2} + Y_{CO}$ |
| $Z, C_{n,2}, h_n$ | $Y_{CO_2} - Y_{O_2}$ |
| $Z, C_{n,3}, h_n$ | $Y_{CO_2}/\mathcal{M}_{CO_2} + Y_{H_2O}/\mathcal{M}_{H_2O} - Y_{O_2}/\mathcal{M}_{O_2}$ |

culate the functional errors as shown in Fig. 12.11b). The analysis shows that the particle clusters significantly affect the functional modelling errors. The overprediction of the progress variable source term inside the particle clusters can be attributed to the limitations of the typically used flamelet tabulated chemistry models in capturing the particle/chemistry interactions. This limitation is especially prominent in the vicinity of particle clusters, which corresponds to a smaller mixing time scale compared to the chemical time scale. In these regions, where the volatiles mix with the oxidiser prior to ignition, the flame behaviour resembles that of a premixed flame, which the non-premixed flamelet approach cannot accurately capture [13].

In the next step, the effects of changing the input parameters, such as excluding enthalpy and altering the progress variable definition in the model performance, are also examined. In order to investigate the effect of heat losses and the progress variable definition in the model, different flamelet tables with various input parameter combinations are generated. The input parameters of these different flamelet tables are summarised in Table 12.1.

From Fig. 12.12a), it is evident that including enthalpy in the flamelet table primarily affects the predictions. In particular, incorporating enthalpy as an input parameter improves the accuracy of the model inside the clusters, especially during ignition. This suggests that the ignition phase is sensitive to changes in enthalpy, especially

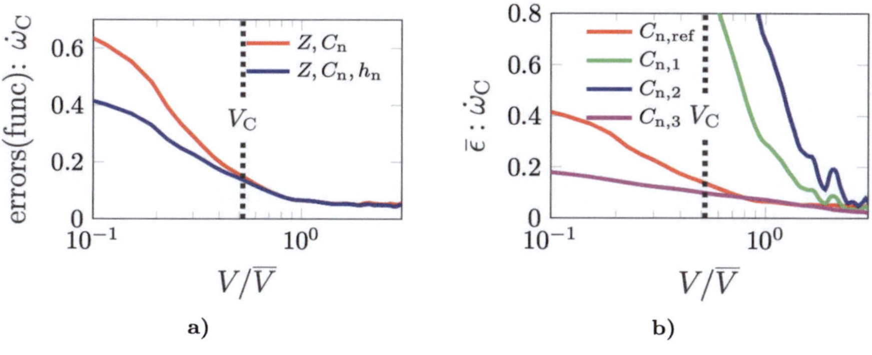


**Fig. 12.12** **a)** Effect of enthalpy and **b)** effect of progress variable definition in the errors of progress variable source term. Solid colour lines show the functional errors, and dashed colour lines correspond to irreducible errors. Volumes larger than $V_C$ represent regions outside of clusters (reprinted and slightly adapted from Farmand et al. [13], licensed under CC BY 4.0).

inside the clusters. During the ignition phase, due to the particle heat-up process and the higher temperature gradients between particles and the gas phase, the heat loss effect is more dominant than at later times [17].

Altering the definition of progress variables can also affect the model's performance. As shown in Fig. 12.12b), excluding CO or O2 from the model significantly amplifies the functional errors of predicting the progress variable source term ($\dot{\omega}_C$), indicating that $C_{n,1}$ and $C_{n,2}$ are not suitable choices for representing the progress variable. On the other hand, replacing CO with $H_2O$ ($C_{n,3}$) resulted in an improvement in predicting the progress variable source term, particularly within the clusters. This shows that the definitions of the input parameters are as important as the combination of the input parameters in the flamelet table in the correct prediction of quantities of interest. More detailed analysis regarding the a priori model assessment of ignition in clustering particle clouds can be found in the study by Farmand et al. [13].

## 12.5 Modelling Pollutant Formation During Solid Fuel Combustion

In solid fuel combustion, controlling hazardous $NO_x$ is one of the major challenges. Unlike gas hydrocarbon combustion, solid fuels contain fuel-bound nitrogen, which is the primary source of $NO_x$. When particles are exposed to heat, the internal bounds inside the particle structure break and the nitrogen element is released as small nitrogen-containing species, which are considered to be the main $NO_x$ precursors. This is a complex and multi-physics process and experiments are limited to providing detailed information regarding the pathways involved. Therefore, numerical simulations with the detailed kinetic models for both solid and gas phases, including the reaction pathways for $NO_x$ can provide additional valuable insights. In this study, solid kinetics is modelled using the CRECK-S model as described in Sect. 8.2.3 and the gas-phase chemistry is modelled with the recently developed Complete-ITV-$NO_x$ chemical kinetic model [22]. A detailed description of these kinetic models can be found in Sects. 8.2.3 and 6.3. This section analyses $NO_x$ formation for single particle and particle group combustion in laminar conditions.

### 12.5.1 $NO_x$ Formation During Ignition and Combustion of Solid Fuels

Since $NO_x$ formation in solid fuels involves kinetic models for both solid and the gas phase, the performance of these models first needs to be assessed compared to available experimental measurements. The drop tube furnace (DTF) configuration, described in Sect. 2.2.3, is typically utilised for measuring $NO_x$ during solid fuel combustion. In the chosen drop-tube furnace, particles with the size of $D_p = 90\,\mu m$ are released from the injector with a specific mass flow rate at the top of a long furnace with high gas-phase temperature. Due to the long residence time, the complete

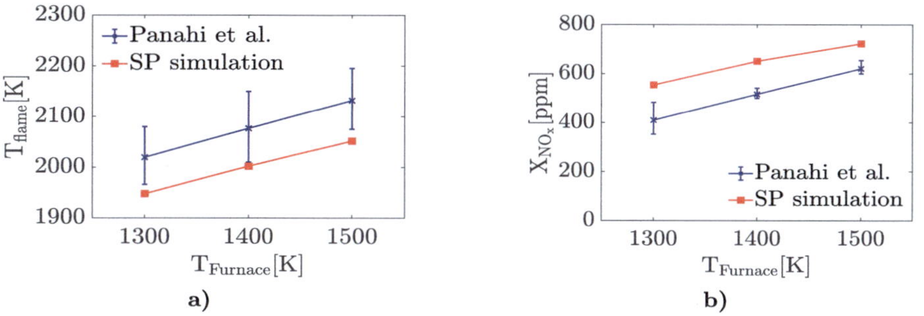


**Fig. 12.13** Comparison of the **a)** flame temperature and **b)** time- and space-averaged model predictions for mole fraction of $NO_x$ for single Pittsburgh coal (USC) particles at different furnace temperatures compared to the experimental data by Panahi et al. [24]

process of ignition, combustion, and $NO_x$ formation can be observed and the species concentrations can be measured at the end of the tube [23].

However, performing direct numerical simulations of a full DTF geometry is numerically expensive. Therefore, these simulations are typically performed in a downscaled inherited DTF geometry, for which the dimensions are chosen such that the domain size is large enough to contain the full flame characteristics from ignition to complete combustion [22]. In this study, $NO_x$ formation during single particle and particle groups combustion is investigated. The furnace operating conditions, particle sizes, and mass flow rate for the particle group configuration is chosen to match that of the experiments [23,24]. Also, to assess the $NO_x$ emissions, simulations were conducted for two flow-through times, which leads to a statistically steady-state for $NO_x$ concentration at the exit plane.

Due to the utilised adopted geometry of the experimental configurations and due to the uncertainties in the numerical models and experimental boundary conditions, only the trends were compared with the measurements. For evaluating the performance of the numerical framework, flame temperature and the time- and space-averaged $NO_x$ at the exit plane for high volatile bituminous Pittsburgh coal (USC) (see Table 1.1) was calculated for comparison with the experiments performed by Panahi et al. [24]. It is worth noting that the simulation results for $NO_x$ mole fractions are scaled by the ratio of the experimental cross-sectional area to that of the simulation so that the volumetric flow rates are the same in both the simulations and experiments. As shown in Fig. 12.13, the trend of the experimental data for flame temperature and $NO_x$ mole fractions can be reproduced, which shows the validity of the numerical framework in capturing the sensitivity of $NO_x$ with respect to different operating parameters. Figure 12.13a) also shows that increasing temperature leads to an increase of $NO_x$ formation, which is due to the increased reactivity in the $NO_x$ formation pathways [22].

After evaluating the numerical framework performance in capturing the $NO_x$ sensitivity with respect to different operating parameters, the comprehensive datasets generated from DNS can be used to study the $NO_x$ production and consumption pathways. To gain a deeper understanding of the $NO_x$ formation process, the reaction pathways are quantitatively analysed using the reaction flux analysis in the next section.

### 12.5.2 $NO_x$ Formation Pathway Analysis

The $NO_x$ formation is inherently a transient process. To obtain a global view of this transient production and consumption pathways, reaction path analysis (RPA) and rate of production analysis (ROPA) for the nitrogen element flux should be performed. The nitrogen element flux $\dot{N}_{AjB}$ from species $A$ to $B$ through reaction $j$ is defined by Farmand et al. [22] and Shamooni et al. [25] as

$$\dot{N}_{AjB} = \frac{n_{N,A}\, n_{N,B}}{n_{N_R,j}} \dot{r}_j, \tag{12.15}$$

where $n_{N,A}$ and $n_{N,B}$ are the numbers of nitrogen atoms in species $A$ and $B$ considering the corresponding stoichiometric coefficients. The term $n_{N_R,j}$ represents the sum nitrogen atoms on either side of reaction $j$, while $\dot{r}_j$ is the net rate of reaction in a chemical mechanism with $N_\mathrm{s}$ species and $N_\mathrm{r}$ reactions, defined as

$$\dot{r}_j \equiv k_{f,j} \prod_{k=1}^{N_s} \left( \frac{\rho Y_k}{W_k} \right)^{\nu'_{k,j}} - k_{b,j} \prod_{k=1}^{N_s} \left( \frac{\rho Y_k}{W_k} \right)^{\nu''_{k,j}}, \tag{12.16}$$

where $\nu'_{k,j}$ and $\nu''_{k,j}$ are the molar stoichiometric coefficients of species $k$ in forward and backward reactions, with Arrhenius rates of $k_{f,j}$ and $k_{b,j}$, respectively.

Following [26], who introduced the time-integrated element flux analysis, all of the analyses in this study are performed using time- and volume-integrated (averaged) fluxes over the entire numerical domain and the total simulation time.

### 12.5.3 $NO_x$ Formation Pathways for Coal and Biomass

As mentioned in Sect. 12.3, the solid fuel type can significantly impact the overall reactive behaviour of solid pulverised fuels. The solid fuel type can change the volatile mixture composition, which can affect the reactions in the gas phase and consequently impact the $NO_x$ formation pathways. To study the effect of fuel type, $NO_x$ formation pathways during the combustion of the high volatile Pittsburgh coal (USC) particles and torrefied miscanthus (TMG) biomass particles are compared. A detailed characterisation of these fuels can be found in Table 1.1. Single particle simulations for the selected coal and biomass types at $T_\mathrm{g} = 1400\,\mathrm{K}$ with a similar particle size of $D_\mathrm{p} = 90\,\mu\mathrm{m}$ in a drop tube configuration described in Sect. 12.5.1 were performed to study the effect of solid fuel type on $NO_x$ formation pathways. Based on the ultimate analysis comparison of Pittsburgh coal and torrefied miscanthus, higher nitrogen content in the coal particle will lead to a higher $NO_x$ formation compared to miscanthus [22]. In this section, the $NO_x$ formation pathways are discussed in more

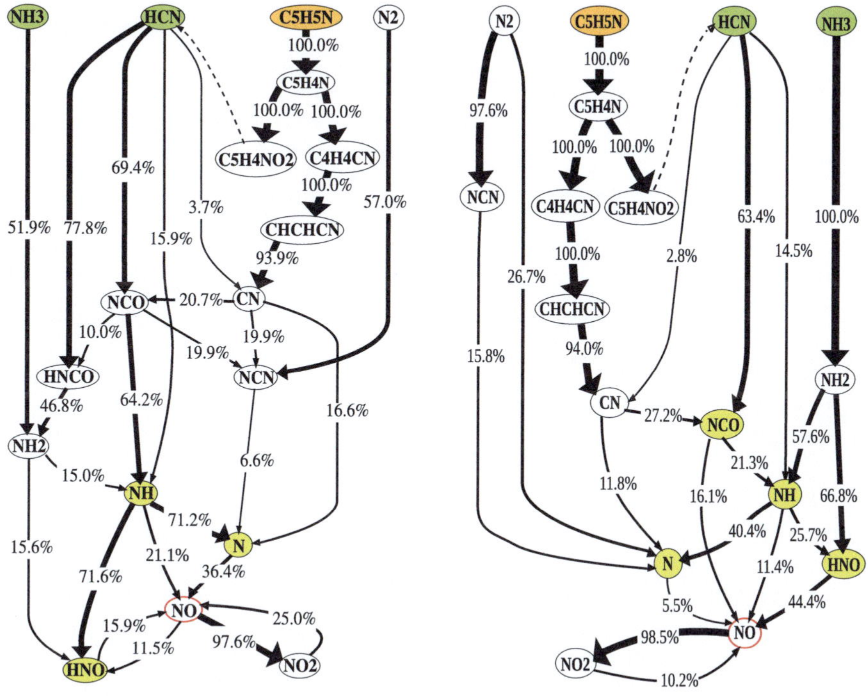


a) USC b) TMG

**Fig. 12.14** The most important time- and space-averaged pathways of N atoms from the released nitrogen-containing volatiles ($NH_3$, HCN, and $C_5H_5N$) and $N_2$, which contribute to NO production for **a)** USC and **b)** TMG biomass. The N-containing light gases ($NH_3$ and HCN) are coloured green and N-containing tar ($C_5H_5N$) is coloured orange. Direct N-containing radicals (N, NH, HNO) are coloured yellow

details. RPA for the nitrogen element flux from the nitrogen-containing volatiles and $N_2$ towards $NO_x$ as illustrated in Fig. 12.14 shows the most important pathways for $NO_x$ formation. The percentage beside each arrow corresponds to the locally normalised flux of N atoms between two species by the total incoming fluxes. The thickness of each arrow is proportional to the N-flux in each path. Some minor pathways are neglected and removed to increase clarity. This approach for presenting the RPA is similar to the method described by Shamooni et al. [25]. It is worth noting that in direct NO production, similar N-containing radicals (NH, N, HNO) participate in both Pittsburgh coal and miscanthus. However, in miscanthus combustion, NCO, which originates from the direct decomposition of HCN, also participates in the direct production of NO. In addition, unlike coal, in which N is the most dominant radical for direct NO production, HNO is the dominant radical in miscanthus combustion. However, both N and HNO originate from the NH radical and it can be concluded that NH is the most important radical for NO production during coal and miscanthus combustion. The most important reactions in NO production and consumption can be studied using the rate of production analysis. Based on the rate of production

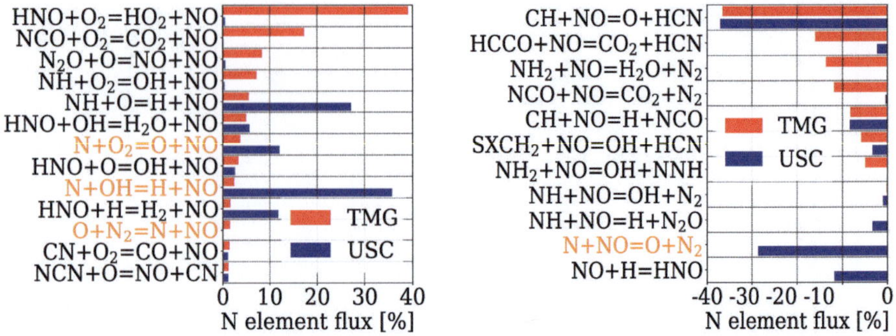


**a)** Rate of production analysis **b)** Rate of destruction analysis

**Fig. 12.15** Rate of NO **a)** production and **b)** destruction analysis for coal and miscanthus. The extended Zeldovich reactions are coloured orange

of NO, as shown in Fig. 12.15a), although similar N-containing radicals participate in direct NO production, the NO formation pathways are completely different in coal compared to biomass. It can be observed that $HNO + O_2 \rightleftharpoons HO_2 + NO$ is the most important reaction for miscanthus. In contrast, during coal combustion, $N + OH \rightleftharpoons NO + H$ is the dominant reaction. This difference originates from different volatile compositions released from the Pittsburgh coal and miscanthus.

The rate of destruction analysis of NO, as shown in Fig. 12.15b), also highlights that NO destruction during miscanthus combustion, occurs primarily in contact with small hydrocarbon radicals such as H, CH, HCCO through the so-called reburn mechanism [22]. However, during coal combustion, the N radical is the most important radical for NO destruction through reaction $N + NO \rightleftharpoons O + N_2$. This reaction is the initiation reaction for the Zeldovich mechanism for thermal-$NO_x$ pathways. The net NO production rate shows that this reaction only contributes to less than 10% of the total NO formation. Therefore, it can be concluded that the thermal-$NO_x$ pathways have a minor impact on the overall NO formation during single-particle coal and miscanthus combustion.

### 12.5.4 Effect of Particle/Particle Interactions on $NO_x$ Formation

After investigating the $NO_x$ formation pathways in single-particle configuration, the particle/particle interaction is studied by comparing the reaction pathways in the particle group configuration and the single particle configuration. Besides varying the total nitrogen mass in the furnace, particle number density can impact the reaction pathways by changing the particle/particle/chemistry interactions. To this end, simulations of a miscanthus particle group with the same boundary conditions and particle size as the single particle configuration were performed to suppress the impact of other operating parameters in the analysis. It should be mentioned that due to the transient nature of the $NO_x$ formation, the particle group simulation was also performed with a long enough residence time ($t = 330\,\text{ms}$) so that $NO_x$ mole fraction

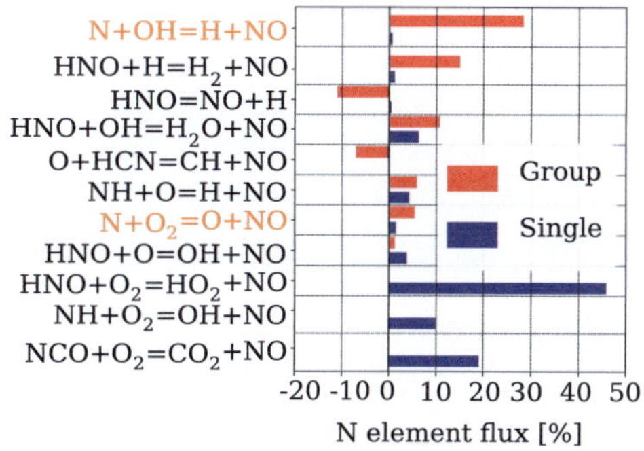


**Fig. 12.16** Effect of particle/particle interactions on the production rate of NO averaged over the entire computational domain and simulation time. The extended Zeldovich reactions are coloured orange

reaches a statistically steady state at the burner exit plane. Rate of production analysis as shown in Fig. 12.16 highlights the fact that the particle group configuration has a completely different reaction pathway for NO formation, which underscores the importance of particle/particle interactions in NO emissions.

In particle group configuration, unlike the single-particle configuration, the reaction $N + OH \rightleftharpoons H + NO$ is the most dominant reaction for miscanthus NO production. It is worth noting that in the particle group configuration, the dense particle group leads to $O_2$ deficiency in the vicinity of the particles. Therefore, in the particle group configuration, this $O_2$ deficiency deactivates the direct reactions with $O_2$ towards NO formation. As a result, the pathways, including the available radicals around the flame, such as OH, become more dominant. Since in the particle group configuration, $NO_x$ emission can change based on the local thermo-chemical conditions, to understand the origin of different pathways in particle group configuration in comparison with single-particle configuration, the flame topology and the local $NO_x$ emissions should be investigated. In the next section, these interactions are discussed in more detail.

### 12.5.5 Flame Topology Effects on $NO_x$ Formation

After investigating the global $NO_x$ formation pathways in particle groups in comparison with the single-particle configuration, the interactions between the local thermo-chemical conditions and $NO_x$ formation is studied. To this end, the same torrefied miscanthus particle group configuration as discussed in Sect. 12.5.4 is considered. Figure 12.17 depicts the opening flame topology at the statistically steady state in the dense particle group configuration, which is illustrated with the OH mole fraction alongside the local $NO_x$ mole fraction in the numerical domain. The local comparison with the flame topology and the $NO_x$ emissions shows that both local maxima of $NO_x$ and OH mole fractions occurs around stoichiometric mixtures.

Interestingly the global maximum of $NO_x$ is observed at the flame tip in the particle stream centre. At this location the nitrogen-containing volatile release has ended and due to high local gas temperature in the vicinity of small radicals shown in Fig. 12.16 the $NO_x$ production is accelerated.

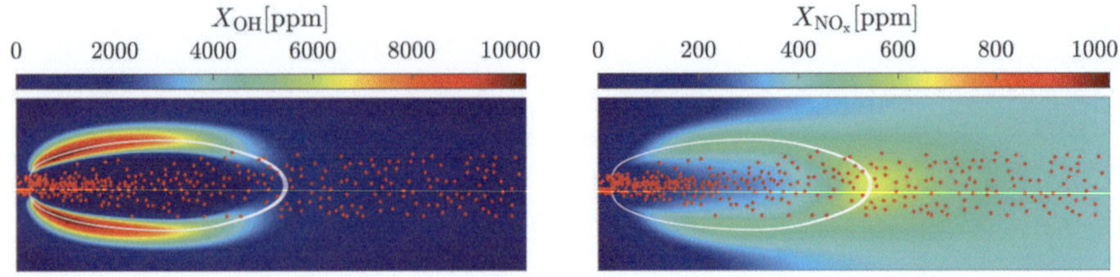


**Fig. 12.17** Snapshots of the OH and $NO_x$ mole fractions in a miscanthus flame at a time with statistically steady-state conditions ($t = 330\,\mathrm{ms}$). Particles are visualised with red points, and the stoichiometric mixture fraction $Z_{\mathrm{st}} \approx 0.15$ is visualised with white isolines

In Fig. 12.18a), a clear demonstration of this behaviour is presented by showing the distribution of $X_{\mathrm{NO}_x}$ and the source terms for N-containing volatiles and $NO_x$ in mixture fraction space coloured by OH mole fraction. The joint correlations with respect to $X_{\mathrm{OH}}$ shows two distinct peaks for the $X_{\mathrm{NO}_x}$ at $Z = Z_{\mathrm{st}}$. By comparing these peaks with the flame topology depicted in Fig. 12.17, it can be concluded that these peaks are related to the high OH region at the flame location and the low OH region at the particle stream centre. As observed in Fig. 12.17, the peak related to the low OH region is the global maximum of $NO_x$. Although the local maximum of the $NO_x$ occurs around stoichiometry, high $NO_x$ mole fractions can also be observed in rich areas. As observed in Fig. 12.18b), the highest production rate of $NO_x$ ($\dot{\omega}_{\mathrm{NO}_x}$), occur in the rich region, which is in the vicinity of the particle stream itself.

The joint correlations with respect to $X_{\mathrm{OH}}$ shows two distinct peaks for the $X_{\mathrm{NO}_x}$ at $Z = Z_{\mathrm{st}}$. By comparing these peaks with the flame topology depicted in Fig. 12.17, it can be concluded that these peaks are related to the high OH region at the flame location and the low OH region at the particle stream centre. As observed in Fig. 12.17, the peak related to the low OH region is the global maximum of $NO_x$. Although the local maximum of the $NO_x$ occurs around stoichiometry, high $NO_x$ mole fractions can also be observed in rich areas. As observed in Fig. 12.18b), the highest production rate of $NO_x$ ($\dot{\omega}_{\mathrm{NO}_x}$), occur in the rich region, which is in the vicinity of the particle stream itself. This observation underlines the importance of fuel-$NO_x$

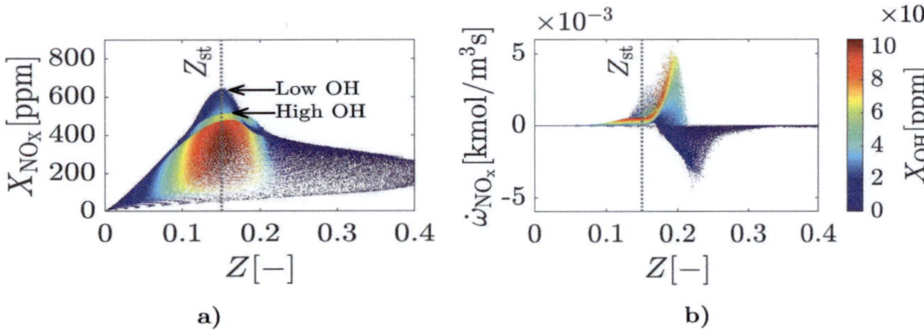


**Fig. 12.18** **a)** $NO_x$ mole fraction ($X_{\mathrm{NO}_x}$), and **b)** $NO_x$ chemical source term ($\dot{\omega}_{\mathrm{NO}_x}$) of the selected torrefied miscanthus flame in mixture fraction space coloured by OH mole fraction

pathways in the overall $NO_x$ formation, which is consistent with previous findings for coal combustion [25].

### 12.5.6 Effect of Operating Conditions on $NO_x$ Formation

Furnace operating conditions can significantly influence $NO_x$ emissions from the combustion of solid pulverised fuels [27]. Examining the effects of various operating conditions on $NO_x$ formation pathways can improve the understanding of $NO_x$ sensitivity to different parameters. Therefore, after investigating the $NO_x$ formation pathways for a reference biomass flame, a parametric study is performed to investigate the sensitivity of $NO_x$ formation with respect to different operating conditions. In Sect. 12.5.1, the effect of furnace temperature on the $NO_x$ emissions was presented. In this section, the effect of atmospheric composition on the $NO_x$ formation is investigated.

To evaluate the atmosphere effects, air and oxy-fuel atmospheres with 21% $O_2$ are considered. Also, the effect of changing oxygen concentration in the oxidiser by increasing the $O_2$ content by 10% (31% $O_2$) is studied. Particle group simulations with the torrefied miscanthus in the DTF configuration were performed. Figure 12.19a) shows the flame illustrated with the OH mole fraction. Lower OH intensity can be observed in oxy-fuel atmosphere, which originates from the higher heat capacity of $CO_2$ compared to $N_2$ and different reaction pathways in $O_2/CO_2$ atmosphere in comparison with $O_2/N_2$ [22].

In an oxygen-rich atmosphere, more H radicals are consumed through the reaction $CO_2 + H \rightarrow CO + OH$, which significantly impacts the overall radical pool. Depletion of H radicals significantly impacts the reaction $O_2 + H \rightarrow OH + O$, the most important chain branching reaction for ignition and combustion. This results in a weaker flame and lower reactivity and, thus, less $NO_x$ formation in oxy-fuel atmosphere as shown in Fig. 12.19b). It is worth noting that increasing the oxygen content in different

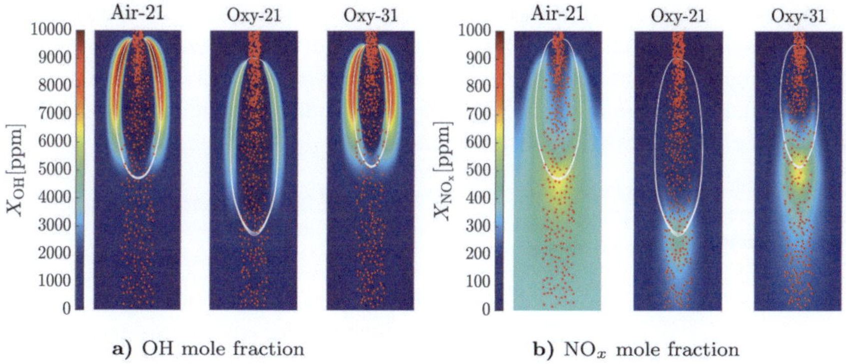


**Fig. 12.19** Effect of different atmospheric compositions on **a)** flame topology visualised by OH mole fraction and **b)** $NO_x$ mole fraction at the time with statistically steady-state

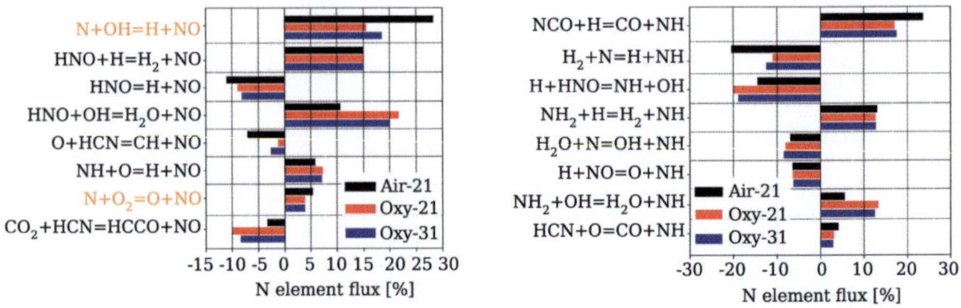


a) ROPA of NO  b) ROPA of NH

**Fig. 12.20** Effect of different atmospheres in ROPA of **a)** NO and **b)** NH, averaged over the entire computational domain and simulation time. The extended Zeldovich reactions are coloured orange

atmospheres increases the reactivity, and, as a result, a higher amount of $NO_x$ values can be observed in those conditions.

To gain a deeper understanding of the effect of atmospheric composition on the $NO_x$ formation mechanism, the rate of production analysis (ROPA) of NO at different atmospheres is compared in Fig. 12.20a). Comparing the rate of production of NO shows that, unlike in air, the HNO pathways in the oxy-fuel atmosphere become more important than N pathways, and $HNO + OH \rightleftharpoons H_2O + NO$ becomes the most important reaction for NO production.

As mentioned in Sect. 12.5.3, direct N-containing radicals such as N and HNO originate from the NH radical. To this end, the rate of production of NH can provide a deeper understanding of the main pathways towards NO production under different atmospheres. Figure 12.20b) highlights the differences in NH production under different atmospheres. It can be observed that in the oxy-fuel atmosphere, a higher NH consumption rate for the reaction $NH + OH \rightleftharpoons H + HNO$ and lower NH consumption rate for reaction $NH + H \rightleftharpoons H2 + N$ is the leading cause for higher direct production of NO from HNO radical in oxy-fuel atmosphere. It is worth noting that increasing oxygen concentration increases the consumption rate of all nitrogen-containing radicals, which results in higher NO production compared to the lower oxygen concentration case in the same atmosphere.

## 12.6 Conclusion

In this chapter, ignition, combustion, and pollutant formation pathways of coal and biomass particles were numerically studied in different configurations and under different conditions. Numerical results were analysed to understand the interactions between particles, flow and chemistry and to explain the ignition, combustion, and pollutant formation pathways of solid fuel particles under different operating conditions. Detailed simulations and discussions for modelling ignition and combustion of single particle and particle groups in laminar conditions are provided in Sect. 12.3, in which, at first, the ignition and combustion of single coal and walnut shells were compared. The observations showed similar combustion characteristics in coal and biomass due to the cancellation of competing effects of particle mass and moisture content. Observations showed that the flame structure is sensitive to different operating conditions, such as particle size and $O_2$ content, and the flame distance is directly related to the competition between $O_2$ diffusion rate and the volatile release rate.

The investigation of the effects of particle/particle interactions on the ignition and flame topology of solid fuels using different particle group configurations shows that increasing the particle number density leads to a delay in the ignition time because of the oxygen deficiency and energy sink due to particle heat-up. This leads to the opening flame behaviour in the co-flow laminar particle jet flames. The simulations employing detailed kinetics for solid and gas phase provide a baseline for assessing the reduced-order models used in large-scale simulations. Therefore, the simplified assumptions in the tabulated flamelet models and the simplified solid kinetic models were compared with the detailed simulations. It was found that well fitted reduced-order solid kinetic models based on the detailed models can predict the particle behaviour correctly. In addition, since the particle/chemistry interactions are neglected in the flamelet tables, these models mispredict the quantities of interest near the particles outside the flammability limit.

In Sect. 12.4, the effect of particle clustering in a particle group configuration on the ignition and combustion of solid fuels under homogeneous isotropic turbulence conditions was studied. It was shown that ignition is more likely to occur outside of particle clusters, where more suitable thermo-chemical conditions for ignition exist. Also, after forming the first auto-ignited regions, the growth and merging of individual ignited regions are the primary mechanisms for flame propagation. Afterwards, the DNS dataset was used to assess the FPV model using the error decomposition and optimal estimator concept. The analyses showed that mixture fraction, progress variable, and enthalpy are the best set of input parameters for the flamelet table. Neglecting the effects of enthalpy and heat loss increases the total modelling errors. In addition, the definition of the input parameters, such as the progress variable, was found to impact the model performance significantly. The analysed reduced order models are utilised for large-scale burner simulations in Chap. 19.

In Sect. 12.5, the $NO_x$ formation during the ignition and combustion of biomass particles under laminar conditions was analysed. Solid miscanthus particles, as a biomass fuel, were employed in a drop tube configuration to study the $NO_x$ formation pathways. Coal and biomass showed similar type of direct nitrogen-containing radicals of N, NH, and HNO for NO formation. However, due to different volatile chemical compositions, completely different reaction pathways in coal and biomass were observed. Additionally, particle/particle interactions were investigated and $O_2$

deficiency was found to be the leading cause of different pathways between single-particle and particle group combustion. In the particle group combustion, the local maximum of $NO_x$ occurs around stoichiometric conditions, and the total maximum of $NO_x$ occurs at the flame tip in the particle stream centre around $Z = Z_{st}$ towards the rich region. In addition, the reaction pathways analysis showed that the main $NO_x$ formation contributors are the fuel-$NO_x$ pathways, and the largest share of $NO_x$ destruction is through the reburn mechanism. Finally, it was also observed that operating conditions such as oxidiser composition significantly changes the $NO_x$ formation pathways and oxy-fuel condition and lower $O_2$ content lead to lower $NO_x$ production.

**Acknowledgements** This work has been funded by the Deutsche Forschungsgemeinschaft (DFG, German Research Foundation)-—215035359-—SFB/TRR 129 'Oxyflame'.

## References

1. Attili, A., Bisetti, F., Mueller, M. E. and Pitsch, H. "Effects of Non-Unity Lewis Number of Gas-Phase Species in Turbulent Nonpremixed Sooting Flames". *Combustion and Flame* 166 (2016), 192–202. https://doi.org/10.1016/j.combustflame.2016.01.018
2. Farazi, S., Hinrichs, J., Davidovic, M., Falkenstein, T., Bode, M., Kang, S., Attili, A. and Pitsch, H. "Numerical Investigation of Coal Particle Stream Ignition in Oxy-Atmosphere". *Fuel* 241 (2019), 477–487. https://doi.org/10.1016/j.fuel.2018.11.108
3. Cai, L., Kruse, S., Felsmann, D. and Pitsch, H. "A Methane Mechanism for Oxy-Fuel Combustion: Extinction Experiments, Model Validation, and Kinetic Analysis". *Flow, Turbulence and Combustion* 106.2 (2021), 499–514. https://doi.org/10.1007/s10494-020-00138-w
4. Miller, R. S. and Bellan, J. "Direct Numerical Simulation of a Confined Three-Dimensional Gas Mixing Layer with One Evaporating Hydrocarbon-Droplet-Laden Stream". *Journal of Fluid Mechanics* 384 (1999), 293–338. https://doi.org/10.1017/S0022112098004042
5. Farazi, S., Attili, A., Kang, S. and Pitsch, H. "Numerical Study of Coal Particle Ignition in Air and Oxy-Atmosphere". *Proceedings of the Combustion Institute* 37.3 (2019), 2867–2874. https://doi.org/10.1016/j.proci.2018.07.002
6. Li, T., Farmand, P., Chen, H., Boehme, C., Nicolai, H., Hasse, C., Pitsch, H. and Böhm, B. "Homogeneous Ignition and Volatile Flame Structure of Single Bituminous Coal and Walnut Shell Particles: Effects of Particle Size and Gas Atmosphere". *Fuel* 371 (2024), 131955. https://doi.org/10.1016/j.fuel.2024.131955
7. Li, T., Farmand, P., Geschwindner, C., Greifenstein, M., Köser, J., Schumann, C., Attili, A., Pitsch, H., Dreizler, A. and Böhm, B. "Homogeneous Ignition and Volatile Combustion of Single Solid Fuel Particles in Air and Oxy-Fuel Conditions". *Fuel* 291 (2021). https://doi.org/10.1016/j.fuel.2020.120101
8. Li, T., Geschwindner, C., Köser, J., Schiemann, M., Dreizler, A. and Böhm, B. "Investigation of the Transition from Single to Group Coal Particle Combustion Using High-Speed Scanning OH-LIF and Diffuse Backlight-Illumination". *Proceedings of the Combustion Institute* 38 (2021), 4101–4109. https://doi.org/10.1016/j.proci.2020.06.314
9. Farmand, P., Nicolai, H., Schumann, C., Attili, A., Berger, L., Li, T., Geschwindner, C., di Mare, F., Hasse, C., Böhm, B., Janicka, J. and Pitsch, H. "Numerical Investigation and Assessment of Flamelet-Based Models for the Prediction of Pulverized Solid Fuel Homogeneous Ignition and Combustion". *Combustion and Flame* 235 (2022), 111693. https://doi.org/10.1016/j.combustflame.2021.111693
10. Attili, A., Farmand, P., Schumann, C., Farazi, S., Böhm, B., Li, T., Geschwindner, C., Köser, J., Dreizler, A. and Pitsch, H. "Numerical Simulations and Experiments of Ignition of Solid

Particles in a Laminar Burner: Effects of Slip Velocity and Particle Swelling". *Flow, Turbulence and Combustion* 106 (2020), 515–531. https://doi.org/10.1007/s10494-020-00222-1

11. Li, T., Schiemann, M., Köser, J., Dreizler, A. and Böhm, B. "Experimental Investigations of Single Particle and Particle Group Combustion in a Laminar Flow Reactor Using Simultaneous Volumetric OH-LIF Imaging and Diffuse Backlight-Illumination". *Renewable and Sustainable Energy Reviews* 136 (2021), 110377. https://doi.org/10.1016/j.rser.2020.110377
12. Palmore, J. A. and Desjardins, O. "Technique for Forcing High Reynolds Number Isotropic Turbulence in Physical Space". *Physical Review Fluids* 3.3 (2018). https://doi.org/10.1103/PhysRevFluids.3.034605
13. Farmand, P., Nicolai, H., Usman, M., Berger, L., Attili, A., Gauding, M., Hasse, C. and Pitsch, H. "Modeling homogeneous ignition processes of clustering solid particle clouds in isotropic turbulence". *Fuel*, Elsevier 371 (2024), 132054. https://doi.org/10.1016/j.fuel.2024.132054
14. Lau, T. C. W. and Nathan, G. J. "The Influence of Stokes Number on Particle Clustering within a Two-Phase Turbulent Jet". *7$^{th}$ Australian Conference on Laser Diagnostics in Fluid Mechanics and Combustion*. Melbourne, Australia, 2015.
15. Nicolai, H., Debiagi, P., Wen, X., Dressler, L., Maßmeyer, A., Janicka, J. and Hasse, C. "Flamelet LES of Swirl-Stabilized Oxy-Fuel Flames Using Directly Coupled Multi-Step Solid Fuel Kinetics". *Combustion and Flame* 241 (2022), 112062. https://doi.org/10.1016/j.combustflame.2022.112062
16. Rieth, M., Clements, A. G., Rabaçal, M., Proch, F., Stein, O. T. and Kempf, A. M. "Flamelet LES Modeling of Coal Combustion with Detailed Devolatilization by Directly Coupled CPD". *Proceedings of the Combustion Institute* 36.2 (2017), 2181–2189. https://doi.org/10.1016/j.proci.2016.06.077
17. Nicolai, H., Li, T., Geschwindner, C., di Mare, F., Hasse, C., Böhm, B. and Janicka, J. "Numerical Investigation of Pulverized Coal Particle Group Combustion Using Tabulated Chemistry". *Proceedings of the Combustion Institute* 38.3 (2021), 4033–4041. https://doi.org/10.1016/j.proci.2020.06.081
18. Monchaux, R., Bourgoin, M. and Cartellier, A. "Analyzing Preferential Concentration and Clustering of Inertial Particles in Turbulence". *International Journal of Multiphase Flow* 40 (2012), 1–18. https://doi.org/10.1016/j.ijmultiphaseflow.2011.12.001
19. Wen, X., Rieth, M., Han, W., Chen, J. H. and Hasse, C. "Investigation of the Ignition Processes of a Multi-Injection Flame in a Diesel Engine Environment Using the Flamelet Model". *Proceedings of the Combustion Institute* 38.4 (2021), 5605–5613. https://doi.org/10.1016/j.proci.2020.05.057
20. Krüger, J., Haugen, N. E., Mitra, D. and Løvås, T. "The Effect of Turbulent Clustering on Particle Reactivity". *Proceedings of the Combustion Institute* 36.2 (2017), 2333–2340. https://doi.org/10.1016/j.proci.2016.06.187
21. Berger, L., Kleinheinz, K., Attili, A., Bisetti, F., Pitsch, H. and Mueller, M. E. "Numerically Accurate Computational Techniques for Optimal Estimator Analyses of Multi-Parameter Models". *Combustion Theory and Modelling* 22.3 (2018), 480–504. https://doi.org/10.1080/13647830.2018.1424353
22. Farmand, P., Boehme, C., Steffens, P., Nicolai, H., Loffredo, F., Debiagi, P., Girhe, S., Chu, H., Gauding, M., and Hasse, C.,"Numerical investigation and modeling of NOx formation in pulverized biomass flames under air and oxyfuel conditions". *Combustion and Flame*. Elsevier 279 (2025) 114284. https://doi.org/10.1016/j.combustflame.2025.114284
23. Ren, X., Sun, R., Meng, X., Vorobiev, N., Schiemann, M. and Levendis, Y. A. "Carbon, Sulfur and Nitrogen Oxide Emissions from Combustion of Pulverized Raw and Torrefied Biomass". *Fuel* 188 (2017), 310–323. https://doi.org/10.1016/j.fuel.2016.10.017
24. Panahi, A., Sirumalla, S. K., West, R. H. and Levendis, Y. A. "Temperature and Oxygen Partial Pressure Dependencies of the Coal-Bound Nitrogen to $NO_x$ Conversion in $O_2/CO_2$ Environments". *Combustion and Flame* 206 (2019), 98–111. https://doi.org/10.1016/j.combustflame.2019.04.015
25. Shamooni, A., Debiagi, P., Wang, B., Luu, T. D., Stein, O. T., Kronenburg, A., Bagheri, G., Stagni, A., Frassoldati, A., Faravelli, T., Kempf, A. M., Wen, X. and Hasse, C. "Carrier-Phase

DNS of Detailed $NO_x$ Formation in Early-Stage Pulverized Coal Combustion with Fuel-Bound Nitrogen". *Fuel* 291 (2021). https://doi.org/10.1016/j.fuel.2020.119998
26. Androulakis, I. P., Grenda, J. M. and Bozzelli, J. W. "Time–Integrated Pointers for Enabling the Analysis of Detailed Reaction Mechanisms". *AIChE Journal* 50.11 (2004), 2956–2970. https://doi.org/10.1002/aic.10263
27. Shamooni, A., Stein, O. T., Kronenburg, A., Kempf, A. M., Debiagi, P., Li, T., Dreizler, A., Böhm, B. and Hasse, C. "Fully-Resolved Simulations of Volatile Combustion and $NO_x$ Formation from Single Coal Particles in Recycled Flue Gas Environments". *Proceedings of the Combustion Institute* 39.4 (2023), 4529–4539. https://doi.org/10.1016/j.proci.2022.07.034

# Experimental Flame Studies

# Laboratory Scale Pulverised Solid Fuel Combustion Experiments

# 13

Janik Hebel, Henrik Schneider, Lukas Becker, Benjamin Böhm and Andreas Dreizler

## Contents



**Abstract**

This chapter addresses the gap in the current understanding and validation of mathematical models for pulverised solid fuel flames, particularly under oxy-

J. Hebel · H. Schneider · L. Becker · B. Böhm · A. Dreizler (✉)
Institute of Reactive Flows and Diagnostics, Technische Universität Darmstadt, Darmstadt, Germany
e-mail: dreizler@rsm.tu-darmstadt.de

J. Hebel
e-mail: hebel@rsm.tu-darmstadt.de

H. Schneider
e-mail: hschneider@rsm.tu-darmstadt.de

L. Becker
e-mail: becker@rsm.tu-darmstadt.de

B. Böhm
e-mail: boehm@rsm.tu-darmstadt.de


R. Kneer et al. (eds.), *Clean Energy Conversion*,
https://doi.org/10.1007/978-3-031-98956-8_13

fuel conditions. Traditional invasive and optical measurement techniques have been insufficient for capturing the coupled processes of fluid mechanics, particle dynamics, and chemical reactions. Therefore, this study employs advanced, non-invasive optical methods in a combustion chamber with large optical access in gas-assisted solid fuel flames, enabling detailed measurements and well-defined boundary conditions. Experiments gradually increased complexity from single-phase, non-reacting flows to two-phase, reacting flows. Parametric studies examined the influence of the fuel (lignite, bituminous coal and biomass) and oxidiser composition on the flame. Inert particles were also used for fundamental studies of particle-flow interaction. Key findings indicate that flame stabilisation of biomass and coal in oxy-fuel combustion is dominated by the combustion of volatiles and that the transition of boilers from air to oxy-fuel operation requires an increase in the oxygen content in the oxidiser.

## 13.1 Introduction

Understanding the combustion of pulverised solid fuels requires knowledge of the fluid mechanics, particle dynamics, and chemical processes involved and how they are coupled. In the past, solid fuel flames were mainly investigated using invasive and, to a lesser extent, optical measurement methods, which are, however, a prerequisite for capturing the coupled processes in detail. Experiments using non-invasive optical measurement techniques can essentially be divided into three categories: Studies on single isolated coal particles in laminar flows [1], gas stabilised burners with higher coal particle loadings in turbulent flows [1] and combustors with stable solid fuel combustion ranging from small laboratory flames of 6 $kW_{th}$ to large scale of up to 600 $MW_{th}$ [2–5]. This work fills the gap in the validation hierarchy in the literature between open laboratory and pilot-scale solid fuel combustion chambers. As Li et al. [6] summarise in their review, detailed investigations of flame stabilisation, particle and gas temperatures, soot formation and the influence of turbulence on flame temperature and species concentrations are required for the validation and development of mathematical models, especially under oxy-fuel conditions.

Therefore, this study aims to contribute to a better understanding in general and to improved physico-chemical models in particular by providing validation data. This requires experimental data with a high level of detail and experiments with clearly defined and well-controlled boundary conditions. Although open laboratory flames allow experiments with the required level of detail on the one hand, they may suffer from secondary air entrainment and have less well-defined boundary conditions. On the other hand, in pilot-scale combustion chambers, experiments with the required level of detail are difficult due to limited optical access, extinction effects and contamination by particle deposits. The combustion chamber used in this study overcomes these limitations by providing large-scale optical access for detailed measurements and well-defined boundary conditions. These reproducible boundary conditions allow parameter studies to investigate the influence of $O_2$ concentration and solid particle variation. The particle variation included inert particles, Rhenish lignite, German bituminous coal (Prosper-Haniel), walnut shells and beech wood. The experiments were stepwise increased in complexity from single-phase non-reacting flows to two-phase reacting flows [7]. A variety of optical, non-invasive in-situ methods were used to characterise key parameters of the different operating conditions: Particle image and particle tracking velocimetry (PIV/PTV) for gas- and particle-phase velocities [7,8], chemiluminescence imaging and planar laser-induced fluorescence (PLIF) of OH for reaction zone characterisation [9,10], PLIF of polycyclic aromatic hydrocarbons (PAH) and laser-induced incandescence (LII) for characterisation of soot and its precursors [9], two-colour pyrometry for particle temperatures [11], coherent anti-Stokes Raman spectroscopy (CARS) for gas temperatures [12] and $O_2$ and $CO_2$ concentrations [13], tunable diode laser absorption spectroscopy (TDLAS) to determine residence times [14,15] and tomographic laser absorption spectroscopy for gas temperatures [12,16] and phosphor thermometry for wall temperatures [12]. Overall, measurement techniques for use in oxy-fuel combustion systems were further developed as part of the experimental investigations presented and applied for the first time in such complex environments.

In the following, selected results are presented and discussed. The burner and combustion chamber are presented first, followed by the operating conditions and optical measurement techniques used.

## 13.2 Methods

### 13.2.1 Solid Fuel Combustor and Operation Conditions

The original geometry of the burner was designed and optimised by Toporov et al. [17] for the study of combustion phenomena in the near-nozzle region. The design has been scaled and adapted for use with laser optical diagnostics. Due to the heat loss caused by the large optical accesses, it is not possible to operate self-sustaining solid fuel flames and gas assistance is required. Gas-assisted solid fuel flames can be operated up to 70 $kW_{th}$.

The geometry of the solid fuel combustor including a sketch of the characteristic flow field of the stabilised swirl flame is shown in Fig. 13.2. The test rig consists of the combustion chamber and the burner head including a diffuser, which is placed on top of the chamber. The down-fired combustor is enclosed by planar walls that form a rectangular chamber cross-section with chamfered corners and exhibits well-known boundary conditions [10,11]. The combustor walls and the diffuser are made of fused silica to provide access for laser optical diagnostics as needed to study combustion processes in the near-nozzle region.

Three different flows enter the combustor. The primary flow (I) and the secondary flow (II) are injected through separate annular orifices of the burner head, while the tertiary flow (III) enters the combustion chamber near the walls. The primary flow consists of partially premixed fuel (methane, solid fuel, oxidiser), the secondary and tertiary flows consist of oxidiser only. The swirl is controlled by dividing the secondary mass flow between four inclined tangential channels arranged at an angle

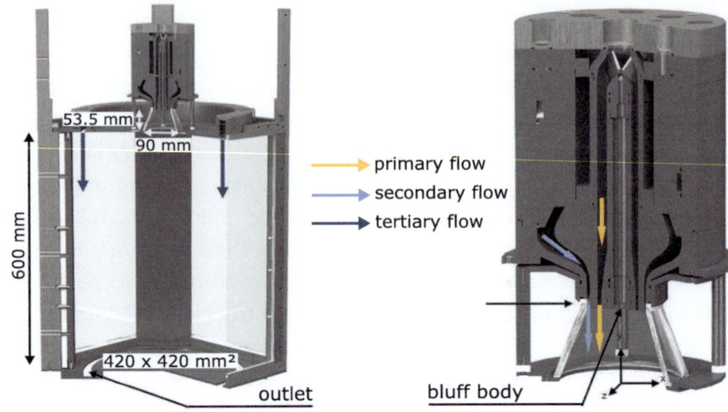


**Fig. 13.1** Optically accessible combustion chamber (left) and schematic of burner assembly consisting of nozzle and diffuser (right). (Reprinted from Schneider et al. [11], Copyright (2021), with permission from Elsevier and minor editing)

of 45% and four straight channels. In addition, a bluff body is located in the centre of the burner head (Fig. 13.1).

The swirl flame stabilisation mechanism is shown on the right side of Fig. 13.2. Hot flue gas is recirculated to the burner head through the internal recirculation zone (IRZ), which heats the fresh gas and solid fuel particles. Inside the chamber, fresh oxidiser from the tertiary flow is transported to the flame by an outer recirculation zone (ORZ).

In a sequence of increasing complexity, the experimental investigations include single-phase gas flows without and with chemical reactions as well as two-phase gas-particle flows without and with chemical reactions. This chapter presents exemplary studies of the chemically non-reacting two-phase flows [7] and gas-assisted combustion of pulverised solid fuels under chemically reacting oxy-fuel conditions [9]. The individual operating conditions are presented separately for these two studies.

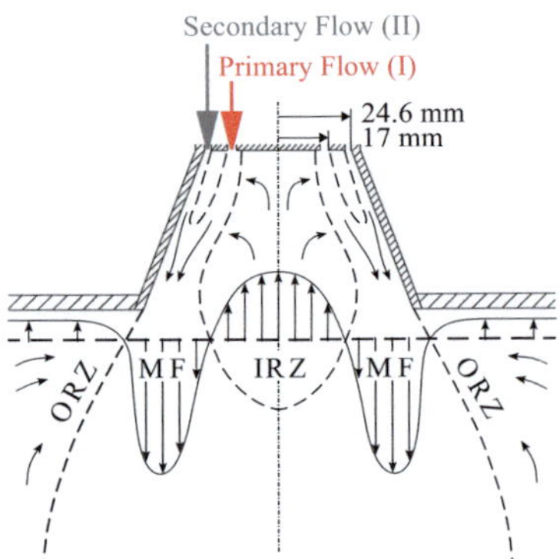


**Fig. 13.2** Schematic of the stabilisation of the swirl flame. (Reprinted from Schneider et al. [13], Copyright (2024), with permission from Elsevier and minor editing)

**Table 13.1** Volume flows of operating conditions NR30 (30 vol.% $O_2$ and 70 vol.% $CO_2$) and NRAir (air) under normal conditions.

| Operating condition | | NRAir | NR30 |
|---|---|---|---|
| Air/$O_2$/$CO_2$ | vol.% | 100/0/0 | 0/30/70 |
| I oxidiser | $m^3\ h^{-1}$ | 13.55 | 8.16 |
| II straight oxidiser | $m^3\ h^{-1}$ | 5.97 | 3.76 |
| II inclined oxidiser | $m^3\ h^{-1}$ | 12.02 | 7.27 |
| III oxidiser | $m^3\ h^{-1}$ | 69.95 | 42.6 |

**Non-Reacting Two-Phase Flows**

For a better understanding of pulverised solid fuel flames, it is necessary to first investigate the particle dynamics without chemical reactions. Due to the particle sizes in the range of 50–200 μm and the associated Stokes numbers greater than one, there is a slip between the particle phase and the surrounding gas phase. This slip significantly influences the momentum, heat, and mass transfer. The objective of this study was therefore to first develop a suitable measurement method for the simultaneous measurement of the velocities of the gas and particle phases, and then to apply this method to selected operating conditions.

Two non-reacting cases were selected that comprise flow conditions relevant for oxy-fuel combustion (NR30) and conventional air combustion (NRAir). Volume flow rates for both cases are listed in Table 13.1. For both cases the swirl number at the nozzle exit for the combined primary and secondary flows was 0.47. The inert particles substituting ground coal were spherical and consisted of a compound of 75% $Al_2O_3$ and 25% $SiO_2$ with a density of $\rho_p = 3.6$ g cm$^{-3}$. The particle size distribution exhibited an arithmetic mean diameter of $d_p = 55$ μm. Compared to typical solid fuel such as coal, the selected particles had a three times larger density such that their aerodynamic behaviour according to the response time [18] corresponds to coal particles with a mean diameter of approximately $d_p = 95$ μm. For each atmosphere, three different particle loadings were investigated: A high particle load of $\dot{m}_p = 2.0\,\mathrm{kg\,h^{-1}}$, a medium load of $\dot{m}_p = 0.5\,\mathrm{kg\,h^{-1}}$, and a low load of below $\dot{m}_p = 0.1\,\mathrm{kg\,h^{-1}}$. The relative uncertainty was 3.6%, 14%, 72%, respectively, due to the resolution of the balance used to measure the mass flow rate. For the low particle load the exact value could therefore not be determined accurately such that it is labelled "low" in this work.

**Chemically Reacting Two-Phase Flow**

For the chemically reacting two-phase flow, four oxy-fuel operating conditions and one comparable air operating condition were selected. The oxygen concentration was varied from 27 to 36 vol.% to study its effect on combustion and to obtain adiabatic flame temperatures as observed under air atmosphere [10]. Table 13.2 summarises all operating conditions. It includes single-phase (methane combustion) and two-phase (gas-assisted solid fuel combustion) operating conditions.

In order to keep the stoichiometric ratio constant at the same volume flow rates, the thermal power is increased at higher oxygen concentrations. Under two-phase

**Table 13.2** Volume flow rates and particle loads for single-phase and the corresponding two-phase operating conditions for AIR, OXY27P, OXY30P, OXY33P, and OXY36P. I: primary flow, II: secondary flow, III: tertiary flow. The solid fuel is indicated by the suffix "-WNS" for walnut shell (e.g. OXY33P-WNS). Two-phase combustion values are listed in parentheses. The inlet temperatures were set at 20 °C. The local stoichiometric ratio is calculated from the oxygen supply through the burner head (primary and secondary flow). Swirl numbers are estimated from Becker et al. [10]. The adiabatic flame temperatures ($T_{ad}$) are calculated using Cantera [19] for single-phase operating conditions, taking into account the primary and secondary flows

| Operating condition | | OXY27P | OXY30P | OXY33P | OXY36P | AIR |
|---|---|---|---|---|---|---|
| Oxidiser: air/$O_2$/$CO_2$ | vol.% | 0/27/73 | 0/30/70 | 0/33/67 | 0/36/64 | 100/0/0 |
| I methane | $m^3\ h^{-1}$ | 2.59 | 2.87 | 3.17 | 3.45 | 2.01 |
| I particle mass flow | | | | | | |
| ...-WNS | $kg\ h^{-1}$ | (4.93) | (5.47) | (6.02) | (6.56) | (3.82) |
| I oxidiser | $m^3\ h^{-1}$ | 12.10 | 12.10 | 12.10 | 12.10 | 12.10 |
| II oxidiser | $m^3\ h^{-1}$ | 11.09 | 11.09 | 11.09 | 11.09 | 11.09 |
| III oxidiser | $m^3\ h^{-1}$ | 24.02 | 24.02 | 24.02 | 24.02 | 24.02 |
| Thermal power | $kW_{th}$ | 25.8 (51.6) | 28.65 (57.3) | 31.5 (63) | 34.35 (68.7) | 20 (40) |
| Primary stoich. ratio | – | 0.63 (0.32) | 0.63 (0.32) | 0.63 (0.32) | 0.63 (0.32) | 0.63 (0.32) |
| Local stoich. ratio | – | 1.21 (0.61) | 1.21 (0.61) | 1.21 (0.61) | 1.21 (0.61) | 1.21 (0.61) |
| Global stoich. ratio | – | 2.47 (1.25) | 2.47 (1.25) | 2.47 (1.25) | 2.47 (1.25) | 2.47 (1.25) |
| Swirl number | – | 0.44 | 0.44 | 0.43 | 0.42 | 0.46 |
| $T_{ad}$ | K | 1905 | 2028 | 2138 | 2228 | 2053 |

conditions, methane and solid fuel combustion contribute equally to the total energy output. Walnut shells (WNS) were selected as a representative solid fuel for biomass combustion. Properties of the solid fuels, including the proximate and ultimate analysis are listed in Table 1.1, while Fig. 1.2 depicts the particle size distribution. This analysis yields $d_{10} = 101.5\ \mu m$, $d_{50} = 140.6\ \mu m$, and $d_{90} = 178.8\ \mu m$.

### 13.2.2 Measurement Techniques

The fluid dynamic, particle dynamic, and chemical processes have been studied using optical measurement techniques, which are reviewed below.

**Combined Particle Image and Particle Tracking Velocimetry**

Simultaneous quantification of gas flow velocities and particle trajectories was achieved using a two-phase PIV/PTV methodology introduced by Becker et al. [7]

and Schneider et al. [9]. This allowed to determine instantaneous slip velocities. The implementation of the two-phase PIV/PTV technique required the acquisition of double-frame images of Mie scattering of both small gas tracer particles and the larger solid fuel particles, which needs a large dynamic range of the camera system. The gas flows were seeded with $Al_2O_2$ particles (*MARTOXID MR-70*, $d_{50} = 0.5$–$0.8\ \mu$m). A separation algorithm was developed and used to distinguish the small gas tracer particles from the larger fuel particles. This algorithm separates the particles based on their size and Mie scattering intensity. A conventional PIV algorithm was then applied to the dual-frame images where only small gas tracer particles were present to obtain gas flow velocities. Simultaneously, the particle trajectories were calculated with a PTV algorithm using dual-frame images with the large particles only.

The particles were illuminated at 10 Hz with a double-pulsed, frequency-doubled Nd:YAG laser formed into a thin light sheet. Dependent on the specific setup used in the studies by Becker et al. [7] and Schneider et al. [9], the light sheet thickness (full width half minimum, FWHM) was between 300 and 700 $\mu$m. The laser pulses were separated in a range of $\Delta t = 25$–$35\ \mu$s, depending on the local flow velocities. The scattering images were recorded with a high-resolution *sCMOS* camera (*LaVision*) coupled either to a 180 mm *Sigma* macro lens or a 105 mm *Sigma* macro lens. The luminosity of the flame was suppressed with a set of filters. The images were pre-processed with a subtract sliding minimum filter prior to phase separation. This pre-processing step effectively minimised the influence of scattered light from the glass diffuser and residual flame luminosity. This was followed by an iterative application of the filter introduced by Khalitov and Longmire [20] to reduce high frequency noise. Large particles were then identified and separated by applying a threshold for pixel number and scattering intensity. The optimal size and intensity thresholds were determined from a sensitivity analysis that examined the impact of misclassified large particles. The sensitivity analysis focused on the quantification of the mean velocity magnitude differences between separated and raw image pairs, with the hypothesis that once all large particles have been removed, additional particle detection would not significantly change the mean velocity magnitude differences. Errors in this procedure arose primarily from the non-uniform illumination of the particles and are discussed in the study by Becker et al. [7].

PIV and PTV processing were performed using *DaVis* software (*LaVision*). For PIV an adaptive weighting multi-pass vector computation was used. Vectors with low peak ratios were eliminated and spurious vectors were removed. Depending on the measurement location, PIV processing resulted in vector spacings ranging from 0.18 to 0.36 mm. A hybrid algorithm was used for PTV. First, an initial velocity estimate was obtained using PIV processing as described above for the small tracer particles. The large particles were then tracked separately using a PTV algorithm. To reduce the likelihood of multiple particles appearing in a single PTV correlation window, the window size was chosen to be slightly larger than the largest particles detected. Vectors with a ratio of first to second correlation peak less than a user-defined threshold were discarded to ensure that correlation windows contained only one particle.

**Flame Imaging by CH* Chemiluminescence**
Chemiluminescence imaging of the CH* radical was used to visualise heat release areas near the nozzle. An *sCMOS* camera (*LaVision*) with a 180 mm or 105 mm *Sigma* macro lens and a 434 nm bandpass filter (*Semrock, 434/17 nm BrightLine*) was used. The exposure time for all measurements was 80 ms.

**Temperature Measurements by Tomographic Laser Absorption Spectroscopy**
The gas phase temperature field in the combustion chamber was measured by tomographic tunable diode laser absorption spectroscopy (TDLAS) [9]. Absorption spectra were obtained in the wavenumber ranges 7178.0–7183.4 $cm^{-1}$ and 4997.7–5001.8 $cm^{-1}$, providing sufficient information for temperature reconstruction in both $H_2O$-rich and $CO_2$-rich regions over the entire expected temperature range. The laser light from two diode lasers was guided via optical fibres to the measurement system. This consists of eight scanning probes with rotating mirrors that were set up around the combustion chamber. The laser light was reflected from the opposite wall of the combustion chamber by means of retroreflectors via an off-centre parabolic mirror onto photodiodes. To enable measurements at different heights, the combustion chamber was traversed vertically in steps of 5 mm. Details on the setup and the tomographic reconstruction can be found in Chap. 16.

**Combined Planar Laser Induced Fluorescence of PAH and Laser Incandescence Measurements**
To study soot formation, LII and PLIF of intermediate hydrocarbons were combined simultaneously. Excitation at 266 nm was chosen for LIF because it is expected to produce a fluorescence signal that is predominantly due to the PAH. The fluorescence signal was recorded using a *CMOS* camera (*LaVision*) with an image intensifier (*LaVision, Low-Speed IRO*) and a 100 mm UV objective (*Cerco*). A filter configuration with two long pass filters (*Asahi, ZUL0275*, cut-off wavelength: 275 nm) to attenuate laser reflections, one short pass filter (*Asahi, ZUV0325*, cut-off wavelength: 325 nm) and an additional band pass filter (*Thorlabs, FGUV11S*, cut-off wavelength: 275–375 nm) to suppress near infrared emissions were used to limit the spectral data acquisition to the range from 275 to 325 nm.

LII was excited at 1064 nm to avoid possible interference with the fluorescence emission of PAH. A laser fluence of $\sim$0.2 $Jcm^{-2}$ was used to increase the LII signal strength while reducing the sublimation of particles [21]. An *sCMOS* camera (*LaVision, Imager sCMOS*) with an image intensifier (*LaVision, Low-Speed IRO X*) and a 180 mm or 105 mm *Sigma* macro lens was used for detection. The image intensifier was controlled temporally for a duration of 50 ns, starting with the laser pulse. A bandpass filter with a centre wavelength of 600 nm (*Edmund Optics, 86-668* with 25 nm FWHM) was used to effectively eliminate potential interference from emission in the $C_2$ Swan band or $C_2$ Swing band [21]. In addition, an infrared cut filter (*Edmund Optics, 54-516*) was placed in front of the lens to reduce the risk of damage to the experimental setup. The LIF and LII light layers were superimposed before entering the combustion chamber and separated in time by 1 $\mu$s to ensure signal separation.

**Temperature Measurements by Coherent Anti-Stokes Raman Spectroscopy**
For gas phase temperature measurements, a dual-band CARS approach was adopted from Roy et al. [22], where $O_2$ and $CO_2$ were measured simultaneously. The laser wavelengths and pulse energies for the pump, Stokes, and probe lasers were 561.5, 612.3, and 532 nm, and 25, 23.5, and 30 mJ, respectively. A broadband Stokes laser was used to simultaneously record the Q-branch transitions of $O_2$ (1556 $cm^{-1}$) and the 2v2 system of $CO_2$ (1388 $cm^{-1}$). For this purpose, a Q-switched, frequency-doubled Nd:YAG laser with two Nd:YAG cavities was modified for individual operation of both cavities. One cavity was used to pump a narrowband dye laser (rhodamine 590 in ethanol) to produce radiation at 561.5 nm. The radiation from the other laser cavity was split into two paths, one serving as a probe beam at 532 nm and one to optically pump a custom-built broadband dye laser (rhodamine 640 in ethanol) to produce the Stokes laser beam at 612.3 nm with a full width at half maximum (FWHM) of approximately 5 nm. In all cases, pulse energy and divergence were controlled for optimal overlap in the probe volume. The pump, Stokes, and probe beams were synchronised by a delay line and focused into the probe volume by a spherical lens with a focal length of $f = 300$ mm in a planar BOXCARS arrangement. The BOXCARS phase matching geometry allowed measurements close to the combustor dump plane. The probe volume was approximately 0.7 mm long. The generated CARS signal was collimated with a spherical lens of $f = 300$ mm. The pump and CARS beams were separated by a set of mirrors, including a notch filter to remove scattered radiation from 532 nm. The polarisation of the CARS signal was optimised by rotating an achromatic half-wave plate. The CARS signal was recorded with a spectrometer (*Jobin Yvon S.A.S., THR 1000 Monochromator*, focal length 1 m, 2400 lines/mm grating) equipped with a high dispersion grating (2400 lines/mm) and a low-noise back-illuminated CCD camera. Using hardware binning in the vertical direction, the camera frame rate was 20 Hz, twice the laser repetition rate, to record background spectra between the CARS spectra. Non-resonant spectra for normalisation were recorded in pure argon.

The `carsfwsc` algorithm from Cutler et al. [23] was used to fit the CARS spectra. The bounds for the fit variables were set wide to avoid any influence of the bounds on the precision of the measurement. $O_2$ and $CO_2$ Raman transitions with Raman shifts between 1372 and 1572 $cm^{-1}$ were fitted. The spectral range 1444.5–1450.5 $cm^{-1}$ had to be blocked from fitting as the Raman active rotational line S(5) of $H_2$ at 1447.28 $cm^{-1}$ [24] interfered with the measured spectra. The third-order non-resonant susceptibility $\chi_{NR}$ was set to $19.74 \times 10^{-18}$ $cm^3$ $erg^{-1}$ $amg^{-1}$ [25]. This is because $H_2O$ is assumed to make up the majority of the buffer gas.

## 13.3 Results and Discussion

Only a few selected results from a larger set of studies are presented in this chapter to illustrate the sequence of data from isothermal flow to pollutant emission. First, the velocity fields of the gas and particle phases are presented for a chemically non-reacting two-phase flow [7]. The resulting instantaneous slip velocities are discussed.

This is followed by investigations of chemically reacting single-phase and selected two-phase oxy-fuel combustion processes, with emphasis on flow fields, PAH, and soot visualisations [9]. Finally, CARS measurements for two operating conditions are presented, with the discussion focusing on temperature, although the method also quantifies mole fractions of $O_2$ and $CO_2$ [13].

### 13.3.1 Flow Field Measurements in Chemically Non-Reacting Two-Phase Flows

The discussion of selected two-phase flow characteristics is limited here to the NR30 operating condition with a particle load of $\dot{m}_p = 0.5$ kg h$^{-1}$. The gas phase flow field is shown in the left part of Fig. 13.3. In general, it has similar characteristics to the single-phase flow discussed in the study by Becker et al. [10]. A large recirculation zone with low velocity magnitudes forms in the centre of the diffuser. Downstream of the dump plane, the recirculation zone first widens but then narrows due to vortex breakdown. The recirculation zone is surrounded by primary and secondary flows, which merge into a main flow about 10–15 mm downstream of the nozzle exit. The interaction with the central recirculation zone pushes the main flow towards the walls of the diffuser, but without significant dissipation of its initial momentum. Decreasing velocity magnitudes of the main flow are observed only as the vortex breakdown region is approached.

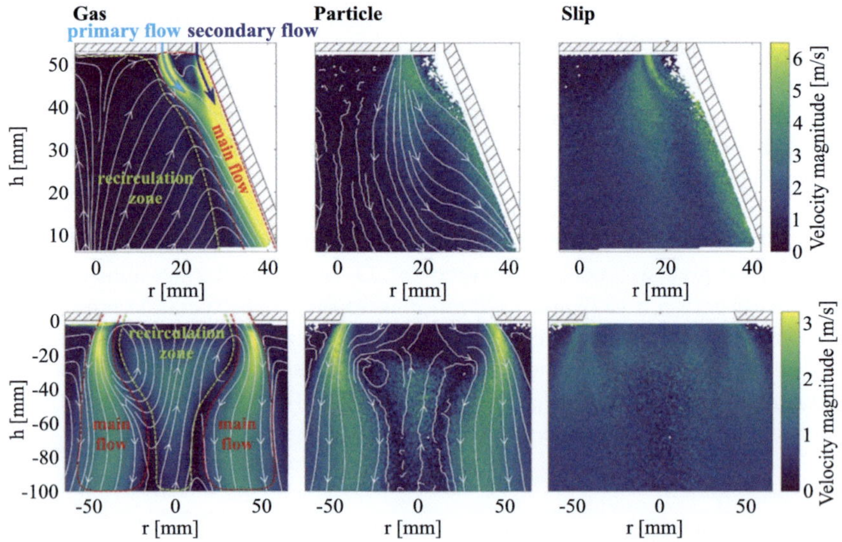


**Fig. 13.3** Gas, particle and slip velocity magnitude of the mean velocity field. Two-phase flow of NR30 atmosphere with particle load $\dot{m}_p = 0.5$ kg h$^{-1}$. Top row: within the diffuser. Bottom row: downstream of the dump plane. Note the different velocity range for the top and bottom rows. (Reprinted from Becker et al. [7], Copyright (2019), with permission from Elsevier and minor editing)

This is evident from the probability density function (PDF) of the particle phase velocity, which shows bimodal behaviour (detailed in the study by Becker et al. [7]). The bimodality indicates that the particles have different residence times in the diffuser. Due to their high momentum at the nozzle exit and relatively high mass inertia, they do not follow the streamlines of the gas phase but enter the central recirculation zone. The other part of the observed particles has already lost the high momentum and is carried back to the burner at a low speed by the central recirculation zone.

This bimodality is further investigated by conditioned mean velocity fields of the particle phase. In the middle plot of Fig. 13.4 particles moving downwards (axial velocity component $< 0$) are averaged, while in the right plot only particle velocities of particles moving upwards are averaged. Obviously, magnitude and direction vary significantly.

Finally, the instantaneous slip velocities are determined by the difference of the simultaneously measured gas and particle velocities. The magnitudes of the averaged slip velocities are shown on the right side of Fig. 13.3.

Within the diffuser, three characteristic regions are observed that have significant slip for different reasons. Directly at the nozzle exit of the primary flow carrying the particles, high slip is observed because a contraction upstream of the nozzle exit

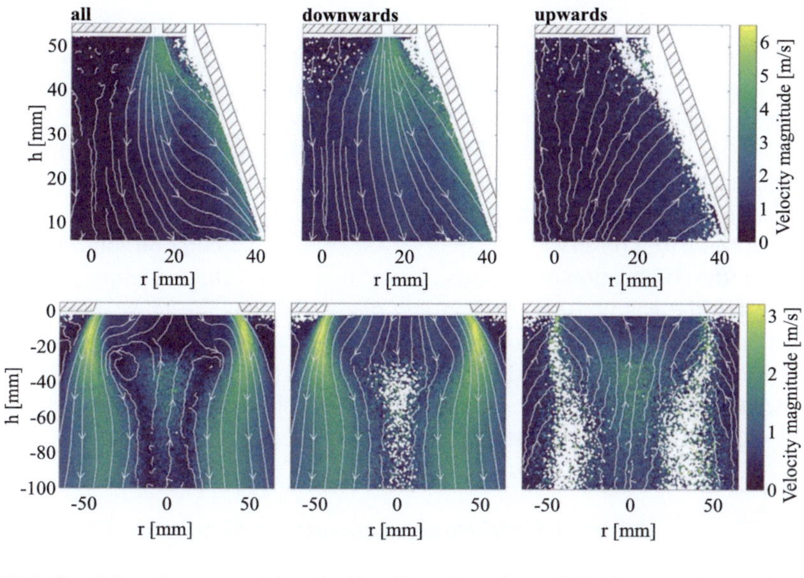


**Fig. 13.4** Conditioned mean particle velocities. Two-phase flow of NR30 atmosphere with particle loading $\dot{m}_\mathrm{p} = 0.5$ kg h$^{-1}$. Top row: inner diffuser. Bottom row: downstream of the dump plane. Velocity magnitude in columns from left to right: all particles, particles moving downstream, particles moving upstream. Note the different velocity range for the top and bottom rows. The white areas indicate areas in which too few velocity vectors could be obtained from the measurement. These areas were discarded. (Reprinted from Becker et al. [7], Copyright (2019), with permission from Elsevier and minor editing)

accelerates the gas flow by a factor of 2.3. The inertia of the particles slows their rapid acceleration, causing the slip near the nozzle.

The second region of high slip is associated with high-momentum particles emanating directly from the nozzle, moving downwards in the opposite direction of the recirculation zone, and penetrating it. The time required to equilibrate the particle velocities to those of the surrounding gas phase in the recirculation zone depends on the turbulence and particle size. Depending on the instantaneous gas velocity field and particle inertia, particles will either be stopped within the recirculation zone or cross it. Particles that cross the recirculation zone move outward and join the main flow between $y = 0 - 40$ mm. Particles stopped in the recirculation zone are carried upward and outward and join the main flow further upstream inside the diffuser. Inspection of the particle and gas velocity fields in Fig. 13.3 shows that most particles equilibrate to the local gas phase velocity at about $y = 25$ mm.

The third region within the diffuser that shows significant slip is located on the wall in the main flow after the primary flow merges with the secondary flow downstream $y = 40$ mm. The strong shear in the main flow, associated with slow moving particles entering from the recirculation zone, is responsible for significant slip in this region.

Using a maximum particle diameter of 87 $\mu$m and a maximum slip velocity between the gas and particle phases, a particle Reynolds number of approximately 75 was obtained, meaning that this two-phase flow is in a regime where turbulence production is not expected to occur. Since particle Reynolds numbers of up to 250 are expected for pulverised solid fuel combustion systems the chemically non-reactive two-phase flow investigated here is in a range relevant to oxy-fuel combustion.

Stokes numbers are estimated to vary between 0.2 and 460 depending on the position either in the diffuser or downstream in the dump plane. Thus, while some of the particles follow the flow, the other part of the particles does not. This estimate is consistent with the behaviour discussed in Fig. 13.3.

The slip velocities observed, especially in the central recirculation zone, are advantageous for the combustion of pulverised solid fuels. High slip velocities favour heat transfer from the gas phase to the particle phase and thus the release of volatiles. In addition, the particle trajectories show that there are sometimes long residence times of the solid particles in the hot recirculation zone, up to 80 ms, which also promotes the release of volatiles. The release of volatiles in the immediate vicinity of the burner stabilises the flame in the diffuser.

### 13.3.2 Flow and Scalar Field Measurements in Single and Two-Phase Oxy-Fuel Combustion

This section presents investigations of the single-phase methane flame and the gas-assisted pulverised solid fuel flame under oxy-fuel conditions. Using two-dimensional imaging techniques such as chemiluminescence imaging, PIV/PTV, PAH-PLIF and LII, the specific flow and reaction characteristics and the dependence of the flow and reaction fields on the oxygen content are presented. The temperature fields shown below have been recorded with the tomographic TDLAS approach. For the variation of the particle type, which was also carried out but not discussed here, reference is made to the study by Schneider et al. [9].

**Single-Phase Measurements**

The investigations of single-phase methane flames under oxy-fuel conditions aim to explore the fundamental properties of the turbulent swirl flame. The same oxy-fuel conditions as in the gas-assisted pulverised solid fuel combustion experiments were set with varying oxygen content between 27 and 36 vol.% (Table 13.2). In addition, an air operating condition was investigated for comparison.

Figure 13.5 shows time-averaged images of the spectrally broadband chemiluminescence of single-phase methane flames. In all cases, the flame stabilises within the diffuser. Apart from the OXY27P operating condition, side flames form in the immediate vicinity of the burner head between the primary and secondary streams. The length of these side flames increases as the oxygen content increases. The visible global flame lengths vary widely between a height of $h = -80$ mm for the OXY27P case and approximately $h = -300$ mm for the OXY36P case.

In the next step, the chemiluminescence of the CH* radicals, the velocity fields and the temperature fields are examined. The summary plot in Fig. 13.6 shows the time-averaged results.

The CH* chemiluminescence images provide information about the location of the heat release. The differences between the zones of high local heat release appear clearer than in the spectral broadband images. As the oxygen concentration increases, the heat release becomes more localised and shifts upstream closer to the burner head. The side flames favour early heat release. The intensification of the heat release associated with the higher oxygen content is related to the associated increase in thermal output: Compared to the OXY27P case, the thermal output in the OXY36P case increased by 33% and compared to the AIR case by 72%. This increase in thermal power was necessary to maintain a constant equivalence ratio.

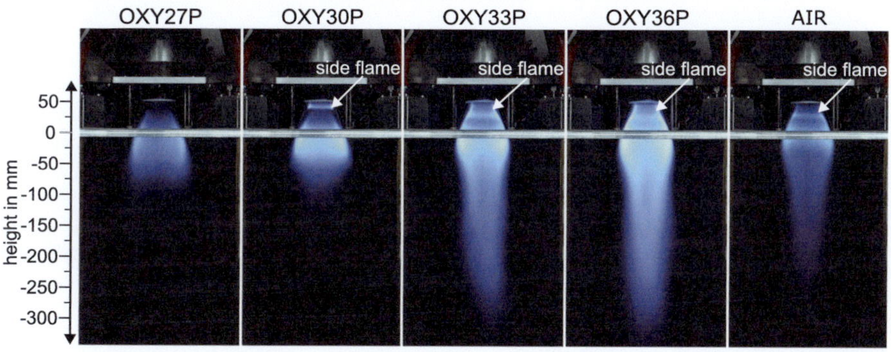


**Fig. 13.5** Time-averaged images of the single-phase methane flames with increasing oxygen content according to Table 13.2. The operating condition of air is shown on the far right for comparison. The camera settings (exposure time, aperture, ISO) were identical for all flames. Side flames are indicated by arrows. (Reprinted from Schneider et al. [9], Copyright (2024), with permission from Elsevier and minor editing)

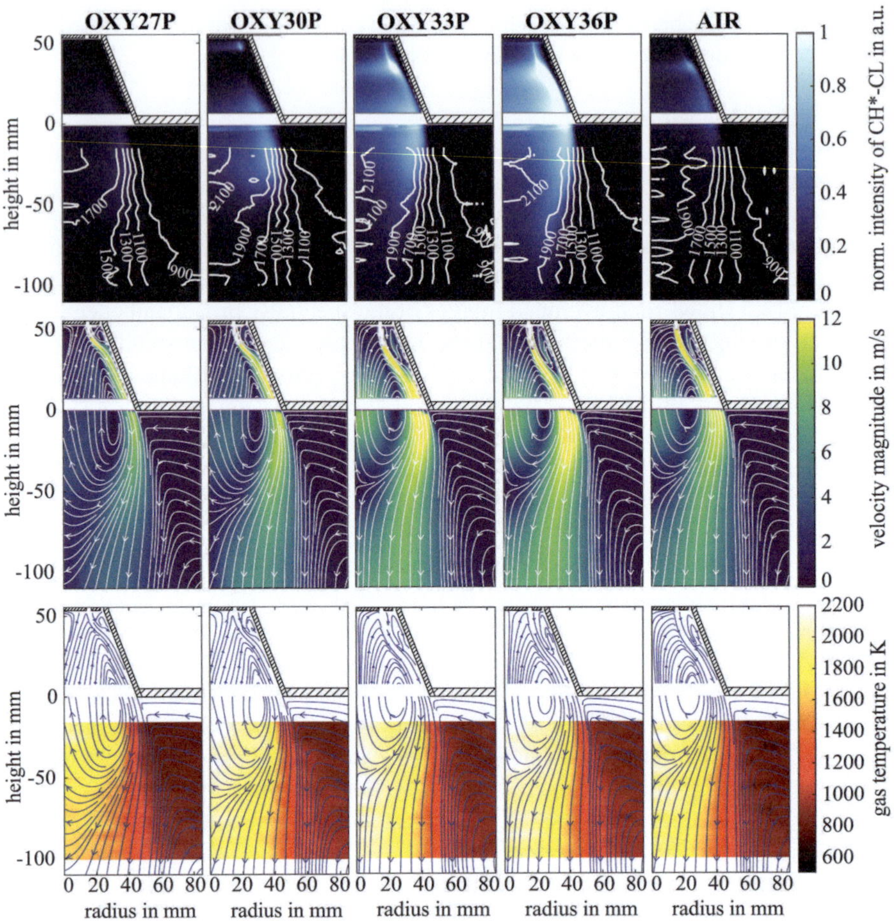


**Fig. 13.6** Comparison of different single-phase methane flames with different oxidant compositions. Top: CH* chemiluminescence with isothermal lines obtained from TDLAS measurements (in Kelvin). The same camera settings were used to capture the images. The intensity values are normalised to the same value. Middle: two components of the mean velocity field and streamlines. Bottom: mean gas temperatures with mean streamline distribution. (Reprinted from Schneider et al. [9], Copyright (2024), with permission from Elsevier and minor editing)

As in the chemically non-reacting case in Sect. 13.3.1, the averaged flow fields show an intense inner recirculation zone. The axial extent of the inner recirculation zone decreases with increasing oxygen content, while the velocity magnitudes increase. This is especially true for the flow velocities in the main flow zone, as defined in Fig. 13.3, top left. These higher velocity magnitudes are due to more intense heat release and thus more intense thermal expansion. This is consistent with the higher mean temperatures observed at higher oxygen concentrations.

The AIR operating condition shows similar flow characteristics to the OXY33P case, although the velocity magnitudes in the main flow zone are slightly lower. In contrast, the temperature fields of the AIR case are closer to the OXY30P case. This

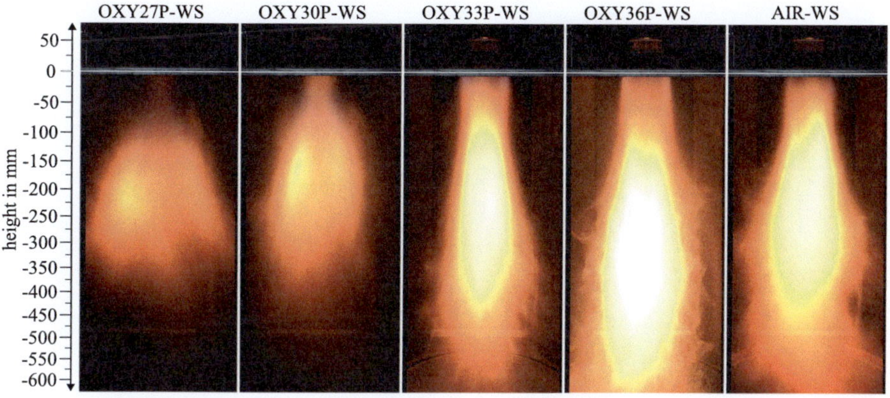


**Fig. 13.7** Time-averaged images of the gas-assisted flames operated with walnut shells (WNS). The camera settings (exposure time, aperture, ISO) were identical for all flames. The exposure time was set to 0.5 s. For the oxy-fuel cases, the oxygen content and thermal output increase from left to right. The operating condition with air is shown on the far right for comparison. In contrast to the single-phase methane flames, the diffuser is not optically accessible because particles are deposited on the walls. (Reprinted from Schneider et al. [9], Copyright (2024), with permission from Elsevier and minor editing)

shows that it is possible to maintain similar flow and temperature fields, which is a prerequisite for converting boilers from air to oxy-fuel operation, by increasing the oxygen content in the oxidant to compensate for the physically different properties of $CO_2$ compared to $N_2$, such as the higher heat capacity.

**Two-Phase Measurements**

To investigate chemically reacting two-phase flows, biomass was added to the methane flames presented in the previous section. The biomass was added to the primary flow, resulting in a gas-assisted pulverised solid fuel flame. The thermal power of the gas and particle phases were the same. In the following, the investigations with walnut shells (WNS) as a representative biomass are discussed.

Figure 13.7 shows the time averaged broadband emission spectra of the oxy-fuel and air operating conditions. Note that no optically accessible diffuser could be used as it became opaque within a short time due to particle deposition. The influence of the particle phase on the combustion is considerable, in particular the flames OXY27P-WNS and OXY30P-WNS are displaced downstream and separated from the diffuser wall. This can be explained by the significantly lower temperatures in the single-phase case, which results in a low heating rate of the particles in the two-phase case and thus a delayed onset of particle combustion. Despite the flame lift-off, temporally stable reaction zones are observed for these two operating conditions. This indicates that the inner recirculation zone is maintained and extends axially beyond the flame stabilisation point.

The OXY27P-WNS flame is radially wider than for the other operating conditions. These flames with higher oxygen concentrations have a stabilisation point upstream in the diffuser and have radially less extended high temperature regions. The flame

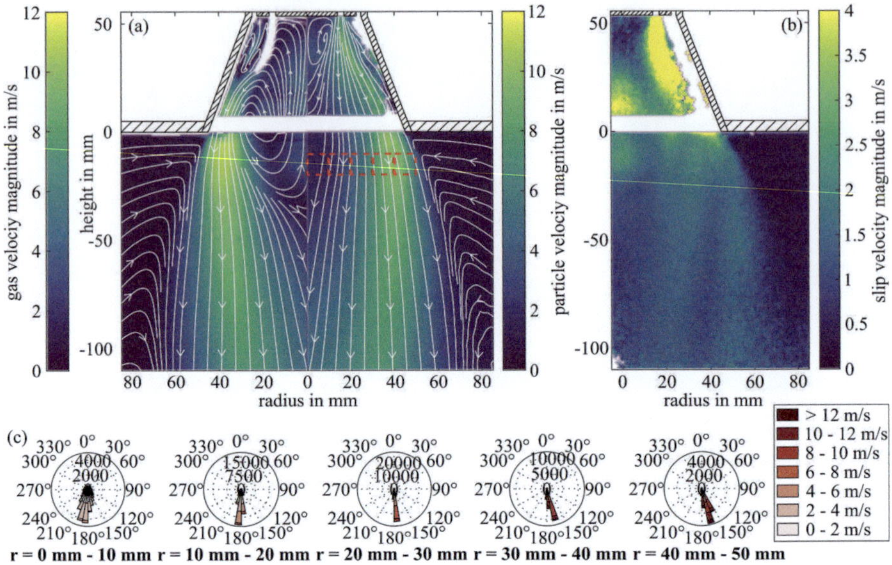

**Fig. 13.8** Mean gas, particle, and slip velocity values of the OXY33P-WNS operating condition. Only the regions with more than 30 vectors are shown in Fig. 13.8. **a**) Gas velocity magnitude (left) and particle velocity magnitude (right). To show them on the same plot, the gas velocity plot is mirror-transformed from right to left side. **b**) Slip velocity magnitude and **c**) polar histograms of particle velocities at different radial positions averaged at a height between $h = -10$ mm and $-20$ mm. The colour indicates the velocity of the particles. The orientation within the plot shows the direction of movement of the particles. The number of particles is marked on the radial axis of the plot, binned by velocity and direction of movement. Areas used for the histograms are marked as red dashed squares in Fig. 13.8a). (Reprinted from Schneider et al. [9], Copyright (2024), with permission from Elsevier and minor editing)

length correlates positively with the oxygen content. This is due to the earlier thermal expansion upstream and the associated acceleration of the gas phase, as well as the higher thermal power associated with maintaining a constant equivalence ratio.

As an example of the other cases, the velocity field for the OXY33P-WNS operating condition is analysed. Figure 13.8 shows the mean gas velocity, the mean particle velocity (Fig. 13.8a)) and the slip velocity (Fig. 13.8b)). The mean particle velocities and velocity directions averaged over selected radial positions are shown as polar histograms in Fig. 13.8 (Fig. 13.8c)). The differences in streamlines and velocity magnitudes between the gas and particle phases are considerable and particularly high inside the diffuser. While the gas phase has an intense inner recirculation zone, this is not observed for the particle phase. The particles penetrate the inner recirculation zone of the gas phase, resulting in high slip velocities between the two phases, especially in the diffuser. The slip velocities are very high, particularly within the recirculation zone at the transition to the dump plane and in the main flow zone. Although the particles are slowed down in the recirculation zone, there is no flow reversal on average. In the area of the main flow zone, the streamlines of the gas phase have relatively high curvatures, which the particles cannot follow.

The polar histograms in Fig. 13.8c) give a more detailed insight into the flow dynamics of the particle phase. In the range $r = 10 - 20$ mm, the particles are directed downwards and slightly towards the centre. Almost no particles are transported towards the burner. As the radial position increases, the number of particles moving upwards continues to decrease and the main particle motion points radially outwards. For radial positions between $r = 40 - 50$ mm, the velocity magnitudes are particularly high due to the fast main flow.

The highest particle Reynolds number is estimated to be $Re_p = 3$. This is derived from a particle diameter of $d_{p,90} = 179$ µm ($d_{90}$ value), a maximum slip velocity of $u_{s,max} = 4.5$ ms$^{-1}$ (calculated as the sum of the maximum mean slip velocity of $u_{s,mean} = 2.5$ ms$^{-1}$ and the maximum standard deviation of the slip velocity of $u_{std} = 2$ ms$^{-1}$), a mean gas temperature within the internal recirculation zone of $T_{mean} = 1700$ K, and an exhaust gas composition of 71 vol.% $CO_2$, 24 vol.% $H_2O$, and 5 vol.% $O_2$. This estimated $Re_p$ is well below 280, which is the threshold for particle-induced turbulence generation by vortex shedding [26]. This analysis suggests that turbulence in the solid fuel combustor under study is attenuated rather than enhanced by the presence of particles, although individual particle Reynolds numbers can span a wide range.

Figure 13.9 shows the probability distribution for the detection of large particles. This probability map was obtained using Mie scattering and separating the large from the small particles, as described in Sect. 13.2.2. The streamlines of the gas phase are superimposed. The highest number density of large particles is located just below the outlet of the primary flow, where the particle flow splits. One part flows through the inner recirculation zone, the other part is directed towards the main flow. As the large particles slow down, they become spatially dispersed and the number density decreases rapidly. The high slip velocities in the diffuser favour rapid heating of the particles and release of volatiles. The number density of large particles is particularly high in these zones. Consistent with this, the intense CH* chemiluminescence signals shown in Fig. 13.11 are also observed here. This indicates that the transition from volatile to char combustion also occurs during the transition to the dump plane.

Polycyclic aromatic hydrocarbon (PAH) formation during secondary pyrolysis of tars released during primary pyrolysis can lead to soot nanoparticle formation through collision and clustering [27,28]. In addition to the formation of PAH during pyrolysis (petrogenic PAH), PAH can also be formed during the combustion process (pyrogenic PAH) [29], which is potentially conducive to soot formation and soot growth. The process of soot formation is particularly sensitive to the local atmosphere. The reaction pathways for PAH and soot particles are strongly dependent on the oxygen concentration. When oxygen levels are high, PAH and soot particles are oxidised to $CO_2$ and $H_2O$. To study this process, PAH-PLIF and LII imaging were performed simultaneously.

Figure 13.10 compares the average PAH-PLIF signals (left) and the average LII signals (right). The gas phase flow streamlines are superimposed. In the diffuser, the zones of high PAH-PLIF intensity correlate with high volatile concentrations. These do not spatially overlap in the diffuser with the zones where high nanoparticle number densities are observed using LII. The release of volatiles occurs mainly

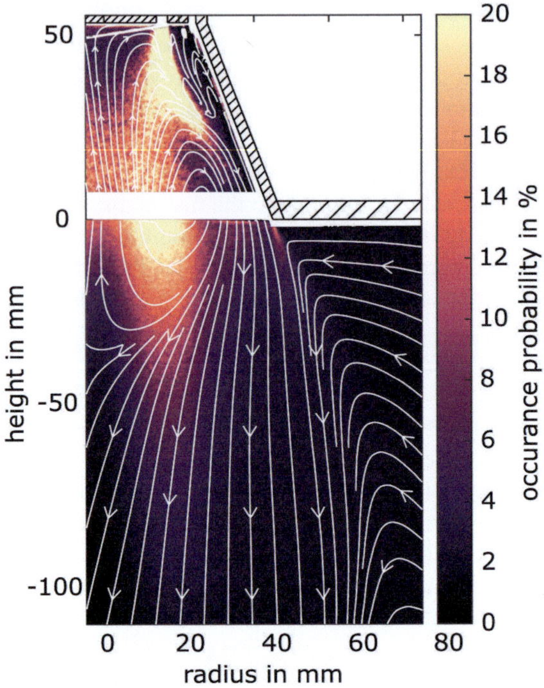


**Fig. 13.9** Probability of occurrence of large particles. This was determined using Mie scattering and the separation algorithm presented in Sect. 13.2.2. The streamlines of the gas phase flow are superimposed. (Reprinted from Schneider et al. [9], Copyright (2024), with permission from Elsevier and minor editing)

between the main flow and the inner recirculation zone. The number density of the biomass particle phase is particularly high in this area. The strongest PAH-PLIF signal occurs at the transition from the diffuser to the dump plane. In contrast, the LII signals in the diffuser are particularly strong in the inner recirculation zone, indicating that the nanoparticles formed during combustion are transported upstream in the recirculation zone towards the burner. Once they reach the main flow zone, they are oxidised and disappear.

Further downstream, in the region between $h = -70$ mm and $h = -300$ mm, the mean PAH-PLIF and LII signals overlap. The cause of this overlap remains uncertain but may be due either to the formation of pyrogenic PAH that coexist with soot in soot-rich areas, or to an incandescence signal generated by the LIF setup. However, laser excitation energy is much less, and the signal was recorded in the region between 275 and 325 nm, such that the latter hypothesis appears unlikely.

In analogy to the single-phase methane flames, the oxygen content was varied between 27–36 vol.% in two-phase flames. Figure 13.11 compares the operating conditions listed in Table 13.2. The mean CH* chemiluminescence images, mean gas phase velocities, large particle occurrence probabilities, mean gas temperatures, normalised PAH-PLIF intensity, and normalised LII signal intensity are shown. Measurements are limited to regions below the dump plane. The reason for this is that the optical access to the diffuser becomes heavily contaminated by particle deposits during long measurement times.

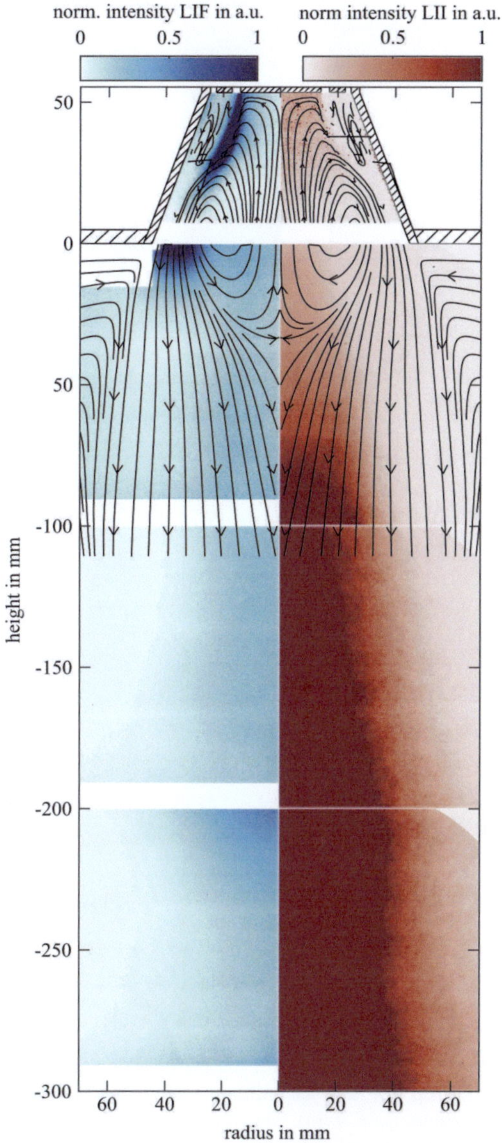


**Fig. 13.10** Average PAH-PLIF and LII signals for the OXY33P-WNS operating condition. The streamlines of the gas phase flow are superimposed. The image consists of four superimposed measurement windows that have been recorded sequentially. For visualisation purposes, the LIF intensity plot is mirrored on the centre line to show LIF and LII results on the same plot. (Reprinted from Schneider et al. [9], Copyright (2024), with permission from Elsevier and minor editing)

The chemiluminescence images in Fig. 13.11 confirm the observations from the spectrally broadband flame images in Fig. 13.7 that the OXY27P-WNS and OXY30P-WNS flames are clearly shifted downstream. This is consistently confirmed by the mean temperature fields, which are particularly low for the OXY27P-WNS case, sometimes falling below 1000 K in the main flow, but again well above 1000 K near the symmetry line in the centre of the combustor. The OXY30P-WNS case shows a much more intense CH* chemiluminescence than the OXY27P-WNS case. This is probably in a zone of intense methane combustion. In contrast, the OXY33P-WNS

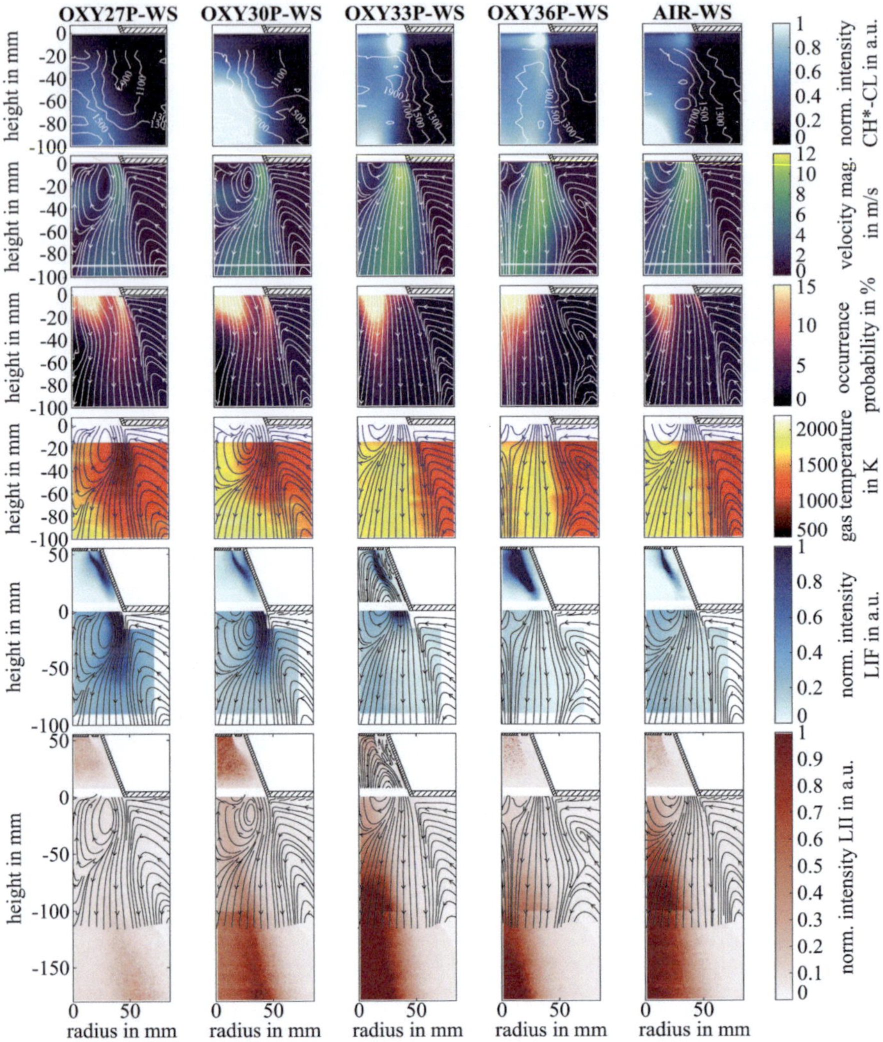


**Fig. 13.11** Comparison of the operating conditions for the two-phase operation with the walnut shell. Top row: CH* chemiluminescence images with isothermal lines obtained from TDLAS measurements (in Kelvin). The same camera settings were used to capture the images. The intensity values are normalised to the same value. Second row: mean gas phase velocity magnitude with gas streamlines. Third row: large particle probability distribution with mean particle trajectories. Fourth row: mean gas phase temperature with gas streamlines. Fifth row: mean LIF signal of PAH with gas streamlines. Sixth row: average LII signal from soot particles with gas streamlines. (Reprinted from Schneider et al. [9], Copyright (2024), with permission from Elsevier and minor editing)

and OXY36P-WNS cases and the AIR-WNS case already burn in the diffuser. For these cases, the CH* chemiluminescence intensities in the main stream are particularly high. This could be due to the intense combustion of volatiles, which is supported by relatively high PAH-PLIF signal intensities. LII signal intensities are particularly pronounced for the OXY33P-WNS and AIR-WNS cases below $h = -80$ mm and $h = -60$ mm, respectively. This indicates zones of intense char burning.

Like single-phase combustion, the inner recirculation zones also become more compact with increasing oxygen content in two-phase combustion. However, the overall axial expansion of the inner recirculation zone is smaller in the presence of particles and the mean gas velocities are lower. This indicates that the thermal expansion of the gas phase is reduced by heat transfer to the particles.

The probability of occurrence of large particles shows significant discrepancies between the different operating conditions. While for the OXY27P-WNS and OXY30P-WNS operating conditions a significant number of particles are observed in the main flow downwards, for the other operating conditions only a few particles are observed in this region. This effect may be related to the smaller recirculation zone between the primary and secondary flows, the area containing the side flames, as previously shown in the single-phase combustion analyses. Mean particle trajectories show that most particles pass through the IRZ without returning to the nozzle for all operating conditions. The OXY36P-WNS case also has a significant number of particles in the centre of the chamber, although peak particle densities are predominantly located in the radial range of $r = 5$–$25$ mm for most operating conditions. The overall lower particle loading is responsible for the reduced signals in the air operation case.

The mean gas temperatures for the two-phase combustion cases in the inner recirculation zone are lower than for the single-phase cases. They are all below 2000 K. In the outer recirculation zone, however, they are higher than in the corresponding single-phase cases. Consistent with the chemiluminescence images, the temperature distributions for the OXY27P-WNS and OXY30P-WNS cases are shifted downstream and differ significantly from those of the other three cases. The low mean temperatures of the OXY27P-WNS and OXY30P-WNS cases naturally lead to a lower particle heating rate. Accordingly, the release of volatiles is delayed, which is confirmed by the PAH-PLIF signal distributions. As the oxygen concentration increases at the OXY33P-WNS and OXY36P-WNS operating conditions, the volatiles are released further upstream. In the case of OXY36P-WNS and AIR-WNS, this is almost complete in the diffuser.

Soot formation, which can be monitored qualitatively from the LII images, increases with increasing oxygen concentration up to the OXY33P-WNS operating condition. However, at the highest oxygen concentration, OXY36P-WNS, it starts farthest downstream. This is due to the high particle loading, which results in delayed heating of the particles and lower mean gas temperatures at the corresponding heights. The occurrence of the soot-rich regions correlates with the flow and temperature fields. While for the two cases with the highest oxygen contents, OXY33P-WNS and OXY36P-WNS, as well as for the AIR-WNS case, the soot concentrations are highest in the central region, the highest LII signals are observed for the OXY27P-WNS and OXY30P-WNS cases with a radial outward shift. This different behaviour cannot be conclusively explained but is probably related to the small recirculation zones between the primary and secondary flow close to the burner in the diffuser, which have a significant influence on the overall flame topology.

As with single-phase operation, the air case AIR-WNS is best represented by the OXY33P-WNS operating condition. The differences of a slightly axially displaced char combustion zone and reduced volatile release in the OXY33P-WNS case are due to the different particle loading required to maintain the equivalence ratio.

The combination of different imaging techniques provides complementary information on flow dynamics and reaction zones in single-phase and corresponding two-phase flames. These investigations as a whole are unique in the literature to date. These data provide new insights into oxy-fuel combustion. In addition, these data support the validation of mathematical models for the numerical simulation of complex pulverised solid fuel combustion processes.

### 13.3.3 CARS Temperature Measurements in Biomass Oxy-Fuel Combustion

The temperature measurements presented in Sect. 13.3.2 were based on the tomographic TDLAS method. This method has a relatively low spatial resolution. Due to the long measurement times of a spatial scan, only mean values can be determined directly. In contrast, the CARS method has a much higher spatial resolution and instantaneous values can be measured. In addition to the mean value, the variance of the temperature can be estimated from these values without the need for a model. Selected results of the CARS measurements are presented in Figs. 13.12 and 13.13, where it should be noted that the mean temperatures between the tomographic TDLAS and CARS methods are in very good agreement [30].

The CARS measurements shown are limited to the OXY33P (single-phase) and OXY33P-WNS (two-phase) operating conditions introduced in Sect. 13.2.1, Table 13.2. The results of the single-phase measurement are presented first, followed by the two-phase measurements.

**Single-Phase Measurements**

CARS measurements were performed at the heights $h = -24.6$ mm, $-49.2$ mm, and $-98.4$ mm. At each height, the combustor was traversed to record radial CARS spectra in 3 mm increments between $r = -15$–$60$ mm and in 5 mm increments between $r = 60$–$90$ mm. The outer radial limit at $r = 90$ mm was due to the design of the corners of the combustion chamber. For each position, 500 CARS spectra were recorded .

The radial gas temperature profiles of the OXY33P operating condition in Fig. 13.12 shows a high temperature plateau region above 2100 K in the centre of the combustor with a steep temperature gradient to an outer region of temperatures around 750 K. Compared to the axial velocity, the high temperature plateau region at $h = -24.6$ mm is within the internal recirculation zone of the swirl flame.

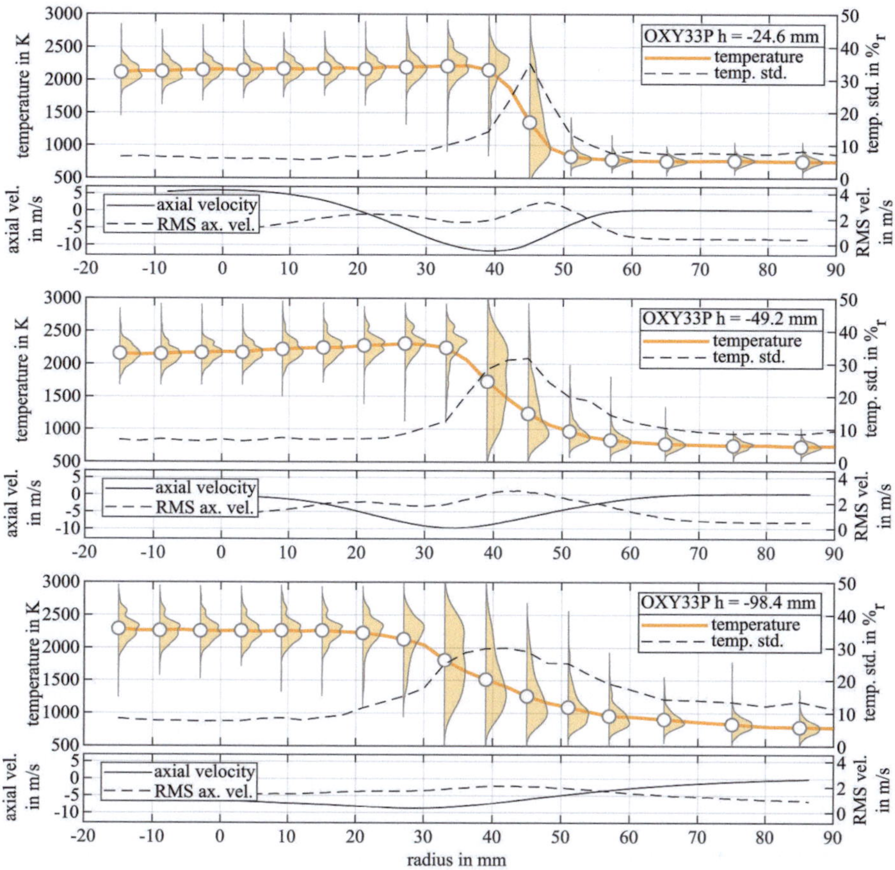


**Fig. 13.12** Gas temperature profiles of single-phase OXY33P operating condition at heights $h = -24.6$ mm, $-49.2$ mm, and $-98.4$ mm. The mean temperature is shown as a solid line. The probability density function (PDF) of every second measurement point is shown. The PDF are normalised for clear representation. The relative standard deviation of the gas temperature is shown as a dashed line. Axial velocity profiles and the corresponding RMS values are shown for comparison and are taken from Sect. 13.3.2 [9]. (Reprinted from Schneider et al. [13], Copyright (2024), with permission from Elsevier and minor editing)

The steep temperature gradient is formed in the shear layer towards the outer recirculation zone. This is indicated by the high fluctuation values of the axial velocity. The height $h = -49.2$ mm is located approximately at the stagnation point of the flow field and the height $h = -98.4$ mm is located downstream of the inner recirculation zone.

The highest gas temperature of 2310 K is measured at $h = -49.2$ mm and $r = 27$ mm. This is about 150 K higher than the gas temperature in the centre. This temperature is close to the maximum adiabatic flame temperature under slightly fuel-rich conditions in comparison with equilibrium calculations performed with *Cantera* [19] and given the measurement accuracy obtained. Within the inner and outer recirculation zone, and in the centre downstream of the inner recirculation zone, the probability density functions (PDF) of the measurement points show rather narrow distributions. Here the relative standard deviations of the gas temperature are close to the relative precision of the $O_2$-$CO_2$-CARS setup. This indicates locally constant temperature conditions.

The PDF become broad at locations where mixing of different flows occurs. The peaks of the PDF on the inner side of the region characterised by the steep temperature gradient show a shift towards higher temperatures than the mean temperatures of the PDF peaks. In contrast, at the lower end of the steep temperature gradient, the PDF peaks shift towards lower temperatures than the mean temperatures of the PDF peaks. For fluid mixing in a shear layer, these observations are typical. As shown by the radial profiles of the relative temperature standard deviation, the steep temperature gradient decreases progressively in the downstream direction. This agrees with the profiles of the axial velocity.

**Two-Phase Measurements**

For the two-phase operating condition OXY33P-WNS, up to 1000 individual CARS single shots were recorded for each measurement point. This increased number compared to the single-phase flame was necessary because the number of optical breakdowns was increased and a minimum number of 300 undisturbed individual measurements is required for statistical evaluation.

The radial gas temperature profiles are shown in Fig. 13.13 for the heights $h = -24.6$ mm, $-49.2$ mm, and $-98.4$ mm. In addition, the fraction of measurements not affected by significant optical breakdown is quantified. In the zone of the internal recirculation, the radial gas temperature profile at $h = -24.6$ mm shows a temperature drop to about 1700 K. In the zone of the main flow downwards, surrounding the inner recirculation zone, the gas temperature rises to the same level as the gas temperature of the single-phase OXY33P operating condition. The fraction of spectra not suffering from optical breakdown is reduced due to elevated probabilities of particles present within the probe volume. This occurs mostly within a 20 mm radius around the central axis. In this region, 20 to 35% of the total number of spectra at each position were affected by optical breakdowns. These observations suggest that the inner recirculation zone is predominantly influenced by the solid fuel particles undergoing heating and releasing volatiles, while the main downward flow is dominated by methane and volatile combustion with rather low particle number density. The temperature gradient towards the outer recirculation zone is flattened for gas-assisted solid fuel combustion compared to single-phase combustion. The temperature level within the outer recirculation zone is about 250 K higher, caused by the increase in total thermal power. Further downstream, at $h = -49.2$ mm and $h = -98.4$ mm, the central zone gas temperature is between 1500 and 2000 K. The highest radial gas temperatures are obtained at the radial position of about $r = 40$ mm. Compared to OXY33P, the steep temperature gradient towards the outer recirculation zone is extended radially outwards, flattened, and shows a reduced absolute temperature difference. The gas temperature distribution of the individual measurements is like that of OXY33P when operating single-phase.

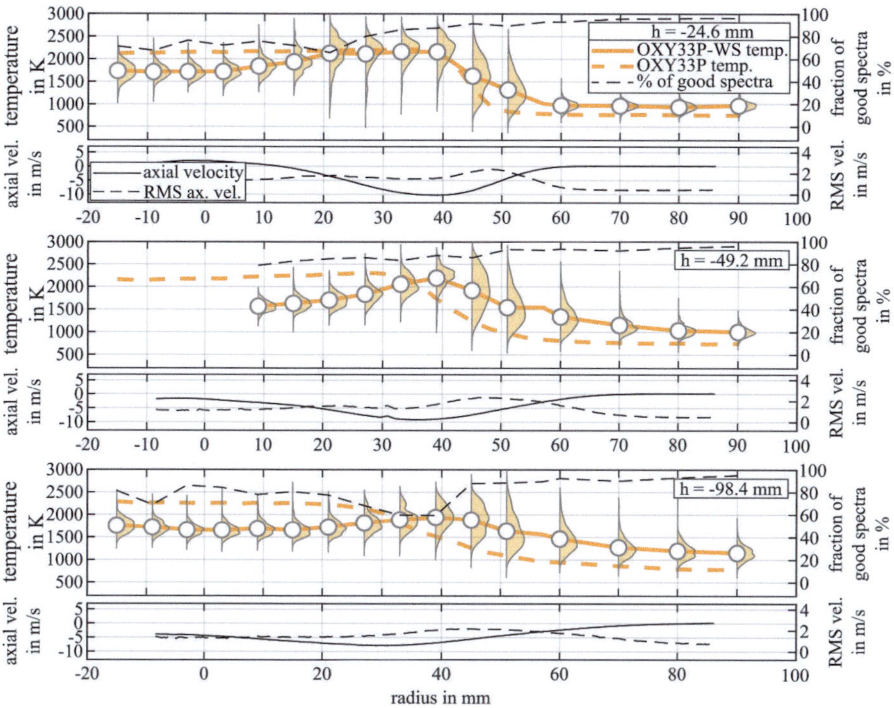


**Fig. 13.13** Gas temperature profiles of the two-phase operating condition OXY33P-WNS at $h = -24.6$ mm; $-49.2$ mm, and $-98.4$ mm. The mean temperature is shown as a solid line and the PDFs are shown as well. For better visibility, the PDFs have been normalised. For comparison, the mean gas temperature of OXY33P is shown. The fraction of spectra without significant interference from optical breakdowns is shown as a black dashed line. Due to low signal-to-noise ratio, measurements in the centre at $h = -49.2$ mm were omitted. (Reprinted from Schneider et al. [13], Copyright (2024), with permission from Elsevier and minor editing)

These gas temperature measurements and the quantitative measurements of $O_2$ and $CO_2$ mole fractions, which are not shown in this chapter, go far beyond previous experimental investigations of pulverised solid fuel flames. They provide new insights and quantitative scalar data for the validation of numerical simulation models. The measurements showed that the inner recirculation zone is characterised by relatively homogeneous distributions of thermochemical states. This inner recirculation zone is dominated by the presence of particles for two-phase operation. In contrast, the downward main flow zone is dominated by volatiles and methane. Here the particle number density is significantly reduced compared to the central flow.

## 13.4 Conclusion

A deeper understanding of the fluid mechanics, particle dynamics, and chemical processes underlying the combustion of pulverised solid fuels under oxy-fuel conditions in the near-nozzle region is essential, as these coupled processes influence devolatilisation, mixing, ignition, and flame stabilisation. The focus was on gas-assisted solid fuel flames in the power range up to 63 $kW_{th}$, which closes the gap in the literature between unconfined laboratory flames and pilot-scale combustion chambers.

Unique features were the complete optical access to the diffuser and the entire combustion chamber as well as the well-controlled boundary conditions. The optical access enabled the use of various advanced laser diagnostic techniques. The quartz glass diffuser provided the opportunity to study the mechanisms that contribute to the stabilisation of the flame. In a series of experiments, the complexity was gradually increased from single-phase, non-reacting flows to two-phase, reacting flows. Parametric studies were performed to investigate the sensitivities of Reynolds numbers, fuel composition, and oxidiser composition. The fuel variation included Rhenish lignite, Prosper-Haniel bituminous coal, walnut shells, and beech wood. In addition, inert particles were used for fundamental investigations of particle-flow interaction.

Overall, existing optical measurement techniques for use in oxy-fuel combustion systems have been further developed and applied for the first time in such complex environments. Combined particle imaging and particle tracking velocimetry (PIV/PTV), laser-induced phosphor thermometry, coherent anti-Stokes Raman spectroscopy (CARS) and pyrometry were applied to investigate the flow field, particle dynamics, slip velocities and wall, gas and particle temperatures. Residence times, gas temperatures and species concentrations were measured using absorption spectroscopy and soot formation employing laser-induced fluorescence of polycyclic aromatic hydrocarbons and laser-induced incandescence. Especially for the oxy-fuel conditions the dual-pump CARS approach was adopted to measure gas temperatures and $CO_2$ and $O_2$ concentrations.

The investigations revealed that flame stabilisation is dominated by the combustion of volatiles. Direct comparison of oxy-fuel with air operating conditions demonstrated that it is possible to maintain similar combustion characteristics by increasing the oxygen content in the oxidiser to compensate for the physically different properties of $CO_2$ compared to $N_2$, such as the higher heat capacity. This is a critical requirement when converting boilers from air to oxy-fuel operation. Furthermore, these data support the validation of mathematical models for the numerical simulation of complex combustion processes with pulverised solid fuels. The comprehensive data set has been well received in the community and is used for the validation of numerical simulations.

**Acknowledgement** This work has been funded by the Deutsche Forschungsgemeinschaft (DFG, German Research Foundation)—215035359—SFB/TRR 129 'Oxyflame'.

## References


1. Li, T., Geschwindner, C., Dreizler, A. and Böhm, B. "Particle-Resolved Optical Diagnostics of Solid Fuel Combustion for Clean Power Generation: A Review". *Measurement Science and Technology* 34.12 (2023), 122001. https://doi.org/10.1088/1361-6501/acef49
2. Bonin, M. P. and Queiroz, M. "Local Particle Velocity, Size, and Concentration Measurements in an Industrial-Scale Pulverized Coal-Fired Boiler". *Combustion and Flame* 85.1–2 (1991), 121–133. https://doi.org/10.1016/0010-2180(91)90181-A
3. Most, J.-M., Trouillet, P., Jallais, S., Mandin, P., Marchand, F., Le-Masson, C., Witwicki, P. and Rampelberg, D. "Development of a LDV Probe for Velocity Measurements in a 600 MW Pulverized Coal Power Plant". *10th International Symposium on Applications of Laser Techniques to Fluid Mechanics*. Lisbon, Portugal, 2000
4. Pickett, L. M., Jackson, R. E. and Tree, D. R. "LDA Measurements in a Pulverized Coal Flame at Three Swirl Ratios". *Combustion Science and Technology* 143.1–6 (1999), 79–107. https://doi.org/10.1080/00102209908924194
5. Sabel, T., Risio, B., Unterberger, S., Schnell, U., Hein, K. R. G., Käß, M., Priesmeier, U. and Thierbach, H.-U. "Full Scale Measurements and Mathematical Modelling Studies for the Investigation of the Combustion Behaviour of a Modern Bituminous Coal-Fired Boiler". *IFRF Combustion Journal* (2001), 200102
6. Li, S., Xu, Y. and Gao, Q. "Measurements and Modelling of Oxy-Fuel Coal Combustion". *Proceedings of the Combustion Institute* 37.3 (2019), 2643–2661. https://doi.org/10.1016/j.proci.2018.08.054
7. Becker, L., Langenthal, T. von, Pielsticker, S., Böhm, B., Kneer, R. and Dreizler, A. "Experimental Investigation of Particle-Laden Flows in an Oxy-Coal Combustion Chamber for Non-Reacting Conditions". *Fuel* 235 (2019), 753–762. https://doi.org/10.1016/j.fuel.2018.08.076
8. Becker, L., Pielsticker, S., Böhm, B., Kneer, R. and Dreizler, A. "Particle Dynamics in a Gas Assisted Coal Combustion Chamber Using Advanced Laser Diagnostics". *Fuel* 269 (2020), 117188. https://doi.org/10.1016/j.fuel.2020.117188
9. Schneider, H., Bonarens, M., Hebel, J., Hamel, H., Emmert, J., Böhm, B., Wagner, S., Kneer, R. and Dreizler, A. "Combined Flow, Temperature and Soot Investigation in Oxy-Fuel Biomass Combustion Under Varying Oxygen Concentrations Using Laser-Optical Diagnostics". *Fuel* 362 (2024), 130771. https://doi.org/10.1016/j.fuel.2023.130771
10. Becker, L., Kosaka, H., Böhm, B., Doost, S., Knappstein, R., Habermehl, M., Kneer, R., Janicka, J. and Dreizler, A. "Experimental Investigation of Flame Stabilization Inside the Quarl of an Oxyfuel Swirl Burner". *Fuel* 201 (2017), 124–135. https://doi.org/10.1016/j.fuel.2016.09.002
11. Schneider, H., Valentiner, S., Vorobiev, N., Böhm, B., Schiemann, M., Scherer, V., Kneer, R. and Dreizler, A. "Investigation on Flow Dynamics and Temperatures of Solid Fuel Particles in a Gas-Assisted Oxy-Fuel Combustion Chamber". *Fuel* 286 (2021), 119424. https://doi.org/10.1016/j.fuel.2020.119424
12. Meißner, C., Schneider, H., Sidiropoulos, E., Hölzer, J. I., Heckmann, T., Böhm, B., Dreizler, A. and Seeger, T. "Investigation on Wall and Gas Temperatures Inside a Swirled Oxy-Fuel Combustion Chamber Using Thermographic Phosphors, $O_2$ Rotational and Vibrational CARS". *Fuel* 289 (2021), 119787. https://doi.org/10.1016/j.fuel.2020.119787
13. Schneider, H., Hebel, J., Böhm, B., Kneer, R. and Dreizler, A. "Rotational-Vibrational $O_2$-$CO_2$ Coherent Anti-Stokes Raman Spectroscopy for Determination of Thermochemical States in Oxy-Fuel Biomass Combustion". *Proceedings of the Combustion Institute* 40.1–4 (2024), 105457. https://doi.org/10.1016/j.proci.2024.105457
14. Bürkle, S., Becker, L., Dreizler, A. and Wagner, S. "Experimental Investigation of the Flue Gas Thermochemical Composition of an Oxy-Fuel Swirl Burner". *Fuel* 231 (2018), 61–72. https://doi.org/10.1016/j.fuel.2018.05.039
15. Bürkle, S., Becker, L., Agizza, M. A., Dreizler, A., Ebert, V. and Wagner, S. "In-Situ Measurement of Residence Time Distributions in a Turbulent Oxy-Fuel Gas-Flame Combustor". *Experiments in Fluids* 58.7 (2017), 647. https://doi.org/10.1007/s00348-017-2366-2

16. Emmert, J., Schneider, H., Böhm, B., Dreizler, A. and Wagner, S. "Phase-Locked Absorption Tomography for Retrieving 5 kHz Time-Resolved Tracer Profiles in Solid Fuel Combustion". *Applications in Energy and Combustion Science* 12 (2022), 100093. https://doi.org/10.1016/j.jaecs.2022.100093
17. Toporov, D., Bocian, P., Heil, P., Kellermann, A., Stadler, H., Tschunko, S., Förster, M. and Kneer, R. "Detailed Investigation of a Pulverized Fuel Swirl Flame in $CO_2/O_2$ Atmosphere". *Combustion and Flame* 155.4 (2008), 605–618. https://doi.org/10.1016/j.combustflame.2008.05.008
18. Raffel, M., Willert, C. E., Scarano, F., Kähler, C. J., Wereley, S. T. and Kompenhans, J. *Particle Image Velocimetry*. Cham: Springer International Publishing, 2018. https://doi.org/10.1007/978-3-319-68852-7
19. Goodwin, D. G., Moffat, H. K., Schoegl, I., Speth, R. L. and Weber, B. W. *Cantera: An Object-Oriented Software Toolkit for Chemical Kinetics, Thermodynamics, and Transport Processes*. Zenodo. 2023. https://doi.org/10.5281/zenodo.8137090
20. Khalitov, D. A. and Longmire, E. K. "Simultaneous Two-Phase PIV by Two-Parameter Phase Discrimination". *Experiments in Fluids* 32.2 (2002), 252–268. https://doi.org/10.1007/s003480100356
21. Michelsen, H. A., Schulz, C., Smallwood, G. J. and Will, S. "Laser-Induced Incandescence: Particulate Diagnostics for Combustion, Atmospheric, and Industrial Applications". *Progress in Energy and Combustion Science* 51 (2015), 2–48. https://doi.org/10.1016/j.pecs.2015.07.001
22. Roy, S., Gord, J. R. and Patnaik, A. K. "Recent Advances in Coherent Anti-Stokes Raman Scattering Spectroscopy: Fundamental Developments and Applications in Reacting Flows". *Progress in Energy and Combustion Science* 36.2 (2010), 280–306. https://doi.org/10.1016/j.pecs.2009.11.001
23. Cutler, A. D., Gallo, E. C. A. and Cantu, L. M. L. "WIDECARS Spectra Fitting in a Premixed Ethylene–Air Flame". *Journal of Raman Spectroscopy* 47.4 (2016), 416–424. https://doi.org/10.1002/jrs.4829
24. Jennings, D. E. and Brault, J. W. "The Ground State of Molecular Hydrogen". *Journal of Molecular Spectroscopy* 102.2 (1983), 265–272. https://doi.org/10.1016/0022-2852(83)90039-5
25. Magnotti, G. "Dual-Pump CARS Development and Application to Supersonic Combustion". PhD thesis. George Washington University, 2012.
26. Subramaniam, S. and Balachandar, S., eds. *Modeling Approaches and Computational Methods for Particle-Laden Turbulent Flows*. Elsevier, 2023. https://doi.org/10.1016/C2020-0-03264-0
27. Wen, X., Ferraro, F., Nicolai, H., Hashimoto, N., Hayashi, J., Nakatsuka, N., Tainaka, K. and Hasse, C. "Flamelet LES of a Turbulent Pulverized Solid Fuel Flame Using a Detailed Phenomenological Soot Model". *Proceedings of the Combustion Institute* 39.3 (2023), 3429–3438. https://doi.org/10.1016/j.proci.2022.07.190
28. Fletcher, T. H., Ma, J., Rigby, J. R., Brown, A. L. and Webb, B. W. "Soot in Coal Combustion Systems". *Progress in Energy and Combustion Science* 23.3 (1997), 283–301. https://doi.org/10.1016/S0360-1285(97)00009-9
29. Desgroux, P., Mercier, X. and Thomson, K. A. "Study of the Formation of Soot and its Precursors in Flames Using Optical Diagnostics". *Proceedings of the Combustion Institute* 34.1 (2013), 1713–1738. https://doi.org/10.1016/j.proci.2012.09.004
30. Emmert, J., Schneider, H., Meißner, C., Sidiropoulos, E., Hölzer, J. I., Seeger, T., Böhm, B., Dreizler, A. and Wagner, S. "Characterization of Temperature Distributions in a Swirled Oxy-Fuel Coal Combustor Using Tomographic Absorption Spectroscopy with Fluctuation Modelling". *Applications in Energy and Combustion Science* 6 (2021), 100025. https://doi.org/10.1016/j.jaecs.2021.100025

# Pilot Scale Pulverised Solid Fuel Combustion Experiments

14

Burak Özer, Reinhold Kneer and Anna Maßmeyer

## Contents



**Abstract**

The experimental characterisation of solid fuel flames lacks systematic analysis regarding the influence of inherent boundary conditions. Heat losses to the environment often require continuous gas co-firing, complicating the separate investigation of this effect in existing literature. Additionally, studies on scaling effects

B. Özer (✉) · R. Kneer · A. Maßmeyer
Institute of Heat and Mass Transfer, RWTH Aachen University, Aachen, Germany
e-mail: burak.oezer@rwth-aachen.de

R. Kneer
e-mail: kneer@wsa.rwth-aachen.de

A. Maßmeyer
e-mail: massmeyer@wsa.rwth-aachen.de





R. Kneer et al. (eds.), *Clean Energy Conversion*,
https://doi.org/10.1007/978-3-031-98956-8_14

from laboratory to semi-industrial scales are limited. This chapter addresses the transition from methane co-fired flames to self-sustained flames and examines the transition from thermal loads typically investigated in laboratory experiments (up to 50 $kW_{th}$) to semi-industrial scales (100 $kW_{th}$). Further on, flame stabilisation in oxy-fuel combustion is explored by comparing different oxygen contents in the oxidiser. Experiments are conducted in a brick-lined combustion chamber with constant wall heating, enabling self-sustained solid fuel combustion and gas co-firing across various thermal loads and fuel compositions. Rhenish lignite and walnut shell flames are studied with and without methane co-firing. Measurements include solid fuel particle velocities and narrow-band imaging to assess how particle flow field and flame shape are affected. The findings reveal that flame stabilisation is sensitive to variations in parameters such as added gas content, volatile content of the fuel, local flame temperature, and local gas composition.

## 14.1 Introduction

The superior optical access of most lab-scale combustion experiments requires the necessity of gas-assistance to volatilise and ignite the solid fuel particles and to prevent flame and particles to extinguish. This concerns single or group particle combustion experiments in flat flame burners (e.g. [1,2]), but also more complex swirl flame experiments [3–7]. In contrast, industrial-scale combustion experiments (e.g. [8–10]) with self-sustained solid fuel combustion (i.e. that do not require gas-assistance) generally provide a very limited optical access for the measurement techniques.

This study, therefore, focuses on pilot-scale experiments, as an intermediate level between lab-scale and industrial-scale, in a combustion chamber with a high operational flexibility and access for both—optical and probe-based—measurement techniques. The operational flexibility encompasses, on the one hand, a wide range of thermal loads (ranging from 10 $kW_{th}$ to more than 100 $kW_{th}$), a large operating range for the oxidiser (air or different oxy-fuel conditions with oxygen contents between 19–36 vol.%), and different degrees of swirl. On the other hand, the configuration provides the option to operate purely self-sustained flames as well as pure gas flames or a mixture of both, and thus to systematically investigate the influence of gas co-firing on flame properties, such as their stabilisation.

In the first study on this configuration, the applicability of Reynolds-averaged Navier-Stokes (RANS) simulations as a burner-design tool under oxy-fuel conditions was the central focus [11]. In the following years, various burner designs with different modifications were employed in diverse studies [5,12–16] (as also discussed in Chaps. 13 and 16) and the experimental findings were used as validation data for different models in RANS or large eddy simulations (LES), e.g. [17–20]. In recent experimental studies, flames with different fossil fuels [12–15,21,22] as well as one torrefied biomass [23] have been thoroughly studied. Earlier studies focused on the limits of applicability of the selected measurement techniques to investigate Rhenish lignite flames [12,14,22] as well as their general flame characteristics when changing from air to oxy-fuel conditions [13,21]. Recently, the effect of variations in oxidiser composition for different fuel types (Rhenish lignite (RL), German bituminous coal (GBC), walnut shells (WNS) and a torrefied biomass) under different inflow conditions [14,15], and the influence of methane-co-firing [24,25] were studied. A summary of the particle velocity determination for lignite flames is given in the study by Zabrodiec [26]. In the lab-scale methane-assisted experiments using a similar burner design, the effects of different oxygen concentrations in the oxidiser stream [5,27], and different thermal loads [28] have been comprehensively investigated for pulverised coal combustion regarding flame structure, flow field, particle and gas temperatures, as well as locally resolved species concentrations.

Flow field characteristics of 40 $kW_{th}$ self-sustained Rhenish lignite flames under oxy-fuel atmosphere with 23–33 vol.% $O_2$ concentrations have been investigated by Zabrodiec et al. [14]. The flow rates of the oxidiser streams were kept constant for all investigated flames and thus, with increasing $O_2$-content, the air-fuel ratio rose. Increasing the air-fuel ratio leads to more intense expansion of the product gases, which is reflected in higher velocities in the main flow and lower velocities in the internal recirculation zone. The same flames have been investigated by Hees et al. [13] to characterise flame length and combustion zones via spectrally narrow- band imaging, showing that with increasing $O_2$-content the combustion close to the burner is intensified and the flames stabilised even closer to the diffuser, although the general flame structure remains unaffected. More recently, Schneider et al. [29] conducted a systematic study for methane-assisted solid fuel flames under various oxy-fuel atmospheres ranging from 27 to 36 vol.% $O_2$, in comparison to a conventional air atmosphere. Unlike in the studies by Hees et al. [13] and Zabrodiec et al. [14], the flow rates and air-fuel ratios of the operated flames were kept constant, but the thermal load was varied to keep the air-fuel ratio constant. Thus, the flames under oxy-fuel atmosphere with higher $O_2$ exhibit higher thermal loads among the selected flames. The flow field measurements show that as the oxygen content in oxy-fuel atmosphere increases, the inner recirculation zone of the flame shrinks and the measured velocities of the main flow increase. Moreover, a shift of the reactive zone towards the burner is observed as the $O_2$ content increases, similar to observations in the study by Hees et al. [13].

In several studies [15,22,30], pulverised coal flames under oxy-fuel with 21 and 25 vol.% $O_2$ were investigated and compared to operation in air atmosphere. In these studies, the 100 $kW_{th}$ burner (detailed in Sect. 14.2.1) was employed. Laser Doppler velocimetry (LDV) measurements [22] show that both oxy-fuel flames have overall higher velocities as well as much stronger recirculation zones than the reference air case, which can be explained by the changes in momentum flow of the oxidiser at the burner exit: The momentum flows are increasing when switching from air to the 21 vol.% $O_2$-case and so are the peak velocities, and decreasing when increasing the $O_2$-content to 25 vol.%, and likewise are the velocities. This mechanism is analysed by Maßmeyer et al. [15] in detail, also for different fuels. This change is similarly observed in the narrow-band images in the preceding study by Hees et al. [30]. The reactive devolatilisation region in the oxy-fuel cases retracts towards the burner, indicated by the OH* emission that increases in intensity close to the burner when increasing the momentum of the swirled flow, thus intensifying the recirculation. The air-flame with a smaller and less strong recirculation zone is more elongated. Hence, the relatively higher velocities in the near burner region are intensified in the oxy-fuel cases, due to the compact combustion zones that lead to a strong gas phase expansion. The vertical wall-incident radiative heat flux profiles [15] recorded below the burner have very similar characteristics for air and oxy-fuel flames (25 vol.% $O_2$). Although the absolute heat flux values are higher in the air flame at each measured distance below the burner, both air and oxy-fuel flames share the same qualitative profile in which the radiative heat flux intensity increases up to the same axial position, and then decreases further downstream. Thus, despite the distinct flow fields [22] and narrow-band images [30] of oxy-fuel and air-flames, the wall-incident radiative heat flux profiles are similar and only their magnitude is affected. The variations in fuel and oxidiser [14, 15] yield another insight: The ratio of momentum flows at the burner is another crucial parameter to influence the flame-attachment to the burner.

The choice to investigate the influence of single parameters on the combustion dynamics, i.e. regions where reactions take place and the flow field as well as the interaction of these two, is always a trade-off, as the variation of one parameter necessarily implies the variation of other parameters and a decision must be made as to which parameter is kept constant. Besides the different approaches in the previous studies (keeping either the air-fuel ratio or the flow rates constant), the results present a clear trend that increasing either the oxygen content or the proportion of the swirled momentum flow both act to reinforce the flame stabilisation at the diffuser.

In most previous studies coal was investigated, such as Rhenish lignite. Despite the urgent need to reduce fossil fuel combustion, e.g. by burning biomass as a substitute and subsequently capturing the $CO_2$, studies for pulverised biomass flames remain scarce, especially under oxy-fuel conditions with varying $O_2$ content. To fill the research gap, in the present chapter, an assortment of recent results on biomass combustion (walnut shells) in the pilot scale facility is presented to bridge between the different experimental scales, with the aim to elucidate the influences of i) the methane assistance needed on the laboratory scale, ii) different oxygen-concentrations in the oxidiser streams, as well as iii) higher thermal load. The focus is on the flame stabilisation in terms of shape and positions of reactive zones and the particles' velocity profiles in the combustion chamber. By comparing the influence of methane combustion for Rhenish lignite as well as ground walnut shells, this overview is appended with a discussion on the influence of the fuel-choice. The results are discussed in the context of the previous investigations on Rhenish lignite in the same experimental configuration, and new insights on the influence of methane assistance by a comparison to the findings from the laboratory-scale experiments discussed in Chap. 13.

## 14.2 Experimental Setup

### 14.2.1 The Pilot-Scale Test Facility

The experiments have been conducted in a pilot-scale test facility illustrated in Fig. 14.1a), consisting of a cylindrical brick-lined combustion chamber with an embedded electrical heating. The burner port, holding the swirl burner, is movable along the axis of the cylindrical chamber. Hence, the flame can be set in a desired vertical distance from the fixed measurement level, where three measurement ports (cf. Fig. 14.1b)) provide access to the flame, either for probe-based or purely optical (and thus non-intrusive) measurements.

The swirl burners employed (cf. Figs. 14.1c) and d)) consist each of a central bluff-body surrounded by two annular nozzles, both used to supply oxidiser. The inner annulus is denoted as "primary", and the outer one as "secondary". The primary stream is fed straight into the combustion chamber and carries the fuel. The swirl of the flame is controlled by the secondary stream via the proportion of the stream that flows straight or at an angle of 45° into a mixing chamber inside the burner. The larger burner (cf. Fig. 14.1d)) features a third stream, denoted "tertiary", feeding a straight oxidiser stream through the gap between burner and burner port, whereas the "staging stream" between burner port and wall is identical for any employed configuration.

The oxidiser can be varied between air or a mixture of $CO_2$ and $O_2$ in the range of 20–35 vol.% $O_2$. The oxidiser streams are adjusted via thermal mass flow controllers, adapted for the respective oxidiser composition. The pulverised solid fuel is dosed by a screw feeder mounted on a scale, and dispersed into the primary air by a Venturi type ejector. Moreover, the solid fuel seeder is flushed with a constant stream of $N_2$ under air or $CO_2$ under oxy-fuel conditions, hence the primary air oxidiser content is slightly reduced to prevent ignition in the fuel feeding system. Each burner can be operated with pulverised solid fuel with diameters up to a few hundred micrometre or likewise with combustible gases like methane, or a mixture of these fuels. The heated walls of the combustion chamber along with slightly lean operating conditions ensure the total burnout of the solid fuels. Thus, the feeding rate of solid fuel can be cross-checked by the monitored oxygen content in the flue gas according to the stoichiometric calculation.

### 14.2.2 Measurement Techniques

To characterise the flames in terms of flame shape, areas of heat-release and local particle velocity, two different measurement techniques have been employed. Measurements are conducted at selected distances below the burner, whereby the lower edge of the burner is the reference point $h = 0\,d$, with $d$ being the outer diameter of the secondary annulus of each burner, i.e. $d_{40\,\text{kW}} = 49.2$ mm and $d_{100\,\text{kW}} = 64$ mm as highlighted in Figs. 14.1c) and d).

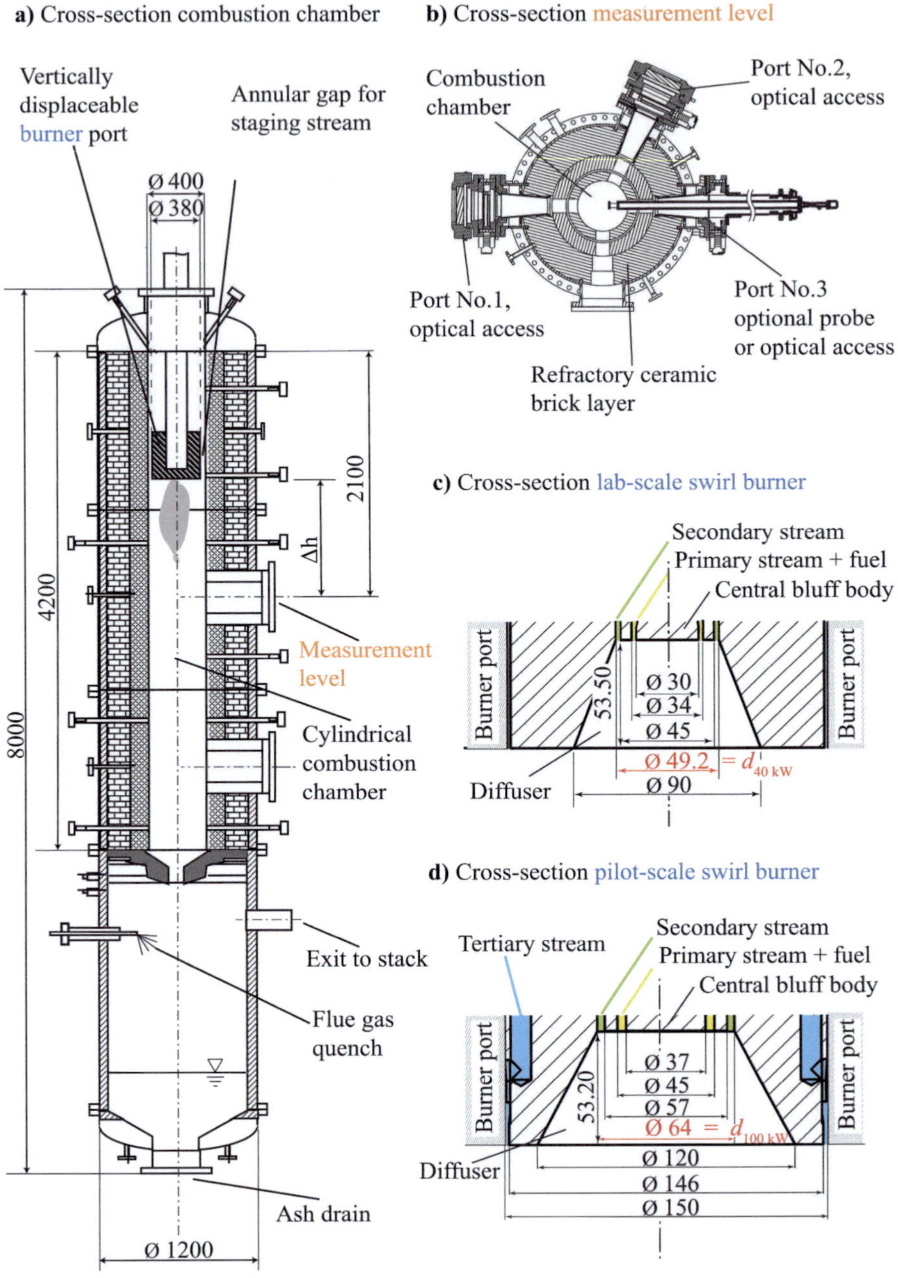


**Fig. 14.1** Illustration of **a**) the experimental test facility, **b**) the measurement level, and the **c**) laboratory-scale and **d**) pilot-scale swirl burners employed different thermal loads. All dimensions are given in millimetre

**Determination of Flame Shape and Reactive Zones: Narrow-Band Imaging**
The conversion process of solid fuel particles involves volatile matter (e.g. hydrocarbons) release, and their combustion. Like in methane combustion, intermediate radicals form, e.g. OH*, $C_2$*, CH* and emit radiation that can be optically detected at specific wavelengths. In solid fuel combustion, this radiation can be superimposed by the blackbody radiation originating from the surface temperature of the solid fuel particles. As discussed in the study by Zabrodiec et al. [22], in the visible part of the spectrum, the blackbody radiation outshines the one originating from radical emission. However, in the UV wavelength range, where blackbody radiation is low in intensity, radical emission can still be detected. In the present work, an image intensifier (intensified relay optics: IRO, *LaVision*) equipped with notch filters (transmitting light in a narrow-band wavelength range), and connected to a CCD camera is employed. Filters were chosen with central wavelengths of 308 and 430 nm, each with a full-width half minimum (FWHM) of 10 nm. At 308 nm, the blackbody radiation of the particles is expected to be low, and thus this wavelength is ideal to determine OH* radical release whose spontaneous emission peaks around this wavelength [31]. For particles in laminar flat-flame burners, 430 nm filters are usually employed to detect the emission occuring from CH* radical release (e.g. Schneider et al. [29]). However, in larger flames, the spectrum in this range is strongly dominated by the blackbody radiation of the particles (e.g. [22]). Thus, from those narrow-band images primarily the char combustion and the related intensity of heat release in different regions in the combustion chamber can be assessed.

Images are collected at constant gain and variable gate times of image intensifier at constant frame rate of 1.2 Hz to account for the different intensities of the different flames. As the recorded intensity scales linearly with the gate time, comparison between different recording settings is enabled in post-processing, by normalising each image based on the applied gate time. Flame images taken at different axial positions are combined consecutively to obtain one single narrow-band image.

The results from narrow-band imaging are represented in terms of radial and axial position below the burner (cf. Figs. 14.2, 14.5 and 14.8). On the left hand side below the burner, the results obtained for the 308 nm filter are depicted, also referenced as OH* chemiluminescence, characterising the regions of devolatilisation and early combustion. On the right hand side below the burner, images obtained with the 430 nm filter are shown, whereby this wavelength is used to identify regions of char burnout.

**Axial and Tangential Particle Velocities: Laser Doppler Velocimetry**
To determine local particle velocities without perturbing the flow, i.e. non-intrusive laser Doppler velocimetry (LDV) is applied. This technique relies on the scattering of monochromatic incident light (i.e. a laser beam) on a tracer particle, here on a fuel particle. Although Rhenish lignite provides proper flow tracers [22], walnut shell particles have high Stokes numbers due to their particle size range of 100–250 μm. Thus, LDV does not capture each detail of the gas flow field of walnut shell flames. Slip velocity measurements of walnut shell particles can be found in Sect. 13.3.2. In the present setup, two optically pumped semiconductor lasers

with a single longitudinal mode (*Coherent Genesis CX-SLM series*) are used in a custom-made long-range (working distance of 1600 mm) LDV system. The detection optics are operated in backscatter mode. For the determination of the axial and tangential velocity components, lasers with wavelenghts of 488 and 514 nm are employed, respectively. Radial profiles of axial and tangential velocity are obtained (by radially traversing the measurement system) at selected distances below the burner (by changing the distance of the burner to the observation port). The radial resolution of the measurements is adjusted based on the spatial gradients in the velocity profiles. Measurements at each point are conducted for at least 60 s. The ellipsoidal measurement volumes of 488 nm and 514 nm lasers have a width of $\approx$ 135–150 $\mu$m and a length of $\approx$ 1500–1800 $\mu$m, respectively. A comprehensive description of this technique in the context of the present configuration is given by Zabrodiec [26].

The radial profiles of axial and tangential velocity are shown at defined distances below the burner and compared for different boundary conditions (e.g. Fig. 14.3, where the level of methane enrichment is changed). Positive axial velocities represent downward flow, i.e. away from the burner. The clockwise rotation of the flow (when looking from above) is described by the negative tangential velocity.

## 14.3 Influences of Methane Co-Firing

### 14.3.1 Operating Conditions

The influence of methane assistance on the flame characteristics has been investigated for Rhenish lignite as well as walnut shell flames with the 40 $kW_{th}$ burner. The thermal load was kept constant at 40 $kW_{th}$ while the fraction of methane in the flames (cf. Table 14.1) was varied. For each solid fuel, the chamber was operated with three different conditions: one with a 50% contribution of methane to the thermal load, i.e. 20 $kW_{th}$ solid fuel (L: lignite, W: walnut shells) and 20 $kW_{th}$ methane (M), hence denoted L/W20M20. For the other two flames the methane fraction was reduced to 25% (L/W30M10) and 0% (L/W40M00). The flames were operated with air as oxidiser, and the oxidisers' volume flows were kept constant for all flames.

### 14.3.2 Results and Discussion

The results of narrow-band imaging are presented in Fig. 14.2 in two columns for Rhenish lignite (left) and walnut shells (right). From top to bottom of the Fig. the solid fuel contribution to the thermal load of the flame is increased from 50 over 75 to 100%, i.e. for lignite (L) and methane (M) L20M20, L30M10, L40M00 and likewise for walnut shells (W) and methane (M) W20M20, W30M10, and W40M00.

For both fuels, the highest OH* radiation intensities are observed in methane-assisted flames. This is attributed to a combination of both, the emission from the methane combustion itself, along with the smaller and denser devolatilisation zones

**Table 14.1** Operating conditions for methane (M)-enriched Rhenish lignite (L) or biomass (walnut shell, W) flames. As the only difference between L- and W-flames is the fuel and its feeding flow rate (due to the different heating values of the two fuels) all six flames are summarised in Table 14.1. Numbers in flame names indicate the contribution to the thermal load (in $kW_{th}$) of the respective fuel (L or W and M)

| Parameter | Units | L/W20M20 | L/W30M10 | L/W40M00 |
|---|---|---|---|---|
| Methane volume flow rate | [$m^3$ $h^{-1}$] | 2.0 | 1.0 | 0.0 |
| Lignite/Biomass mass flow rate | [kg $h^{-1}$] | 3.3/3.8 | 5/5.7 | 6.7/7.6 |
| Thermal load of biomass/methane | [$kW_{th}$/$kW_{th}$] | 20/20 | 30/10 | 40/00 |
| $O_2$ fraction of primary stream | [vol.%] | 19.5 | 19.5 | 19.5 |
| $O_2$/$N_2$ secondary and staging streams | [vol.%] | 21/79 | 21/79 | 21/79 |
| Volume flow rate of primary stream | [$m^3$ $h^{-1}$] | 12.1 | 12.1 | 12.1 |
| Volume flow rate of secondary stream | [$m^3$ $h^{-1}$] | 11.1 | 11.1 | 11.1 |
| Volume flow rate of staging stream | [$m^3$ $h^{-1}$] | 24.0 | 24.0 | 24.0 |

of these flames due to the increased heating rates in the near burner region. Furthermore, the higher volatile content of the walnut shells compared to the lignite (cf. Table 1.1) leads to OH* radiation intensity differences between lignite and walnut shell flames. The highest OH* radical emission in lignite flames varies between 700 and 2500, whereas for walnut shells the range is 1600 to 6000, as described in Sect. 14.4. Intensity values are normalised with their gate times and filter transmissivities to enable comparable results.

For the methane-assisted lignite flames with 50 and 75% (i.e. L20M20 and L30M10) solid fuel fraction, the devolatilisation region is limited to the near burner region. However, in the self-sustained flame without methane assistance (L40M00), the maximum is also located in the near burner region, but while in the other flames radiation is limited to this region, the devolatilisation process continues further downstream. This shift can also be observed in the walnut shell flames. Without methane-assistance (W40M00), the devolatilisation in self-sustained walnut flames predominantly takes place in the far field of the burner. This indicates, that in the case of co-firing, combustion in the near burner region is strongly dominated by methane combustion as it provides a high heating rate to the solid fuel particles. Thus, both, the methane combustion and the solid fuel particle size, determine where devolatilisation takes place, i.e. by continuously decreasing the methane share in the flame, the

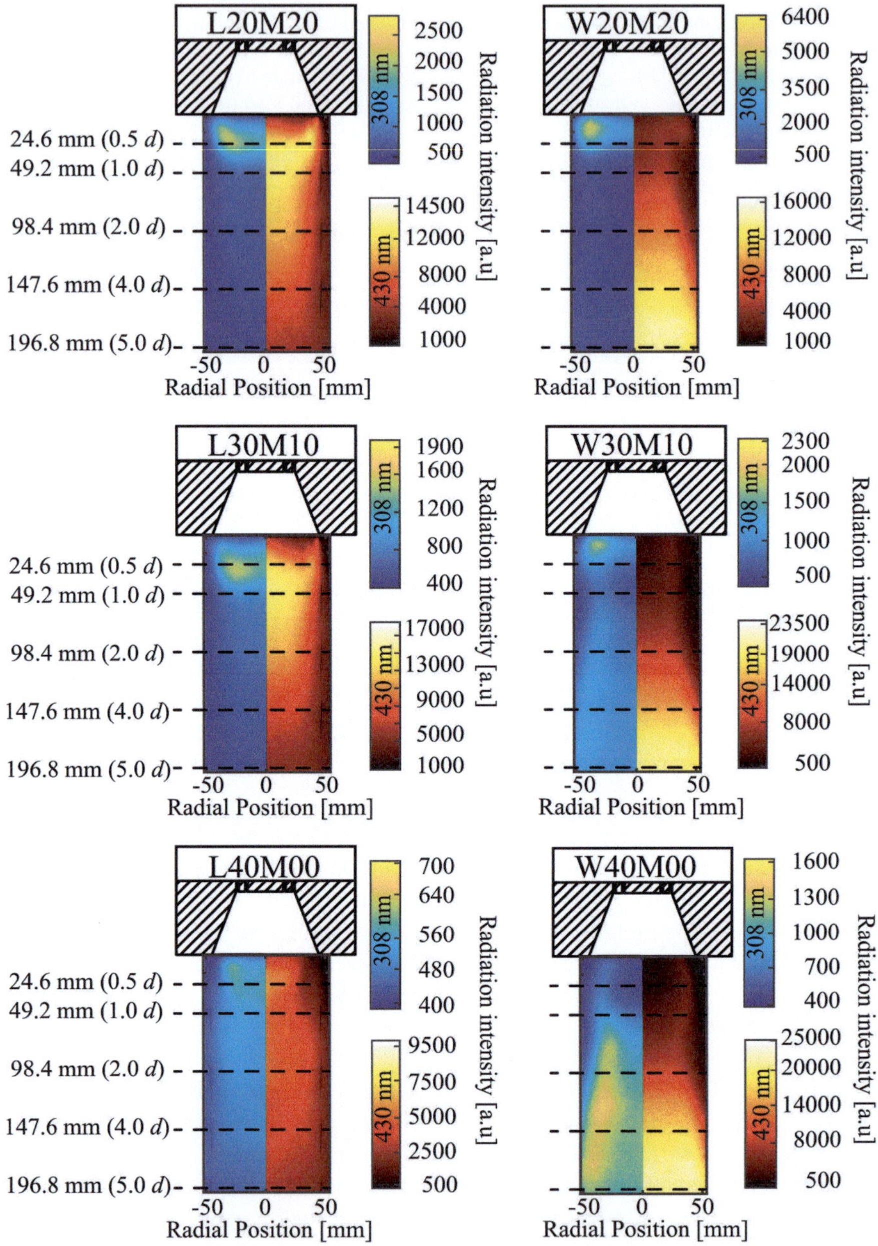


**Fig. 14.2** Average narrow-band emission images of pulverised Rhenish lignite (L) flames (left) and pulverised walnut shell (W) flames (right). The wavelength 308 nm corresponds to OH* radical emission (volatile release and homogeneous combustion) and 430 nm indicates regions of solid fuel heat release (heterogeneous char combustion)

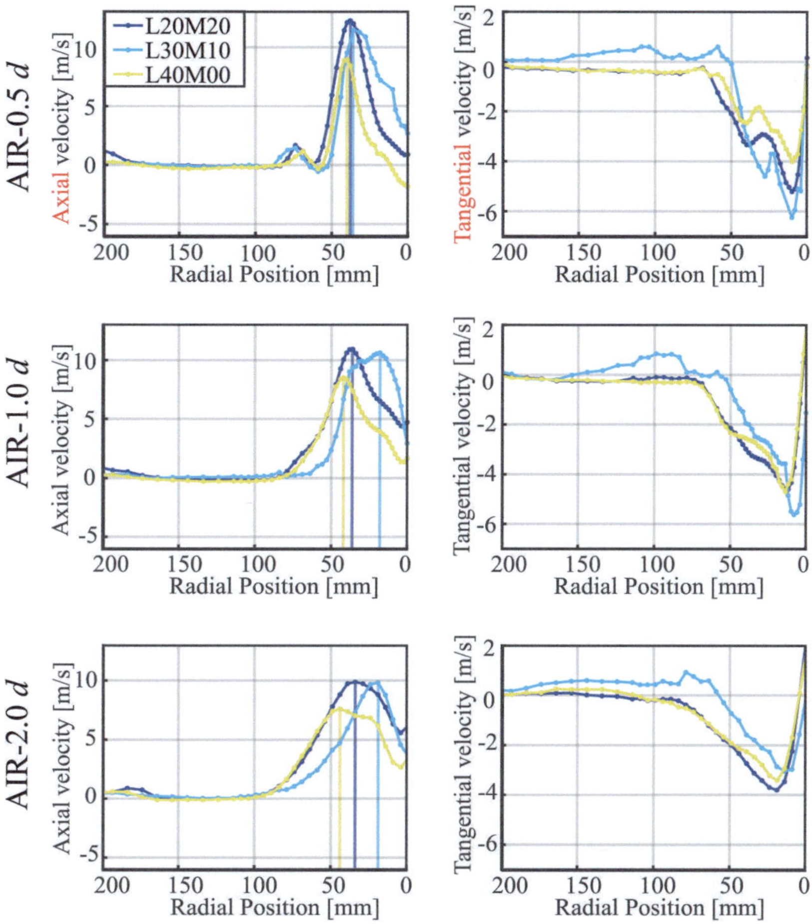


**Fig. 14.3** Mean velocity profiles obtained from the LDV measurements of the axial and tangential velocity components for Rhenish lignite flames with different methane enrichment. Vertical lines on the radial axes of the axial velocity indicate the position of the peak of the main flow for each profile. Positive axial velocities represent downward flow

particle-based combustion is notably delayed for the larger walnut shell particles. A similar mechanism was also reported by Schneider et al. [29]. The influence of particle size on the regions where char combustion takes place (as seen in the narrow-band images at 430 nm), is more pronounced. For all lignite flames, the highest thermal radiation appears in the near burner region. In contrast, in the walnut shell flames the main char combustion occurs further downstream. The large distance between regions of strong devolatilisation and char burnout in the methane-assisted walnut shell flames can again be linked to the larger particle size (cf. Fig. 1.2), as the bigger walnut shell particles need longer to heat up and ignite the char later than the smaller lignite particles.

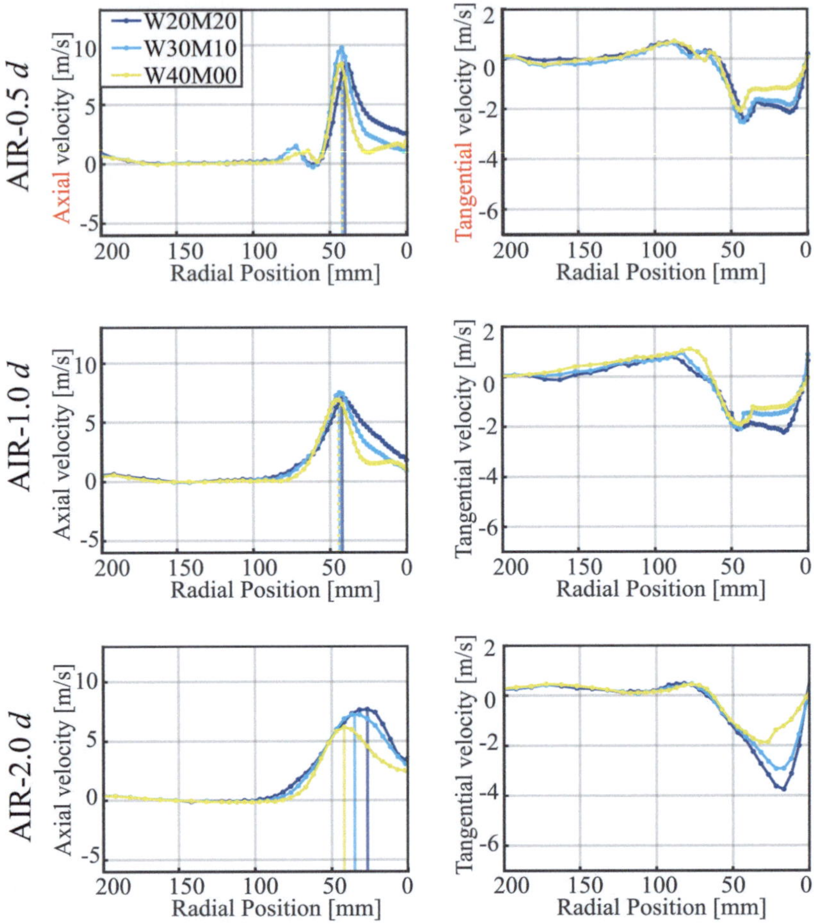


**Fig. 14.4** Mean velocity profiles obtained from the LDV measurements of the axial and tangential velocity components for walnut shell flames with different methane enrichment. Vertical lines on the radial axes of the axial velocity indicate the position of the peak of the main flow for each profile. Positive axial velocities represent downward flow

The corresponding flow field measurements for Rhenish lignite flames are shown in Fig. 14.3 and for walnut shell flames in Fig. 14.4. Comparing the axial and tangential components of the main flow, it is observed that the flames with methane-assistance have relatively higher velocities than the self-sustained pulverised flames for both fuels. This is attributed to the strong gas phase expansion in the near burner region, which is associated to the higher heating rate of the gas phase due to methane combustion. The tangential velocities generally exhibit two peaks in the vicinity of the burner, which unify further downstream into only one peak that progresses, along with the main flow, towards the axis of the chamber.

Similar to the narrow-band images, the velocity profiles of the walnut shell flames have a gradual transition from a methane-assisted to a self-sustained pulverised solid fuel flame: As the methane fraction increases, the particle velocities increase, especially close to the chamber axis due to the thermal expansion of the gas combustion. While the walnut shell flames exhibit this gradual change, the differences in the lignite flames' flow fields are more distinct: Despite a similar location of the main flow for the self-sustained L40M00 and the methane-assisted L30M10 and L20M20 flames, an inner recirculation zone (IRZ, i.e. negative axial velocities at or around the chamber axis) develops only in the self-sustained flame.

Comparing walnut shell flames with lignite flames side by side, the velocities measured in the lignite flames are higher than in the walnut shell flames, despite the identical inflow conditions. This can likely be linked to the above-described different areas of heat release: While for the lignite flames particles ignite close to the burner, walnut shell particles do not face a flame's gas expansion until $H = 2.0\,d$. As a result of this ignition delay, the velocity magnitudes of the walnut shell flames are significantly lower than those of lignite flames.

Among all six flames, an internal recirculation zone (IRZ) is only observed in the L40M00 flame. Although the velocity profiles of the three lignite flames are similar at $H = 0.5\,d$, only the L40M00 flame has an IRZ. The reason is presumably that the recirculation zones in these flames are comparably shorter than that of L40M00. Thus, an IRZ develops, and is either only located inside the quarl, and hence not observable in this configuration, or the mean particle velocity derived from the measurements of the larger walnut shell particles (i.e. high Stokes number) prohibits its detection.

The results show that the presence of methane in the near burner region can act as an anchor of the flame to the burner. This is well in line with the conclusions in Sect. 13.4, that "flame stabilisation is dominated by the combustion of volatiles", whereby the methane in cofired flames has a similar effect as the volatile release. The heating rate imposed on the solid fuel particles, originating from the gas-phase combustion in these cases, largely influences in turn their volatile release and as a consequence also the location of heat release. In this regard, also the fuel particle size is a crucial parameter. The char combustion of the larger walnut shell particles is spatially delayed compared to that of Rhenish lignite.

In conclusion, this investigation shows that the availability of gaseous combustibles—be it methane or the release of volatiles from the solid fuel—has a huge impact on the evolving flow field as well as the regions of heat release. The impact of volatile release has been previously reported by Schneider et al. [29], where also the influence of particle size is highlighted.

## 14.4 Influence of Oxygen Content

### 14.4.1 Operating Conditions

The influence of the oxygen content in the oxidiser stream is investigated for a set of four 60 $kW_{th}$ pulverised walnut shell flames (cf. Table 14.2), using the 100 $kW_{th}$ burner. The reference flame is the 60 $kW_{th}$ flame with 25 vol.% oxygen in the oxidiser stream that has already been thoroughly investigated for Rhenish lignite [15,22,30].

**Table 14.2** Operating conditions for different oxygen contents in the oxidiser streams for walnut shell flames with 60 $kW_{th}$

| Parameter | Units | OXY25 | OXY27 | OXY30 | OXY33 |
|---|---|---|---|---|---|
| Biomass flow rate | [kg $h^{-1}$] | 11.4 | 11.4 | 11.4 | 11.4 |
| Thermal load | [kW] | 60 | 60 | 60 | 60 |
| $O_2$ fraction of primary stream | [vol.%] | 22.6 | 24.4 | 27.1 | 29.8 |
| $O_2/CO_2$ secondary and staging streams | [vol.%] | 25/75 | 27/73 | 30/70 | 33/67 |
| Volume flow rate of | | | | | |
| Primary stream | [$m^3$ $h^{-1}$] | 8.5 | 8.5 | 8.5 | 8.5 |
| Secondary stream | [$m^3$ $h^{-1}$] | 23.7 | 21.4 | 18.5 | 16.2 |
| Tertiary stream | [$m^3$ $h^{-1}$] | 4.2 | 3.8 | 3.3 | 2.9 |
| Staging stream | [$m^3$ $h^{-1}$] | 22.7 | 21.0 | 18.9 | 17.2 |
| Momentum flow rate of | [kg m $s^{-2}$] | | | | |
| Fuel | | 57.9 | 57.9 | 57.9 | 57.9 |
| Secondary stream | $\times 10^{-3}$ | 120.02 | 97.3 | 72.0 | 54.7 |
| Momentum ratio[a] | $\times 10^{-3}$ | 2.1 | 1.7 | 1.2 | 0.9 |

[a]Ratio of the momentum flow rate of secondary stream over momentum flow rate of fuel

Here it is investigated for walnut shell particles. The oxygen content varies from 25 to 33 vol.% and the flames are denoted accordingly as OXY25, OXY27, OXY30, and OXY33. As the air-fuel ratio is kept constant, the total flow rate of the oxidiser streams decreases with increasing oxygen content from 25 to 33% $O_2$. As this changes the momentum flows at the burner, the resulting values are calculated and added to Table 14.2 for the discussion of the results.

### 14.4.2 Results and Discussion

Among all narrow-band images (cf. Fig. 14.5), the OXY25 flame is the only one stabilised downstream from the diffuser, i.e. OH* radical and thermal radiation start at $H = 0.5\,d$. With increasing oxygen content, the OXY27 flame stabilises at the diffuser, and for the OXY30 flame, the peak intensities decrease and OH* radical

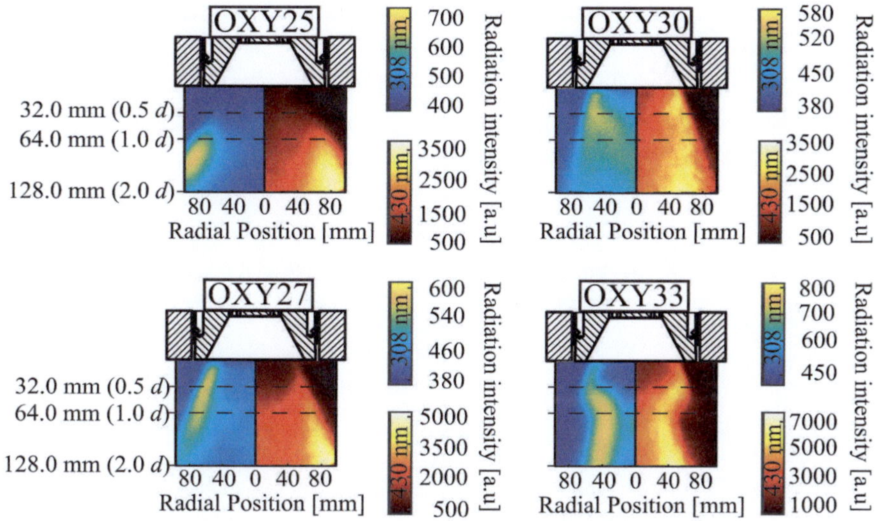


**Fig. 14.5** Average narrow-band emission images of four walnut shell flames with 60 $kW_{th}$ thermal loads under different oxy-fuel atmospheres: OXY25, OXY27, OXY30, and OXY33

emission as well as thermal radiation are more homogeneously distributed, especially at the flame axis. An initial necking is observed which turns into a neck-like shape when further increasing the oxygen content (OXY33). From OXY30 to OXY33 the peak intensities in the narrow-band images increase substantially. This is attributed to the significantly lower diameter of the reactive zone of OXY33 flame compared to other flames. As the diameter of the reactive zone decreases, devolatilisation and thermal radiation zones become denser, so a higher radiative emission is observed. It should be noted that the thermal radiation of the particles is not representative for the main char combustion region, which can be assumed to take place below $H = 2.0\,d$ according to walnut shell flame images shown in Fig. 14.2. Here only the initial devolatilisation stage is observed, which is why at 430 nm the intensities are far lower than the maximum intensities observed in Fig. 14.2.

The velocity profiles are shown in Fig. 14.6. As seen in the OH* images, the OXY25 flame is the only flame not stabilised at the diffuser, consequently leading to the fact, that the thermally induced expansion of the gases also occurs further downstream. Thus, at $H = 1.0\,d$, the axial velocity of the main flow, which spreads towards the wall, increases. In the OXY27 flame the peak velocities remain almost constant from $H = 0.5\,d$ to $H = 1.0\,d$, while the main flow diffuses and also spreads towards the wall. For the OXY30 flame, the main flow progresses towards the centre with decreasing peak velocity. This transition from 25–30 vol.% $O_2$ content can be attributed to two reasons. One is that the higher oxygen content causes higher availability of $O_2$, so the ignition takes place closer to burner. The second one is the momentum ratio between secondary stream and pulverised fuel flow. With increasing oxygen content, the momentum ratio of the swirling flame decreases. Thus, OXY25 and OXY27 flames shows the characteristics of a highly swirling flow which con-

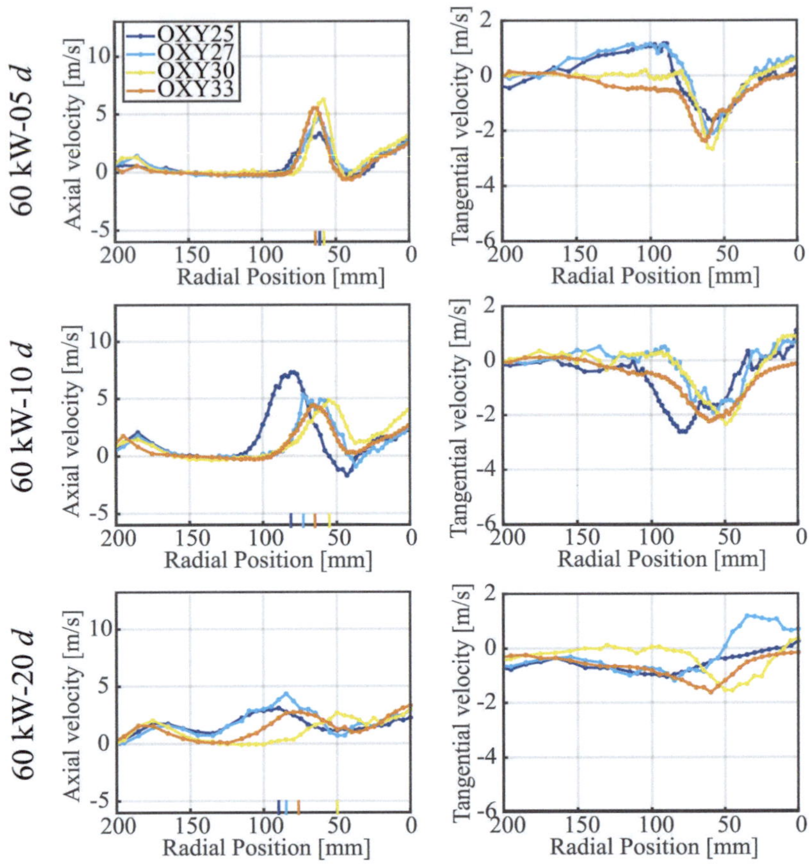


**Fig. 14.6** Mean velocity profiles of four 60 $kW_{th}$ walnut shell flames (OXY25, OXY27, OXY30, and OXY33) obtained from the LDV measurements of the axial and tangential velocity components. Vertical lines on the radial axes of the axial velocity indicate the position of the peak of the main flow for each profile. Positive axial velocities represent downward flow

tinuously moves away from the axis towards the wall of the chamber while the main flow of OXY30 flame keeps its main flow position at $R = 50$ mm. With a further increase in $O_2$ content, the OXY33 flame's velocity profiles are similar to those of OXY30, but the main flow spreads towards the wall below the necking of the flame at $H = 1.0\,d$.

The local and global air-fuel ratios and the thermal load were kept constant for the present experiments. For operational purposes, also the primary flow is kept constant. This inevitably leads to a decrease of the secondary (swirled) flow rate as the $O_2$ oxidiser content increases. Despite the associated reduction of the momentum flux of the swirled air, the flame stabilisation at the diffuser improves. It can therefore be concluded that increasing the locally available oxygen content increases the local combustion temperatures involving also higher particle heating rate, which thus accelerates the volatile release, whose combustion in turn improves the flame stabili-

**Table 14.3** Operating conditions for different thermal loads for walnut shell flames. The 60 $kW_{th}$ flame corresponds to the OXY25 flame in Table 14.2

| Parameter | Units | 60 $kW_{th}$ | 80 $kW_{th}$ | 100 $kW_{th}$ |
|---|---|---|---|---|
| Biomass flow rate | [$kg\ h^{-1}$] | 11.4 | 15.3 | 19.1 |
| Thermal load | [$kW_{th}$] | 60 | 60 | 60 |
| $O_2$ fraction of primary stream | [vol.%] | 22.6 | 22.6 | 22.6 |
| $O_2/CO_2$ secondary and staging streams | [vol.%] | 25/75 | 25/75 | 25/75 |
| Volume flow rate of | | | | |
| Primary stream | [$m^3\ h^{-1}$] | 8.5 | 8.5 | 8.5 |
| Secondary stream | [$m^3\ h^{-1}$] | 23.7 | 34.0 | 44.3 |
| Tertiary stream | [$m^3\ h^{-1}$] | 4.2 | 6.0 | 7.8 |
| Staging stream | [$m^3\ h^{-1}$] | 22.7 | 30.3 | 37.9 |
| Momentum flow rate of | [$kg\ m\ s^{-2}$] | | | |
| Fuel | | 57.9 | 77.8 | 97.1 |
| Secondary stream | $\times 10^{-3}$ | 120.02 | 247.0 | 419.4 |
| Momentum ratio[a] | $\times 10^{-3}$ | 2.1 | 3.2 | 4.3 |

[a]Ratio of the momentum flow rate of secondary stream over momentum flow rate of fuel

sation at the diffuser. A similar effect is observed by Schneider et al. [29], whereby in that study also a more pronounced effect onto the peak velocities is observed, which increase along with an increasing $O_2$ content. This is not observed in the present results, where the thermal load of the flame is kept constant, and thus the momentum flows reduce with increasing $O_2$ content, unlike in the study by Schneider et al. [29].

## 14.5 Influence of Thermal Load

### 14.5.1 Operating Conditions

Starting from the OXY25 flame at 60 $kW_{th}$, the thermal load is increased in steps of 20 $kW_{th}$ at a constant oxidiser's oxygen content of 25 vol.% as illustrated in Table 14.3. To ease the discussion, the flames are referred to as 60, 80, or 100 $kW_{th}$ flames, whereby the 60 $kW_{th}$ is identical to the OXY25 flame. As the air-fuel ratio and primary air flow rate remain constant, the swirl at the burner increases with increasing thermal load, leading to an increase in the momentum flow of the secondary stream compared to the primary stream (both given as reference for discussion).

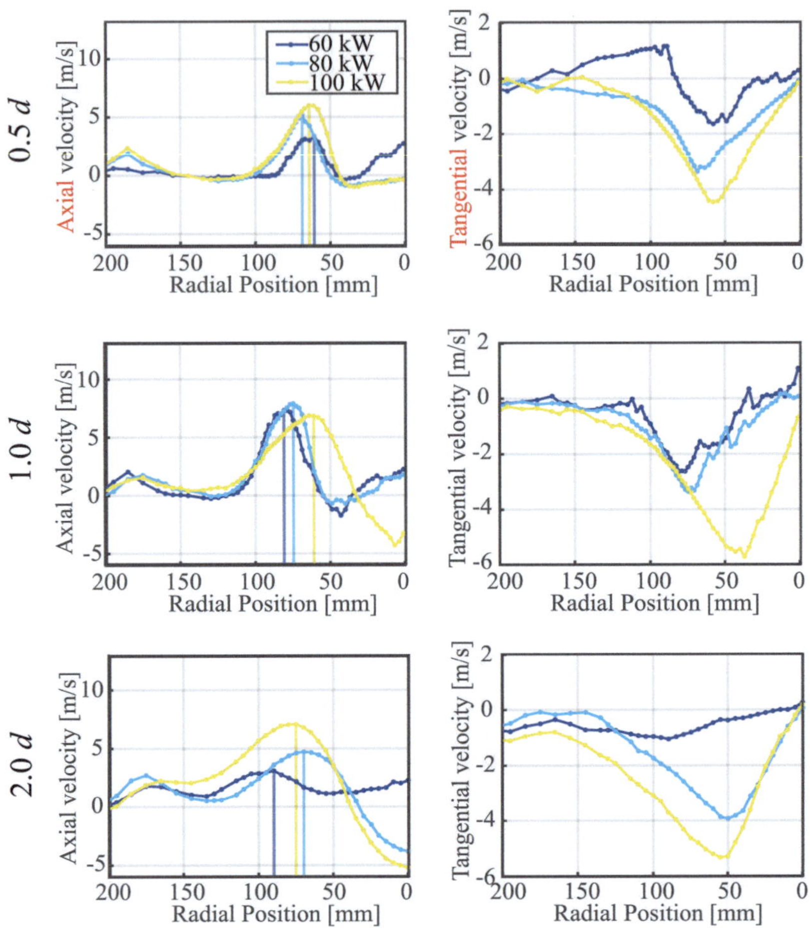


**Fig. 14.7** Mean velocity profiles of three walnut shell flames with three different thermal loads (60, 80, and 100 $kW_{th}$) obtained from the LDV measurements of the axial and tangential velocity components. Vertical lines on the radial axes of the axial velocity indicate the position of the peak of the main flow for each profile. Positive axial velocities represent downward flow

### 14.5.2 Results and Discussion

The velocity profiles of the three flames (cf. Fig. 14.7) exhibit two different flame patterns regarding the existence of an IRZ. The 60 $kW_{th}$ flame exhibits no IRZ and the main flow expands with increasing distance to the burner. In the other two flames (80 and 100 $kW_{th}$), an IRZ establishes, increasing in size with increasing thermal load. The IRZ in the 80 $kW_{th}$ flame is seen first in the velocity profile at $H = 2.0\,d$, while the 100 $kW_{th}$ flame has a stagnation point at $H = 0.5\,d$ and an IRZ at $H = 1.0\,d$. Associated with the IRZ, high tangential velocities in the shear zones, i.e. where the axial velocity changes direction, are reached, as well as a retraction of the main flow towards the centre in these regions. For the 60 $kW_{th}$ flame, the highest velocities

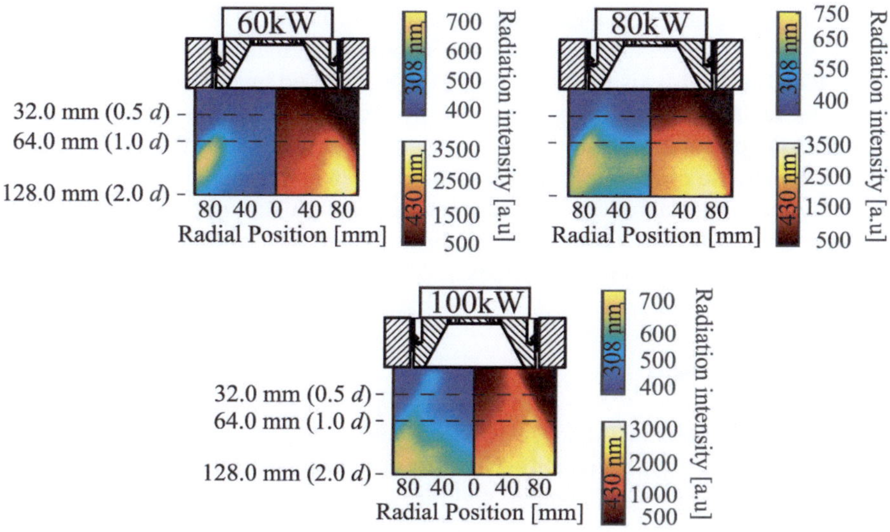


**Fig. 14.8** Average narrow-band emission images of three walnut shell flames with three different thermal loads (60, 80, and 100 $kW_{th}$) under same oxy-fuel atmosphere with 25 vol.% $O_2$ content (OXY25)

are not measured in the vicinity of the burner, but at $1.0\,d$. Further downstream, the velocities decrease as the main flow expands. For the flames with IRZ, the axial velocities of the main flow show no significant increase or decrease, whereas the tangential velocity increases with the presence of an IRZ.

The narrow-band images (cf. Fig. 14.8) yield a similar picture. The 60 $kW_{th}$ flame has a dense devolatilisation zone and exhibits high thermal radiation in the main flow, while in the 80 and 100 $kW_{th}$ flame, radiation increases on both wavelengths, especially at the flame axis between $H = 1.0$ and $2.0\,d$. The recirculation of hot combustion gases and hot particles in the 80 and 100 $kW_{th}$ flames yield higher radiation intensities in the IRZ. Thereby, the recirculation also provides heat to the near burner region, leading to an earlier release of volatiles. Thus, the increase in thermal load enhances the stabilisation at the diffuser, as can be seen below the edge of the diffuser at the position where OH* radical emission as well as thermal radiation occurs. The OH* radical emission starts closer to the burner with increasing thermal load, indicating an improvement of stabilisation at the diffuser, despite the increased flow rates. The position of the reaction zone in the main flow goes along with increasing axial velocities, where a strong increase can be observed in the measurements of the 80 and 100 $kW_{th}$ flames, as a result of the thermal expansion inside the reaction zone. Since the 60 $kW_{th}$ flame lacks an IRZ, its reactive zone is restricted to its main flow only. The ignition of volatiles inside the diffuser of the 100 $kW_{th}$ flame also affects the velocity field. Although there is no significant increase in the peak axial velocity, the main flow diffuses in radial direction at constant maximum velocity, i.e. the bulk velocity increases due to the thermal expansion.

To sum up, the flow field measurements show that, due to the increased momentum of the swirled oxidiser, thermal load has a significant influence on the development of an IRZ. This is also visible in the flame shape recorded by narrow-band imaging. Thus, the recirculated particles play an active role in the combustion process. A change in the position of the reaction zones and the flow field of a flame can be seen with each increase of 20 $kW_{th}$ thermal load. This can be explained by the variations in the momentum ratio of the secondary stream and pulverised fuel flow between the flame conditions (cf. Table 14.3). As the ratio increases with thermal power, the recirculation zone becomes more intense.

## 14.6 Conclusion

In the present chapter, the influence of various operating parameters on the solid fuel combustion is investigated, namely the effect of methane-enrichment, the solid fuel type, the oxidiser's oxygen content, and the thermal power. Effects on the flames are discussed regarding the particle velocities and the stabilisation mechanism at the burner as well as the regions in the combustion chamber where volatile and heat release take place. Particle velocities are determined using backscattered 2-D laser Doppler velocimetry. Flame visualisation is accomplished using a CCD camera coupled with an optical intensifier. Two optical filters are used to narrow the flame images to wavelengths of 308 and 430 nm, each with FWHM of 10 nm.

For both fuels (Rhenish lignite and walnut shell) the methane-cofiring improves the stabilisation of the flames at the diffuser, and the measured particle velocities increase with the methane enrichment. Also the fuel type affects the flame shape, i.e. for the methane-cofired Rhenish lignite char combustion occurs closer to the burner than for the walnut shell flames. The effect of methane enrichment on the starting position of walnut shell char combustion is less pronounced, which is due to the bigger particle sizes and thus the slower heat-up for the walnut shell particles.

For the investigated oxygen range (25–33 vol.% $O_2$), the particle velocities are hardly affected, despite the different inflow conditions and thus the different momentum flows. However, with increasing oxygen content, volatile-release and combustion occur closer to the burner and with a further increase, a neck-like flame shape evolves.

Increasing the thermal load has an immediate effect on the particle flow field, as an inner recirculation zone evolves, which is also reflected in the narrow band images. This can mainly be attributed to the related increase of the momentum ratio of the swirled secondary compared to the purely axial pulverised fuel flow of the primary stream.

To summarise, the results append to what is discussed in the study by Schneider et al. [29] and Chap. 13, i.e. studies with methane assistance, as well as previous results obtained mostly for Rhenish lignite in the same combustion chamber as the one discussed in the present chapter [12, 15, 30]. A comparison of the results with respect to the different parameters that are kept constant or varied throughout the different studies thereby highlights different mechanisms that acton the flame stabilisation:

- Methane co-firing has a similar effect as early volatile release, their combustion imposes a significant heat-flux onto the particles which influences the flow field and the regions of heat release.
- The particle size determines how fast the particles heat up and thus where the ignition of volatiles is observed and where char combustion starts.
- The availability of oxygen in the vicinity of the burner affects the combustion intensity, in turn influencing the regions of volatile and heat release.
- With increasing thermal load, the flame stabilisation at the diffuser improves, whose dependency on the momentum flow ratio between secondary stream and pulverised fuel flow needs further investigation.

The present chapter yields a comprehensive review of recent findings on biomass combustion. Two different routes can be taken to achieve flame stabilisation at the diffuser, namely a high availability of homogeneous combustibles (be it methane-enrichment or an early volatile release) and an aerodynamic stabilisation of a swirl flame whereby the recirculation zone establishes with an increasing ratio of the secondary swirled compared to purely axial pulverised fuel. The results yield a broader view on the findings from literature, i.e. experiments that have been performed with methane-assisted solid fuel combustion as well as earlier studies on the present experimental configuration.

**Acknowledgement** This work has been funded by the Deutsche Forschungsgemeinschaft (DFG, German Research Foundation)—215035359—SFB/TRR 129 'Oxyflame'.

## References

1. Ma, P., Huang, Q., Gao, Q. and Li, S. "Effects of Na and Fe on the Formation of Coal-Derived Soot in a Two-Stage Flat-Flame Burner". *Fuel* 265 (2020), 116914. https://doi.org/10.1016/j.fuel.2019.116914
2. Li, T., Geschwindner, C., Dreizler, A. and Böhm, B. "Particle-Resolved Optical Diagnostics of Solid Fuel Combustion for Clean Power Generation: A Review". *Measurement Science and Technology* 34.12 (2023), 122001. https://doi.org/10.1088/1361-6501/acef49
3. Balusamy, S., Kamal, M. M., Lowe, S. M., Tian, B., Gao, Y. and Hochgreb, S. "Laser Diagnostics of Pulverized Coal Combustion in $O_2/N_2$ and $O_2/CO_2$ Conditions: Velocity and Scalar Field Measurements". *Experiments in Fluids* 56.5 (2015). https://doi.org/10.1007/s00348-015-1965-z
4. Li, T., Köser, J., Dreizler, A., Schiemann, M. and Böhm, B. "Time-Resolved Volumetric Measurement of OH Radicals on Single Particle and Particle Group Combustion". *$14^{th}$ International Conference on Sustainable Development of Energy, Water and Environment Systems*. Dubrovnik, Croatia, 2019
5. Becker, L., Kosaka, H., Böhm, B., Doost, S., Knappstein, R., Habermehl, M., Kneer, R., Janicka, J. and Dreizler, A. "Experimental Investigation of Flame Stabilization Inside the Quarl of an Oxyfuel Swirl Burner". *Fuel* 201 (2017), 124–135. https://doi.org/10.1016/j.fuel.2016.09.002
6. Ribeirete, A. and Costa, M. "Detailed Measurements in a Pulverized-Coal-Fired Large-Scale Laboratory Furnace with Air Staging". *Fuel* 88.1 (2009), 40–45. https://doi.org/10.1016/j.fuel.2008.07.033
7. Schneider, H., Hebel, J., Böhm, B., Kneer, R. and Dreizler, A. "Rotational-Vibrational $O_2$-$CO_2$ Coherent Anti-Stokes Raman Spectroscopy for Determination of Thermochemical States in

Oxy-Fuel Biomass Combustion". *Proceedings of the Combustion Institute* 40.1-4 (2024), 105457. https://doi.org/10.1016/j.proci.2024.105457

8. Andersson, K., Johansson, R., Hjärtstam, S., Johnsson, F. and Leckner, B. "Radiation Intensity of Lignite-Fired Oxy-Fuel Flames". *Experimental Thermal and Fluid Science* 33.1 (2008), 67–76. https://doi.org/10.1016/j.expthermflusci.2008.07.010
9. Ballester, J., Barroso, J., Cerecedo, L. M. and Ichaso, R. "Comparative Study of Semi-Industrial-Scale Flames of Pulverized Coals and Biomass". *Combustion and Flame* 141.3 (2005), 204–215. https://doi.org/10.1016/j.combustflame.2005.01.005
10. Damstedt, B., Pederson, J. M., Hansen, D., Knighton, T., Jones, J., Christensen, C., Baxter, L. L. and Tree, D. R. "Biomass Cofiring Impacts on Flame Structure and Emissions". *Proceedings of the Combustion Institute* 31.2 (2007), 2813–2820. https://doi.org/10.1016/j.proci.2006.07.155
11. Toporov, D., Bocian, P., Heil, P., Kellermann, A., Stadler, H., Tschunko, S., Förster, M. and Kneer, R. "Detailed Investigation of a Pulverized Fuel Swirl Flame in $CO_2/O_2$ Atmosphere". *Combustion and Flame* 155.4 (2008), 605–618. https://doi.org/10.1016/j.combustflame.2008.05.008
12. Zabrodiec, D., Becker, L., Hees, J., Maßmeyer, A., Habermehl, M., Hatzfeld, O., Dreizler, A. and Kneer, R. "Detailed Analysis of the Velocity Fields from 60 kW Swirl-Stabilized Coal Flames in $CO_2/O_2$- and $N_2/O_2$-Atmospheres by Means of Laser Doppler Velocimetry and Particle Image Velocimetry". *Combustion Science and Technology* 189.10 (2017), 1751–1775. https://doi.org/10.1080/00102202.2017.1332598
13. Hees, J., Zabrodiec, D., Maßmeyer, A., Hatzfeld, O. and Kneer, R. "Experimental Investigation into the Influence of the Oxygen Concentration on a Pulverized Coal Swirl Flame in Oxy-Fuel Atmosphere". *Fuel* 240 (2019), 64–74. https://doi.org/10.1016/j.fuel.2018.11.111
14. Zabrodiec, D., Maßmeyer, A., Hees, J., Hatzfeld, O. and Kneer, R. "Flow Pattern and Behavior of 40 $kW_{th}$ Pulverized Torrefied Biomass Flames Under Atmospheric and Oxy-Fuel Conditions". *$14^{th}$ International Conference on Sustainable Development of Energy, Water and Environment Systems*. Dubrovnik, Croatia, 2019
15. Maßmeyer, A., Zabrodiec, D., Hees, J., Kreitzberg, T., Hatzfeld, O. and Kneer, R. "Flame Pattern Analysis for 60 $kW_{th}$ Flames Under Conventional Air-Fired and Oxy-Fuel Conditions for Two Different Types of Coal". *Fuel* 271 (2020), 117457. https://doi.org/10.1016/j.fuel.2020.117457
16. Becker, L., Pielsticker, S., Böhm, B., Kneer, R. and Dreizler, A. "Particle Dynamics in a Gas Assisted Coal Combustion Chamber Using Advanced Laser Diagnostics". *Fuel* 269 (2020), 117188. https://doi.org/10.1016/j.fuel.2020.117188
17. Sadiki, A., Agrebi, S., Chrigui, M., Doost, S., Knappstein, R., di Mare, F., Janicka, J., Maßmeyer, A., Zabrodiec, D., Hees, J. and Kneer, R. "Analyzing the Effects of Turbulence and Multiphase Treatments on Oxy-Coal Combustion Process Predictions Using LES and RANS". *Chemical Engineering Science* 166 (2017), 283–302. https://doi.org/10.1016/j.ces.2017.03.015
18. Nicolai, H., Debiagi, P., Wen, X., Dressler, L., Maßmeyer, A., Janicka, J. and Hasse, C. "Flamelet LES of Swirl-Stabilized Oxy-Fuel Flames Using Directly Coupled Multi-Step Solid Fuel Kinetics". *Combustion and Flame* 241 (2022), 112062. https://doi.org/10.1016/j.combustflame.2022.112062
19. Wen, X., Nicolai, H., Debiagi, P., Zabrodiec, D., Maßmeyer, A., Kneer, R. and Hasse, C. "Flamelet LES of a 40 $kW_{th}$ Pulverized Torrefied Biomass Furnace in Air and Oxy-Fuel Atmospheres". *Proceedings of the Combustion Institute* 39.4 (2023), 4563–4572. https://doi.org/10.1016/j.proci.2022.07.135
20. Askarizadeh, H., Pielsticker, S., Nicolai, H., Kneer, R., Hasse, C., Özer, B. and Maßmeyer, A. "Numerical Study on the Influence of Devolatilisation Kinetics on Pulverised Solid Fuel Turbulent Swirling Flames Under Oxyfuel Conditions". *$14^{th}$ European Conference on Industrial Furnaces and Boilers*. Algarve, Portugal, 2024
21. Hees, J., Zabrodiec, D., Maßmeyer, A., Pielsticker, S., Gövert, B., Habermehl, M., Hatzfeld, O. and Kneer, R. "Detailed Analyzes of Pulverized Coal Swirl Flames in Oxy-Fuel Atmospheres".

*Combustion and Flame* 172 (2016), 289–301. https://doi.org/10.1016/j.combustflame.2016.07.028

22. Zabrodiec, D., Hees, J., Maßmeyer, A., Lehn, F. vom, Habermehl, M., Hatzfeld, O. and Kneer, R. "Experimental Investigation of Pulverized Coal Flames in $CO_2/O_2$- and $N_2/O_2$-Atmospheres: Comparison of Solid Particle Radiative Characteristics". *Fuel* 201 (2017), 136–147. https://doi.org/10.1016/j.fuel.2016.11.097
23. Zabrodiec, D., Maßmeyer, A., Hees, J., Hatzfeld, O. and Kneer, R. "Flow Pattern and Behavior of 40 $kW_{th}$ Pulverized Torrefied Biomass Flames Under Atmospheric and Oxy-Fuel Conditions". *Renewable and Sustainable Energy Reviews* 138 (2021), 110493. https://doi.org/10.1016/j.rser.2020.110493
24. Özer, B., Maßmeyer, A., Zabrodiec, D., Hatzfeld, O. and Kneer, R. "Experimental Study on the Influence of Methane-Cofiring on Pulverized Coal Flames Under Air and Oxy-Fuel Conditions". *30. Deutscher Flammentag*. Hannover, Germany, 2021. https://doi.org/10.18154/RWTH-2022-02000
25. Özer, B., Zabrodiec, D., Kneer, R. and Maßmeyer, A. "Experimental Investigation of 40 $kW_{th}$ Methane-Assisted and Self-Sustained Pulverized Biomass Flames". *Proceedings of the Combustion Institute* 39.3 (2023), 3343–3351. https://doi.org/10.1016/j.proci.2022.07.112
26. Zabrodiec, D. "Experimental Characterization of Turbulent Particle Laden Reactive Flows Under Air and Oxy-Fuel Conditions". PhD thesis. RWTH Aachen University, 2021. https://doi.org/10.18154/RWTH-2022-02974
27. Bürkle, S., Becker, L., Dreizler, A. and Wagner, S. "Experimental Investigation of the Flue Gas Thermochemical Composition of an Oxy-Fuel Swirl Burner". *Fuel* 231 (2018), 61–72. https://doi.org/10.1016/j.fuel.2018.05.039
28. Schneider, H., Valentiner, S., Vorobiev, N., Böhm, B., Schiemann, M., Scherer, V., Kneer, R. and Dreizler, A. "Investigation on Flow Dynamics and Temperatures of Solid Fuel Particles in a Gas-Assisted Oxy-Fuel Combustion Chamber". *Fuel* 286 (2021), 119424. https://doi.org/10.1016/j.fuel.2020.119424
29. Schneider, H., Bonarens, M., Hebel, J., Hamel, H., Emmert, J., Böhm, B., Wagner, S., Kneer, R. and Dreizler, A. "Combined Flow, Temperature and Soot Investigation in Oxy-Fuel Biomass Combustion Under Varying Oxygen Concentrations Using Laser-Optical Diagnostics". *Fuel* 362 (2024), 130771. https://doi.org/10.1016/j.fuel.2023.130771
30. Hees, J., Zabrodiec, D., Maßmeyer, A., Habermehl, M. and Kneer, R. "Experimental Investigation and Comparison of Pulverized Coal Combustion in $CO_2/O_2$- and $N_2/O_2$-Atmospheres". *Flow, Turbulence and Combustion* 96.2 (2016), 417–431. https://doi.org/10.1007/s10494-015-9662-9
31. Gaydon, A. G. *The Spectroscopy of Flames*. London: Chapman and Hall, 1974. https://doi.org/10.1007/978-94-009-5720-6

# Semi-Industrial Scale Pulverised Solid Fuel Combustion Experiments

15

Dominik König, Janik Hebel, Jochen Ströhle, Andreas Dreizler and Bernd Epple

## Contents




D. König (✉) · J. Ströhle · B. Epple
Institute for Energy Systems and Technology, Technische Universität Darmstadt, Darmstadt, Germany
e-mail: dominik.koenig@est.tu-darmstadt.de

J. Ströhle
e-mail: jochen.stroehle@est.tu-darmstadt.de

B. Epple
e-mail: bernd.epple@est.tu-darmstadt.de

J. Hebel · A. Dreizler
Institute of Reactive Flows and Diagnostics, Technische Universität Darmstadt, Darmstadt, Germany
e-mail: hebel@rsm.tu-darmstadt.de

A. Dreizler
e-mail: dreizler@rsm.tu-darmstadt.de






R. Kneer et al. (eds.), *Clean Energy Conversion*,
https://doi.org/10.1007/978-3-031-98956-8_15

**Abstract**

The main objective of this chapter is to identify the effect and interaction of flame stabilising phenomena in a furnace operating close to industrial conditions. Compared to the laboratory-scale combustion chamber outlined in Chap. 13 and the pilot-scale chamber in Chap. 14, the focus is on the influence of upscaling on the combustion properties such as flame stabilisation, which will allow assessing to what extent results from smaller furnaces can be transferred to an industrial scale. The experiments are conducted in a water-cooled top-fired furnace with a maximum thermal power of 1 $MW_{th}$. Using a variety of complementary measurement technologies at various ports along the furnace axis, mean values and fluctuations of temperatures, particle velocities, gas composition, and radiative heat transfer are determined. Various biomass flames are investigated to provide a comprehensive study of combustion phenomena in a semi-industrial-scale furnace.

## 15.1 Introduction

Different phenomena in a flame have various effects depending on the size of the flame and thus the overall thermal power. Using small laboratory-scale combustion chambers helps analysing the occurring phenomena in the flame, due to good accessibility for measurement techniques. But with increasing thermal power and thus flame size, the mutual interactions and combustion phenomena change. Therefore the flame behaviour has also to be investigated in a semi-industrial combustion chamber to understand which characteristics can be transferred from a lab-scale to an industrial-scale combustor. For this purpose, a 1 $MW_{th}$ burner is developed, based on the burner geometries used at laboratory scale. The upscaling of thermal power from a lab-scale combustion chamber to industrial dimensions is accompanied by different boundary conditions influencing the overall structure and stability of the flame. Especially the radiative heat loss and the influence of the different volume-to-surface ratios may lead to local flame extinction and formation of pollutants. The interaction between these different physical and chemical mechanisms on the flame has to be analysed. With this knowledge, the experimental results of laboratory-scale furnaces discussed in the Chaps. 13 and 14 with lower thermal power can be interpreted and the possibility of transfer to industrial scales can be assessed.

In literature, different oxy-fuel combustion systems are investigated. For laboratory scale, often a synthetic flue gas is used. Depending on the recirculation path, the flue gas in the oxidiser line contains also water, nitrogen, and other combustion products. The various components lead to more complex chemistry. Hees et al. [1] and Habermehl et al. [2] have analysed the flame stability of a 60 $kW_{th}$ burner using

CO and flame images as an indicator. They have shown that unstable flames do not anchor at the diffuser outlet. Toporov et al. [3] extracted the influence of the oxygen content on flame stability showing that for an oxidiser consisting of $CO_2$ and $O_2$ the oxygen content has to rise to 27% to reach similar flame temperatures as for the air case. All the experiments of Hees et al. [1], Habermehl et al. [2] and Toporov et al. [3] are performed with a mixture of $O_2$ and $CO_2$ in a refractory lined combustion chamber with heated walls. Analysis with recirculation of flue gas was performed by Smart et al. [4] on a test facility with a thermal power of 500 $kW_{th}$. According to this study, the recycling ratio in the oxidiser stream influences the flame stability, which was quantified via the temperature profile and the oscillation frequency of the flame. Another experimental investigation of oxy-fuel combustion using recycled flue gas was performed by Hjärtstam et al. [5] in a 100 $kW_{th}$ combustion chamber at *Chalmers University*. Different measurement techniques indicated that an earlier ignition and more intense combustion take place in oxy-fuel combustion due to the higher oxygen content in the oxidiser stream.

For classical air combustion and oxy-fuel combustion with a synthetic $O_2/CO_2$ mixture, the dominating chemical reaction determining the flame speed and the promotion of chain branching is known [6,7]. Hjärtstam et al. [5] analysed the CO content in the flame at *Chalmers University*, using a top fired 100 $kW_{th}$ oxy-fuel burner with flue gas recirculation. The oxygen content was varied between 25 and 29%. As fuel, pulverised lignite was used. These previous experiments revealed that compared to air combustion, with increasing oxygen content the CO content in the inner recirculation zone increases for oxy-fuel combustion. However, the overall exit CO was similar for all cases.

It is concluded that previous studies have focused on lignite as a fuel, and measurements of CO in oxy-fuel flames have been made only with $O_2/CO_2$ mixtures or with flue gas recirculation. The influence of recirculated flue gas on the combustion, compared to a mixture of $CO_2$ and $O_2$ is not investigated yet. In the last few years scientific experiments were focused on coal combustion or were performed in pilot plants with low thermal power. Thus experimental data from a semi-industrial combustion chamber with the possibility of flue gas recirculation and a mixture of $O_2/CO_2$, water-cooled walls, and biomass as fuel are missing.

In this chapter, this knowledge gap is reduced, using different measurement techniques, to quantify as many parameters as possible. At first, a new burner is uspcaled, based on the geometry of the burners outlined in the Chaps. 13 and 14. Secondly, the experimental setup is introduced. The main focus is on measuring the flow field, temperature and species distribution in the flame, and radiative heat flux from the flame to the wall. The sum of these measurement results serves the validation of the simulation of this combustion chamber as described in Chap. 19.

## 15.2 Development and Experimental Setup

The laboratory scale burners introduced in Sects. 13.2.1 and 14.2.1 operated with a thermal power of 40 and 60 $kW_{th}$ are based on a design originally introduced by

Toporov et al. [3] and advanced by Heil et al. [8] and Habermehl et al. [9]. For consistency, the geometry of the 1 $MW_{th}$ burner aligns with the characteristics of these burners. Based on this requirement, a scaling correlation, dimensional analysis, and CFD simulations are applied.

Mathematical correlations are often used to reduce the effort involved in the up-scaling process. The aim of these correlations is to obtain similar values for emissions, heat transfer rate, and flame stability for the up-scaled burner as for the reference burner [10]). Weber and Mancini [11] used a constant velocity (CV) and a constant residence time (CRT) approach for scaling in the range from 0.176 to 50 $MW_{th}$. For penetration flames, both approaches are suitable for the entire thermal power range investigated by Weber and Mancini [11]. The idea of the CRT approach is to keep the ratio of the burner diameter and the inlet velocity constant, which is a proxy for the macro turbulent mixing time scale for fuel and air. Intense mixing of the fuel and the oxidiser ensures stable combustion [12]). Carrasco-Maldonado et al. [13] used the CV and CRT approaches to downscale an oxy-fuel-fired clinker burner from 47–0.5 $MW_{th}$. Pramanik and Ravikrishna [14] introduced two approaches—constant volume to jet momentum ratio (CM) and constant volume to jet kinetic energy ratio (CK)—for the up-scaling of a 3.3 $kW_{th}$ combustor to 25 $kW_{th}$. Their investigation showed that the CK and CM criterion corresponds closely to the CV approach and is therefore suitable for the scaling of combustors with reverse-flow.

Based on these options documented in literature, different approaches are compared to select the most suitable one. Afterwards the new burner is simulated and compared with the two laboratory-scale burners.

### 15.2.1 Scaling Correlation and Dimensional Analysis

The two pilot-scale burners at the *Institute of Heat and Mass Transfer* (WSA) are both top-fired and enable operation with an air or an oxy-fuel atmosphere with a variable mixture of $O_2/CO_2$. Both are mounted in a combustion chamber with an inner height of 4.2 m and an inner diameter of 0.4 m. The walls are refractory lined and heated during operation. Detailed information is provided in Sect. 14.2.1. The upscaled burner at the *Institute for Energy Systems and Technology* is mounted in a combustion chamber with a height of 7 m and an inner diameter of 1.1 m. The wall has a hexagonal cross-section and is water cooled. It can be operated with a maximum thermal load of 1 $MW_{th}$.

For the upscaling process the correlation between the maximum thermal power $\dot{Q}_0$ and the burner diffuser diameter is applied, using (15.1) for $\dot{Q}_0$:

$$\dot{Q}_0 = \dot{m}_{0,\text{fuel}}\, \Delta H_\text{u} = u_0\, \rho_0\, \frac{\pi}{4}\, d_0^2\, \Delta H_\text{u} = K\, u_0\, \rho_0\, d_0^2 \tag{15.1}$$

The thermal power $\dot{Q}_0$ is calculated via the lower heating value $\Delta H_\text{u}$ of the fuel and the fuel mass flow rate $\dot{m}_{0,\text{fuel}}$. With the assumption of a constant lower heating value, the geometric factor $\pi/4$, and the reference density $\rho_0$, the scaling parameter $K$ is deduced. To calculate the upscaled diameter of the diffuser, the pursued thermal load

**Table 15.1** Results using correlations from the literature for up-scaling WSA40 to 500 $kW_{th}$ termed EST500 [15]. Geometric values are illustrated in Fig. 15.3. For detailed information of the final geometry, see also Sect. 15.2.3

| | Thermal load | Flow cross-section area | | Diffuser | |
|---|---|---|---|---|---|
| | $\dot{Q}$ [$kW_{th}$] | $A_{prim}$ [$mm^2$] | $A_{sec}$ [$mm^2$] | $L$ [mm] | $d$ [mm] |
| WSA40 | 40 | 292.2 | 310.7 | 53 | 90 |
| EST500 | 500 | 3782.0 | 10307.6 | 235 | 438 |
| CV | 500 | 9833 | 11157 | 187.4 | 318.2 |
| CRT | 500 | 1547.4 | 1645.7 | 122 | 207.1 |
| CM | 500 | 1033.0 | 1098.6 | 99.7 | 169.2 |
| CK | 500 | 2564.3 | 2727.3 | 157 | 266.6 |

of the upscaled burner $\dot{Q}_{scaled}$ is related to the reference burner $\dot{Q}_0$ and the diameter of the reference burner $d_0$ as detailed in (15.2):

$$d_{scaled} = d_0 \left( \frac{\dot{Q}_{scaled}}{\dot{Q}_0} \right)^C , \tag{15.2}$$

whereby the values for the exponent $C$ differ between the various upscaling approaches. From (15.2), all other dimensions of the burner can be derived, because the geometric similarity is maintained. Table 15.1 summarises the dimensions of the different upscaling methods used as well as the inlet areas of the laboratory-scale burner (WSA40) and the semi-industrial burner (EST500). The differences between the theoretical geometric dimension of the upscaling approaches and the actual dimension of the burner EST500 refer to the geometrical boundary condition in the combustion chamber.

With the scaling approach used, only the diffuser and burner diameter can be estimated. Based on the inlet streams, a dimensionless analysis is conducted. For this purpose Buckingham's $\Pi$-theorem is used, which is a theorem of similarity theory and dimensional analysis [16]. The parameter with the highest influence is extracted. In accordance to the study by Tong et al. [17], this is the inner recirculation zone (IRZ), which is used as characteristic value. For a swirl-stabilised burner, the size, the opening angle and the velocities at the inlet have the strongest influence on the IRZ. The dimension of the IRZ marked by the axis of a comparable ellipsoid ($a$ and $c$) as well as the opening angle $\alpha$ are shown schematically in Fig. 15.1. For the upscaling process, the ratio of the IRZ surface, normalised by the diffuser surface, is used and held constant for similarity.

Detailed information regarding the dimensional analysis are documented in the studies performed by Richter et al. [10, 15]. The exact geometric dimension of the new burner, which is detailed in Fig. 15.3 and Table 15.2 in Sect. 15.2.3, is determined by the spatial dimensions of the existing combustion chamber. The scaled position of the primary and secondary duct is the same as for the two laboratory-scale burners. The velocity and thus the mass flow rate of the gas in the primary stream is designed in such a way, that the momentum ensures the transport of the pulverised fuel. Following

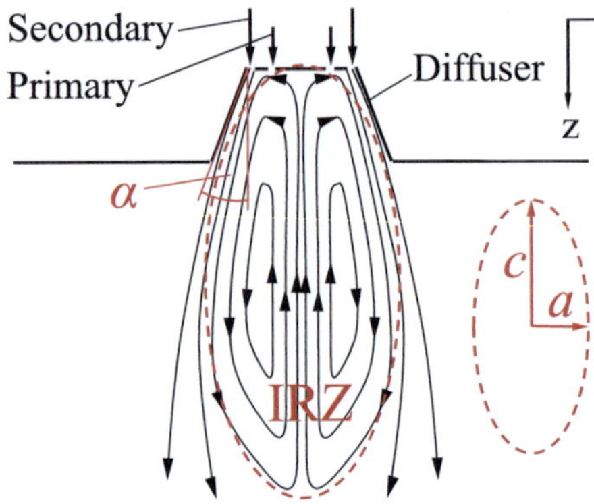


**Fig. 15.1** Qualitative sketch of the IRZ at the diffuser of a burner with primary and secondary oxidant annual channels. The approximation of the IRZ with an ellipsoid is marked by red dashed lines [10, 17] (reprinted from Richter et al. [10], licensed under 5995901264073)

empirical experience, the transport gas mass flow has to be twice the maximum fuel mass flow. Based on this experience, a primary air velocity of 8.8 $\mathrm{ms^{-1}}$ is chosen. As the ratio of secondary to primary velocity is 2.4 for the WSA40 burner, the velocity of the secondary flow for the upscaled burner is set to 20 $\mathrm{ms^{-1}}$ [10].

### 15.2.2 Numerical Simulations

To validate the theoretical upscaling, multiple numerical simulations are performed to optimise the burner geometry. For these simulations, the open-source code OpenFOAM® v2006 is applied considering the flow, particles, chemical reactions as well as heat and mass transfer. The solver is termed `simpleCoalParcelFoam` and is based on an Euler-Lagrange method where the gas phase is modeled with a RANS approach. All equations for the continuous phase are Favre-averaged. The effective dynamic viscosity is calculated by the turbulence model $k$-$\epsilon$. In the OpenFOAM® standard codes, the assumption is made that the Schmidt number Sc is equal to one and thus the diffusivity corresponds to the viscosity. Detailed information regarding the simulation of the disperse phase and the combustion chemistry is provided in the study by Richter et al. [15]. The numerical model is validated using experimental data of the WSA40 and WSA60 reference burners. The WSA60 burner has an increased power of 60 $\mathrm{kW_{th}}$ but geometrical features similar to the WSA40 burner.

Summarising key features of the simulation, the velocity of the three different burners, WSA40, WSA60 and EST500 are compared in Fig. 15.2. It can be seen that at the centre line, the new 500 $\mathrm{kW_{th}}$ burner shows a higher similarity compared to the WSA60 burner, despite the fact that the diffuser opening angle is similar to WSA40. Due to the different location of tertiary air injection, the outer recirculation zone (ERZ) is located at a higher position compared to the two WSA burners. The overall maximum velocity in the diffuser is higher for the WSA burners, compared to the EST burner. However, the inlet velocity of the EST burner is much higher with values of 18 $\mathrm{ms^{-1}}$

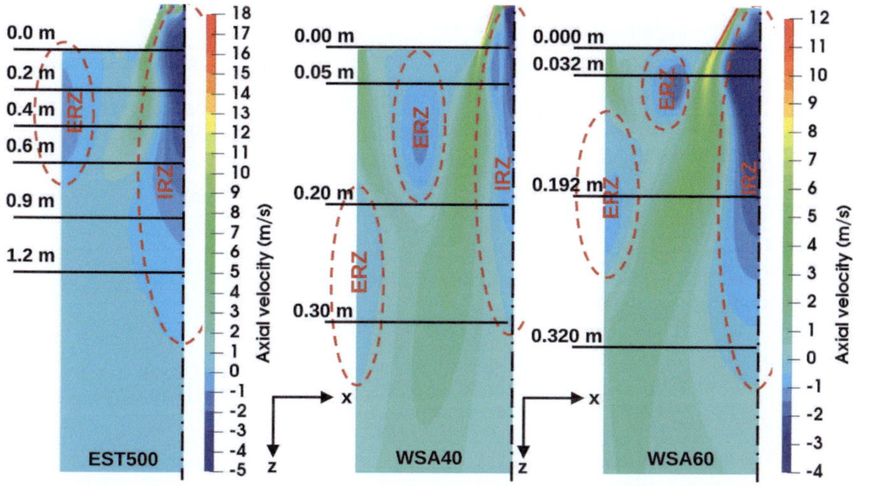


**Fig. 15.2** Velocity ($z$-direction) field results from CFD simulations for EST500 (left), WSA40 (middle), and WSA60 (right) (reprinted from Richter et al. [10], licensed under 5995901264073)

Through the upscaling approach the burners show a similarity for the inner recirculation zone in the upper part close to the burner. However, the shape of the recirculation zone in the middle and lower area of the flame changes due to the altered injection of the tertiary air. The outer recirculation zone also has a different stabilisation point in the combustion chamber for the EST500 burner compared to the configuration WSA60.

### 15.2.3 Geometry

The geometry of the burner is based on the upscaling, presented in Sect. 15.2. The different streams and dimensions are shown in Fig. 15.3 and Table 15.2. The burner is mounted in a top down fired furnace with a height of 7 m and a diameter of 1.1 m.

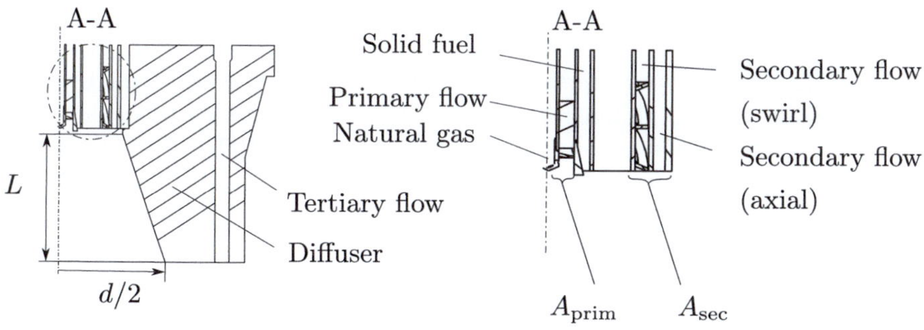


**Fig. 15.3** The different inlet streams for natural gas, solid fuel, and the four oxidiser channels of the burner (reprinted and slightly adapted from König et al. [18], licensed under CC BY 4.0)

**Table 15.2** Geometric dimensions of the burner

| Channel | Mean diameter [mm] | Width [mm] | Angle of incidence [°] |
|---|---|---|---|
| Solid fuel | 45.9 | 9.8 | 0.0 |
| Primary flow | 24.25 | 15.5 | 45.0 |
| Secondary flow (swirl) | 98.25 | 12.5 | 45.0 |
| Secondary flow (axial) | 112.75 | 6.4 | 0.0 |
| Tertiary flow | 332.5 | 28.0[1] | 0.0 |

[1] Diameter of one tertiary stream bore

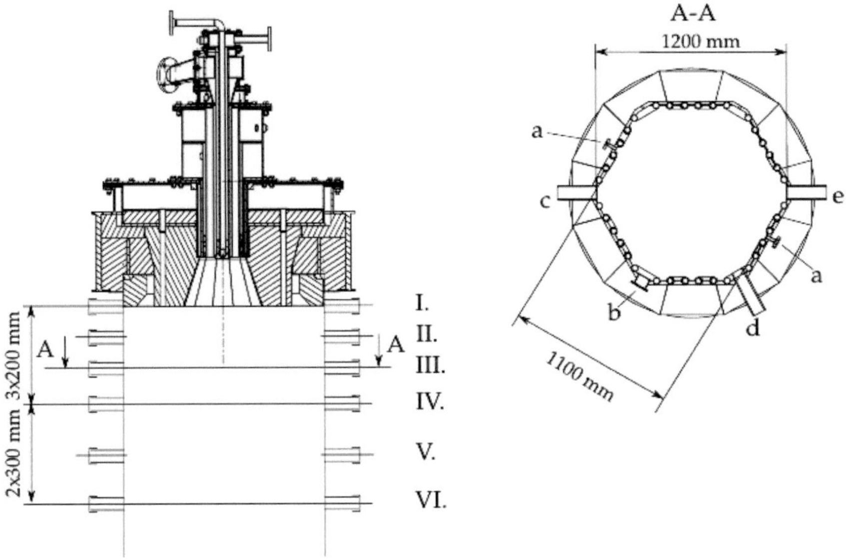


**Fig. 15.4** Vertical cut on the combustion chamber setup with measurement ports (left) and horizontal cut on the different measurement ports (right) (reprinted from König et al. [18], licensed under CC BY 4.0)

Additionally, different measurement ports are placed on the water cooled wall to enable measurements at different heights. A sketch of the combustion chamber is shown in Fig. 15.4. The two ports "a" have both a height of 100 mm and a width of 20 mm. Port "b" has a quadratic cross-section with an edge length of 100 mm. The three ports "c"–"e" exhibit a cylindrical shape with a diameter of 80 mm.

## 15.3 Measurement Techniques

For a comprehensive characterisation of the flame as needed for investigating parametric sensitivities and for providing data for validation of numerical simulations. For this purpose multiple measurement techniques are applied for flow field measurements, flame position, flame stability, temperature distribution, and species distribution in the flame.

### 15.3.1 Flue Gas Analysis

The investigation of the stability of the different operation conditions is primary based on the CO content in the flue gas. The flue gas is analysed using two gas analyser units from *ABB*. For oxygen analysis in the off-gas, an *ABB Magnos* and for the other combustion flue gas species like CO and $CO_2$ an *ABB Uras 26* is used. The moisture is measured using a *HYDROPHIL H2430-11*.

### 15.3.2 Flame Images

Flame luminosity is recorded in the visible spectral range through port “b”. For this purpose two CMOS cameras (2560×2160 px) are used that are equipped with a *Nikon Nikkor 50 mm f1.2* lens each. Both are operated with 50 Hz. The cameras capture a field of view of 200×250 $mm^2$ in the combustion chamber and are aligned to image one side of the combustion chamber including the geometric centre line. As the flame is supposed to be statistically axisymmetric, following this approach the flame can be analysed with a higher spatial resolution. The software *LaVision Davis 10.2.1* is used for image post-processing. Different operating conditions are investigated.

### 15.3.3 Particle Tracing Velocimetry

Particle tracking velocimetry (PTV) is based on the temporal tracking of individual tracer particles, which are illuminated with a laser and imaged onto a camera. The PTV method is a non-intrusive measurement that does not interfere with the combustion process. Using PTV, the velocity of solid fuel particles (pulverised walnut shells) is measured at various heights within the combustion chamber. Different operating conditions are examined to assess the influence of swirl and oxidising gas composition on particle velocity. These have Stokes numbers Stk $= 1$–$9$ and therefore do not follow the local gas phase flow. As the reaction progresses, the fuel particles lose mass and disappear because the walnut shell particles used have a low ash content. As a result, the local particle number density decreases with increasing axial distance from the burner. Therefore, statistically reliable particle velocity measurements are only possible up to a height of $-700$ mm.

The particles are illuminated using a double-pulsed frequency-doubled Nd:YAG laser (*InnoLas PIV 600*) with a repetition rate of 10 Hz and a pulse separation of $\Delta t = 60\,\mu s$. To illuminate the field of view of approximately $200 \times 250\,mm^2$ in the centre of the combustion chamber, and taking into account the considerable radiance of the flame and particles, a high laser energy of 360 mJ per pulse is employed. A laser sheet is formed using the optical setup illustrated in Fig. 15.5. A laser sheet with a thickness of approximately 900 µm full width at half maximum (FWHM) is generated using a cylindrical lens with a focal length of 2000 mm. To illuminate the entire height of the field of view monitored by the camera, the beam is expanded using a cylindrical lens with a focal length of −30 mm. The laser sheet is introduced into the combustion chamber at different heights. This is achieved by using a periscope to couple the laser beam into the combustion chamber through the two ports labelled "a" in Fig. 15.4 at the heights II–IV. Two additional mirrors are required to guide the laser beam around the heat exchanger (Fig. 15.5). The area, in which the laser is housed, is secured by optically impermeable textiles and cooled with an air conditioning system to protect the laser from dust and high temperatures during combustion operation, and to comply with laser safety standards.

Mie scattering images are recorded using a double frame CMOS camera (*LaVision Imager sCMOS*, $2560 \times 2160$ px, 16 bit) equipped with a 50 mm lens (*Nikon Nikkor 50 mm f1.2*). To mitigate the incandescence of the flame and the particles, two band-pass filters with a central wavelength of 532 nm and a 10 nm FWHM are employed (2× *Edmund Optics 65-216*). The cameras are mounted inside a water-cooled housing, which is connected to port "b" of the combustion chamber (see Fig. 15.4). A quartz window inside the housing seals the camera from the combustion chamber while maintaining optical accessibility. To prevent solid fuel particles from adhering to the window, a constant purge gas flow is employed in front of the window. Depending on the gas atmosphere, either air or $CO_2$ is utilised for purging.

The recorded Mie scattering images are processed to eliminate background noise resulting from the high radiance of the flame, which is particularly evident under oxy-fuel combustion conditions. In the next step, a multipass cross-correlation approach with a final interrogation window size of $48 \times 48$ px is employed to calculate particle image velocimetry (PIV) vector fields. The vector fields are post-processed based on peak ratio, correlation value, and an outlier detection algorithm. The PIV results are then used to track individual particles by the PTV methodology, with intensity and particle size thresholds adjusted to ensure accurate detection of solid fuel particles. The resulting vectors are binned onto a grid of $48 \times 48$ px with an overlap of 75%, resulting in a vector spacing of 1.14 mm. The processing is conducted using *LaVision Davis 10.2.1*.

### 15.3.4 Fourier-Transform Infrared Spectroscopy

With the Fourier-transform infrared spectroscopy (FTIR) measurements, radial profiles of the local gas composition are measured with an air-cooled extraction probe. The extracted gas is transferred to a filter unit and the FTIR analyser. Heated tubes are

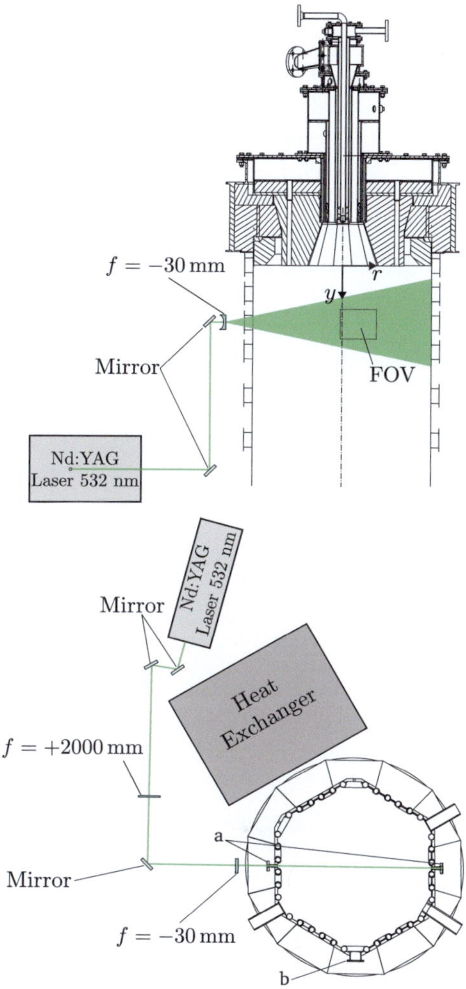


**Fig. 15.5** Schematic of the optical setup. In the upper part, a side view is presented. The periscope allows for adjustment of the measurement height within the combustor and includes a spherical lens with a focal length of −30 mm to expand the laser sheet. In the lower part, a top view of the setup is depicted

used to prevent the condensation of water. A sketch of the setup is shown in Fig. 15.6. Measurement data are collected at radial intervals of 5 cm. The data is averaged over 5 min. The instrument used is a *Ansyco Gasmet CX4000* FTIR analyser with *Calcmet* evaluation software.

### 15.3.5 Radiative Heat Flux Measurement

A Gardon gauge heat flux sensor type *GG01-250-SW* from *HUKSEFLUX* is used to measure wall heat fluxes up to 250 kW m$^{-2}$. The sensor is schematically shown in Fig. 15.7. The sensor is mounted on a lance that has purge nozzles to protect the

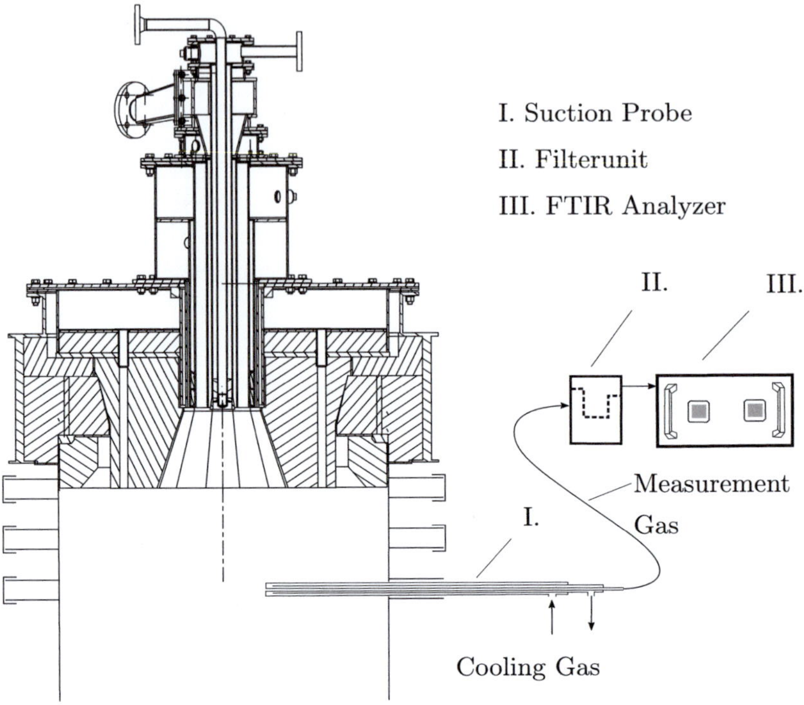


**Fig. 15.6** Setup of the FTIR measurements

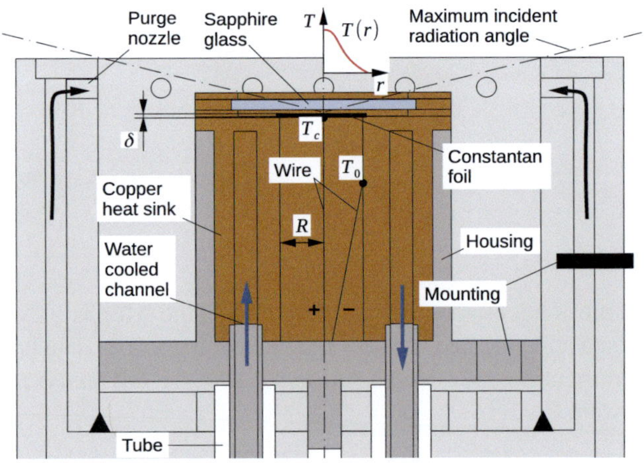


**Fig. 15.7** Sketch of the measurement lance and Gardon gauge sensor with the temperature profile of the absorption foil (reprinted from Richter et al. [19], licensed under CC BY 4.0)

sensor window from slagging, and consists of a water-cooled copper block, which serves as a heat sink.

At the upper part of the copper housing, a thermocouple is connected to a blackened foil. The radiative heat is absorbed by the black foil, which is made of constantan, and subsequently dissipates the heat along its radius to the surrounding heat sink. The heat flux $\dot{q}$, which is assumed to be transferred via radiation, is calculated by (15.3)

$$\dot{q} = \frac{4\,\lambda\,\delta}{R^2}\,(T_c - T_0) \tag{15.3}$$

where $\lambda$ is the thermal conductivity, $\delta$ the thickness, and $R$ the radius of the foil. $T_c$ is the centre temperature of the black foil and $T_0$ the temperature of the heat sink [20]. To exclude convective heat transfer from the measurement, a sapphire window is attached to the front of the sensor, which has a nominal transmissivity of 86% at a wavelength of 1 μm. This transmissivity is almost constant for the spectral range from 0.2 to 5.5 μm. As the sensor measures the relative change of incoming radiation, the error due to the sapphire glass is negligible. Together with the protective window, the sensor has an incidence angle limited to 150°, which means that not the entire hemisphere is covered. According to the manufacturer, the resulting error is also negligible, since the part of the field of view that is cut off has a very small angle of incidence to the sensor surface and has approximately the same temperature as the sensor. The calibration error of the sensor is specified as approximately 6%. The calibration is carried out in accordance with the ISO 14934-3 standard. The uncertainty caused by the transmitter and the control unit is low. In addition to the calibration uncertainties, there are other uncertainties, which are difficult to determine, such as spectral properties of the flame and are not considered here. Measurements are conducted at the ports III–VI for stable flame conditions for 5 min [19].

## 15.4 Operating Conditions

The analysed operation conditions are based on the stability analysis discussed in Sect. 15.5.1 and ranges for key parameters are specified in Table 15.3.

First, three different air operating conditions are analysed. Afterwards corresponding oxy-fuel operating conditions are developed. These operating conditions have different objectives and thus different parameters. They are summarised in Table 15.4.

**Table 15.3** Range of the varied operating conditions for the stability analysis

| Parameter | Operation range |
|---|---|
| Global air ratio [–] | 1.4–2.0 |
| Local air ratio [–] | 0.7–1.2 |
| Swirl number [–] | 0.7–1.0 |
| Thermal power [$kW_{th}$] | 400–750 |

**Table 15.4** Analysed air operating conditions (OP)

| Operating condition | | Staging $Air_{AS}$ | High mom. $Air_{HM}$ | Low power $Air_{LP}$ |
|---|---|---|---|---|
| Global air ratio $\lambda_{glob}$ | [–] | 1.65 | 1.7 | 1.75 |
| Local air ratio $\lambda_{loc}$ | [–] | 0.95 | 1.45 | 1.3 |
| Swirl number $S$ | [–] | 0.95 | 0.965 | 0.95 |
| Thermal power $P$ | [$kW_{th}$] | 520 | 440 | 420 |
| Fuel mass flow $\dot{m}_{fuel}$ | [$kg\,h^{-1}$] | 112 | 95 | 90 |
| Sec. flow (axial) $\dot{V}_{axial}$ | [$Nm^3\,h^{-1}$] | 50 | 50 | 45 |
| Sec. flow (swirl) $\dot{V}_{swirl}$ | $Nm^3\,h^{-1}$] | 345 | 450 | 350 |
| Tertiary flow $\dot{V}_{ter}$ | $Nm^3\,h^{-1}$] | 360 | 90 | 170 |

The $Air_{AS}$ operating condition corresponds to a classical air staging configuration. The high air momentum operating condition $Air_{HM}$ features a low fuel to air momentum flux ratio and thus a strong recirculation zone, but a local air ratio higher than one. The low power air operating condition $Air_{LP}$ has a similar fuel to air momentum flux ratio compared to the standard air operation, but has a lower thermal load. This is due to the fact that with this operating condition the differences between synthetic flue gas and recirculated flue gas are analysed. Due to technical restriction of the pilot plant, the thermal power has to be reduced for this experiments. Due to the low power, also an higher local air ratio is chosen. For each operation condition, a natural gas assistance of 2 $Nm^3h^{-1}$ is needed. The primary air is not used during pulverised fuel combustion. The air ratio $\lambda$, the swirl number $S$, and the air to fuel momentum ratio ($MR$) are calculated via (15.4)–(15.7):

$$\lambda_{loc} = \frac{\dot{V}_{O_2,TG} + \dot{V}_{O_2,Sec}}{\dot{V}_{O_2,stoi}}, \tag{15.4}$$

$$\lambda_{glob} = \frac{\dot{V}_{O_2,TG} + \dot{V}_{O_2,Sec} + \dot{V}_{O_2,Ter}}{\dot{V}_{O_2,stoi}}, \tag{15.5}$$

$$S = \frac{L_{swirl}}{\sum r\, I_{axial}} = \frac{r_{swirl}\,\dot{V}_{swirl}\,u_{swirl}\,\cos(\alpha)}{r_{swirl}\,\dot{V}_{swirl}\,u_{swirl}\,\sin(\alpha) + r_{axial}\,\dot{V}_{axial}\,u_{axial}}, \tag{15.6}$$

$$MR = \frac{\dot{m}_{fuel}\,u_{fuel}^2\,\frac{1}{\dot{V}_{fuel}}}{\dot{m}_{Swirl}\,u_{Swirl}^2\,\frac{1}{\dot{V}_{Swirl}}} \tag{15.7}$$

with

$$\dot{m}_{\text{fuel}} = \dot{m}_{\text{solid}} + \dot{m}_{\text{TG}}. \tag{15.8}$$

For the local air ratio, only the transport gas, primary and secondary oxidiser inlet flows are taken into account. The global air ratio $\lambda_{\text{glob}}$ is calculated considering all oxidiser streams entering the combustion chamber. The calculation of the swirl number is based on the geometrical swirl number of the burner, where $L_{swirl}$ is the angular momentum, $r$ the radius of the axial momentum components, and $I_{axial}$ is the momentum of the axial oxidiser flows. From the geometry, it follows that a maximum swirl number of $S = 1$ can be achieved. Claypole and Syred [21] showed that the geometric swirl number and the real swirl number correlate linearly. The transport of the solid fuel from the feeding unit to the combustor is performed with air for the air combustion and with a mixture of $CO_2$ and $O_2$ for the oxy-fuel combustion. The amount of transport gas is defined by the maximum solid fuel mass flow and held constant for each operating condition. Additionally, the fuel momentum flux into the combustion chamber has to be considered. To secure a similar entrance momentum of the transport gas into the combustion chamber, the transport gas is set to 110 $\text{Nm}^3\,\text{h}^{-1}$ for air combustion and to 95 $\text{Nm}^3\,\text{h}^{-1}$ for oxy-fuel combustion. As other investigations have shown that a small amount of oxygen in the transport gas supports the combustion stability [22], the oxygen content in the oxy-fuel transport gas is set to 15 vol.%. For the first analysis, different parameter are varied in the test run. The focus is on a thermal power of 500 $\text{kW}_{\text{th}}$, but also operation with 750 and 400 $\text{kW}_{\text{th}}$ have been analysed. The definition of the air to fuel momentum ratio, is a measure of the swirl generating momentum of the secondary stream and the swirl prohibiting momentum of the transport gas and the fuel [23].

For the following experiments, walnut shells are used as fuel. A detailed characterisation of the fuel can be found in Table 1.1.

Corresponding to the shown air operating conditions, various parameters are varied or held constant to ensure corresponding oxy-fuel operating conditions. In the oxidiser stream, the oxygen concentration for the different oxy-fuel conditions is set to 27, 30 and 33 vol.%. With varying oxidiser content in the inlet stream, one of the three other parameters, air ratio, thermal power or the volume flows and thus the velocity field at the burner inlet, varies too. Additionally the diluent is varied between a synthetic flue gas with a mixture of $CO_2/O_2$ and an enrichment of recirculated flue gas with $O_2$. The variation for the different oxy-fuel combustion conditions are presented in Table 15.5.

**Table 15.5** Overview over the analysed oxy-fuel operating conditions. The $X$ symbolises the oxygen content in the oxidiser line for the different oxy-fuel conditions. The indices describe the varied parameter of the operating condition. Detailed information of the analysed air and oxy-fuel operating conditions can be found in Tables 15.4, 15.6, and 15.7

| Operating cond. | OXY-$X_{P}$ | OXY-$X_{Vol}$ | OXY-$X_{AR}$ | OXY-$X_{Dry}$ |
|---|---|---|---|---|
| $O_2$ cont. [vol.%] | 27, 30, 33 | 27, 30, 33 | 27, 30, 33 | 33 |
| Constant par. | $\lambda$, $\dot{V}$ | $\lambda$, $\dot{Q}$ | $\dot{Q}$, $\dot{V}$ | $\dot{Q}$, $\dot{V}$ |
| Variable par. | $P$ | $\dot{V}$ | $\lambda$ | $\lambda$ |
| Dilutant | flue gas | flue gas | flue gas | $CO_2$ |
| Corr. air OP | $Air_{AS}$, $Air_{HM}$ | $Air_{AS}$, $Air_{LP}$ | $Air_{AS}$ | $Air_{LP}$ |

**Table 15.6** Parameters for the low swirl (LS) operating conditions

| Parameter | $Air_{LS}$ | $OXY_{LS}$-$27_{P}$ | $OXY_{LS}$-$30_{P}$ | $OXY_{LS}$-$33_{P}$ |
|---|---|---|---|---|
| Global AR $\lambda_{glob}$ [–] | 1.65 | 1.75 | 1.75 | 1.75 |
| Local AR $\lambda_{loc}$ [–] | 0.95 | 0.95 | 0.95 | 0.95 |
| Swirl number $S$ [–] | 0.95 | 0.95 | 0.95 | 0.95 |
| Power $P$ [$kW_{th}$] | 500 | 570 | 610 | 650 |
| $O_2$ cont. [vol.%] | 21 | 27 | 30 | 33 |

**Table 15.7** Parameters of the low power (LP) operating conditions

| Parameter | $Air_{LP}$ | $OXY_{LP}$-$33_{Vol}$ | $OXY_{LP}$-$33_{Dry}$ |
|---|---|---|---|
| Global AR $\lambda_{glob}$ [–] | 1.73 | 1.73 | 1.7 |
| Local AR $\lambda_{loc}$ [–] | 1.31 | 1.43 | 1.42 |
| Power $P$ [$kW_{th}$] | 420 | 420 | 415 |
| $O_2$ cont. [vol.%] | 21 | 33 | 33 |

## 15.5 Experimental Results

This section starts with a discussion of the the stability analysis to provide an initial overview of the operating range of the combustor. The operating conditions selected from within this range are then examined in detail with the previously described measurement techniques, to analyse flow field, local gas composition and radiation properties.

### 15.5.1 Stability Analysis

For the detailed stability analysis the CO concentration is used as an indicator of whether an operating condition is stable or not. The CO content is measured without moisture in the flue gas. Figure 15.8 shows the different operating conditions according to their local and global air ratio.

The operating conditions correspond to a classical air staging with a local air ratio lower than one and a global air ratio higher than one. In the upper left hand area a region is identified in which some operating points with low CO emissions are observed. For better visibility this area is highlighted by three dashed orange lines. On the right hand side, the lean blow off limit is marked with a purple dashed line at a global air ratio of approximately $\lambda_{glob} = 2.0$. The high CO concentration in the flue gas, even for the most stable operating conditions is due to the higher particle size of the walnut shells compared to other pulverised fuel types with a $D_{50} = 200\,\mu m$ for walnut shells, and $D_{50} = 130\,\mu m$ for the coal [10] (cf. Fig. 1.2). Thus the burnout needs more time and the probability of quenching the flame increases. The most stable operating conditions are located at a global air ratio $\lambda_{glob} = 1.57$–$1.65$ and a local air ratio $\lambda_{loc} > 0.8$. In particular, a local air ratio of less than one must be preferred in order to limit $NO_x$ emissions.

For a better understanding of the different operating conditions and their stability, the swirl number is considered in a next step.

The CO emissions for different air operating conditions are shown in dependence of the swirl number and the global air ratio in Fig. 15.9. Figure 15.9 shows the same operating conditions as Fig. 15.8. Again, an area with decreased CO emissions of the flame is located between $\lambda_{glob} = 1.57$–$1.65$ and $S = 0.7$–$0.95$, and all four limits are marked with dashed magenta lines. The lean blow off limit is shown in red on the right-hand-side. It can be seen, that for higher power, a lower swirl is sufficient to reach the same stabilisation of the flame, compared to the 500 $kW_{th}$ flame. Due to the same transport gas volume flow and the nearly constant fuel mass flow, the fuel momentum is mostly unchanged. With increasing thermal power, the volume flow through the secondary channel increases. This ratio can be calculated via (15.7) in Sect. 15.4. It is obvious that the fuel mass flow increases as a result of the increasing

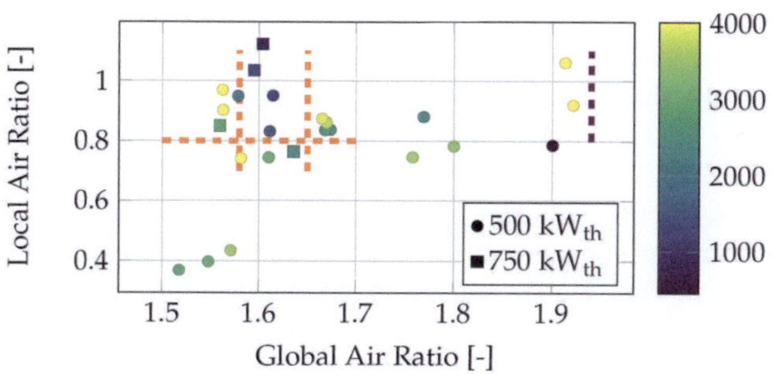


**Fig. 15.8** Emissions of CO for different air operating conditions for 500 and 750 $kW_{th}$, in dependence of the global and local air ratio. Operating conditions marked with a dot have a thermal power between 520 and 560 $kW_{th}$, those marked with a square have a thermal power of 750 $kW_{th}$. The CO content in the flue gas varies between 500 adn 6000 ppm, all operating conditions with an higher CO content than 4000 ppm are marked in yellow. The area of low CO content in the flue gas is marked as well as the lean blow of limit with orange and purple dashed lines, respectively (reprinted from König et al. [18], licensed under CC BY 4.0)

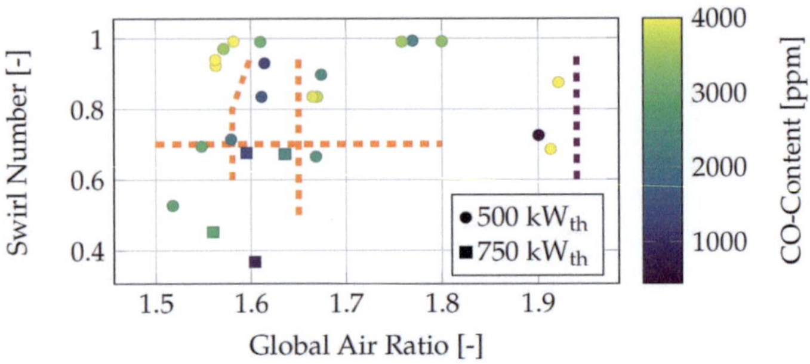


**Fig. 15.9** Emissions of CO at different air operating conditions for 500 and 750 $kW_{th}$ in dependence of the global air ratio and the swirl number. Operating conditions marked with a dot have a thermal power between 520 and 560 $kW_{th}$, the one marked with a square have a thermal power of 750 $kW_{th}$. The CO content in the flue gas is between 500 and 6000 ppm, all operating conditions with a higher CO content than 4000 ppm are marked in yellow. The area of low CO content in the flue gas is marked as well as the lean blow of limit with orange and purple dashed lines, respectively (reprinted from König et al. [18], licensed under CC BY 4.0)

power, but at the same time the air mass flow increases as well. Due to the constant transport gas volume flow, the inlet velocity of the fuel does not change significantly. This leads to a decrease of the fuel to air momentum flux ratio. Thus, a lower swirl is needed to develop a proper recirculation zone. The additionally tested operating condition with 400 $kW_{th}$ can not be stabilised with a classical air staging approach. This is due to the low flow velocity and an insufficient development of the recirculation zone. For this operating condition, a higher local air ratio is needed to generate enough swirl momentum for flame stabilisation.

### 15.5.2 Flame Structure

The flame images are used for a rough assessment of the flame position in the flame and together with the flow field, the anchoring of the flame can be analysed. The images of the three different air operating conditions $Air_{AS}$, $Air_{HM}$, and $Air_{LP}$ are shown in Fig. 15.10.

The images show the normalised intensity signal of the CMOS camera recordings for the different operating conditions at different heights. For the classical air staging operating condition in Fig. 15.10a), it is observed that the highest intensity and thus the main position of the flame is orientated vertically for the axial positions 0.2 and 0.4 m. Further downstream, at 0.6 m, the flame spreads horizontally and gets wider. On the last image at a distance of 0.9 m from the burner plane, the luminous part of the flame ends. The high swirl operating condition in Fig. 15.10b) shows a wider spreading in horizontal direction of the intensity on the first two heights at 0.2-0.4 m. This is due to the high swirl intensity. The fuel to air momentum flux ratio is lower and thus a strongly recirculating flame is formed, as previously discussed by Chen et al. [23]. The end of the flame is reached at a distance of 0.6 m from the burner plane.

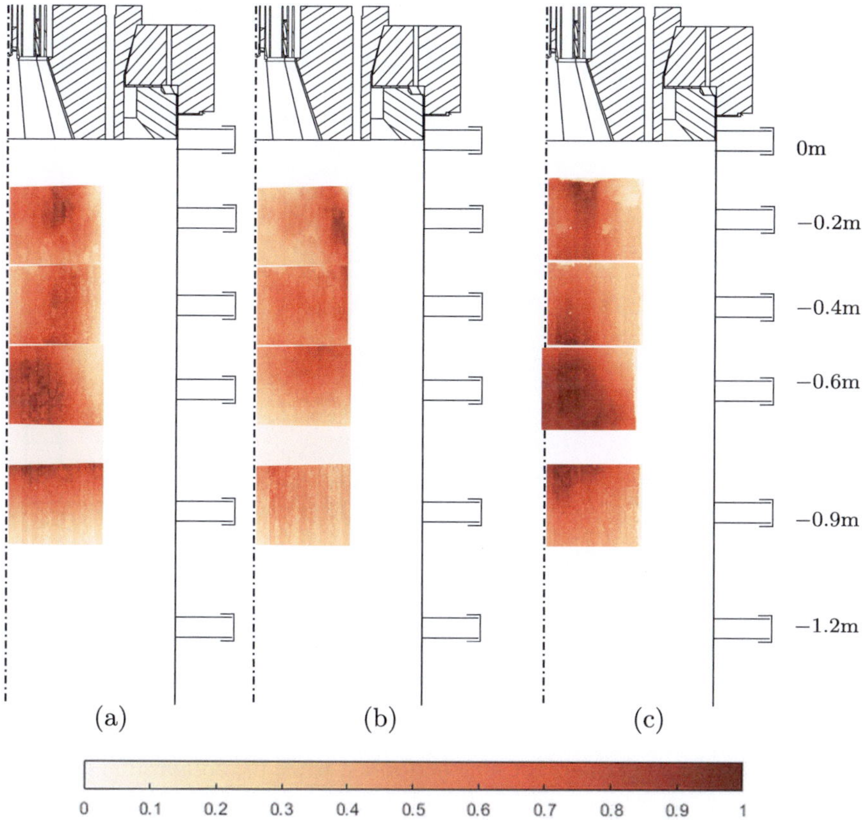


**Fig. 15.10** Normalised intensity images of the flame in the combustion chamber for **a)** classical air staging operating condition $Air_{AS}$, **b)** high momentum air $Air_{HM}$, and **c)** low power air $Air_{LP}$ operating condition (reprinted from König et al. [18], licensed under CC BY 4.0)

The low power operating condition in Fig. 15.10c) shows a similar shape compared to the air staging operating condition in Fig. 15.10a). The flame is oriented vertically in the first two heights and spreads on the fourth height in radial direction, which is caused by the similar fuel to air momentum flux ratio.

### 15.5.3 Particle Velocity Field

The results of the particle velocity measurements for the operating conditions $Air_{AS}$ and $OXY_{AS}$-$33_P$ are shown in Fig. 15.11. Both measurements use the same field of view to the right of the burner's central axis (see Fig. 15.5). For a better comparison of the mean velocity fields of the two operating conditions, the velocity fields measured for $Air_{AS}$ are mirrored on the centre line. It is also important to note that the different heights are measured on different days, as the traversing in height and adjustment of

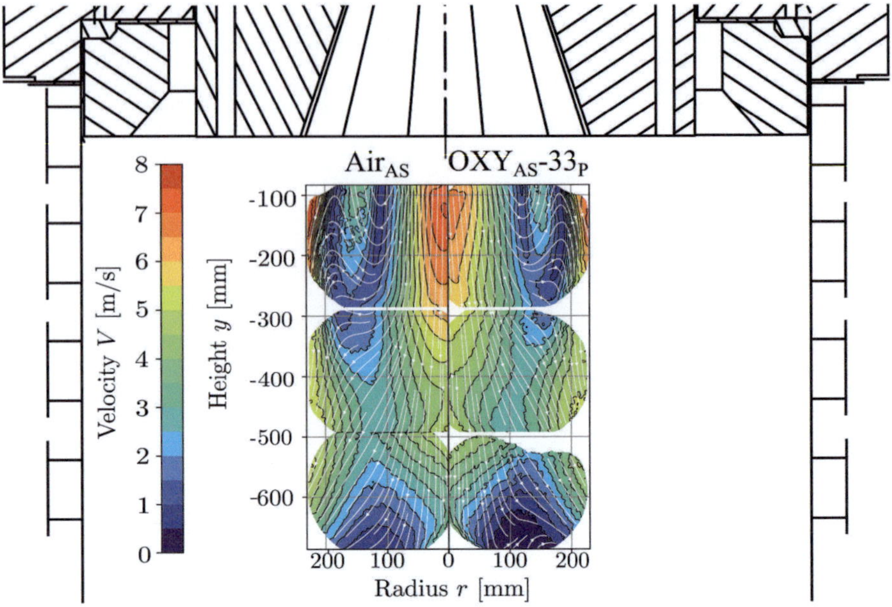


**Fig. 15.11** Mean particle velocity fields of the operating conditions Air$_{AS}$ (left) and OXY$_{AS}$-33$_{P}$ (right). To visualise both operating conditions in the same plot, Air$_{AS}$ is mirrored at the centre line of the burner. Note that the different heights were not measured simultaneously. The technical drawing is scaled to fit the measured area

the laser sheet and the camera must be performed while the combustion chamber is shut down.

At different heights of the same operating condition the velocity fields show good consistency with no significant offset from one height to another. This indicates good reproducibility of the operating conditions as the measurements were taken sequentially.

Both operating conditions show similar characteristics in the velocity field. Particles enter the combustion chamber close to the centre line of the burner in a jet-like manner. This can also be observed in the Mie scattering images, which indicate that the majority of the particles are located along this particle jet. At a height of $y = -100$ and $-200$ mm, particles at the outer edge of the central jet zone are accelerated upward towards the burner. Following this zone further outwards, particles are affected by the secondary flow and are strongly accelerated downwards. The area below $y = -300$ mm is dominated by the expansion of the central jet and the secondary flow.

In the optically accessible combustion chamber discussed in Sect. 13.2.2, both particle and gas phase measurements are performed simultaneously. There it is observed that the highest particle velocity is in the main jet, located at the outer radial position of the diffuser opening. Due to an inner gas phase recirculation zone, the particle velocity in the centre line is reduced. In contrast, the particle velocities measured in the present combustion chamber are highest in the central jet. Although gas phase velocities cannot be measured, the particle velocities indicate that there is no inner gas phase recirculation zone.

The comparison between combustion in an air atmosphere and in an oxy-fuel atmosphere with 33% $O_2$ shows lower particle velocities in the oxy-fuel operating condition. In general, replacing $N_2$ with $CO_2$ increases the heat capacity of the oxidising gas. This leads to a lower temperature [24], which in turn causes a lower expansion and therefore acceleration of the gas phase and subsequently the particle phase. This is compensated for by increasing the oxygen volume fraction of the oxidising gas for operation in an oxy-fuel atmosphere. Referring to Table 15.5, increasing the oxygen volume fraction also increases the thermal power in the case of $OXY_{AS}$-$33_P$. Although the velocity field of $OXY_{AS}$-$33_P$ is overall very close to the velocities in the air operating condition compared to the oxy-fuel operating conditions with lower oxygen volume, the difference is still apparent.

Comparing the flow field and the images of the flame, it is noticeable, that the main combustion and thus the highest intensity is located in the main shear layer of the secondary stream. In this area, the fuel, which comes together with the carrier gas from the centre, and the secondary flow, which is attached to the diffuser, are mixed.

### 15.5.4 Spatially Resolved Species Concentrations

For the experiments, three low power operating conditions are selected, introduced in Sect. 15.4. This operation conditions are chosen due to limits of the amount of synthetic flue gas. Together with the air operating condition $Air_{LP}$ in Fig. 15.12a) and the two oxy-fuel operating conditions $OXY_{LP}$-$33_{Vol}$ with flue gas recirculation and a mixture of $O_2/CO_2$ in Fig. 15.12b) and $OXY_{LP}$-$33_{Dry}$ in Fig. 15.12c) are analysed. The radial CO concentration profiles for the three flames are shown in Fig. 15.12.

It is noticeable that the CO content decreases with increasing distance from the centre line of the combustion chamber at the three measurement positions ranging from 0.4 to 0.9 m from the burner plane. For every operating condition, the highest CO concentration is measured in the flame on the fourth port height at 0.6 m with an amount between 18 and 23%, respectively. With increasing distance from the burner plane, the decreasing trend of the CO concentration is retarded. Following the approach of Habermehl et al. [2] that CO can be used as a marker for the flame, it can be concluded that the flame spreads in radial distance. However, at 1.2 m the profiles do not show a consistent trend such that a distinct reaction zone cannot be identified. Comparing the air and the oxy-fuel conditions, it is apparent that the CO concentration is significantly increased for oxy-fuel combustion. This is consistent with previous studies and is caused by the increased dominance of $CO_2$-based reactions in the gas mixture, according to (15.9) [5,6,25,26]:

$$H + CO_2 \leftrightarrow CO + OH. \tag{15.9}$$

This chemical reaction is of similar importance as (15.10)

$$H_2 + O_2 \leftrightarrow H_2O + O \tag{15.10}$$

for air combustion. The $CO_2$ competes with $O_2$ for molecular hydrogen. With an increasing moisture content in the flame, due to a wet flue gas recirculation, the water-gas shift reaction

$$H_2O + CO \leftrightarrow CO_2 + H_2 \tag{15.11}$$

has an increased influence for oxy-fuel combustion. For all operating conditions, however, the exit CO concentration of the flue gas is similar, as summarised in Table 15.8. The change from recirculated flue gas to pure carbon dioxide leads to a slightly decreased carbon monoxide concentration in the flue gas. This can be attributed to the higher proportion of water in the combustion process as shown in (15.11). As a result, the change from technical $CO_2$ to recirculated flue gas causes a locally increased CO content in the flame, but with only small variations in the exit carbon monoxide concentrations in the flue gas.

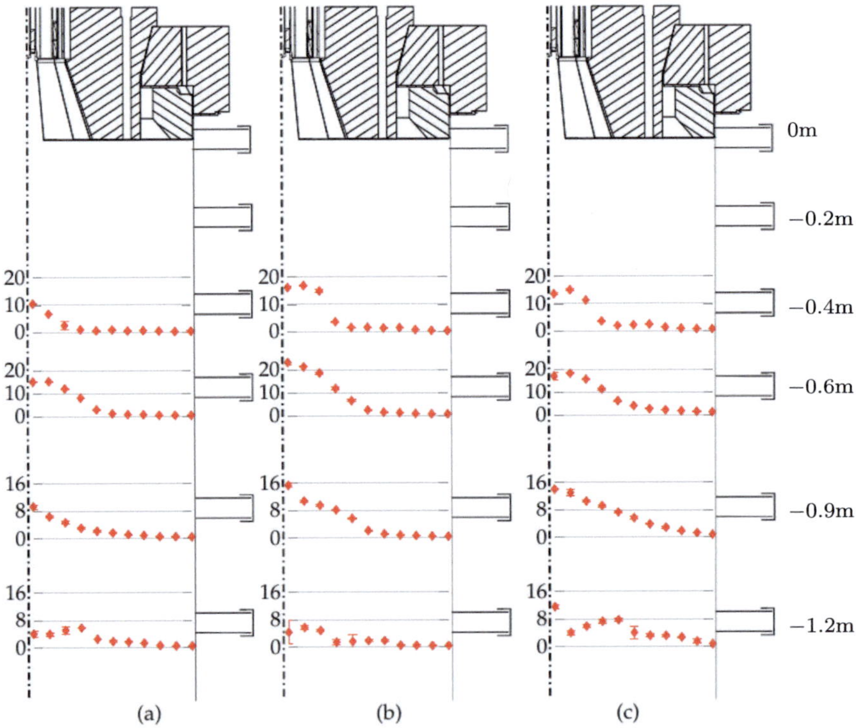


**Fig. 15.12** Dry CO concentration in ppm of the three operating conditions **a)** $Air_{LP}$, **b)** $OXY_{LP}$-$33_{Vol}$, and **c)** $OXY_{LP}$-$33_{Dry}$

**Table 15.8** Concentrations of CO and $H_2O$ in the flue gas

| | $Air_{LP}$ | $OXY_{LP}$-$33_{AR}$ | $OXY_{LP}$-$33_{Dry}$ |
|---|---|---|---|
| CO [ppm (dry)] | 780 | 1100 | 800 |
| $H_2O$ [vol.%] | 7.8 | 15.4 | 9 |

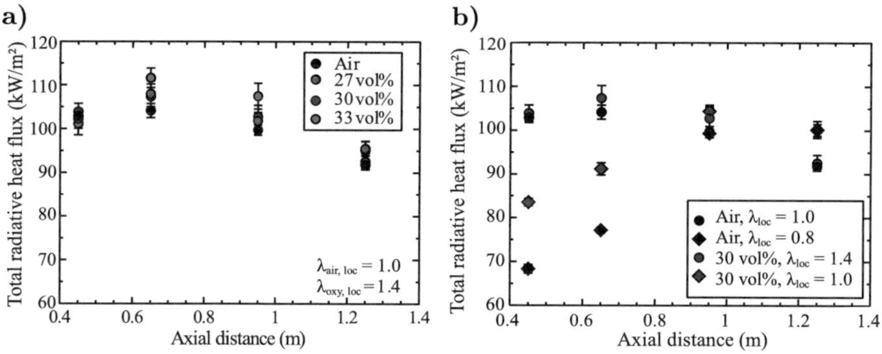


**Fig. 15.13** Total radiative heat flux at the combustion chamber wall in axial direction for walnut shells: comparison of **a)** air and oxy-fuel combustion at three different oxygen concentrations and **b)** different local air ratios for air and oxy-fuel combustion (reprinted from Richter et al. [19], licensed under CC BY 4.0)

### 15.5.5 Radiative Heat Flux Measurements

For the heat flux measurements, the high momentum operating condition $Air_{HM}$ with a thermal power of 500 $kW_{th}$, together with different air and oxy-fuel operating conditions are assessed. In a second step, the local air ratio has been changed for the $Air_{HM}$ and the $OXY_{HM}$-30 from 1.0 to 0.8 and from 1.4 to 1.0, respectively.

Figure 15.13a) shows the results of the radiative wall heat flux measurements plotted against the axial distance to the quarl for air and oxy-fuel combustion.

It is immediately noticeable that the axial progression does not decrease continuously but exhibits a local maximum. The highest incident radiation of 112 kW $m^{-2}$ is measured at the 0.6 m position for $OXY_{HM}$-30, due to the highest $O_2$ concentration and correspondingly increased adiabatic flame temperatures. The deviations from the air operating condition increase with increasing oxygen content in the oxidiser stream. Figure 15.13b) shows the comparison of an air point and $OXY_{HM}$-30 with low and high local air ratios. The global air ratio of $\lambda_{glob} = 1.65$ is constant for all operating conditions shown. The results for the corresponding air case from Fig. 15.13a) are included for comparison. For locally substoichiometric conditions, a lower radiative heat flux and lower temperatures are measured compared to the stoichiometric case. Furthermore, a long flame is generated, which distributes the heat radiation over a larger area.

## 15.6 Conclusion

In this chapter, the development and investigation of a semi-industrial oxy-fuel burner is presented. The combustor is based on a pilot-scale burner, described in Sect. 14.2.1. With increasing the size of a burner, the influence of different phenomena changes due to a change in boundary conditions, like volume-to-surface ratio and wall temperature. To analyse oxy-fuel combustion on a semi-industrial scale and to get a detailed insight into the flame, multiple measurement techniques are applied, analysing the combustion process. These are flame luminosity, flow field, temperature, radiative heat flux, and species concentration measurements. Together with selected operation conditions, the differences between air and oxy-fuel combustion is analysed. In this chapter two different operation conditions are discussed in detail. The influence of different oxygen contents in the oxidiser stream on flame position, flow field, and temperature profile in the flame shows that the flame position and the flow field have the highest similarity with the air flame for an oxygen content of 33 vol.% in the oxidiser stream. The flame temperature is more similar for the lowest analysed oxygen content of 27 vol.%. It is therefore not possible to optimise both the flow field and the temperature distribution in the flame to achieve similar values to the air case. The differences between a synthetic oxy-fuel process using a $CO_2/O_2$ mixture and a flue gas recirculation process also have been analysed. Changing the oxy-fuel mode from synthetic to flue gas recirculation, the moisture content in the oxidiser line increases from zero to 8%. Thus a higher amount of water is present in the flame. This leads to a slight increase of carbon monoxide in the flame zone and a slight increase of CO emissions measured in the flue gas. In total the CO concentrations in the oxy-fuel case are higher, compared to air combustion.

**Acknowledgements** This work has been funded by the Deutsche Forschungsgemeinschaft (DFG, German Research Foundation)—215035359—SFB/TRR 129 'Oxyflame'.

## References

1. Hees, J., Zabrodiec, D., Maßmeyer, A., Habermehl, M. and Kneer, R. "Experimental Investigation and Comparison of Pulverized Coal Combustion in $CO_2/O_2$- and $N_2/O_2$-Atmospheres". *Flow, Turbulence and Combustion* 96.2 (2016), 417–431. https://doi.org/10.1007/s10494-015-9662-9
2. Habermehl, M., Hees, J., Maßmeyer, A., Zabrodiec, D., Hatzfeld, O. and Kneer, R. "Comparison of Flame Stability Under Air and Oxy-Fuel Conditions for an Aerodynamically Stabilized Pulverized Coal Swirl Flame". *Journal of Energy Resources Technology* 138.4 (2016), 042209. https://doi.org/10.1115/1.4032940
3. Toporov, D., Bocian, P., Heil, P., Kellermann, A., Stadler, H., Tschunko, S., Förster, M. and Kneer, R. "Detailed Investigation of a Pulverized Fuel Swirl Flame in $CO_2/O_2$ Atmosphere". *Combustion and Flame* 155.4 (2008), 605–618. https://doi.org/10.1016/j.combustflame.2008.05.008
4. Smart, J. P., Lu, G., Yan, Y. and Riley, G. S. "Characterisation of an Oxy-Coal Flame Through Digital Imaging". *Combustion and Flame* 157.6 (2010), 1132–1139. https://doi.org/10.1016/j.combustflame.2009.10.017

5. Hjärtstam, S., Andersson, K., Johnsson, F. and Leckner, B. "Combustion Characteristics of Lignite-Fired Oxy-Fuel Flames". *Fuel* 88.11 (2009), 2216–2224. https://doi.org/10.1016/j.fuel.2009.05.011
6. Scheffknecht, G., Al-Makhadmeh, L., Schnell, U. and Maier, J. "Oxy-Fuel Coal Combustion—A Review of the Current State-of-the-Art". *International Journal of Greenhouse Gas Control* 5 (2011), S16–S35. https://doi.org/10.1016/j.ijggc.2011.05.020
7. Glarborg, P. and Bentzen, L. L. B. "Chemical Effects of a High $CO_2$ Concentration in Oxy-Fuel Combustion of Methane". *Energy & Fuels* 22.1 (2008), 291–296. https://doi.org/10.1021/ef7005854
8. Heil, P., Toporov, D., Stadler, H., Tschunko, S., Förster, M. and Kneer, R. "Development of an Oxycoal Swirl Burner Operating at Low $O_2$ Concentrations". *Fuel* 88.7 (2009), 1269–1274. https://doi.org/10.1016/j.fuel.2008.12.025
9. Habermehl, M., Erfurth, J., Toporov, D., Förster, M. and Kneer, R. "Experimental and Numerical Investigations on a Swirl Oxycoal Flame". *Applied Thermal Engineering* 49 (2012), 161–169. https://doi.org/10.1016/j.applthermaleng.2011.07.055
10. Richter, M., Ströhle, J. and Epple, B. "Up-Scaling of a Laboratory-Scale Pulverised Oxy-fuel Burner to Semi-Industrial-Scale Through a Flow Similarity Approach". *Fuel* 371 (2024), 131809. https://doi.org/10.1016/j.fuel.2024.131809
11. Weber, R. and Mancini, M. "On Scaling and Mathematical Modelling of Large Scale Industrial Flames". *Journal of the Energy Institute* 93.1 (2020), 43–51. https://doi.org/10.1016/j.joei.2019.04.010
12. Smart, J. P. and Morgan, D. J. "Exploring the Effects of Employing Different Scaling Criteria on Swirl Stabilised Pulverised Coal Burner Performance". *Combustion Science and Technology* 100.1-6 (1994), 331–343. https://doi.org/10.1080/00102209408935459
13. Carrasco-Maldonado, F., Spörl, R., Fleiger, K., Hoenig, V., Maier, J. and Scheffknecht, G. "Oxy-Fuel Combustion Technology for Cement Production – State of the Art Research and Technology Development". *International Journal of Greenhouse Gas Control* 45 (2016), 189–199. https://doi.org/10.1016/j.ijggc.2015.12.014
14. Pramanik, S. and Ravikrishna, R. V. "Investigation of Novel Scaling Criteria on a Reverse-Flow Combustor". *Energy* 206 (2020), 118156. https://doi.org/10.1016/j.energy.2020.118156
15. Richter, M., Maßmeyer, A., Askarizadeh, H., Ströhle, J. and Epple, B. "Up-Scaling of a Lab-Scale Oxyfuel Burner to a 1 $MW_{th}$ Semi-Industrial Oxyfuel Combustion Chamber (Poster)". $4^{th}$ *International Workshop on Oxy-Fuel Combustion – Book of Abstracts*. Ed. by Pielsticker, S., Maßmeyer, A., Scherer, V., Pitsch, H., Dreizler, A., Senneca, O. and Kneer, R. Naples, Italy, 2023. https://doi.org/10.18154/RWTH-2023-06259
16. Buckingham, E. "On Physically Similar Systems: Illustrations of the Use of Dimensional Equations". *Physical Review* 4.4 (1914), 345–376. https://doi.org/10.1103/PhysRev.4.345
17. Tong, Y., Chen, S., Li, M., Li, Z. and Klingmann, J. "Experimental Study on Bluff-Body Stabilized Premixed Flame with a Central Air/Fuel Jet". *Energies* 10.12 (2017), 2011. https://doi.org/10.3390/en10122011
18. König, D., Richter, M., Ströhle, J. and Epple, B. "Stability and Flame Structure Analysis of a Semi-Industrial Swirl-Stabilized Oxy-Fuel Combustion Chamber System for Biomass". *Energies* 18.6 (2025), 1513. https://doi.org/10.3390/en18061513
19. Richter, M., König, D., Ströhle, J. and Epple, B. "Radiative Heat Flux Measurement in a Semi-Industrial Oxyfuel Combustion Chamber with Biomass and Coal". *Energies* 17.11 (2024), 2735. https://doi.org/10.3390/en17112735
20. Gardon, R. "An Instrument for the Direct Measurement of Intense Thermal Radiation". *The Review of Scientific Instruments* 24.5 (1953), 366–370. https://doi.org/10.1063/1.1770712
21. Claypole, T. C. and Syred, N. "The Effect of Swirl Burner Aerodynamics on $NO_x$ Formation". *Symposium (International) on Combustion* 18.1 (1981), 81–89. https://doi.org/10.1016/S0082-0784(81)80013-6
22. Guo, J., Zhang, T., Huang, X., Luo, W., Hu, F., Luo, Z., Li, P. and Liu, Z. "Oxy-Fuel Combustion Characteristics of Pulverized Coal in a 3 MW Pilot-Scale Furnace". *Energy & Fuels* 32.10 (2018), 10522–10529. https://doi.org/10.1021/acs.energyfuels.8b02275

23. Chen, R.-H., Driscoll, J. F., Kelly, J., Namazian, M. and Schefer, R. W. "A Comparison of Bluff-Body and Swirl-Stabilized Flames". *Combustion Science and Technology* 71.4-6 (1990), 197–217. https://doi.org/10.1080/00102209008951632
24. Meißner, C., Schneider, H., Sidiropoulos, E., Hölzer, J. I., Heckmann, T., Böhm, B., Dreizler, A. and Seeger, T. "Investigation on Wall and Gas Temperatures Inside a Swirled Oxy-Fuel Combustion Chamber Using Thermographic Phosphors, $O_2$ Rotational and Vibrational CARS". *Fuel* 289 (2021), 119787. https://doi.org/10.1016/j.fuel.2020.119787
25. Ontyd, C., Pielsticker, S., Yildiz, C., Schiemann, M., Hatzfeld, O., Ströhle, J., Epple, B., Kneer, R. and Scherer, V. "Experimental Determination of Walnut Shell Pyrolysis Kinetics in $N_2$ and $CO_2$ via Thermogravimetric Analysis, Fluidized Bed and Drop Tube Reactors". *Fuel* 287 (2021), 119313. https://doi.org/10.1016/j.fuel.2020.119313
26. Pielsticker, S., Heuer, S., Senneca, O., Cerciello, F., Salatino, P., Cortese, L., Gövert, B., Hatzfeld, O., Schiemann, M., Scherer, V. and Kneer, R. "Comparison of Pyrolysis Test Rigs for Oxy-Fuel Conditions". *Fuel Processing Technology* 156 (2017), 461–472. https://doi.org/10.1016/j.fuproc.2016.10.010

# Robust Laser Diagnostics for Quantitative Process Characterisation

16

Matthias Bonarens, Johannes Emmert and Steven Wagner

## Contents



**Abstract**

The scientific investigation of biomass combustion strongly relies on the ability of measurement methods to resolve temporal and spatial scales in high-temperature process environments. Specifically, linking generic flames and the physicochemical knowledge gained from them to industrial-scale combustion

M. Bonarens · J. Emmert · S. Wagner (✉)
Institute of Reactive Flows and Diagnostics, Technische Universität Darmstadt, Darmstadt, Germany
e-mail: wagner@rsm.tu-darmstadt.de

M. Bonarens
e-mail: bonarens@rsm.tu-darmstadt.de

J. Emmert
e-mail: emmert@rsm.tu-darmstadt.de



R. Kneer et al. (eds.), *Clean Energy Conversion*,
https://doi.org/10.1007/978-3-031-98956-8_16

chambers requires measurement techniques that span a wide range of scales. Tunable diode laser absorption spectroscopy is used as a robust and quantitative measurement method to investigate gas species mole fractions, residence times, and temperature distributions in several combustion systems. Besides different measurement approaches (line-of-sight, 1-D and 2-D) to resolve spatial distributions, this method was enhanced for investigating combustion systems with increasing complexity, including a one-dimensional laboratory-scale gas flame, a single particle combustor, a gas-assisted pulverised biomass combustor, and a semi-industrial scale swirl combustor. Selected results are discussed, which provide validation data for models and numerical simulations.

## 16.1 Introduction

The combustion of biomass in oxy-fuel atmospheres presents a multitude of challenges for the modelling of reaction kinetics, the physico-chemical environment, and, as a consequence, also the flow conditions. To understand the coupled chemical reactions, pollutant generation, and flame stability, it is useful to follow a structured experimental approach where the complexity of the investigated systems is systematically increased, from generic to large-scale combustion. Experimental data on numerous parameters, including the gas composition, temperature, and velocity field, is required for the development and validation of sophisticated numerical models.

In order to establish a data basis that encompasses multiple scales, ranging from generic laboratory flames to semi-industrial-scale combustors of approximately 1 $MW_{th}$, it is beneficial to perform measurements using a single measurement method. This eliminates systematic uncertainties that exist between measurement methods, allowing unbiased comparisons between different combustion systems.

This demand implies several challenges, as the different combustion scenarios range from

- laminar to highly turbulent flows,
- homogeneous to inhomogeneous distribution of gases,
- single-phase to multiphase flows,
- from stationary flow conditions to turbulent flows,
- start-up of the combustion up to stationary operation,
- small laboratory devices up to industrial-scale combustors.

Laser spectroscopic measurement methods have been demonstrated to be well suited for probing various molecular species within combustion systems [1]. Apart from a suitable optical access, the flow is not disturbed by the measurement process, making laser spectroscopic measurements minimally invasive. Moreover, with many methods, temporal resolutions below 1 ms can be achieved. As such, they outperform extractive measurements or intrusive sensors placed in the flow, like thermocouples, and therefore allows for the analysis of highly dynamic processes.

Laser spectroscopy of chemical species can be classified into resonant and Raman-based measurement methods. Raman techniques are based either on spontaneous Raman scattering [2–4] or on non-linear coherent anti-Stokes Raman spectroscopy (CARS) [5]. For time-resolved quasi-instantaneous measurements, spontaneous Raman scattering requires high laser pulse energies because the scattering cross-sections are small. This causes several challenges, such as plasma formation, scattering by particles in two-phase biomass combustion, and damage to optical windows. Furthermore, extensive optical access is required for efficient detection. These limitations make spontaneous Raman scattering unsuitable for measurements in industrial-scale burners in general and in particle-laden flows in particular. CARS, in contrast, is generally capable of simultaneously detecting different chemical species [6]. However, the equipment required is complex, and the sensitivity is limited, especially for low species concentrations. Moreover, the optical access must be designed for several laser beam directions in order to achieve phase matching. The laser pulse energies required are also high, although they are smaller than for spontaneous Raman scattering. This makes the application of CARS to particle-laden two-phase flows of different scales and widely varying compositions extremely challenging.

Resonant measurement techniques make use of absorption of the laser radiation by targeted molecules. Following a single-photon absorption process, molecules enter an energetically excited state. In combustion diagnostics, electronic and vibrational excitations are commonly used. In absorption spectroscopy, the attenuation of laser radiation is utilised directly to identify the molecular species and quantify their thermochemical state [1,7]. If the spontaneous emission of light, known as fluorescence, following the energetic excitation is used instead of absorption, a high spatial resolution can be achieved, for instance with a detection direction perpendicular to the direction of laser light propagation. This measurement technique, commonly referred to as laser-induced fluorescence (LIF) [2–5,8], is primarily utilised for qualitative concentration measurements. A major impediment to quantitative concentration measurements is quenching, which depends on temperature, pressure, and the collision partners within the measurement volume. Similar to spontaneous Raman scattering, extensive optical access is required for effective detection. The majority of the species targeted by LIF are intermediates with a low concentration, requiring UV radiation for electronic excitation. The required equipment is complex, especially if several species are to be measured quasi-simultaneously. Furthermore, volatiles released during the early phase of biomass combustion also absorb ultraviolet radiation, which makes quantitative LIF concentration measurements impossible [9]. Thus, laser absorption spectroscopy with tunable diode lasers emitting in the infrared spectral region is the most appropriate laser spectroscopic measurement method for investigating pulverised biomass combustion in combustors of several scales. It is robust, allows for quantitative measurements of multiple process parameters, and is significantly more cost-effective than the alternatives discussed above. Furthermore, it can be applied to burners of varying sizes, ranging from generic laboratory setups to semi-industrial test rigs, without altering the setup or optical access design.

The tunable diode laser absorption spectroscopy (TDLAS) technique is widely recognised for its robust measurement of quantitative gas species mole fractions, gas temperatures, and residence times with high temporal resolution. In order to achieve improved sensitivity, robustness, and spatial resolution, several modifications were required in order to enable measurements on scales from generic flames to a 1 $MW_{th}$ combustor. The methodological extensions concern improved spectroscopic models for the evaluation of absorption spectra, as well as advanced tomographic evaluation routines for the two-dimensional reconstruction of thermochemical states. This chapter presents these methodological advancements and their application.

## 16.2 Methods

In this section, the fundamentals of TDLAS are first introduced concisely (Sect. 16.2.1), before an efficient tomographic approach for the reconstruction of two-dimensional parameter distributions from TDLAS measurements along multiple paths is introduced (Sect. 16.2.2). Furthermore, a novel method that merges TDLAS with time-of-flight lidar to enable single-ended, spatially resolved measurements along a line in particle-laden flows is discussed (Sect. 16.2.3).

### 16.2.1 Tunable Diode Laser Absorption Spectroscopy

In contrast to many common techniques, TDLAS provides quantitative information regarding thermochemical states without the need for calibration. As stated in the introduction, in TDLAS the molecule-specific and wavelength-dependent absorption is probed using the laser light emitted by a tunable diode laser. The measurement principle is based on the Beer-Lambert law [10]:

$$I(\nu) = I_0(\nu)\, Tr(t)\, \exp[-S(T)\, g(\nu - \nu_0)\, N\, L] + E(t) \tag{16.1}$$

In (16.1), $I_0(\nu)$ is the intensity entering the probe volume and $I(\nu)$ the measured intensity after passing the probe volume. The absorption is described by the temperature-dependent line strength of the transition $S(T)$, the number density of the molecule $N$, the absorption path length in the volume $L$, and the normalised transition line shape function $g(\nu - \nu_0)$, which is centred at the wavelength $\nu_0$. Parasitic broadband losses, e.g. due to the interaction with biomass and soot particles, beam steering or absorption by optical elements, can disturb the transmission spectra. They are considered by the term $Tr(t)$. In combustion systems, strong background radiation occurs, which is compensated by the introduction of an emission term $E(t)$. Probing the transition of a molecule by spectrally scanning over multiple wavenumbers $\nu$, the number density is determined by resolving the Beer-Lambert law for

$$N = -\frac{1}{S(T)\, L} \int \ln\left(\frac{I(\nu) - E(t)}{I_0(\nu)\, T(t)}\right) \frac{\partial \nu}{\partial t}. \tag{16.2}$$

The tuning coefficient of $\frac{\partial \nu}{\partial t}$ is an intrinsic characteristic for a diode laser and only needs to be determined once for a given modulation scheme. The mole fraction can be calculated using the ideal gas law if the absorption path length, the gas pressure, and the gas temperature are known. The geometric configuration of the investigated system determines the absorption path length. The line strength $S$ is transition-specific and is governed by the temperature dependent Boltzmann distribution and the lower state energy. Hence, the gas temperature has to be determined by a complementary measurement method (thermocouple) or extracted from absorption spectra as well. By measuring two distinct transitions of a molecule, the ratio of line strengths is directly correlated with temperature through the Boltzmann distribution (two-line thermometry [11, 12]). If more than two absorption lines are present in the measured spectrum, this method can be extended by utilising multiple lines. This stabilises the temperature calculation and reduces uncertainties significantly. The pressure in the combustion systems discussed here is known to be at ambient pressure and does not need to be determined from the line shape function.

The diode lasers used for TDLAS are mostly fibre coupled. This reduces the size of the optical access to the combustion system to small windows (e.g. diameter 25 mm). By using fibre splitters, the laser light can be divided into multiple beams and subsequently guided through the combustion chamber in different directions simultaneously. Measurements along single beams provide line-of-sight (LOS) information, and do not allow characterising the spatial distribution of the measured parameters along the path. The use of multiple beams with distinct projections enables tomographic reconstruction approaches for chemical species and temperature distributions.

Biomass combustion systems are characterised by spatially inhomogeneous distributions of gas species mole fractions and temperature. To measure this locally varying thermochemical variables, two methodologies were developed. Using a tomographic approach, the spatial distribution of temperature and mole fractions was reconstructed by utilising various projection angles, which requires large optical access (see Sect. 16.2.2). Furthermore, a new method is being developed for industrial-scale combustors, where optical access is very limited. Here, the parameter distribution along the beam (only a single window as optical access) is reconstructed employing the time-of-flight principle (see Sect. 16.2.3).

### 16.2.2 Linear Hyperspectral Tomography

The single projection line-of-sight approach fails to measure thermochemical states accurately in spatially inhomogeneous flows due to the non-linear dependence of the absorption spectrum on temperature. In order to overcome this limitation, linear hyperspectral absorption tomography (LHAT) was proposed, as discussed in the studies by Emmert et al. [7, 13, 15–18] and Grauer et al. [14]. The method is briefly outlined in the following.

A tomographic reconstruction is based on a large number of transmission spectra along different paths through the area of interest. The higher the number of projections, the more information is available and the less ill-posed the reconstruction. For the reconstruction, the probed area is triangularised [14], as depicted schematically in Fig. 16.1. Each beam within a specific projection only intersects a small portion

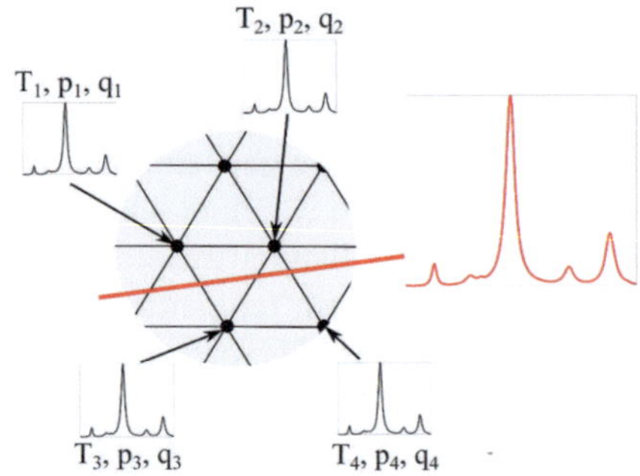


**Fig. 16.1** Mesh design of the LHAT approach. The knots are connected in triangles to define small regions, and the spectra (black) for each triangle are recalculated from the LOS integrated spectrum (red)

of the triangular regions. A single photodiode used for measuring $I(\nu)$ in a specific line-of-sight (LOS) detects the superposition of the absorbance of the spectra from all the triangles intersected by the beam. This corresponds to the LOS signal in the classical approach. However, in the tomographic reconstruction, the contribution from each triangle is reconstructed using prior knowledge of the process.

The nonlinear relations between each local absorbance spectrum and the local thermochemical properties in combination with the high number of discretisation elements (triangles) makes the nonlinear tomographic problem numerically expensive [14, 16]. Directly conducting this nonlinear tomographic reconstruction requires specialised optimisation methods [19], but still is numerically prohibitive for practical applications on large datasets. Linear hyperspectral absorption tomography resolves this issue by dividing the problem into a linear tomographic reconstruction of local absorption coefficients and a multitude of small nonlinear fits to determine the local thermochemical properties from the local absorption coefficients. This subdivision of the problem reduces the numerical costs by orders of magnitude [14, 16], and thus makes the method practically applicable for the experiments described in Sect. 16.3.4 (Fig. 16.2).

In the tomographic measurements described in the following sections, numerous LOS spectra were recorded by scanning laser beams across the area of interest. Each

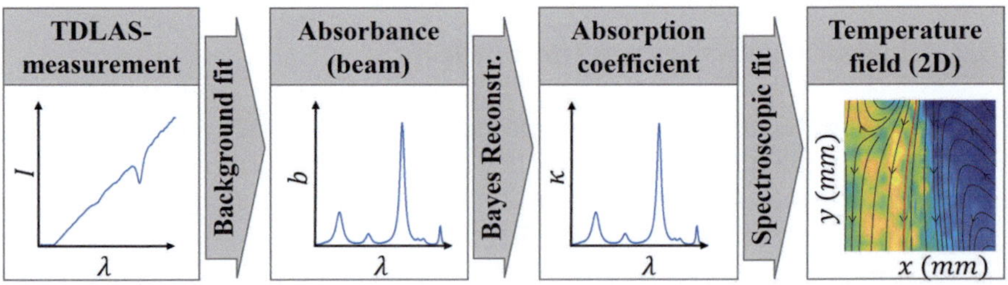


**Fig. 16.2** Block diagram of the LHAT procedure. The raw intensity signal from the measurement is treated by subtracting the background signal (i.e. laser ramp, background losses or emission fluctuation) and transferring it into an absorbance spectrum. These are subsequently used to infer local spectra (within the triangles) by employing a Bayesian reconstruction approach. Each triangle spectrum is then evaluated by fitted model spectra to them. All triangle results together will create a 2-D temperature distribution

scanning position was probed for several seconds, which is much longer than the time scales of the turbulent fluctuations. These turbulent fluctuations impacted the spectrum of each triangular area. In order to account for these temporal fluctuations, a normal distribution of the state variables was assumed within each triangular region. By incorporating a fluctuation model into the tomographic reconstruction, the influence of turbulence was considered [7].

### 16.2.3 Hyperspectral Time-of-Flight Absorption Spectroscopy

The aforementioned tomographic approach necessitates extensive optical access to the combustion chamber in order to comprehensively scan the entire area of interest. As such, it is not applicable to industrial-scale systems where the optical access is severely limited. Hence, a TDLAS spectrometer that allows to perform spatially resolved measurements in particle-laden flows and requires only a single optical access is currently being developed [20]. In this new approach, the information on the spatial distribution of the quantities of interest is not inferred from the information on the absorption along multiple paths, but from the temporal delay between signal components scattered by particles at different locations. As depicted in Fig. 16.3, a laser beam travelling through a particle-laden flow, like those found in pulverised biomass combustion, will experience several effects along its path, including:

- Reflection at the window used for optical access,
- Reflection at the wall opposing the window,
- Scattering, reflection and absorption by particles,
- Absorption by the gas phase.

The detected transmittance $T(r, \nu)$, can be described as a function of the distance from the entrance window to the position of reflection $r$ and the emitted wavelength of the laser $\nu$, and consist of three components: The molecular absorption coefficient of the gas phase $k_g$, the effective absorption coefficient of the particular matter $k_p$, and the scattering coefficient $\sigma_s$ [20]:

$$T(r, \nu) = exp\left(-2\int_0^r k_g(\tilde{r}, \nu) + k_p(\tilde{r}) + \sigma_s(\tilde{r})d\tilde{r}\right) \tag{16.3}$$

Absorption and scattering by particles have a spectrally smooth influence on $T(r, \nu)$ compared to gas absorption, allowing to isolate the latter. Thus, by launching short laser pulses that feature sharp edges and deconvoluting the backscattered signal with the shape of the emitted pulse, the attenuation of the light along its path can be reconstructed spatially [21]. If such measurements are performed at multiple wavelengths, absorption spectra that characterise the local state of the gas along the beam path are obtained, which can each be evaluated using common algorithms.

Of course, this approach requires a fast photodiode and data acquisition unit, as the time between two intensity measurements needs to be small compared to the time

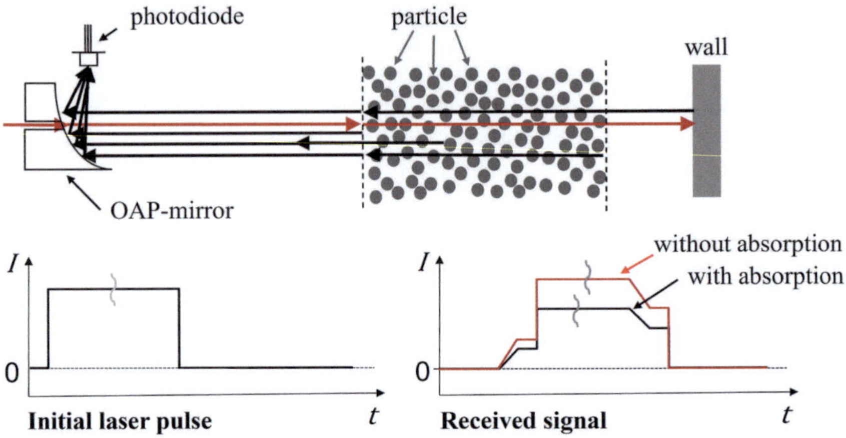


**Fig. 16.3** Setup of the hyperspectral time-of-flight absorption spectroscopic measurement procedure. The laser beam passes through a hole in an off-axis parabolic mirror and enters the probe volume, which contains particulate matter dispersed in a gas matrix. The temporal profile of the laser pulse is ideally rectangular (bottom graph left). As a result of absorption by particles and gas and scattering and reflection by particles, the back-scattered signal contains information regarding the concentrations of local species (as depicted in the bottom right graph). The information on the gas along the absorption path is reconstructed from the complex pulse shape

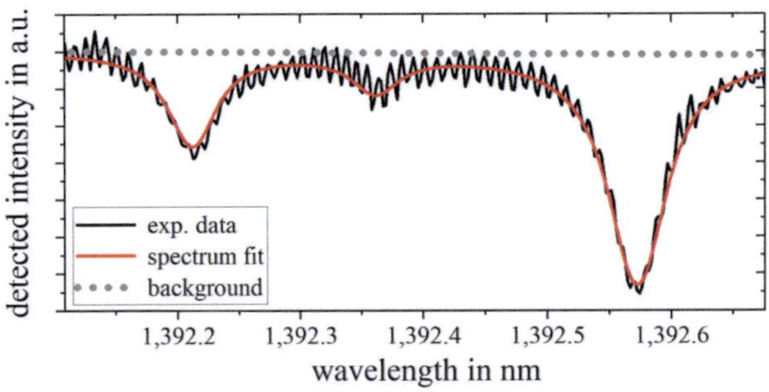


**Fig. 16.4** Example of a reconstructed spectrum of a single reflecting metal element, superimposed by an etalon structure caused by the optical setup (black) and the result of the spectroscopic fit (red)

light needs to pass through the probed volume. Moreover, a suitably large number of reflecting particles along the beam path, and a high-power laser source, are needed. Furthermore, advanced numerical evaluation approaches including regularisation are necessary to interpret the acquired signals (Fig. 16.4).

## 16.3 Results and Discussion

In the following sections, the application of TDLAS to various oxy-fuel combustion systems is discussed. Starting with a generic gas flame, the complexity of the combustion system is continuously increased, accompanied by increasing challenges for the TDLAS diagnostics.

### 16.3.1 Generic—Wolfhard-Parker Burner

The adaptation of the TDLAS methodology to oxy-fuel combustion in a generic flame requires appropriate lasers to probe the species of interest, as well as extending the spectroscopic model to these species. Additional species may cause spectral interference and line broadening. Both effects must be considered when performing a spectral fitting procedure.

For the development, testing, and validation of the TDLAS spectrometer, a Wolfhard-Parker burner (WHP) [22] was used. It creates a two-dimensional laminar $CH_4$/air non-premixed flame, which is well suited for LOS absorption spectroscopy. It offers a long absorption path within the flame, is stationary, and easily accessible with a laser. There is a database available that contains 27 different flame parameter (such as mole fraction, temperatures, velocities and many more), previously measured by various methods using a burner in the laboratory of Kermit Smyth [23]. As such, the burner can be used as a benchmark to verify the accuracy of the species mole fraction and temperature measurements of the new spectrometer. In addition to the $CH_4$/air operation for this verification, it was switched to oxy-fuel conditions ($O_2$, $N_2$, $CO_2$ and $CH_4$) for the transfer of the TDLAS methodology to the new fuel configuration.

A modified line-of-sight setup was used to measure the distribution of the combustion species (see Fig. 16.5) during operation at $CH_4$/air and $CH_4$/oxy-fuel conditions [22]. The comparison of flame structure at various fuels provides an initial indication of fundamental modifications in the physicochemical boundary conditions. The comparison with the reference values from the study by Smyth et al. [23] revealed a satisfactory agreement in the shape of the mole fraction profiles and the absolute values. A minor discrepancy exists between the two slots in the setup from Bürkle et al. [22]. Small differences are expected due to uncertainties in the gas mixtures, the manufactured burner setup (modified based on the reference design as described by Smyth et al. [23]), and the flow condition around the burner in laboratory air. As can be observed from the change of the stoichiometry from 21 vol.% $O_2$ in air to 30 vol.% $O_2$, the peak temperature decreases slightly due to increased heat capacity and heat transfer. Furthermore, higher mole fractions of the major species CO and $H_2O$ are observed, which implies a strong impact on the main reaction path.

Based on this data, it was shown that the modified TDLAS spectrometer provides accurate mole fraction and temperature measurements.

### 16.3.2 Generic—Counter-Flow-Burner

After the transfer to the oxy-fuel environment and verification of the line-of-sight approach in a generic diffusion flame, the linear hyperspectral absorption spec-

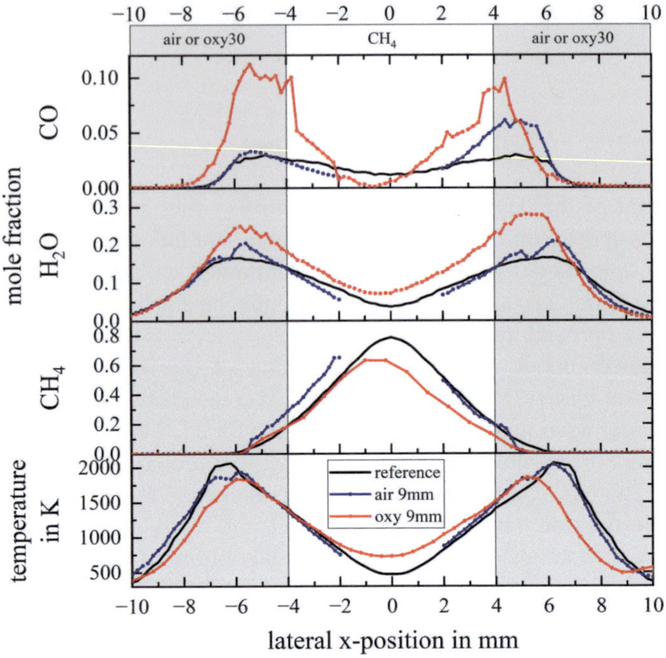


**Fig. 16.5** Comparison of the temperature and mole fraction profiles measured at the Wolfhard-Parker burner for methan-air-flame and oxy-fuel operation

troscopy methodology is demonstrated next. The method has been developed to spatially resolve the distribution of gas species and temperature in complex combustion environments. Prior to its application in turbulent combustion systems, it is demonstrated in a generic setting. To this end, the counter-flow burner [18] at the *Institute for Combustion Technology* (ITV) at *RWTH Aachen University* was utilised as a testing platform. It provides a suitable environment with defined boundary conditions for method development and comparisons with complementary measurement methods.

The burner (see Fig. 16.6) was operated at stationary conditions. For the LHAT evaluation, the assumption of an axisymmetric flame structure allows applying an onion peeling algorithm in a Bayesian framework. Due to its axisymmetry, traversing the laser along the $x$-axis is sufficient to reconstruct the parameter distribution for the entire $x$-$y$-plane. By repeating this procedure while translating in the vertical direction ($z$), a parameter map of the distribution is extracted (see Fig. 16.7). These maps contain quantities of different species and temperatures, as well as changes in flame curvature due to different flow conditions for various operating points (see Fig. 16.8).

As the burner can also be operated under oxy-fuel conditions, it was used:

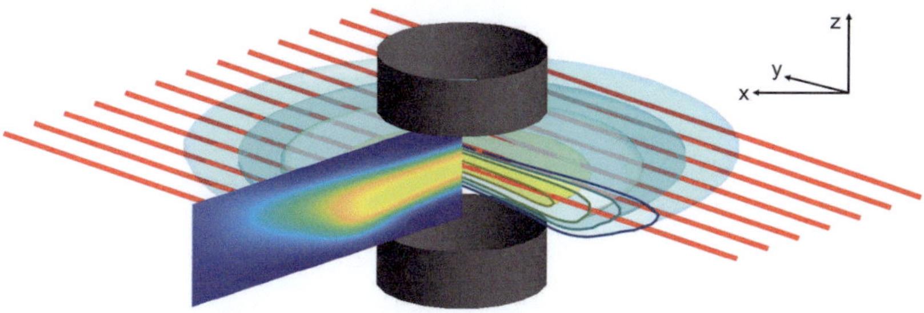


**Fig. 16.6** Scanning scheme for the measurement of the mole fraction and temperature distribution in the counter flow flame. Horizontal traversing of the LOS provides parallel transmission spectra in the probe plane, which can independently be adjusted. From the single planes, the entire volume is reconstructed, as depicted in the coloured inlet picture located to the right of the centre axis (reprinted from Emmert [24], licensed under CC BY-SA 4.0)

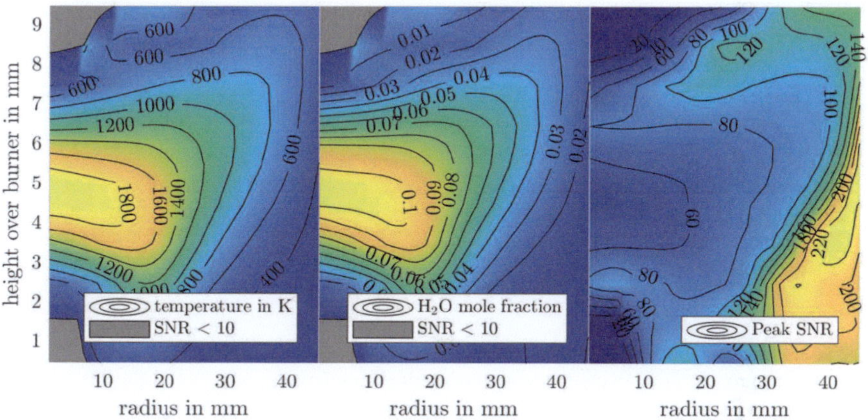


**Fig. 16.7** Reconstructed distributions of temperature and $H_2O$ mole fraction in the counter-flow burner (reprinted from Emmert et al. [18], licensed under CC BY 4.0)

1. to investigate the changes in flame chemistry due to fuel changes. The measurements provided valuable data for chemical model development.
2. for comparisons with the results of the numerical simulations (see Fig. 16.8) and model validation.
3. for comparisons of complementary measurement methods (i.e. TDLAS and extractive mass spectrometry).

The application of LHAT to the generic counter flow burner shows good agreement with reference data from complementary methods and is subsequently used to study oxy-fuel conditions. First comparisons to numerical simulations also demonstrated good agreement in shape and position [18].

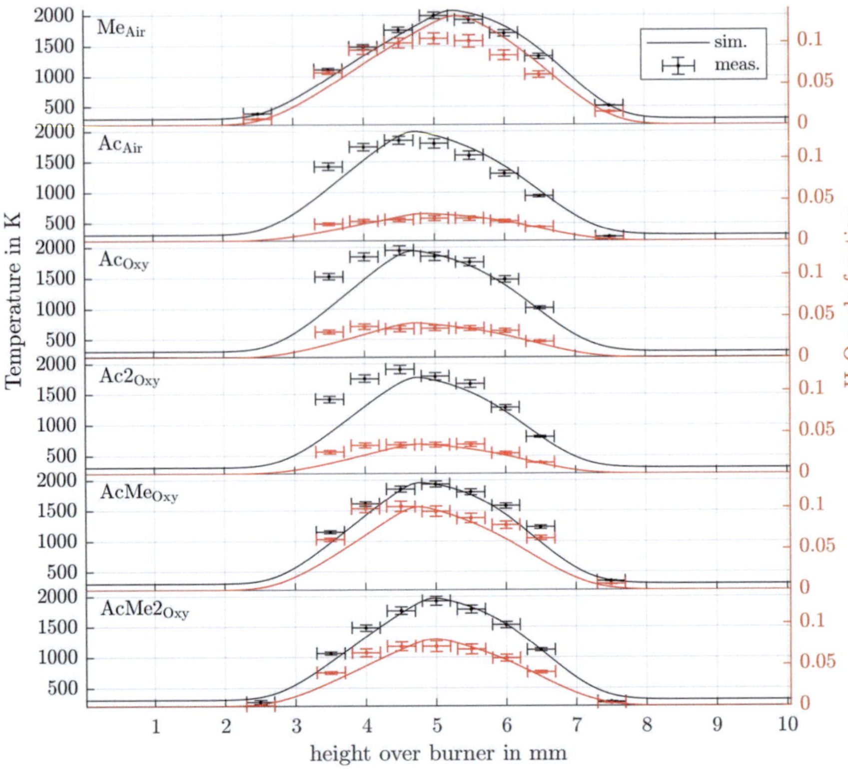


**Fig. 16.8** Temperature and $H_2O$ mole fraction along the burner centre axis reconstructed by using the LHAT approach (dots) in comparison to numerical simulation results (line). Results are shown for 6 different fuel configurations (reprinted from Emmert et al. [18], licensed under CC BY 4.0)

### 16.3.3 Generic—Single Particle Burner

Next, a microwave-heated, turbulent jet is characterised in Sect. 10.2.1 [24]. The focus of the experiment is on the devolatilisation of a biomass particle and its ignition, which are of great importance for developing numerical models. LHAT was adapted and applied to determine the temperature distribution of the single-phase flow around a centred fuel nozzle. Specifically, a $CO_2$ co-flow (67% $N_2$, 21% $O_2$, and 12% $CO_2$) was characterised. The information on the temperature distribution is important, as small asymmetries in the co-flow could result in disturbances of the solid particle ignition.

In order to resolve the 3 mm inner nozzle that is surrounded by the co-flow, a dense beam pattern is required. Hence, it is imperative to traverse the single-ended sensor head in precise increments around the outlet of the test stand. It was rotated around the central axis of the burner and simultaneously translated in step widths ranging from 50 to 100 μm. The triangular mesh used for the tomographic reconstruction was also designed to be dense near the inner nozzle outlet and coarse in outer regions. The

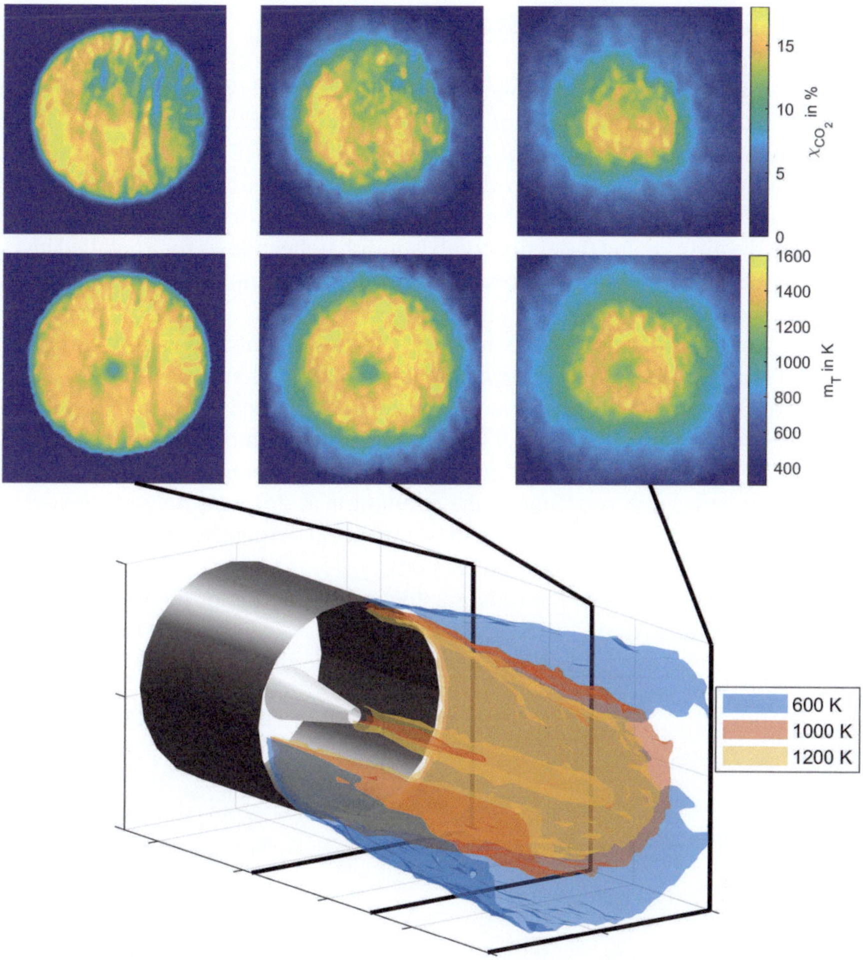


**Fig. 16.9** Reconstructed temperature and $CO_2$ mole fraction distributions for planes at three distances to the fuel nozzle exit (reprinted from Emmert [24], licensed under CC BY-SA 4.0). The cold flow in the middle and the shape of the hot co-flow are clearly visible [24]

fluctuation model [7] was applied, and the absolute temperature values on individual points in the flow were compared with thermocouple measurements.

The tomographic measurements were performed at different distances from the nozzle exit (5, 75, and 145 mm). Figure 16.9 shows the three-dimensional results for an exemplary operating point. As can be seen, the inner fuel nozzle was not resolved, but the co-flow temperature topography was reconstructed. An asymmetry in the distribution is observed, which may be related to imperfect mixing of cold $CO_2$ in the co-flow nozzle chamber.

### 16.3.4 Small Scale—Swirl-Stabilised Burner

The LHAT methodology was applied to a small-scale and optically accessible combustion chamber at the *Institute of Reactive Flows and Diagnostics* (RSM) of the *Technical University of Darmstadt* (Fig. 16.10) as described in detail in Sect. 13.2.1. This chamber is equipped with a diffuser of realistic geometry, yet features large optical access from four sides. The optical accessibility of the combustion system including the diffuser and the entire combustion chamber allows for a detailed investigation of the thermochemical processes using different complementary measurement methods.

Since the two-phase combustion process in this chamber is highly turbulent, a LOS approach is not sufficient for elucidating spatial inhomogeneities [10, 25–27]. Therefore, a tomographic setup was developed to be applied to the optically accessible combustion chamber (Fig. 16.11) in combination with the LHAT approach (see Sect. 16.2.2). Eight scanning devices operated simultaneously guided laser beams through the entire cross-section of the combustor at different heights downstream the dump plane. Since the chamber can be translated vertically, the chamber volume can be scanned stepwise in three dimensions.

For validation purposes, the tomographic reconstruction of the temperature was compared to results from CARS measurements [7]. As can be seen from Fig. 16.12, the comparison shows a good agreement regarding absolute values as well as the

**Fig. 16.10** Photograph of the combustion chamber in full operation. On the bottom front and the sides, the scanning probes of the tomographic measurement system are visible (reprinted from Emmert [24], licensed under CC BY-SA 4.0). The blue tubes are used for purging and cooling of the optics and detection units [24]

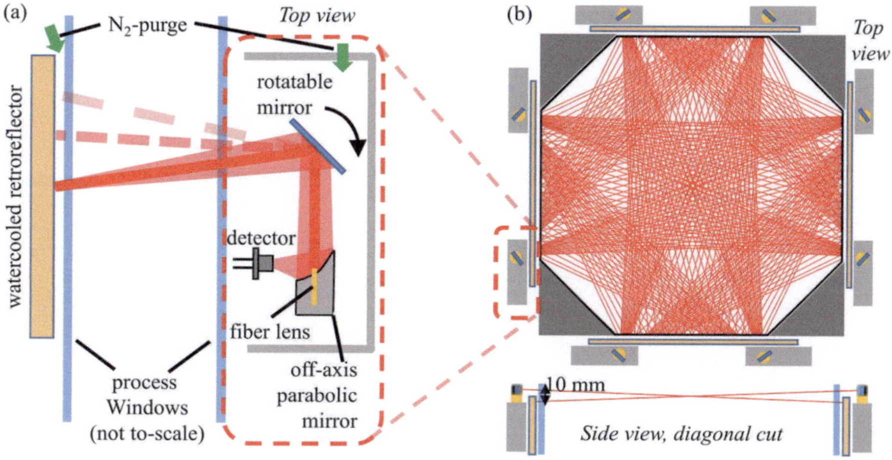


**Fig. 16.11** Setup of a single probe (left) and the overview of the positioning of all probes around the combustion chamber (right) (reprinted from Emmert et al. [13], licensed under CC BY-NC-ND 4.0)

shape of the temperature profile. The tomographic measurements, particularly when evaluated using the fluctuation model, yielded useful results across the entire cross-section of the combustor, whereas CARS was only applicable in the outer regions of the combustor.

Using the LHAT approach in combination with the tomographic spectrometer, the residence times of fluid elements in the combustor were also determined in a spatially resolved manner. Residence time distributions (RTD) are important for understanding combustion processes in complex combustors and an inherent property of the flow within a combustion volume. The experimental measurement of RTD is very helpful for the development and validation of models implemented in numerical simulation tools. RTD characterises the pulse response to a time-dependent gas mole fraction profile entering the combustor at $t = 0$. Here, the RTD is modelled as a normal distribution $E(t)$ that integrates to unity and has to be strictly non-negative in the case of gas concentrations. The exiting mole fraction profile is therefore described by the convolution of $E(t)$ with the input mole fraction $\chi_{\text{inlet}}$:

$$\chi_{\text{exit}}(t) = \chi_{\text{inlet}}(t) \circledast E(t) = \int_{-\infty}^{\infty} \chi_{\text{inlet}}(t)\, E(t-\tau)\, d\tau \tag{16.4}$$

The mole fraction $\chi_{inlet}$ is a function of the turbulent flow field and the mixing processes downstream of the nozzle. According to (16.4), the measurement of the RTD can be accomplished by injecting HCl as a chemically inert gas in a short pulse into the combustor and detecting the time resolved mole fraction at various distances from the nozzle using the TDLAS setup described above. The laser spectrometer was fixed in position while the combustion chamber was translated vertically for

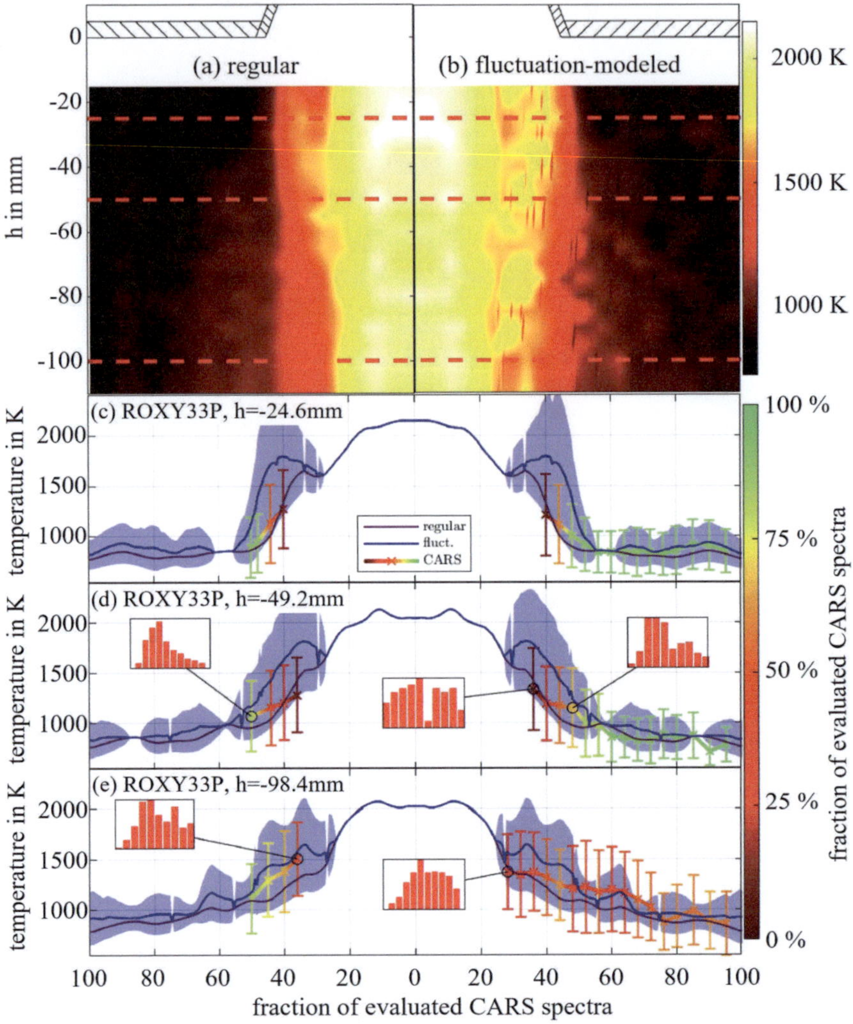


**Fig. 16.12** Reconstruction of the temperature map (top) for both the regular approach (**a**) and the fluctuation model (**b**). For the comparison with $O_2$-CARS measurements, measured profiles at distances of $h = -24.6$ mm, $-49.2$ mm, and $-98.4$ mm are shown. The CARS data are colour-coded regarding the number of evaluated spectra at each position. The blue-shaded area and error bars of the CARS measurements correspond to a standard deviation of the determined temperatures. Histograms of selected CARS measurements are depicted to clarify the quality of the results (reprinted and fonts adapted from Emmert et al. [7], licensed under CC BY-NC-ND 4.0)

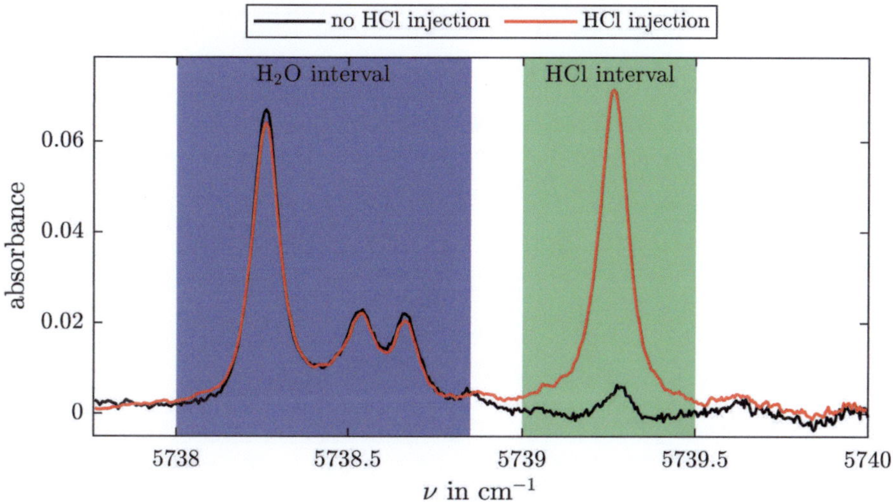


**Fig. 16.13** Comparison of two measured spectra with and without HCl injection. While the $H_2O$ spectrum does not change significantly (small changes due to fluctuations in the process), the HCl absorption is clearly visible. Additionally, a small $H_2O$ background absorption below the HCl spectrum can be seen if no HCl is present in the probe volume (reprinted from Emmert et al. [7], licensed under CC BY-NC-ND 4.0)

measurements at different axial positions. In this way, the RTD was measured in a spatially resolved manner [13].

To deduce reliable RTD, the local mole fraction of an inert tracer needs to be measured temporally resolved and with high accuracy. This poses a significant challenge in a strongly fluctuating gas flow. Hydrogen chloride was chosen as a tracer because its spectrum comprises suitable strong absorption lines. Moreover, the acquired spectra were phase-averaged. A typical spectrum with and without HCl injection in the gas flow is shown in Fig. 16.13. The probes HCl line overlaps with water transitions, which are compensated for in the evaluation. A typical reconstructed time series is shown in Fig. 16.14.

For a selected operation condition (ROXY33P: 33% $O_2$, 67% $CO_2$, and 31.5 $kW_{th}$), RTD moments are shown in Fig. 16.15 for varying axial and radial positions within the combustion chamber. This enables a quasi-3-D representation of the RTD within the combustion chamber, thereby enabling a more realistic reconstruction of the combustion flow fields' structure by numerical simulations.

A detailed discussion of the measurement and evaluation of laser absorption spectra for the determination of RTD in a combustion environment is given in the studies by Emmert et al. [13] and Bürkle et al. [25,26].

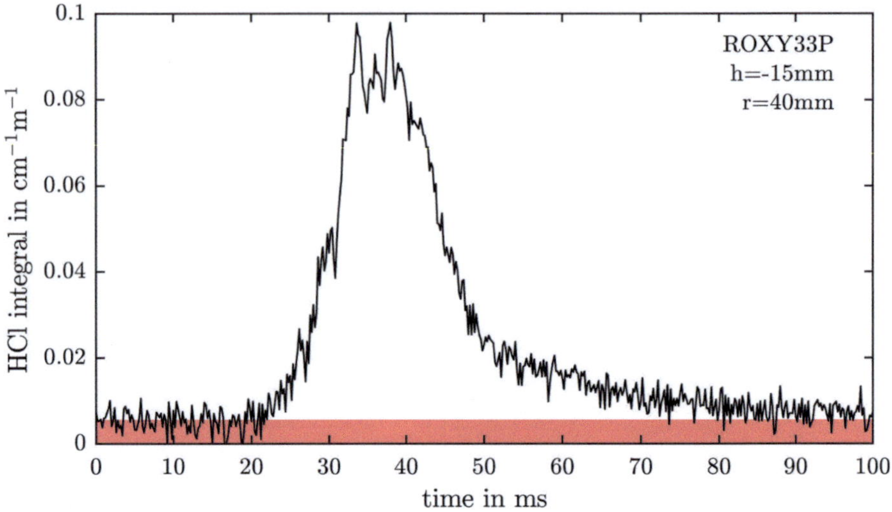


**Fig. 16.14** Plot of an exemplary HCl time series taken from the tomographic reconstruction at a certain radius and height below the diffuser (reprinted from Emmert et al. [13], licensed under CC BY-NC-ND 4.0)

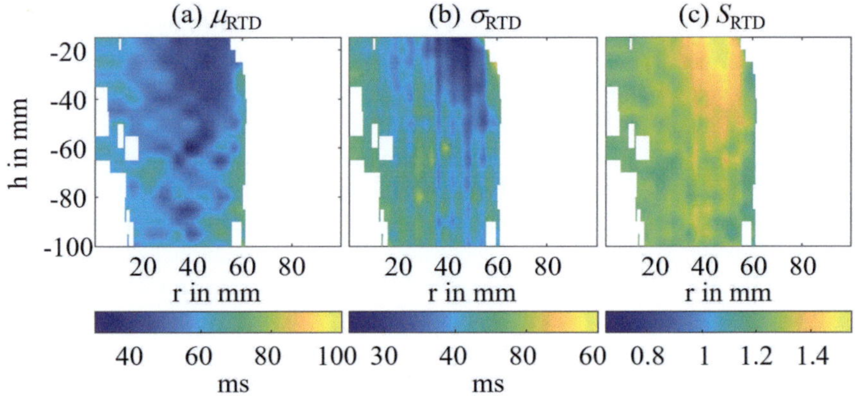


**Fig. 16.15** Reconstructed RTD moments for a single operating point (ROXY33P). Mean ($\mu_{\mathrm{RTD}}$), standard deviation ($\sigma_{\mathrm{RTD}}$) and signal-to-noise ratio ($S_{\mathrm{RTD}}$) are only shown for $S_{\mathrm{RTD}} > 8$ (reprinted and fonts adapted from Emmert et al. [13], licensed under CC BY-NC-ND 4.0)

### 16.3.5 Semi-Industrial Scale—1 $\mathrm{MW_{th}}$ Combustor

The investigation of biomass combustion in an industrial-scale combustor of 1 $\mathrm{MW_{th}}$ was conducted at the furnace of the *Institute for Energy Systems and Technology* (EST) located at the *Technical University of Darmstadt* (as described in detail in Sect. 15.3 [28]. The combustor shares a geometrically similar diffuser geometry as the burner in Sect. 16.3.4. In contrast to the optically fully accessible combustor,

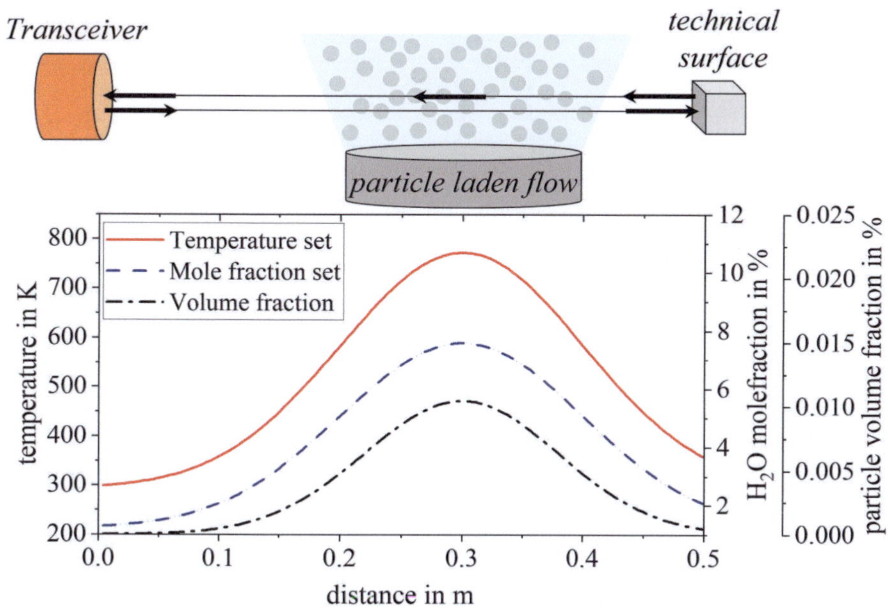


**Fig. 16.16** Overview of the simulated setup and the assumed distribution of water mole fraction, temperature, and particle density (reprinted from Hansemann et al. [20], licensed under CC BY 4.0)

this close-to-reality furnace provides only limited optical access to the combustion volume. Opposing ports on different heights and three horizontal angles are available. Since this is a highly inhomogeneous and turbulent system, the distribution of temperature and species mole fraction is of special interest. Due to the limitations of the ports, measurements along a sufficient number of beam paths are unfeasible. Therefore, the conventional method of tomography to reconstruct the temperature field fails under these conditions.

Using established absorption-based measurement approaches, only path-integrated measurements are possible. Such measurements were performed to infer path-integrated temperatures and mole fractions of $H_2O$, $CO_2$, CO, and $CH_4$. However, they only provide limited insights into spatial structures within the combustor.

To overcome the need for extensive optical access for the application of existing spatially resolving methods, a single-ended hyperspectral time-of-flight method exploiting particle reflection is being developed (see Sect. 16.2.3). In the future, the technique may enable single-ended measurements of spatially resolved temperature and species profiles in particle-laden flows.

The ToF method is being developed, intending to utilise it at the 1 $MW_{th}$ furnace for the investigation of oxy-fuel combustion. However, many challenges like developing a detector system sensible enough to measure the weak reflected signals from solid particles while achieving the necessary temporal resolution to separate the back-scattered signal components by their origin had to be addressed first in the laboratory. Hence, a generic experimental setup was constructed with the objective of

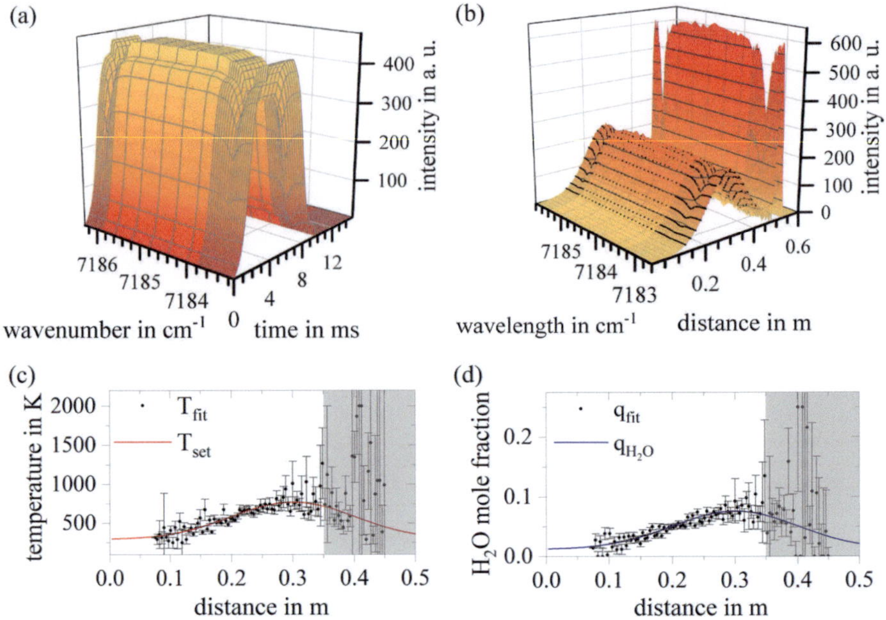


**Fig. 16.17** Visualised results of the finite volume simulation of the particle flow. From the **a)** simulated signal, the **b)** detected signal can be deconvoluted. From these spectra, the **c)** local temperatures and **d)** $H_2O$ mole fractions are extracted using the spectroscopic fit procedure (reprinted and fonts adapted from Hansemann et al. [20], licensed under CC BY 4.0)

enhancing the hardware components and developing a suitable reconstruction algorithm. The uncertainties due to the characteristics of the experimental setup as well as the reconstruction algorithm significantly impact the achievable spatial resolution and were investigated based on simulations [20]. In these simulations, a generic particle-laden flow served as a test (Fig. 16.16). Based on this artificial case, the system response during the measurement and the uncertainty of the reconstructed values are calculated (Fig. 16.17).

According to the simulation results, the uncertainty increases significantly with increasing distance from the emitting laser. From a certain point on, the signal becomes so weak that the uncertainty in the reconstructed parameters is excessively high. However, the simulation suggests that more than half of the travelled distance within the volume of interest may be reconstructed. While many practical problems remain, the simulations show that the developed principle may provide valuable insight into two-phase combustion processes in the future. In future experimental works, the detector system developed by Hansemann et al. [20] used for a lab-based proof-of-concept needs to be transferred to more realistic combustion environments and finally to the industrial-scale furnace.

## 16.4 Conclusion

The investigation of physicochemical processes during biomass combustion necessitates sophisticated measurement technology to provide experimental insights on the characteristics of species distribution, temperatures, and residence time distributions. For the accurate determination of combustion parameters in different combustion systems of varying scales and complexity, tunable diode laser absorption spectroscopy was employed. Three tools based on TDLAS (LOS, LHAT, ToF) were presented here. All of them were either modified and adapted or newly developed for the application under oxy-fuel combustion boundary conditions. The application of these methods to various combustors and the results obtained at multiple scales have demonstrated the efficacy of absorption spectroscopy in providing fundamental insights into biomass combustion for the development and validation of models. The verification of the measurement method at a generic burner setup (WHP) under oxy-fuel conditions provided the foundation for its application in various combustion systems. The increasing complexity of the necessary measurements at a counter flow burner and a single-particle burner allowed further development regarding the robustness of the TDLAS approaches. By utilising complementary techniques, a comprehensive parameter set for the swirl-stabilised burner was obtained. Especially, the quasi-volume reconstructions allow comprehensive insight into the flame structures within this particular combustion chamber. The newly developed technique of ToF-based TDLAS may facilitate future investigations of industrial-scale combustors with severely restricted optical access.

**Acknowledgements** This work has been funded by the Deutsche Forschungsgemeinschaft (DFG, German Research Foundation)—215035359—SFB/TRR 129 'Oxyflame'.

## References

1. Hanson, R. K., Spearrin, R. M. and Goldenstein, C. S. *Spectroscopy and Optical Diagnostics for Gases*. Cham: Springer International Publishing, 2016. https://doi.org/10.1007/978-3-319-23252-2
2. Duan, X. R., Meier, W., Weigand, P. and Lehmann, B. "Phase-Resolved Laser Raman Scattering and Laser Doppler Velocimetry Applied to Periodic Instabilities in a Gas Turbine Model Combustor". *Applied Physics B* 80.3 (2005), 389–396. https://doi.org/10.1007/s00340-004-1722-0
3. Wehr, L., Meier, W., Kutne, P. and Hassa, C. "Single-Pulse 1D Laser Raman Scattering Applied in a Gas Turbine Model Combustor at Elevated Pressure". *Proceedings of the Combustion Institute* 31.2 (2007), 3099–3106. https://doi.org/10.1016/j.proci.2006.07.148
4. Weigand, P., Meier, W., Duan, X. R., Giezendanner-Thoben, R. and Meier, U. "Laser Diagnostic Study of the Mechanism of a Periodic Combustion Instability in a Gas Turbine Model Combustor". *Flow, Turbulence and Combustion* 75.1–4 (2005), 275–292. https://doi.org/10.1007/s10494-005-8585-2
5. Greifenstein, M., Hermann, J., Böhm, B. and Dreizler, A. "Flame–Cooling Air Interaction in an Effusion-Cooled Model Gas Turbine Combustor at Elevated Pressure". *Experiments in Fluids* 60.1 (2019). https://doi.org/10.1007/s00348-018-2656-3

6. Schneider, H., Hebel, J., Böhm, B., Kneer, R. and Dreizler, A. "Rotational-Vibrational $O_2$-$CO_2$ Coherent Anti-Stokes Raman Spectroscopy for Determination of Thermochemical States in Oxy-Fuel Biomass Combustion". *Proceedings of the Combustion Institute* 40.1–4 (2024), 105457. https://doi.org/10.1016/j.proci.2024.105457
7. Emmert, J., Schneider, H., Meißner, C., Sidiropoulos, E., Hölzer, J. I., Seeger, T., Böhm, B., Dreizler, A. and Wagner, S. "Characterization of Temperature Distributions in a Swirled Oxy-Fuel Coal Combustor Using Tomographic Absorption Spectroscopy with Fluctuation Modelling". *Applications in Energy and Combustion Science* 6 (2021), 100025. https://doi.org/10.1016/j.jaecs.2021.100025
8. Meier, U. E., Wolff-Gaßmann, D. and Stricker, W. "LIF Imaging and 2D Temperature Mapping in a Model Combustor at Elevated Pressure". *Aerospace Science and Technology* 4.6 (2000), 403–414. https://doi.org/10.1016/S1270-9638(00)00142-5
9. Becker, L., Pielsticker, S., Böhm, B., Kneer, R. and Dreizler, A. "Particle Dynamics in a Gas Assisted Coal Combustion Chamber Using Advanced Laser Diagnostics". *Fuel* 269 (2020), 117188. https://doi.org/10.1016/j.fuel.2020.117188
10. Bürkle, S., Becker, L., Dreizler, A. and Wagner, S. "Experimental Investigation of the Flue Gas Thermochemical Composition of an Oxy-Fuel Swirl Burner". *Fuel* 231 (2018), 61–72. https://doi.org/10.1016/j.fuel.2018.05.039
11. Liu, X., Jeffries, J. B. and Hanson, R. K. "Measurement of Non-Uniform Temperature Distributions Using Line-of-Sight Absorption Spectroscopy". *AIAA Journal* 45.2 (2007), 411–419. https://doi.org/10.2514/1.26708
12. Ebert, V., Fernholz, T., Giesemann, C., Pitz, H., Teichert, H., Wolfrum, J. and Jaritz, H. "Simultaneous Diode-Laser-Based in-Situ-Detection of Multiple Species and Temperature in a Gas-Fired Power-Plant". *$28^{th}$ International Symposium on Combustion*. Edinburgh, Scotland, 2000.
13. Emmert, J., Schneider, H., Böhm, B., Dreizler, A. and Wagner, S. "Phase-Locked Absorption Tomography for Retrieving 5 kHz Time-Resolved Tracer Profiles in Solid Fuel Combustion". *Applications in Energy and Combustion Science* 12 (2022), 100093. https://doi.org/10.1016/j.jaecs.2022.100093
14. Grauer, S. J., Emmert, J., Sanders, S. T., Wagner, S. and Daun, K. J. "Multiparameter Gas Sensing with Linear Hyperspectral Absorption Tomography". *Measurement Science and Technology* 30.10 (2019), 105401. https://doi.org/10.1088/1361-6501/ab274b
15. Emmert, J., Wagner, S. and Daun, K. J. "Quantifying the Spatial Resolution of the Maximum a Posteriori Estimate in Linear, Rank-Deficient, Bayesian Hard Field Tomography". *Measurement Science and Technology* 32 (2021), 025403. https://doi.org/10.1088/1361-6501/abb550
16. Emmert, J., Grauer, S. J., Wagner, S. and Daun, K. J. "Efficient Bayesian Inference of Absorbance Spectra from Transmitted Intensity Spectra". *Optics Express* 27.19 (2019), 26893–26909. https://doi.org/10.1364/OE.27.026893
17. Emmert, J., Blume, N. G., Dreizler, A. and Wagner, S. "PolySpec: Polynomial Spectrum Models for Fast and Light-Weight Spectroscopic Evaluation". *Applied Optics* 57.34 (2018), 9907. https://doi.org/10.1364/AO.57.009907
18. Emmert, J., Baroncelli, M., van der Kley, S., Pitsch, H. and Wagner, S. "Axisymmetric Linear Hyperspectral Absorption Spectroscopy and Residuum-Based Parameter Selection on a Counter Flow Burner". *Energies* 12.14 (2019), 2786. https://doi.org/10.3390/en12142786
19. Cai, W. and Kaminski, C. F. "A Tomographic Technique for the Simultaneous Imaging of Temperature, Chemical Species, and Pressure in Reactive Flows Using Absorption Spectroscopy with Frequency-Agile Lasers". *Applied Physics Letters* 104.3 (2014), 034101. https://doi.org/10.1063/1.4862754
20. Hansemann, C., Bonarens, M., Emmert, J., Daun, K. J. and Wagner, S. "Towards a Spatially Resolved, Single-Ended TDLAS System for Characterizing the Distribution of Gaseous Species". *Scientific Reports* 14.1 (2024), 11708. https://doi.org/10.1038/s41598-024-61644-9
21. Kovalev, V. A. and Eichinger, W. E. *Elastic Lidar: Theory, Practice, and Analysis Methods*. Hoboken N.J.: John Wiley, 2004.

22. Bürkle, S., Dreizler, A., Ebert, V. and Wagner, S. "Experimental Comparison of a 2D Laminar Diffusion Flame Under Oxy-Fuel and Air Atmosphere". *Fuel* 212 (2018), 302–308. https://doi.org/10.1016/j.fuel.2017.10.067
23. Smyth, K. C., Miller, J., Dorfman, R. C., Mallard, W. and Santoro, R. J. "Soot Inception in a Methane/Air Diffusion Flame as Characterized by Detailed Species Profiles". *Combustion and Flame* 62.2 (1985), 157–181. https://doi.org/10.1016/0010-2180(85)90143-9
24. Emmert, J. "Spectrally Resolved Absorption Tomography for Reacting, Turbulent Gas Phase Systems: Theory and Application". PhD thesis. Technische Universität Darmstadt, 2022. https://doi.org/10.26083/tuprints-00020750
25. Bürkle, S., Becker, L., Agizza, M. A., Dreizler, A., Ebert, V. and Wagner, S. "In-Situ Measurement of Residence Time Distributions in a Turbulent Oxy-Fuel Gas-Flame Combustor". *Experiments in Fluids* 58.7 (2017), 647. https://doi.org/10.1007/s00348-017-2366-2
26. Bürkle, S., Becker, L., Agizza, M. A., Dreizler, A. and Wagner, S. "Comparison of Two Measurement Strategies to Obtain the Residence Time Distribution in Combustion Chambers Using Tunable Diode Laser Absorption Spectroscopy". *Applied Optics* 58.10 (2019), C36. https://doi.org/10.1364/AO.58.000C36
27. Agizza, M. A., Bagheri, G., Bürkle, S., Faravelli, T., Wagner, S. and Dreizler, A. "Investigation of Oxy-Fuel Combustion Through Reactor Network and Residence Time Data". *Energies* 15.1 (2022), 252. https://doi.org/10.3390/en15010252
28. Richter, M., König, D., Ströhle, J. and Epple, B. "Radiative Heat Flux Measurement in a Semi-Industrial Oxyfuel Combustion Chamber with Biomass and Coal". *Energies* 17.11 (2024), 2735. https://doi.org/10.3390/en17112735
29. Schneider, H., Bonarens, M., Hebel, J., Hamel, H., Emmert, J., Böhm, B., Wagner, S., Kneer, R. and Dreizler, A. "Combined Flow, Temperature and Soot Investigation in Oxy-Fuel Biomass Combustion under Varying Oxygen Concentrations Using Laser-Optical Diagnostics". *Fuel* 362 (2024), 130771. https://doi.org/10.1016/j.fuel.2023.130771

# Radiation

# 17 Radiation Modelling

Matthias Koch, Hossein Askarizadeh, Stefan Pielsticker, Lukas Pörtner and Reinhold Kneer

## Contents



M. Koch · H. Askarizadeh · S. Pielsticker · R. Kneer (✉)
Institute of Heat and Mass Transfer, RWTH Aachen University, Aachen, Germany
e-mail: kneer@wsa.rwth-aachen.de

M. Koch
e-mail: koch@wsa.rwth-aachen.de

H. Askarizadeh
e-mail: askarizadeh@wsa.rwth-aachen.de

S. Pielsticker
e-mail: pielsticker@wsa.rwth-aachen.de

L. Pörtner
Department of Energy Plant Technology, Ruhr-Universität Bochum, Bochum, Germany
e-mail: poertner@leat.rub.de



R. Kneer et al. (eds.), *Clean Energy Conversion*,
https://doi.org/10.1007/978-3-031-98956-8_17

**Abstract**

Due to the presence of high-temperatures within a combustion chamber, radiation is an essential heat transfer mechanism. Therefore, this chapter investigates the interaction of thermal radiation with gas and solid fuel particles and utilises these findings to develop a novel, computationally efficient model for coupled radiation/computational fluid dynamics (CFD) simulations. The gas radiation model is briefly introduced, while the development of the particle radiation model is explained in detail. In the development of this radiation model, the absorption and scattering coefficients, as well as the scattering phase function, are considered on a single-particle level with subsequent transfer to particle clouds. For this purpose, Maxwell's equations for a single particle are solved analytically and numerically depending on the problem's complexity. The final model considers shape and structure influences on the radiative properties depending on the burnout and the type of solid fuel. Furthermore, the influence of each single particle position in a particle group is considered without resolving the particles. The result is a very detailed and efficient particle model.

## 17.1 Introduction

As outlined in Chap. 1, modelling and simulation of the combustion process requires consideration of several physical phenomena. The quality of the simulation depends on the chosen submodels and their coupling. A crucial phenomenon for the accurate simulation of any combustion process is radiation. Due to the high temperature and high temperature differences in a combustion chamber, radiation is the dominant heat transfer mechanism [1], which requires different modelling approaches for particle and gas phases.

Several studies analysed the contributions of particle and gas radiation in semi-industrial burners. Guo et al. [2] investigated the effects of gas and particle radiation on the temperature distribution of the 2.5 $MW_{th}$ horizontal swirling combustion furnace from the *International Flame Research Foundation* (IFRF). They highlighted that particle radiation dominates gas radiation near the burner due to the high particle number density. Johansson et al. [3] investigated the influence of particle and gas radiation on the radiative transfer for a cylindrical chamber with varying radii. They concluded that particle radiation is dominant in the vicinity of the burner and also dominates the radiation in the burnout region. Similar conclusions were drawn in studies by Kez et al. [4] and Askarizadeh et al. [5]. Nevertheless, gas radiation remains crucial for accurately modelling radiative heat transfer.

In an oxy-fuel flame, the most important radiating gas species are $H_2O$ and $CO_2$. Their absorption properties are strongly varying throughout the wavelength spectrum. For accurate modelling of gas radiation, this wavelength dependency has to be addressed correctly [6]. Line-by-line (LBL) integration of the spectral lines provide accurate solutions, however, in this case about $10^6$ equations need to be solved. Solving the wavelength dependency efficiently is an important subject, and many different models were developed for air combustion, and later on adjusted for oxy-fuel conditions.

In a numerical study by Kez et al. [4], various models were compared, with the weighted-sum-of-grey-gases model (WSGGM) [7,8] and the full-spectrum correlated-$k$ model (FSCKM) [9] indicating the best results considering the trade-off between accuracy and efficiency. These so-called global models first integrate the wavelength-dependent gas properties over the entire spectrum and then apply these integrated properties to solve the radiative transfer equation. The radiative transfer equation describes the radiation interaction with gas and particles. Due to its simple implementation and extensibility, the WSGGM has been chosen to consider wavelength dependency in the current radiation model. Its exact formulation is briefly explained in Sect. 17.2.1.

Kez et al. [10] presented a method to incorporate the wavelength dependency of particle radiation. However, this study also showed that incorporating the particle wavelength dependency only improves the accuracy slightly compared to Planck-mean values [11,12]. Therefore, while the gas properties are wavelength-dependent, the particle properties are averaged over the entire spectrum using Planck-mean values. Hence, the goal of the following sections is not to present an improvement of the wavelength dependency but to accurately represent the radiative properties of particle groups in the radiation model without oversimplifying the internal and external structures of the single particles.

The Mie theory [13] is the basis for modelling the radiation properties of single particles. Here, the radiative properties are the absorption and scattering efficiency—a measure to define how much radiation is absorbed or scattered—and the phase function—a measure to determine in which direction radiation is scattered. Although the Mie theory is an analytical solution to Maxwell's equations and, thus, an exact solution, its drawback is the restriction to spherical and homogeneous particles. Approximating the shape of coal particles by a sphere was justified by Gronarz et al. [14]. In their study, coal particles were approximated by axisymmetric shapes with axis ratios up to two. The resulting radiative properties differed negligibly compared to the properties of spheres. In a consecutive study [15], the homogeneity assumption was investigated by comparing a locally resolved particle with different indices of refraction (resampling different materials) to a homogeneous particle characterised by one effective index of refraction. No distinct differences between the inhomogeneous and homogeneous modelling approaches were noticed. However, the various refractive indices of the inhomogeneous material only considered solid materials. The influence of porosity was not evaluated.

Therefore, the influence of porosity on the radiative properties is investigated in Sect. 17.3.2. Here, the focus is the difference between unstructured and structured porosity and the change of the particle composition during burnout. Structured porosity is connected to long porous ducts in biomasses, e.g. beech wood. Biomasses like beech wood, poplar, and pine also indicate much larger axis ratios than two, and thus, they were not covered in the study by Gronarz et al. [14]. Riaza et al. [16] also showed that these biomass species keep their non-spherical shapes for up to 50%

burnout. Consequently, the influence of non-spherical particles with aspect ratios up to six is investigated in Sect. 17.3.3.

The final result is the efficient modelling of the radiation properties of a single particle with variable shape and composition. Corrections for non-spherical and highly porous particles are still based on Mie theory, and thus—with minimal additional computational effort—a significant increase in accuracy is achieved.

This detailed single-particle model is then used to investigate the particle-to-particle radiation interaction within particle groups. Since calculating the radiative properties of all individual particles within a combustion chamber is not computationally feasible, the continuum approach for particle clouds is used [12]. Here, effective radiative properties of a particle group are calculated, and only the particle number density and the particle size distribution need to be specified. Typically, the models assume a uniform spatial particle distribution within the group. However, within realistic particle groups, the individual particles are distributed unevenly. Thus, a new but equally efficient method is introduced in Sect. 17.4 to account for this phenomenon.

## 17.2 Radiative Transfer Through a Participating Medium

The radiative heat transfer within a combustion chamber is affected by the chamber geometry and various physical phenomena. Radiation is emitted, absorbed and scattered by solid fuel, gases and chamber walls. All of the effects are described by the radiative transfer equation (RTE). The RTE is derived either by the conservation of energy of a focused beam passing through an absorbing, scattering, and emitting medium [17] or by simplifying Maxwell's equations [18]. Equation (17.1) outlines the radiative transfer equation for a single wavelength $\lambda$ and describes the change of intensity $I$ along an infinitesimal path d$s$ in the direction of the solid angle $\Omega$. Properties of the gas are marked by the index "g" and the particles by "p":

$$\begin{aligned}\frac{\mathrm{d}I_\lambda^{\Omega}}{\mathrm{d}s} = & -\underbrace{\left(\kappa_{\mathrm{g},\lambda} + \kappa_{\mathrm{p},\lambda}\right) I_\lambda^{\Omega}}_{\text{Absorption}} - \underbrace{\sigma_{\mathrm{p},\lambda}\, I_\lambda^{\Omega}}_{\text{Outscattering}} \\ & + \underbrace{\frac{\sigma_{\mathrm{p},\lambda}}{4\pi}\int_{4\pi} I_\lambda^{\Omega'}\, \Phi_{\mathrm{p},\lambda}^{\Omega\Omega'}\, \mathrm{d}\Omega'}_{\text{Inscattering}} + \underbrace{\epsilon_{\mathrm{g},\lambda}\, I_{\mathrm{b},\lambda}(T_\mathrm{g}) + \epsilon_{\mathrm{p},\lambda}\, I_{\mathrm{b},\lambda}(T_\mathrm{p})}_{\text{Emission}} \qquad (17.1)\end{aligned}$$

The right-hand side is divided into four terms: The first term describes the radiation attenuation due to absorption including the absorption coefficients $\kappa$ for gas and particles. The outscattered radiation in the second term is only relevant for particles and is quantified by the scattering coefficient $\sigma$. The third term accounts for increased radiation intensity due to radiation inscattering from other solid angles $\Omega'$ into the current solid angle $\Omega$. The total amount of inscattered radiation depends also on the scattering coefficient $\sigma$ and the phase function $\Phi$, which controls the redistribution in each solid angle. The fourth group of terms considers the radiation intensity increase due to radiation emission. Here, the radiation emission depends on the temperature-dependent blackbody intensity $I_{b,\lambda}$ and the emission coefficient $\epsilon$. Kirchhoff's law [19] is applied, if the emission coefficient is unknown, and the absorption coefficient is used instead. However, a measured emission coefficient $\epsilon_p$ is used resulting from the measurements described in Chap. 18.

### 17.2.1 Weighted-Sum-Of-Grey-Gases Model

To consider the wavelength dependency efficiently, the weighted-sum-of-grey-gases model (WSGGM) is used, which was first presented in the study by Hottel and Sarofim [20]. The underlying idea is the approximation of the spectral properties of a gas (non-grey) with a pre-defined number of wavelength-independent gases (grey gases). The resulting WSGG absorption coefficient $k$ is constant with respect to the wavelength for each grey gas $i$ and covers a fixed portion of the spectrum. In addition, the weighting factor $a$ defines the percentage of the total spectrum for which a corresponding $k$ is determined. With these two parameters, the RTE is reformulated into the WSGG form for each grey gas $i$:

$$\begin{aligned}\frac{\mathrm{d}I_\mathrm{i}}{\mathrm{d}s} = & -\underbrace{\left(k_{\mathrm{g,i}} + \overline{\kappa}_\mathrm{p}\right) I_\mathrm{i}^{\Omega}}_{\text{Absorption}} - \underbrace{\overline{\sigma}_\mathrm{p}\, I_\mathrm{i}^{\Omega}}_{\text{Outscattering}} \\ & + \underbrace{\frac{\overline{\sigma}_\mathrm{p}}{4\pi}\int_{4\pi} I_\mathrm{i}^{\Omega'}\, \overline{\Phi}_\mathrm{p}^{\Omega\Omega'}\, \mathrm{d}\Omega'}_{\text{Inscattering}} + \underbrace{a_{\mathrm{g,i}}\left(k_{\mathrm{g,i}}\, I_\mathrm{b}(T_\mathrm{g}) + \overline{\epsilon}_\mathrm{p}\, I_\mathrm{b}(T_\mathrm{p})\right)}_{\text{Emission}}\end{aligned} \tag{17.2}$$

The WSGGM derived by Bordbar et al. [21] is used in the current radiation model, whereby the parameters $a_{g,i}$ and $k_{g,i}$ are expressed for five grey gases with polynomials depending on the temperature and the concentrations of $CO_2$ and $H_2O$. Here, the parameters only describe an oxy-fuel atmosphere.

For particles, the phase function $\overline{\Phi}_p$, the scattering coefficient $\overline{\sigma}_p$, and the absorption coefficient $\overline{\kappa}_p$ are spectrally averaged (grey) via Planck-mean values [12] and, thus, the same for each grey gas. In (17.3), the Planck-mean value of the scattering coefficient is exemplarily calculated:

$$\overline{\sigma}_\mathrm{p} = \frac{\int \sigma_{\mathrm{p},\lambda}\, I_{\mathrm{b},\lambda}\, \mathrm{d}\lambda}{\int I_{\mathrm{b},\lambda}\, \mathrm{d}\lambda} \tag{17.3}$$

The modelling methods for the non-averaged properties ($\sigma_{p,\lambda}$, $\kappa_{p,\lambda}$, $\Phi_{p,\lambda}$) are introduced in the next sections.

## 17.3 Radiation Interaction with a Single Particle

If an electromagnetic wave interacts with a single particle, two phenomena are observed [22]:

- radiation scattering and
- radiation absorption.

For the quantification of an electromagnetic wave, the direction of propagation, the wavelength $\lambda$, and its incident intensity $I_0$ must be specified. The scattered wave is described by the scattering angles $\theta$ ranging from 0 to 180° and $\varphi$ ranging from 0 to 360°. Therefore, $\theta = 0°$ is the forward direction and $\theta = 180°$ the backward direction. With these angles, the scattered intensity $I_{\mathrm{Sca}}$ in a point $r$ is defined by (17.4) [23]:

$$I_{\mathrm{Sca}} = I_0 \frac{F(\theta, \varphi)\, \lambda^2}{4\pi^2 r^2} = I_0 \frac{\Phi(\theta, \varphi)\, C_{\mathrm{Sca}}}{r^2} \tag{17.4}$$

The dimensionless function $F$ additionally depends on the scattering angles, the orientation of the particle relative to the incident wave, the shape of the particle, the composition of the particle, and the wavelength. However, it is independent of $r$. The function $F$ can be replaced by the phase function $\Phi$ with the following relation:

$$F = \left(\frac{2\pi}{\lambda}\right)^2 \Phi\, C_{\mathrm{Sca}} \tag{17.5}$$

The phase function $\Phi$ is interpreted as a redistribution function that gives the proportion of intensity $I_0$ scattered into all possible directions, where its sum is normed to $4\pi$. The scattering cross-section $C_{\mathrm{Sca}}$ is a theoretical area that describes the area within which the incident electromagnetic wave is influenced due to scattering. Like the scattering cross-section, an absorption cross-section $C_{\mathrm{Abs}}$ is defined as the area within which the incident electromagnetic wave is influenced due to absorption.

All in all, scattering and absorption cross-section $C$ and the phase function $\Phi$ represent the radiative properties of a single particle. These properties depend on the radiation wavelength, particle shape, size, composition, and orientation of the particle and, in the case of the phase function, on the scattering angles. The following section presents the modelling methods and suitable simplifications for solid fuel particles.

### 17.3.1 Modelling Methods

Maxwell's equations are solved to determine the radiative properties of a single particle interacting with radiation. Different methods are used depending on the complexity of the particle shape and internal structure. Furthermore, to define the application range of some methods, the size parameter $x$ is defined as the ratio of the particle perimeter to the wavelength.

Therefore, the discrecte dipole approximation, the Mie theory and some extensions of the Mie theory are presented in the following sections. These methods are used to derive the final radiation model.

**Discrete Dipole Approximation**

The discrete dipole approximation (DDA) is a numerically exact method to solve Maxwell's equation [24]. Theoretically, it has no restrictions, and thus, it can be applied to any particle shape and composition. The idea of this method is to mimic the microscopic behaviour of the electric charges within the particle. For this purpose, the particle is divided into small regions, each represented by a single dipole. An example of such a complex particle is illustrated in Fig. 17.1a).

The incident electromagnetic wave induces a dipole moment, and due to the oscillations of the dipoles, radiation is scattered in different directions. When the number of dipoles is increased, and thus the distance between dipoles is decreased, a solid particle is simulated. With an infinite number of dipoles, the method yields an exact solution. However, it is also sufficiently accurate as long as the distances between adjacent dipoles is smaller than the critical distance $d$ [25],

$$d \leq \frac{\lambda}{10\,|\mathrm{IOR}|} \tag{17.6}$$

with complex index of refraction IOR describing the particle composition. With increasing size parameter, the number of dipoles increases exponentially, and thus, the computational costs (simulation time and required memory).

All in all, the DDA is a benchmark model to derive correction terms for more efficient methods in developing the current radiation model. For this purpose, the open-source code A-DDA [25] is used, and particles with a size parameter $x < 100$ are investigated.

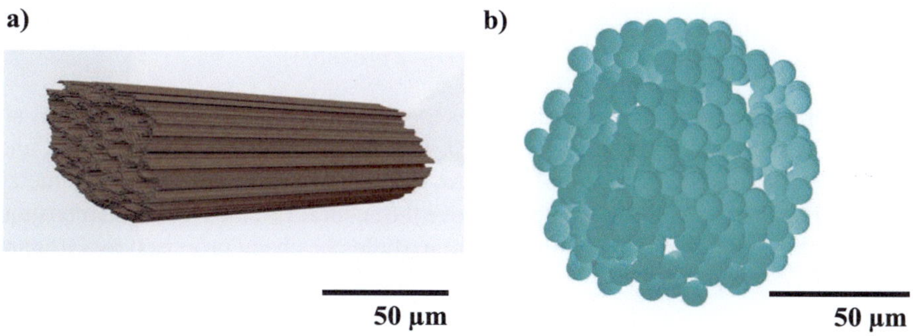


**Fig. 17.1** **a)** Complex particle modelled by approximately $1.8 \times 10^7$ dipoles for the discrete-dipole approximation (DDA) and **b)** particle modelled by generalised multiparticle Mie (GMM) method

**Mie Theory**

The Mie theory yields an analytical solution of Maxwell's equations for a homogeneous sphere interacting with radiation. For solid fuels, this simplification is only valid for selected cases, e.g. coal particles.

For its application, the index of refraction IOR of the homogeneous material and the size parameter with

$$x = \frac{\pi\, D}{\lambda} \tag{17.7}$$

are required. Mie theory has no restriction, considering the size parameter. However, approximations of the Mie theory with less computational costs exist for $x \ll 1$ (Rayleigh scattering [26]) and for $x \gg 1$ (geometric optics [27]). For pulverised solid fuel particles, the size parameter covers the range $1 \leq x \leq 500$. Therefore, Mie theory serves as basis model, and the DDA is used to develop corrections terms to consider shape and material influences. A detailed description of the Mie theory is given by Bohren and Huffman [22], and the Mie code used for the investigations in the following sections is based on the implementations by Mätzler [28].

When Mie theory is applied to calculate the radiative properties, determination of the phase function $\Phi$ is computationally the most expensive step. To save this computational time, the radiative properties can be pre-calculated and stored in tabulated format. An alternative is to determine the radiative cross-sections $C$ utilising the Mie theory and using mathematical approximations of the Mie phase functions. In several numerical investigations [4,29,30], the $\delta$-Eddington approximation [31] and the Henyey-Greenstein phase function [32] yield appropriate approximations of the Mie phase function. In the $\delta$-Eddington approximation, a part of the scattered radiation is treated as purely forward scattering (transmission), while the remaining part is modelled by an isotropic phase function. The Henyey-Greenstein phase function is a more complex approximation represented by an analytical expression that depends on the asymmetry parameter $g$, which is defined for $-1 \leq g \leq 1$,

$$\Phi_{\mathrm{HG}} = \frac{1}{4\,\pi} \frac{1 - g^2}{\left(1 + g^2 - 2\,g\,\cos(\theta)\right)^{1.5}} \tag{17.8}$$

The asymmetry parameter solely describes the scattering direction: For $g = -1$, all radiation is scattered backward, for $g = 0$, radiation is scattered isotropically, and for $g = 1$, all radiation is scattered forward. For solid fuel particles, the asymmetry parameter equals $g \geq 0.6$. To further improve the accuracy of the Henyey-Greenstein phase function, a modified version for coal particles has been proposed by Gronarz et al. [33].

**Extensions of the Mie Theory**

As the Mie theory is limited to homogeneous spheres, three modifications were used to consider inhomogeneity: effective material approximations, the coated Mie theory, and the generalised multiparticle Mie solution.

The simplest method to consider the index of refraction of different materials simultaneously is to calculate an effective index of refraction $\mathrm{IOR}_{\mathrm{eff}}$. In (17.9), the influence of each material $i$ on the effective IOR is considered by multiplying the corresponding IOR by the volume fraction $f$.

$$\mathrm{IOR}_{\mathrm{eff}} = \sum f_i \, \mathrm{IOR}_i \tag{17.9}$$

The Maxwell-Garnett equation [34] is also an effective medium approximation. Initially, it was used to calculate an effective electric permittivity, but in the context of particle radiation, it calculates an effective complex index of refraction:

$$\frac{\mathrm{IOR}_{\mathrm{eff}}^2 - \mathrm{IOR}_{\mathrm{m}}^2}{\mathrm{IOR}_{\mathrm{eff}}^2 + 2\,\mathrm{IOR}_{\mathrm{m}}^2} = \sum_{\mathrm{inc}} f_{\mathrm{inc}} \frac{\mathrm{IOR}_{\mathrm{inc}}^2 - \mathrm{IOR}_{\mathrm{m}}^2}{\mathrm{IOR}_{\mathrm{inc}}^2 + 2\,\mathrm{IOR}_{\mathrm{m}}^2} \tag{17.10}$$

In (17.10), $\mathrm{IOR}_{\mathrm{eff}}$ is the resulting effective index of refraction representing a sphere with a medium (index "m") with inclusions (index "inc"). Therein, $f_{\mathrm{inc}}$ denotes the volume fraction of the inclusions. In the investigation of radiative properties of single particles, the Maxwell-Garnett approximation is mainly used to consider porosity or to model inhomogeneous compositions. The drawback of effective medium approximations is the assumption of a perfect, even distribution of the materials.

As the name suggests, the coated Mie theory was originally derived for coated particles [35]. This extension models the radiative properties of a sphere, consisting of a homogeneous core with a homogeneous coating and, thus, considering two different IOR simultaneously. Additionally, the Maxwell-Garnett approximation can be used to calculate an effective IOR for the core and the coating to consider even more different IOR effectively. In the context of solid fuel combustion, this extension is used to calculate the radiative properties of solid fuel particles during the burnout. The core is a porous carbon core, and a porous ash film represents the coating. The results are discussed in Sect. 17.3.2.

The generalised multiparticle Mie solution (GMM) [36] calculates simultaneously the radiative properties of multiple particles that are closely packed within a sphere. The spheres can have different sizes and different IOR. Here, it is used to construct a porous particle. An example of such a particle is given in Fig. 17.1b).

**Representation of the Phase Function**

The mentioned methods calculate the scattering cross-section $C_{\mathrm{Sca}}$, absorption cross-section $C_{\mathrm{Abs}}$, and the phase function $\Phi$. The cross-sections are scalar values, which are directly transferred to coupled CFD/radiation simulation. However, the phase function additionally depends on the scattering angles and must be discretised for simulation application. Figure 17.2a) illustrates a continuous phase function and Fig. 17.2b) the corresponding discretised phase function. Here, the scattering phase function is divided into eight constant segments over the scattering angle $\theta$ and into eight constant segments over $\varphi$. This discretisation leads to 64 segments. However, this simple discretisation scheme is only used in the following section to analyse the phase function. The actual phase function applied in the CFD/radiation simulation is divided into 80 segments applying the FTn scheme [37].

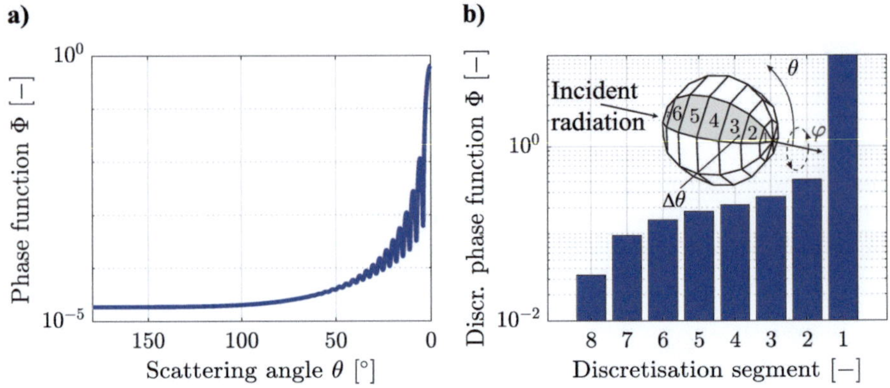


**Fig. 17.2** **a)** Continuous phase function and **b)** corresponding discretised phase function dividing $\Phi$ in 8 constant segments over $\theta$ and 8 constant segments over $\varphi$

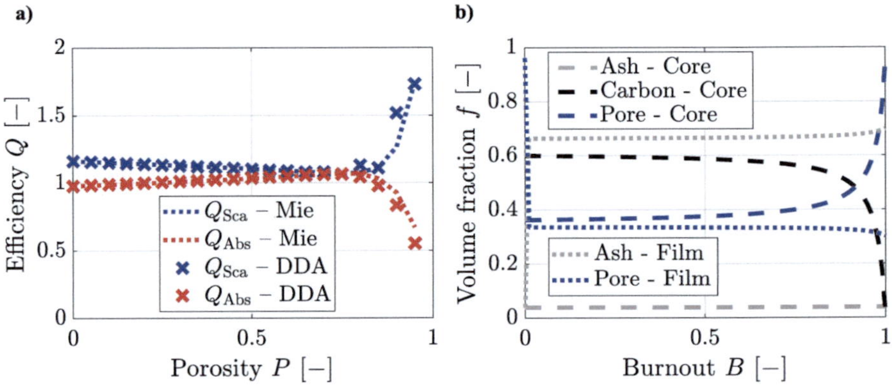


**Fig. 17.3** **a)** The efficiencies depending on the porosity calculated by the DDA and by utilising the Mie theory with an effective index of refraction and **b)** utilising the CCK model, the volume fractions of ash, carbon and pores are calculated depending on the burnout $B$: Dotted line represents the volume fractions of the core and dashed lines represent the volume fraction of the ash film (based on the publication by Koch et al. [38])

## 17.3.2 Influence of Porosity and Burnout

During the conversion process, first, the volatiles are released from the particle, while the minerals are transformed into ash. Depending on the reaction conditions and the oxygen transport within the particle pores, effects like ash film formation, particle swelling, and particle shrinking occur. A more detailed description is given in Sect. 3.1. The pore size, the porosity fraction, and thus the total particle morphology change. Considering the particle radiation, the question arises as to what degree of accuracy the change in morphology should be accounted for, i.e. is it necessary to develop correction terms with the DDA?

Figure 17.3a) illustrates a spherical particle's absorption and scattering efficiency depending on the porosity with a size parameter $x \approx 52$. The efficiency is the ratio of radiation cross-section $C$ to geometrical cross-section $G$, and it is often used to compare the radiative properties of different size parameters. To calculate these efficiencies, the radiation interaction is modelled and solved by the DDA, where the IOR of the dipoles equal either IOR $= 1.4 + 0.1i$ (IOR of solid fuel [39]) or IOR $= 1.0 + 0.0i$ (vacuum), and the efficiencies are calculated by using the Mie theory with an effective IOR applying the Maxwell-Garnett approximation (MGA).

For a pore volume fraction with $P < 0.8$, the MGA, in combination with the Mie theory, matches almost perfectly the results calculated by the DDA. For $P \leq 0.8$, the differences increase, but they are still below 10% and are thus negligible. The increased deviation from the DDA results is due to the assumption of a spherical particle shape. For $P \leq 0.8$, too many dipoles have the same IOR as the surrounding gas. Thus, the particle loses its spherical shape with increasing porosity. Overall, an effective material property can account for the pores, and no correction term for the Mie theory is necessary.

As the pore volume fraction increases and the minerals are transformed into ash during the conversion process, the volume fraction of the pores must be specified in addition to the volume fraction of ash and carbon content of the burning particles. To account for the influences of these fractions on the radiative properties, two approaches are proposed in the current radiation model:

1. a one-way coupling of the radiative properties to the char conversion kinetics (CCK) model and
2. applying an effective material property in a simplified approach.

A detailed description of the CCK model is given in Sect. 8.2.2.

To quantify the effects of structural changes during the char conversion process on the radiative properties, the CCK model determines among others the particle size, ash, carbon and pore volume fraction. These output data are then used to build artificial particles at different conversion degrees. These particles comprise a porous ash film and a porous carbon core with a time-independent ash mass fraction. The resulting volume fractions $f$ for a particle with diameter $D_p = 100\,\mu m$ are displayed in Fig. 17.3b) depending on the burnout $B$. Burnout $B$ is a measure that describes the progress of the reaction and equals $B = 1 - w_c$ with the carbon mass fraction $w_c$. Therefore, $B = 0$ indicates the state after pyrolysis, and $B = 1$ means no carbon is left within the particle, and only a porous ash particle remains. The exact calculation of the volume fractions $f$ utilising the CCK data are described in the study by Koch et al. [38].

During the conversion process, the overall particle size and relative core size decrease, as the relative ash film size increases. Finally, the radiation interaction is calculated by applying Mie theory for coated particles, with the following IOR for the coating ($IOR_{film}$) and the core:

$$\mathrm{IOR}_{\mathrm{film}} = f_{\mathrm{ash}}\,\mathrm{IOR}_{\mathrm{ash}} + f_{\mathrm{pore}} \tag{17.11}$$

$$\mathrm{IOR}_{\mathrm{core}} = f_{\mathrm{ash}}\,\mathrm{IOR}_{\mathrm{ash}} + f_{\mathrm{carbon}}\,\mathrm{IOR}_{\mathrm{coal}} + f_{\mathrm{pore}} \tag{17.12}$$

An alternative to the one-way coupling procedure is applying a simplified approach. Here, standard Mie theory for homogeneous spheres is applied, and an effective $\mathrm{IOR}_{\mathrm{eff}}$ is calculated by applying the following equation, which was derived in the study by Koch et al. [38]:

$$\mathrm{IOR}_{\mathrm{eff}} = \left(\mathrm{IOR}_{\mathrm{SF}}\,(1 - B^{12.4}) + \mathrm{IOR}_{\mathrm{ash}}\,B^{12.4}\right)(1 - f_{\mathrm{pore}}) + f_{\mathrm{pore}} \tag{17.13}$$

The index "SF" denotes the solid fuel. However, this approach still needs the specification of the pore volume fraction $f_{\mathrm{pore}}$. In previous studies, another approach was developed that neglects porosity, is not coupled to the CCK model, and depends on the Mie theory for coated particles [40]. The absorption and scattering efficiency results depending on the burnout are illustrated in Fig. 17.4a) for the two mentioned approaches, the one-way coupling process and an approach from literature by Guo et al. [41]. The one-way coupling process is the most accurate approach. Thus, it is treated as a benchmark. For burnout $B \leq 0.8$, all simplified approaches result in appropriate approximations of the efficiencies, whereas the approaches developed by the authors are more accurate than the one from literature. However, for $B > 0.9$, considering the influence of porosity is mandatory. Only the recently developed approach by Koch et al. [38] agrees with the benchmark. The porosity volume fraction is essential, especially for ash particles ($B = 1$). Figure 17.4a) illustrates the discretised phase function exemplarily for a burnout $B = 0.9$. Both approaches by the authors yield equally accurate results.

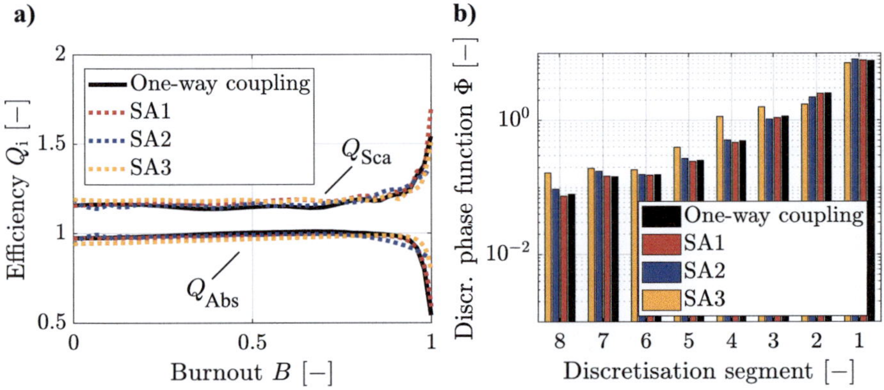


**Fig. 17.4** The radiative properties for four different approaches: The one-way coupling with the CCK model represents the most accurate solution. Three simplified approaches (SA) are developed by Koch et al. [38] (SA1), Gronarz et al. [40] (SA2, neglection of porosity), and Guo et al. [41] (SA3): **a)** absorption and scattering efficiency (based on the publication by Koch et al. [38]) and **b)** phase function for a burnout $B = 0.9$

These investigations were also conducted for different particle sizes, and it was noticed that the correct porosity volume fraction has an even more significant influence on the accuracy of particles with decreasing diameters.

All in all, the correct pore volume fraction is mandatory to calculate the efficiencies accurately. However, it is not necessary to derive a correction term from the DDA using exact structure representations of the particle. It is sufficient to apply an effective IOR for spherical solid fuel particles. The influence of non-sphericity and structured porosity is investigated in the next section.

### 17.3.3 Influence of Highly Non-Spherical Shapes

As seen in the previous section, the particle shape deviates from spheres with increasing porosity. These shape differences result in divergent radiative properties to the spherical shape. Furthermore, only random inclusions of the pores have been investigated so far. For coal and walnut shell (WNS) particles, these assumptions are accurate. However, when biomass species like pine, poplar, or beech wood are compared to coal or WNS, two significant differences are noticed: These biomass samples are distinctly more non-spherical and have an internal, structured porosity. Riaza et al. [16] investigated the shape of non-spherical biomasses during conversion, observing almost spherical particle shapes for $B > 90\%$. However, pyrolysis did not strongly influence the shape, as the particles stayed non-spherical for most of the conversion process. For example, the aspect ratio of pine biomass was still four for 50% burnout, whereas the starting aspect ratio was six. To investigate the influence of these aspects, pulverised beech wood particles were modelled as realistically as possible, considering the asymmetric three-dimensional shape and tube-like pores. Then, the shape and internal structure were simplified stepwise [42]. The resulting particles are illustrated in Fig. 17.5.

Case A in Fig. 17.5 resembles the artificial beech wood particle. Its base area is an ellipse with a long axis $D_{\text{long}} = 50\,\mu\text{m}$ and a short axis $D_{\text{short}} = 37.5\,\mu\text{m}$, where the curvature is disturbed by noise. The resulting average axis length is $\overline{D}_{\text{base}} = 43.75\,\mu\text{m}$. The base area is extruded with varying heights $H = 50–250\,\mu\text{m}$. Furthermore, 60 ducts with a diameter $D_{\text{duct}} = 3.75\,\mu\text{m}$ are evenly distributed, which represent the internal structured porosity. Finally, the end faces with random fractures are added. As each end face has a maximal height of $25\,\mu\text{m}$, the total height varies from $100–300\,\mu\text{m}$, resulting in aspect ratios of 2.2–6.8. In case B, the fractal end pieces are removed, but the internal structure is kept. In case C, the internal structure is removed, and an effective IOR is calculated. The last case D displays a homogeneous sphere with an equivalent volume compared to the particle in case C. The radiative properties were calculated by the DDA for cases A–C, and by the Mie theory for case D. Due to the non-spherical shapes, the relation between the direction of propagation of the radiation and the orientations of the particles in cases A–C has to be specified. For the calculation, the direction of propagation is constant with (0/0/1), and the orientation angle $\beta$ is introduced. This orientation angle describes the rotation of the particle around the $y$-axis.

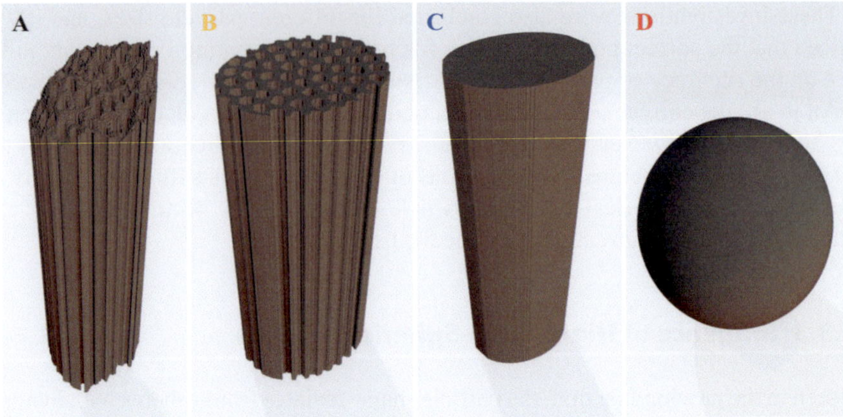


**Fig. 17.5** **a)** Artificially generated beech wood particle, **b)** approximation by an extruded ellipse, **c)** approximation by an extruded ellipse with an effective material property, and **d)** approximation by a volume-equivalent sphere

The mean relative deviation $E$ was calculated to quantify the accuracy of the simplified particles considering the particle radiation interaction:

$$E_1(H) = \frac{1}{N_\beta} \sum_{\beta=0^\circ}^{\beta=180^\circ} \frac{|C^\beta_{\mathrm{Abs,i}}(H) - C^\beta_{\mathrm{Abs,j}}(H)|}{C^\beta_{\mathrm{Abs,i}}(H)} \tag{17.14}$$

where $C$ denotes the radiation cross-section, and the indices $i$ and $j$ are abbreviations for the cases. This means that when the mean deviation of case B to case A is calculated, $C_{\mathrm{Abs,i}}$ is replaced by the calculated cross-sections of case B and $C_{\mathrm{Abs,j}}$ by the ones of case A. The resulting mean deviations are plotted in Fig. 17.6a).

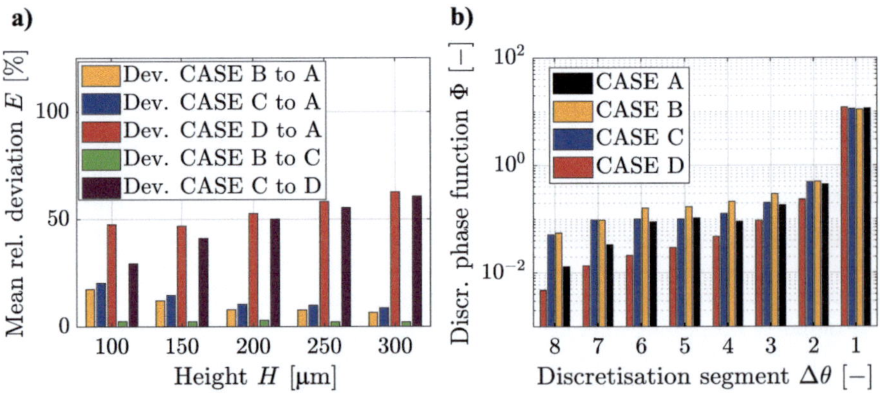


**Fig. 17.6** **a)** Mean relative deviations between the cases depending on the height $H$ applying (17.14) and **b)** discretised phase function averaged over the orientation angles of the particle with a total height $H = 100\,\mu\mathrm{m}$ (based on the publication by Koch et al. [42])

These results indicate that it is sufficient to consider even the structured porosity by an effective material property. The only change between cases B and C is the approximation of the porosity by an effective material property, and the deviation of these two approaches is always $E < 3\%$. Furthermore, with increasing height, and thus, with increasing axis ratio, the deviation of the sphere (case D) to the remaining cases increases. Comparing the artificial beech wood particle (case A) to the sphere, the deviation is around 40–60%, while the assumption of an extruded ellipse (case B) led to deviations of 10–20% between the radiation cross-sections of the artificial beech wood particle (case A). These deviations highlight that a good shape approximation of the particle is mandatory, and a volume-equivalent sphere leads to more significant deviations. The same conclusions are drawn for the scattering cross-section $C_{\text{Sca}}$ and the phase function $\Phi$. Figure 17.6b) illustrates the discretised phase function for the four cases. The deviations are indeed the largest for the sphere but negligible and in a similar order to the extruded ellipse. Most of the scattered radiation is scattered forward. Thus, for highly non-spherical particles, the same $\delta$-Eddington phase function is advised as for volume-equivalent spheres.

These investigations were also done for varying indices of refractions, where the real part of the IOR is changed to $n \in \{1.2, 1.4, 1.6, 1.8\}$ and the imaginary part is changed to $k \in \{0.001, 0.01, 0.1\}$. Similar results were noticed, and thus, these conclusions are not limited to a specific index of refraction.

All in all, the performed numerical investigation showed that an effective material property can approximate inhomogeneity/porosity, but the shape approximation of the particle should be as close as possible. A shape approximation by a cylinder or an ellipsoid is recommended for highly non-spherical particles like beech wood. As discussed in Sect. 17.3.1, calculating the radiative properties of non-spherical particles requires methods like DDA and, thus, a lot of calculation time. Therefore, a correction term for the Mie theory considering non-spherical shapes is derived from exact data calculated by the DDA.

### Corrected Mie Theory for Non-Spherical Particles

To develop a correction term for the Mie theory that considers non-sphericity, a general shape simplification must be done. Here, the non-spherical particles are approximated by prolate ellipsoids (cf. Fig. 17.7a), whereas the diameter of the base area $D_{\text{Base}}$ and the aspect ratio $\zeta = D_{\text{Long}}/D_{\text{Base}}$ define the shape. For $\zeta = 1$, the Mie theory is used with size parameter $x = \pi\, D_{\text{Base}}/\lambda$. Furthermore, the orientation angle $\beta$ is additionally required. Figure 17.7b) illustrates the absorption efficiency $Q_{\text{Abs}}$ for non-spherical particles with a base area $D_{\text{Base}} \in (20, 30)\ \mu\text{m}$ and aspect ratios $\zeta \in (2, 3)$ depending on the orientation angle $\beta$.

The profiles of all $Q^{\beta}_{\text{Abs}}$ are similar: The minimum value is at $\beta = 0°$, the maximum is at $\beta = 90°$, and trigonometric functions can describe the overall profile. When the aspect ratio is kept constant, and the base area is increased (continuous line to dotted one), the course of $Q^{\beta}_{\text{Abs}}$ is the same, with a shift of $Q_{\text{Abs}}$ to a lower value. Increasing the aspect ratio (red to blue) decreases the value of $Q_{\text{Abs}}$ at $\beta = 0°$,

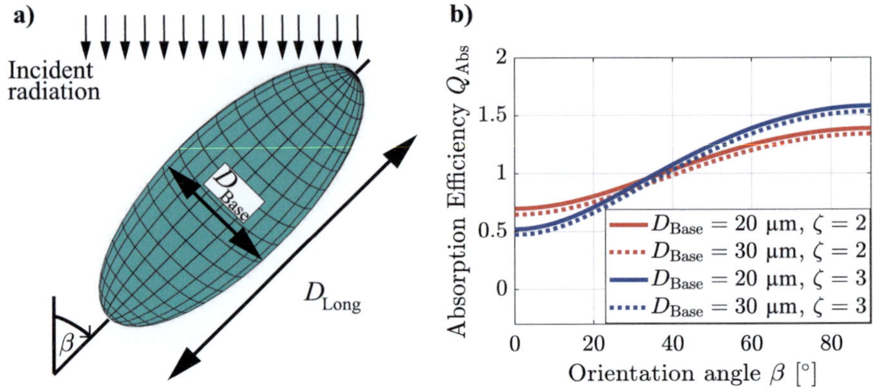


**Fig. 17.7** Influence of the aspect ratio $\zeta$ and the orientation angle $\beta$ on the absorption efficiency $Q_{\mathrm{Abs}}$: **a)** definition of the geometry and **b)** simulation results

and increases the value of $Q_{\mathrm{Abs}}$ at $\beta = 90°$. This similarity is used to derive a Mie theory correction term depending on $\zeta$ and $\beta$ for a wide parameter range.

For a reference case ($D_{\mathrm{Base}} = 30\,\mu\mathrm{m}$, IOR $= 1.4 + 0.1\mathrm{i}$, $\lambda = 3\,\mu\mathrm{m}$), the absorption efficiencies at $\beta \in (0°, 45°, 90°)$ are fitted depending on the aspect ratio $\zeta$. For $\zeta = 1$, the Mie solution $Q_{\mathrm{M}}$ is used. For other base diameters $D_{\mathrm{Base}}$, wavelengths $\lambda$ or indices of refraction IOR, a shift of the fit function is calculated based on the differences between the Mie solution for $\zeta = 1$. Figure 17.8a illustrates the absorption efficiency $Q_{\mathrm{Abs}}$ depending on the aspect ratio $\zeta$ for the three orientation angles $\beta \in (0°, 45°, 90°)$. The crosses mark the calculated values by the DDA, and the dotted line represents the fit applying the following functions

$$Q_{\mathrm{Abs}}^{0°} = Q_{\mathrm{R}}\, \zeta^{-0.7} + Q_{\mathrm{D}} \tag{17.15}$$

$$Q_{\mathrm{Abs}}^{90°} = Q_{\mathrm{R}}\, \zeta^{0.32} + Q_{\mathrm{D}} \tag{17.16}$$

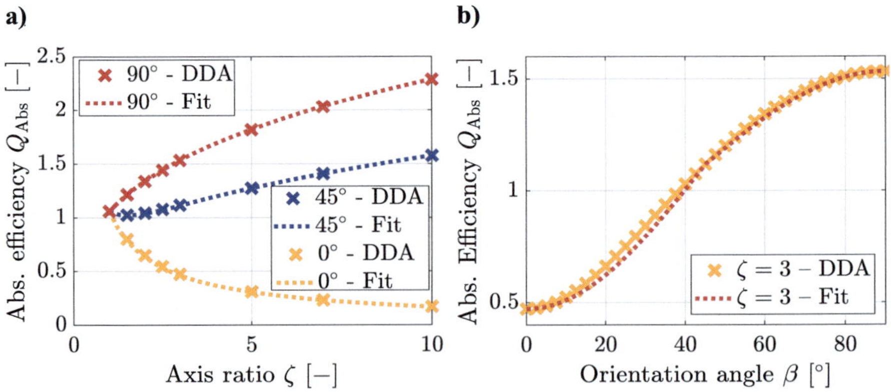


**Fig. 17.8** DDA results and the resulting fits for the absorptions efficiency $Q_{\mathrm{Abs}}$ **a)** depending on the aspect ratio $\zeta$ for $\beta \in (0°, 45°, 90°)$ and **b)** depending on the orientation angle $\beta$ for $\zeta = 3$

$$Q^{45°}_{\mathrm{Abs}} = \begin{cases} 1.14\exp(-2.37\,\zeta) + 0.94\exp(0.07\,\zeta) + Q_{\mathrm{D}} & \text{for } \zeta \leq 3 \\ 0.31\,\zeta^{0.51} + 0.62 + Q_{\mathrm{D}} & \text{for } \zeta > 3 \end{cases} \tag{17.17}$$

where $Q_{\mathrm{R}} = 1.11$ denotes the Mie solution of the reference case and the shift to a higher or lower value depending is considered by $Q_{\mathrm{D}} = Q_{\mathrm{M}} - Q_{\mathrm{R}}$. Similar fit functions are derived for the scattering efficiency $Q_{\mathrm{Sca}}$:

$$Q^{0°}_{\mathrm{Sca}} = 0.823\,\zeta^{-0.661} + 0.318 + Q_{\mathrm{D}} \tag{17.18}$$

$$Q^{90°}_{\mathrm{Sca}} = 0.741\,\zeta^{0.406} + 0.4 + Q_{\mathrm{D}} \tag{17.19}$$

$$Q^{45°}_{\mathrm{Sca}} = \begin{cases} 1.4\exp(-2.57\,\zeta) + 0.96\exp(0.08\,\zeta) + Q_{\mathrm{D}} & \text{for } \zeta \leq 3 \\ 0.37\,\zeta^{0.485} + 0.576 + Q_{\mathrm{D}} & \text{for } \zeta > 3 \end{cases} \tag{17.20}$$

With increasing aspect ratio, the three absorption efficiencies converge to a constant value. Furthermore, $Q_{\mathrm{Abs}}$ at $\beta = 45°$ is not the arithmetic mean of $Q_{\mathrm{Abs}}$ at $\beta = 0°$ and at $\beta = 90°$. Therefore, it is not sufficient to fit the course of $Q^{\beta}_{\mathrm{Abs}}$ in Fig. 17.7b by a single trigonometric function, and instead the following functions are used:

$$Q^{\beta}_{\mathrm{i}} = \begin{cases} \left(2\cos(\beta + 90°)^2 - 1\right)\,(Q^{45°}_{\mathrm{i}} - Q^{0°}_{\mathrm{i}}) + Q^{45°}_{\mathrm{i}} & \text{for } \beta \leq 45° \\ \left(2\cos(\beta + 90°)^2 - 1\right)\,(Q^{90°}_{\mathrm{i}} - Q^{45°}_{\mathrm{i}}) + Q^{45°}_{\mathrm{i}} & \text{for } \beta > 45° \end{cases} \tag{17.21}$$

with the index i for scattering ("Sca") and absorption ("Abs"). The absorption efficiency calculated by the DDA and the resulting trigonometric function are illustrated in Fig. 17.8b. For $\beta \geq 45°$ the deviation of the fit is smaller than for $\beta < 45°$, but still with a mean relative deviation lower than 3.5%.

**Validation and Alternative of the Corrected Mie Theory for Non-Spherical Particles**

The previous section proposed an efficient method to calculate the efficiencies for non-spherical particles. In the following, the accuracy of this new model is tested by applying (17.22),

$$E(\zeta) = \frac{1}{N_{\beta}} \sum_{\beta=0°}^{\beta=90°} \frac{|Q^{\beta}_{\mathrm{Abs,Model}}(\zeta) - Q^{\beta}_{\mathrm{Abs,DDA}}(\zeta)|}{Q^{\beta}_{\mathrm{Abs,DDA}}(\zeta)} \tag{17.22}$$

where $Q^{\beta}_{\mathrm{Abs,DDA}}$ is the benchmark absorption efficiency calculated by the DDA, and $Q^{\beta}_{\mathrm{Abs,Model}}$ represents the absorption efficiency calculated by applying the proposed non-spherical particle (NSP) model or the volume-equivalent sphere (VES) model. Therefore, $E$ represents the deviation of a model with respect to the benchmark solution averaged over the orientation angle $\beta$ depending on the aspect ratio $\zeta$. The deviation $E$ is calculated for varying indices of refraction, where the real part of the IOR is changed to $n \in \{1.2, 1.5, 1.8, 2.1\}$ and the imaginary part is changed to $k \in \{0.005, 0.01, 0.05, 0.1, 0.5\}$. Furthermore, the base area varies for $D_{\mathrm{Base}} \in$

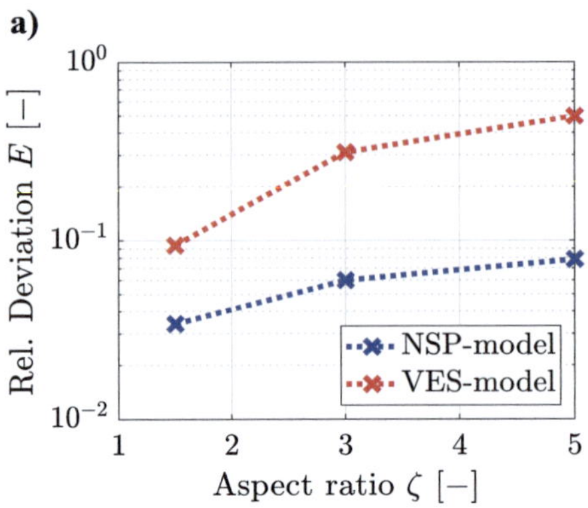


**Fig. 17.9** Mean relative deviations between the non-spherical particle (NSP) model and volume-equivalent sphere (VES) model to the benchmark depending on the aspect ratio $\zeta$ applying (17.22) for **a)** IOR $= 1.2 + 0.5\text{i}$ and $D_{\text{Base}} = 5\,\mu\text{m}$, and **b)** IOR $= 1.2 + 0.005\text{i}$ and $D_{\text{Base}} = 5\,\mu\text{m}$

$\{5, 10, 25, 50\}\,\mu\text{m}$ and the aspect ratio for $\zeta \in \{1.0, 1.5, 3, 5\}$. The wavelength is constant with $\lambda = 3\,\mu\text{m}$.

Figure 17.9 displays the resulting deviations for IOR $= 1.2 + 0.005\text{i}$ and IOR $= 1.2 + 0.5\text{i}$ with $D_{\text{Base}} = 5\,\mu\text{m}$. In almost all cases, the deviation $E$ increases with increasing aspect ratio. To evaluate all validation cases' accuracy graphically, each validation case is categorised according to the deviation $E$ at $\zeta = 5$, which is illustrated in Fig. 17.10. From these results, the following statements are concluded:

- With increasing base diameter $D_{\text{Base}}$, increasing refractive index $n$ or increasing absorptivity index $k$, the accuracy of the proposed NSP model increases, while the accuracy of the VES model decreases.
- The NSP model for base diameter $D_{\text{Base}} > 25\,\mu\text{m}$ is sufficiently accurate for $k \geq 0.05$. For lower values of $k$, the (VES) model performs better, but the deviation is always between $E = 0.5$–$1.0$.
- For large base diameter $D_{\text{Base}} > 25\,\mu\text{m}$ and for $k \geq 0.05$, the NSP model is always more accurate than the volume equivalent sphere. Deviation $E$ reaches its maximum at $\zeta = 5$, while the deviation of the VES model to the benchmark increases with increasing $\zeta$ (not illustrated in Fig. 17.10).
- The non-spherical model for $D_{\text{Base}} = 10$–$25\,\mu\text{m}$ is sufficiently accurate for $k \geq 0.1$. A value of $0.05 < k < 0.1$ already yields deviations of $E < 0.2$. For lower values of $k$, the volume equivalent model is advised. However, deviations of $E = 0.5$ are possible.
- For $D_{\text{Base}} = 5$–$10\,\mu\text{m}$, the non-spherical model only generates accurate results for large absorptivity indices with $k \geq 0.5$. Due to the small base diameter, it is advised to use the DDA for the calculation, as the required simulation resources are affordable.

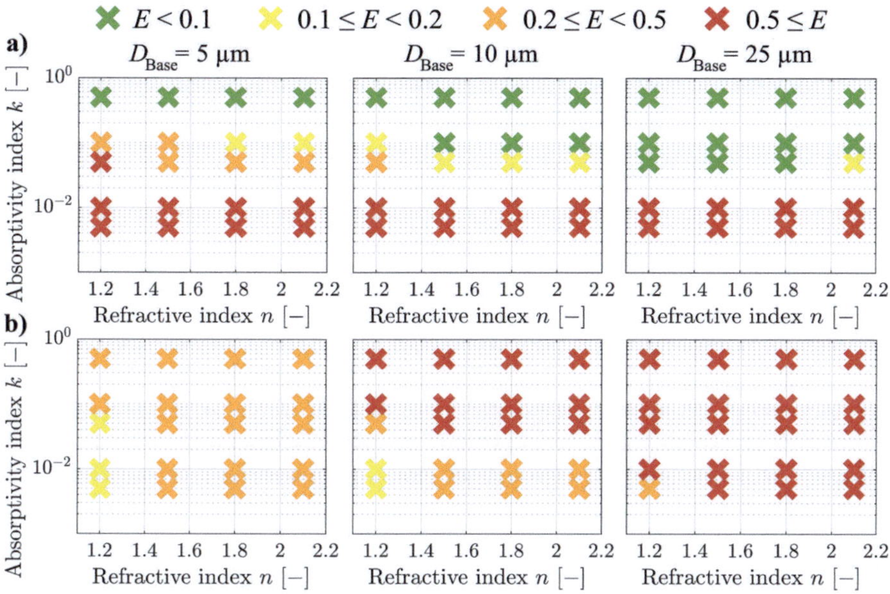


**Fig. 17.10** Mean relative deviation $E$ for varying IOR, and base diameter: Validation results for **a)** the proposed non-spherical particle (NSP) model and **b)** the volume-equivalent sphere (VES) model

## 17.4 Radiation Interaction with a Particle Group

In the previous sections, modelling the radiative properties of a single particle was presented. However, tracking all particles and calculating the radiation interaction individually is computationally too expensive in a combustion chamber, and thus, the principle of particle clouds is used. A particle cloud defines a mixture of particles within a gas atmosphere. A single absorption coefficient, a single scattering coefficient, and a single scattering phase function describe the radiative properties of all particles simultaneously. The overall absorption coefficient of the particle clouds is calculated by summing the gas absorption coefficient and the particle absorption coefficient. Scattering by the gas molecules is neglected. These properties apply to the entire volume of the particle cloud, which means that the cloud is not described by a few solids with spaces filled by the gas atmosphere but by a homogeneous mass. Therefore, this approach is also called the continuum approach.

In order to calculate these properties, the volume of the cloud $V_{PC}$, the number of particles $N_p$, and a particle size distribution (PSD) are required. The particle size distribution yields a Sauter mean diameter, which is used to calculate the absorption and scattering cross-section for a single particle representing all particles on average. Additionally, for non-spherical particles, a mean base diameter $D_{Base}$ and mean aspect ratio are required. The number of particles and the particle cloud volume are applied to calculate a particle number density $n$, which is then linearly scaled with the absorption and scattering cross-section $C$. Since the radiation cross-section $C$ equals the efficiencies $Q$ multiplied with the geometrical cross-section $G$, (17.23) and (17.24) express absorption coefficient

$$\kappa_{p,\lambda} = n\, Q_{Abs,\lambda}\, G \tag{17.23}$$

and scattering coefficient

$$\sigma_{p,\lambda} = n\, Q_{Sca,\lambda}\, G. \tag{17.24}$$

However, this linear scaling is only accurate, if three criteria are fulfilled:

1. All waves interacting with the particles are assumed to be planar, since this assumption was also made for the single particle radiation interaction.
2. No wave travels unhindered through the cloud volume, while other waves interact with particles.
3. No multiple scattering occurs.

When the first criterion is not fulfilled, dependent scattering has to be modelled [43]. Dependent scattering means that the particles are too close to each other. Each scattered or emitted wave of a particle is a spherical wave. However, this wave can be assumed to be planar in the far distance to its origin. Therefore, for low particle volume fractions, dependent scattering events are low compared to all scattering events, and the modelling can be neglected. Furthermore, Drolen and Tien [44] stated dependent scattering is only relevant for volume fractions above 1%, which is higher than volume fractions of typical solid fuel combustion.

The second criterion is only fulfilled for perfectly uniformly distributed particles. Figure 17.11 illustrates three different particle clouds, each with a particle volume fraction of 0.05%. Figure 17.11a) illustrates a perfectly uniform particle distribution. In comparison, Figs. 17.11b) and c) depict non-uniform particle distributions. For the particle distribution in Fig. 17.11b), the minimum distance between the particles is set to 20 mean diameters to avoid dependent scattering, while the particle distribution in Fig. 17.11c) is random. These non-uniform particle distributions result in the radiation being able to pass unhindered through areas with low particle volume

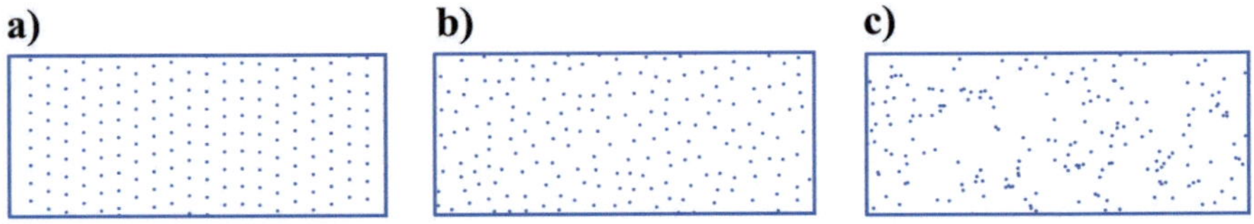


**Fig. 17.11** Illustration of three particle clouds each with a particle volume fraction of 0.05% and 400 monodisperse particles: **a)** a uniform particle distribution, **b)** random non-uniform particle distribution with a minimal distance of each particle set to 20 diameters, and **c)** random non-uniform particle distribution without any restrictions

fraction, while areas with high particle volume fraction are exposed to multiple scattering [23,45]. Multiple scattering increases the radiation interactions with particles in the cloud, while the first phenomenon decreases the interactions.

As particle distributions depicted in Figs. 17.11b) and c) are more realistic, the second and third criteria are not fulfilled. How these phenomena influence the resulting absorption and scattering coefficient of a particle cloud and how these phenomena can be modelled efficiently was shown in numerical studies [46–48]. The methodology and the most important results are briefly presented in the following sections.

### 17.4.1 Fully Resolved Monodisperse Particle Clouds

Since the focus of the following numerical investigation was to investigate the influence of the spatial particle distribution—either uniform or non-uniform—within a particle cloud on the radiative properties, the following simplifications were made:

- Influence of gas radiation is neglected, filling the empty spaces between the particles with vacuum.
- No wavelength dependency is considered.
- Particle emission is not considered in the investigations, but the results are directly transferable to particle emission [47].
- Cartesian meshes are used to approximate the particle shape.

In this section, particle clouds with monodisperse, spherical particles ranging from 10 to 100 $\mu$m are investigated. Section 17.4.2 presents the results for polydisperse particle clouds. For the numerical investigation, a Monte-Carlo ray-tracing method was used. On one side of the particle clouds, rays enter the cloud perpendicularly, and on the other side, the intensities of the rays leaving the cloud are summed. Here, the particle cloud properties are modelled by two approaches. One is the standard approach stated in (17.23) and (17.24), which is called the linear scaling approach (LSA). Here, it is assumed that the absorption and scattering properties of the cloud are identical at all positions across the cloud. The particle positions are lost and the particle cloud is interpreted as black box. Equations (17.23) and (17.24) calculate effective absorption and scattering coefficients. The particle number density

$$n = \frac{N_{\mathrm{P}}}{V_{\mathrm{PC}}} \tag{17.25}$$

describes the number of particles $N_{\mathrm{P}}$ per particle cloud volume $V_{\mathrm{PC}}$, and thus, the resulting unit of the absorption and scattering coefficients is m$^{-1}$. Therefore, these coefficients express the statistical interaction of radiation with particles and how much radiation is absorbed or scattered on average when passing through the particle cloud. The advantage of this approach is low computational costs due to a coarse grid.

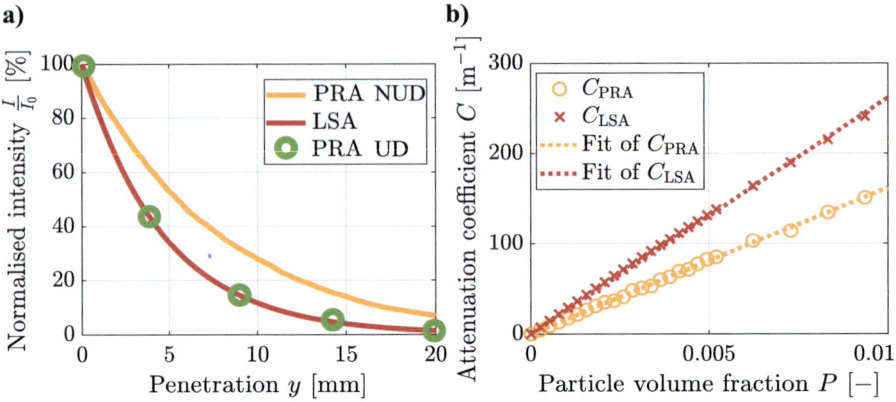


**Fig. 17.12** **a)** Normalised intensity as a function of penetration $y$ for $Q_{\text{Abs}} = 1$, $D_{\text{p}} = 80\,\mu\text{m}$ and $P = 0.01$ and **b)** attenuation coefficient $C$ depending on the particle volume fraction $P$ for $Q_{\text{Abs}} = 1$ and $D_{\text{p}} = 80\,\mu\text{m}$

The other method serves as benchmark, where each particle is resolved, and thus, this method is called particle-resolved approach (PRA). Since each particle is resolved, a fine mesh is necessary with a mesh size equal to the particle diameter $D_{\text{p}}$. For the cells filled with a particle, (17.23) and 17.24 are used, but $n$ is replaced by $n^*$,

$$n^* = \frac{1}{V_{\text{p}}} \tag{17.26}$$

with the particle volume $V_{\text{p}}$. As a consequence of the non-uniform particle distribution, the number of particle interactions of one ray depend on its entering points. Multiple interactions lead to high absorption and strong scattering in different directions. On the contrary, when no particle is in the path of the radiation illustrated by rays, the ray leaves the cloud with the starting intensity. This effect results in increased local transmission compared to the mean transmission of the cloud.

Finally, the intensity leaving the cloud was summed and compared to the intensity leaving the cloud modelled by the LSA. Due to the random particle distribution in the PRA, the simulations were repeated 15 times with a new random distribution each time and averaged. Figure 17.12a) depicts the normalised intensities $I/I_0$ as a function of penetration $y$. The yellow curve depicts the intensity distribution of the averaged PRA, and the red curve illustrates the intensity distribution applying the LSA. The green circles mark the intensity distribution for applying the PRA with a uniform particle distribution, which fulfills the assumptions made for the LSA (no multiple scattering, all interacting waves are planar, and no partial transmission). This results in a distribution similar to the one illustrated in Fig. 17.11a). Such particle distributions are constrained to the particle cloud geometry. Thus, it is only possible for specific penetration $y$.

These curves were fitted by the Beer-Lambert law [49]:

$$I = I_0 \exp(C\, y) \tag{17.27}$$

The exponent $C$ is the fit parameter and was determined for particle clouds modelled by the LSA and PRA for varying parameters. Here, the particle volume fraction was varied within $P = 0$–$1\%$, the particle mean diameter within $D_\mathrm{p} = 10$–$100\,\mu\mathrm{m}$, and the absorption efficiency for $Q_\mathrm{Abs} = 0.6$–$1.6\,\mu\mathrm{m}$. The scattering efficiency is set constant to $Q_\mathrm{Sca} = 1$, and the simulations were performed in a pseudo-2-D particle cloud with a geometry of $4 \times 20\,\mathrm{mm}^2$ since the influence of scattering efficiency $Q_\mathrm{Sca}$ and the particle cloud geometry are negligible for the exponent $C$. Figure 17.12b) illustrates the resulting $C_\mathrm{LSA}$ and $C_\mathrm{PRA}$ for a constant $Q_\mathrm{Abs} = 1$, $D_\mathrm{p} = 80\,\mu\mathrm{m}$ depending on the particle volume fraction $P$. The data were fitted by $C_\mathrm{i}(Q_\mathrm{Abs}) = m_\mathrm{i}(Q_\mathrm{Abs})\,P$ with index $i$ equal to "LSA" or "PRA". Correlations for the slopes $m$ and the values of its parameters are given in Koch et al. [47].

Based on this slope, a correction term for the LSA was determined, enabling the PRA accuracy while maintaining the LSA efficiency. The absorption and scattering coefficient determined by the LSA are scaled by the weighting factor $W$

$$W(D_\mathrm{p}, Q_\mathrm{Abs}) = \frac{C_\mathrm{PRA}}{C_\mathrm{LSA}} = \frac{m_\mathrm{PRA}(D_\mathrm{p}, Q_\mathrm{Abs})}{m_\mathrm{LSA}(D_\mathrm{p}, Q_\mathrm{Abs})}, \tag{17.28}$$

and thus, this new method is called weighted scaling approach (WSA). It only depends on the particle diameter and absorption efficiency, not on the particle volume fraction $P$ as long as $P < 1\%$. In general, the weighting factor is always smaller than one. Therefore, the LSA overpredicts the particle radiation interaction in a cloud, i.e. in reality, fewer particle radiation interactions are noticed. It is also applied to the emission coefficient of the particle cloud, assuming less emitted radiation per metre, which was also verified by simulations.

The scaling approach did not remodel the phase function. Most of the radiation is scattered forward, and the phase function of the particle cloud matches that of the representative particle. Due to multiple scattering, the sidewards scattered radiation of a cloud is more significant than for a single particle. However, as long as the numerical cell (representing the particle cloud) has the same size as the numerical cell in the fluid simulation, this approximation of the phase function should give appropriate solutions.

### 17.4.2 Fully Resolved Polydisperse Particle Clouds

The WSA was determined for monodisperse particle groups applying the accuracy of the PRA. Due to the implementation of the computational grid, only complete shadowing was possible (i.e. a particle could either be placed entirely in the shadow of another particle or not). However, a polydisperse particle group describes reality more accurately. Therefore, the effect of polydisperse particle groups and spatial distributions enabling partial shadowing were also investigated [48].

Figure 17.13a) illustrates a Gaussian particle size distribution (PSD) for a particle volume fraction of $P = 0.01$ with mean diameter $\overline{D}_\mathrm{p} = 50\,\mu\mathrm{m}$ and standard deviation of $\sigma = 13.3\,\mu\mathrm{m}$. For monodisperse particle clouds, only the specification of

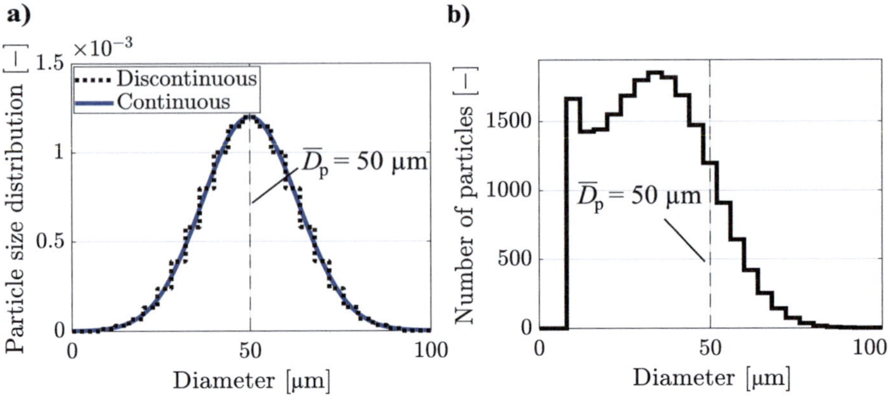


**Fig. 17.13** **a)** Continuous and discontinuous particle size distribution (PSD) with a mean diameter $\overline{D}_p = 50\,\mu m$ and a standard deviation of $\sigma = 13.3\,\mu m$ and **b)** corresponding numbers of particles within a particle cloud with dimensions $x \times y \times z = 4 \times 20 \times 4\,mm^3$ and particle volume fraction of $P = 0.01$

the mean diameter $\overline{D}_p$ is necessary. For polydisperse particle clouds, the continuous PSD is discretised specifying different particle size classes. Figure 17.13b) depicts the resulting number of particles in each size class for a particle cloud volume of $V_{PC} = 320\,mm^3$.

Again, the Beer-Lambert law from (17.27) was used to characterise the particle cloud with the attenuation coefficient $C$, showing that the LSA overestimates the particle radiation interactions again. However, the deviation of $C_{LSA}$ to the polydisperse particle resolved approach (PDA) was $\approx 32\%$, while the deviation of the LSA to the monodisperse particle resolved approach (MDA) was $\approx 37\%$. Increasing the standard deviation further decreases the deviation of $C_{LSA}$ to $C_{PDA}$. This decrease is related to the projected area of all particles $\sum A_p$. With increasing $\sum A_p$, more radiation interacts with the particles in the cloud. Since increasing the standard deviation $\sigma$ also increases $\sum A_p$, more radiation interactions happen compared to the monodisperse particle group. To account for the influence of the polydisperse particle group, it is suggested to multiply the weighting factor $W$ by 1.1.

## 17.5 Influence of the New Radiation Model on the Coupled CFD/Radiation Simulation

As mentioned in the introduction of this chapter, particle radiation is essential for accurately modelling the heat transfer in a combustion chamber. Therefore, several coupled CFD/radiation simulations of a 60 $kW_{th}$ flame under oxy-fuel conditions in the chamber presented in Sect. 14.2.1 are performed. Since the combustion chamber is cylindrical (cf. Fig. 14.1), the radiation evaluation is carried out depending on the chamber radius $r$ in the following. Simulations are carried out in Ansys Fluent 17.1, applying a RANS model for the fluid dynamics [50]. As this chapter focuses

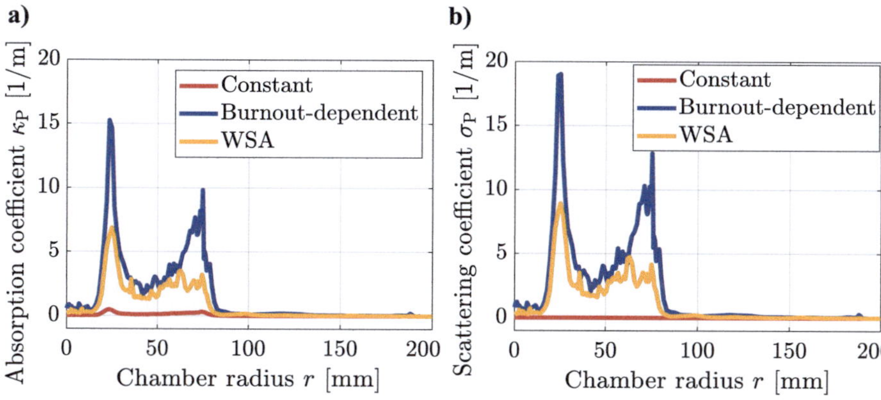


**Fig. 17.14** Radiative properties of a particle cloud including **a)** absorption coefficient $\kappa_P$ and **b)** scattering coefficient $\sigma_P$ at 32 mm below the burner depending on the chamber radius $r$ (reprinted and slightly adapted from Askarizadeh et al. [50], licensed under CC BY 4.0). More details about the chamber geometry are described in Sect. 14.2.1

on radiation modelling, the details of turbulence modelling, the reaction kinetics, the char conversion process, and other applied models are not listed here. However, they are described by Askarizadeh et al. [50]. The only difference in the simulations originates in the particle radiation modelling, where three cases are investigated:

- Constant: The radiative properties (efficiencies $Q_{\mathrm{Abs}}$ and $Q_{\mathrm{Sca}}$) of the particles are constant according to parameters such as wavelength, temperature, burnout degree, and particle diameter.
- Burnout-dependent: The radiative properties of the particle are calculated applying the Mie-theory considering burnout-dependent particles as outlined in Sect. 17.3.2.
- WSA: The burnout-dependet model is applied, and the correction term for the properties of particle clouds (see Sect. 17.4) is considered.

Figure 17.14 illustrates the resulting absorption and scattering coefficients depending on the chamber radius, which are evaluated at a distance of 32 mm below the burner. Applying the burnout-dependent Mie model results in significantly larger coefficients compared to the constant particle properties. Additionally, Fig. 17.15a) illustrates the net contribution of the particles in the radiation exchange. For a chamber radius $r =$ 20–50 mm, much larger source terms are observed, and for $r =$ 50–80 mm, larger sinks compared to the remaining parts of the chamber. Neglecting the more detailed modelling of the particle radiative properties and assuming constant efficiencies finally results in local temperature differences up to 200 K, cf. Fig. 17.15b). More details about this investigations are also published by Askarizadeh et al. [50].

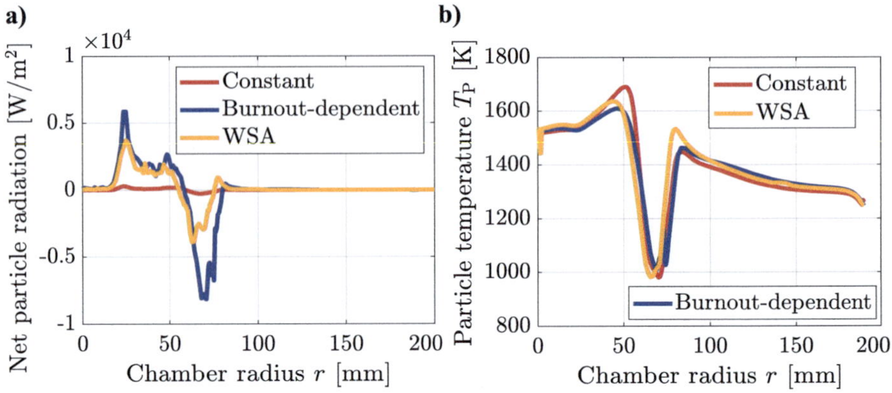


**Fig. 17.15** **a)** Net contribution of the particle radiation exchange which is calculated by substracting the absorbed from the emitted amount of radiation in each numerical cell and **b)** resulting particle temperatures (reprinted and slightly adapted from Askarizadeh et al. [50], licensed under CC BY 4.0)

## 17.6 Conclusion

In this chapter the radiation model was presented. The highly wavelength-dependent properties of the gas are considered by a weighted-sum-of-grey-gases model, where the coefficients were derived by Bordbar et al. [21]. The particle properties are averaged over the entire spectrum by Planck-mean values. However, a novel, computationally efficient model was developed that considers the following aspects:

- internal structure changes of the particles during conversion,
- highly non-spherical shapes, and
- influence of individual particle positions in particle groups on the radiative properties.

The influence of this model on the temperature was tested in simulations of a 60 $kW_{th}$ flame in an oxy-fuel atmosphere. As the investigated combustion chamber was small, the role of particle radiation was especially important in the near-burner region. Further downstream, the differences between the resulting temperatures in different models were negligible. More considerable influences of the particles were expected for larger chambers, mainly due to scattering.

However, it is always advised to use more detailed modelling approaches for the radiative properties of the particles. In the end, the presented particle radiation models slightly increase the computational time in the simulations but considerably affects the simulation results in the near-burner region. This region plays the main role in the formation of the flame and, thus, using a more accurate model for particle radiative properties can increase the accuracy in the predictions of flame behaviour.

**Acknowledgements** This work has been funded by the Deutsche Forschungsgemeinschaft (DFG, German Research Foundation)—215035359—SFB/TRR 129 'Oxyflame'.

## References


1. Bordbar, M. H. and Hyppänen, T. "Modeling of Radiation Heat Transfer in a Boiler Furnace". *Advanced Studies in Theoretical Physics* 1.12 (2007), 571–584.
2. Guo, J., Hu, F., Jiang, X., Li, P. and Liu, Z. "Effects of Gas and Particle Radiation on IFRF 2.5 MW Swirling Flame Under Oxy-Fuel Combustion". *Fuel* 263 (2020), 116634. https://doi.org/10.1016/j.fuel.2019.116634
3. Johansson, R., Leckner, B., Andersson, K. and Johnsson, F. "Influence of Particle and Gas Radiation in Oxy-Fuel Combustion". *International Journal of Heat and Mass Transfer* 65 (2013), 143–152. https://doi.org/10.1016/j.ijheatmasstransfer.2013.05.073
4. Kez, V., Consalvi, J.-L., Liu, F., Gronarz, T., Ströhle, J., Kneer, R. and Epple, B. "Investigation of Gas and Particle Radiation Modelling in Wet Oxy-Coal Combustion Atmospheres". *International Journal of Heat and Mass Transfer* 133 (2019), 1026–1040. https://doi.org/10.1016/j.ijheatmasstransfer.2019.01.013
5. Askarizadeh, H., Koch, M., Nicolai, H., Pielsticker, S., Kneer, R., Hasse, C. and Maßmeyer, A. "Influence of Gas Radiation, Particle Radiation Interactions, and Conversion-Dependent Particle Radiative Properties on Pulverized Solid Fuel Combustion". *31. Deutscher Flammentag*. Berlin, Germany, 2023.
6. Kez, V., Consalvi, J.-L., Liu, F., Ströhle, J. and Epple, B. "Assessment of Several Gas Radiation Models for Radiative Heat Transfer Calculations in a Three-Dimensional Oxy-Fuel Furnace Under Coal-Fired Conditions". *International Journal of Thermal Sciences* 120 (2017), 289–302. https://doi.org/10.1016/j.ijthermalsci.2017.06.017
7. Denison, M. K. and Webb, B. W. "A Spectral Line-Based Weighted-Sum-of-Gray-Gases Model for Arbitrary RTE Solvers". *Journal of Heat Transfer* 115.4 (1993), 1004–1012. https://doi.org/10.1115/1.2911354
8. Denison, M. K. and Webb, B. W. "The Spectral Line-Based Weighted-Sum-of-Gray-Gases Model in Nonisothermal Nonhomogeneous Media". *Journal of Heat Transfer* 117.2 (1995), 359–365. https://doi.org/10.1115/1.2822530
9. Modest, M. F. and Zhang, H. "The Full-Spectrum Correlated-$k$ Distribution for Thermal Radiation from Molecular Gas-Particulate Mixtures". *Journal of Heat Transfer* 124.1 (2002), 30–38. https://doi.org/10.1115/1.1418697
10. Kez, V., Consalvi, J.-L., Liu, F., Koch, M., Hatzfeld, O. and Kneer, R. "Development of a Weighted-Sum-of-Gray-Gases Model for Modeling Radiative Heat Transfer in Coal-Fired Oxy-Fuel Boilers". *9$^{th}$ International Symposium on Radiative Transfer*. Athens, Greece, 2019.
11. Rohsenow, W. M., Hartnett, J. P. and Cho, Y. I., eds. *Handbook of Heat Transfer*. New York: McGraw-Hill, 1998.
12. Modest, M. F. *Radiative Heat Transfer*. New York: Academic Press, 2013.
13. Mie, G. "Beiträge zur Optik trüber Medien, speziell kolloidaler Metallösungen". *Annalen der Physik* 330.3 (1908), 377–445. https://doi.org/10.1002/andp.19083300302
14. Gronarz, T., Schnell, M., Siewert, C., Schneiders, L., Schröder, W. and Kneer, R. "Comparison of Scattering Behaviour for Spherical and Non-Spherical Particles in Pulverized Coal Combustion". *International Journal of Thermal Sciences* 111 (2017), 116–128. https://doi.org/10.1016/j.ijthermalsci.2016.08.014
15. Gronarz, T., Höges, C., Kez, V., Habermehl, M. and Kneer, R. "Comparison of Scattering Behaviour for Inhomogeneous Particles in Pulverized Coal Combustion". *International Journal of Thermal Sciences* 140 (2019), 1–7. https://doi.org/10.1016/j.ijthermalsci.2019.02.034

16. Riaza, J., Mason, P. E., Jones, J. M., Williams, A., Gibbins, J. and Chalmers, H. "Shape and Size Transformations of Biomass Particles During Combustion". *Fuel* 261 (2020), 116334. https://doi.org/10.1016/j.fuel.2019.116334
17. Chandrasekhar, S. *Radiative Transfer*. New York: Dover Publications, 1960.
18. Ripoll, J. "Derivation of the Scalar Radiative Transfer Equation from Energy Conservation of Maxwell's Equations in the Far Field". *Journal of the Optical Society of America. A, Optics, image science, and vision* 28.8 (2011), 1765–1775. https://doi.org/10.1364/JOSAA.28.001765
19. Kirchhoff, G. "Ueber das Verhältniss zwischen DEM Emissionsvermögen und DEM Absorptionsvermögen der Körper für Wärme und Licht". *Annalen der Physik* 185.2 (1860), 275–301. https://doi.org/10.1002/andp.18601850205
20. Hottel, H. C. and Sarofim, A. F. *Radiative Transfer*. New York: McGraw–Hill, 1967.
21. Bordbar, M. H., Węcel, G. and Hyppänen, T. "A Line by Line Based Weighted Sum of Gray Gases Model for Inhomogeneous $CO_2$-$H_2O$ Mixture in Oxy-Fired Combustion". *Combustion and Flame* 161.9 (2014), 2435–2445. https://doi.org/10.1016/j.combustflame.2014.03.013
22. Bohren, C. F. and Huffman, D. R. *Absorption and Scattering of Light by Small Particles*. New York: John Wiley & Sons, 1983.
23. van de Hulst, H. C. *Light Scattering by Small Particles*. New York: Dover Publications, 1981.
24. Purcell, E. M. and Pennypacker, C. R. "Scattering and Absorption of Light by Nonspherical Dielectric Grains". *The Astrophysical Journal* 186 (1973), 705. https://doi.org/10.1086/152538
25. Yurkin, M. A. and Hoekstra, A. G. "The Discrete-Dipole-Approximation Code ADDA: Capabilities and Known Limitations". *Journal of Quantitative Spectroscopy and Radiative Transfer* 112.13 (2011), 2234–2247. https://doi.org/10.1016/j.jqsrt.2011.01.031
26. Rayleigh, F. R. S. "On the Transmission of Light Through an Atmosphere Containing Small Particles in Suspension, and on the Origin of the Blue of the Sky". *The London, Edinburgh, and Dublin Philosophical Magazine and Journal of Science* 47.287 (1899), 375–384. https://doi.org/10.1080/14786449908621276
27. Schuster, A. and Nicholson, J. W. *An Introduction to the Theory of Optics*. London: Edward Arnold & Co, 1924.
28. Mätzler, C. *MATLAB Functions for Mie Scattering and Absorption – Version 2*. 2002.
29. Gronarz, T., Habermehl, M., Hatzfeld, O. and Kneer, R. "Investigation of Modelling Approaches for Scattering and Absorption Properties of Particles in Pulverized Coal Combustion". *Industrial Combustion* (2016), 201602.
30. Gronarz, T., Johansson, R. and Kneer, R. "Modeling of Anisotropic Scattering of Thermal Radiation in Pulverized Coal Combustion". *Journal of Heat Transfer* 140.6 (2018), 062701. https://doi.org/10.1115/1.4038912
31. Joseph, J. H., Wiscombe, W. J. and Weinman, J. A. "The Delta-Eddington Approximation for Radiative Flux Transfer". *Journal of the Atmospheric Sciences* 33.12 (1976), 2452–2459. https://doi.org/10.1175/1520-0469(1976)033%3C2452:TDEAFR%3E2.0.CO;2
32. Henyey, L. C. and Greenstein, J. L. "Diffuse Radiation in the Galaxy". *The Astrophysical Journal* 93 (1941), 70. https://doi.org/10.1086/144246
33. Gronarz, T., Habermehl, M. and Kneer, R. "Approximations for Particle Scattering Phase Functions in Solid Fuel Combustion". *41$^{st}$ International Technical Conference on Clean Coal & Fuel Systems*. Clearwater, Florida, USA, 2016.
34. Chang, H.-M. and Liao, C. "A Parallel Derivation to the Maxwell-Garnett Formula for the Magnetic Permeability of Mixed Materials". *World Journal of Condensed Matter Physics* 01.02 (2011), 55–58. https://doi.org/10.4236/wjcmp.2011.12009
35. Yang, W. "Improved Recursive Algorithm for Light Scattering by a Multilayered Sphere". *Applied Optics* 42.9 (2003), 1710–1720. https://doi.org/10.1364/AO.42.001710
36. Xu, Y.-l. and Gustafson, B. Å. "A Generalized Multiparticle Mie-Solution: Further Experimental Verification". *Journal of Quantitative Spectroscopy and Radiative Transfer* 70.4-6 (2001), 395–419. https://doi.org/10.1016/S0022-4073(01)00019-X

37. Kim, S. H. and Huh, K. Y. "A New Angular Discretization Scheme of the Finite Volume Method for 3-D Radiative Heat Transfer in Absorbing, Emitting and Anisotropically Scattering Media". *International Journal of Heat and Mass Transfer* 43.7 (2000), 1233–1242. https://doi.org/10.1016/S0017-9310(99)00211-2
38. Koch, M., Pielsticker, S., Tarlinski, D., Scherer, V. and Kneer, R. "Coupling Radiative Properties with Detailed Char Conversion Kinetics". *Fuel* 363 (2024), 130973. https://doi.org/10.1016/j.fuel.2024.130973
39. Koch, M., Pörtner, L., Gu, Y., Schiemann, M., Rohlfs, W. and Kneer, R. "Comparison of Scattering Phase Functions of Reacting and Non-Reacting Pulverised Fuel Particles". *Fuel* 287 (2021), 119415. https://doi.org/10.1016/j.fuel.2020.119415
40. Gronarz, T., Habermehl, M. and Kneer, R. "Modeling of Particle Radiative Properties in Coal Combustion Depending on Burnout". *Heat and Mass Transfer* 53.4 (2017), 1225–1235. https://doi.org/10.1007/s00231-016-1896-0
41. Guo, J., Hu, F., Luo, W., Li, P. and Liu, Z. "A Full Spectrum k-Distribution Based Non-Gray Radiative Property Model for Unburnt Char". *Proceedings of the Combustion Institute* 37.3 (2019), 3081–3089. https://doi.org/10.1016/j.proci.2018.06.009
42. Koch, M., Pielsticker, S. and Kneer, R. "Biomass Particle-Radiation-Interaction and the Effect of Shape and Structure Simplifications". *Fuel* 361 (2024). https://doi.org/10.1016/j.fuel.2023.130657
43. Mishchenko, M. I. ""Independent" and "Dependent" Scattering by Particles in a Multi-Particle Group". *OSA Continuum* 1.1 (2018), 243. https://doi.org/10.1364/OSAC.1.000243
44. Drolen, B. L. and Tien, C. L. "Independent and Dependent Scattering in Packed-Sphere Systems". *Journal of Thermophysics and Heat Transfer* 1.1 (1987), 63–68. https://doi.org/10.2514/3.8
45. Yashchuk, V. P., Tikhonov, E., Prygodjuk, O., Levandovska, O. and Zhuravsky, M. "Additional Absorption in the Multiply Scattering Absorbing Media". *Solid State Phenomena* 106 (2005), 109–116. https://doi.org/10.4028/www.scientific.net/SSP.106.109
46. Koch, M., Pielsticker, S. and Kneer, R. "Einfluss inhomogener Partikelverteilungen innerhalb dichter Partikelwolken auf die Wärmestrahlung bei der Feststoffverbrennung". *30. Deutscher Flammentag*. Hannover, Germany, 2021. https://doi.org/10.18154/RWTH-2021-09142
47. Koch, M., Pielsticker, S. and Kneer, R. "Numerical Investigation of the Radiative Properties of a Fully Resolved Particle Cloud". *International Journal of Thermal Sciences* 185 (2023), 108102. https://doi.org/10.1016/j.ijthermalsci.2022.108102
48. Koch, M., Pielsticker, S., Ströhle, J. and Kneer, R. "Numerical Investigation of the Radiative Properties of Fully Resolved Polydisperse Particle Clouds". *17. International Heat Transfer Conference*. Cape Town, South Africa, 2023. https://doi.org/10.1615/IHTC17.380-30
49. Beer, A. "Bestimmung der Absorption des rothen Lichts in farbigen Flüssigkeiten". *Annalen der Physik* 162.5 (1852), 78–88. https://doi.org/10.1002/andp.18521620505
50. Askarizadeh, H., Pielsticker, S., Nicolai, H., Koch, M., Kneer, R., Hasse, C. and Maßmeyer, A. "Radiation Modelling Considering Burnout-Dependent Properties and Cellwise Non-Uniform Particle Distributions in the Numerical Simulation of Pulverised Solid Fuel Combustion". *Fuel* 381 (2025), 133338. https://doi.org/10.1016/j.fuel.2024.133338

# Experimental Radiation Investigations 18

Lukas Pörtner, Philipp Graeser, Jeanette Gorewoda, Matthias Koch, Martin Schiemann, Viktor Scherer and Reinhold Kneer

## Contents



L. Pörtner (✉) · P. Graeser · J. Gorewoda · M. Schiemann · V. Scherer
Department of Energy Plant Technology, Ruhr-Universität Bochum, Bochum, Germany
e-mail: poertner@leat.rub.de

P. Graeser
e-mail: graeser@leat.rub.de

J. Gorewoda
e-mail: gorewoda@leat.rub.de

M. Schiemann
e-mail: schiemann@leat.rub.de

V. Scherer
e-mail: scherer@leat.rub.de

M. Koch · R. Kneer
Institute of Heat and Mass Transfer, RWTH Aachen University, Aachen, Germany
e-mail: koch@wsa.rwth-aachen.de

R. Kneer
e-mail: kneer@wsa.rwth-aachen.de



R. Kneer et al. (eds.), *Clean Energy Conversion*,
https://doi.org/10.1007/978-3-031-98956-8_18

**Abstract**

The experimental investigation of radiation in large combustion chambers is challenging due to the harsh environment and various radiation effects. This chapter presents experimental findings from several laboratory-scale studies that examined these effects separately. One goal was to determine the emissivity of the solid fuels and of ash utilising two setups: (1) observing a particle streak in a laminar flow reactor (LFR) and (2) studying laser-ignited fuel particles. The emissivity of both investigated solid fuels (Colombian bituminous coal and walnut shells) decreased with increasing temperature and progressing conversion, where mineral content has also an influence. For particles at higher levels of conversion, a final increase in emissivity was noticed. A spectroscopic experiment examined the emissivity of typical ash components and their mixtures. Carbonates and sulphates exhibited characteristic temperature- and wavelength-dependent emission trends, with specific emission bands attributed to $CO_3$ and S–O functional groups. Another goal was to determine the index of refraction of the solid fuels experimentally. It was measured at room temperature and for burning particles applying inverse Mie-theory.

## 18.1 Introduction

In pulverised fuel combustion chambers, radiative heat transfer is one of the major heat transfer mechanisms. Especially in large systems, the influence of radiation exceeds the heat transfer contributions of convection and conduction. Considering all possible radiation sources in such a chamber, the radiation can be separated into three main parts. These parts are particle radiation, gas radiation, and the radiation emitted by the chamber walls. The main source of thermal radiation in combustion systems are solid materials, namely the fuel particles and the chamber walls. Consequently, the individual solid fuel particle and wall material radiative parameters have a decisive influence on the behaviour of the combustion process. In addition to the emission of radiation from varying sources, also the interaction of solid materials with thermal radiation inside the chamber needs to be considered. The interaction of single fuel particles with thermal radiation is of special interest, as the typical particle sizes lead to complex radiation scattering effects.

## 18.2 Emissivity of Solid Fuel Particles

### 18.2.1 State of Research

From the study by Chen et al. [1] it is known that the contribution of particle radiation to the overall radiation in an oxy-fuel combustion system can rise to 70%. This confirms the high interest in a well-defined emissivity of the applied single fuel particles visible in the series of studies which aimed for this parameter.

First, the influence of iron oxide $Fe_2O_3$ and $Fe_3O_4$ on the emissivity of fuels was investigated and proven by Boow and Goard [2]. Following, this result was confirmed by Wall et al. [3] and Brewster and Kunitomo [4]. For $Fe_2O_3$, these results were extended by the results of Greffrath et al. [5], who showed a dependence of fuel ash emissivity from the ash's iron content.

With the known influence of different fuel compositions, Baxter et al. [6] measured the emissivity of single coal particle layers of various degrees of carbonisation at temperatures below 200 °C between $\epsilon = 0.6$ and 0.8. Besides the influence of carbonisation degree, they also found a decreasing trend of the emissivity with higher temperatures. Via emission/transmission spectroscopy at 300–782K, Solomon et al. [7,8] confirmed the influence of the fuel's degree of carbonisation and found an influence of the fuel degree of pyrolysis, resulting in the observation of a non-grey, and therefore wavelength dependent emission behaviour. These results were further confirmed and extended by Bhattacharya and Wall [9], where the investigations of thin layers of coal chars at temperatures up to 1000 °C showed a dependence of the emissivity from the coal type, temperature, particle diameter and conversion progress. Furthermore, in the wavelength range from 2 to 5 μm the non-grey behaviour was confirmed with emissivity variations between $\epsilon = 0.6$ and 0.85. Single particle emissivities in combustion systems were first measured by Rego-Barcena et al. [10]. At a wavelength range of 3.95 μm the particle emissivity was proven to be temperature dependent. Furthermore, Bäckström et al. [11] measured the emissivities at different flame locations, varying from $\epsilon = 0.15$ to 0.45. Overviewing all these studies and their individual findings, Yin [12] concluded that the available literature data of fuel emissivities were still incomplete, even though they are highly mandatory for simulations.

### 18.2.2 Emissivity Definition

The wavelength-dependent total emissivity of a single fuel particle describes its ability to emit radiation at a specific wavelength $\lambda$. The total radiation intensity of the particle emitted at this wavelength $I_{P_\lambda}$ is compared to the respective intensity $I_{BB_\lambda}$ emitted by a theoretically identical blackbody radiator, given by:

$$\epsilon_{P_\lambda} = \frac{I_{P_\lambda}}{I_{BB_\lambda}} \tag{18.1}$$

The emitted radiation intensity of the blackbody depends on the particle surface area $A$ and temperature $T$, as given by the Stefan-Boltzmann law for a blackbody radiator:

$$I_{BB} = \sigma \, A \, T^4 \tag{18.2}$$

As shown in Chaps. 2 and 3, the composition of an individual fuel particle changes during burnout, raising the question how composition and structural changes affect the particle emissivity as the particle reacts. Consequently, the total emitted radiation as well as the individual temperature and surface area of single reacting particles during their burnout need to be determined experimentally. In the following, the received particle parameters are compared to reference radiation intensities received via calibration measurements of a blackbody reference radiator.

### 18.2.3 Experimental Setups

For the experimental determination of the particle size, temperature, and total emitted radiation intensity during burnout, two different setups were applied. While the first setup by Graeser and Schiemann [13] is used for investigations of particles in a laminar flow reactor (LFR), the second setup by Pörtner et al. [14] observes laser-induced burnout of single particles in a single particle reactor (SPR). The laminar flow reactor setup is used to determine particle parameters at different, but limited, particle reaction times, while the single particle setup provides results for the entire reaction time of a single particle. Both setups are schematically illustrated in Fig. 18.1.

In the LFR setup of Graeser and Schiemann [13], the radiation emitted by the burning particles in the laminar flow reactor is captured by a 90° off-axis parabolic mirror and focused onto a coded aperture. This aperture is used to measure the particle size, as it has several slits of defined size that allow to determine the average diameter of the particle from the shape of the detector signals. The radiation passing this aperture is collimated by a second 90° off-axis parabolic mirror and guided to a polka-dot filter. As this filter has an array of reflecting and transmitting areas, the particle radiation is splitted into two identical parts. The reflected part of the radiation is focused by a $CaF_2$ lens onto an IR detector, which records the total emitted radiation intensity. Either an InSb detector with an operation wavelength range in the mid infrared region (MIR) from 2.4–5.5 μm or a spectrometer with an operation wavelength range in the near infrared region (NIR) from 1.25–2.25 μm is applied. Radiation that is not reflected by the polka-dot filter passes through it and hits a beam splitter cube, which splits the incident radiation into three different parts by multiple surface reflections. Two of the resulting beams are spectrally filtered, using narrow bandpass filters with central wavelengths of 550 ± 10 nm and 700 ± 10 nm. Each of the beams is focused onto a photomultiplier tube (PMT), which records the intensity of the incident particle radiation in the respective spectral range. The intensities measured by the PMT are then used to determine the particle temperature through two-colour pyrometry.

When measuring the total emitted radiation, it is essential to ensure that a measured particle is located in the centre of the probe volume. Therefore, a He-Ne laser is

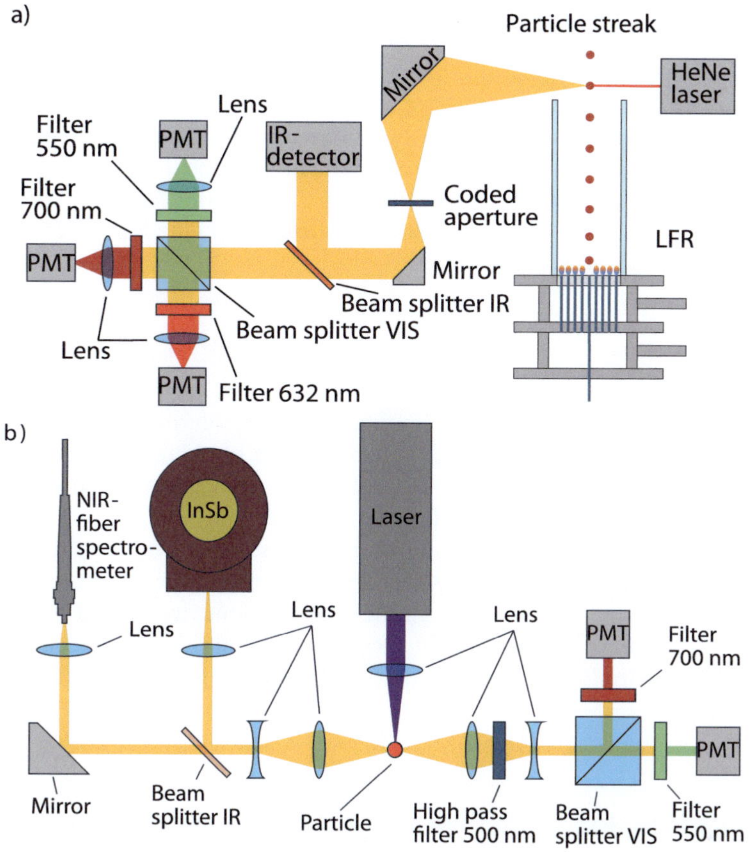


**Fig. 18.1** Experimental setups for the determination of fuel particle emissivity **a)** in a laminar flow reactor (reprinted from Graeser and Schiemann [13], Copyright (2016), with permission from Elsevier) and **b)** for the determination of fuel particle emissivity via laser-induced single particle burnout [14]

focused on the centre of the beam using a 90° off-axis parabolic mirror. The laser radiation scattered by the particle is also captured by the previously described setup and directed to a third PMT. This PMT is equipped with a narrow band filter with a central wavelength of $632 \pm 10$ nm, which ensures the detection of laser radiation only.

A laser-induced ignition setup is used for single fuel particle investigations in the SPR. A single particle is positioned on a mechanical support at the centre of the setup and ignited by a diode laser. The diode laser emits radiation with a total power of 1.2 W at 450 nm, which is focused onto the particle via a biconvex lens. After the particle is ignited, the radiation emitted by the particle is analysed in different spectral ranges.

Similar to the optical setup on the LFR, the emitted particle radiation is collected by a lens system and directed onto a 50:50 beam splitter for particle temperature determination by two-colour pyrometry. A high-pass filter with a cutoff wavelength of 500 nm suppresses scattered laser radiation from the particle surface. Prior to detection by two PMT, the particle radiation is spectrally filtered using narrow band-pass filters with central wavelengths of $550 \pm 10$ nm and $700 \pm 10$ nm.

In contrast to the LFR setup, the two infrared detectors of the SPR setup are used simultaneously to determine the total emitted radiation intensity of the particle. The emitted particle radiation is collected by a $CaF_2$ lens system and directed to an infrared 50:50 beam splitter for simultaneous detection in the NIR (1.25–2.25 μm) and MIR (2.4–5.5 μm) spectral ranges.

### 18.2.4 Methods

The determination of the individual particle parameters during burnout is based on reference calibration measurements performed on a blackbody reference radiator. The respective reference data sets contain temperature, diameter, and total emitted radiation calibrations, which are outlined in the following.

As discussed, in the laminar flow reactor (LFR) setup of Graeser and Schiemann [13], the determination of the individual particle diameter is performed by the coded aperture. As a particle moves through the probe volume, two signal intensities $I_1$ and $I_2$ are recorded using the specific aperture shown in Fig. 18.2. It can be shown, that this aperture leads to an intensity ratio $I_1/I_2$ which is proportional to the particle diameter $D_{\mathrm{P}}$.

$$I_{\mathrm{r}} = \frac{I_1}{I_2} \sim D_{\mathrm{P}} \tag{18.3}$$

To determine the diameter for various intensity ratios, the system was calibrated simulating radiating moving particles of different sizes. Radiating moving particles were simulated by a backlit rotating disc with various precision pinhole apertures. The different pinhole diameters cause varying intensity ratios detected with the PMT. An illustration of the coded aperture with a single particle (yellow dot) and the resulting ratios of the reference signal peaks, together with a linear regression function, are shown in Fig. 18.2.

In contrast to this procedure, in the single particle experiments of Pörtner et al. [14] in the SPR, the particle diameter was assumed to be constant. Therefore, the results presented later must be treated with caution, as the variation in particle diameter has a large effect on the emissivity determined.

Two-colour pyrometry is used to determine the particle temperature in both setups, the LFR and SPR, using the radiation intensities $I_{550}$ and $I_{700}$ at $550 \pm 10$ nm and $700 \pm 10$ nm, and calculating their ratio $I_{\mathrm{r}}$

$$I_{\mathrm{r}} = \frac{I_{550}}{I_{700}}. \tag{18.4}$$

To assign an intensity ratio obtained from the particle experiments to the respective temperature, calibration measurements using a blackbody reference were performed

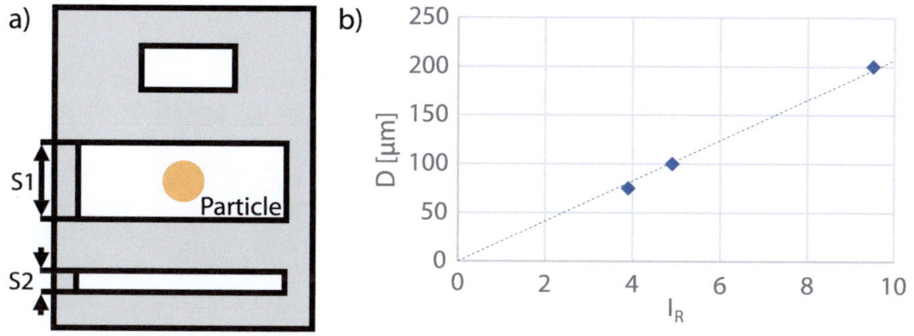


**Fig. 18.2** **a)** Schematic illustration of the coded aperture with the defined slit sizes S1 and S2. (reprinted from Graeser and Schiemann [13], Copyright (2016), with permission from Elsevier) and **b)** diameter $D$ calibration curve determined by precision pinhole experiments [15]

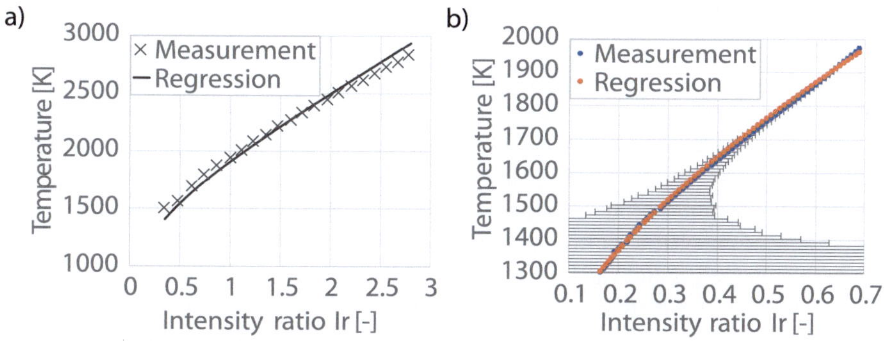


**Fig. 18.3** **a)** Temperature calibration curve of the laminar flow reactor setup [16] and **b)** temperature calibration curve of the single particle setup [14]

for both experiments. For varying radiator temperatures, the radiation intensities $I_{550}$ and $I_{700}$ of the reference radiation were recorded by the PMT, resulting in temperature dependent intensity ratios $I_r$. Via the definition of $I_\mathrm{r}$ and Wien's radiation law, the relation between the measured intensity ratio and the temperature of the observed blackbody radiator is given in the study by Graeser et al. [16] as

$$T = \frac{h\, c_0\, (\lambda_1 - \lambda_2)}{k_\mathrm{B}\, \lambda_1\, \lambda_2} \left[ \ln \left( \frac{1}{\kappa} \left( \frac{\lambda_1}{\lambda_2} \right)^5 \right) + \ln (I_\mathrm{r}) \right]^{-1} \tag{18.5}$$

A regression of (18.5) on the measured intensity ratios via the fit coefficient $\kappa$ leads to the calibration curves of both setups shown in Fig. 18.3. This allows to determine the particle temperature for a wide range of particle radiation intensity ratios. Note that the fit coefficient $\kappa$ is different for the two optical setups, as it takes into account all individual setup radiation intensity losses. Caused by the measurement uncertainties, this method is only valid for temperatures above 1550 K.

For each possible particle temperature and size, a respective blackbody radiation intensity is mandatory to determine the particle emissivity. For the determination

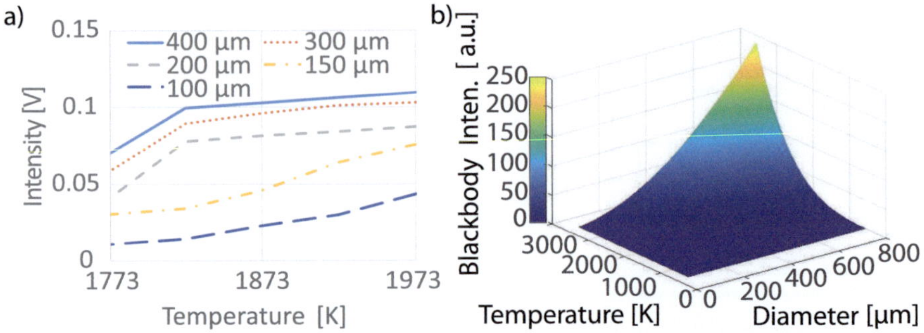


**Fig. 18.4** **a)** Measured blackbody reference data for the laminar flow reactor setup. The different lines denote different diameters of the pinholes as specified in the legend (reprinted from Graeser and Schiemann [13], Copyright (2016), with permission from Elsevier). **b)** Reference surface determined from the calibration measurements of the single particle setup [14]

of this reference radiator intensity $I_{BB}$, the optical setup needs another calibration. The respective optical part of each setup is positioned in front of a blackbody reference radiator, which was equipped with precision pinhole apertures of varying size. These apertures were positioned between the optical setup and the blackbody reference radiator in the probe volume of each individual setup. The signal intensities received from the NIR-spectrometer and the InSb detector (MIR) were recorded for all pinhole diameters and varying temperatures and thus, form the intensity reference data. These data sets were interpolated to provide reference intensities also for particle temperature and diameter combinations not represented by the chosen reference radiator temperatures and pinhole diameters. Illustrations of the reference data sets in the MIR are presented for both setups in Fig. 18.4.

## 18.2.5 Results

### Colombian Coal Particle Emissivity with Carbon Conversion Evolution

The carbon conversion evolution of the single particle emissivity of Colombian coal bituminous (CBC) particles was investigated by Graeser et al. [17, 18]. The particle emissivity was measured in the laminar flow reactor at five different heights of the burning particle streak, covering early and late char burnout regions. Conversion identification was performed by taking particle samples from the streak at the corresponding emissivity measurement heights, applying a water-cooled, nitrogen-purged sampling probe. The ultimate analysis results are shown in Table 18.1 while the measurement heights as well as the determined emissivity and conversion is displayed in Fig. 18.5. With increasing ash content, the carbon conversion continuously rises from 49% at a height of 12.5 cm to 62% at a height of 22.5 cm. At the same time, as the conversion of the particles increases, the emissivity in both spectral regions decreases. In the NIR, the emissivity decreases almost linearly from 0.53 at a height

**Table 18.1** Results of the ultimate analysis (dry basis), ash yield $a_{Sample}$ (dry basis), and carbon related conversion $c$ at different heights of the flat flame burner combustion chamber from Graeser and Schiemann [17]

| Height [cm] | C (± 1.6) [wt.%] | H (± 0.1) [wt.%] | N (± 0.1) [wt.%] | $a_{Sample}$ (± 0.3) [wt.%] | c [%] |
|---|---|---|---|---|---|
| 0.0 | 72.4 | 4.6 | 2.0 | 5.9 | 0 |
| 12.5 | 82.3 | 1.0 | 2.2 | 12.4 | 49±3.3 |
| 15.0 | 85.1 | 0.4 | 1.8 | 13.1 | 50±3.2 |
| 17.5 | 83.7 | 0.2 | 1.5 | 13.9 | 54±2.9 |
| 20.0 | 82.2 | 0.2 | 1.4 | 14.7 | 57±2.7 |
| 22.5 | 81.1 | 0.3 | 1.3 | 16.3 | 62±2.4 |

**Table 18.2** Results of the ultimate analysis (dry basis), ash yield $a_{dry}$ (dry basis) and carbon related conversion $c$ of the initial fuel samples and samples taken from the particle streak of the LFR [19]

| | Initial fuel sample | | Taken from LFR | |
|---|---|---|---|---|
| | $SiO_2$-HTC [%] | $Fe_2O_3$-HTC [%] | $SiO_2$-HTC [%] | $Fe_2O_3$-HTC [%] |
| C | 66.4 | 67.7 | 70.4±1.6 | 69.6±1.6 |
| H | 3.9 | 4.5 | 3.6±0.1 | 3.6±0.1 |
| N | 0.3 | 0.8 | 0.6±0.1 | 0.8±0.1 |
| O | 29.4 | 27.0 | 25.5±1.8 | 26.0±1.8 |
| $a_{dry}$ | 3.2 | 23.2 | 5.4±0.3 | 33.7±0.3 |
| c | 0.0 | 0.0 | 39.0±13.0 | 40.0±3.0 |

of 12.5 cm to 0.41 at 22.5 cm. Confirming this trend, the emissivity data in the MIR decreases from 0.70 at 12.5 cm to 0.63 at 22.5 cm.

### Influence of Fuel Mineral Content on the Emissivity of Synthetic Carbonaceous Fuel

To investigate the influence of the fuel mineral content on the total particle emissivity, Graeser et al. [19] performed emissivity measurements in the LFR on two samples of a synthetic carbonaceous fuel (HTC). Further details on HTC generation can be found in the study by Graeser et al. [19] and Sect. 4.2. As $SiO_2$ and $Fe_2O_3$ are important coal ash constituents, both samples were doped with 5% $SiO_2$ and one of the samples with additional 25% $Fe_2O_3$. The sample's emissivities were investigated in the previously described setup at a height of 7.5 cm above the burner surface in an oxy-fuel atmosphere containing 25% $O_2$, 56% $CO_2$, and 19% $H_2O$. Fuel samples were taken at this burner height to provide information on particle conversion. The ultimate analysis results of the two initial HTC samples and fuel samples taken at a burner height of 7.5 cm are presented in Table 18.2.

A composition comparison of the raw fuel and the samples taken at a burner height of 7.5 cm show an increasing carbon and decreasing hydrogen content with burnout. Additionally, the ash yields increase and the resulting conversions are comparable for both samples. The results of the emissivity measurements are presented in Fig. 18.6.

In both spectral ranges (NIR and MIR), the $Fe_2O_3$-doped sample exhibits significantly higher mean and median emissivity compared to the sample doped solely with $SiO_2$. With a difference of 0.13 in the NIR, this effect is slightly more pronounced in this spectral range compared to the MIR, where the difference is 0.10. Possible explanations include the additional doping process with iron oxide, which could induce changes in particle structure and/or porosity, that are known influencing factors [20]. Additionally, catalytic effects of $Fe_2O_3$ and heterogeneous mineral distributions within the fuel may have influenced the fuel structure. Furthermore, determined single particle emissivities with values above 1 are caused by measurement method uncertainties.

**Temperature Dependence of Colombian Bituminous Coal Particle Emissivity**
The influence of the particle temperature on single bituminous Colombian coal particles (CBC) was investigated in the LFR by Graeser and Schiemann [21]. An analysis of the raw coal sample with a sieve fraction of 160–200 μm showed a carbon content of 72.4%, hydrogen content of 4.6%, and nitrogen content of 2.0%. The ash yield of the samples was determined to be 5.9%. For the measurements in the laminar flow reactor, the oxygen concentration of the reaction atmosphere was varied to achieve different particle temperatures. Since the char oxidation rate is directly linked to the oxygen partial pressure, this adjustment increased the reaction temperature and consequently the particle temperature. To cover a broad spectrum of particle temperatures, the oxygen content ranged from 15 to 40%. Table 18.3 presents a list of all examined reaction atmospheres, their respective average temperatures, and the subsequently determined average particle temperatures from two-colour pyrometry measurements.

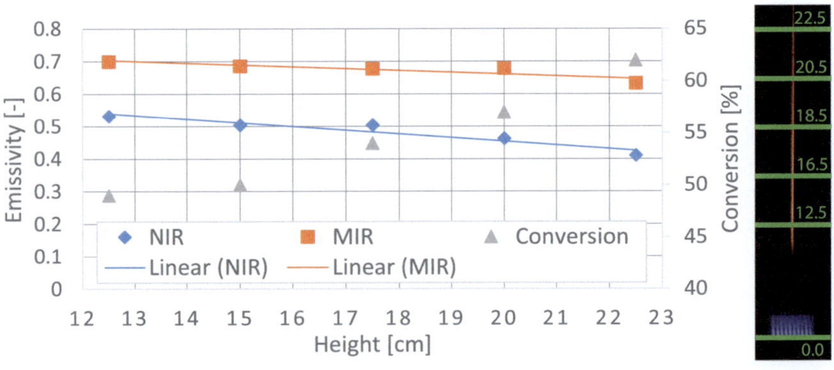


**Fig. 18.5** Average total emissivity (left ordinate) in the spectral range from 1.25–2.25 μm and 2.4–5.5 μm, and the carbon conversion (right ordinate) at each level (left) (reprinted from Graeser and Schiemann [18], Copyright (2018), with permission from Elsevier) and image of a particle streak including lines for measurement level height illustration (right)

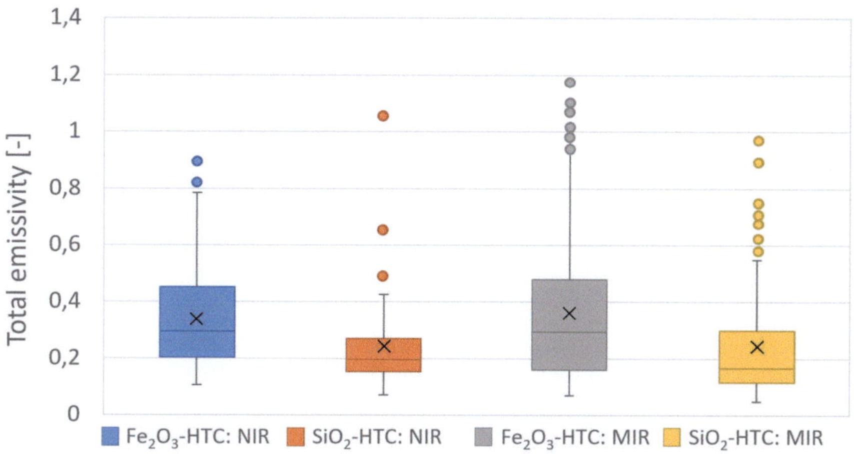


**Fig. 18.6** Boxplot of the average total emissivity in two spectral ranges from 1.25–2.25 μm and 2.4–5.5 μm of both char samples at a burner height of 7.5 cm [19]

**Table 18.3** Composition of the reaction atmospheres and their influence on the particle temperature for the investigation of the emissivity from Graeser and Schiemann [21]

| Nr. | Atmosphere composition | | | Gas temperature [K] | Particle temperature [K] |
|---|---|---|---|---|---|
| | $O_2$ [vol.%] | $CO_2$ [vol.%] | $H_2O$ [vol.%] | | |
| 1 | 15.0 | 66.0 | 19.0 | 1767 | 1956 |
| 2 | 17.0 | 64.0 | 19.0 | 1776 | 2028 |
| 3 | 19.0 | 62.0 | 19.0 | 1788 | 2071 |
| 4 | 21.0 | 60.0 | 19.0 | 1796 | 2109 |
| 5 | 23.0 | 58.0 | 19.0 | 1808 | 2149 |
| 6 | 25.0 | 56.0 | 19.0 | 1819 | 2189 |
| 7 | 30.0 | 51.0 | 19.0 | 1844 | 2378 |
| 8 | 35.0 | 46.0 | 19.0 | 1873 | 2458 |
| 9 | 40.0 | 41.0 | 19.0 | 1902 | 2548 |

The measurements were conducted at a fixed elevation of 15 cm above the burner surface. At this level, devolatilisation had already completed and pure char burnout was observed. Emissivity was measured in both the NIR and MIR spectral regions. The averaged emissivity of both spectral ranges, dependent on the corresponding average particle temperature of each atmosphere, is illustrated in Fig. 18.7.

In both spectral ranges, the predominant trend is a decrease in emissivity with increasing particle temperature. In the NIR, emissivity steadily declined from 0.33 at 1972 K to 0.23 at 2180 K. For higher temperatures, values remained relatively constant at around 0.21. In the MIR, emissivity decreased linearly across the entire temperature range, from 0.54 at 1950 K to 0.26 at 2545 K.

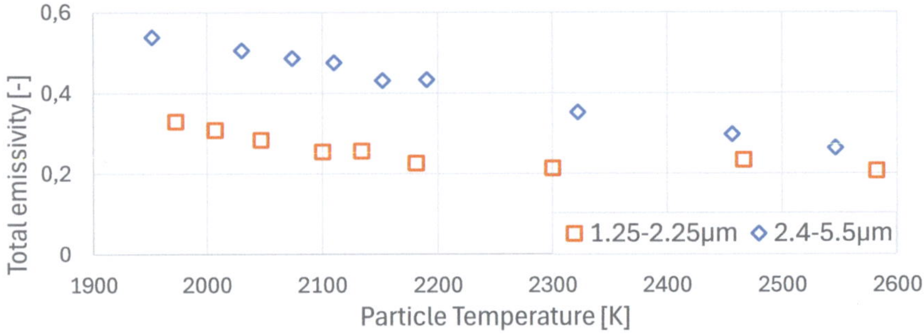


**Fig. 18.7** Spectrally averaged emissivity in dependence on the particle temperature for both wavelength ranges (reprinted from Graeser and Schiemann [21], Copyright (2020), with permission from Elsevier)

**Time-Resolved Single Coal and Biomass Particle Emissivity**
Time resolved single fuel particle emissivity analysis were performed by Pörtner et al. [14] in the SPR setup. Within this study, Colombian bituminous coal (CBC) and walnut shell (WNS) particles were investigated. Based on the raw sample size fraction limits, the coal particles were assumed to have an average diameter of 180 μm and the walnut shell particles an average diameter of 325 μm. Ultimate and proximate analysis of the two fuels is reported in Table 1.1.

Pörtner et al. [14] investigated five different coal and six walnut shell particles. The results of each particle were restricted to a reliable region, excluding low signal intensities and the devolatilisation process. The resulting pairs of particle temperature and respective emissivity were sorted into temperature groups of 100 K width. The resulting temperature dependent distribution of the fuel particle emissivities is presented in Fig. 18.8.

For both fuels, the dominant trend is a decrease in emissivity with increasing particle temperature, in agreement with results from the study by Graeser and Schiemann [21] using the same coal. The overall emissivity of walnut shells decreases from 0.17 at 2100 K to 0.04 at 2900 K. In comparison, the emissivity of coal is consistently higher, decreasing from 0.4 at 2300 K to 0.01 at 3400 K. This difference between coal and walnut shell particles can be assigned to the significantly higher volatile content of the walnut shells compared to the coal, as therefore, the char burnout of the walnut shells is inhibited. Consequently, the char burnout of the coal particles starts earlier, which leads to slightly higher burnout temperatures.

## 18.3 Normal Spectral Radiative Emittance of Ash Layers Under Oxy-Fuel Conditions

Ashes from solid fuel combustion form deposits on heat exchanger surfaces in boilers. These deposits influence the heat transfer from the flame to the heating surfaces. This is particularly critical in the furnace section, where radiation dominates and the

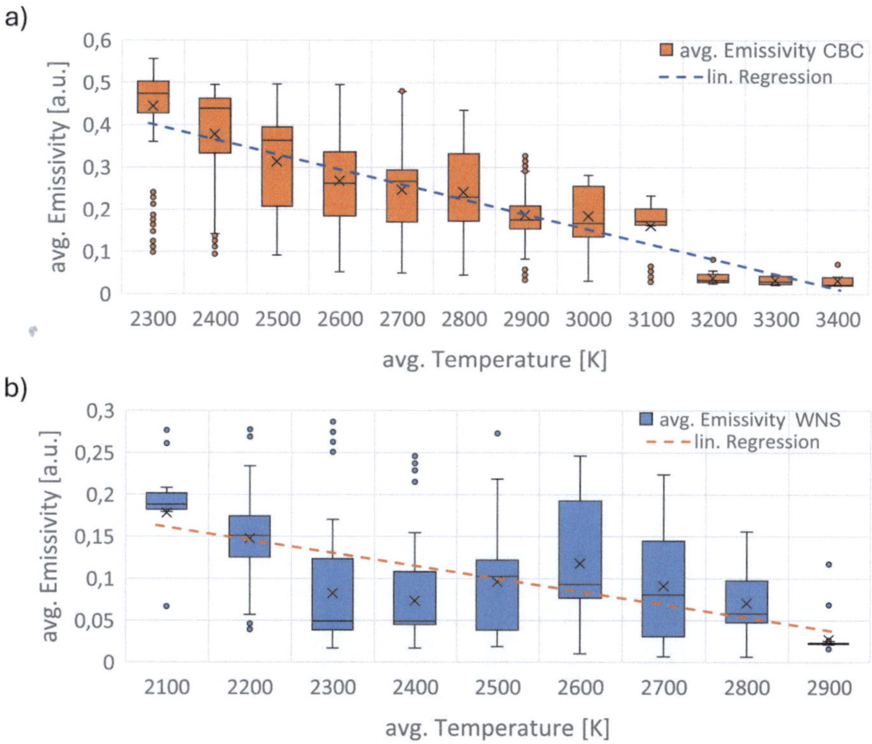


**Fig. 18.8** Temperature-dependent emissivity of **a)** Colombian bituminous coal (CBC) and **b)** walnut shell (WNS) particles [14]

emittance properties of the ash layer play a key role. Consequently, spectral dependence of radiative emittance is important for boiler layout. Coal ashes are strongly non-grey, which has to be considered in spectrally resolved CFD simulations, or if non-spectrally resolved radiation balances are carried out, appropriate averaging of the spectral emittance has to be accounted for (e.g. a weighted averaging procedure with Planck's law as weighting function [22]). Please note, even if this section concentrates on coal ashes, it is expected that the general results can be transferred to other fuels as biomass.

Already in 1969, Boow and Goard [2] concluded that the emittance is influenced by the structure of the surface (mean particle size, roughness of the layer) and by the mineral composition of the ashes. In the studies by Shimogori et al. [23] and Saljnikov et al. [24], it is shown that spatially averaged total emittance is decreasing with increasing ash temperature. Correlations of emittance as a function of particle diameter are reported for synthetic ash layers by Boow and Goard [2] and Greffrath et al. [20] with decreasing emittance for smaller particle diameter. This dependency was confirmed, e.g. by Wall et al. [25], based on Mie theory and for real coal ashes by measurements of Markham et al. [26]. Surface sintering at elevated temperature increases emittance as well as surface fusion as reported by Boow and Goard [2].

Greffrath et al. [5] showed that fusion—depending on the ash composition—can either increase emittance (formation of glass-like layers) or can decrease emittance (formation of a mirror-like surface). The $Fe_2O_3$ acts as a so-called colouring agent that increases emittance of ash layers. Shimogori et al. [2] and Shimogori et al. [27] presented a correlation of the spectral emittance of coal ashes as a function of $Fe_2O_3$ content. Boow and Goard [28] verified the influence of $Fe_2O_3$ and the measured complex refractive index of slags as a function of composition. All these studies mentioned-above are not related to the specifics of oxy-fuel combustion. Oxy-fuel firing systems—compared to air fired systems—have significantly higher $CO_2$ concentrations in the combustion zone but also show an increased concentration of minor species like $SO_2$ due to intensive flue gas recirculation. Consequently, oxy-fuel ashes might contain more carbonates and sulphates than ashes from air-fired systems. However, the statements in literature whether carbonates or sulphates dominate in oxy-fuel ashes are contradictory.

Against this background, the spectral characteristics of the emittance for typical carbonates and sulphates found in coal ashes were investigated. Due to space limitations, this chapter will focus on $CaCO_3$ and $CaSO_4$, as calcium is commonly present in significant quantities in coal ash and can form both carbonates and sulphates. For comparison, $K_2SO_4$ is also included in the analysis. Note that further detailed studies on the emittance of Ca-, Mg-, and Sr-carbonates as pure substances and mixed with $SiO_2$ have been conducted [22,29]. The minerals $CaCO_3$, $CaSO_4$, and $K_2SO_4$ were prepared in powder form and sieved into two particle size fractions. Results are presented for $CaCO_3$ and $CaSO_4$ with a particle size fraction of $x = 125–160\,\mu m$, and for $K_2SO_4$, results are also shown for the finer fraction of $x < 32\,\mu m$.

### 18.3.1 Experimental Setup

The measurements of thermal radiation were carried out in a facility, where samples are heated to temperatures ranging from 500 to 1000 °C. The test rig consists of a Fourier-transform infrared (FTIR) spectrometer and an electrical heating unit. The heating unit contains both, a sample holder and a blackbody as reference radiator. The radiation emitted by the heated sample (22 mm diameter, 2 mm thickness) and the blackbody is directed to the FTIR spectrometer using a gold-coated off-axis parabolic mirror. Thermal radiation is detected in the range from 1.6–12 μm. The sample surface temperature is determined by two thermocouples (type K, 0.25 mm diameter) installed 1 and 2 mm below the sample surface (in the central axis of the disc-shaped sample). A schematic illustration of the setup is presented in Fig. 18.9.

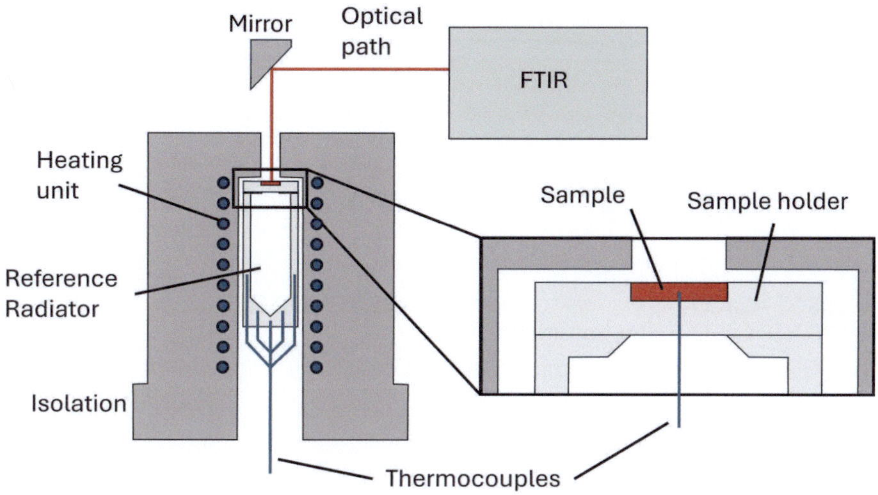


**Fig. 18.9** Experimental setup for material emissivity determination

## 18.3.2 Methods

In order to obtain the spectral normal emittance in (18.6), the sample radiation $S_{\text{sample}}(\lambda, T)$ is compared with the radiation of the blackbody $S_{\text{BB}}(\lambda, T)$:

$$\epsilon(\lambda, T) = \frac{S_{\text{sample}}(\lambda, T)}{S_{\text{BB}}(\lambda, T)} \tag{18.6}$$

The total average emissivity in a spectral range between $\lambda_1$ and $\lambda_2$ results from:

$$\overline{\epsilon}(T) = \frac{\int_{\lambda_1}^{\lambda_2} S_{\text{sample}}}{\int_{\lambda_1}^{\lambda_2} S_{\text{BB}}} \tag{18.7}$$

The sample surface temperature is calculated by assuming a linear temperature profile across the sample's thickness. Consequently, the probe surface temperature is determined via the measured temperatures of the thermocouples located at the bottom and the centre of the sample.

## 18.3.3 Results

As examples for carbonates and sulphates, the spectral emittances of $CaCO_3$, $CaSO_4$, and $K_2SO_4$ are shown in Fig. 18.10.

The general trend of the spectral emittance is similar and typical for minerals, i.e. also for coal ashes. The spectral emittance is small at low temperatures and increases with wavelength to values in the order of 0.8–0.9 above 10 $\mu$m. The spectral

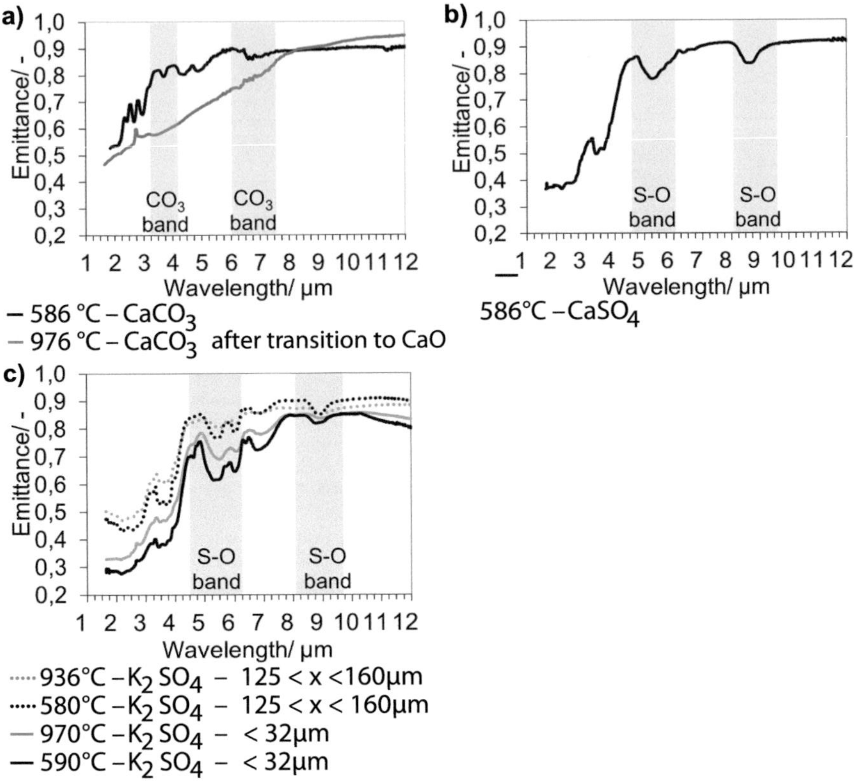


**Fig. 18.10** Spectral emittance of **a)** $CaCO_3$, particle size fraction $x = 125$–$160\,\mu$m, **b)** $CaSO_4$, particle size fraction $x = 125$–$160\,\mu$m, and **c)** $K_2SO_4$, particle size fractions $x < 32\,\mu$m and $x = 125$–$160\,\mu$m

emittance of the carbonates and sulphates show characteristic spectral bands due to the presence of the $CO_3$ and the S–O functional groups. Note that the specific bands around 3 μm in Fig. 18.10 are to the presence of $CO_2$ in the surrounding air.

Figure 18.10a) presents the spectral emittance for $CaCO_3$ for the particle size fraction $x = 125$–$160\,\mu$m. Spectral bands of the characteristic $CO_3$ group are shaded in grey. Thermodynamic calculation carried out with *FactSage* predict a decomposition at 775 °C for $CaCO_3$ to CaO. Thus, $CaCO_3$ is completely decomposed at 976 °C as confirmed by X-ray diffraction measurements. The phase transition from $CaCO_3$ to CaO is clearly visible in Fig. 18.10a), with a noticeable reduction in emittance during the transformation and the characteristic spectral peaks of $CaCO_3$ disappear.

Comparing Figs. 18.10a) and b), it is evident that the spectral emittance of calcium carbonate (Fig. 18.10a)) is higher than that of the corresponding sulphates (Fig. 18.10b)) up to a wavelength of approximately 4.5 μm. Above 4.5 μm, the spectral emittance of $CaCO_3$ and $CaSO_4$ are similar in the order of 0.9. Also $CaSO_4$ (Fig. 18.10b)) exhibits characteristic bands but they occur at different wavelength ranges compared to $CaCO_3$. Note that $CaSO_4$ remains stable at the typical temperatures of coals ashes in boilers, indicating that the S–O bands can also been detected in real coals ashes, which have been exposed to very high temperatures.

Figure 18.10c) shows the characteristic S–O bands as well for $K_2SO_4$. Instead, $K_2SO_4$ shows no phase transformation even at elevated temperature similar to $CaSO_4$. Figure 18.10c) clearly shows that larger particle sizes result in higher emittance, a well-documented effect for coal ashes in general [23].

This section about spectral emittance of coal ashes with respect to oxy-fuel conditions cannot cover all aspects. For example, values for total emittance as a function of temperature for carbonates and sulphates can be found in the study by Gorewoda and Scherer [22]. Detailed studies on the influences of particle size on emittance are presented by Gorewoda et al. [30] and Gorewoda [31]. Furthermore, it has been shown that the characteristic emittance bands of carbonates and sulphates can be found in real coal ashes [31].

## 18.4 Scattering Experiment

The complex index of refraction IOR is a mandatory material property to determine the radiative characteristics of solid fuel particles. In this section, two experimental setups are presented for the measurement of the IOR. These experiments aim to measure the scattered intensity in multiple spatial directions and then analyse the scattering pattern to determine the IOR. A crucial condition is to apply non-intrusive methods to avoid falsification of the radiation interaction with the particle. Both setups achieve this goal differently. The first setup utilises an acoustic levitator (setup A). The second setup achieves this condition using a flat flame burner (setup B). In the following section, these two setups are described and measurement results are presented.

### 18.4.1 Setup A: Acoustic Levtiator

Figure 18.11 displays setup A. As the complex index of refraction is wavelength-dependent, a broadband source is required. Therefore, a silicon carbide light source was used as a source for scattering radiation from particles, with radiation in the range of 0.5–9 $\mu$m. The temperature of the silicon carbide light source is kept constant to ensure constant emission intensity. An ellipsoid reflector and a $CaF_2$ lens are used to focus the emitted radiation on the particle.

The particle is kept contactless in position by the acoustic levitator. The levitator consists of 2 shells, each with four speaker rings equipped with 4, 8, 16, and 16 sound sources. Each sound source points towards the centre of the levitator. The emitted sound waves superimpose in the levitator, creating an interference pattern and resulting in a standing sound wave. The sound pressure creates a force on the particle, which compensates the gravitational force. Further details and explanations of the basic principle are explained by Marzo et al. [32].

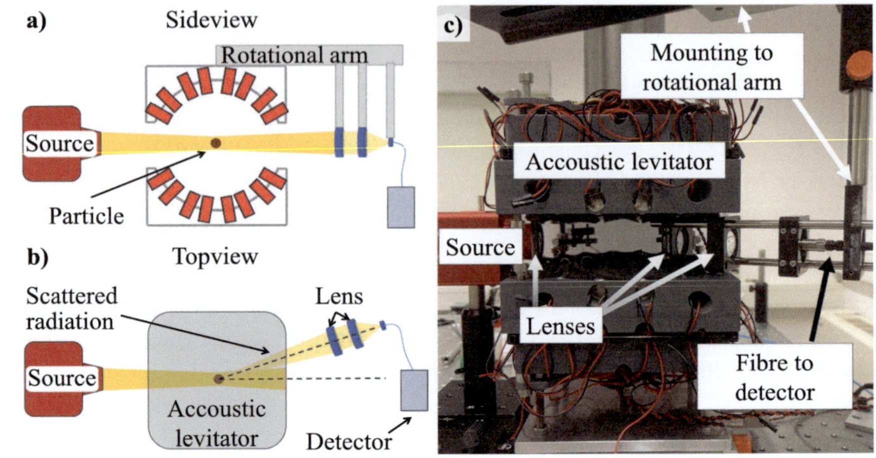


**Fig. 18.11** Experimental setup A: Radiation is focused on a levitating particle and the scattered radiation is collected by lenses: schematic illustration of the setup in **a)** sideview and **b)** topview, as well as **c)** a picture of the setup

The radiation interacts with the levitated particle and is finally scattered in different spatial directions. This scattered radiation is collected by lens optics, which are mounted to an angle adjustable arm. The centre of the rotational axis coincides with the $z$-axis and thus with the position of the particle. The lens optics consists of two $CaF_2$ lenses, both with a lens diameter $D_L = 25.4$ mm and a focal length $f_L = 40$ mm. The first lens collects the scattered radiation and collimates it. The second lens focuses the collimated radiation onto the end of an indium fluoride glass fibre.

During the experiments, scattering angles $\theta = 21$–$114°$ were investigated. For $\theta \geq 114°$, the radiation collecting lenses block the path between particle and source. For $\theta \leq 21°$ the fraction of emitted radiation that radiates directly from the source into the lens optics falsified the measurement.

For the evaluation of scattered radiation, either an FTIR spectrometer from *ARCoptix* was used (supporting wavelengths ranging from $\lambda = 2000$ to 6000 nm) or a spectrometer for the visible and near infrared range from *Oceanview* (supporting wavelengths ranging from $\lambda = 200$ to 1100 nm).

### 18.4.2 Setup B: Flat Flame Burner

The distribution of radiation scattered by single particles of a particle streak was measured by the setup illustrated in Fig. 18.12.

The particle streak in this setup is either provided by a flat flame burner (burning particles) or a feeding system doping cold particles through the centre of the setup. To provide radiation of a wide wavelength range, a 1 kW Xe-band lamp is focused via two parabolic 90° off-axis mirrors onto the streak. While transmitted lamp radiation is reflected by another mirror into a beam trap, the scattered radiation

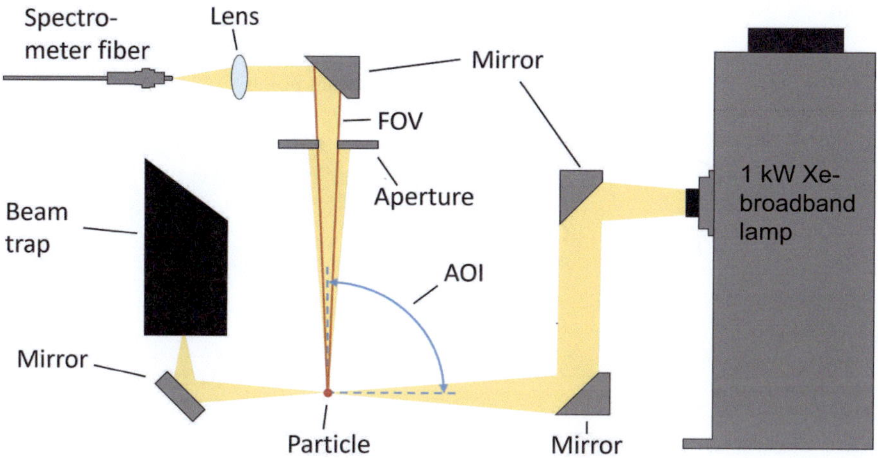


**Fig. 18.12** Schematic illustration of the experimental setup B for investigations of the radiation scattering phase function of a particle streak

distribution is measured by a detection unit. The radiation redirected by the particle passes an aperture of 25.4 mm, which defines the field of view (FOV). The radiation is collimated by a 90° off-axis parabolic mirror and focused by a $CaF_2$ lens into the fibre of a spectrometer sensitive to a wavelength range of 200–1000 nm. For the determination of the radiation distribution, the angle of incidence (AOI) of the lamp radiation is varied via a rotation stage.

### 18.4.3 Assessment of the Phase Function

In both experimental setups, a circular optic is used to collect the scattered radiation. To evaluate the phase function, it is integrated over the circular area of the collection optics

$$\Phi_{\Delta\theta,i} = 2 \int_{\theta^*-\frac{\Delta\theta}{2}}^{\theta^*+\frac{\Delta\theta}{2}} \Phi \sin(\theta) \sin(\varphi^*)\, d\theta, \tag{18.8}$$

where $\theta^*$ is the angular position of the collecting optic. The integration width $\Delta\theta$ is related to the optics diameter $D_{\mathrm{L}}$ and its focal length $f_{\mathrm{L}}$:

$$\Delta\theta = \arctan\left(\frac{D_{\mathrm{L}}}{f_{\mathrm{L}}}\right) \tag{18.9}$$

The factor $\sin(\varphi^*)$ considers the integration over a circular element. These angles and parameters are illustrated in Fig. 18.13.

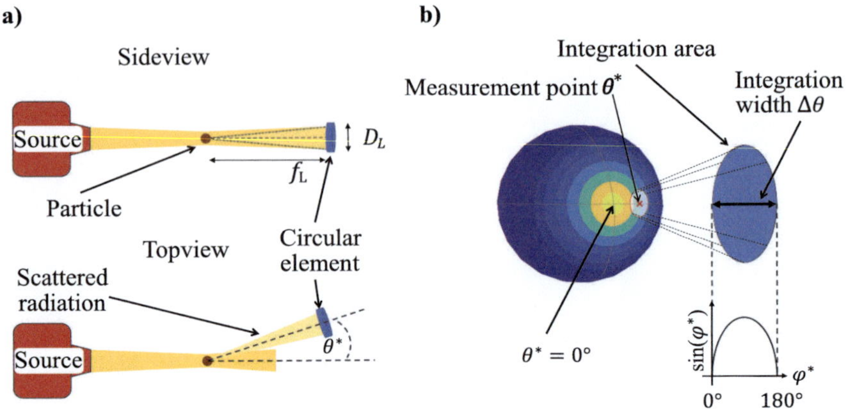


**Fig. 18.13** **a)** Side- and topview of the scattered radiation, which is collected by a circular element and **b)** phase function plotted on a particle and the relation of $\theta^*$ and $\sin(\varphi^*)$ in (18.8)

### 18.4.4 Resulting Complex Index of Refraction

To determine the complex index of refraction, an inverse problem is solved, which is transferred into an minimisation problem

$$\min\left[\Delta\Phi(\mathrm{IOR})\right] \tag{18.10}$$

with the functional $\Delta\Phi(\mathrm{IOR})$,

$$\Delta\Phi = \sum_{i=1}^{N_{\theta*}} \frac{\mathrm{abs}\left(\log(\Phi_{\Delta\theta_i,\mathrm{Calc}}) - \log(\Phi_{\Delta\theta_i,\mathrm{Meas}})\right)}{\mathrm{abs}\left(\log(\Phi_{\Delta\theta_i,\mathrm{Calc}})\right)} \tag{18.11}$$

The indices "Meas" and "Calc" are short for measured and calculated. This function resembles the relative difference between measured and calculated integrated phase function. This difference is minimised by changing the complex index of refraction IOR.

Figure 18.14 exemplary illustrates the fit of the measured index of refraction for walnut shells (WNS) utilising setup A (Fig. 18.14a)) and the wavelength-dependent IOR of cold and burning Colombian bituminous coal (CBC) particles from setup B (Fig. 18.14b)).

Comparing the real and imaginary parts of reacting and non-reacting coal particles, the received values show an approximately constant behaviour in the visible spectral range. In contrast to this, the walnut shells showed a clear increase of both IOR parts with wavelength. Please note, the received IOR parts of the walnut shells were determined for a spectral range of 2–5.5 µm, differing from the coal results.

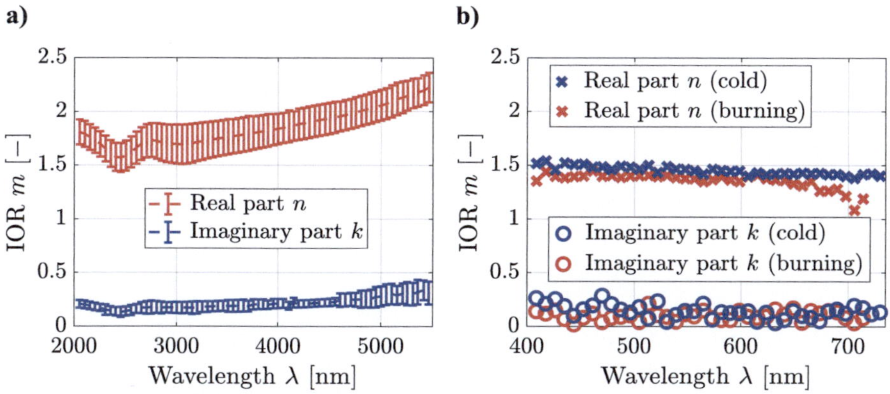


**Fig. 18.14** Wavelength dependent complex index of refraction (IOR) of **a)** walnut shell particles determined by setup A (reprinted and slightly adapted from Koch et al. [33], licensed under CC BY 4.0) and **b)** reacting and non-reacting coal particles determined by setup B (reprinted from Koch et al. [34], Copyright (2021), with permission from . and minor editing)

## 18.5 Conclusion

This chapter has presented several different approaches to the experimental investigation of the radiative effects that contribute to the radiative heat transfer mechanisms within combustion facilities. The emissivity of single fuel particles, the emissivity of typical ash materials, and the interaction of single particles with incident radiation were addressed, providing a novel base of experimental data for combustion system simulations.

The received results of fuel particle emissivities extended the previously available data, which mainly consisted of results from non-reacting fuel samples by reacting single particle data. Within this, a decrease of the emissivity with increasing particle conversion and the effect of the mineral content of the fuel on the emissivity has been investigated by experiments with doped fuels. A doping with $Fe_2O_3$ showed an increased emissivity compared to $SiO_2$. Single particle emissivity was shown to be temperature dependent, with emissivity decreasing as particle temperature increased.

Addressing the emissivity of ashes, typical ash components as well as their mixtures were investigated spectroscopically. For carbonates and sulphates, the spectral emittance trend is similar and typical for minerals. The emittance is small for low temperatures and increases with wavelength. Characteristic emission bands were identified, caused by $CO_3$ and the S–O functional groups.

At last, the interaction of radiation with single fuel particles was investigated by determining a representative complex index of refraction. Walnut shell and coal particles have been studied by measuring the scattering phase function, which provides the refractive index by applying an inverse Mie theory approach. While the real and imaginary part of reacting and non-reacting coal are approximately constant, the walnut shell showed a clear increase of both IOR parts with wavelength.

**Acknowledgements** This work has been funded by the Deutsche Forschungsgemeinschaft (DFG, German Research Foundation)—215035359—SFB/TRR 129 'Oxyflame'.

## References

1. Chen, L., Yong, S. Z. and Ghoniem, A. F. "Oxy-Fuel Combustion of Pulverized Coal: Characterization, Fundamentals, Stabilization and CFD Modeling". *Progress in Energy and Combustion Science* 38.2 (2012), 156–214. https://doi.org/10.1016/j.pecs.2011.09.003
2. Boow, J. and Goard, P. R. C. "Fireside Deposits and Their Effect on Heat Transfer in a Pulverized-Fuel-Fired Burner: Part III: The Influence of the Physical Characteristics of the Deposit on its Radiant Emittance and Effective Thermal Conductance". *Journal of the Institute of Fuel* 42 (1969), 412–419.
3. Wall, T. F., Lowe, A., Wibberley, L. J. and McC. Stewart, I. "Mineral Matter in Coal and the Thermal Performance of Large Boilers". *Progress in Energy and Combustion Science* 5.1 (1979), 1–29. https://doi.org/10.1016/0360-1285(79)90017-0
4. Brewster, M. Q. and Kunitomo, T. "The Optical Constants of Coal, Char, and Limestone". *Journal of Heat Transfer* 106.4 (1984), 678–683. https://doi.org/10.1115/1.3246738
5. Greffrath, F., Schiemann, M., Scherer, V. and Gwosdz, A. "Spectral Emittance Characteristics of Powdery and Molten Coal Ashes". *World of Coal Ash Conference*. Denver, Colorado,USA, 2011.
6. Baxter, L. L., Fletcher, T. H. and Ottesen, D. K. "Spectral Emittance Measurements of Coal Particles". *Energy & Fuels* 2 (1988), 423–430. https://doi.org/10.1021/ef00010a007
7. Solomon, P. R., Carangelo, R. M., Best, P. E., Markham, J. R. and Hamblen, D. G. "The Spectral Emittance of Pulverized Coal and Char". *Symposium (International) on Combustion* 21.1 (1988), 437–446. https://doi.org/10.1016/S0082-0784(88)80271-6
8. Solomon, P. R., Chien, P. L., Carangelo, R. M., Best, P. E. and Markham, J. R. "Application of FT-IR Emission/Transmission (E/T) Spectroscopy to Study Coal Combustion Phenomena". *Symposium (International) on Combustion* 22.1 (1989), 211–221. https://doi.org/10.1016/S0082-0784(89)80027-X
9. Bhattacharya, S. and Wall, T. F. "Development of Emittance of Coal Particles During Devolatilisation and Burnoff". *Fuel* 78.5 (1999), 511–519. https://doi.org/10.1016/S0016-2361(98)00178-1
10. Rego-Barcena, S., Saari, R., Mani, R., El-Batroukh, S. and Thomson, M. J. "Real Time, Non-Intrusive Measurement of Particle Emissivity and Gas Temperature in Coal-Fired Power Plants". *Measurement Science and Technology* 18.11 (2007), 3479–3488. https://doi.org/10.1088/0957-0233/18/11/031
11. Bäckström, D., Johansson, R., Andersson, K., Johnsson, F., Clausen, S. and Fateev, A. "Measurement and Modeling of Particle Radiation in Coal Flames". *Energy & Fuels* 28.3 (2014), 2199–2210. https://doi.org/10.1021/ef402271g
12. Yin, C. "On Gas and Particle Radiation in Pulverized Fuel Combustion Furnaces". *Applied Energy* 157 (2015), 554–561. https://doi.org/10.1016/j.apenergy.2015.01.142
13. Graeser, P. and Schiemann, M. "Char Particle Emissivity of Two Coal Chars in Oxy-Fuel Atmospheres". *Fuel* 183 (2016), 405–413. https://doi.org/10.1016/j.fuel.2016.06.116
14. Pörtner, L., Gu, Y. and Schiemann, M. "Investigation of Pulverized Biomass and Coal Char Emissivity". *Energies* 13.18 (2020), 4620. https://doi.org/10.3390/en13184620
15. Graeser, P. and Schiemann, M. "An Attempt to Measure the Emissivity of Burning Coal Particles in Oxy-Fuel Atmospheres: Test Rig Calibration and First Results". *40th International Technical Conference on Clean Coal & Fuel Systems*. Clearwater, Florida, USA, 2015.
16. Graeser, P., Wütscher, A., Cerciello, F., Muhler, M. and Schiemann, M. "Emissivity Comparison Between Chars and Demineralized Coal Chars Under Oxycombustion Conditions". *Chemical Engineering & Technology* 41.7 (2018), 1490–1496. https://doi.org/10.1002/ceat.201700040

17. Graeser, P. and Schiemann, M. "Emissivity of Burning Bituminous Coal Char Particles – Burnout Effects". *Fuel* 196 (2017), 336–343. https://doi.org/10.1016/j.fuel.2017.01.110
18. Graeser, P. and Schiemann, M. "Corrigendum to "Emissivity of Burning Bituminous Coal Char Particles – Burnout Effects" [Fuel 196 (2017) 336–343] (-)". *Fuel* 234 (2018), 225–226. https://doi.org/10.1016/j.fuel.2018.07.027
19. Graeser, P., Pörtner, L. and Schiemann, M. "Influence of the Mineral Content on the Emissivity of Hydrothermal Char". *Chemical Engineering & Technology* 43.5 (2020), 1016–1020. https://doi.org/10.1002/ceat.201900623
20. Greffrath, F., Gorewoda, J., Schiemann, M. and Scherer, V. "Influence of Chemical Composition and Physical Structure on Normal Radiant Emittance Characteristics of Ash Deposits". *Fuel* 134 (2014), 307–314. https://doi.org/10.1016/j.fuel.2014.05.047
21. Graeser, P. and Schiemann, M. "Investigations on the Emissivity of Burning Coal Char Particles: Influence of Particle Temperature and Composition of Reaction Atmosphere". *Fuel* 263 (2020), 116714. https://doi.org/10.1016/j.fuel.2019.116714
22. Gorewoda, J. and Scherer, V. "Influence of Carbonate Decomposition on Normal Spectral Radiative Emittance in the Context of Oxyfuel Combustion". *Energy & Fuels* 30.11 (2016), 9752–9760. https://doi.org/10.1021/acs.energyfuels.6b01398
23. Shimogori, M., Yoshizako, H., Shimogori, Y., Richardson, M. F., Shimogori, M., Yoshizako, H. and Shimogori, Y. "Characterization of Coal Ash Emissivity in High Temperature Atmospheres". *JSME International Journal Series B* 49.2 (2006), 265–270. https://doi.org/10.1299/jsmeb.49.265
24. Saljnikov, A., Gojak, M., Trifunović, M., Andrejević, S. and Dobrnjac, M. "Research on Infrared Emission Spectra of Pulverized Coal Ash Deposits". *FME Transactions* 41.1 (2013), 41–48.
25. Wall, T., Bhattacharya, S., Zhang, D. K., Gupta, R. P. and He, X. "The Properties and Thermal Effects of Ash Deposits in Coal-Fired Furnaces". *Progress in Energy and Combustion Science* 19.6 (1993), 487–504. https://doi.org/10.1016/0360-1285(93)90002-V
26. Markham, J. R., Best, P. E., Solomon, P. R. and Yu, Z. Z. "Measurement of Radiative Properties of Ash and Slag by FT-IR Emission and Reflection Spectroscopy". *Journal of Heat Transfer* 114.2 (1992), 458–464. https://doi.org/10.1115/1.2911295
27. Shimogori, M., Yoshizako, H. and Matsumura, Y. "Determination of Coal Ash Emissivity Using Simplified Equation for Thermal Design of Coal-Fired Boilers". *Fuel* 95 (2012), 241–246. https://doi.org/10.1016/j.fuel.2011.11.005
28. Goodwin, D. G. and Mitchner, M. "Infrared Optical Constants of Coal Slags: Dependence on Chemical Composition". *Journal of Thermophysics and Heat Transfer* 3.1 (1989), 53–60. https://doi.org/10.2514/3.125
29. Gorewoda, J. and Scherer, V. "Normal Radiative Emittance of Coal Ash Sulfates in the Context of Oxyfuel Combustion". *Energy & Fuels* 31.4 (2017), 4400–4406. https://doi.org/10.1021/acs.energyfuels.6b02866
30. Gorewoda, J., Maas, P. and Scherer, V. "Emittance of Typical Ash Minerals: Influence of Particle Size and Mixtures of Carbonates and Sulfates". *Energy Procedia* 158 (2019), 1780–1785. https://doi.org/10.1016/j.egypro.2019.01.420
31. Gorewoda, J. "Bestimmung der temperaturabhängigen spektralen Emissionsgrade von Aschen und Aschebestandteilen unter Oxyfuel-Bedingungen". PhD thesis. Ruhr-Universität Bochum, 2018.
32. Marzo, A., Seah, S. A., Drinkwater, B. W., Sahoo, D. R., Long, B. and Subramanian, S. "Holographic Acoustic Elements for Manipulation of Levitated Objects". *Nature communications* 6 (2015), 8661. https://doi.org/10.1038/ncomms9661
33. Koch, M., Pielsticker, S., Ströhle, J. and Kneer, R. "Experimental Determination of the Optical Properties of Walnut Shell Particles". *Journal of Quantitative Spectroscopy and Radiative Transfer* 329 (2024), 109202. https://doi.org/10.1016/j.jqsrt.2024.109202
34. Koch, M., Pörtner, L., Gu, Y., Schiemann, M., Rohlfs, W. and Kneer, R. "Comparison of Scattering Phase Functions of Reacting and Non-Reacting Pulverised Fuel Particles". *Fuel* 287 (2021), 119415. https://doi.org/10.1016/j.fuel.2020.119415

# Full-Scale Modelling

# OxySim-129 Simulation Framework

# 19

Pascal Steffens, Antje Vahl, Leon Loni Berkel, Hendrik Nicolai and Christian Hasse

## Contents



P. Steffens (✉) · A. Vahl · L. L. Berkel · H. Nicolai · C. Hasse
Institute for Simulation of Reactive Thermo-Fluid Systems,Technische Universität Darmstadt, Darmstadt, Germany
e-mail: steffens@stfs.tu-darmstadt.de

A. Vahl
e-mail: vahl@stfs.tu-darmstadt.de

L. L. Berkel
e-mail: berkel@stfs.tu-darmstadt.de

H. Nicolai
e-mail: nicolai@stfs.tu-darmstadt.de

C. Hasse
e-mail: hasse@stfs.tu-darmstadt.de


R. Kneer et al. (eds.), *Clean Energy Conversion*,
https://doi.org/10.1007/978-3-031-98956-8_19

**Abstract**

This chapter explores the unsteady modelling and simulation of turbulent pulverised solid fuel combustion, transitioning models from coal to biomass in oxy-fuel chambers. The first section introduces the models implemented in *OxySim-129*, structured around three pillars: (1) solid fuel conversion, (2) gas-phase turbulence-chemistry interaction, and (3) turbulent flow, mixing, and particle movement. Solid fuel conversion covers devolatilisation and char burnout. Gas-phase combustion of complex volatiles is handled using an extended tabulated flamelet approach, balancing fidelity and computational efficiency. Turbulent flow is modeled with large eddy simulations (LES), while particles are treated as Lagrangian point particles. Computational reactor networks (CRN) enable detailed chemistry integration with a reduced-order fluid dynamics model. The second section applies this framework to combustion systems of varying scales, focusing on model validation and key physical processes observed in research burners.

## 19.1 OxySim-129 Framework

The complexity of the physical and chemical processes involved in oxy-fuel combustion makes it difficult to characterise all the features of the processes using experimental techniques only. High-fidelity simulations provide a thorough understanding of oxy-fuel combustion by giving insight into properties that are not accessible in experiments. The *OxySim-129* framework is a comprehensive 3-D computational fluid dynamics (CFD) framework developed for the combustion of pulverised solid fuels at various scales, with a focus on oxy-fuel conditions. This chapter presents the framework, starting with an introduction to the implemented models, focusing on solid fuel conversion, gas phase turbulence, and reaction modelling. It also introduces computational reactor networks (CRN) as a viable option to use detailed chemistry with a simplified fluid dynamics model, allowing, for example, more accurate determination of minor species such as pollutants. The second part of this chapter highlights the applicability of the framework, showing the development towards more advanced models, increasing combustion chamber sizes and biomass as a solid fuel, and discussing certain findings and results in more detail.

### 19.1.1 Overview of Simulation Code and Utilised Models

The *OxySim-129* framework is an extension of the open-source CFD code OpenFOAM®. The governing equations for the gas phase are solved using the finite volume method (FVM). Various numerical schemes are available including a second-order central differencing scheme for spatial and a second-order backward differentiation scheme for temporal integration. Parallel processing across multiple cores is facilitated by the message passing interface (MPI). The multiphase problem of solid fuel combustion requires the combination of several physical models. Within the *OxySim-129* framework, the particle and gas phases are modelled with an Euler-Lagrange approach. The particles are modelled as Lagrangian point particles using the particle-in-cell (PIC) method. The conversion models for carbonaceous solid fuels such as coal and biomass are introduced in Sect. 19.1.2. Turbulence is treated by means of large eddy simulation (LES), which is presented in Sect. 9.1.3. The gas phase chemical reactions are modelled using a tabulated flamelet manifold approach, which is discussed in Sect. 9.1.3. Furthermore, gas phase and particle radiation modelling is briefly explained in Sect. 19.1.4.

### 19.1.2 Particle Modelling

Particles are incorporated into the simulation using a Lagrangian approach, where their interaction with the gas phase is accounted for by particle source terms in the momentum equation, energy equation, and mass transfer according to the particle-source-in cell (PSI) method [1]. Particle motion is tracked by integrating the force balance of inertial, drag, and gravitational forces. The drag force couples the particle motion to the gas phase momentum equation. The drag coefficient is based on the particle Reynolds number and takes into account the flow regime of the particle. Buoyancy effects are included in the gravity term, and particle-wall interactions are treated as fully elastic reflections. The particle temperature is derived by balancing the heat flux across the particle surface and changes in the internal particle energy. The heat transfer coefficient is calculated for a given Nusselt number based on the relationship proposed by Ranz and Marshall [2]. For calculating mass transfer between particle and gas phase, additional particle conversion models are introduced in the following section.

**Conversion Models**

Solid fuel conversion is described by two submodels for the main fuel conversion processes. One determines the devolatilisation rate and the other the char burnout. Modelling approaches for both are presented below.

Two reduced order models for devolatilisation are implemented in *OxySim-129* to limit the computational cost. Detailed solid fuel conversion simulations at relevant conditions are used to tune the model parameters. The simplest approach to modelling devolatilisation is a single step model with a first order reaction [3]. More complex devolatilisation behaviour is described with a competing two-step model (C2SM) proposed by Kobayashi et al. [4]. The kinetics of both the single first order reaction of the SFOR and the two competing single order reactions of the C2SM are described by the Arrhenius law, the parameters of which are fuel specific and therefore need to be determined in advance. This is further explained in Chap. 8.

The input parameters of the reduced order models are obtained using the pyrolysis kinetic preprocessor developed by Vascellari et al. [5]. It adapts the Arrhenius kinetic parameters ($A$ and $E_{\mathrm{A}}$) and the final volatile yield of the models to the specific fuel and combustor conditions using a generic algorithm. In the *OxySim-129* framework,

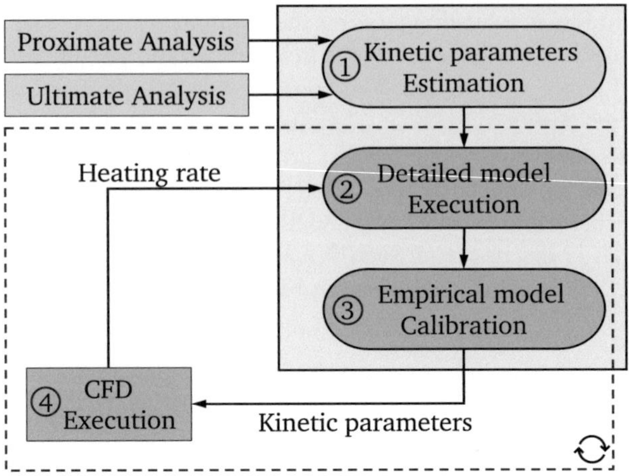


**Fig. 19.1** Schematic workflow of the pyrolysis kinetic preprocessor. (Based on the publication by Vascellari et al. [5])

the CRECK-S and CRECK-S-B models for coal and biomass are used as detailed reaction mechanisms. They are described in detail in Sect. 8.2.3. The workflow to obtain the parameters for the reduced models is an iterative optimisation process as shown in Fig. 19.1. It starts with (1) estimation of the kinetic parameters based on the solid fuel composition characterised by proximate and ultimate analysis data. With the kinetic parameters and an initial estimate of the heating rate, (2) the detailed CRECK-S or CRECK-S-B model is applied to obtain a dynamic release of the volatile composition over time. The release rate is then in turn used to (3) fit the empirical reduced order model coefficients. The final volatile composition is assumed to be released homogeneously over time and used as a fuel in the gas phase chemistry model. Finally, (4) the resulting kinetic parameters are fed into the CFD simulation, providing the actual particle heating rates with which the calibration process can be repeated until the solution converges.

Char conversion in the *OxySim-129* framework is implemented according to the model developed by Baum and Street [6].

### 19.1.3 Gas-Phase Modelling

The gas phase requires dedicated models to account for turbulent flow and chemistry. Turbulence is simulated using the large eddy simulation (LES) approach, which resolves the large turbulent scales while applying models for the subgrid fluctuations. The chemical reactions in the gas phase and the resulting changes in temperature and transport properties are addressed by a tabulated flamelet manifold approach.

**Turbulence**

Turbulence plays a major role in industrially-relevant applications and strongly influences the species and heat transport, as well as particle movement and distribution. It is characterised by the irregular, transient behaviour of the fluid, which leads to strong fluctuations of velocity, strong mixing, and eddy structures on a range of length scales. Since the energy transport from the largest to the smallest scales is essential for the dissipation of kinetic energy, models are needed to describe the scales that cannot be resolved on the computational grid.

Two methods are commonly used for turbulence modelling: Reynolds-averaged Navier-Stokes models (RANS) and large eddy simulation (LES). Modelling with RANS is based on decomposing the variables in the Navier-Stokes equations into a time-averaged and fluctuating part, and time averaging over the decomposed equations. As a result, the Reynolds stress tensor appears, which is unknown and has to be modelled to account for turbulence. In LES, the velocity field is filtered spatially. Therefore, the scales of the turbulent flow that are larger than the filter size—the large eddies—are resolved. Filtering the Navier-Stokes equations gives rise to the subgrid-scale stress tensor. Similar to the Reynolds-stress tensor, the subgrid-scale stress tensor is unknown and needs to be modelled for closure of the equations. Due to the increased availability of computational resources, LES have become more common for pulverised solid fuels in the scientific literature. Furthermore, comparisons show LES to be far superior to RANS simulations [7–9]. However, due to the lower computational effort, RANS simulations can be used to evaluate the necessary degree of modelling of different submodels or the influence of individual parameters in advance [10–12].

To model the subgrid-scale stress tensor and close the LES equations, an eddy-viscosity approach is selected. In the *OxySim-129*, different models with eddy-viscosity approaches are available. In the validation studies, eddy viscosities were obtained using two different models, namely the dynamic Smagorinsky model [13] and the $\sigma$-model [14].

**Tabulated Flamelet Manifold Methods**

Solid fuel volatiles consist of various chemical species. Therefore, the chemical reaction mechanisms that describe the volatile combustion kinetics are usually large, encompassing numerous species and reactions. To efficiently and accurately capture combustion, appropriate models are needed.

In the 2000s, reduced order models for premixed [15] and non-premixed [16] combustion were developed based on the flamelet model [17].

The main assumption of these models is that the chemical time scales are much smaller than the flow time scales (fast chemistry assumption). A turbulent flame can hence be decomposed into many one-dimensional flamelets, which can be calculated in advance and stored in a flamelet database. Based on the dominant kind of flame structure in the three-dimensional simulation, the flamelets are calculated as either simplified premixed or non-premixed flames as shown in Fig. 19.2. The database can then be accessed during the simulation to retrieve the full thermochemical state, including temperature, species mass fractions, and transport quantities. For simple

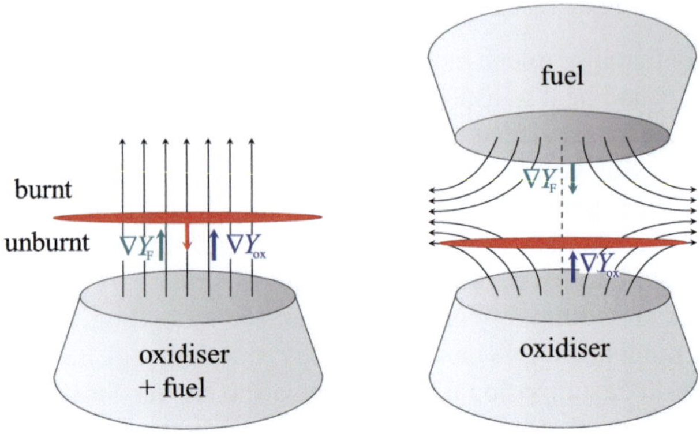


**Fig. 19.2** One dimensional flames used for pre-calculation of flamelets: premixed flame (left) and non-premixed flame (right)

single-phase combustion cases these variables are the fuel mixture fraction $Z$ and the progress variable $PV$.

Over the past few years, several advancements have been made to improve tabulated flamelet methods for solid fuel combustion [18–20]. The principal advantage of the flamelet model is its capability to integrate detailed gas-phase chemistry (which comprises many species and reactions—especially for biomass) at a manageable computational expense.

Before applying the flamelet model, the underlying flame structure must be defined as either premixed or non-premixed. While a non-premixed flame structure is expected for the homogeneous combustion of single coal or biomass particles, the choice becomes more intricate in the context of a real burner.

Assessing and validating combustion models exemplifies the necessity of hierarchical modelling approaches. Employing detailed chemical kinetics in a mixing layer, CP-DNS (carrier phase-direct numerical simulation) [21] allows for a comprehensive evaluation of flame structures in turbulent multi-phase combustion. It is found that turbulent mixing, coupled with the continuous recirculation of flue gases, introduces complexities resulting in both premixed and non-premixed combustion interactions [22,23]. While both premixed and non-premixed flame structures can be observed in solid fuel flames, the overall error in the premixed table was found to be greater than in the non-premixed table.

Prolonged mixing times between fuel and oxidiser can result in partially-premixed combustion. While there are continuous advancements in the formulation of multi-regime flamelet approaches [24], these are currently only under investigation in the context of pure gas-phase combustion. Currently, the choice of flamelet structure should be made with careful consideration of the configuration under investigation. For instance, laboratory-scale configurations with assisting premixed flames are more suitable for premixed flames, whereas self-sustained solid combustion is predominantly a non-premixed process.

In turbulent flows, not only velocity fluctuations can occur below the grid scale, but also fluctuations of chemical species that lead to wrinkling of the flame. This is called turbulence-chemistry interaction. Like the velocity, all transported quantities are implicitly filtered by the computational grid. Therefore, a model must be applied to describe the unresolved subgrid distributions of the transported scalars, which determines the overall progress variable source term in a cell, as well as the average transport quantities, temperature, and density.

Two approaches are used in the context of the present studies to address turbulent chemistry interaction. On the one hand, the artificially thickened flame approach adapts the scalar transport equations in a way that results in more gradual gradients, while maintaining flame speed and modelling subgrid wrinkling [25]. In this way, source terms can be resolved more smoothly.

On the other hand, the presumed $\beta$-PDF approach [26] is based on a probability density function derived from direct numerical simulations that relates the subgrid distribution of the mixture fraction sum $Z$ with the mean and variance of the filtered $\tilde{Z}$ and $\widetilde{Z''^2}$. The remaining scalars as well turbulence-radiation-interaction are modelled with a $\delta$-function, i.e. the filtered value is assumed to be constant in a computational cell. Conveniently, all tabulated quantities can be integrated in advance for a range of $\widetilde{Z''^2}$.

**Flamelet Modelling for Solid Fuel Combustion**

In addition to determining the flame regime, solid fuels have a significant impact on the chemical processes in the gas phase. Therefore, extensions to the classical flamelet model are required. Furthermore, the obtained flamelet solutions have to be diligently assembled to achieve the best possible results.

**Flamelet Model Extensions**

On the one hand, particles affect the local enthalpy balance. During heat-up, fuel particles act as heat sinks, while during char burnout on the particle surface, they release heat to the gas phase. The enthalpy exchange is modelled with a enthalpy source term in the enthalpy transport equation. On the other hand, the heterogeneity of the released species (e.g. volatiles and char products) or pilot flames leads to a spectrum of fuel compositions. To reflect these influences on the gas phase, additional dimensions and accessing variables must be added to a conventional tabulated flamelet manifold, which is normally based on a single fuel mass fraction and progress variable.

An increased enthalpy $h$ can be accomplished by preheating the fuel and oxidiser of the flamelets [19]. Enthalpy losses must be treated differently depending on the flamelet structure. In premixed combustion, lower enthalpy flamelets can be achieved by exhaust gas recirculation (EGR)—adding differing amounts of cooled combustion products to the fresh mixture while keeping the overall elemental composition and temperature constant [15,19,27]. Physically, this would correspond to fresh gases mixing with burned gases that have been cooled in the vicinity of burner walls or particles. Figure 19.3 shows the effect on the affected premixed flamelets.

For non-premixed flamelets, two methods have been proposed to reduce the enthalpy [28]. In the first method, an artificial radiation loss term is introduced in the

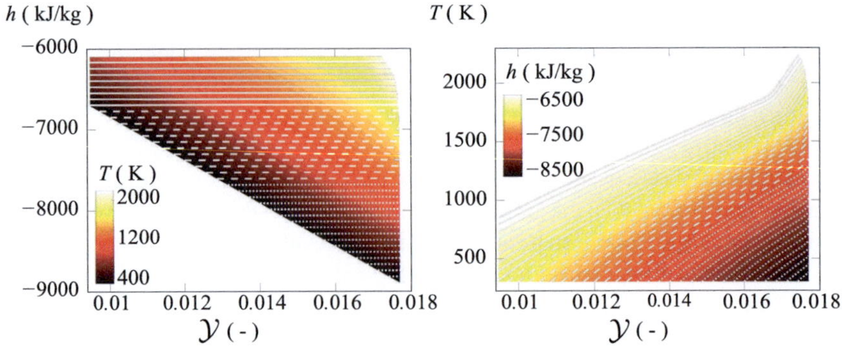


**Fig. 19.3** Premixed flamelets with enthalpy variation for a given mixture fraction $Z$. Enthalpy $h$ and temperature $T$ are shown over the progress variable $\mathcal{Y}$. Enthalpy is increased by raising the temperature of the fresh mixture (solid lines). A decrease in enthalpy is achieved by exhaust gas recirculation, which leads to a higher $\mathcal{Y}$ at a lower temperature (dashed lines). An interpolation is performed to a fully recirculated and cooled mixture (dotted lines)

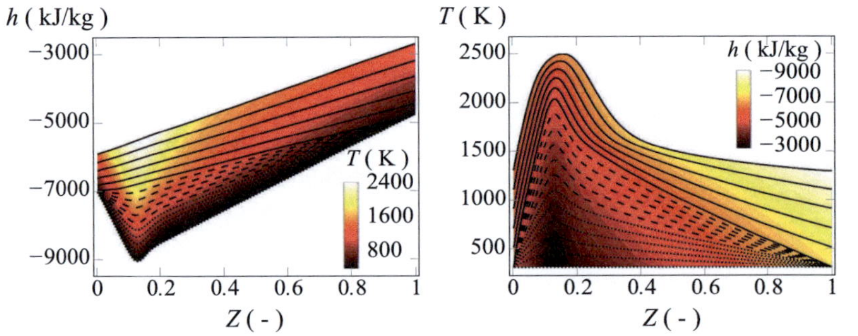


**Fig. 19.4** Enthalpy $h$ and temperature $T$ over mixture fraction $Z$ of non-premixed flamelets with a enthalpy variation for a given scalar dissipation rate $\chi$. Enthalpy is increased by preheating (solid lines). A decrease in enthalpy is achieved by adding an artificial radiation term or by damping the heat release—as was done here (dashed lines). The last burning solution (before a further enthalpy reduction would cause quenching) is cooled while keeping the species composition constant. Solutions are interpolated between the last burning solution and the cooled one (dotted lines)

energy equation. In the second method, the heat release from the chemical reactions is reduced by a damping factor. The effect of the former is shown in Fig. 19.4.

The variation in fuel composition can be obtained by using a spectrum of fuel blends in the flamelet tabulation process for premixed or non-premixed flamelets. The mixture fraction $Z$ then becomes the sum of all fuel mass fractions $Z_i$. The blending factor, denoted as $Y = Z_2/(Z_1 + Z_2)$ represents the fraction of fuel 2 in the overall fuel mixture [29]. The transformation is shown in Fig. 19.5. This approach can be further extended for more fuels [20].

The definition of the char-off gases has to account for the oxygen consumption. It is therefore not simply defined as CO, but as CO + M, where M is the oxidiser without oxygen. For air, M is N2 and for oxy-fuel atmospheres it is $CO_2$. According

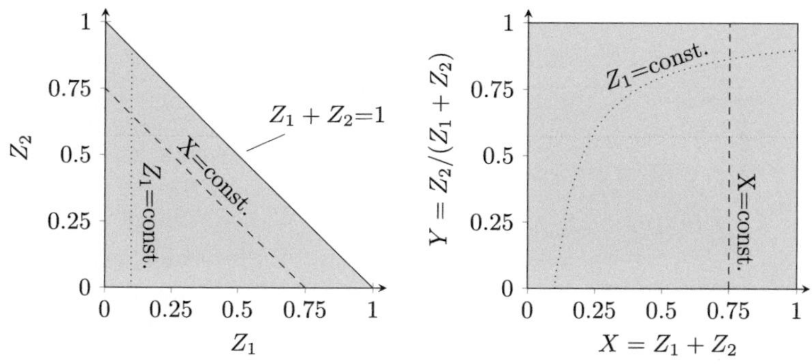


**Fig. 19.5** To capture the mixing of different fuels (e.g. volatiles and char off-gases) in the combustion chamber with a tabulated flamelet manifold, the mixture fractions $Z_1$ and $Z_2$ are transformed from $Z_1$-$Z_2$-space into $X$-$Y$-space, i.e. the mixture fraction sum and a blending factor. This representation has the advantage of being rectangular and allows for more efficient data storage and access

to the study by Knappstein et al. [18], the chemical reaction can be written as

$$2\mathrm{C(s)} + \mathrm{O_2} + \alpha\mathrm{M} \rightarrow 2\mathrm{CO} + \alpha\mathrm{M}. \tag{19.1}$$

For air, the molar fraction of M and $O_2$, $\alpha$, is 3.7619. For oxy-fuel atmospheres, $\alpha$ depends on the specific ratio. For example, for Oxy33 it would be $\alpha = X_{CO_2}/X_{O_2} = 0.67/0.33 = 2.03$. The molar fractions of the char-off gases can then be easily calculated.

**Assembly of the Flamelet Lookup Table (FLUT)**
The specific steps for assembling the flamelet solutions into a flamelet lookup table depend on the flame structure of the solution basis, i.e. whether the flamelets are premixed or non-premixed.

Premixed flamelets can only be calculated within the flammable mixture fraction range [19]. A reduction of enthalpy through EGR will eventually also lead to non-flammable conditions. Nevertheless, the simulations may require thermophysical properties in these non-flammable regions. Therefore, multiple interpolation steps must be performed between flamelets at the flammability limits and synthetic solutions at the edge of the parameter space, e.g. for $Z = 0$ (pure oxidiser), $Z = 1$ (pure fuel), or EGR $= 1$ (pure cooled combustion products). These synthetic solutions correspond to the fresh mixture under present conditions. The basis for these interpolations is the assumption that the species and temperature mix linearly between the flammability limits and the edges of the parameter space. After interpolation two corrections have to be made. (1) The density must be recalculated as it is non-linearly related to the temperature [19]. (2) The progress variable source term $\omega_{\mathcal{Y}}$ should be set to zero outside the flammability limits to avoid spurious reactivity (Fig. 19.6).

For non-premixed flamelets, no interpolation in mixture-fraction direction is needed, as the whole mixture fraction range is covered by design. However, the range of scalar dissipation rates $\chi$, which are a measure of mixing intensity, should

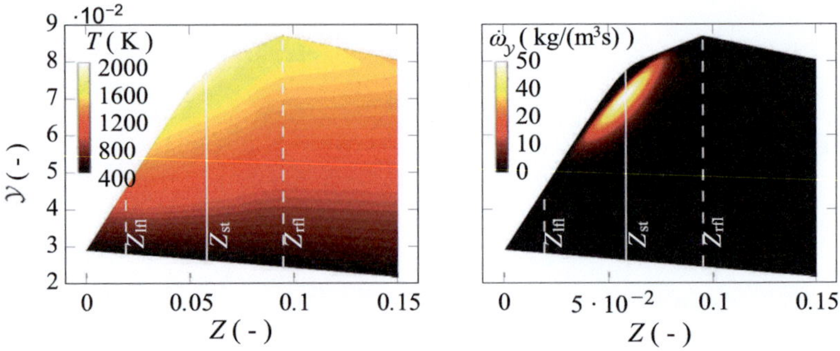


**Fig. 19.6** Manifold of premixed flamelets. Premixed flamelet solutions can only be calculated between the lean flammability limit $Z_{\mathrm{lfl}}$ and the rich flammability limit $Z_{\mathrm{rfl}}$. Thermophysical quantities between the flammability limits and the edges of the manifold must be interpolated

be wide enough to include a solution where mixing is so strong that the flame cannot burn, and a pure mixed solution is obtained instead.

When the entire manifold is filled with data points, it has to be parameterised to be accessible in the simulation, i.e. pre-heating and EGR or heat release parameters need to be mapped to the enthalpy. The spatial coordinate $x$ or the scalar dissipation rate $\chi$—for premixed or non-premixed flamelets, respectively—need to be mapped to the progress variable $\mathcal{Y}$ [19].

The progress variable should be defined in a way that it increases monotonically. This can be challenging when assembling a complex multi-dimensional flamelet table. An approach to facilitate the choice of the progress variable was proposed by Prüfert et al. [30]. They used an evolutionary algorithm to find an optimal progress variable definition. Ultimately, the final table should always be checked against the original solutions.

### 19.1.4 Radiation Modelling

The radiative heat transfer is computed using the discrete ordinates method (DOM) with a finite volume approach, employing a total of 96 ray directions to discretise the radiation field. Gas-phase absorption is modeled with the weighted sum of grey gases (WSGG) approach, allowing for an efficient representation of radiative properties across different spectral bands. In this framework, the radiative transport equation (RTE) is solved for the gaseous components within the WSGG model to ensure accurate coupling with the DOM solver. This has been shown to be a suitable approximation for coal particles [31] and are assumed to be also valid for biomass. Then, the final form of the RTE becomes

$$\frac{\mathrm{d}I_j(\mathbf{r},\mathbf{s})}{\mathrm{d}s} = \kappa_{j,\mathrm{g}}\, a_{j,\mathrm{g}}\, I_{\mathrm{b,g}} + E_{j,\mathrm{p}} - (\kappa_{j,\mathrm{g}} + \kappa_{\mathrm{p}})\, I_j(\mathbf{r},\mathbf{s}), \tag{19.2}$$

where $\mathbf{r}$ and $\mathbf{s}$ represent position and direction of rays, $j$ is the index of gases in the WSGG model, $I_{\mathrm{b}}$ is the blackbody intensity, $E_{j,\mathrm{p}}$ is the particle emissive power, $\kappa_{\mathrm{p}}$

is the particle absorption coefficient, $\kappa_{j,\mathrm{g}}$ and $a_{j,\mathrm{g}}$ are the absorption coefficient and weight factor of the gas phase for the WSGG model.

WSGG coefficients by Dorigon et al. [32] are employed for air and by Bordbar et al. [33] for oxy-fuel conditions.

### 19.1.5 Chemical Reactor Networks

#### Fundamentals of Chemical Reactor Networks

CFD simulations—and LES in particular—are capable of describing the flow field and general combustion characteristics with great detail and high accuracy. Nevertheless, the computational cost of LES with detailed chemical kinetics is punitive due to the stiff equation systems required for large chemical mechanisms such as those in Chap. 6. Therefore, direct coupling of large chemical mechanisms to LES in a finite-rate chemistry approach is not feasible. As an alternative, reduced-order chemistry representations, such as flamelet-based models presented earlier, have been extensively used in LES. However, the slow formation of minor chemical species and pollutants poses a challenge for accurate predictions using flamelet-based models. Another challenge is the high computational cost of the flamelet-LES, which can easily surpass thousands of CPU hours. This makes the method impractical for repetitive engineering calculations and parameter studies. Chemical reactor networks (CRN) overcome both challenges by directly exploiting large chemical mechanisms with significantly lower computational cost (which may be as low as CPU seconds). This is achieved by simplifying the aerodynamic/hydrodynamic representations of the system rather than the chemistry.

A CRN is a combination of ideal chemical reactors, and auxiliary units (mixers and splitters) to direct the flow between them. The ideal reactors are a fundamental tool for studying reactive systems in chemical engineering and they represent simple flow patterns that are easy to calculate and potentially desirable to achieve in real systems [34]. The two ideal continuous-flow reactor types are the perfectly stirred reactor (PSR) and the plug flow reactor (PFR). A generic CRN example is shown in Fig. 19.7.

The PSR assumes infinitely fast mixing, resulting in a 0-dimensional model where diffusion is neglected due to the absence of concentration gradients. At steady-state conditions, the PSR is described by a set of algebraic equations

$$\frac{\omega_{i,\mathrm{in}} - \omega_{i,\mathrm{out}}}{\tau} + \frac{R_i\, M_i}{\rho_{\mathrm{out}}} = 0\,, \tag{19.3}$$

for each species $i$. Here, $\omega_{i,\mathrm{in}}$, $\omega_{i,\mathrm{out}}$, $R_i$, and $M_i$ are the inlet mass fraction, outlet mass fraction, formation rate and molecular mass of species $i$, respectively. Residence time and mixture density are given as $\tau$ and $\rho_{\mathrm{out}}$.

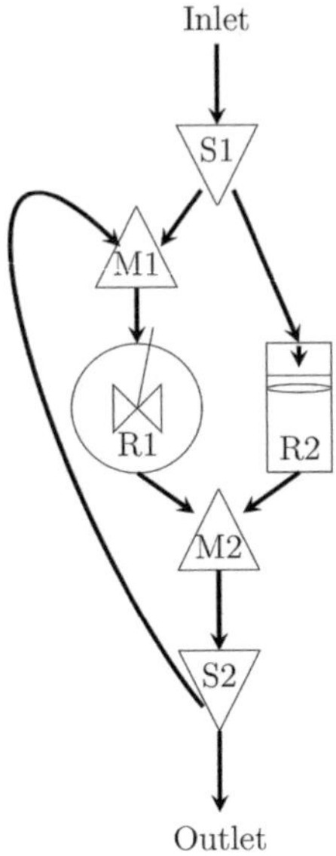


**Fig. 19.7** A simple CRN example containing a PSR (R1), a PFR (R2), flow mixers (M1, M2) and splitters (S1, S2). (Reprinted with permission from Trespi et al. [35], copyright 2020 American Chemical Society)

The PFR assumes fast, unidirectional convective flow, resulting in a 1-dimensional model where diffusion is neglected due to dominating convective flow. At steady-state conditions, the PFR is described by a set of ordinary differential equations

$$\frac{\mathrm{d}\omega_i}{\mathrm{d}\tau} = \frac{R_i\, M_i}{\rho}, \tag{19.4}$$

for each species $i$.

In addition to the inlet flows, either the volume ($V$) or the residence time ($\tau$) of the individual reactors are required to close the equations. The relation between volume, residence time, and volumetric flow rates ($\dot{V}$) are described by

$$\tau = \frac{V}{\dot{V}}. \tag{19.5}$$

The mixer and splitter are simple linear operators with no reactions, conserving the mass flows of individual chemical species.

For representing real systems, usually multiple ideal reactor units are required. A difficulty for the CRN approach is the creation of a reasonably representative network of reactors for the system to be simulated. This involves the identification of (1) zones where the flow assumptions of PFR and PSR are valid, and (2) the mass flows between these zones. An early example of a CRN with multiple units is by Touchton [36], where serially connected reactors are used to calculate $NO_x$ emissions in a gas turbine. In their work, the CRN is created by manually identifying the flame, post-flame, and quenching zones, and the residence times of the reactors are fitted based on experimental data. The crucial flow information to create the representative CRN can also be extracted from a CFD result. In this case, the part of the chemistry that potentially changes the flow field should be captured by the CFD (i.e. combustion reactions). This flow field can then be used to generate the CRN for calculating minor species (i.e. pollutants), assuming they have a negligible impact on the flow field. An initial example of a CFD based CRN is proposed by Ehrhardt et al. [37] for predicting $NO_x$ in a pilot-scale furnace. In their work, a simple global chemical mechanism is employed in CFD to obtain the flow field, which is then used to generate the CRN. Afterwards, a detailed mechanism is employed in the CRN for $NO_x$ prediction. This CFD post-processing approach has become a common method for generating representative CRN. A recent and significant example is the kinetic post processor (KPP) by Cuoci et al. [38,39] and Stagni et al. [40], which clusters the CFD cells to be represented by (up to thousands of) PSR to construct a representative CRN. The work described in the following sections uses the simulation framework NetSMOKEpp [35,41,42], which is based on OpenSMOKEpp [43].

## 19.2 Simulation of Turbulent Solid Fuel Combustion

Numerous investigations were conducted to vigorously validate the models that are implemented in *OxySim-129* at different scales and for different operating conditions. In this process, valuable understanding of the underlying physics could be gathered. Figure 19.8 shows an overview of work that was published in this context. Apart from the shift from coal to biomass, these studies were characterised by three major transitions.

1. Over time, burners of increasing application relevance and complexity have been investigated.
2. Studies focused increasingly on oxy-fuel combustion.

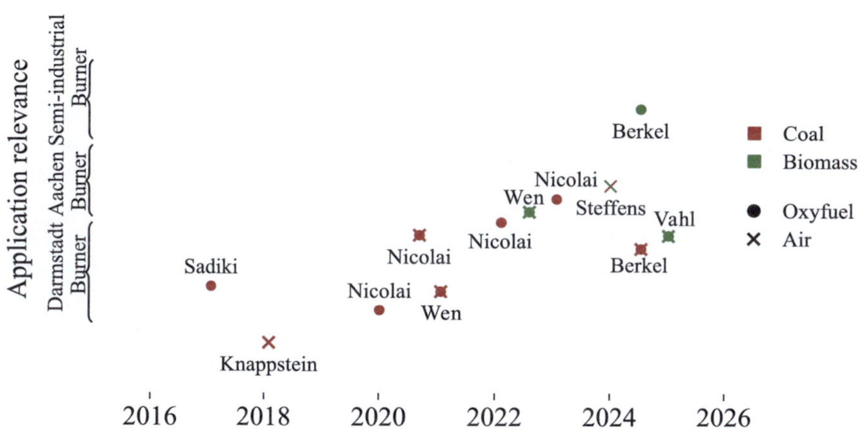


**Fig. 19.8** Simulations of turbulent combustion of pulverised coal and biomass in air and oxy-fuel atmospheres in the last ten years at different burner scales

3. The models were transferred from the research code FASTEST and implemented in the popular open-source code OpenFOAM® as the *OxySim-129* framework.

Both the Darmstadt and Aachen burners, introduced in Chaps. 13 and 14, mimic mechanisms seen in industrial combustors, e.g. relying on swirl-induced recirculation zones for stabilisation. The Aachen burner, with heated walls, achieves self-sustaining flames of up to 120 $kW_{th}$. Complementary to this, the Darmstadt burner, which reaches 40 $kW_{th}$, offers ample optical access but requires gas assistance to compensate for heat losses. Both the Aachen and Darmstadt burners are equipped with two concentric annular nozzles surrounding a central bluff body. The nozzles are positioned atop a diffuser. The central primary nozzle releases oxidiser, solid fuel, and, if necessary, methane. Surrounding this, the outer secondary nozzle supplies a swirling oxidiser flow. Inside the diffuser, the primary and secondary flows mix and enter the combustion chamber. Additionally, a tertiary staged stream enters near the outer walls of the combustion chamber. Although originally designed for coal, recent numerical and experimental studies have focused on biomass.

The semi-industrial burner, which is also located in Darmstadt, aims to bridge the gap between laboratory-scale burners and industrial combustion chambers. Analogous to the laboratory-scale burners, the burner features a swirl-induced recirculation zone with two concentric inlets.

**Sadiki et al.** [7] **(2017)** were the first to simulate coal under oxy-fuel conditions in the Aachen burner. Their work from 2017 focuses on the differences between RANS and LES and particle dispersion, while using simplified gas-phase kinetics. They showed that LES results were more accurate.

The first simulation of the newly constructed Darmstadt optically accessible solid fuel combustor was performed by **Knappstein et al.** [18] **(2018)**. Compared to previous studies, they increased the fidelity of the simulations by using an LES and tabulated flamelet approach (flamelet-LES) with a methane/volatile and a char mass fraction, and used the artificially thickened flame (ATF) model to account for turbulence-chemistry interactions. Examining single-phase non-reactive, single-phase reactive ($CH_4$+air) and two-phase reactive ($CH_4$+air+coal) operating conditions, they showed good agreement with the available experimental data.

Also using flamelet-LES, **Nicolai et al.** [19] **(2020)** simulated coal combustion in the Darmstadt burner in an oxy-fuel atmosphere. The study analysed gas-assisted coal combustion with two different coal loads and analysed the effects on the reaction zones, showing that the impact of coal scales with the amount that is fed.

A follow-up study of **Nicolai et al.** [44] **(2021)** investigated self-sustained coal combustion in air and oxy-fuel conditions with 21-25 vol.% $O_2$ content in the Aachen burner using tabulated chemistry, ATF, and LES. The results showed good agreement with experimental data that encompassed particle velocity and temperature measurements, and exhaust gas composition. A correlation between the increased gas-phase temperatures and initial particle heating and conversion was found for the oxy-fuel operating conditions.

**Wen et al.** [20] **(2021)** examined pollutant formation, namely $SO_x$ and $NO_x$, inside the Darmstadt burner for air and oxy-fuel conditions by means of flamelet-

LES. To resolve the assisting methane, coal volatiles, and char off-gases, the tabulated flamelet manifold was constructed with three mixture fractions. The turbulence-chemistry interaction was modelled with a presumed $\beta$-PDF approach. The results showed some differences between combustion with oxy-fuel and air atmospheres. The $NO_x$ formation moved closer to the diffuser under oxy-fuel conditions. For the 25 vol.% $O_2$ condition, a higher amount of $SO_x$ was reported.

To analyse the influence of fuel conversion models, **Nicolai et al.** [45] **(2022)** performed a flamelet-LES of oxy-fuel coal combustion in the Aachen combustion chamber. The simplified single first-order kinetics, and the slightly more sophisticated competing two-step model were compared with the detailed CRECK-S model. The models and results are discussed in more detail in Sect. 19.2.2.

The first biomass flamelet-LES included in this chapter was carried out by **Wen et al.** [46] **(2023)**. The self-sustained combustion of torrefied biomass combustion was examined in the Aachen burner for both air and oxy-fuel conditions. The results showed good agreement with experimental results. To achieve constant equivalence ratios, the flow rates were changed when switching from air to oxy-fuel. This was also reflected in the results where a longer and wider particle flame was obtained due to suppression of the inner recirculation zone.

**Nicolai et al.** [47] **(2023)** analysed the differences between multiple oxy-fuel operating conditions in coal combustion with different oxygen contents ranging from 23 to 33 vol.%. The results are discussed more thoroughly in Sect. 19.2.2.

**Steffens et al.** [48] **(2024)** thoroughly examines the difference between turbulent coal and biomass combustion in the Aachen burner using flamelet-LES. The most important results are discussed in Sect. 19.2.2.

**Berkel et al.** [49] **(2024)** investigated different oxy-fuel and air operating conditions of the Darmstadt chamber focusing on the effects of atmosphere on particle behaviour.

Further, the combustion of biomass in the Darmstadt semi-industrial combustion chamber is examined, the results are presented in Sect. 19.2.3.

### 19.2.1 Gas-Assisted Solid Fuel Combustion

LES of gas-assisted solid fuel combustion in the optically accessible Darmstadt burner have been carried out for air-fired coal combustion [18] and oxy-fuel coal combustion [19,20] with the aim of validating the OxySim-129 framework for laboratory-scale configurations. In addition, the studies investigate how variations of both oxidiser and thermal power affect the flame structure. The complexity of the system is approached by gradually increasing the number of reacting phases, starting from single-phase non-reactive (SPNR) conditions. The configuration is then extended to single-phase reactive (SPR) and finally to two-phase reactive (TPR) conditions. The oxidiser composition as well as the thermal power of the conditions of increasing complexity are shown in Table 19.1, sorted by the publication in which they are discussed.

**Table 19.1** Oxidiser composition and thermal power of operating conditions investigated by Knappstein et al. [18] and Nicolai et al. [19]. Values without brackets and in brackets belong to methane and coal respectively

| Publication | Knappstein et al. [18] | Nicolai et al. [19] |
|---|---|---|
| Oxidiser air/$O_2$/$CO_2$ [vol.%] | 100/0/0 | 0/30/70 |
| Operating conditions | Thermal power [$kW_{th}$] | |
| Single-phase non-reactive | 0 | – |
| Single-phase reactive | 20 | 20 |
| Two-phase reactive | 20 (20) | 20 (7), 20 (20) |

For single-phase non-reactive conditions, the thermophysical state is constant and no flamelet lookup is required. To extend the model to single-phase reactive conditions, one mixture fraction is sufficient. However, to represent the state under two-phase reactive conditions, two mixture fractions $f_i$, the enthalpy $h$ and a reaction progress variable, here $Y_{CO_2}$, are required. Physically there are three gaseous fuels, the volatile matter and the char from the coal particles, and methane as a supporting gas. To simplify the flamelet table, the volatile composition has been approximated to pure methane. Therefore, methane and volatiles can be treated as one species and only two mixture fractions are required to account for two of the gaseous fuels. The first mixture fraction treats the methane, while the second is the carbon monoxide formed in the particle surface reaction, which represents the gaseous char products.

**Stepwise Validation Approach**

The single-phase non-reactive configuration analyses the flow field of pure cold air [18]. This configuration allows a stand-alone assessment of the appropriate boundary conditions and reproduction of the velocity field. Of particular interest are the regions within and below the diffuser as these are important for flame stabilisation. The velocity profiles below the diffuser of such a simulation are visualised in Fig. 19.9 and generally show good agreement with the experimental data. In particular, the simulated vorticity and the extent of the large recirculation zone, as indicated by the tangential and axial velocity components below the diffuser, are well reproduced by the simulation. The velocity profiles in the diffuser show discrepancies, which may be related to the difficulties of measurements close to walls due to reflections.

In the single-phase reactive configuration, a fuel stream of pure methane is mixed with oxygen in the combustion chamber. The flow provides a thermal power of 20 $kW_{th}$. The comparison of the single-phase reactive simulations with experimental data allows an extended assessment of the performance of the numerical simulation in both air and oxy-fuel conditions. Since the methane flame contributes significantly to the heat release, its understanding is also crucial for the two-phase reaction.

The experiments show that, in addition to the main flame, small side flames stabilise between the primary and secondary flow, for both air and oxy-fuel operating conditions [50]. There are several observations regarding the side flame that can be applied to both operating conditions due to similarities. The formation of a stable side flame depends on the initial conditions [18]. When a numerical side flame is

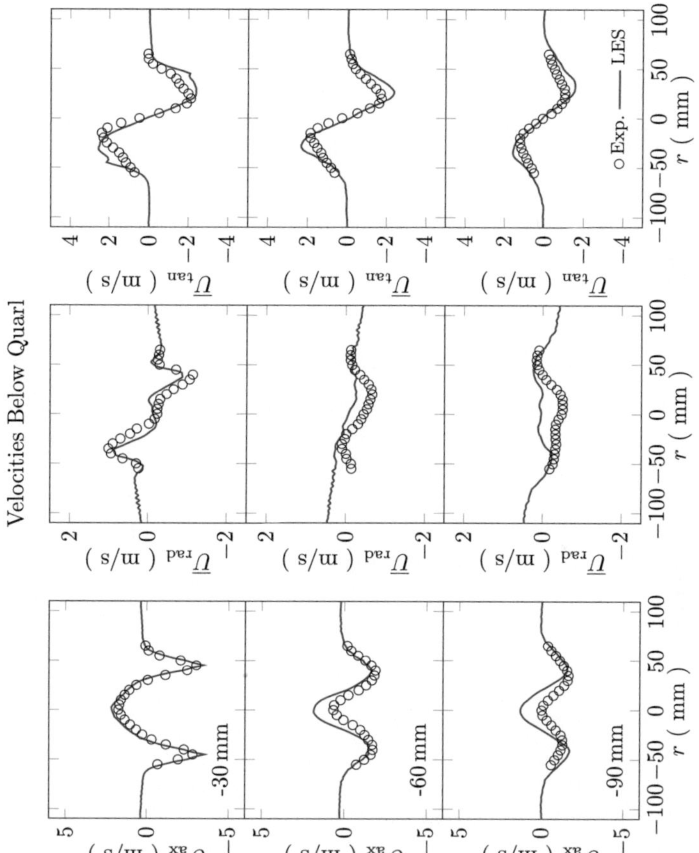


**Fig. 19.9** Comparison of experimental and numerical mean velocity data below the diffuser for a single-phase non-reactive oxy-fuel operating condition

initialised, it is larger and therefore merges with the main flame, whereas in the experiments the side flame is detached from the main flame. As the flow field inside the diffuser is affected by the shape of the side flame, the combustion mode without side flame shows a different velocity field. The velocity field is closer to the experimental case and will therefore be referred to in the following. Additional information about the reaction zone with air as oxidiser can be obtained by comparing the non-reactive and reactive experimental velocity data [18]. Changes in the flow field induced by density changes due to heat release during chemical reactions can be identified. The good agreement between the numerical and experimental results indicates that the chemical reaction is generally well reproduced by the simulation. To verify the reaction zone under oxy-fuel conditions, the numerically obtained mean OH mass fraction fields are qualitatively compared with the mean laser-induced OH fluorescence signal from the experiment [19]. Both simulations and experiments show symmetrical fields in the diffuser with a slightly narrower and longer reaction zone in the experiments. While the flow field measurements become asymmetric below the diffuser and are therefore constantly shifted away from the numerical flow field, within the diffuser there is good agreement between the velocities of the simulation and the measurement results.

The single-phase reactive configuration is further extended by injecting coal particles with an additional thermal power of 20 kW using air as oxidiser, and 7 and 20 kW under oxy-fuel conditions. The performance of the model is evaluated by comparing the numerical and experimental gas-phase velocity fields, again focusing on the diffuser-zone where flame stabilisation takes place. In addition, the simulation allows the analysis of particle histories, devolatilisation processes, char burnout, and particle-gas phase interaction. The two different particle loads give an indication of their influence on the flow combustion characteristics. The applied model is able to reproduce the experimental gas-phase velocity field with air as oxidiser, particularly with regard to the axial component [18]. When further analysing the particle histories, a rapid heat-up is detected that induces fast devolatilisation in the diffuser, and hence, a richer mixture in the stabilised main gas flame. The short residence times of the particles in the high-temperature regions result in only small fractions of the overall char conversion within the chamber. Analysing the numerical results at oxy-fuel conditions, the essential flow characteristics of both operating points are well reproduced by the models. However, there are deviations in regions of high particle density, which can be explained by the challenging separation process necessary for evaluating the measured PIV data. A comparison of the operating points with different coal mass flows, namely 7 $kW_{th}$ (TPR7) and 20 $kW_{th}$ (TPR20), reveals several differences [19]. Comparing the TPR7 results with the SPR reaction zones, the flame burns much further below the burner inlet, close to the walls at the exit of the diffuser. The TPR20 results show even more significant changes as the increased coal mass flow shifts the reaction zone further downwards to be outside the diffuser, as shown in the two temperature fields in Fig. 19.10a).

To estimate the effect of the shifted and larger areas of higher temperature on particle conversion, the amount of char gases in the gas phase is shown by the second mixture fraction in Fig. 19.10b). As the particles pass through the combustion

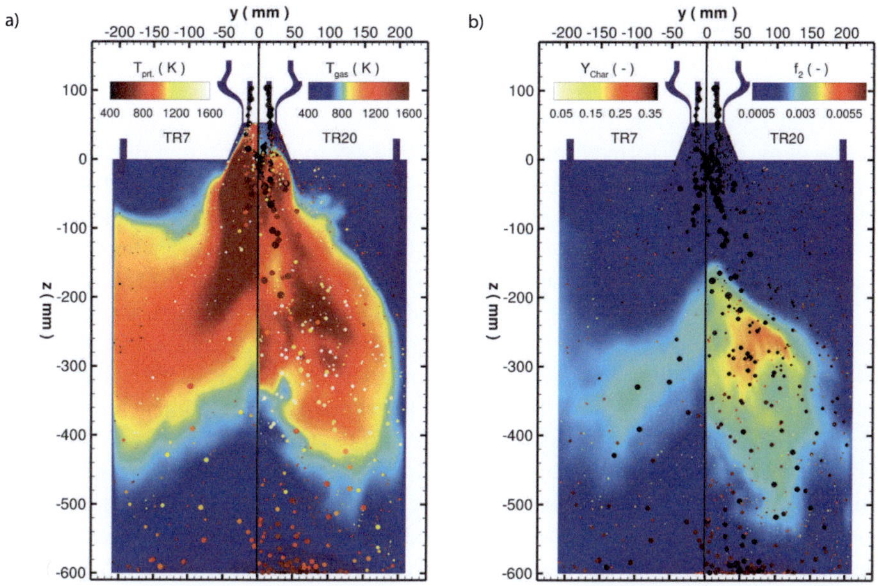


**Fig. 19.10** **a)** Mean gas temperature field with instantaneous particle temperatures and **b)** mean second mixture fraction and particle char content on the right hand side for both oxy-fuel operating points [19]. The second mixture fraction indicates the amount of char gases. (Reprinted from Nicolai et al. [19], Copyright (2020), with permission from Elsevier)

chamber, larger zones of high temperature are shifted into regions of lower particle velocities, resulting in longer particle residence times in the hot gases. This effect is amplified for small particles, as they decelerate rapidly after leaving the diffuser where the highest particle velocities occur. If sufficient oxygen is available, this in turn increases the char reaction. As the char conversion correlates very strongly with the temperature regions, and the tertiary flow provides sufficient amounts of oxygen, it can be concluded that the high temperature in this combustion chamber is the driving factor for the char conversion [19].

The simulations of gas-assisted solid fuel combustion were able to reproduce the flow characteristics observed in the experiments. The models implemented in the OxySim-129 simulation framework are applicable to air and oxy-fuel conditions with coal combustion with different thermal powers and particle loads at laboratory scales.

### 19.2.2 Self-Sustained Solid Fuel Combustion

In large industrial-scale combustion chambers, solid fuel combustion is usually self-sustaining without the need for a pilot flame. This self-sustained combustion mode can be investigated in the Aachen combustion chamber at a laboratory scale due to its heated walls. In the following, (1) the influence of the fuel conversion models, (2) the differences of different oxygen contents in self-sustained coal combustion, and finally (2) the differences between coal and carbon-neutral biomass fuels are discussed.

**Fidelity of Combustion Models**

Nicolai et al. [45] performed multiple flamelet-LES to compare two reduced-order devolatilisation models combined with the char combustion model by Baum and Street [6] with the detailed CRECK-S model. Since simulations with the detailed model were shown to use approximately six times more computational resources, it is crucial to understand the capabilities and limitations of simpler models.

The reduced-order models used in this study were the single first-order model and the competing two-step model. The former describes volatile release with a single Arrhenius-type equation. It therefore always results in the same volatile yield, which has to has to be prescribed in advance with the remainder being char. It represents the solid fuel conversion process with two reactions. One converts the solid fuel to volatiles and the other to char. This model is therefore closer to the realistic physics of solid fuel conversion, and the total volatile yield is dependent on the heating rate. The model coefficients of both reduced-order models are fitted to CRECK-S simulations that are run with representative heating rates of the simulation, and are thus tuned to the present conditions.

The study was performed with an oxy-fuel atmosphere containing 27 vol.% $O_2$ and 73 vol.% $CO_2$ and a thermal output of 40 $kW_{th}$. The volume-based median particle diameter of the coal powder sample used in multiple studies is $D_{50} = 29\ \mu m$, with $D_{10} = 5\ \mu m$ and $D_{90} = 132\ \mu m$. A coarse and a fine mesh were tested to evaluate grid dependency. The fine mesh with $3 \times 10^6$ cells and a finest resolution of $1 \approx 0.5$ mm was also used in the later studies in the same configuration.

The fidelity of the conversion models was assessed after a validation with experimental reference data, which showed good agreement for all three models. The particle conversion analyses show some differences between the models, e.g. a greater variety of the volatile and char reaction rates using the CRECK-S model. Moreover, a temporary slowing of the mass release can be identified in the CRECK-S simulations that cannot be captured by the simpler models.

Despite these differences in mass release, only very minor differences can be observed when comparing particle histories over the whole residence time in the burner, such as particle temperature or particle burnout, as shown in Fig. 19.11. Furthermore, little differences can be seen in the gas phase quantities. In conclusion, it appears that the present reduced-order models can be used successfully in a complex burner configuration if they are correctly tuned to the prevailing conditions.

**Oxy-Fuel Operating Conditions**

In the work from Nicolai et al. [47], the same experimental configuration was investigated. In this study, the focus was less on model validation and more on the analysis of the combustion physics of three different oxy-fuel atmospheres with oxygen volume fractions of 23, 27 and 33 vol.%. In order to keep the velocity profiles comparable, the volume flow rates were kept constant, leading to small deviations in the global equivalence ratio. Even though the previous study showed minor differences between

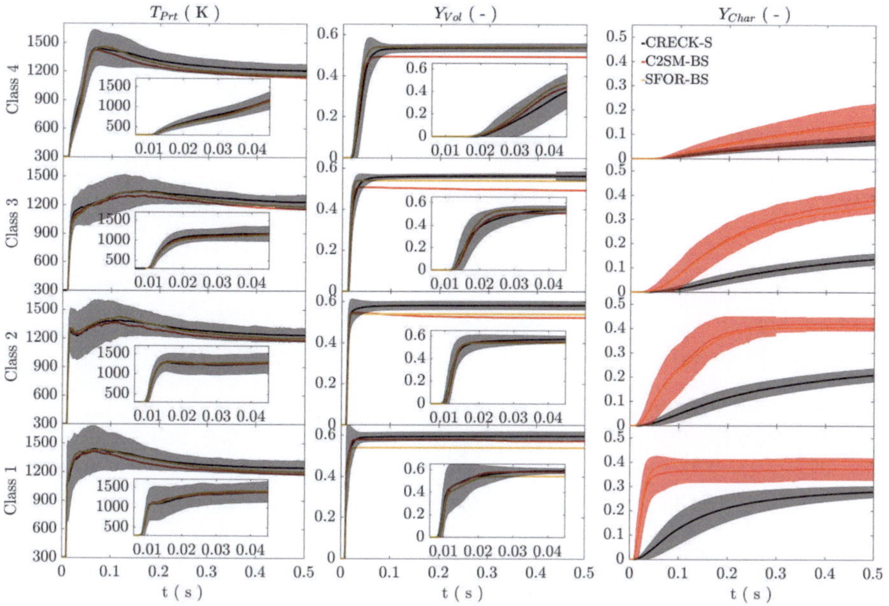


**Fig. 19.11** Particle histories with different fuel conversion models. (Reprinted from Nicolai et al. [45], Copyright (2022), with permission from Elsevier)

detailed and simplified solid fuel conversion models, the detailed CRECK-S model was selected for maximum fidelity.

Velocity and temperature fields from this study are displayed in Fig. 19.12. The key findings of the study were the following: As the volume flow rates are kept constant, there are only minor differences in particle velocities, even though the local combustion characteristics, which can affect the flow, change. High $O_2$ to $CO_2$ ratios increase the flame temperature independent of the global equivalence ratio. This leads to faster particle heat-up and mass release. Therefore, higher $O_2$ to $CO_2$ ratios lead to a greater particle burnout at the bottom of the chamber, even though particle residence times decrease. Finally, the stability and anchoring of the flame inside the diffuser are not affected by the increase in oxygen.

**Coal vs. Biomass**

In the study by Steffens et al. [48], self-sustained coal and biomass flames were compared directly in the same burner in an air atmosphere using flamelet-LES. Both coal and biomass share similar combustion characteristics, such as the initial devolatilisation and gas phase reactions, and the subsequent char burnout.

However, there are significant differences in properties that affect the combustion process. Biomass usually has a higher volatile content, contains more moisture, and is richer in oxidated volatile species, which are less energy-dense. Therefore, a slightly higher solid fuel mass flow is needed to retain the overall power output when

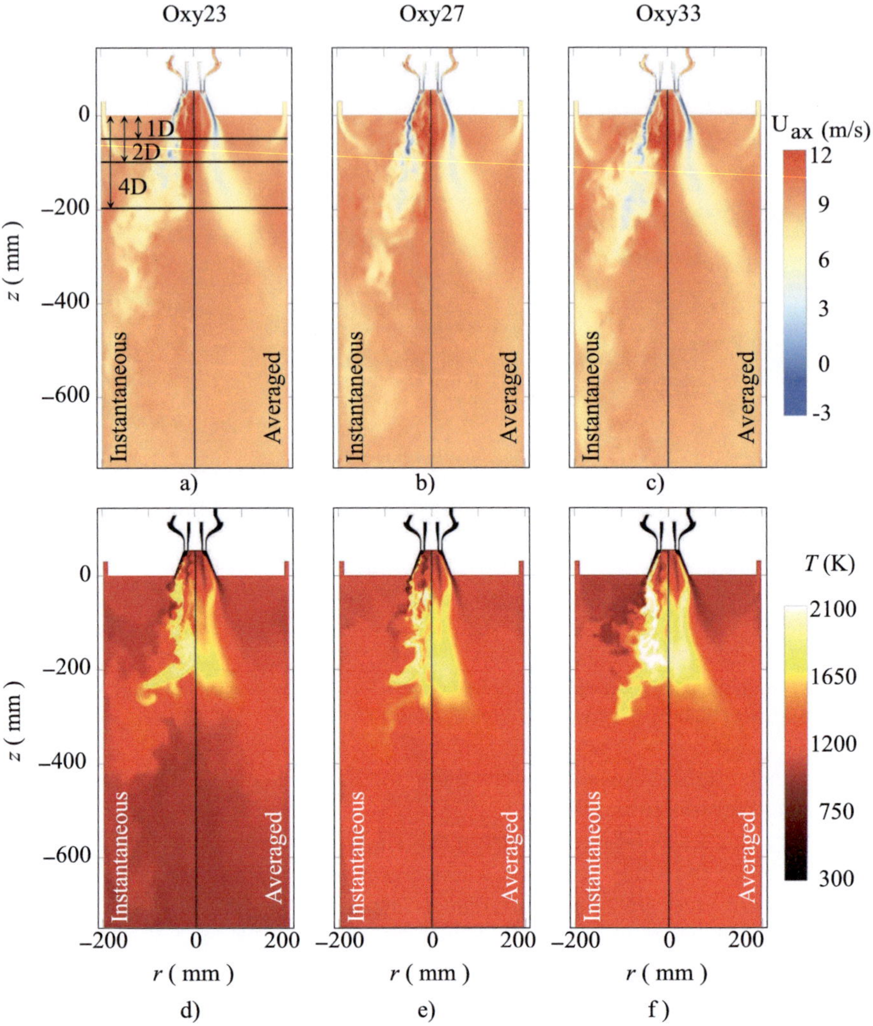


**Fig. 19.12** Velocity and temperature fields for different oxy-fuel conditions. (Reprinted from Nicolai et al. [47], Copyright (2023), with permission from Elsevier and minor editing)

switching from coal to biomass. In addition to releasing more volatiles, biomass releases them more quickly for the same heating rate. Finally, particle sizes can vary considerably. In the present study, biomass particles are significantly larger with a volume-based median of $D_{50} = 140\ \mu\text{m}$, $D_{10} = 101.5\ \mu\text{m}$ and $D_{90} = 178.8\ \mu\text{m}$.

In this study, Rhenish lignite (RL) and walnut shells (WNS) were selected as representatives of coal and biomass. For both, the volatile release was calculated with a competing two-step model, whose parameters were obtained from a fit to the results of the CRECK-S and CRECK-S-B models for lignite and walnut shells, respectively. The detailed simulations were run with the actual heating rates from

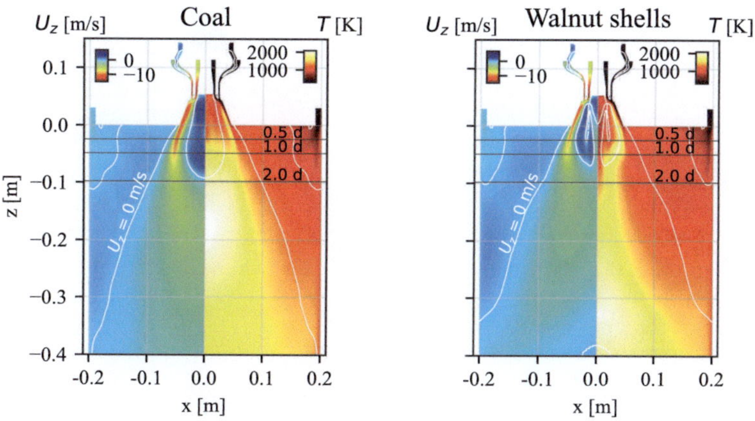


**Fig. 19.13** Averaged axial velocity and temperature fields of Rhenish lignite (RL) and walnut shell (WNS) flames. (Reprinted from Steffens et al. [48], Copyright (2024), with permission from Elsevier)

the burner. The model by Baum and Street [6] was used for char burnout for both fuels.

Gas phase reactions were modelled using a tabulated flamelet method, with detailed kinetics suitable for describing reactions of complex coal and biomass volatile species. Including two fuel mixture fractions for volatiles and char off-gases, an enthalpy dimension, the progress variable, and the mixture fraction variance to model the turbulence-chemistry interaction, both flamelet tables are 5-dimensional.

The results show significant differences between the two fuels. Considering the axial velocity and temperature contours in Fig. 19.13, large deviations can be observed in the velocity field which are also reflected in the temperature fields. The most striking differences in the walnut shell case are the partially suppressed recirculation zone in the diffuser and the longer flame shape.

To identify the reasons for this behaviour, a detailed analysis of particle trajectories and volatile release areas was conducted, e.g. particle trajectories were examined by grouping similar particles together as shown in Figs. 19.14 and 19.15. This revealed that most of the coal particles follow the gas flow more closely, directing them towards the flame inside the diffuser where they devolatilise very quickly. In contrast, the larger walnut shell particles tend to gather in the centre of the diffuser, locally suppressing the recirculation zone and heating slowly. They therefore deliver their volatiles at a lower position inside the burner, which causes the lengthening of the flame.

Even though the biomass particles release volatiles at a lower height inside the burner and modify the shape of the recirculation zone, the flame stabilisation mechanism is the same as for the coal flame, with the flame anchoring in the top of the diffuser in the shear layer between the primary and secondary inflows.

The studies of self-sustained solid fuel combustion establish the applicability of the OxySim-129 framework for different parameters. On the one hand, it is shown that simplified models can efficiently predict fuel conversion when tuned to the

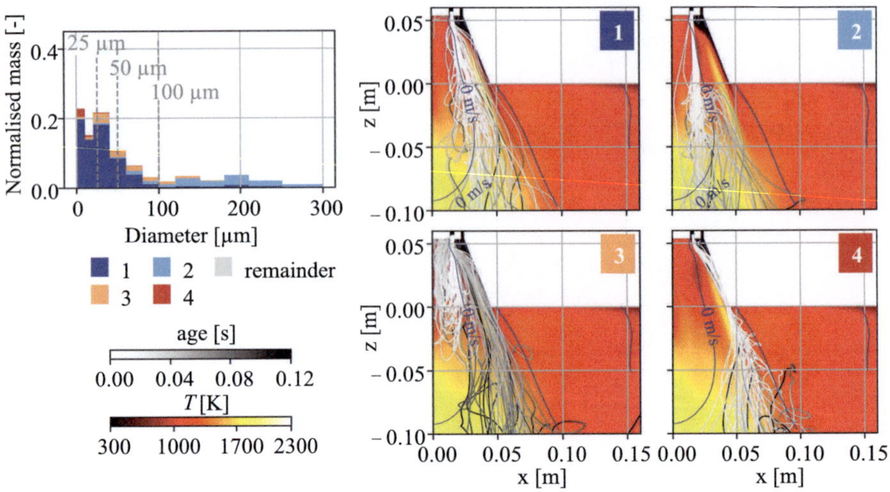


**Fig. 19.14** Coal particles grouped by similar trajectories. (Reprinted from Steffens et al. [48], Copyright (2024), with permission from Elsevier and minor editing)

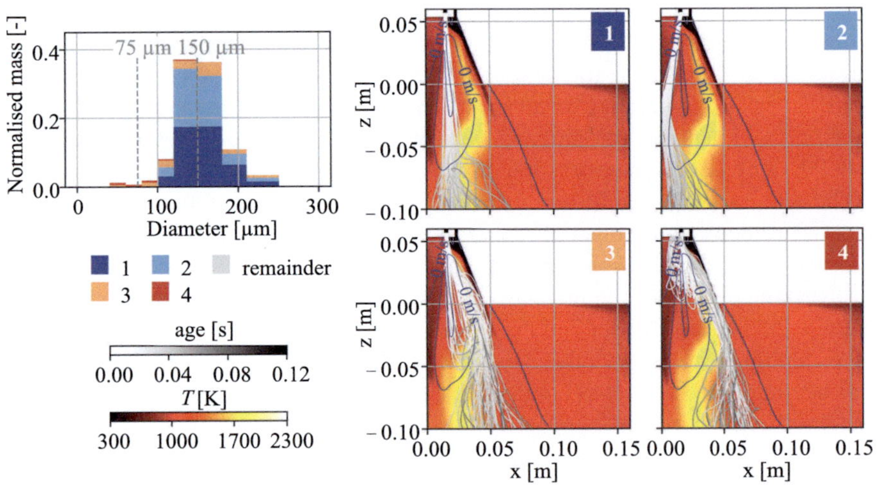


**Fig. 19.15** Walnut particles grouped by similar trajectories. (Reprinted from Steffens et al. [48], Copyright (2024), with permission from Elsevier and minor editing)

present conditions in the combustion system. On the other hand, the performance of the models was demonstrated for a spectrum of oxy-fuel operating conditions and for two fuels. The physical analyses indicate that the variation of oxygen in oxy-fuel coal combustion leads to faster fuel burnout. The comparison of coal and biomass highlighted the importance of particle size. Large biomass particles can affect the flow field and increase the flame length.

### 19.2.3 Solid Fuel Combustion in a Semi-Industrial Combustion Chamber

The semi-industrial Darmstadt combustion chamber is the scaled-up version of the self-sustained Aachen burner described in the previous section. The burner is designed to operate up to 1 $MW_{th}$ and is significantly larger than the Aachen burner. The burner head in comparison with the Aachen burner is shown in Fig. 19.16.

In the work by Berkel et al. [49], the effect of scaling-up is analysed by comparing the CFD results of the Aachen and Darmstadt burners, both operated with pulverised walnut shells (WNS). The selected preliminary operating condition for simulating the semi-industrial Darmstadt burner corresponds to a thermal load of 500 $kW_{th}$. The inlet flow rates (Fig. 19.16) for the simulated conditions for the Aachen and Darmstadt burners are compared in Table 19.2. The simulation setup is similar to that of Steffens et al. [48], using 5-dimensional flamelet tables.

The results are shown in Fig. 19.17, including the contour plots for velocity, temperature and mixing fields. Isolines for the velocity ($U_z = 0\,\mathrm{m\,s^{-1}}$, blue) and OH mass fraction ($0.15\,\mathrm{OH_{max}}$, red) as well as particle locations with their temperatures are given. The velocity and OH mass fraction isolines are used as indicators for the internal recirculation zones (IRZ) and reaction zones, respectively. Significant differences are observed as a result of scaling up.

The first column of Fig. 19.17 shows that the shape of the recirculation zone remains intact in the larger combustor and the velocities are higher, resulting in stronger swirl stabilisation. The particle trajectories are also significantly different. In the smaller chamber, the cold particles are able to penetrate the IRZ, whereas in the larger burner the particles go around the IRZ and heat up while still located in the diffuser. The second column in Fig. 19.17 shows that the IRZ temperatures are higher in the larger burner, partly because of the smaller number of particles in it. Inspecting the volatile mixture fractions ($Z_1$), very rich volatile mixtures form in the IRZ of the smaller burner, and reactions continue below the dump plane ($z = 0$ m).

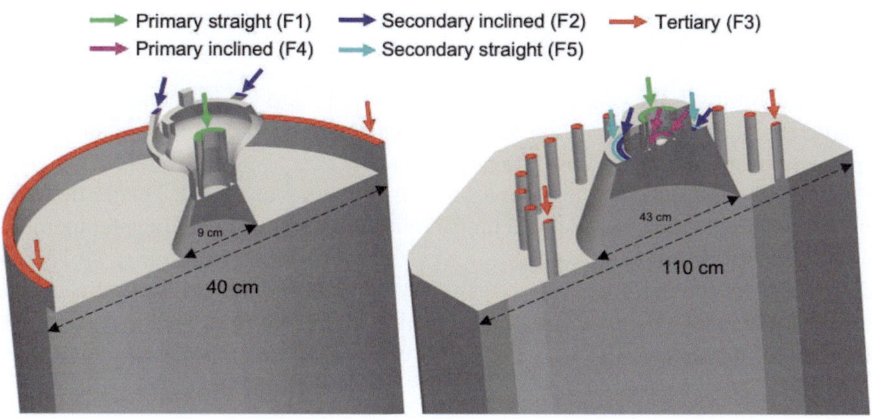


**Fig. 19.16** Self-sustained Aachen burner (left) and semi-industrial Darmstadt burner (right)

**Table 19.2** Feed flows of the two burners for the compared simulations

| Burner<br>Power | | | Aachen<br>40 $kW_{th}$ | Darmstadt<br>500 $kW_{th}$ |
|---|---|---|---|---|
| Primary solid (WNS) | $F1_{solid}$ | [kg $h^{-1}$] | 6.5 | 81.3 |
| Primary gas (straight) | $F1_{gas}$ | [m $s^{-1}$] | 6.22 | 4.34 |
| Secondary gas (inclined) | F2 | [m $s^{-1}$] | 19.9 | 32.5 |
| Tertiary gas | F3 | [m $s^{-1}$] | 0.6 | 1.04 |
| Primary gas (inclined) | F4 | [m $s^{-1}$] | - | 14.1 |
| Secondary gas (straight) | F5 | [m $s^{-1}$] | - | 10.3 |

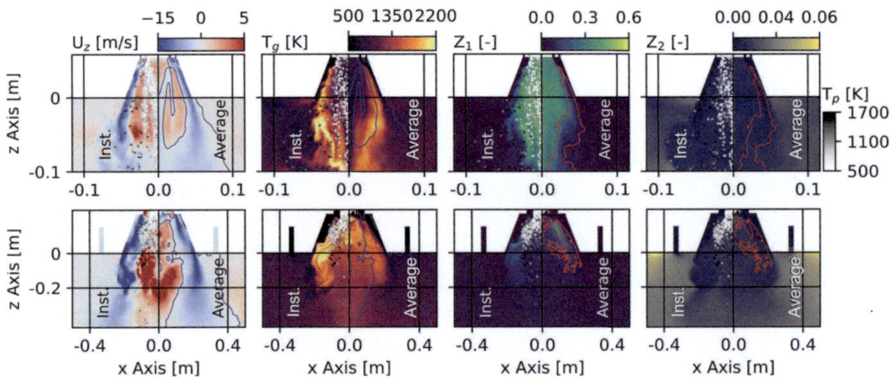


**Fig. 19.17** Simulation results for the Aachen (upper row) and Darmstadt (bottom row) burners. From left to right the four columns show the contours for axial velocity ($U_z$), temperature ($T_g$), volatile mixture fraction ($Z_1$) and char off-gas mixture fraction ($Z_2$). Blue isolines and red isolines show $U_z = 0$ m $s^{-1}$ and OH $= 0.15$ $OH_{max}$, respectively

In the larger burner, the reaction zone is mostly limited to the inside of the diffuser. Finally, the char reactions in the larger burner have a significantly higher impact, especially near the tertiary flow (staging flow).

Finally, the evolution of the particles in both burners is analysed. The particles are divided into size groups based on their different trajectories. The average temperature and conversion evolution of four different size groups are shown in Fig. 19.18. The heating rates of the particles decrease slightly in the larger burner. Nevertheless, the conversion of small particles is generally similar in both burners. As the particle size increases, both the heating rate and the conversion differences increase.

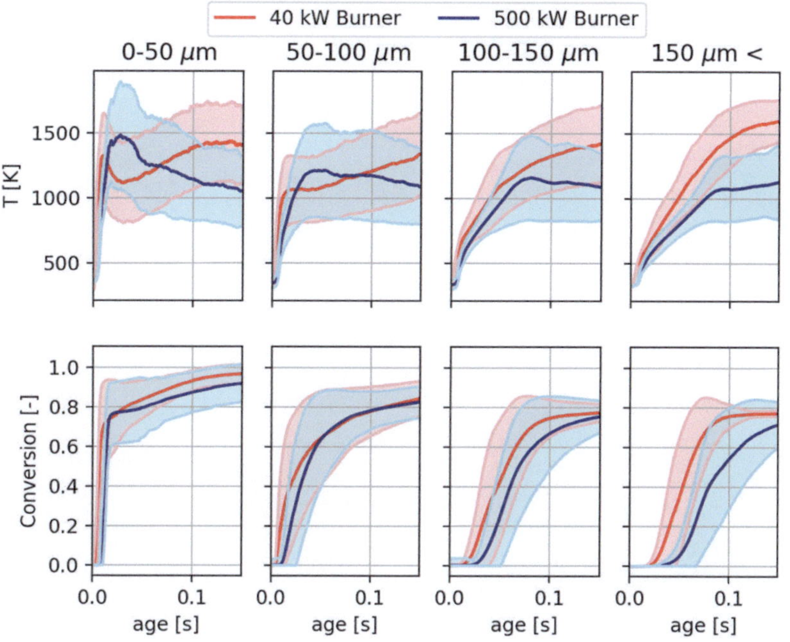


**Fig. 19.18** Particle temperature and conversion histories for different particle size groups

## 19.2.4 Analysis of Minor Species with CRN

### CFD-Based CRN for Oxy-Fuel Combustion

In the context of oxy-fuel combustion, Trespi et al. [35] introduced an iteratively enhanced CFD-based CRN for the swirled lab-scale Darmstadt burner operated with methane under oxy-fuel conditions. The CRN is based on the flow and temperature fields obtained from a flamelet-LES by Nicolai et al. [19]. First, the recirculation regions are identified, and are represented by PSR. The remaining regions with straight flow are then filled with PFR. The identified zones and three CRN constructed with increasing complexity (RN1, RN2, and RN3) based on these zones are shown in Fig. 19.19.

Once the zones have been identified, the mass flows between the reactors (i.e. the splitting ratios of the splitters) are adjusted so that the residence time distribution (RTD) of the CRN matches the experimental RTD measurements. This optimisation procedure ensures that the CRN actually represents the flow field of the combustor.

The final CRN results, together with experimental measurements and CFD results are shown in Table 19.3. The main species including $H_2O$, $O_2$, and $CO_2$ are well captured by both the CFD and CRN approaches as expected. However, the CO concentrations can be calculated more accurately with CRN, especially with increasing the complexity based on the target species.

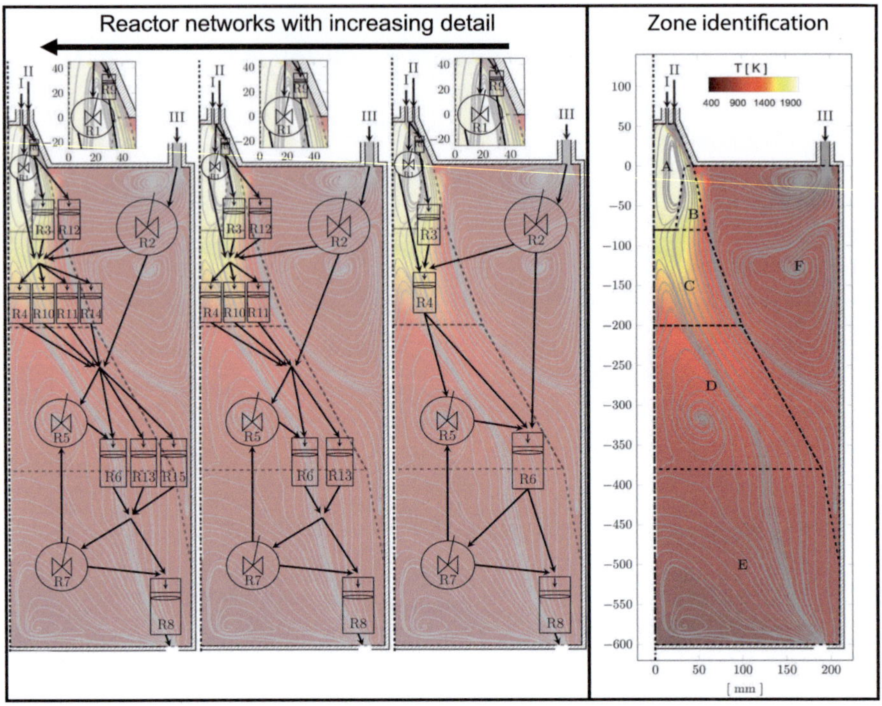


**Fig. 19.19** Zones identified through CFD and resulting reactor networks with increasing complexity. Adapted from Trespi et al. [35]

**Table 19.3** Comparison of experimental measurements against CFD and CRN results

| | $O_2$ [vol.%] | $H_2O$ [vol.%] | $CO_2$ [vol.%] | CO [ppm] |
|---|---|---|---|---|
| Exp. | $22.4 \pm 1.5$ | $6.1 \pm 0.4$ | $72.5 \pm 0.7$ | 40 |
| CFD | 22.88 | 6.63 | 70.48 | 621.4 |
| RN1 | 22.62 | 6.42 | 70.96 | 0.3235 |
| RN2 | 22.62 | 6.42 | 70.94 | 131.47 |
| RN3 | 22.62 | 6.42 | 70.95 | 71.6 |

### Extending CRN to Multiphase Systems

With promising results for gas-phase cases, the CRN approach is recently extended to multiphase systems by Berkel et al. [41]. For this purpose, the species balance equations are derived for a multiphase, solid-gas PSR (sPSR). The perfect mixing assumption applies within each phase, while the gas and the solid parts can exchange mass between them (depending on the kinetic mechanism) inside the sPSR. The governing equations and their derivation can be found in the study by Berkel et al. [41].

The sPSR is applied for several different fluidised bed reactor experiments using biomass as the fuel. For consistent representation of varying fluidised bed

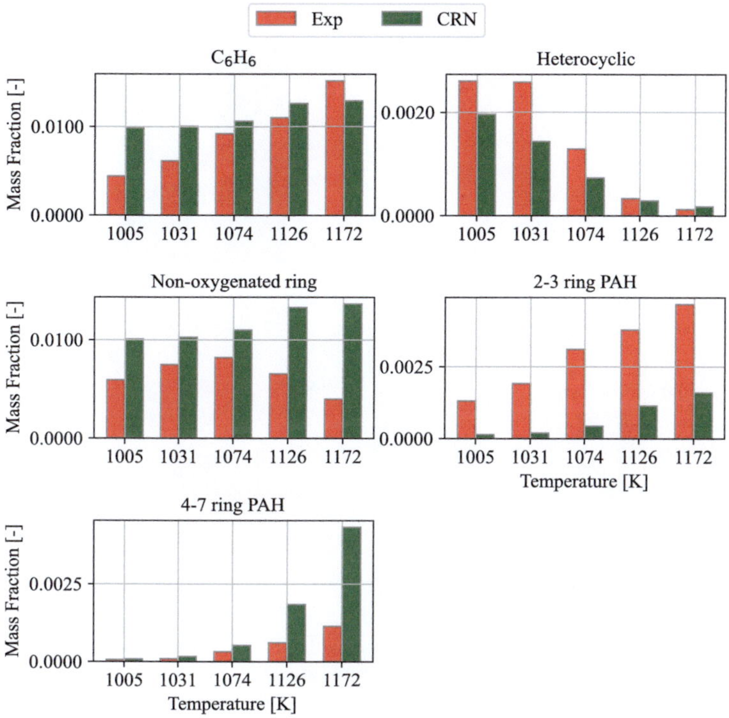


**Fig. 19.20** Comparison of tar measurements and CRN calculations at different fluidised bed operating temperatures. Experimental data from van Paasen and Kiel [55]. (Based on the publication by Berkel et al. [41])

experiments, an automatic generation algorithm is implemented and used [41], based on empirical correlations describing the bubbling behaviour in fluidised beds [51].

The employed sensitivity analysis in the study allowed to identify important aspects during biomass gasification in fluidised beds, such as the effect of mild combustion, water-gas shift reaction catalysis and the reactivity in different regions of the fluidised bed. The full (gas and solid) CRECK-S-B kinetic model [52–54] is used for the CRN calculations, which includes more than 20,000 reactions and 600 species (gas and solid phase combined). As a result, a very detailed chemical description of the produced gas is obtained. The tars from the simulation results are categorised for comparison with experimental measurements. The predictions of the model compared with experimental measurements from van Paasen and Kiel [55] are shown in Fig. 19.20. The trends of the different tar classes are satisfactorily captured by the CRN model of the fluidised bed.

### CFD-Based Multiphase CRN

Dübal et al. [42] introduced the solid-gas PFR (sPFR), similar to the sPSR. The governing equations and their derivation can be found in the study by Dübal et al. [42].

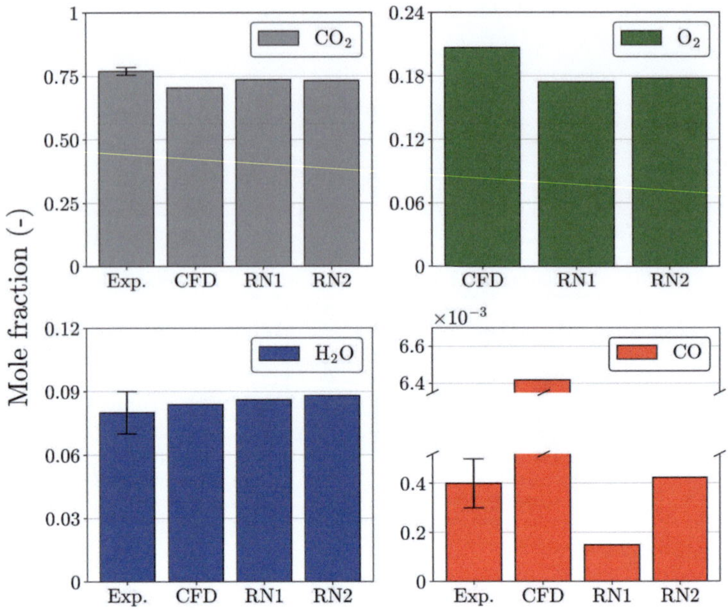


**Fig. 19.21** Calculations using CRN in comparison with CFD and experimental results. (Reprinted from Dübal et al. [42], Copyright (2024), with permission from Elsevier)

With the addition of sPFR, the CRN approach can be extended to more complex multi-phase systems. The CRN simulation of the methane-assisted pulverised coal flames in the swirled lab-scale Darmstadt burner is employed by Dübal et al. [42]. The CRN are generated by using the flamelet-LES of Nicolai et al. [19]. Similar to the study by Trespi et al. [35], a sensitivity study is carried out on the initial CRN. The temperatures of individual reactors are changed and the CO concentrations are tracked depending on the changes. The reactors with high sensitivity for CO production are divided into multiple ideal reactors to refine the CRN. The results of the initial (RN1) and improved (RN2) networks, as well as experimental and CFD data, are shown in Fig. 19.21. For CO, the CRN resulted in more accurate calculations than the CFD. Additionally, significant improvements in CO prediction are observed with the refined network compared to the initial one. The same refinement approach is also applied to the calculation of minor species including $NO_x$, $SO_x$, $C_6H_6$, and $C_{10}H_8$ by using detailed kinetics. However, validation was not possible due to lack of measured data.

## 19.3 Conclusion

In this chapter, the most important models of the OxySim-129 simulation framework were introduced and their applicability to experimental burners at different scales was demonstrated. Three aspects need to be addressed for the successful simulation of application-relevant solid fuel combustion systems, namely turbulent flow and mixing, turbulence-chemistry interaction and solid fuel conversion. In the OxySim-129 framework, turbulence is modelled using a LES approach, which yields the highest fidelity at a reasonable computational cost and resolves the large turbulent structures important for solid fuel conversion and gas phase combustion. For the turbulence-chemistry interaction, tabulated flamelet manifold methods are used—reduced-order models that are based on detailed chemistry but require only few transported quantities. Finally, the solid fuel conversion models are needed for both volatile release and char burnout. Especially for the volatile release, simple models—reducing the process to one or two reactions—can perform well if the reaction coefficients are tuned to the present conditions.

The applicability of these models was demonstrated in experimental combustion systems of different sizes. The Darmstadt burner allows for ample optical access and can provide abundant experimental data. The Aachen burner is still laboratory scale, but its heated walls allow for self-sustained solid fuel combustion. Therefore, experimental and numerical studies can be performed under more industrially-relevant conditions. Finally, the simulation framework was tested in the Darmstadt semi-industrial combustion chamber with a thermal power of 1 $MW_{th}$. Apart from varying scales, the framework was applied to two different fuels (coal and biomass) and two different atmospheres (air and multiple oxy-fuel atmospheres). The studies carried out under this variety of conditions showed good agreement with experimental results and yielded valuable insights into solid fuel combustion.

**Acknowledgements** This work has been funded by the Deutsche Forschungsgemeinschaft (DFG, German Research Foundation)—215035359—SFB/TRR 129 'Oxyflame'.

## References

1. Crowe, C. T., Sharma, M. P. and Stock, D. E. "The Particle-Source-in Cell (PSI-Cell) Model for Gas-Droplet Flows". *Journal of Fluids Engineering* 99.2 (1977), 325–332. https://doi.org/10.1115/1.3448756
2. Ranz, W. E. and Marshall, W. R. "Evaporation from Drops: Part I". *Chemical Engineering Progress* 48.3 (1952), 141–146.
3. Anthony, D. B. and Howard, J. B. "Coal Devolatilization and Hydrogasification". *AIChE Journal* 22.4 (1976), 625–656. https://doi.org/10.1002/aic.690220403
4. Kobayashi, H., Howard, J. B. and Sarofim, A. F. "Coal Devolatilization at High Temperatures". *Symposium (International) on Combustion* 16.1 (1977), 411–425. https://doi.org/10.1016/S0082-0784(77)80341-X
5. Vascellari, M., Arora, R., Pollack, M. and Hasse, C. "Simulation of Entrained Flow Gasification with Advanced Coal Conversion Submodels. Part 1: Pyrolysis". *Fuel* 113 (2013), 654–669. https://doi.org/10.1016/j.fuel.2013.06.014

6. Baum, M. M. and Street, P. J. "Predicting the Combustion Behaviour of Coal Particles". *Combustion Science and Technology* 3.5 (1971), 231–243. https://doi.org/10.1080/00102207108952290
7. Sadiki, A., Agrebi, S., Chrigui, M., Doost, S., Knappstein, R., di Mare, F., Janicka, J., Maßmeyer, A., Zabrodiec, D., Hees, J. and Kneer, R. "Analyzing the Effects of Turbulence and Multiphase Treatments on Oxy-Coal Combustion Process Predictions Using LES and RANS". *Chemical Engineering Science* 166 (2017), 283–302. https://doi.org/10.1016/j.ces.2017.03.015
8. Mehdizadeh, A., Doost, S., Sadiki, A., Janicka, J. and Karimi, N. "Assessment of Predictive Capability of Hybrid URANS/LES Methods in Residence Time Calculation". *Chemical Engineering Science* 183 (2018), 47–59. https://doi.org/10.1016/j.ces.2018.02.035
9. Cai, R., Luo, K., Watanabe, H., Kurose, R. and Fan, J. "Recent Advances in High-Fidelity Simulations of Pulverized Coal Combustion". *Advanced Powder Technology* 31.7 (2020), 3062–3079. https://doi.org/10.1016/j.apt.2020.05.001
10. Askarizadeh, H., Pielsticker, S., Nicolai, H., Kneer, R., Hasse, C. and Maßmeyer, A. "Investigation of the Coupling Schemes Between the Discrete and the Continuous Phase in the Numerical Simulation of a 60 $kW_{th}$ Swirling Pulverised Solid Fuel Flame Under Oxyfuel Conditions". *Fire* 7.6 (2024), 185. https://doi.org/10.3390/fire7060185
11. Askarizadeh, H., Pielsticker, S., Nicolai, H., Koch, M., Kneer, R., Hasse, C. and Maßmeyer, A. "Radiation Modelling Considering Burnout-Dependent Properties and Cellwise Non-Uniform Particle Distributions in the Numerical Simulation of Pulverised Solid Fuel Combustion". *Fuel* 381 (2025), 133338. https://doi.org/10.1016/j.fuel.2024.133338
12. Askarizadeh, H., Pielsticker, S., Nicolai, H., Özer, B., Kneer, R., Hasse, C. and Maßmeyer, A. "Influence of the Devolatilisation Kinetics on the Numerical Simulation of Pulverised Fuel Swirl Flames Under Oxyfuel Conditions". *Fuel* (2025), 135177. https://doi.org/10.1016/j.fuel.2025.135177
13. Smagorinsky, J. "General Circulation Experiments with the Primitive Equations". *Monthly Weather Review* 91.3 (1963), 99–164. https://doi.org/10.1175/1520-0493(1963)091%3C0099:GCEWTP%3E2.3.CO;2
14. Nicoud, F., Toda, H. B., Cabrit, O., Bose, S. and Lee, J. "Using Singular Values to Build a Subgrid-Scale Model for Large Eddy Simulations". *Physics of Fluids* 23.8 (2011). https://doi.org/10.1063/1.3623274
15. van Oijen, J. A. and Goey, L. de. "Modelling of Premixed Laminar Flames Using Flamelet-Generated Manifolds". *Combustion Science and Technology* 161.1 (2000), 113–137. https://doi.org/10.1080/00102200008935814
16. Pierce, C. D. and Moin, P. "Progress-Variable Approach for Large-Eddy Simulation of Non-Premixed Turbulent Combustion". *Journal of Fluid Mechanics* 504 (2004), 73–97. https://doi.org/10.1017/S0022112004008213
17. Peters, N. "Laminar Diffusion Flamelet Models in Non-Premixed Turbulent Combustion". *Progress in Energy and Combustion Science* 10.3 (1984), 319–339. https://doi.org/10.1016/0360-1285%2884%2990114-X
18. Knappstein, R., Künne, G., Becker, L., di Mare, F., Sadiki, A., Dreizler, A. and Janicka, J. "Large Eddy Simulation of a Novel Gas-Assisted Coal Combustion Chamber". *Flow, Turbulence and Combustion* 101.3 (2018), 895–926. https://doi.org/10.1007/s10494-018-9910-x
19. Nicolai, H., Künne, G., Knappstein, R., Schneider, H., Becker, L., Hasse, C., di Mare, F., Dreizler, A. and Janicka, J. "Large Eddy Simulation of a Laboratory-Scale Gas-Assisted Pulverized Coal Combustion Chamber Under Oxy-Fuel Atmospheres Using Tabulated Chemistry". *Fuel* 272 (2020), 117683. https://doi.org/10.1016/j.fuel.2020.117683
20. Wen, X., Nicolai, H., Schneider, H., Cai, L., Janicka, J., Pitsch, H. and Hasse, C. "Flamelet LES of a Swirl-Stabilized Multi-Stream Pulverized Coal Burner in Air and Oxy-Fuel Atmospheres with Pollutant Formation". *Proceedings of the Combustion Institute* 38.3 (2021), 4141–4149. https://doi.org/10.1016/j.proci.2020.05.061

21. Rieth, M., Kempf, A. M., Kronenburg, A. and Stein, O. T. "Carrier-Phase DNS of Pulverized Coal Particle Ignition and Volatile Burning in a Turbulent Mixing Layer". *Fuel* 212 (2018), 364–374. https://doi.org/10.1016/j.fuel.2017.09.096
22. Wen, X., Rieth, M., Scholtissek, A., Stein, O. T., Wang, H., Luo, K., Kempf, A. M., Kronenburg, A., Fan, J. and Hasse, C. "A Comprehensive Study of Flamelet Tabulation Methods for Pulverized Coal Combustion in a Turbulent Mixing Layer—Part I: A Priori and Budget Analyses". *Combustion and Flame* 216 (2020), 439–452. https://doi.org/10.1016/j.combustflame.2019.05.046
23. Wen, X., Rieth, M., Scholtissek, A., Stein, O. T., Wang, H., Luo, K., Kronenburg, A., Fan, J. and Hasse, C. "A Comprehensive Study of Flamelet Tabulation Methods for Pulverized Coal Combustion in a Turbulent Mixing Layer—Part II: Strong Heat Losses and Multi-Mode Combustion". *Combustion and Flame* 216 (2020), 453–467. https://doi.org/10.1016/j.combustflame.2019.12.028
24. Scholtissek, A., Popp, S., Hartl, S., Olguin, H., Domingo, P., Vervisch, L. and Hasse, C. "Derivation and Analysis of Two-Dimensional Composition Space Equations for Multi-Regime Combustion Using Orthogonal Coordinates". *Combustion and Flame* 218 (2020), 205–217. https://doi.org/10.1016/j.combustflame.2020.05.011
25. O'Rourke, P. and Bracco, F. "Two Scaling Transformations for the Numerical Computation of Multidimensional Unsteady Laminar Flames". *Journal of Computational Physics* 33.2 (1979), 185–203. https://doi.org/10.1016/0021-9991(79)90015-9
26. Pierce, C. D. "Progress-Variable Approach for Large-Eddy Simulation of Turbulent Combustion". PhD thesis. Stanford University, 2001.
27. Ketelheun, A., Kuenne, G. and Janicka, J. "Heat Transfer Modeling in the Context of Large Eddy Simulation of Premixed Combustion with Tabulated Chemistry". *Flow, Turbulence and Combustion* 91.4 (2013), 867–893. https://doi.org/10.1007/s10494-013-9492-6
28. Wollny, P., Rogg, B. and Kempf, A. M. "Modelling Heat Loss Effects in High Temperature Oxy-Fuel Flames with an Efficient and Robust Non-Premixed Flamelet Approach". *Fuel* 216 (2018), 44–52. https://doi.org/10.1016/j.fuel.2017.11.127
29. Hasse, C. and Peters, N. "A Two Mixture Fraction Flamelet Model Applied to Split Injections in a DI Diesel Engine". *Proceedings of the Combustion Institute* 30.2 (2005), 2755–2762. https://doi.org/10.1016/j.proci.2004.08.166
30. Prüfert, U., Hartl, S., Hunger, F., Messig, D., Eiermann, M. and Hasse, C. "A Constrained Control Approach for the Automated Choice of an Optimal Progress Variable for Chemistry Tabulation". *Flow, Turbulence and Combustion* 94.3 (2015), 593–617. https://doi.org/10.1007/s10494-015-9595-3
31. Gronarz, T., Schulze, J., Laemmerhold, M., Graeser, P., Gorewoda, J., Kez, V., Habermehl, M., Schiemann, M., Ströhle, J., Epple, B., Scherer, V. and Kneer, R. "Quantification of the Influence of Parameters Determining Radiative Heat Transfer in an Oxy-Fuel Operated Boiler". *Fuel Processing Technology* 157 (2017), 76–89. https://doi.org/10.1016/j.fuproc.2016.11.012
32. Dorigon, L. J., Duciak, G., Brittes, R., Cassol, F., Galarça, M. and França, F. H. "WSGG Correlations Based on HITEMP2010 for Computation of Thermal Radiation in Non-Isothermal, Non-Homogeneous $H_2O/CO_2$ Mixtures". *International Journal of Heat and Mass Transfer* 64 (2013), 863–873. https://doi.org/10.1016/j.ijheatmasstransfer.2013.05.010
33. Bordbar, M. H., Węcel, G. and Hyppänen, T. "A Line by Line Based Weighted Sum of Gray Gases Model for Inhomogeneous $CO_2$-$H_2O$ Mixture in Oxy-Fired Combustion". *Combustion and Flame* 161.9 (2014), 2435–2445. https://doi.org/10.1016/j.combustflame.2014.03.013
34. Levenspiel, O. *Chemical Reaction Engineering*. New York: John Wiley & Sons, 1999.
35. Trespi, S., Nicolai, H., Debiagi, P., Janicka, J., Dreizler, A., Hasse, C. and Faravelli, T. "Development and Application of an Efficient Chemical Reactor Network Model for Oxy-Fuel Combustion". *Energy & Fuels* 35.9 (2021). https://doi.org/10.1021/acs.energyfuels.0c03560
36. Touchton, G. L. "An Experimentally Verified $NO_x$ Prediction Algorithm Incorporating the Effects of Steam Injection". *Journal of Engineering for Gas Turbines and Power* 106.4 (1984), 833–840. https://doi.org/10.1115/1.3239647

37. Ehrhardt, K., Toqan, P., Jansohn, P., Teare, J. D., Beer, J. M., Sybon, G. and Leuckel, W. "Modeling of $NO_x$ Reburning in a Pilot Scale Furnace Using Detailed Reaction Kinetics". *Combustion Science and Technology* 131.1-6 (1998), 131–146. https://doi.org/10.1080/00102209808935758
38. Cuoci, A., Frassoldati, A., Buzzi-Ferraris, G., Faravelli, T. and Ranzi, E. "The Ignition, Combustion and Flame Structure of Carbon Monoxide/Hydrogen Mixtures. Note 2: Fluid Dynamics and Kinetic Aspects of Syngas Combustion". *International Journal of Hydrogen Energy* 32.15 (2007), 3486–3500. https://doi.org/10.1016/j.ijhydene.2007.02.026
39. Cuoci, A., Frassoldati, A., Stagni, A., Faravelli, T., Ranzi, E. and Buzzi-Ferraris, G. "Numerical Modeling of $NO_x$ Formation in Turbulent Flames Using a Kinetic Post-Processing Technique". *Energy & Fuels* 27.2 (2013), 1104–1122. https://doi.org/10.1021/ef3016987
40. Stagni, A., Cuoci, A., Frassoldati, A., Faravelli, T. and Ranzi, E. "A Fully Coupled, Parallel Approach for the Post-Processing of CFD Data Through Reactor Network Analysis". *Computers & Chemical Engineering* 60 (2014), 197–212. https://doi.org/10.1016/j.compchemeng.2013.09.002
41. Berkel, L., Debiagi, P., Nicolai, H., Amjed, M. A., Stagni, A., Hasse, C. and Faravelli, T. "Development of a Multiphase Chemical Reactor Network Method as a Tool for Simulating Biomass Gasification in Fluidized Beds". *Fuel* 357 (2024), 129731. https://doi.org/10.1016/j.fuel.2023.129731
42. Dübal, S., Berkel, L., Debiagi, P., Nicolai, H., Faravelli, T., Hasse, C. and Hartl, S. "Chemical Reactor Network Modeling in the Context of Solid Fuel Combustion Under Oxy-Fuel Atmospheres". *Fuel* 364 (2024), 131096. https://doi.org/10.1016/j.fuel.2024.131096
43. Cuoci, A., Frassoldati, A., Faravelli, T. and Ranzi, E. "OpenSMOKE++: An Object-Oriented Framework for the Numerical Modeling of Reactive Systems with Detailed Kinetic Mechanisms". *Computer Physics Communications* 192 (2015), 237–264. https://doi.org/10.1016/j.cpc.2015.02.014
44. Nicolai, H., Wen, X., Miranda, F., Zabrodiec, D., Maßmeyer, A., di Mare, F., Dreizler, A., Hasse, C., Kneer, R. and Janicka, J. "Numerical Investigation of Swirl-Stabilized Pulverized Coal Flames in Air and Oxy-Fuel Atmospheres by Means of Large Eddy Simulation Coupled with Tabulated Chemistry". *Fuel* 287 (2021), 119429. https://doi.org/10.1016/j.fuel.2020.119429
45. Nicolai, H., Debiagi, P., Wen, X., Dressler, L., Maßmeyer, A., Janicka, J. and Hasse, C. "Flamelet LES of Swirl-Stabilized Oxy-Fuel Flames Using Directly Coupled Multi-Step Solid Fuel Kinetics". *Combustion and Flame* 241 (2022), 112062. https://doi.org/10.1016/j.combustflame.2022.112062
46. Wen, X., Nicolai, H., Debiagi, P., Zabrodiec, D., Maßmeyer, A., Kneer, R. and Hasse, C. "Flamelet LES of a 40 $kW_{th}$ Pulverized Torrefied Biomass Furnace in Air and Oxy-Fuel Atmospheres". *Proceedings of the Combustion Institute* 39.4 (2023), 4563–4572. https://doi.org/10.1016/j.proci.2022.07.135
47. Nicolai, H., Debiagi, P., Janicka, J. and Hasse, C. "Flamelet LES of Oxy-Fuel Swirling Flames with Different $O_2/CO_2$ Ratios Using Directly Coupled Seamless Multi-Step Solid Fuel Kinetics". *Fuel* 344 (2023), 128089. https://doi.org/10.1016/j.fuel.2023.128089
48. Steffens, P., Berkel, L., Gierth, S., Debiagi, P., Özer, B., Maßmeyer, A., Nicolai, H. and Hasse, C. "LES of a Swirl-Stabilized 40 $kW_{th}$ Biomass Flame and Comparison to a Coal Flame". *Fuel* 372 (2024), 132098. https://doi.org/10.1016/j.fuel.2024.132098
49. Berkel, L., Steffens, P., Nicolai, H. and Hasse, C. "Towards Using LES for Analysis of Combustion Characteristics in a Semi-Industrial Biomass Burner". *$14^{th}$ European Conference on Industrial Furnaces and Boilers*. Algarve, Portugal, 2024.
50. Becker, L., Kosaka, H., Böhm, B., Doost, S., Knappstein, R., Habermehl, M., Kneer, R., Janicka, J. and Dreizler, A. "Experimental Investigation of Flame Stabilization Inside the Quarl of an Oxyfuel Swirl Burner". *Fuel* 201 (2017), 124–135. https://doi.org/10.1016/j.fuel.2016.09.002
51. Grace, J. R., Leckner, B., Zhu, J. and Cheng, Y. "Fluidized Beds". *Multiphase Flow Handbook*. Ed. by Crowe, C. T. Boca Raton: CRC Press, 2005.

52. Debiagi, P., Pecchi, C., Gentile, G., Frassoldati, A., Cuoci, A., Faravelli, T. and Ranzi, E. "Extractives Extend the Applicability of Multistep Kinetic Scheme of Biomass Pyrolysis". *Energy & Fuels* 29.10 (2015), 6544–6555. https://doi.org/10.1021/acs.energyfuels.5b01753
53. Debiagi, P., Gentile, G., Pelucchi, M., Frassoldati, A., Cuoci, A., Faravelli, T. and Ranzi, E. "Detailed Kinetic Mechanism of Gas-Phase Reactions of Volatiles Released from Biomass Pyrolysis". *Biomass and Bioenergy* 93 (2016), 60–71. https://doi.org/10.1016/j.biombioe.2016.06.015
54. Debiagi, P., Gentile, G., Cuoci, A., Frassoldati, A., Ranzi, E. and Faravelli, T. "A Predictive Model of Biochar Formation and Characterization". *Journal of Analytical and Applied Pyrolysis* 134 (2018), 326–335. https://doi.org/10.1016/j.jaap.2018.06.022
55. van Paasen, S. V. B. and Kiel, J. H. A. *Tar Formation in a Fluidised-Bed Gasifier: Impact of Fuel Properties and Operating Conditions*. 2004.

MIX
Papier aus verantwortungsvollen Quellen
Paper from responsible sources
FSC® C105338

If you have any concerns about our products,
you can contact us on
**ProductSafety@springernature.com**

In case Publisher is established outside the EU,
the EU authorized representative is:
**Springer Nature Customer Service Center GmbH**
**Europaplatz 3, 69115 Heidelberg, Germany**

Printed by Libri Plureos GmbH
in Hamburg, Germany